medEssentials for the USMLE™ Step 1

FOURTH EDITION

ALSO FROM KAPLAN MEDICAL

Books

USMLE® Step 1 Qbook

Flashcards

USMLE® Diagnostic Test Flashcards:

The 200 Questions You Need to Know for the Exam for Steps 2 & 3

USMLE® Examination Flashcards: The 200 "Most Likely Diagnosis" Questions You Will See on the Exam for Steps 2 & 3

USMLE® Pharmacology and Treatment Flashcards: The 200 Questions You're Most Likely to See on Steps 1, 2 & 3

USMLE® Physical Findings Flashcards: The 200 Questions You're Most Likely to See on the Exam

Online

Dr. Conrad Fischer's Comprehensive Cases

Updated USMLE® Step 3 Qbank

medEssentials for the USMLE™ Step 1

FOURTH EDITION

High-Yield Review for First- and Second-Year Medical Students

USMLE™ is a registered trademark of the Federation of State Medical Boards (FSMB) of the United States and the National Board of Medical Examiners® (NBME®), neither of which sponsors or endorses this product.

USMLE® is a registered trademark of the Federation of State Medical Boards (FSMB) of the United States and the National Board of Medical Examiners® (NBME®), neither of which sponsors or endorses this product.

This publication is designed to provide accurate and authoritative information in regard to the subject matter covered. It is sold with the understanding that the publisher is not engaged in rendering legal, accounting, or other professional service. If legal advice or other expert assistance is required, the services of a competent professional should be sought.

© 2012 by Kaplan, Inc.
Published by Kaplan Publishing, a division of Kaplan, Inc.
395 Hudson Street, 4th Floor
New York, NY 10014

All rights reserved. The text of this publication, or any part thereof, may not be reproduced in any manner whatsoever without written permission from the publisher.

Printed in the United States of America

10 9 8 7 6 5

ISBN-13: 978-1-60978-026-5

Kaplan Publishing books are available at special quantity discounts to use for sales promotions, employee premiums, or educational purposes. Please call the Simon & Schuster special sales department at 866-506-1949.

LEAD AUTHORS/EDITORS

Michael S. Manley, M.D.
Senior Director, Step 1 Curriculum
Kaplan Medical

Leslie D. Manley, Ph.D.
Director, Step 1 Curriculum
Kaplan Medical

CONTRIBUTORS

Thomas H. Adair, Ph.D.
Professor
Department of Physiology and Biophysics
University of Mississippi Medical Center

John Barone, M.D.
Anatomic and Clinical Pathologist

Stuart Bentley-Hibbert, M.D., Ph.D.
New York Presbyterian Hospital
Weill Cornell Medical Center

Douglas E. Fitzovich, Ph.D.
Professor of Physiology
DeBusk College of Osteopathic Medicine
Lincoln Memorial University

Beth Forshee, Ph.D.
Assistant Professor of Physiology
Lake Erie College of Osteopathic Medicine

Steven R. Harris, Ph.D
Associate Dean for Academic Affairs
Professor of Pharmacology
Kentucky College of Osteopathic Medicine

Robert F. Kissling III, M.D.

John A. Kriak, Pharm.D.
President/Chief Clinical Consultant, CAMMCO, LLC

Nancy Standler, M.D., Ph.D.
Department of Pathology
Valley View Medical Center
Intermountain Healthcare

Sam Turco, Ph.D.
Professor, Department of Biochemistry
University of Kentucky College of Medicine

James S. White, Ph.D.
Assistant Professor of Cell Biology
School of Osteopathic Medicine
University of Medicine and Dentistry of New Jersey
Adjunct Assistant Professor of Cell and Developmental Biology
University of Pennsylvania School of Medicine

Glenn C. Yiu M.D., Ph.D.
Clinical Fellow in Medicine
Harvard Medical School

Contents

(*A more comprehensive list appears on the first page of each chapter.*)

How to Use *medEssentials*

Doing well on USMLE® Step 1 means mastering over 6,000 different terms, as well as countless concepts and relationships. To accomplish this, most students focus on brute memorization, trying to jam into their heads the information they think they will need and then spitting out relevant facts when exam time comes. This memorization strategy works fairly well for "recall exams," which require you to demonstrate that you know a particular, defined set of information.

Unlike most medical school exams, however, the USMLE is a "thinking exam," requiring you to not just recognize, but apply, use, and reason with the content you have learned. Questions are drawn from the broad array of basic science knowledge, and not confined to some explicit subset. To do well—indeed, simply to pass the USMLE Step 1—you must store up information over weeks and months and then think intelligently about that material. Sheer memorization will not get you to this goal. You need a more organized and intelligent approach.

To master the information you will need for Step 1, you must (1) decide what to focus on and (2) organize it in a way that fosters long-term retention. This means that you have to know what is important, and then link it together so you can think and reason as you solve the exam questions.

MedEssentials for the USMLE Step 1 was created to help you master this learning process. First, it distills the mass of material you must master to emphasize the most important concepts and details. Second, it organizes the material you need to know in a way that facilitates retention and application to clinical problems. *MedEssentials* has already helped thousands gain a better understanding of the basics of medicine and earn better exam results. In this single volume, we have organized the high-yield USMLE information with a conscious eye towards fostering the kind of insight and understanding the current USMLE demands.

DISTILLATION

When you are plowing through the stack of material you need to study for your medical school and wondering what really matters, *medEssentials* will serve as a guide. If it is in this book, you should know it.

The purpose of *medEssentials* is not to cover everything you need to master, but rather to highlight and emphasize what matters most. The material included in *medEssentials* is the most high-yield content, selected by Kaplan Medical's best faculty, drawing on their long years of experience preparing students for the USMLE Step 1. This faculty-selected content is then periodically reviewed by recent USMLE test-takers and rated in terms of exam relevance.

Be aware that your medical school exams may contain other material not in *medEssentials*. We urge you to not neglect that additional content for your medical school exams. But if you are overwhelmed by the sheer volume of what you have to learn, *medEssentials* will ensure that you do not lose your way or neglect what is the most important.

RETENTION AND APPLICATION

Our brains are basically lazy. Old information is continually purged to make way for the new. The only things our brains hold on to for the long haul are things that we *use* and things that are *connected.*

Forget about cramming.

The most common strategy for preparing for a recall exam is to cram the needed information into short-term memory and hope it lingers long enough to be used on the exam. Unfortunately, what is simply memorized is soon forgotten. The long lists of anatomic names, pharmacological substances, and pathological conditions are absorbed one day, but gone the next.

Cramming information into your head may get you past today's exam, but does not prepare you for the more complex exams or the medical practice you will face tomorrow. To solve the problems that are posed to you on USMLE Step 1, you must store up information over weeks and months, and then think intelligently to apply that material.

Don't just sit there, do something!

To be effective, study must be active. The concepts you are learning are like the clothes in your closet: We tend to hold on to clothing that we wear every day. It's the stuff piled in the back of the closet that gets lost or tossed. Likewise, simply rereading the same thing over and over is ineffective. The key to retention is doing something with the content you want to master.

Doing something with the material as you learn activates the hippocampus and amygdala. The hippocampus is a mental map that organizes information, facilitating both general comprehension and active recall. The amygdala puts an emotional tag on knowledge and events, making them stand out from the sensory stream of daily experience. When you do something with the content, it signals your brain that the content is important and activates the mental processes that allow for long-term recall.

Common strategies for active learning involve taking summary notes, making charts, drawing figures or graphs, or talking with others. Each of these techniques can be beneficial, but all take a lot of time. However, recent research tells us that the simple technique of Reading, Reciting, and Reviewing can provide the same benefits of active learning in *about half the time*.

To use this method, you first read a short section of material. Next, you cover it or look away and in your own words, recite for yourself the salient points. Finally, look back at the material and check your understanding. That's it. Nothing complicated. This simple technique breaks you out of the boredom of endless repetition and forces you to do something concrete with the material as you encounter it. And doing something means holding on to what you are learning.

Compare and contrast for long-term retention.

Blocks of content are easier to retain than scattered facts. This means you should learn content in chunks rather than as individual items. When you study, never focus on learning individual things. Rather, learn sets of things, and then learn to differentiate within the set.

For instance, don't try to learn all about a particular drug or bug. Learn classes of drugs and then the key details that let you differentiate among them. Learn groups of pathogens and then the critical differences among them based on symptoms, origins, properties, or treatments. Organize everything you learn into groupings that share common features. Once you have defined what is in the grouping, turn your attention to rehearsing how you will distinguish within the group.

Note that not all details matter. The critical ones are those that help you make distinctions. Everything else is mental decoration. Details that do not help you make decisions make you feel smart, but do nothing to help you answer USMLE questions.

Why study things in groupings? First, things that are connected form clusters that provide a coherent set of information and facilitate associations to other content. The cluster gives you a dual basis for understanding—learning what something is, and by contrast, what it is not. I may understand being "male" by knowing anatomic, hormonal, and developmental characteristics, but a full understanding only arrives when I am able to contrast "male" with "female."

Second, your main task on the USMLE is to say which of the presented options is "most likely," given the information presented in the question stem (the lead-in paragraph of the question). To answer a question correctly, you have to recognize the clues in the paragraph and use these clues to guide your choice of an answer. The options form a cluster of possibilities. The clues in the question stem tell you which of the possible options is the "most likely" answer out of the presented possibilities. When you study by differentiating within groupings, you are *anticipating the options* that you will encounter and *practicing how you will decide* among them.

Mastering this mental task has implications far beyond your USMLE score. Medical practice itself is essentially a series of multiple-choice questions. For each patient, you enter the examination room, assess available information, call to mind a set of options for how to proceed, and then pick one of the options. On the Step 1 and Step 2 CK exams, you do not have to know the options; they will be given to you. All you have to do is pick the correct choice out of the options presented. If you organize your medical knowledge as groupings and know how to differentiate within the grouping, you are organizing your thinking for medical practice as you prepare for the USMLE.

When reviewing *medEssentials* tables, focus on contrasts.

Material in *medEssentials* is presented as a set of tables and diagrams. This organization of the content is the most efficient method for fostering the type of learning that the USMLE requires. Do not simply rote memorize lists of things. Let *medEssentials* give you the organization that allows the intelligent decision-making you will need.

Review the tables in this book always with an eye toward differentiation among terms and concepts. To do this, pick two presented concepts (e.g., different pathogens, drugs, diagnoses) and rehearse how they are the same and how they are different. Once you have mastered that pair, pick another pair and repeat the process, and so on until you have rehearsed how to differentiate all of the items in the table. In other words, you should not be looking at the tables to memorize words, but to rehearse relationships. Memory fades quickly, but relationships are remembered. The key question is not, "How will I remember this?" but "How do I tell things apart?"

When reviewing *medEssentials* diagrams, focus on connections.

Diagrams highlight distinct points of larger processes. Your task is to master not just the named points, but also the process of how you get from one to the next. So, rather than simply memorizing a sequence of steps by rote, think about the transformations or dynamic processes that lead from one step to the next. The key question is not, "What is the next step in the process?" but, "What happens to bring about the next step?"

For instance, when reviewing anatomic diagrams, focus on context, not simply memorizing the name of bodily parts. What is this part connected to? What lies beside it or under it? Questions on the USMLE are likely to require the mastery of such linkages. You are trying not just to learn the pieces, but how they all fit together.

MedEssentials is not a list of things to be memorized, but an organization of knowledge to allow for the retention and use of what you learn. If you use it the right way, you should find that you have a clearer understanding and enhanced confidence when applying core basic science knowledge on the USMLE, medical school exams, or even the rigors of medical practice.

Remember, it's not what you know, but how it's organized in your brain that makes the difference! Push yourself to move beyond passive memorization to active understanding. Focus on mastering contrasts and connections, and you will know what you need to pick the best answer on exam day.

Preface

Welcome to Kaplan Medical. This ***medEssentials*** course includes a comprehensive high-yield review book with interactive online exercises covering both first and second year medical school subjects. These tools are an excellent adjunct to your medical school curriculum and an important resource for your board preparations.

The *medEssentials: High-Yield USMLE® Step 1* Review

The *medEssentials* review book, structured by organ system and in a compact, concise fashion, presents the most relevant and important basic medical principles with reference charts and corresponding images. Our design allows a unique level of integration between the disciplines. For example, rather than sequentially reviewing each discipline within an organ system (anatomy, physiology, pathology, and pharmacology), the book integrates important information across various disciplines for a given subtopic in one place—sometimes on a single page. This way of reviewing allows you to obtain a complete, comprehensive understanding of any given topic.

The first section of the book (General Principles) covers the general principles of pathology, pharmacology, physiology, behavioral science, biostatistics, biochemistry, molecular biology, cell biology, genetics, microbiology, immunology, embryology, and histology. These subjects precede the organ system chapters and serve as a comprehensive foundation for the organ-specific facts that follow.

The second section of the book features the following organ systems: cardiovascular; respiratory; renal and urinary; hematologic and lymphoreticular; nervous; musculoskeletal, skin and connective tissue; gastrointestinal; endocrine; and reproductive. Each chapter includes high-yield information from the disciplines corresponding to the basic science courses taken during the first two years of medical school. Each organ system is viewed from histologic, embryologic, physiologic, pathologic, and pharmacologic perspectives.

We want to hear what you think of this course. Please share with us your feedback by emailing **medfeedback@kaplan.com**.

Study Techniques

Effective study techniques for medical school and ultimately for the board exams are about making choices. At each stage of your medical career, you need to choose study strategies that will lead you to success.

Many medical students fear that not knowing all of the details may affect their clinical performance. It is natural to feel this way, but rest assured: Your ability to treat patients in these coming years will grow from mastery of concepts and by practicing your clinical skills. Your "clinical eye" will grow in time. Understanding basic mechanisms and key principles is crucial to developing the ability to apply what you know to real patients.

One of the most effective strategies for studying the basic sciences is to apply the medical concepts you have learned to your imagined, future patients. For example, when studying muscle groups in anatomy, imagine yourself as a surgeon and how you would find the structures of interest. In physiology, imagine a patient asking you about cortisol, what it is and what it does. Practice explaining to a patient the biochemical differences between lipids, how they differ, and what this means medically. *By doing so, you are simplifying the concepts, making them more memorable and clinically relevant; you are learning actively.*

Active use of the material that you learn increases retention and facilitates recall. Repetition makes memories. Each instance of recall produces a new memory trace, linking concepts and increasing the chance of recall in the future. Recall actually changes neuronal structures. To be truly useful, a piece of information needs to be triangulated, connected to a number of other concepts, or better yet, experienced. In other words, mere memorization is not your goal, but rather the ability to process and apply that information in a fully integrated manner.

Rereading textbooks from cover to cover and underlining—yet again, in a different color—every line on every page is *not* an efficient way to learn. You need to focus on the material most likely to be on your exams and on the material that is considered high-yield.

Begin your studies by following this simple outline:

- Start every study session with a list of specific goals.
- Make your notes richer by color-highlighting, adding notes to the diagrams, and re-summarizing what you have learned. This is your book; personalize it to get the most out of it.
- After reviewing your intended subject for the session, imagine how you'd teach the same concepts to someone else.

Using *medEssentials: High-Yield USMLE™ Step 1*

1. Learn the basic definitions and concepts central to each discipline (Section I: General Principles). The book provides the core vocabulary to understand the content of those disciplines. Terms and definitions are learned by the use of *associational memory.*
2. Learn central concepts for each of the subject areas and how they integrate within an organ system (Section II: Organ System). Integration of concepts and disciplines within each major organ system is the key to success in both medical school and licensing exams. Your basic mental task here is reconstructive memory, learning to recall the concepts in terms of how things fit together within an organ system. At this stage, patterns begin to emerge. The diagrams, tables, and pictures in this book are specifically designed for this stage of learning.
3. Engage in active learning by applying the concepts to scenarios, clinical settings, and mini-case presentations. This is the hardest stage of preparation, and one that most students neglect. Your task at this level is reasoning, comprehension, and deduction.

Get a Mobile Version of This Book

This book comes with a free mobile download. To get your free download, follow these simple steps:

1. Go to **kaptest.com/booksonline**.
2. Follow the on-screen instructions. Have a copy of your book available.
3. Click on the link for directions on how to download the mobile version of *medEssentials for the USMLE Step 1.*

Access to the mobile version is limited to the original owner of this book and is nontransferable. Kaplan is not responsible for providing access to customers who purchase or borrow used copies of this book.

On behalf of Kaplan Medical, we wish you the best of success in your studies and your medical career!

Section I

General Principles

Behavioral Science

Chapter 1

Learning and Behavior Therapy

▶ Conditioning

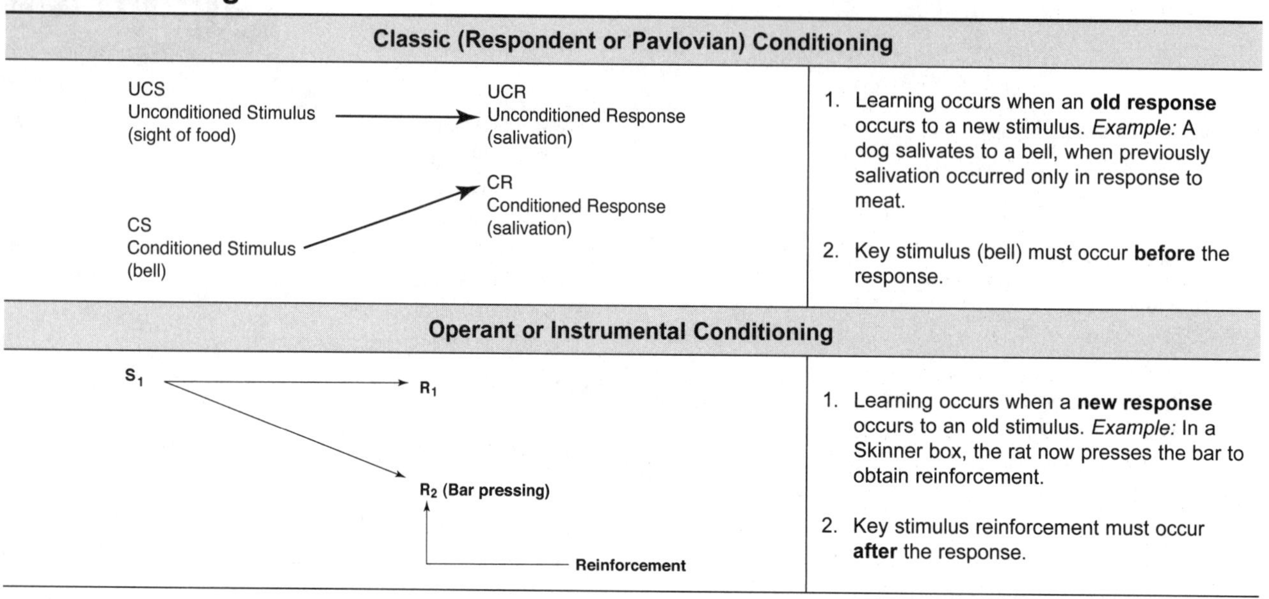

▶ Types of Reinforcement

		Stimulus (S)	
		Add	**Remove**
Behavior (R)	**Stops**	Punishment	Extinction
	Increases	Positive reinforcement	Negative reinforcement

▶ Reinforcement Schedules

		Contingency	
		Time	**Behaviors**
Schedule	**Constant**	Fixed interval (FI)	Fixed ratio (FR)
	Changing	Variable interval (VI)	Variable ratio (VR)

▶ Learning-Based Therapies

On the Basis of Classic Conditioning	
Systematic desensitization	• Often used to treat anxiety and phobias **Step 1:** hierarchy of stimuli (least to most feared) **Step 2:** technique of muscle relaxation taught **Step 3:** patient relaxes in presence of each stimulus on the hierarchy • Works by replacing anxiety with relaxation, an incompatible response
Exposure (also flooding or implosion)	• Simple phobias treated by forced exposure to the feared object • Exposure maintained until fear response is extinguished
Aversive conditioning	• Properties of the original stimulus are changed to produce an aversive response • Can help reduce deviant behaviors
On the Basis of Operant Conditioning	
Shaping	• Achieves target behavior by reinforcing successive approximations of the desired response • Reinforcement gradually modified to move behaviors from general responses to specific responses desired
Extinction	• Discontinuing the reinforcement maintaining an undesired behavior • "Time out" with children or for test anxiety
Stimulus control	• Sometimes stimuli inadvertently acquire control over behavior; when this is true, removal of that stimulus can extinguish the response • *Example:* an insomniac only permitted in bed when he/she is so tired that sleep comes almost at once
Biofeedback	• Using external feedback to modify internal physiologic states; often uses electronic devices to present physiologic information, e.g., heart monitor to show heart rate • Works by means of trial-and-error learning and requires repeated practice to be effective
Fading	• Gradually removing the reinforcement while: *1*) without the subject discerning the difference and *2*) maintaining the desired response • *Example:* gradually replacing postoperative painkiller with a placebo

Child Development

► Apgar Scoring System

Evaluation	0 Points	1 Point	2 Points
Heart rate	0	<100/min	>100/min
Respiration	None	Irregular, shallow gasps	Crying
Color	Blue	Pale, blue extremities	Pink
Tone	None	Weak, passive	Active
Reflex irritability	None	Facial grimace	Active withdrawal

► Infant Development

Evidenced at birth	• Reaching and grasping behavior • Ability to imitate facial expressions • Ability to synchronize limb movements with speech of others • Attachment behaviors, such as crying and clinging
Newborn characteristics	• Prefers: – Large, bright objects with lots of contrast – Moving objects – Curves versus lines – Complex versus simple designs – Facial stimuli (girls more than boys) • Can discriminate between language and nonlanguage stimuli • At 1 week old, the infant responds differently to the smell of mother compared with father
Smiling	• The smile develops from an innate reflex present at birth (endogenous smile) • Shows exogenous smiling in response to a face at 8 weeks • A preferential social smile, e.g., to the mother's rather than another's face, appears about 12 to 16 weeks

► Figures Copied and Approximate Ages

Figure Copied	Approximate Age
Circle	3
Cross	4
Rectangle	4½
Square	5
Triangle	6
Diamond	7

▶ Key Developmental Terms

Brain-growth spurt	• "Critical period" of great vulnerability to environmental influence • Extends from last trimester of pregnancy through first 14 postnatal months • Size of cortical cells and complexity of cell interconnections undergo their most rapid increase
Stranger anxiety	• Distress in the presence of unfamiliar people • Appears at 6 months, reaches peak at 8 months, then disappears after 12 months
Separation anxiety	• Distress of infant after separation from caretaker • Appears at 8 to 12 months; begins to disappear at 20 to 24 months • **Separation anxiety disorder** (school phobia) is failure to resolve separation anxiety • Treatment focuses on child's interaction with parents, not on activities in school

▶ Children's Conceptions of Illness and Death

More than death, a preschool child is more likely to fear:	Separation from parents, punishment, mutilation (Freud's castration anxiety)
When they become ill:	May interpret illness or treatment as punishment Often have all sorts of misconceptions about what is wrong with them
Until age 5:	Children usually have no conception of death as an irreversible process
Only after age 8 or 9:	Child really understands that death is universal, inevitable, and irreversible

▶ Child Development Milestones

Age	Physical and Motor Development	Social Development	Language Development
First year of life	• Puts everything in mouth • Sits with support (4 mo) • Stands with help (8 mo) • Crawls, fear of falling (9 mo) • Pincer grasp (12 mo) • Follows objects to midline (4 wk) • One-handed approach/grasp of toy • Feet in mouth (5 mo) • Bang and rattle stage • Changes hands with toy (6 mo)	• Parental figure central • Issues of trust are key • Stranger anxiety (6 mo) • Play is solitary and exploratory • Pat-a-cake, peek-a-boo (10 mo)	• Laughs aloud (4 mo) • Repetitive responding (8 mo) • "mama, dada" (10 mo)
Year 1	• Walks alone (13 mo) • Climbs stairs alone (18 mo) • Emergence of hand preference (18 mo) • Kicks ball, throws ball • Pats pictures in book • Stacks three cubes (18 mo)	• Separation anxiety (12 mo) • Dependency on parental figure (rapprochement) • Onlooker play	• Great variation in timing of language development • Uses 10 words
Year 2	• High activity level • Walks backward • Can turn doorknob, unscrew lid jar • Scribbles with crayon • Able to aim to throw ball • Stands on tiptoes (30 mo) • Stacks six cubes (24 mo)	• Selfish and self-centered • Imitates mannerisms and activities • May be aggressive • "No" is favorite word • Parallel play	• Use of pronouns • Parents understand most words • Telegraphic sentences • Two-word sentences • Uses 250 words • Identifies body parts by pointing

(Continued)

► Child Development Milestones *(Cont'd.)*

Age	Physical and Motor Development	Social Development	Language Development
Year 3	• Rides tricycle • Stacks 9 cubes (36 mo) • Alternates feet going upstairs • Bowel and bladder control (toilet training) • Draws recognizable figures • Catches ball with arms • Cuts paper with scissors • Unbuttons buttons	• Fixed gender identity • Sex-specific play • Understands "taking turns" • Knows sex and full names	• Completes sentences • Uses 900 words • Understands 4× that • Strangers can understand • Recognizes common objects in pictures • Can answer: "Which block is bigger?"
Year 4	• Alternates feet going down stairs • Hops on one foot • Grooms self (brushes teeth) • Counts fingers on hand	• Imitation of adult roles • Curiosity about sex (playing doctor) • Nightmares and monster fears • Imaginary fears	• Can tell stories • Uses prepositions • Uses plurals • Compound sentences
Year 5	• Complete sphincter control • Brain at 75% of adult weight • Draws recognizable man with head, body, and limbs • Dresses and undresses self • Catches ball with two hands	• Conformity to peers important • Romantic feeling for others • Oedipal phase	• Asks the meaning of words • Abstract words elusive
Years 6 to 12	• Boys heavier than girls • Refined motor skills • Rides bicycle • Gains athletic skill • Coordination increases	• "Rules of the game" are key • Organized sports possible • Being team member focal for many • Separation of the sexes • Demonstrating competence is key	• Shift from egocentric to social speech • Incomplete sentences decline • Vocabulary expands geometrically (50,000 words by age 12)
Years ≥12 (adolescence)	• Adolescent "growth spurt" (girls before boys) • Onset of sexual maturity (≥10 y) • Development of primary and secondary sexual characteristics	• Identity is critical issue • Conformity most important (11 to 12 y) • Organized sports diminish for many • Cross-gender relationships	• Adopts personal speech patterns • Communication becomes focus of relationships

► Tanner Stages of Development

	Female	Female and Male	Male
Stage	**Breast**	**Pubic Hair**	**Genitalia**
1	Preadolescent	None	Childhood size
2	Breast bud	Sparse, long, straight	Enlargement of scrotum/testes
3	Areolar diameter enlarges	Darker, curling, increased amount	Penis grows in length; testes continue to enlarge
4	Secondary mound; separation of contours	Course, curly, adult type	Penis grows in length/breadth, scrotum darkens, testes enlarge
5	Mature female	Adult, extends to thighs	Adult shape/size

► Types of Abuse and Important Issues

	Child Abuse	Elder Abuse	Spousal Abuse
Annual cases	Over 2 million	5 to 10% in population	Over 4 million
Most common type	Physical battery/neglect	Neglect	Physical battery
Likely sex of victim	Before age 5: female After age 5: male	63% female	Female
Likely sex of perpetrator	Female	Male or female	Male
Mandatory reporting?	Yes	Yes	No
Physician's response	Protect and report	Protect and report	Counseling and information

Sleep: Physiology and Disorders

► Sleep Physiology

Types of Sleep

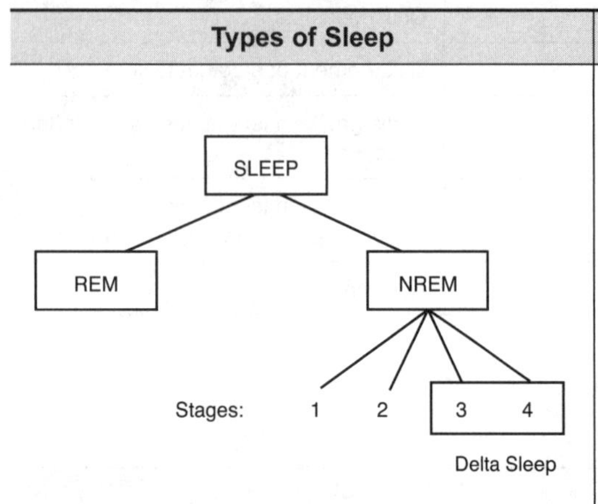

Sleep Architecture Diagram Showing Stages of Sleep in Sequence

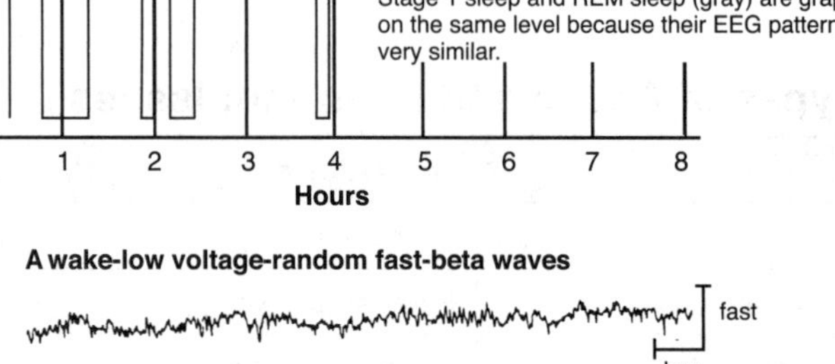

A wake-low voltage-random fast-beta waves

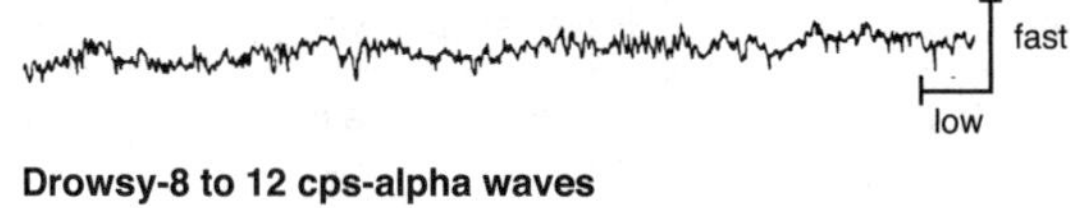

Drowsy-8 to 12 cps-alpha waves

Stage 1-3 to 7 cps theta waves

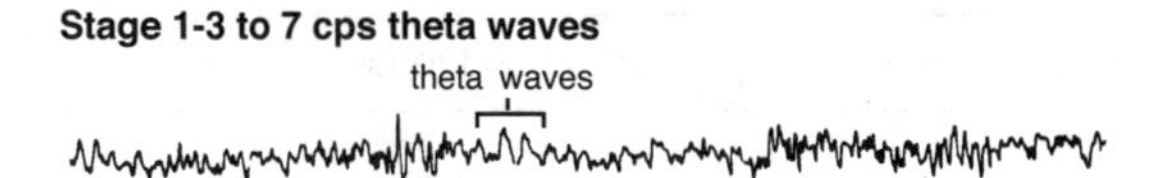

Stage 2-12 to 14 cps-sleep spindles and K complexes

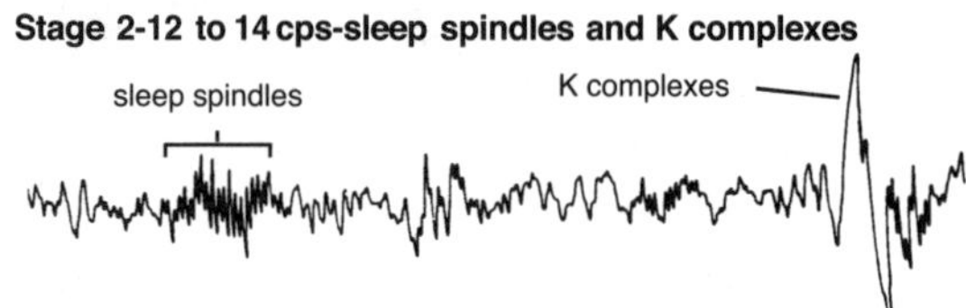

Delta sleep-1/2 to 2 cps-delta waves>75

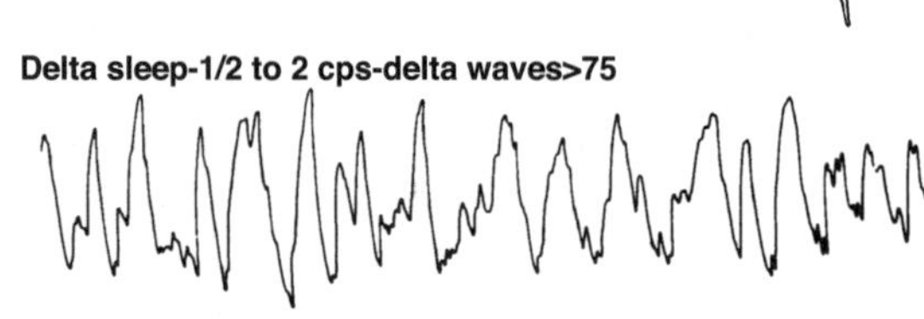

REM sleep-low voltage-random, fast with sawtooth waves

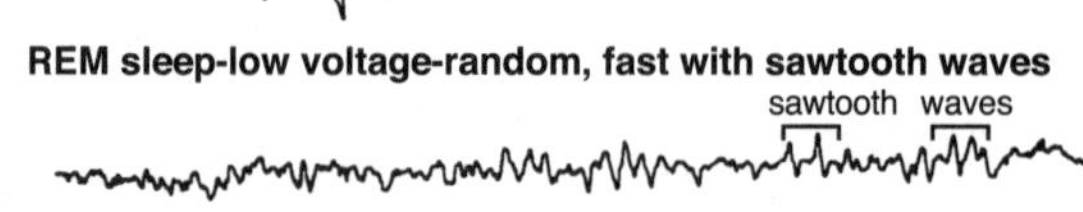

Changes of Daily Sleep Over the Life Cycle

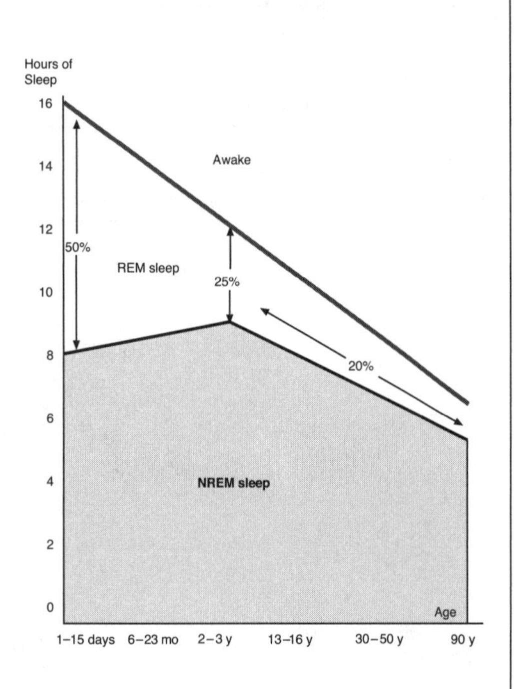

► Changes During the First 3 Hours of Sleep

Human growth hormone (HGH)	↑
Prolactin	↑
Dopamine	↓
Serotonin	↑
Thyroid-stimulating hormone (TSH)	↓

► Sleep Deprivation

Getting 5 or less hours of sleep = functioning at level of someone legally drunk	
Lymphocyte levels	↓
Cortisol levels	↑
Blood pressure	↑
Glucose tolerance	↓
Amygdala activation	↑
Negative mood	↑
Prefrontal cortical activity	↓

► Sleep Disorders

Disorders	Signs, Symptoms, and Issues	Treatments
Narcolepsy	• Sleep attacks • Cataplexy • Hypnagogic hallucinations • Sleep paralysis • REM latency <10 minutes	• **Modafinil:** nonamphetamine alternative to CNS stimulants • Traditionally treated with CNS stimulants (e.g., amphetamines)
Sleep apnea syndromes	• Absence of respiration for extended periods during sleep • Patient often overweight • Risk of sudden death • Obstructive: rasping snoring • Central: Cheyne-Stokes	• Weight loss • Continuous positive airway pressure (CPAP) • Condition so sleeping position not on back • Surgery for severe cases
Sudden infant death syndrome (SIDS)	Unexplained death in children younger than 1 year	• Lay baby to sleep on back • Avoid overstuffed bedding and pillows • Rate higher if household member smokes • Fetal exposure to maternal smoking is strong risk factor
Insomnia	Causes: • Hypnotic medication abuse • Emotional problems • Conditioned poor sleep • Withdrawal from drugs	• Behavioral therapy best treatment • Pharmacology: acute relief by benzodiazepines, zolpidem, zaleplon, eszopiclone (no tolerance)
Somnambulism (sleepwalking)	• Stage 4 sleep • If awakened, person confused and disoriented	Identify anxiety issues
Enuresis (bedwetting)	• Delta sleep • Boys to girls: 2:1 • Defense mechanism of regression	Imipramine acutely Desmopressin chronically
Bruxism (teeth grinding)	• Stage 2 sleep • Patient may be unaware unless told by others	Reduce anxiety, oral devices

► Night Terrors Versus Nightmares

	Night Terrors	Nightmares
Sleep stage	Stage 4 (delta sleep)	REM
Physiologic arousal	Extreme	Elevated
Recall upon waking	No	Yes
Waking time anxiety	Yes, usually unidentified	Yes, often unidentified
Other issues	• Runs in families • More common in boys • Can be a precursor to temporal lobe epilepsy	• Common from ages 3 to 7 • Desensitization behavior therapy provides marked improvement

Substance Abuse

► Alcohol

Key Concepts

- Most costly health problem
- 10% of people are problem drinkers
- Heavy drinking increasing in younger people
- Accounts for 50% of auto accident deaths
- One of the leading reasons for hospitalizations (detox, trauma, injuries, GI problems, etc.)
- Leading known cause of mental retardation (fetal alcohol syndrome [FAS])
- Genetic vulnerability activated by environment and behavior
- Capacity to tolerate alcohol confers greatest risk

Medical Complications

Cirrhosis, alcoholic hepatitis, pancreatitis, gastric or duodenal ulcer, esophageal varices, middle-age onset of diabetes, gastrointestinal cancer, hypertension, peripheral neuropathies, myopathies, cardiomyopathy, cerebral vascular accidents, erectile dysfunction, gout, vitamin deficiencies, pernicious anemia, and brain disorders, including Wernicke-Korsakoff syndrome (mortality rate of untreated Wernicke encephalopathy is 50%; treatment is with thiamine)

Main Treatment: Alcoholics Anonymous

Twelve-step program, consists of meetings and sponsors; least expensive; believed to be most successful

Pharmacologic Treatment

Drug	Description
Disulfiram (Antabuse®)	Inhibits **aldehyde dehydrogenase**; decreases alcohol consumption; produces symptoms of nausea, chest pain, hyperventilation, tachycardia, vomiting; effective for short-term treatment only
Acamprosate	Helps prevent relapse; hypothesized to decrease glutamate receptor sensitivity
Benzodiazepines	Helps prevent alcohol-related seizures
Naltrexone	Opiate-receptor antagonist; reduces cravings; helps patient to stop after first drink

Alcohol Effects and Metabolism

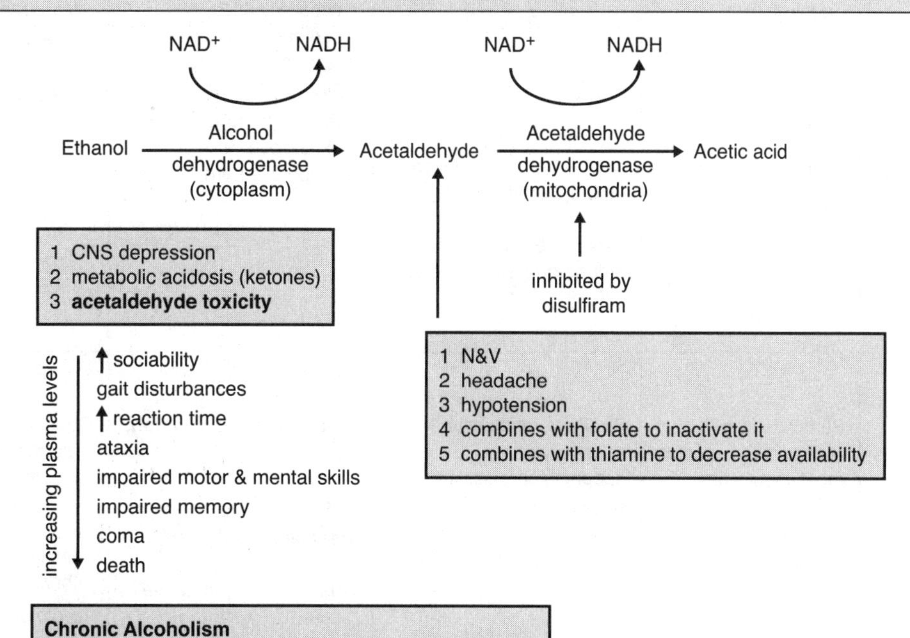

Withdrawal
• The chronic user can experience an alcohol withdrawal syndrome upon discontinuance of ethanol consumption. **Symptoms** include anxiety, tremor, insomnia, and possibly **delirium tremens (DTs)** and **life-threatening seizures**; arrhythmias, nausea/vomiting, and diarrhea may also occur. • **Treatment** includes thiamine, sedative-hypnotics with gradual tapering; clonidine and propranolol are useful to correct the hyperadrenergic state of withdrawal; lorazepam is used for seizures.

▶ Drugs of Abuse

Class/Agent	Mechanisms	Effects	Toxicity	Withdrawal
CNS stimulants				
Cocaine (crack is smokable form)	Blocks DA, NE, 5HT reuptake (also local anesthetic)	Euphoria, hypervigilance, anxiety, stereotyped behavior, grandiosity, tachycardia, pupillary dilation, ↓ appetite	Cardiac arrhythmias, myocardial infarction, stroke, hallucinations, paranoia, hyperthermia, seizures, death *Treatment*: BZs, neuroleptics; control hyperthermia and CV effects, supportive care	Craving, depression, fatigue, ↑ sleep time, ↑ appetite
Amphetamines (speed, crystal meth; ice is smokable form)	Releases DA, NE, 5HT, weak MAO inhibitor **Both:** NE most important for peripheral effects; DA most important for central effects			
CNS depressants				
Benzodiazepines	↑ frequency of $GABA_A$ channel opening	Light to moderate CNS depression	Impaired judgment, slurred speech, uncoordination, unsteady gait, stupor, respiratory depression, death *Treatment*: **flumazenil** for BZ overdose, supportive care	Anxiety, delirium, insomnia, possible life-threatening seizures *Treatment*: long-acting BZ to suppress acute symptoms, taper dose
Barbiturates	↑ duration of $GABA_A$ channel opening	Can produce light to severe CNS depression (EtOH, barbiturates) Cross-tolerance, additivity among drugs of this class		
Ethanol (EtOH) (*see* previous table)				
Opioids				
(heroin, morphine, oxycodone)	Stimulate μ, κ, δ receptors; μ receptor most important in abuse	Euphoria, analgesia, sedation, cough suppression, miosis (except meperidine), constipation	**Respiratory depression**, nausea/vomiting, **miosis**, sedation, **coma**, death *Treatment:* naloxone (short half-life), naltrexone, supportive care	Gooseflesh ("cold turkey"), diarrhea, rhinorrhea, lacrimation, sweating, yawning, muscle jerks ("kicking the habit"), very unpleasant but not life-threatening *Treatment*: methadone, LAAM, buprenorphine, clonidine

(Continued)

▶ Drugs of Abuse *(Cont'd.)*

Class/Agent	Mechanisms	Effects	Toxicity	Withdrawal
Cannabis				
(marijuana ["grass," hashish])	Binds CB_1 and CB_2 cannabinoid receptors; THC is the most active component	Euphoria, disinhibition, perceptual changes, reddened conjunctiva, dry mouth, ↑ appetite, antiemetic effects (**dronabinol** used as antiemetic)	Amotivational syndrome, respiratory effects	Mild irritability/anxiety
Hallucinogens				
(LSD, mescaline, psilocybin)	Interacts with 5HT receptors	Perceptual changes, synesthesias (i.e., hearing a smell), nausea	Panic reaction ("bad trip") possible, flashbacks	Minimal because of lack of physiologic dependence

Miscellaneous	
Phencyclidine (PCP, "angel dust")	• Assaultive, combative, impulsive, agitated, nystagmus, ataxia, muscle rigidity, ↓ response to pain, hyperacusis, paranoia, unpredictable violence, psychosis, hypertension, life-threatening seizures • **Ketamine**, a congener of PCP, also abused
MDMA ("ecstasy")	• 5HT releasers (amphetamine-like mechanism, except releases more 5HT than dopamine) • May cause damage to serotonergic neurons; causes hyperthermia; popular in "raves"
Anticholinergics	Deliriant effects, e.g., Jimson weed, scopolamine; psychotic and anticholinergic effects
Nicotine	• Tobacco use associated with cardiovascular, respiratory, and neoplastic disease • Nicotine patches and gum; bupropion, varenicline, and bromocriptine are used for cessation
Inhalants (glue, solvents)	Belligerence, impaired judgement, uncoordination; causes multiple organ damage

Definition of abbreviations: BZ, benzodiazepine; DA, dopamine; 5HT, serotonin; LAAM, levo-α-acetylmethadol; LSD, lysergic acid diethylamide; MDMA, methylenedioxymethamphetamine; NE, norepinephrine; CV, cardiovascular; THC, tetrahydrocannabinol.

Defense Mechanisms

▶ Common Freudian Defense Mechanisms

Defense Mechanism	Short Definition	Important Associations
Projection	Attributing inner feelings to others	Paranoid behavior
Denial	Saying it is not so	Substance abuse, reaction to death
Splitting	The world composed of polar opposites	Borderline personality; good versus evil
Blocking	Transient inability to remember	Momentary lapse
Regression	Returning to an earlier stage of development	Enuresis, primitive behaviors
Somatization	Physical symptoms for psychological reasons	Somatoform disorders
Introjection	The outside becomes inside	Superego, being like parents
Displacement	Source stays the same, but target changes	Redirected emotion, phobias, scapegoat
Repression	Forgetting so it is nonretrievable	Forget and forget
Isolation of affect	Facts without feeling	Blunted affect, *la belle indifference*
Intellectualization	Affect replaced by academic content	Academic, not emotional, reaction
Acting out	Affect covered up by excessive action or sensation	Substance abuse, fighting, gambling
Rationalization	Why the unacceptable is okay in this instance	Justification, string of reasons
Reaction formation	The unacceptable transformed into its opposite	Manifesting the opposite; feel love but show hate: "Girls have cooties."
Undoing	Action to symbolically reverse the unacceptable	Fixing or repairing, obsessive–compulsive behaviors
Passive-aggressive	Passive nonperformance after promise	Unconscious, indirect hostility
Dissociation	Separating self from one's own experience	Fugue, depersonalization, amnesia, multiple personality
Humor	A pleasant release from anxiety	Laughter hides the pain
Sublimation	Moving an unacceptable impulse into an acceptable channel	Art, literature, mentoring
Suppression	Forgetting, but is retrievable	Forget and remember

Psychopathology

► Five Major Diagnostic Axes

Axis I	**Clinical disorders**	• Includes schizophrenic, affective, anxiety, and somatoform disorders • Also includes anorexia nervosa, bulimia nervosa, sexual disorders, sleep disorders, and autism
Axis II	**Personality disorders and mental retardation**	Personality disorders and mental retardation
Axis III	**Physical conditions and disorders**	Any physical diagnosis
Axis IV	**Psychosocial and environmental problems**	Includes primary support group, social occupation, education, housing, economics, health care services, and legal issues
Axis V	**Global assessment of functioning (GAF)**	Scored on a descending scale of 100 to 1, where 100 represents superior functioning, 50 represents serious symptoms, and 10 represents persistent danger of hurting self or others

► Mental Retardation

Level	IQ	Functioning
Mild	70 to 50	• Self-supporting with some guidance • 85% of retarded persons • Two times as many males as females • Usually diagnosed first year in school
Moderate	49 to 35	• "Trainable" • Benefit from vocational training, but need supervision • Sheltered workshops
Severe	34 to 20	• Training not helpful • Can learn to communicate • Basic habits
Profound	Below 20	• Need highly structured environment, constant nursing care supervision

► Pervasive Developmental Disorders: Autism

General characteristics	• Usually diagnosed during age 2 • Male:female ratio—4:1 • 80% have IQs below 70
Clinical signs	• Problems with reciprocal social interaction • Abnormal/delayed language development, impaired verbal and nonverbal communication • No separation anxiety • Oblivious to external world • Fails to assume anticipatory posture, shrinks from touch • Preference for inanimate objects • Stereotyped behavior; decreased repertoire of activities and interests
Potential causes	• Association with prenatal and perinatal injury, e.g., rubella in first trimester • Possible role of environmental mercury exposure • Proposed mechanism: failure of apoptosis in cortex • Not linked to MMR vaccine
Treatment	• Behavioral techniques (shaping)

▶ Attention Deficit/Hyperactivity Disorder (ADHD)

General comments	• Male to female ratio—10:1 • Overtreatment is common; differentiate from child who is simply "overactive"
Clinical signs	• *Difficulty sustaining attention:* difficulty with organization, easily distracted, often does not listen when spoken to, doesn't complete tasks • *Hyperactivity:* fidgets, often leaves seat in classroom, difficulty playing quietly, talks excessively • *Impulsivity:* interrupts others, blurts out answers, difficulty waiting turn
Treatment	• Behavior therapy • Drug therapy: methylphenidate, dextroamphetamine, atomoxetine

▶ Schizophrenia

Criteria	• Bizarre delusions • Auditory hallucinations (in 75%) • Blunted affect • Loose associations • Deficiency in reality testing, distorted perception; impaired functioning overall • Disturbances in behavior and form and content of language and thought • Changes in psychomotor behavior; loss of prosody • Symptoms for longer than 6 months; <6 months = schizophreniform • Brief psychotic disorder = <30 days, full return to previous function
Epidemiology	• Onset: males, ages 15 to 24; females, ages 25 to 34 • Prevalence: 1% of population cross-culturally • More often in low social economic status • 50% patients attempt suicide; 10% succeed
Neurochemistry	• **"Dopamine hypothesis of schizophrenia"** suggests that symptoms arise because of a functional excess of dopamine activity in the CNS (mesolimbic/mesocortical pathways) • Serotonin and glutamate also play roles
Subtypes	
Paranoid	• Delusions of **persecution** or **grandeur** • Often accompanied by hallucinations (voices)
Catatonic	• **Complete stupor** or pronounced decrease in spontaneous movements • Alternatively, **can be excited** and evidence extreme motor agitation
Disorganized	• Incoherent, primitive, uninhibited • Unorganized behaviors and speech • Active, but aimless • Pronounced thought disorder
Undifferentiated	• Psychotic symptoms but does not fit paranoid, catatonic, or disorganized diagnoses
Residual	• Previous episode, but no prominent psychotic symptoms at evaluation • Some lingering negative symptoms

▶ Antipsychotic Medications: An Overview

Clinical uses	Antipsychotics (neuroleptics) are used in a variety of clinical settings, including schizophrenia, schizoaffective disorders, mania (lithium, initial management), Tourette syndrome (haloperidol, pimozide), sedation (promethazine), drug or radiation emesis (prochlorperazine), and neuroleptic anesthesia (droperidol).
Anatomic targets	Antipsychotics unfortunately affect all dopaminergic tracts in the brain, including the **mesolimbic/ mesocortical** (thought and mood), the **nigrostriatal** (extrapyramidal motor), and **tuberoinfundibular** (dopamine inhibits prolactin release) pathways. The goal is to block the mesolimbic/mesocortical pathways, but side effects result from blocking the nigrostriatal (extrapyramidal side effects **[EPS]**) and tuberoinfundibular (**hyperprolactinemia**) pathways.
Typical antipsychotics	This older group of antipsychotics falls into two general categories: **high potency** and **low potency** (see table below). Their mechanism of action is thought to be related primarily to their ability to block D_2 receptors.
Atypical antipsychotics	Newer group of antipsychotics; in general, weaker D_2 receptor antagonists and stronger **$5HT_2$** antagonists. They also have **less EPS**.

▶ Typical Antipsychotics: "Haloperidol and the -zines"

- The **high-potency** drugs block DA receptors well (cause more EPS), so a low drug dose can be used, minimizing other nonspecific side effects (α **blockade, antimuscarinic effects, sedation**).
- The **low-potency** drugs do not block DA receptors as well as high-potency drugs do, and a higher dose is required. Therefore, there is less EPS and more nonspecific side effects.
- The nonspecific side effects include orthostatic hypotension, male sexual dysfunction (α blockade); constipation, dry mouth, urinary retention, visual problems (muscarinic blockade); and sedation.
- **Hyperprolactinemia** may occur due to D_2-receptor blockade in the pituitary.
- **Neuroleptic malignant syndrome**, which is potentially life-threatening, may occur. Symptoms include extreme muscle rigidity, hyperthermia, and autonomic instability. Treatment may include **dantrolene** and **dopamine agonists**.
- Weight gain and a decrease in the seizure threshold may also occur.

Potency	EPS	Nonspecific Side Effects
High (haloperidol, fluphenazine)	High	Low
Low (chlorpromazine, thioridazine*)	Low	High

Definition of abbreviations: DA, dopamine; EPS, extrapyramidal side effects.

*Thioridazine can cause retinal deposits, leading to visual problems, and conduction defects that may result in fatal ventricular arrhythmias.

▶ Extrapyramidal Side Effects

	Early Onset and Reversible
Dystonia	• Involuntary contraction primarily of the face, neck, tongue, and extraocular muscles • Responds to anticholinergics • Peak: 1 week
Parkinsonism	• Akinesia, muscle rigidity, tremor, shuffling gait (typically appearing in that order) • Peak: akinesia, 2 weeks; rigidity, 3 weeks; tremors, 6 weeks
Akathisia	• Motor restlessness and the urge to move • Give benzodiazepines • Peak: 10 weeks
	Late Onset and Irreversible
Tardive dyskinesia (TD)	• Diagnosis requires exposure to neuroleptics for at least 3 months, but often takes longer • Involuntary repetitive movements of lips, face, tongue, limbs • Try to prevent by using lowest possible dose of antipsychotic medication • Anticholinergics worsen TD • Try to reduce dose or discontinue medication if TD occurs (although increasing dose will temporarily mask symptoms) • Switch to an atypical antipsychotic

▶ Atypical Antipsychotics: "-dones and -pines"

Drugs	Characteristics	Side effects
Clozapine* Risperidone Olanzapine Quetiapine Ziprasidone Aripiprazole	• Generally good **$5HT_2$ antagonists** and weaker D_2 antagonists • Treats positive and negative symptoms of schizophrenia (typicals treat mostly positive) • **Less EPS** • Tend to be more expensive than typicals	• **Clozapine**: **agranulocytosis** • Ziprasidone: prolongs QT interval, may lead to torsades • Risperidone: some EPS

*Clozapine blocks D_4, rather than D_2, receptors.

▶ Mood Disorders: Overview

	Mild	Severe
Stable	Dysthymia	Unipolar (major depression)
Alternating	Cyclothymia	Bipolar (manic depression)

▶ Mood Disorder Subtypes

Types	Characteristics
Dysthymia	• Depressed mood • Loss of interest or pleasure (anhedonia) • Chronic (at least 2 years)
Cyclothymia (nonpsychotic bipolar)	• Alternating states • Chronic • Often not recognized by affected person
Seasonal affective disorder (SAD)	• Depressive symptoms during winter months (shortest days, so least amount of light) • Caused by abnormal melatonin metabolism • Treat with bright light therapy
Unipolar depression (major depression)	• Symptoms for at least 2 weeks • Must be a change from previous functioning • May be associated with anhedonia, no motivation, feelings of worthlessness, decreased concentration, weight loss or gain, depressed mood, recurrent thoughts, insomnia or hypersomnia, psychomotor agitation or retardation, somatic complaints, delusions or hallucinations (mood congruent), loss of sex drive • Diurnal improvement as day progresses • Suicide: 60% of depressed patients have suicidal ideation; 15% die by suicide • Neurochemistry: "biogenic amine theory of depression"—caused by decreased NE/5HT • Sleep: ↑ REM in first half of sleep, ↓ REM latency, ↓ stage 4 sleep, ↑ REM time overall, early morning wakening
Bipolar disorder (manic-depression)	• Symptoms of major depression and symptoms of mania (period of abnormal and persistent elevated, expansive, or irritable mood) • Disability most linked to depressive phase • Subtypes: – **Bipolar I:** mania more prominent – **Bipolar II:** recurrent depressive episodes, plus hypomanic episodes – **"Rapidly cycling bipolar disorder":** if alternates within 48–72 hours • Manic symptoms: ↑ self-esteem or grandiosity, low frustration tolerance, ↓ need for sleep, flight of ideas, excessive involvement in activities, weight loss and anorexia, erratic and uninhibited behavior, ↑ libido

Definition of abbreviations: 5HT, serotonin; NE, norepinephrine; REM, rapid eye movement.

▶ Normal Grief Versus Depression

Normal Grief	Depression
Normal up to 1 year	Longer than 1 year, sooner if symptoms severe
Crying, ↓ libido, weight loss, insomnia	Same
Longing, wish to see loved one, may think they hear or see loved one in a crowd	Abnormal overidentification, personality change
Loss of other	Loss of self
Suicidal ideation rare	Suicidal ideation common
Self-limited, usually <6 months	Symptoms do not stop (may persist for years)
Antidepressants not helpful	Antidepressants helpful

▶ Kübler-Ross Stages of Adjustment

- **Stages:** denial, anger, bargaining, depression, and acceptance
- People can move back and forth through the stages
- Describes people who are dying or people who are dealing with loss or separation

▶ Antidepressants

Class/Agents	Mechanism	Side Effects and Comments
Tricyclic antidepressants (TCAs) (amitriptyline, imipramine, nortriptyline, desipramine)	Block reuptake of NE and 5HT	• Anticholinergic • Alpha blockade • Sedation • ↓ seizure threshold • **Overdose:** triad ("3 Cs"): coma, convulsions, cardiotoxicity • **Drug interactions:** do not mix with SSRIs and MAOIs; potentially fatal
Heterocyclics (amoxapine, bupropion, maprotiline, trazodone, mirtazapine, nefazodone, venlafaxine)	Mechanism varies	• Trazodone causes priapism, sedation • Amoxapine causes EPS (also dopamine receptor blocker) • Maprotiline, amoxapine: seizures, cardiotoxicity • Nefazodone, venlafaxine: P450 inhibitors • Bupropion used in smoking cessation • Mirtazepine used for depression concomitant with anorexia nervosa; causes weight gain
Selective serotonin reuptake inhibitors (SSRIs) (fluoxetine, sertraline, citalopram, fluvoxamine, paroxetine; escitalopram)	Blocks reuptake of 5HT	• Anxiety, agitation, insomnia • Nausea • Sexual dysfunction • **Serotonin syndrome** (muscle rigidity, hyperthermia, myoclonus, ANS instability), seizures occurs with **TCAs**, **MAOIs**, **meperidine**
MAO inhibitors (MAOIs) (phenelzine, tranylcypromine)	Interferes with metabolism of NE and 5HT by blocking monoamine oxidase (MAO) types A and B	• Orthostatic hypotension, weight gain • **Hypertensive crisis** if patient consumes food with **tyramine** and other indirect-acting sympathomimetics • **Serotonin syndrome** when combined with SSRIs

▶ Electroconvulsive Therapy (ECT)

- **Common uses:**
  - Depression (80%)
  - Schizoaffective disorder (10%)
  - Bipolar disorder
- 90% show some immediate improvement
- Usually requires 5 to 10 treatments
- Treats depressive episodes, not for prophylaxis

- **Side effects:**
  - Anesthesia eliminates fractures and anticipatory anxiety
  - Memory loss and headache common, returns to normal in several weeks
  - Serious complications <1:1,000
- Contraindication: ↑ cranial pressure (e.g., tumor)

▶ Lithium

- Used in **bipolar disorder**
- Mechanism not well understood but may inhibit the recycling of neuronal **phosphoinositides**
- Used in conjunction with antidepressants
- Neuroleptics and/or benzodiazepines may be used initially because lithium has very slow onset
- **Very narrow therapeutic index**
- Side effects: sedation, ataxia, tremor, reversible nephrogenic diabetes insipidus, edema, acne, leukocytosis
- Neonatal toxicity if administered to pregnant women
- Other drugs used in bipolar disorder: valproic acid, atypical antipsychotics

▶ Eating Disorders

Characteristics	Anorexia Nervosa	Bulimia Nervosa
Sex	F > M	F > M
Age	Mid-teenage years	Late adolescence/early adulthood
Socioeconomic status (SES)	May be high or low	May be high or low
Weight*	>15% ideal body weight loss	Varies, usually normal or > normal
Neurotransmitters	Serotonin/norepinephrine?	Serotonin/norepinephrine?
Binge/purge	Yes	Yes
Laxative/diuretics	Yes	Yes
Sexual adjustment	Poor	Good
Medical complications	• Amenorrhea • Lanugo • High mortality • Dental cavities • Electrolyte imbalances • Cardiac abnormalities	• Dental cavities • Calluses on hands/fingers • Enlarged parotid glands • Electrolyte imbalances • Cardiac abnormalities

* **Body mass index (BMI)** = weight(kg)/height^2 (meter)^2 ; Underweight: <18.5, Normal: 18.5-24.9; Overweight: 25.0-29.9; Obese: > 30.0; Extremely obese: > 40.0

► Anxiety Disorders

Generalized anxiety disorder	• Symptoms exhibited more days than not for longer than a 6-month period – Motor tension (fidgety, jumpy) – Autonomic hyperactivity (heart pounding, sweating, chest pains), hyperventilation – Apprehension (fear, worry, rumination), difficulty concentrating – Vigilance and scanning (impatient, hyperactive, distracted) – Fatigue and sleep disturbances common, especially insomnia and restlessness • **Treatment:** benzodiazepines, buspirone
Specific phobias (fear of specific object, e.g., spiders, snakes)	• Anxiety when faced with identifiable object • Phobic object avoided • Persistent and disabling fear
Agoraphobia (fear of open spaces or places from which escape is difficult)	• Also sense of helplessness or humiliation • Manifest anxiety, panic-like symptoms • Travel restricted
Social phobia (fear of feeling or being stupid, shameful)	• Leads to dysfunctional circumspect behavior, e.g., inability to urinate in public washrooms • May accompany avoidant personality disorder • Discrete performance anxiety (stage fright): most common phobia; treat with paroxetine (SSRI) or atenolol or propranolol (beta blocker) • **Treatment:** paroxetine (SSRI) or atenolol or propranolol (beta blocker); for generalized social anxiety, use phenelzine (MAO inhibitor) or paroxetine
Obsessive–compulsive disorder	• **Obsession:** focusing on one thought, usually to avoid another • **Compulsion:** repetitive action shields person from thoughts, action "fixes" bad thought • Primary concern of patient is to not lose control • ↑ frontal lobe metabolism, ↑ activity in the caudate nucleus • **Treatment:** fluoxetine, fluvoxamine, or other SSRI, clomipramine
Panic disorder	• Three attacks in 3-week period with no clear circumscribed stimulus • Abrupt onset of symptoms, peak within 10 minutes • **Clinical signs:** – Great apprehension and fear – Palpitations, trembling, sweating – Fear of dying or going crazy – Hyperventilation, "air hunger" – Sense of unreality • **Treatment:** alprazolam, clonazepam, imipramine

► Sedative-Hypnotic Drugs

Class	Notes	
Benzodiazepines	• Used as anxiolytics, hypnotics, anticonvulsants (diazepam, lorazepam, clonazepam), muscle relaxants, for anesthesia (e.g., midazolam) • Binds $GABA_A$ receptor and increases frequency of Cl^- ion channel opening • Dose-dependent CNS depression occurs (not as much as barbiturates when used alone) • Differ in half-life and metabolism • Three BZs are not metabolized in liver: ("Out The Liver: **O**xazepam, **T**emazepam, **L**orazepam)	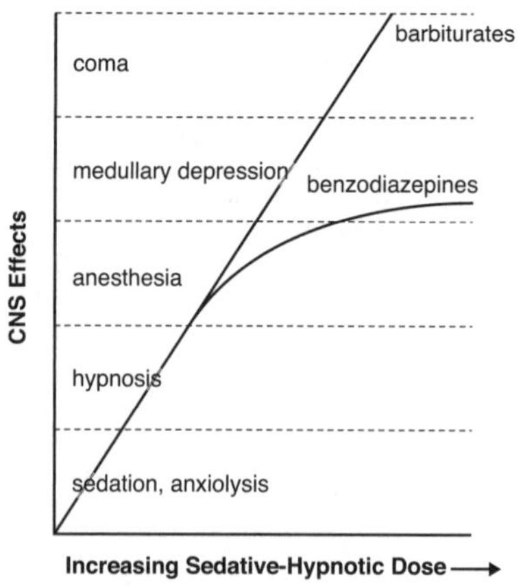 barbiturates coma medullary depression benzodiazepines anesthesia hypnosis sedation, anxiolysis CNS Effects Increasing Sedative-Hypnotic Dose →
Barbiturates	• Used as anticonvulsants (phenobarbital, long-acting), to induce anesthesia (thiopental, short-acting) • Binds $GABA_A$ receptor and increases duration of Cl^- ion channel opening	
Miscellaneous	**Zolpidem**, **zaleplon**, and **eszopiclone**—(all bind BZ_1 receptors) and **ramelteon** (melatonin receptor agonist) used for sleep **Buspirone**—nonbenzodiazepine (anxiolytic)	

▶ Important Benzodiazepines

Generic Name	Common Uses
Alprazolam	Panic, anxiety
Chlordiazepoxide	Alcohol detoxification
Clonazepam	Panic, anxiety, seizures
Diazepam	Anxiety, insomnia, pre-op sedation, muscle relaxation
Flurazepam	Insomnia
Lorazepam	Anxiety, alcohol-related seizures
Midazolam*	Anesthesia
Oxazepam	Alcohol detoxification
Temazepam	Insomnia
Triazolam†	Insomnia

*Shortest acting
†Short acting

▶ Somatoform Disorders

Somatization disorder	• Set of eight or more symptoms (four pain, two gastrointestinal, one sexual, one pseudoneurologic) • Onset before age 30 • Symptoms usually occur over period of years • More common in women than in men (20 to 1)
Conversion disorder	• One or more symptoms • Altering of physical functioning, suggesting physical disorder • Usually skeletal, muscular, sensory, or some peripheral nonautonomic system, e.g., paralysis of the hand, loss of sight • Loss of functioning is real and unfeigned • Look for *la belle indifference*
Hypochondriasis	• Unrealistic interpretation of physical signs as abnormal • Preoccupation with illness or fear of illness when none present • Preoccupation persists in spite of reassurance • At least 6 months' duration • Treat by simple palliative care and fostering relationship
Somatoform pain disorder	• Severe, prolonged pain with no organic cause found • Pain disrupts day-to-day life • Look for secondary gain
Body dysmorphic disorder	• Preoccupation with unrealistic negative evaluation of personal attractiveness • Sees self as ugly or horrific when normal in appearance • Preoccupation disrupts day-to-day life • May seek multiple plastic surgeries or other extreme interventions • May occur in eating disorders

▶ Differentiating Somatoform Disorders from Factitious Disorders and Malingering*

	Somatoform	Factitious	Malingering
Symptom production	Unconscious	Intentional	Intentional
Motivation	Unconscious	Unconscious	Intentional

*All three may present with similar symptom profile. The key to the differential is level of patient awareness.

► Personality Disorders

Types	Definition	Epidemiology	Associated Defenses
Cluster A: Odd or eccentric			
Paranoid	Feelings of persecution; feels that others are conspiring to harm them; suspicious	• Men > women • Increased incidence in families with schizophrenia	Projection
Schizoid	Isolated lifestyle; has no longing for others ("loner")	• Men > women • Increased incidence in families with schizophrenia	—
Schizotypal	Eccentric behavior, thought, and speech	• Prevalence is 3% • Men > women	—
Cluster B: Dramatic and emotional			
Histrionic	Excessive emotion and attention seeking	• Women > men • Underdiagnosed in men	• Regression • Somatization • Conversion • Dissociation
Narcissistic	Grandiose; overconcerned with issues of self-esteem	• Common	Fixation at subphase of separation/individualization
Borderline	Instability of mood, self-image, and relationships	• Women > men • ↑ mood disorders in families	• Splitting • Projective identification • Dissociation • Passive-aggression
Antisocial	Does not recognize the rights of others	Prevalence: 3% in men; 1% in women	Superego lacunae
Cluster C: Anxious and fearful			
Avoidant	Shy or timid; fears rejection	• Common • Possible deforming illness	Avoidance
Dependent	Dependent, submissive	• Common • Women > men • May end up as abused spouse	—
Obsessive-compulsive	Perfectionistic and inflexible, orderly, rigid	• Men > women • ↑ concordance in identical twins	• Isolation • Reaction formation • Undoing • Intellectualization

▶ Delirium Versus Dementia

Characteristics	Delirium	Dementia
History	Acute, identifiable date	Chronic, cannot be dated
Onset	Rapid	Insidious
Duration	Days to weeks	Months to years
Course	Fluctuating	Chronically progressive
Level of consciousness	Fluctuating	Normal
Orientation	Impaired periodically	Disorientation to person
Memory	Recent memory markedly impaired	Remote memories seen as recent
Perception	Visual hallucinations	Hallucinations less common
Sleep	Disrupted sleep-wake cycle	Less sleep disruption
Reversibility	Reversible	Mostly irreversible
Physiologic changes	Prominent	Minimal
Attention span	Very short	Not reduced

▶ Common Abnormalities on Neurologic Examination

Original Drawing	Patient's Drawing	Name	Localization
	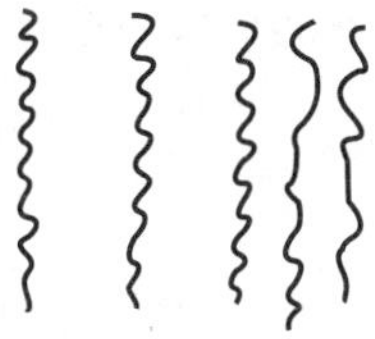	Perseveration	Frontal lobe
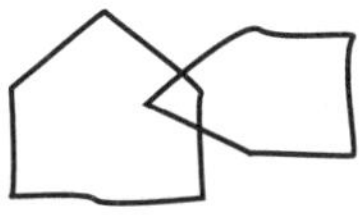		Constructional apraxia	Nondominant (right) parietal lobe
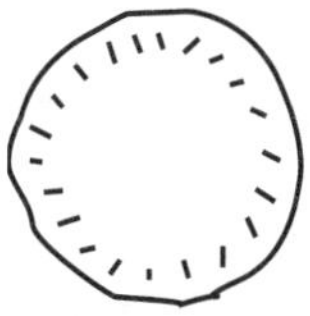		Hemineglect/ hemi-inattention	Right parietal lobe

Sexual Disorders

► Paraphilias

Pedophilia	Sexual urges toward children; most common sexual assault
Exhibitionism	Recurrent desire to expose genitals to strangers
Voyeurism	Sexual pleasure from watching others who are naked, grooming, or having sex; begins early in childhood
Sadism	Sexual pleasure derived from others' pain
Masochism	Sexual pleasure derived from being abused or dominated
Fetishism	Sexual focus on objects, e.g., shoes, stockings *Variant:* transvestite fetishism (fantasies or actual dressing by heterosexual men in women's clothing for sexual arousal)
Frotteurism	Male rubbing of genitals against fully clothed woman to achieve orgasm; subways and buses
Zoophilia	Animals preferred in sexual fantasies or practices
Coprophilia	Combining sex and defecation
Urophilia	Combining sex and urination
Necrophilia	Preferring sex with cadavers
Hypoxyphilia	Altered state of consciousness secondary to hypoxia while experiencing orgasm *Variants:* autoerotic asphyxiation, poppers, amyl nitrate, nitric oxide

► Gender Identity and Preferred Sexual Partner of a Biologic Male

Common Label	Gender Identity	Preferred Sexual Partner
Heterosexual	Male	Female
Transvestite fetishism	Male	Female
Gender identity disorder (transsexual)	Female	Male
Homosexual	Male	Male

► Sexual Disorders

Disorders of Sexual Desire	
Hypoactive	Deficiency or absence of fantasies or desires, 20% of population, more common in women
Sexual aversion	Aversion to all sexual contact
Sexual Arousal Disorders	
Female sexual arousal disorder	• As high as 33% of females; sometimes hormonally related • Antihistamine and anticholinergic medications ↓ vaginal lubrication
Male erectile disorder (impotence)	• *Primary:* never able to achieve erection • *Secondary:* previously able to achieve erection – Up to 20% lifetime prevalence; point prevalence 3% – 50% of men treated for sexual disorders, incidence ↑ with age, more likely in smokers
Orgasm Disorders	
Anorgasmia (inhibited female orgasm)	• 5% of married women older than 35 have never achieved orgasm • Overall prevalence from all causes: 30% • Likelihood to have orgasm ↑ with age
Inhibited male orgasm (retarded ejaculation)	• Usually restricted to inability to orgasm in the vagina • 5% general prevalence • Differentiate from retrograde ejaculation
Premature ejaculation	• Male regularly ejaculates before or immediately after entering vagina • *Treatments:* stop and go technique, squeeze technique, SSRIs
Sexual Pain Disorders	
Dyspareunia	• Recurrent and persistent pain before, during, or after intercourse in either man or woman • More common in women • Chronic pelvic pain is a common complaint of women raped or sexually abused
Vaginismus	• Involuntary muscle constriction of the outer third of the vagina • Prevents penile insertion • *Treatment:* relaxation, Hegar dilators

Physician–Patient Relationships

▶ General Rules

Think what the best physician should do, not necessarily what you have seen in practice.		
Rule #1:	Always place the interests of the patient first.	Make it a point to ask about and know the patient's wishes.
Rule #2:	Nothing should come between you and the patient.	• Get rid of tables and computers. • Ask family members to leave the room if necessary. • Family should not translate for family members.
Rule #3:	Tell the patient everything, even if he or she does not ask.	• The patient should know what you know and when you know it. • Information should flow through the patient to the family, not the reverse.
Rule #4:	Work on long-term relationships with patients, not just short-term problems.	• Every encounter is a chance to develop a better relationship. • Good relationships mean good medical practice. • Make eye contact; both patient and physician should both be sitting, if at all possible. Arrange the setting for comfortable, close communication. If patient is in room, talk to patient, not to colleagues. The patient is always the focus.
Rule #5:	Listening is better than talking.	• When patient talks, you are learning. • Take time to listen to the patient in front of you, even if other patients or colleagues are waiting. • Ask what the patient knows before explaining.
Rule #6:	The patient is the decision-maker.	• Patients make medical decisions, physicians do not. • Negotiate, do not order.
Rule #7:	Solve the problem presented; anticipate future problems.	• Find out what you need to; get the resources you need. • Change initial plans as information changes.
Rule #8:	Admit to the patient when you make a mistake.	Take responsibility; don't blame the nursing staff or a medical student.
Rule #9:	Never "pass off" your patient to someone else.	• Refer to psychiatrist or other specialist only when beyond your expertise. • Provide instruction in aspects of care, e.g., nutrition, use of medications.
Rule #10:	Express empathy, then give control.	• "I'm sorry, what would you like to do?" • Important rule to remember when faced with grieving or angry patient or upset family members.
Rule #11:	Agree on the problem with the patient before moving to the solution.	Informed consent requires the patient to fully understand what is wrong before treatment options are presented.
Rule #12:	Be sure you understand what the patient is talking about before intervening.	Seek information before acting, clarify emotionally loaded words, begin with open-ended questions, then move to closed-ended questions.
Rule #13:	Patients do not get to select inappropriate treatments.	Patients select treatments, but only from presented, appropriate choices.
Rule #14:	Best answers serve multiple goals.	• Think about patient health, relationships, and ethics for each answer. • The best answers solve both long and short term goals.
Rule #15:	Never lie.	• Do not lie to patients, their families, or insurance companies. • Do not deceive to protect a colleague.
Rule #16:	Accept the health beliefs of patients and talk to them in those terms.	• Be accepting of benign folk medicine practices. Expect them. • Diagnoses need to be explained in the way patients can understand, even if not technically precise.

(Continued)

► General Rules for Physician–Patient Relationships (*Cont'd.*)

Rule #17: Accept patients' religious beliefs and participate, if appropriate.	Religion is a source of comfort to many. Ask about a patient's religious beliefs if you are not sure.
Rule #18: Anything that increases communication is good.	• Take the time to talk with patients, even if others are waiting; ask why, not just what. • Seek information about the patient beyond the disease. • Ask about job, family, children, etc. • Ask "Is there anything else?"
Rule #19: Be an advocate for the patient.	• Work to get the patient what he or she needs. • Need, not payment, should decide.
Rule #20: How you do it matters as much as what you do.	• Focus on the process, not just goals. Means, not just the ends. • Do the right thing, the right way. • Treat family members with courtesy and tact, but the wishes and interests of the patient come first.

Ethical and Legal Issues

► General Rules

Rules	Comments
Rule #1: Competent patients have the right to refuse medical treatment.	• Patients have an almost absolute right to refuse. • Patients have almost absolute control over their own bodies.
Rule #2: Assume that the patient is competent unless clear behavioral evidence indicates otherwise.	Competence is a legal, not a medical issue. A diagnosis, by itself, tells you little about a patient's competence. Clear behavioral evidence would be: • Patient attempts suicide. • Patient is grossly psychotic and dysfunctional. • Patient's physical or mental state prevents simple communication.
Rule #3: Avoid going to court. Decision-making should occur in the clinical setting if possible.	Consider going to court only if: • There is intractable disagreement about a patient's competence, who should be the surrogate, or who should make the decision about life support. • You perceive a serious conflict of interest between surrogate and patient's interests. • Court approval of decision to terminate life support is, therefore, rarely required.
Rule #4: When surrogates make decisions for a patient, they should use the following criteria and in this order:	1. Subjective standard • Actual intent, advance directive • What did the patient say in the past? 2. Substituted judgment • Who best represents the patient? • What would patient say if he or she could? 3. Best interests standard • Burdens versus benefits • Interests of patient, not preferences of the decision-maker
Rule #5: If the patient is incompetent, physician may rely on advance directives.	• Advance directives can be oral. • Living will: written document expressing wishes • Health power of attorney: designating the surrogate decision-maker, "speaks with the patient's voice"

(Continued)

► General Rules About Ethical and Legal Issues (*Cont'd.*)

Rule #6:	Feeding tube is a medical treatment and can be withdrawn at the patient's request.	A competent person can refuse even lifesaving hydration and nutrition. This is not considered "killing the patient," but terminating treatment at the patient's request.
Rule #7:	Do nothing to actively assist the patient to die sooner.	• Passive, i.e., allowing to die is okay; active, i.e., killing is not okay • But do all you can to reduce the patient's suffering (e.g., giving pain medication).
Rule #8:	The physician decides when the patient is dead.	• What if there are no more treatment options (the patient is cortically dead), and the family insists on treatment? *If there are no options, there is nothing the physician can do; treatment must stop.* • What if the physician thinks continued treatment is futile (the patient has shown no improvement), but the surrogate insists on continued treatment? *The treatment should continue.*
Rule #9:	Never abandon a patient.	• Lack of financial resources or results are never reasons to stop the treatment of a patient. • An annoying or difficult patient is still your patient.
Rule #10:	Keep the physician-patient relationship within bounds	• Intimate social contact with anyone who is or has been a patient is prohibited. • Do not date family members of patients. • Do not treat family members or write prescriptions for colleagues. • When patients act inappropriately, make clear to them what appropriate behavior would be. • Any gift, beyond small tokens, should be declined.
Rule #11:	Stop harm from happening	• Beyond "do no harm": you must stop anyone from hurting your patient OR your patient from hurting anyone else. • Stopping harm may require breaching confidentiality. • Harm can mean spreading disease, physical assault, abuse, neglect, infliction of pain, etc.
Rule #12:	Always obtain informed consent.	• The patient must receive and understand five pieces of information: 1. Nature of procedure 2. Purpose or rationale 3. Benefits 4. Risks 5. Availability of alternatives • Four exceptions to informed consent: 1. Emergency 2. Waiver by patient 3. Patient is incompetent 4. Therapeutic privilege
Rule #13:	Special rules apply with children.	• Children younger than 18 years are minors and are legally incompetent. • Exceptions: emancipated minors – If patient is older than 13 years and taking care of self, i.e., living alone, treat as an adult. – Marriage makes a child emancipated, as does serving in the military. – Pregnancy or having a child, in most cases, does not. • Partial emancipation – Generally age 14 and older – Consent for certain issues only: Substance drug treatment Prenatal care Sexually transmitted disease treatment Birth control

(Continued)

► General Rules About Ethical and Legal Issues (*Cont'd.*)

Rule #14:	Parents cannot withhold life- or limb-saving treatment from their children.	If parents refuse permission to treat child: 1. If immediate emergency, go ahead and treat. 2. If not immediate, but still critical (e.g., juvenile diabetes), generally the child is declared a ward of the court and the court grants permission. 3. If not life- or limb-threatening (e.g., child needs minor stitches), listen to the parents.
Rule #15:	Organ donation usually requires patient's and family consent.	• The patient's advance directive is key. • Prior discussion with family members eliminates confusion regarding wishes. • If the family refuses, do not cause them stress by insisting.
Rule #16:	Good Samaritan Laws limit liability in nonmedical settings.	• Physician is not required to stop and help. • If help offered, shielded from liability provided: – Actions are within physician's competence. – Only accepted procedures are performed. – Physician remains at scene after starting therapy until relieved by competent personnel. – No compensation changes hands.
Rule #17:	Confidentiality is (almost always) absolute.	• Physicians cannot tell anyone anything about their patient without the patient's permission. • Physician must strive to ensure that others *cannot access* patient information. • Getting a consultation is permitted, as the consultant is bound by confidentiality, too. However, watch the location of the consultation. Be careful not to be overheard (e.g., in elevator or cafeteria). • If you receive a court subpoena, show up in court but do not divulge information about your patient. • If patient is a threat to self or others, the physician *must* break confidentiality. – Duty to warn and to protect (Tarasoff case) – A specific threat to a specific person – Suicide, homicide, and child and elder abuse are obvious threats – Infectious diseases may need to be reported to public officials or an innocent third party – Impaired drivers
Rule #18:	Patients should be given the chance to state DNR (do not resuscitate) orders, and physicians should follow them.	• DNR refers only to cardiopulmonary resuscitation. • Continue with ongoing treatments. • DNR decisions can be made by the patient or surrogate. • Have DNR discussions as part of your first encounter with the patient.
Rule #19:	Committed mentally ill patients retain their rights.	• Committed mentally ill adults are legally entitled to the following: – They must have treatment available. – They can refuse treatment. – They can command a jury trial to determine "sanity." • They lose only the civil liberty to come and go.
Rule #20:	Detain patients to protect them or others.	• Emergency detention can be effected by a physician and/or a law enforcement person for 48 hours, pending a hearing. • A physician can detain; only a judge can commit.
Rule #21:	Remove from patient contact health care professionals who pose risk to patients.	Types of risks: • Infectious disease (e.g., TB) • Substance abuse • Depression (or other psychological issues) • Incompetence
Rule #22:	Focus on what is the best ethical conduct, not simply the letter of the law.	The best conduct is both legal *and* ethical.

Epidemiology and Biostatistics

► Incidence and Prevalence

Incidence	$\text{Incidence rate} = \dfrac{\text{Number of } \textbf{new events} \text{ in a specified period}}{\text{Number of persons exposed to risk of acquiring the condition during this period}} \times \mathbf{10^n}$
Prevalence	$\text{Prevalence rate} = \dfrac{\text{All cases of a disease at a given point/period}}{\text{Total population at risk for having the condition at a given point or period}} \times \mathbf{10^n}$

► Types of Mortality Rates

Crude mortality rate	Deaths ÷ population
Cause-specific mortality rate	Deaths from cause ÷ population
Cause-fatality rate	Deaths from cause ÷ number of persons with the disease/cause
Proportionate mortality rate (PMR)	Deaths from cause ÷ all deaths

► Screening Results in a 2 × 2 Table

		Disease				
		Present		**Absent**		**Totals**
Screening Test Results	**Positive**	TP	60	FP	70	TP + FP
	Negative	FN	40	TN	30	TN + FN
	Totals	TP + FN		TN + FP		TP + TN + FP + FN

Sensitivity = TP/(TP + FN)	Detecting disease in population
Specificity = TN/(TN + FP)	Identifying healthy individuals in population
Positive predictive value = TP/(TP + FP)	What % of positive test results will be correct?
Negative predictive value = TN/(TN + FN)	What % of negative test results will be correct?
Accuracy = (TP + TN)/(TP + TN + FP + FN)	How good is the test overall?

Definition of abbreviations: FN, false negatives; FP, false positives; TN, true negatives; TP, true positives.

► Healthy and Diseased Populations Along a Screening Dimension

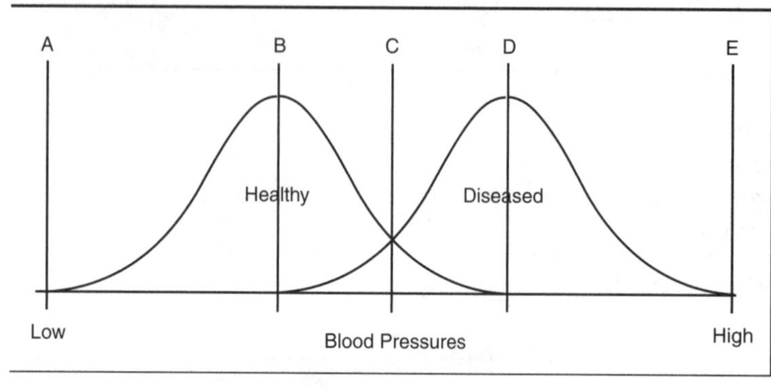

B, optimal sensitivity and optimal negative predictive value

D, optimal specificity and optimal positive predictive value

C, highest accuracy or lowest number of combined falses

Remember, we are looking for a cut-off score that best selects the desired group.

▶ Type of Bias in Research and Important Associations

Type of Bias	Definition	Important Associations	Solutions
Selection	Sample not representative	Berkson's bias, nonrespondent bias	Random, independent sample
Measurement	The process of gathering information distorts it	Hawthorne effect	Control group/placebo group
Experimenter expectancy	Researcher's beliefs affect outcome	Pygmalion effect	Double-blind design
Lead time	Early detection confused with increased survival	Benefits of screening	Measure "back end" survival
Recall	Subjects cannot remember accurately	Retrospective studies	Multiple sources to confirm information
Late-look	Severely diseased individuals are not uncovered	Early mortality	Stratify by severity
Confounding	Unanticipated factors obscure results	Hidden factors affect results	Multiple studies, good research design
Design	Parts of the study do not fit together	Non-comparable control group	Random assignment

▶ Differentiating Observational Studies

Characteristic	Cross-Sectional Studies	Case-Control Studies	Cohort Studies
Time	One time point	Retrospective	Prospective
Incidence	No	No	Yes
Prevalence	Yes	No	No
Causality	No	Yes	Yes
Role of disease	Measure disease	Begin with disease	End with disease
Assesses	Association of risk factor and disease	Many risk factors for single disease	Single risk factor affecting many diseases
Data analysis	Chi-square to assess association	Odds ratio to estimate risk **(Refer to Appendix A for equation.)**	Relative and attributable risk to estimate risk **(Refer to Appendix A for relative and attributable risk equations.)**

▶ Making Decisions Using *p*-Values

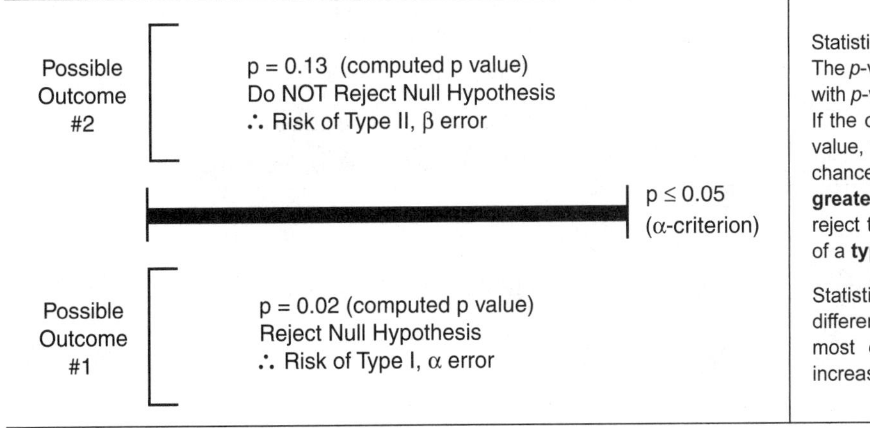

Statistical tests are used for making decisions. The *p*-value generated from the data is compared with *p*-value criterion selected by the investigator. If the computed value is **less than** the criterion value, then reject the null hypothesis, but with a chance of a **type I error**. If the computed value is **greater than** the criterion value, you should not reject the null hypothesis, but there is a chance of a **type II error**.

Statistical **power** is the capacity to detect a difference if it is present. **Power = 1–β.** The most common way to increase power is to increase sample size.

▶ Confidence Intervals

Confidence intervals (CI) estimate the **population** value based on the data from a **sample**. We give up precision, knowing exactly the population number, in exchange for confidence. Confidence intervals tell us that reality is most likely **within** the specified range.

Confidence interval of the mean	$\overline{X} \pm Z\left(\frac{S}{\sqrt{N}}\right)$	Where: $\overline{X}$ = sample mean Z = Z-score* S = standard deviation N = sample size *Z = 1.96 for 95% confidence Z = 2.58 for 99% confidence
Interpretation of Confidence Intervals		
Confidence intervals for the mean	If the CIs for two means overlap, then they could be the same. Therefore, we have no evidence that they are different. If the CIs do not overlap, then we usually assume that they are different (statistical significance). In general, any overlap in CIs indicates no difference.	
Confidence intervals for relative risk (RR) or odds ratios	If the CIs contain the number 1.0, then the population parameters compared in the ratio could be the same. Therefore, we cannot assume that they are different. If 1.0 is not included in the CI, then we assume that they are different (statistical significance). A 1.0 in the CI means that it is not significant.	

▶ Types of Scales in Statistics

Type of Scale	Description	Key Words	Examples
Nominal (categorical)	Different groups	"This" as opposed to "that"	Gender, comparing among treatment interventions
Ordinal	Groups in sequence	Comparative quality, rank order	Olympic medals, class rank in medical school
Interval	Exact differences among groups	Quantity, mean, and standard deviation	Height, weight, blood pressure, drug dosage
Ratio	Interval + true zero point	Zero means zero	Temperature measured in degrees Kelvin

▶ Types of Scales and Basic Statistical Tests

	Variables		
Name of Statistical Test	**Interval**	**Nominal**	**Comment**
Pearson correlation	2	0	Is there a linear relationship?
Chi-square	0	2	Any number of groups
t-test	1	1	Two groups only
One-way ANOVA	1	1	Two or more groups
Matched pairs *t*-test	1	1	Two groups, linked data pairs, before and after
Repeated measures ANOVA	1	1	More than two groups, linked data

Biochemistry

Chapter 2

Biochemistry

► Glycolysis

Glycolysis is a **cytoplasmic** pathway used by all cells to generate energy from glucose. **One** glucose molecule is converted into **2 pyruvate** molecules, generating a net of **2 ATPs** by substrate-level phosphorylation, and **2 NADHs**. When oxygen is present, NADH delivers electrons to the electron transport chain in mitochondria to generate ATP by oxidative phosphorylation. Under **anaerobic** conditions, lactate is generated and NADH is reoxidized to NAD^+.

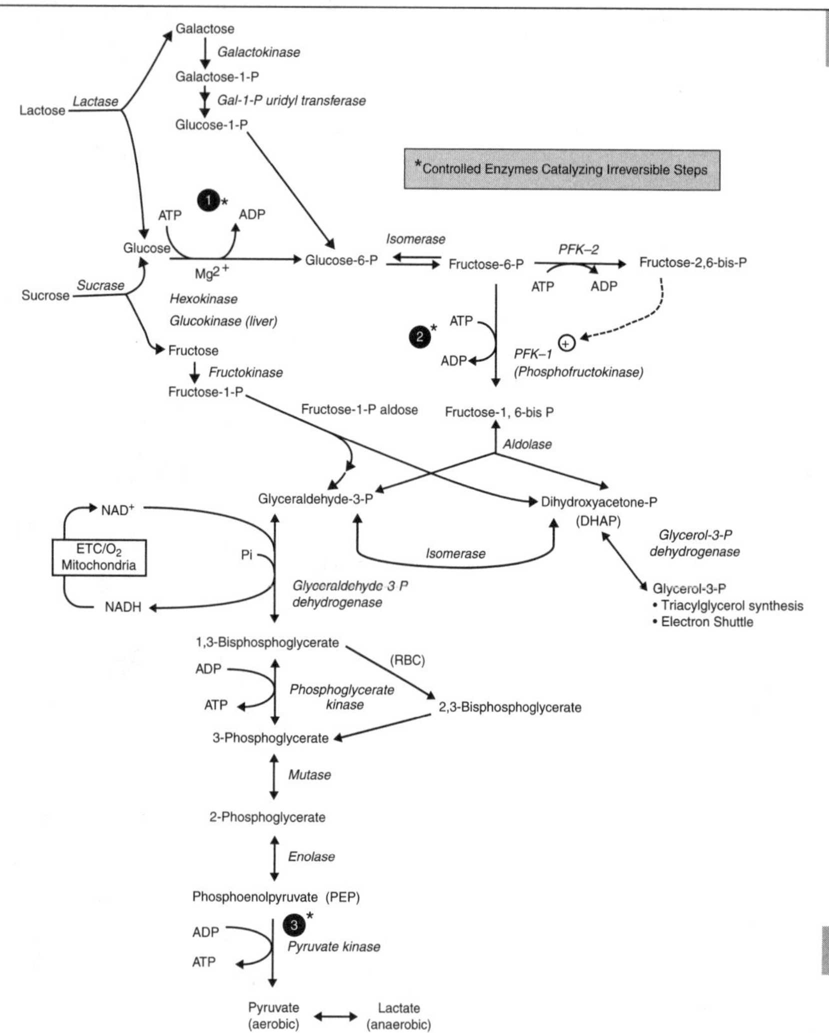

Regulation: Three irreversible steps

Hexokinase	Glucokinase*
Most tissues	Liver, β-islet cells
Low K_m	High K_m
⊖ G-6-P	Induced by insulin in liver

❷

PFK-1	PFK-2
→ F-1,6-BP	→ F-2,6-BP
Rate-limiting step of glycolysis	Glycolysis regulator: ↑ glycolysis ↓ gluconeogenesis
⊕ AMP ⊕ F-2,6-BP† ⊖ ATP ⊖ Citrate	⊕ Insulin ⊖ Glucagon

Pyruvate kinase	
⊕ F-1,6-BP	⊖ ATP ⊖ Alanine†

*Glukokinase mutations may lead to a form of MODY.
†Liver specific

Glucose Transport

GLUT-1 and -3: basal uptake (most cells)

GLUT-2: storage (liver); glucose sensor (β-islet)

GLUT-4: ↑ by insulin (adipose, skeletal muscle); ↑ by exercise (skeletal muscle)

Disease Association	
Galactokinase deficiency	Galactosemia/galactosuria, cataracts in childhood (excess galactose is converted to galactitol via aldose reductase); Tx: no galactose in diet
Gal-1-P uridyl transferase deficiency	Same as above, but more severe with vomiting/diarrhea after milk ingestion, liver disease, lethargy, mental retardation; Tx: no galactose in diet
Fructokinase deficiency	Fructosuria; benign
Fructose-1-P aldolase B deficiency	Fructosuria, liver and proximal renal tubule disorder; Tx: no fructose in diet
Pyruvate kinase deficiency	Chronic hemolysis, ↑ 2,3-BPG and other glycolytic intermediates in the RBC, no Heinz bodies, autosomal recessive

Definition of abbreviations: MODY, mature-onset diabetes of the young; PFK, phosphofructokinase; RBC, red blood cell; Tx, treatment.

► The Citric Acid Cycle

The citric acid cycle **(tricarboxylic acid cycle)** is a **mitochondrial** pathway that occurs **only** under **aerobic conditions**. Each acetyl-CoA generated from pyruvate is used to produce **3 NADH, 1 $FADH_2$**, and **1 GTP**. Both the NADH and $FADH_2$ deliver electrons to the electron transport chain (ETC) to generate ATP by oxidative phosphorylation.

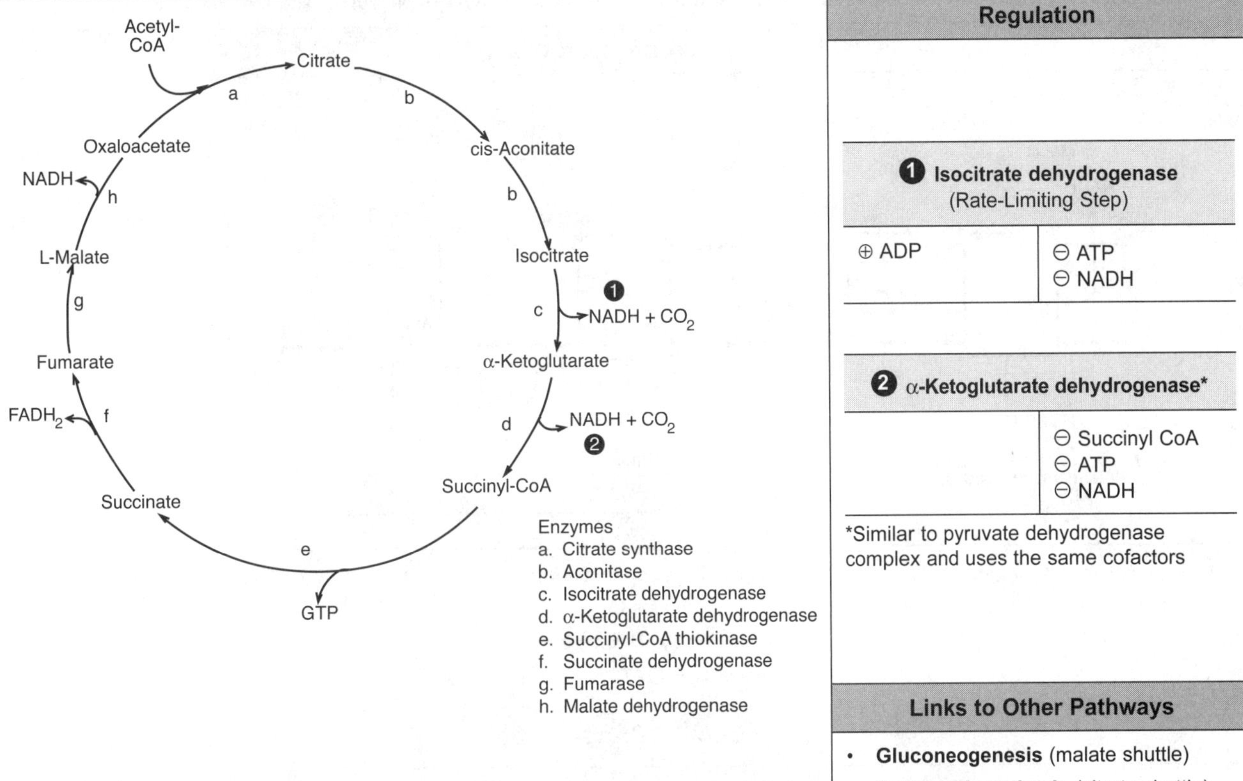

Regulation

❶ **Isocitrate dehydrogenase** (Rate-Limiting Step)	
⊕ ADP	⊖ ATP ⊖ NADH

❷ **α-Ketoglutarate dehydrogenase***	
	⊖ Succinyl CoA ⊖ ATP ⊖ NADH

*Similar to pyruvate dehydrogenase complex and uses the same cofactors

Links to Other Pathways

- **Gluconeogenesis** (malate shuttle)
- **Fatty acid synthesis** (citrate shuttle)
- **Amino acid synthesis** (oxaloacetate and α-ketoglutarate)
- **Heme synthesis** (succinyl CoA)

Stoichiometry of the Citric Acid Cycle

$$\text{Acetyl-CoA} + 3\ NAD^+ + FAD + GDP + P_i \rightarrow 2\ CO_2 + 3\ NADH + FADH_2 + GTP + CoA$$

The only net fate of acetyl-CoA as it proceeds through the citric acid cycle is conversion to CO_2. **The citric acid cycle is NOT a means to convert acetyl groups to glucose.** Humans lack the capacity to form glucose from acetyl-CoA.

▶ Oxidative Phosphorylation

Electron transport and the coupled synthesis of ATP are known as oxidative phosphorylation. The **electron transport chain (ETC)** is a series of carrier enzymes in the **inner mitochondrial membrane** that pass electrons, in a stepwise fashion, from NADH and $FADH_2$ to **oxygen**, the final electron acceptor. These carriers create a proton gradient across the inner membrane, which drives the F_0/F_1 ATP synthase, with a net production of **3 ATPs** per **NADH** and **2 ATPs** per **$FADH_2$**.

Electron Transport Chain

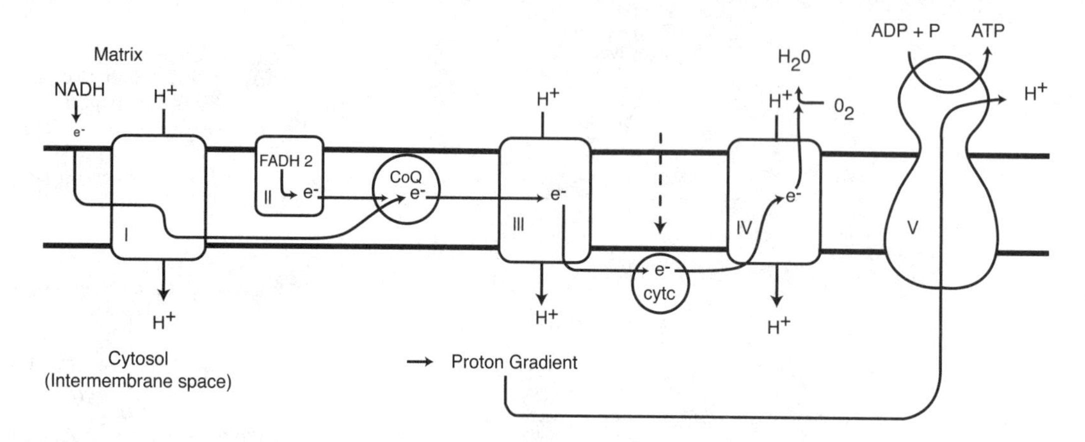

Complex I: NADH dehydrogenase Complex II: succinate dehydrogenase CoQ: coenzyme Q	Complex III: cytochrome b/c_1 Cyt C: cytochrome c Complex IV: cytochrome a/a_3 Complex V: F_oF_1ATP synthase

Clinical Correlation

Cyanide Poisoning

Blocks cytochrome a/a_3; cyanide from burning polyurethane (mattress/furniture stuffing); Tx: nitrites (creates methemoglobin, which binds cyanide)

Inhibitors

Inhibit ETC and O_2 consumption
Inhibit ATP synthesis

- Antimycin A, piscicide (Complex I)
- Cyanide (cyt oxidase)
- Rotenone, broad-spectrum insecticides, other pesticides (Complex 1)
- Oligomycin (F_o)

Uncouplers

Increase ETC and O_2 consumption
Decrease ATP synthesis
Produce heat

- 2,4-dinitrophenol (2,4-DNP)
- Salicylate (metabolite of aspirin)
- Uncoupling proteins (e.g., thermogenin)

▶ Pyruvate Metabolism

❶ **Lactate dehydrogenase:** *Anaerobic tissues:* converts pyruvate to lactate, reoxidizing cytoplasmic NADH to NAD^+.
Liver: converts lactate to pyruvate for gluconeogenesis or for metabolism to acetyl CoA

❷ **Pyruvate dehydrogenase:** generates acetyl-CoA for fatty acid synthesis and the citric acid cycle; complex of 3 enzymes

❸ **Pyruvate carboxylase:** produces oxaloacetate for gluconeogenesis and the citric acid cycle

❹ **Alanine aminotransferase (ALT, GPT):** *Muscle:* converts pyruvate to alanine to transport amino groups to the liver.
Liver: converts alanine to pyruvate for gluconeogenesis and delivers the amino group for urea synthesis

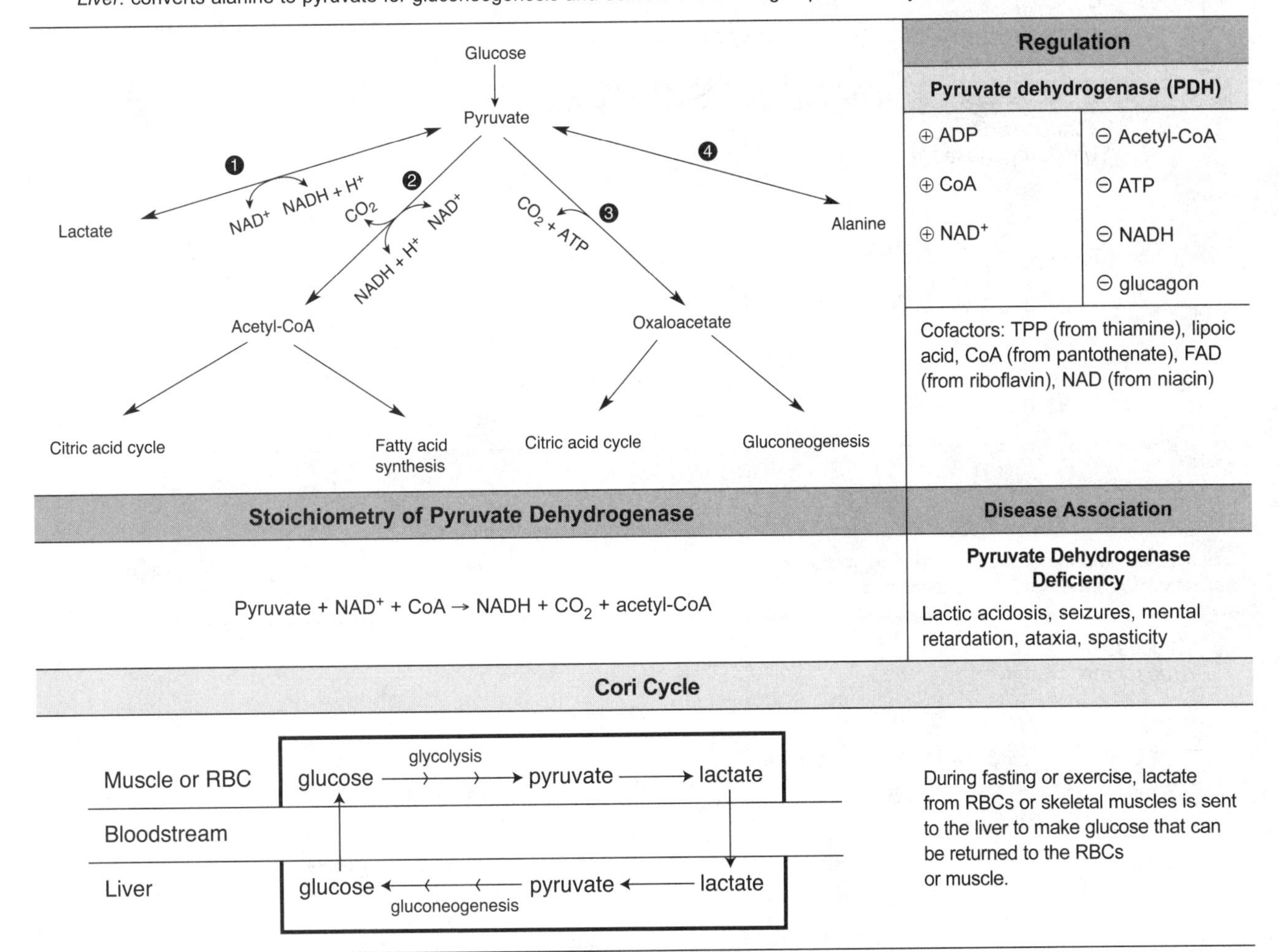

Definition of abbreviations: TCA, tricarboxylic acid; TPP, thiamine pyrophosphate.

▶ Hexose Monophosphate Shunt

The **hexose monophosphate (HMP) shunt** (pentose phosphate pathway) is a **cytosolic** pathway that uses **glucose-6-phosphate** to reduce NADP to NADPH, and synthesize **ribose-5-P**. NADPH is important for fatty acid and steroid biosynthesis, maintenance of reduced glutathione to protect against reactive oxygen species (ROS), and for bactericidal activity in polymorphonuclear leukocytes (PMNs). Ribose-5-P is required for nucleotide synthesis.

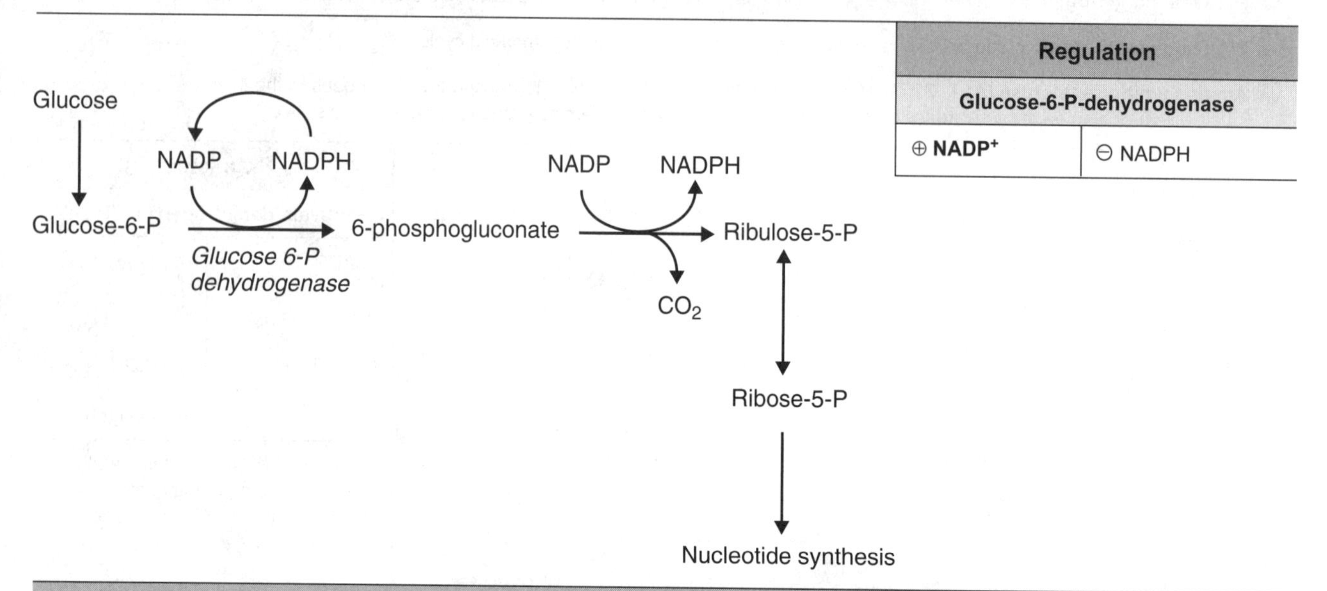

Regulation	
Glucose-6-P-dehydrogenase	
⊕ **NADP⁺**	⊖ NADPH

Disease Association
Glucose-6-Phosphate Dehydrogenase Deficiency
Episodic self-limiting hemolytic anemia induced by infection and drugs (common) or chronic hemolysis (rare); X-linked recessive; female heterozygotes have ↑ resistance to malaria

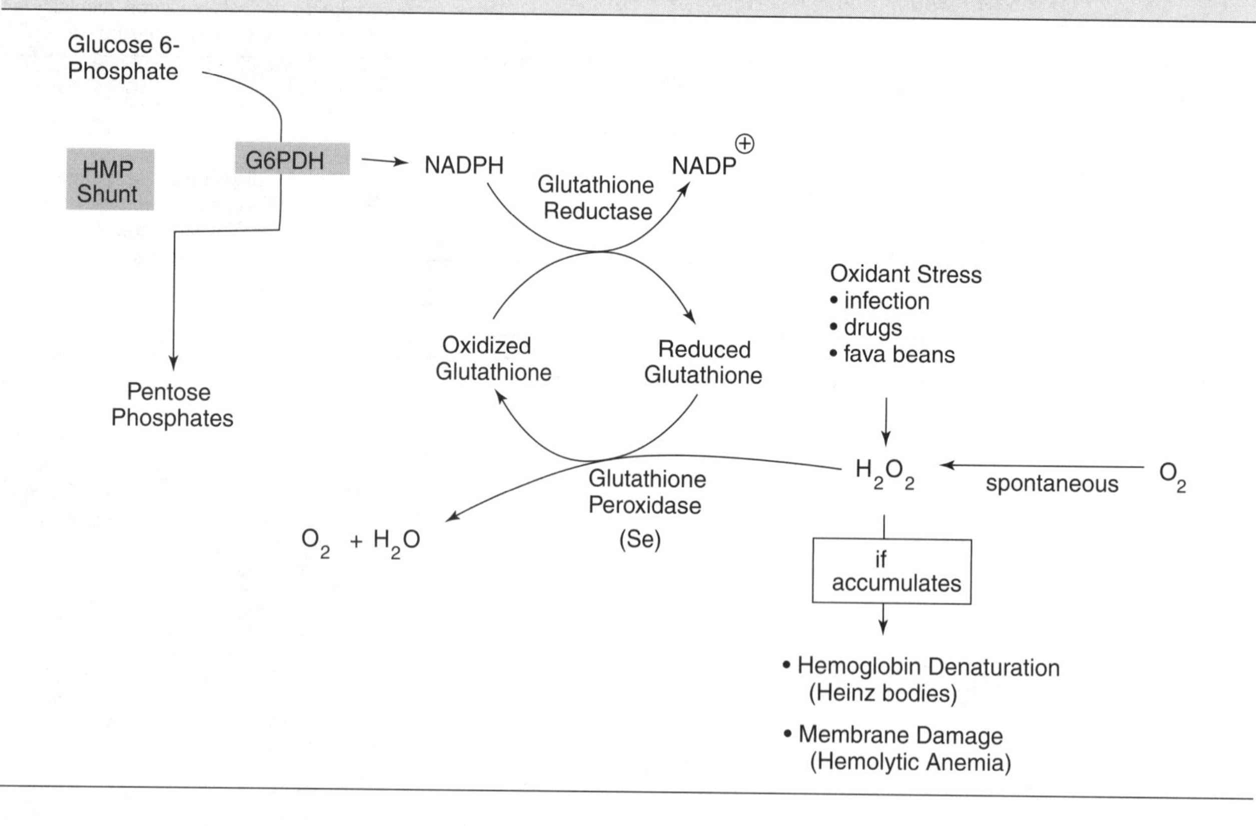

▶ Glycogenesis and Glycogenolysis

Glycogen is a branched polymer of glucose, stored primarily in liver and skeletal muscles, which can be mobilized during hypoglycemia (liver) or muscular contraction (muscles). Synthesis of glycogen (**glycogenesis**) is mediated by **glycogen synthase**, while its breakdown (**glycogenolysis**) is carried out by **glycogen phosphorylase**. Branching of the glycogen polymer occurs via a **branching enzyme**, which breaks an α-1,4-bond and transfers a block of glucosyl residues to create a new α-1,6-bond. This is reversed by a **debranching enzyme**.

Glycogen Metabolism

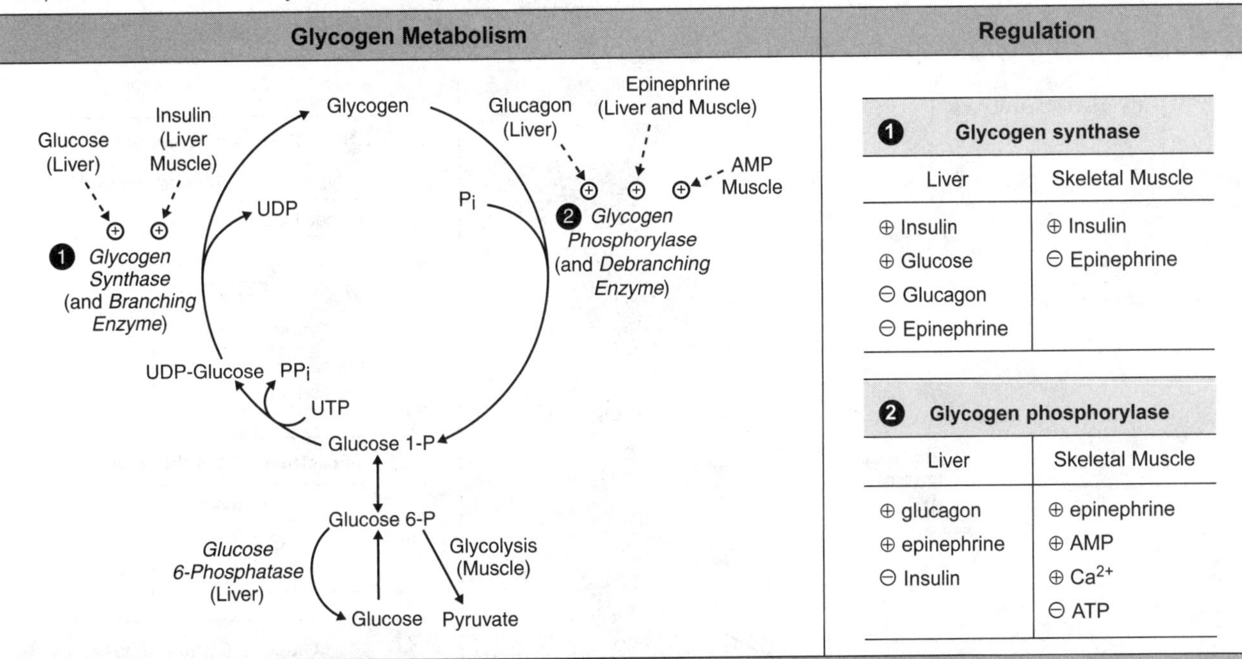

Regulation

❶ Glycogen synthase

Liver	Skeletal Muscle
⊕ Insulin	⊕ Insulin
⊕ Glucose	⊖ Epinephrine
⊖ Glucagon	
⊖ Epinephrine	

❷ Glycogen phosphorylase

Liver	Skeletal Muscle
⊕ glucagon	⊕ epinephrine
⊕ epinephrine	⊕ AMP
⊖ Insulin	⊕ Ca^{2+}
	⊖ ATP

Branching and Debranching Steps

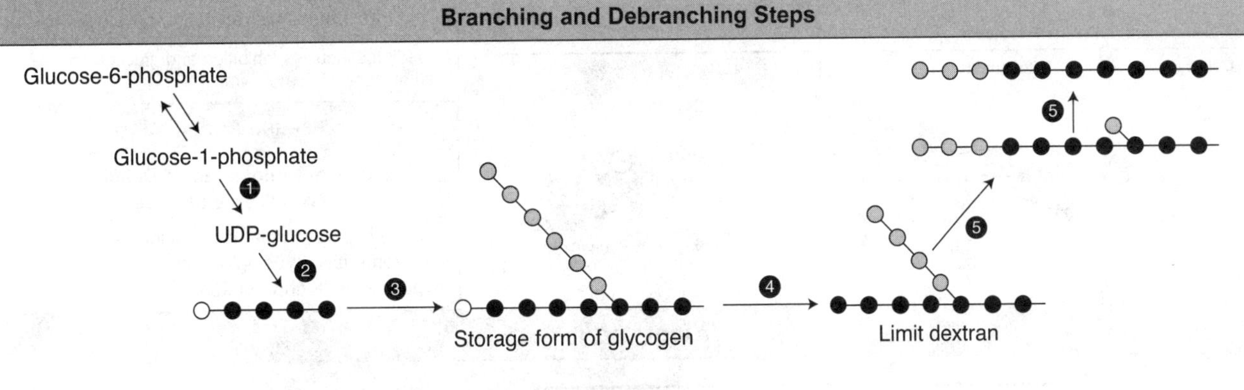

❶ UDP-glucose pyrophosphorylase

❷ Glycogen synthase

❸ Branching enzyme

❹ Glycogen phosphorylase

❺ Debranching enzyme

Glycogen Storage Diseases

Type I: von Gierke disease (↓ glucose-6-phosphatase)	Severe fasting hypoglycemia, lactic acidosis, hepatomegaly, hyperlipidemia, hyperuricemia, short stature
Type II: Pompe disease (↓ lysosomal-α-1,4-glucosidase)	Cardiomegaly, muscle weakness, death by 2 years
Type III: Cori disease (↓ glycogen debranching enzyme)	Mild hypoglycemia; liver enlargement
Type IV: Andersen disease (↓ branching enzyme)	Infantile hypotonia, cirrhosis, death by 2 years
Type V: McArdle disease (↓ muscle glycogen phosphorylase*)	Muscle cramps/weakness during initial phase of exercise, possible rhabdomyolysis and myoglobinuria
Type VI: Hers disease (↓ hepatic glycogen phosphorylase)	Mild fasting hypoglycemia, hepatomegaly, cirrhosis

*Also known as myophosphorylase.

▶ Gluconeogenesis

Gluconeogenesis is a pathway for de novo synthesis of **glucose** from **C3 and C4 precursors** using both **mitochondrial** and **cytosolic** enzymes. Occurring only in liver, kidney, and intestinal epithelium, this pathway functions to provide glucose for the body, especially the brain and RBCs, which require glucose for energy (the brain can also use ketone bodies during fasting conditions). Gluconeogenesis occurs during fasting, as glycogen stores become depleted. Important substrates for gluconeogenesis are gluconeogenic **amino acids** (protein from muscle), **lactate** (from RBCs and muscle during anaerobic exercise), and **glycerol-3-P** (from triacylglycerol from adipose tissues).

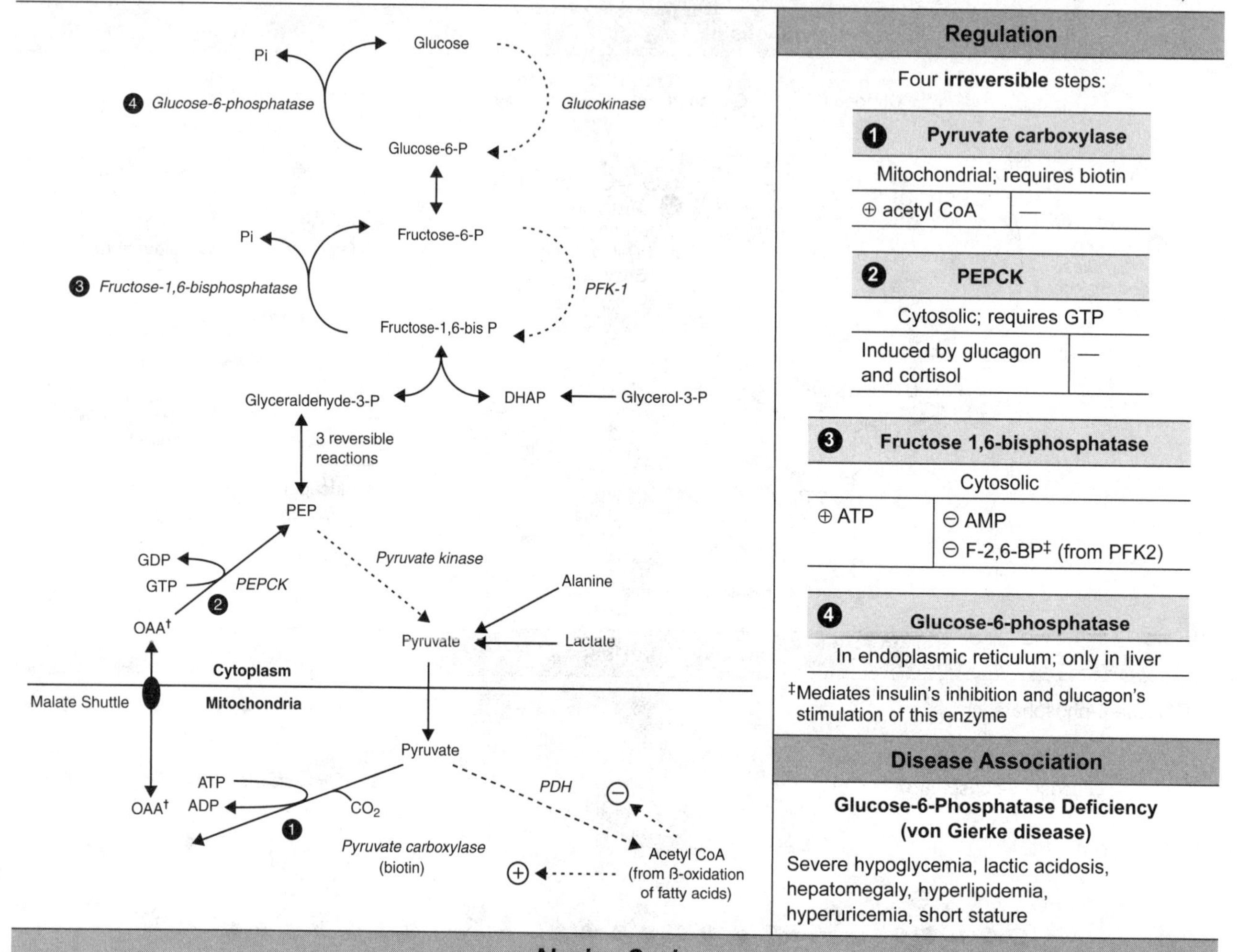

Regulation

Four **irreversible** steps:

❶ **Pyruvate carboxylase**	
Mitochondrial; requires biotin	
⊕ acetyl CoA	—

❷ **PEPCK**	
Cytosolic; requires GTP	
Induced by glucagon and cortisol	—

❸ **Fructose 1,6-bisphosphatase**	
Cytosolic	
⊕ ATP	⊖ AMP ⊖ F-2,6-BP‡ (from PFK2)

❹ **Glucose-6-phosphatase**
In endoplasmic reticulum; only in liver

‡Mediates insulin's inhibition and glucagon's stimulation of this enzyme

Disease Association

Glucose-6-Phosphatase Deficiency (von Gierke disease)

Severe hypoglycemia, lactic acidosis, hepatomegaly, hyperlipidemia, hyperuricemia, short stature

Alanine Cycle

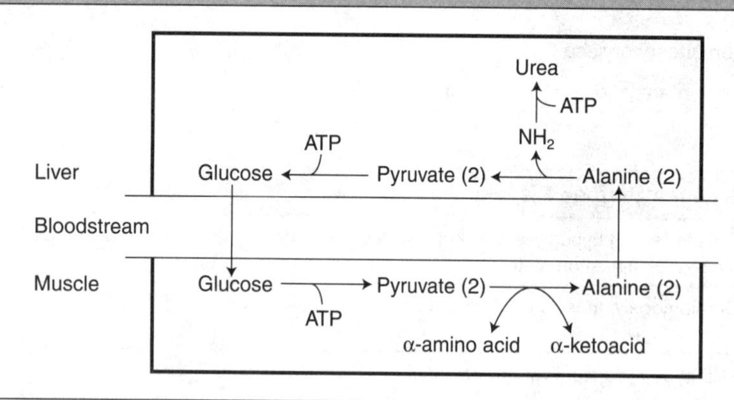

A pathway by which muscles release alanine to the liver, delivering both a gluconeogenic substrate (pyruvate) and an amino group for urea synthesis

Definition of abbreviations: PEPCK, phosphoenolpyruvate carboxykinase; PFK2, phosphofructokinase 2; RBC, red blood cell.

† OAA is not transported across the membrane directly. Instead, it is transported as malate in exchange for asparate via the malate shuttle.

▶ Amino Acid Structures

Hydrophobic Amino Acids

Nonpolar, Aliphatic Side Chains

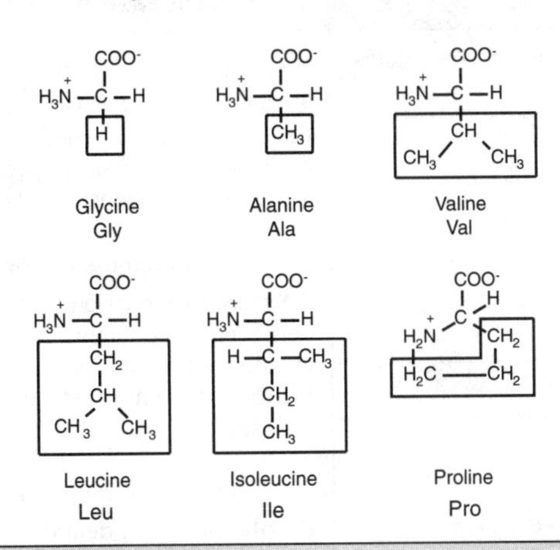

Aromatic Side Chains

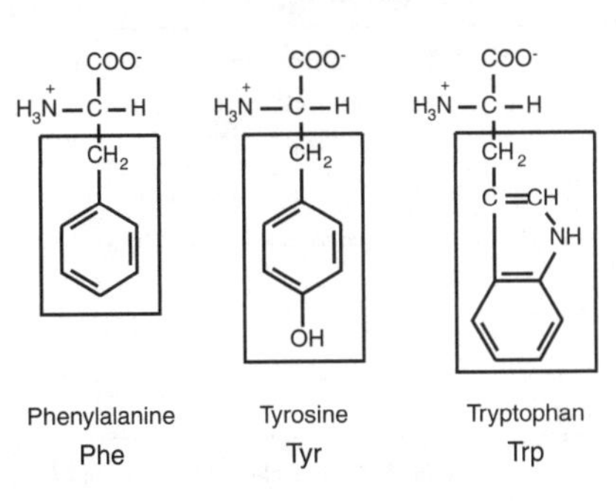

Hydrophilic Amino Acids

Positively Charged R Groups

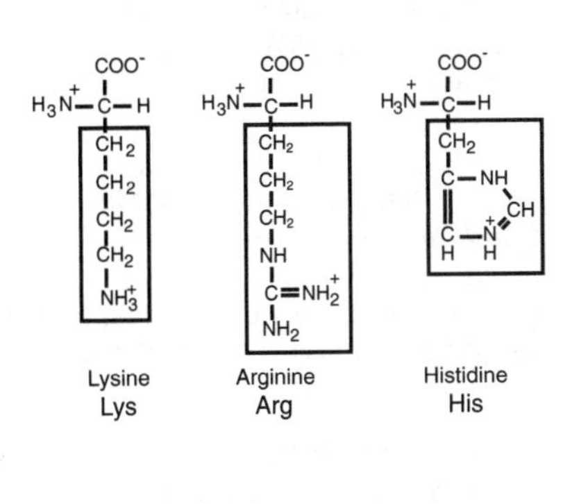

Polar, Uncharged R Groups

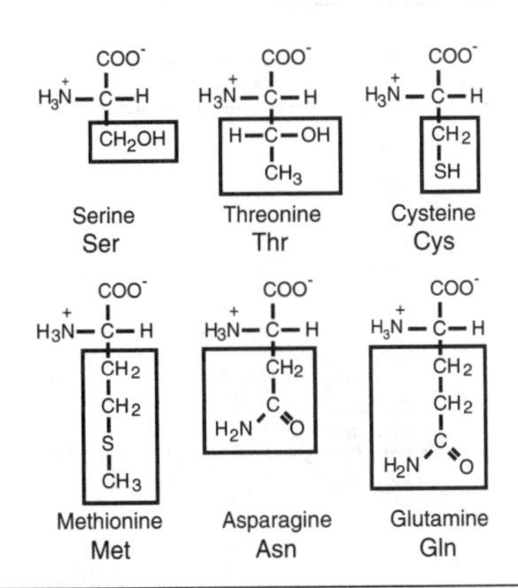

Negatively Charged R Groups

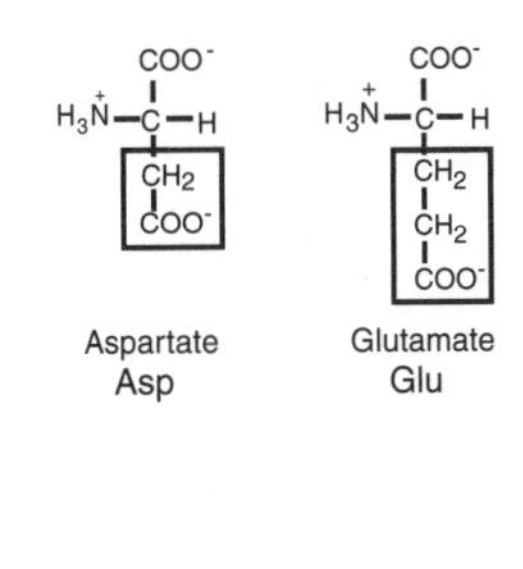

▶ Amino Acid Derivatives

Besides being the building blocks of proteins, amino acids are also precursors for various chemicals, such as **hormones**, **neurotransmitters**, and other small molecules.

Amino Acid	Product
Tyrosine	Thyroid hormones (T_3, T_4); melanin; catecholamines (dopamine, epinephrine)

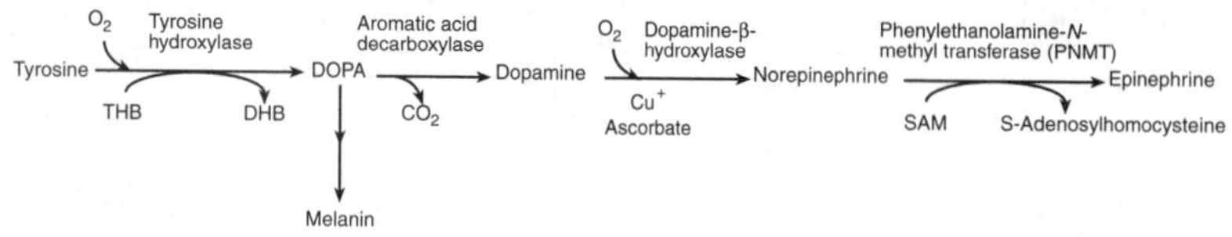

Tyrosine —(Tyrosine hydroxylase; O_2, THB → DHB)→ DOPA —(Aromatic acid decarboxylase; → CO_2)→ Dopamine —(Dopamine-β-hydroxylase; O_2, Cu^+, Ascorbate)→ Norepinephrine —(Phenylethanolamine-*N*-methyl transferase (PNMT); SAM → S-Adenosylhomocysteine)→ Epinephrine

DOPA → Melanin

Amino Acid	Product
Tryptophan	Serotonin (5-HT); melatonin; NAD; NADP

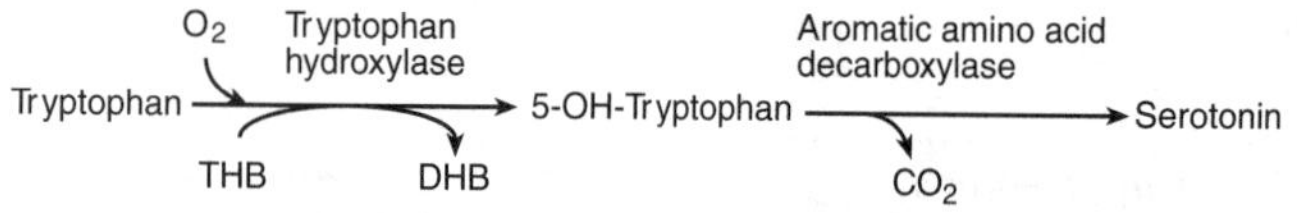

Tryptophan —(Tryptophan hydroxylase; O_2, THB → DHB)→ 5-OH-Tryptophan —(Aromatic amino acid decarboxylase; → CO_2)→ Serotonin

Amino Acid	Product
Glycine	Heme

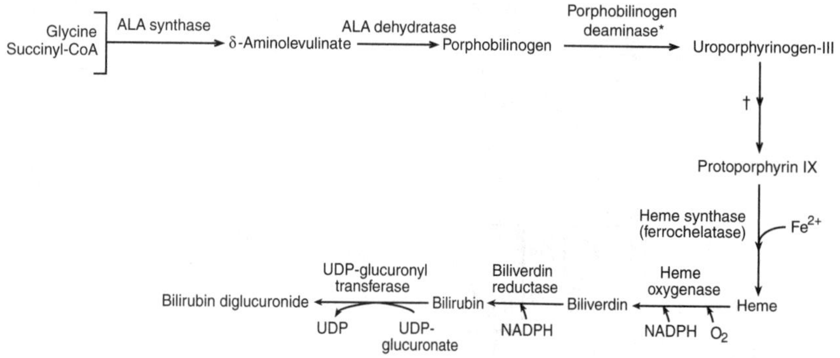

Glycine + Succinyl-CoA —(ALA synthase)→ δ-Aminolevulinate —(ALA dehydratase)→ Porphobilinogen —(Porphobilinogen deaminase*)→ Uroporphyrinogen-III —(†)→ Protoporphyrin IX —(Heme synthase (ferrochelatase); Fe^{2+})→ Heme —(Heme oxygenase; NADPH, O_2)→ Biliverdin —(Biliverdin reductase; NADPH)→ Bilirubin —(UDP-glucuronyl transferase; UDP-glucuronate → UDP)→ Bilirubin diglucuronide

Amino Acid	Product
Glutamate	γ-aminobutyric acid (GABA)

Glutamate —(Glutamate decarboxylase)→ γ-Aminobutyric acid (GABA)

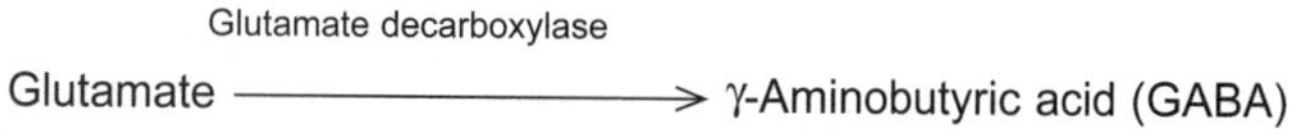

Amino Acid	Product
Arginine	Nitric oxide (NO)

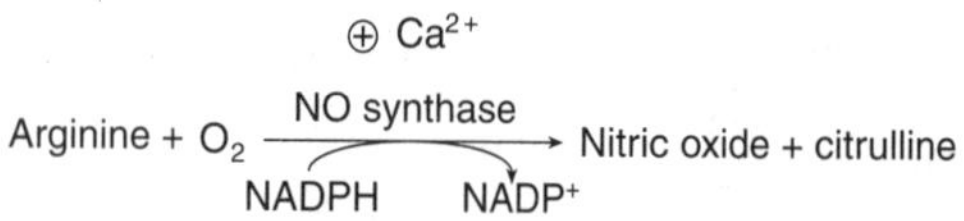

Arginine + O_2 —(NO synthase; ⊕ Ca^{2+}; NADPH → $NADP^+$)→ Nitric oxide + citrulline

Amino Acid	Product
Histidine	Histamine

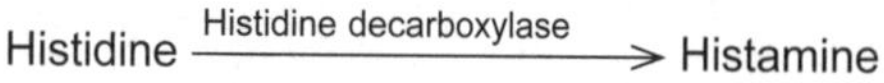

Histidine —(Histidine decarboxylase)→ Histamine

Amino Acid	Product
Methionine	S-adenosylmethionine (SAM; methylating agent)
Arginine, glycine, SAM	Creatine

Disease Association

Albinism

Tyrosine hydroxylase (type I) or tyrosine transporter (type II) deficiency; ↓ pigmentation of skin, eyes, and hair, ↑ risk of skin cancer, visual defects

Carcinoid Syndrome

↑ **Serotonin** excretion from gastrointestinal neuroendocrine tumors (carcinoid tumors); cutaneous flushing, venous telangiectasia, diarrhea, bronchospasm, cardiac valvular lesions

Acute Intermittent Porphyria

Porphobilinogen deaminase* deficiency; episodic expression, acute abdominal pain, anxiety, confusion, paranoia, muscle weakness, no photosensitivity, port-wine urine in some patients, urine excretion of ALA and PBG; autosomal dominant; onset at puberty, 15% penetrance, variable expression; more common in women

Porphyria Cutanea Tarda

Uroporphyrinogen decarboxylase† deficiency; photosensitivity, skin inflammation, and blistering; cirrhosis often associated; autosomal dominant; late onset

Lead Poisoning

Inhibits **ALA dehydratase** and **ferrochelatase**; microcytic sideroblastic anemia; basophilic stippling of erythrocytes; headache, nausea, memory loss, abdominal pain, diarrhea (lead colic), lead lines in gums, neuropathy (claw hand, wrist-drop), ↑ urine excretion of ALA; Tx: dimercaprol and EDTA

Hemolytic Crisis

Jaundice due to ↑ bilirubin from severe hemolysis; ↓ hemoglobin; ↑ reticulocytes; may result from:

(1) G6PD deficiency hemolysis
(2) Sickle cell crisis
(3) Rh disease of newborn

UDP-Glucuronyl Transferase Deficiency

Jaundice due to low bilirubin conjugation; may result from:

(1) Crigler-Najjar syndromes
(2) Gilbert syndrome
(3) Physiologic jaundice of newborn, especially premature infants

* Also known as hydroxymethylbilane synthase; †an enzyme in the pathway between Uroporphyrinogen-III and Protoporphyrin IX.

▶ Amino Acid Synthesis and Metabolism

Amino acids are required for protein synthesis. Although some amino acids can be synthesized de novo (**nonessential**), others (**essential**) must be obtained from the digestion of dietary proteins. Nonessential amino acids are synthesized from intermediates of glycolysis and the citric acid cycle or from other amino acids. Degradation of amino acids occurs by transamination of the amino group to **glutamate**, while the remaining carbon skeletons of the amino acids may be oxidized to $CO_2 + H_2O$, or reverted to citric acid cycle intermediates for conversion to glucose (**glucogenic**) or ketones (**ketogenic**).

Genetic Deficiencies of Amino Acid Metabolism

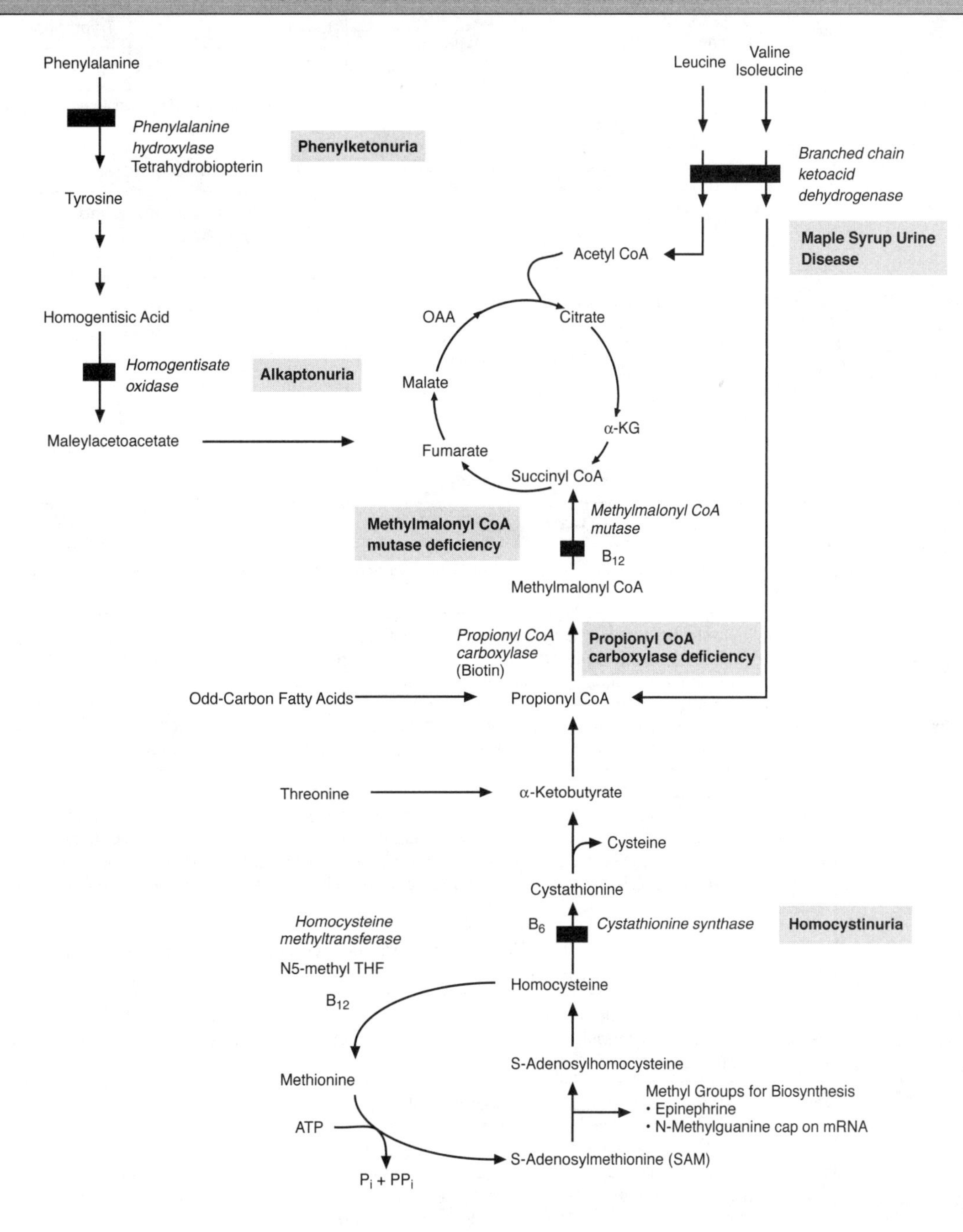

(Continued)

► Amino Acid Synthesis and Metabolism (*Cont'd.*)

Precursors for Nonessential Amino Acids

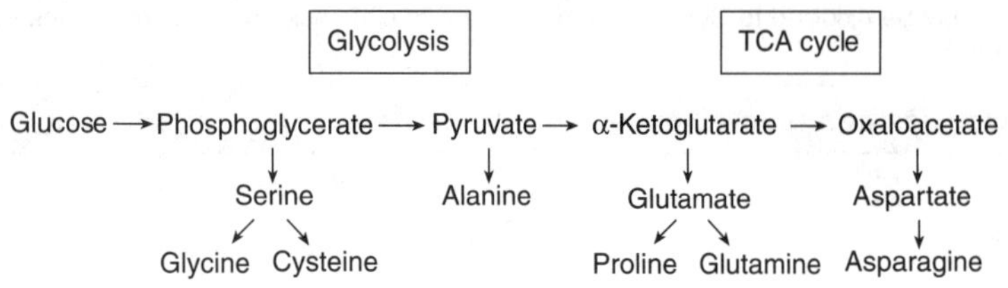

Glycolysis: Glucose → Phosphoglycerate → Pyruvate

TCA cycle: α-Ketoglutarate → Oxaloacetate

- Phosphoglycerate → Serine → Glycine, Cysteine
- Pyruvate → Alanine
- α-Ketoglutarate → Glutamate → Proline, Glutamine
- Oxaloacetate → Aspartate → Asparagine

Essential Amino Acids*

Arginine†	Methionine
Histidine	Phenylalanine
Isoleucine	Threonine
Leucine	Tryptophan
Lysine	Valine

***Mnemonic:** PVT. TIM HALL; †essential during periods of growth and pregnancy

Transfer of α-Amino Groups to α-Ketoglutarate

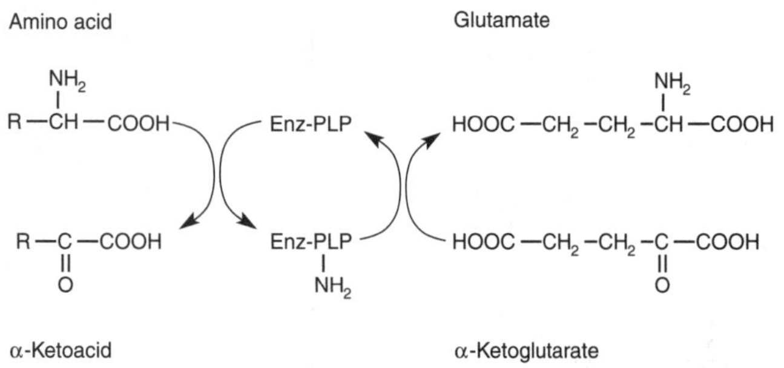

Amino acid ($R-CH(NH_2)-COOH$) + Enz-PLP → α-Ketoacid ($R-C(=O)-COOH$) + Enz-PLP-NH_2

Enz-PLP-NH_2 + α-Ketoglutarate ($HOOC-CH_2-CH_2-C(=O)-COOH$) → Enz-PLP + Glutamate ($HOOC-CH_2-CH_2-CH(NH_2)-COOH$)

Glucogenic and Ketogenic Amino Acids

Ketogenic	Ketogenic and Glucogenic	Glucogenic
Leucine Lysine	Phenylalanine Tyrosine Tryptophan Isoleucine Threonine	All others

Disease Association

Disease	Description
Hartnup disease	Transport protein defect with ↑ excretion of neutral amino acids; symptoms similar to pellagra; autosomal recessive
Phenylketonuria	Phenylalanine hydroxylase or dihydrobiopterin reductase deficiency → buildup of phenylalanine; tyrosine becomes essential; musty body odor, mental retardation, microcephaly, autosomal recessive; Tx: ↓ phenylalanine in diet; avoid aspartame (Nutrasweet®)
Alkaptonuria	Homogentisate oxidase deficiency (for tyrosine degradation); ↑ homogentisic acid in blood and urine (darkens when exposed to air), ochronosis (dark pigment in cartilage), arthritis in adulthood
Homocystinuria	↑ homocystine in urine. Classic homocystinuria, caused by a deficiency in cystathionine synthase, is associated with dislocated lens, deep venous thrombosis, stroke, atherosclerosis, mental retardation, and Marfan-like features. Deficiency of pyridoxine, folate, or vitamin B_{12} can produce a mild homocystinemia with elevated risk of atherosclerosis (previously listed symptoms absent). Methionine synthase (homocysteine methyltransferase) deficiency is extremely rare and is associated with megaloblastic anemia and mental retardation.
Cystinuria	Transport protein defect with ↑ excretion of lysine, arginine, cystine, and ornithine; excess cystine precipitates as kidney stones; Tx: acetazolamide
Maple syrup urine disease	Branched-chain ketoacid dehydrogenase deficiency; branched-chain ketoacidosis from infancy; weight loss, lethargy, alternating hypertonia/hypotonia, maple syrup odor of urine; ketosis/coma/death if untreated; Tx: ↓ valine, leucine, isoleucine in diet
Propionyl-CoA carboxylase deficiency **Methylmalonyl-CoA mutase deficiency**	Neonatal ketoacidosis from blocked degradation of valine, isoleucine, methionine, threonine, and odd-carbon fatty acids; Tx: ↓ these amino acids in diet *Propionyl-CoA carboxylase deficiency:* neonatal metabolic acidosis; hyperammonemia; elevated propionic acid, hydroxypropionic acid, and methylcitrate; poor feeding, vomiting, lethargy, coma *Methylmalonyl-CoA mutase deficiency:* symptoms similar to propionyl CoA carboxylase deficiency, but accumulating metabolites differ (↑ methylmalonic acid)

Definition of abbreviation: PLP, pyridoxal-phosphate, formed from vitamin B_6.

► Urea Cycle

Amino acids transported to the liver are transaminated to glutamate, which undergoes deamination to produce **NH_4^+** or transamination to make **aspartate**. Both of these are used for synthesis of urea in the liver for excretion via the **urea cycle**.

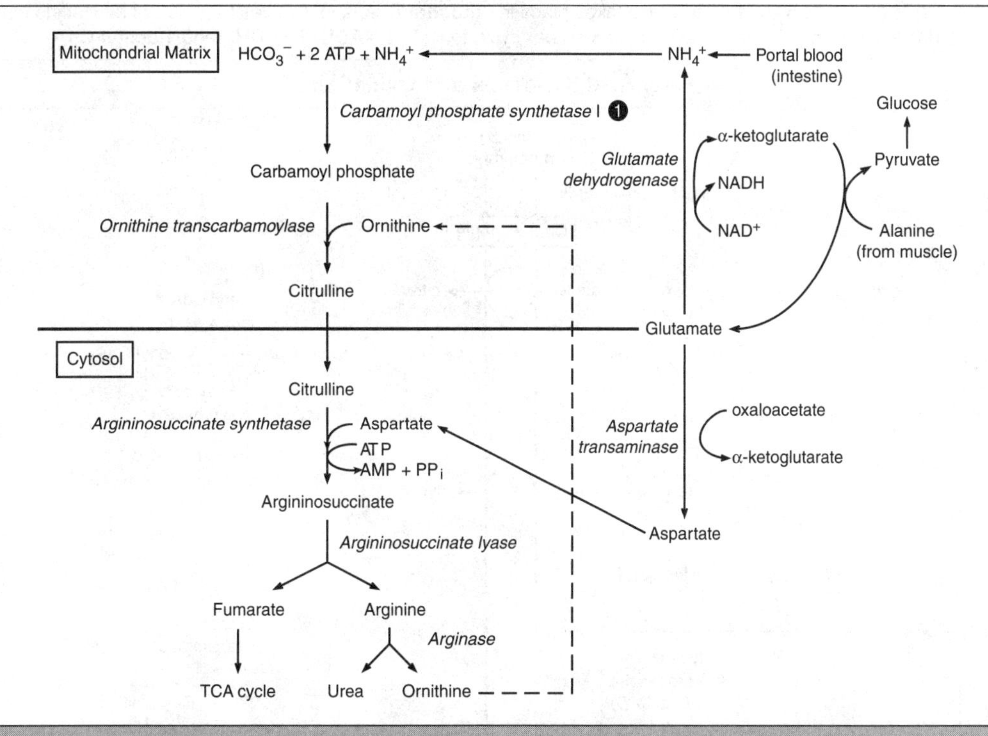

Regulation
❶ **Carbamoyl phosphate synthase I**

⊕ *N*-acetylglutamate*

*High protein diet → ↑ glutamate in mitochondria → ↑ *N*-acetylglutamate

Disease Association	
Carbamoyl Phosphate Synthetase Deficiency	**Ornithine Transcarbamoylase Deficiency**
↑ [NH_4^+]; hyperammonemia	↑ [NH_4^+]; hyperammonemia
↑ blood glutamine	↑ blood glutamine
↓ BUN	↓ BUN
No increase in uracil or orotic acid	Uracil and orotic acid ↑ in blood and urine*
Cerebral edema	Cerebral edema
Lethargy, convulsions, coma, death	Lethargy, convulsions, coma, death

*OTC deficiency: ↑ carbamoyl-P stimulates pyrimidine synthesis, causing ↑ orotic acid and uracil

► Lipid Synthesis and Metabolism

Fatty acids are synthesized from excess glucose in the liver and transported to adipose tissues for storage. Fatty acid **synthesis** occurs in the **cytosol** and involves the transport of **acetyl-CoA** from the mitochondria via the **citrate shuttle**, carboxylation to **malonyl CoA**, and linking together 2 carbons per cycle to form long fatty acid chains. Synthesis stops at **C_{16} palmitoyl-CoA**, requiring **7 ATP** and **14 NADPH**. Metabolism of fatty acids occurs by **β-oxidation**, which takes place in **mitochondria**, and involves transport of fatty acids from the cytosol via the **carnitine shuttle**, then oxidative removal of 2 carbons per cycle to yield **1 NADH**, **1 $FADH_2$**, and **1 acetyl-CoA**.

Fatty Acid Synthesis and Oxidation

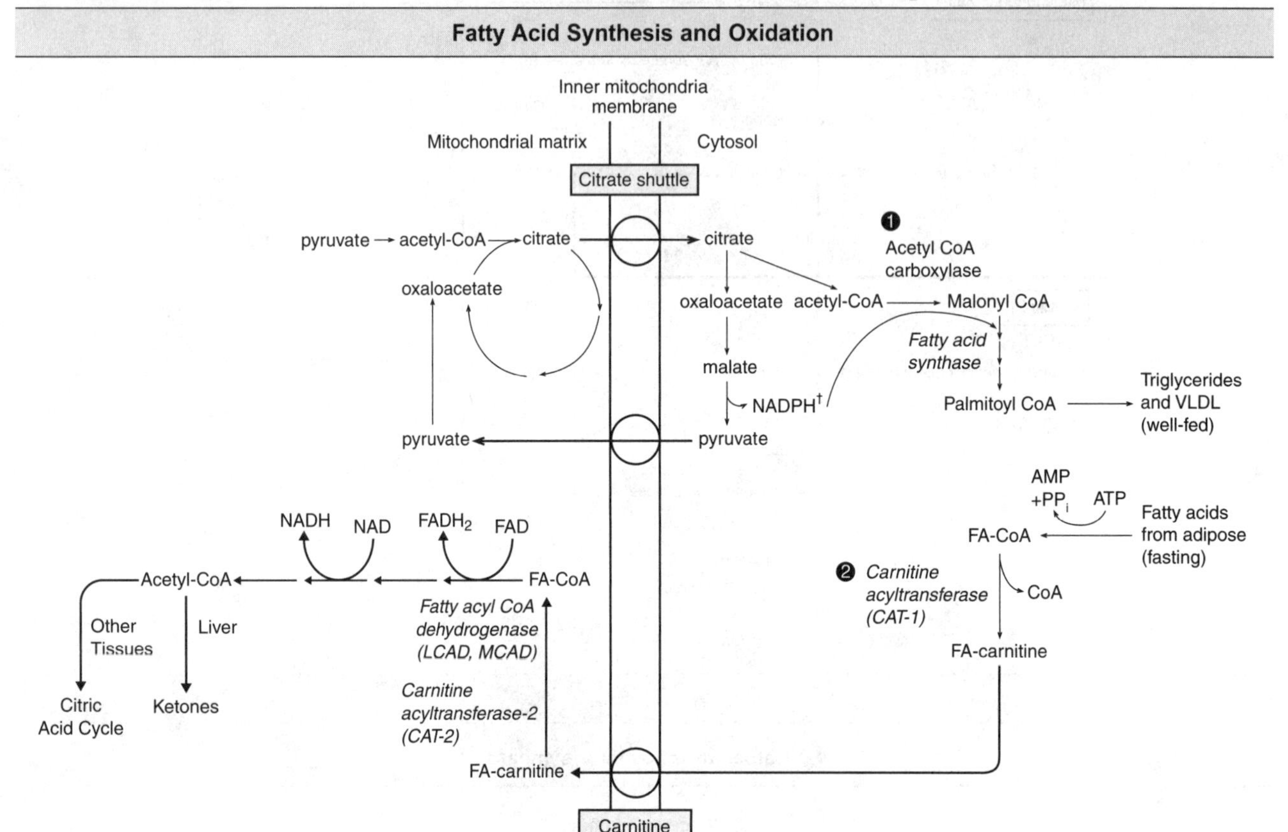

†Another important source of NADPH is the HMP shunt.

Triacylglycerols (triglycerides), the storage form of fatty acids, are formed primarily in the liver and adipose tissues by attaching **3 fatty acids** to a **glycerol-3-P**. Triacylglycerols are transported from liver to adipose as VLDL. Fatty acids from the diet are transported as chylomicrons. Both are digested by lipoprotein lipase (induced by insulin) in the capillaries of adipose and muscle. Fatty acids may be mobilized from triacylglycerols in adipose by hormone-sensitive lipase. Free fatty acids are delivered to tissues for beta oxidation.

Regulation		
❶ Acetyl-CoA carboxylase		**❷ Carnitine acyltransferase-1 (CAT-1)**
Rate-limiting for fatty acid synthesis; requires biotin		**Rate-limiting for fatty acid oxidation**
⊕ insulin ⊕ citrate	⊖ glucagon ⊖ palmitoyl-CoA	⊖ malonyl-CoA

Disease Association	
Myopathic CAT-2/CPT-2 Deficiency	**Medium Chain Acyl-Dehydrogenase (MCAD) Deficiency**
Muscle aches/weakness, myoglobulinuria provoked by prolonged exercise, ↑ muscle triacylglycerols	Fasting hypoglycemia, no ketone bodies, dicarboxylic acidemia, C8–C10 acyl carnitines in blood, vomiting, coma, death; Tx: give IV glucose, avoid fasting, maintain high carb/low fat diet, including short chain FAs, which can be metabolized

(Continued)

▶ Lipid Synthesis and Metabolism (*Cont'd.*)

Triacylglycerol (Triglyceride) Synthesis

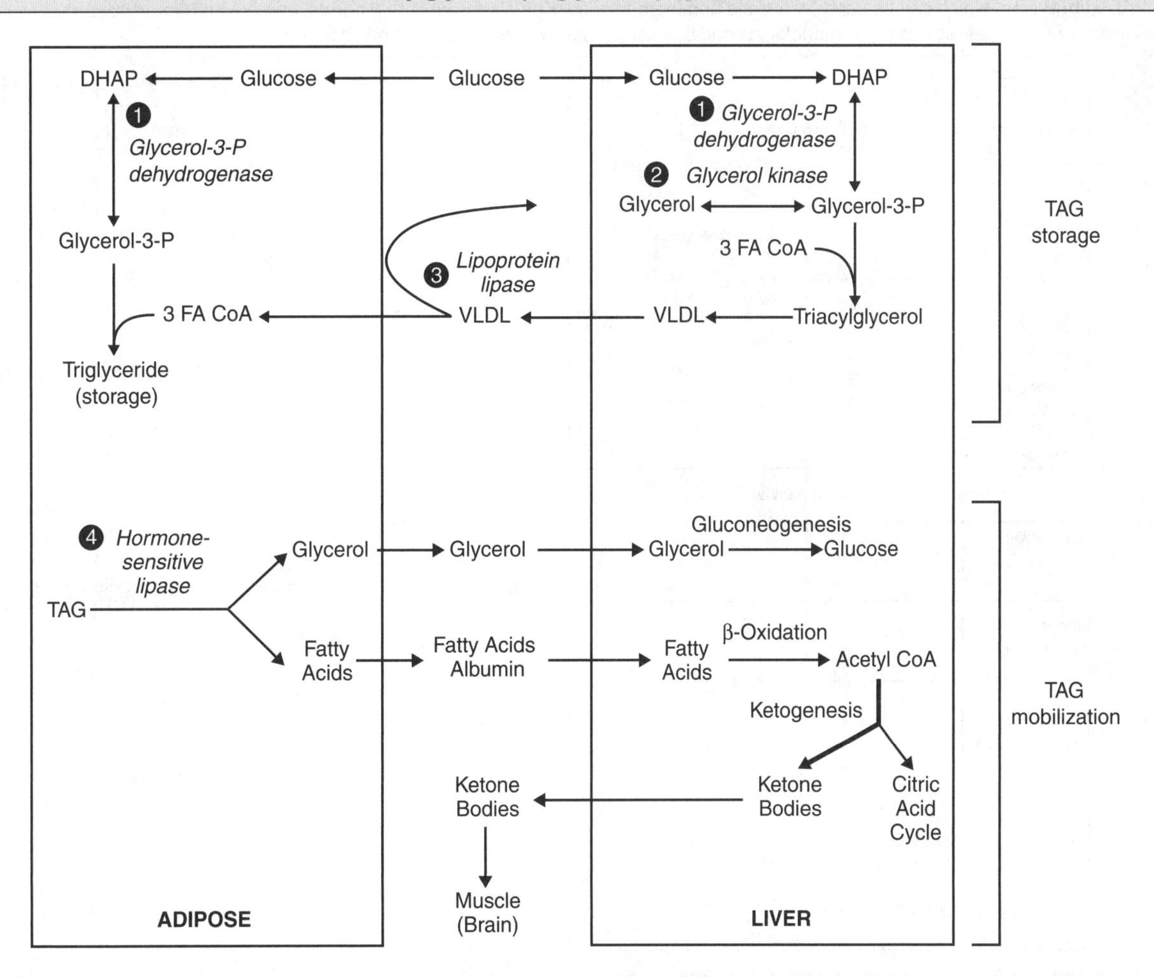

Notes	
❶ Glycerol-3P-dehydrogenase (adipose, liver) **❷ Glycerol kinase (liver only)**	Triacylglycerol synthesis from fatty acids
❸ Lipoprotein lipase	Located on luminal membrane of endothelial cells in adipose tissue

Regulation			
❸ Lipoprotein Lipase		**❹ Hormone sensitive lipase**	
Digests TGL in VLDL and chylomicrons. Fatty acids enter adipose		Mobilizes fatty acids from triacylglycerols	
Induced by insulin ⊕ ApoC-II	Repressed by ↓ insulin	⊕ Epinephrine Induced by cortisol	⊖ insulin

Definition of abbreviations: CAT, carnitine acyltransferase (a.k.a. CPT, carnitine palmitoyl transferase); L/MCAD, long/medium chain acyl-dehydrogenase; TAG, triacylglycerols.
Diabetic ketoacidosis results from overactive hormone-sensitive lipase often in the context of stress, trauma, or infection.

▶ Ketone Body Metabolism

During fasting, the liver converts excess acetyl-CoA from beta-oxidation of fatty acids into two ketone bodies, **acetoacetate** and **β-hydroxybutyrate**, which can be used by muscle and brain tissues. Ketosis represents a normal and advantageous response to fasting/starvation, whereas ketoacidosis is a pathologic condition associated with diabetes and other diseases.

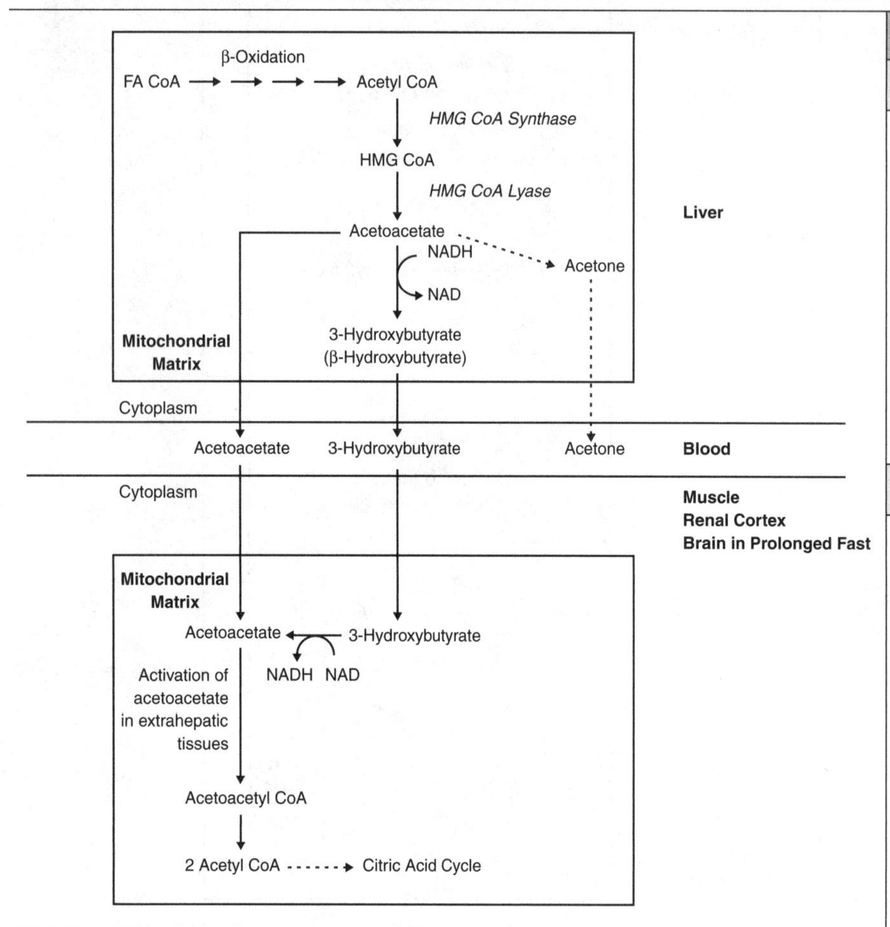

Disease Association

Diabetic Ketoacidosis

Excess ketone bodies in blood associated with type 1 diabetes mellitus not adequately managed with insulin, or precipitated by infection or trauma. Characterized by polyuria, dehydration, CNS depression and coma, sweet fruity breath (acetone).

With the prevalence of obesity and stressful environments, ketoacidosis is now becoming more prevalent in type 2 diabetics, e.g., a diabetic in ketoacidosis cannot be assumed to be type 1.

Alcoholic Ketoacidosis

Excess ketone bodies due to high NADH/NAD ratio in liver; symptoms same as above

Note: In either type of ketoacidosis, 3-hydroxybutyrate (β-hydroxybutyrate) is the predominant ketone body formed (not detected by the urine test). Measure 3-hydroxybutyrate to more accurately evaluate ketoacidosis.

▶ Cholesterol Synthesis

Cholesterol is obtained from diet (about 20%) or synthesized de novo (about 80%). Synthesis occurs primarily in the liver for storage and bile acid synthesis, but also in adrenal cortex, ovaries, and testes for steroid hormone synthesis. Cholesterol may also be esterified into **cholesterol esters** by acyl-cholesterol acyl-transferase (**ACAT**) in cells for storage.

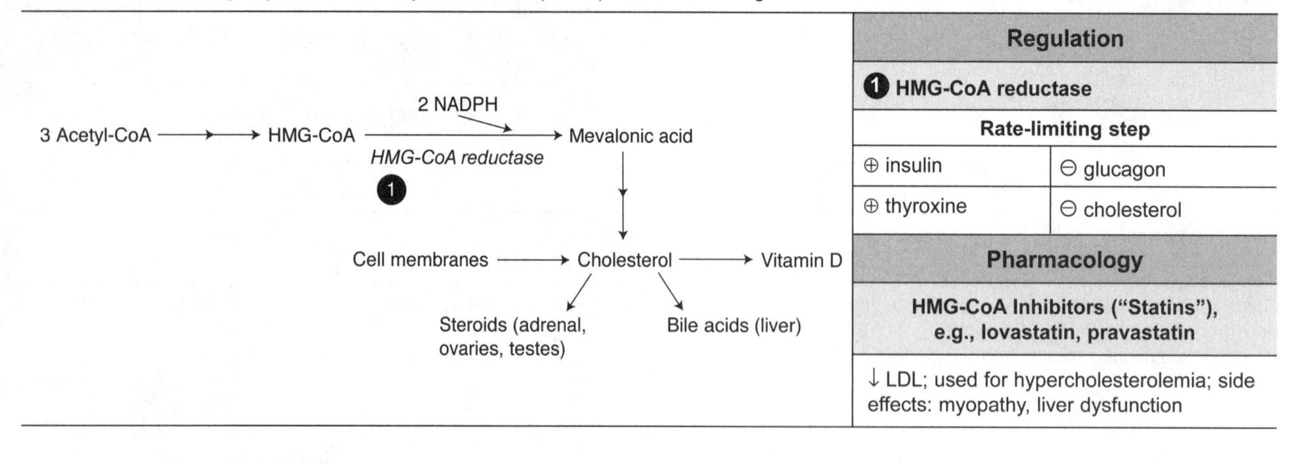

Regulation

1 HMG-CoA reductase

Rate-limiting step	
⊕ insulin	⊖ glucagon
⊕ thyroxine	⊖ cholesterol

Pharmacology

HMG-CoA Inhibitors ("Statins"), e.g., lovastatin, pravastatin

↓ LDL; used for hypercholesterolemia; side effects: myopathy, liver dysfunction

► Lipoprotein Transport and Metabolism

Free fatty acids are transported by serum albumin, whereas neutral lipids (triacylglycerols and cholesterol esters) are transported by **lipoproteins**. Lipoproteins consist of a hydrophilic shell and a hydrophobic core and are classified by their density into **chylomicrons**, **VLDL**, **LDL**, and **HDL**.

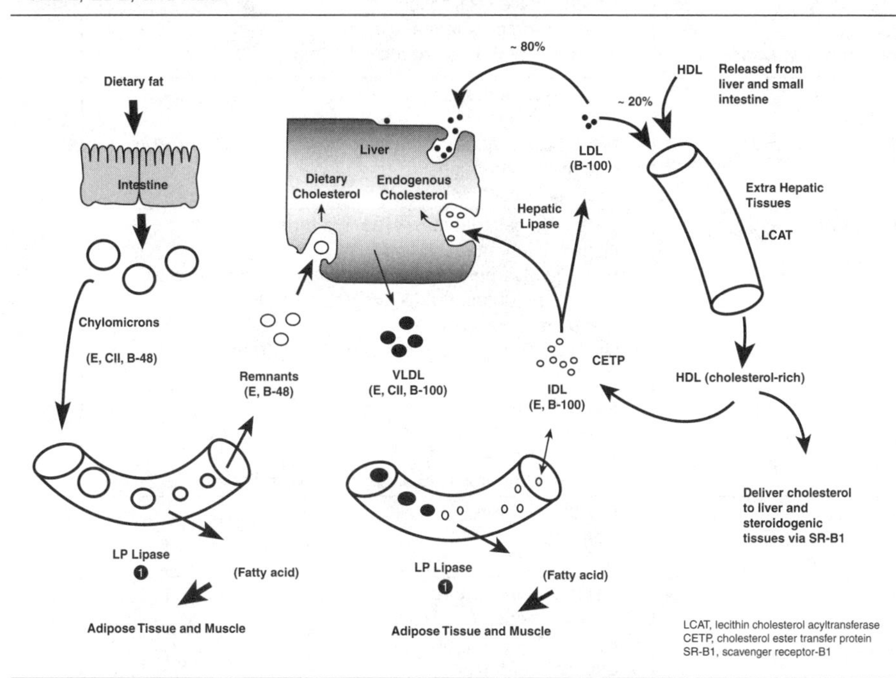

Classes of Lipoproteins and Important Apoproteins

Lipoprotein	Functions	Apoproteins	Functions
Chylomicrons	Transport dietary triglyceride and cholesterol from intestine to tissues	apoB-48 apoC-II apoE	Secreted by epithelial cells Activates lipoprotein lipase Uptake by liver
VLDL	Transports triglyceride from liver to tissues	apoB-100 apoC-II apoE	Secreted by liver Activates lipoprotein lipase Uptake of remnants by liver
LDL	Delivers cholesterol into cells	apoB-100	Uptake by liver and other tissues via LDL receptor (apoB-100 receptor)
IDL (VLDL remnants)	Picks up cholesterol from HDL to become LDL Picked up by liver	apoE	Uptake by liver
HDL	Picks up cholesterol accumulating in blood vessels Delivers cholesterol to liver and steroidogenic tissues via scavenger receptor (SR-B1) Shuttles apoC-II and apoE in blood	apoA-1	Activates LCAT to produce cholesterol esters

Regulation

❶ Lipoprotein lipase

Hydrolyzes fatty acids from triacylglycerols from chylomicrons and VLDL

Induced by insulin
⊕ ApoC-II

Hyperlipidemias

Type I Hypertriglyceridemia

Lipoprotein lipase deficiency; ↑ triacylglycerols and chylomicrons; orange-red eruptive xanthomas, fatty liver, acute pancreatitis, abdominal pain after fatty meal; autosomal recessive

Type II Hypercholesterolemia

LDL receptor deficiency; ↑ risk of atherosclerosis and CAD, xanthomas of Achilles tendon, tuberous xanthomas on elbows, xanthelasma (lipid in eyelid), corneal arcus, homozygotes die <20 years; autosomal dominant

Pharmacology

Cholestyramine/Colestipol

↑ Elimination of bile salts leads to ↑ LDL receptor expression, leading to ↓ LDL; for hypercholesterolemia; side effect: GI discomfort

Gemfibrozil/Clofibrate ("Fibrates")

↑ elimination of VLDL leads to ↓ triacylglycerols and ↑ HDL; for hypertriglyceridemia; side effect: muscle toxicity ; acts via PPAR-α to induce LP lipase gene

Nicotinic Acid

↓ VLDL synthesis leads to ↓ LDL; for hypercholesterolemia and hypertriglyceridemia; side effects: GI irritation; hyperuricemia, hyperglycemia, flushing, pruritus

Definition of abbreviations: CAD, coronary artery disease; CETP, cholesterol ester transfer protein; HDL, high-density lipoprotein; LCAT, lecithin-cholesterol acyl transferase; LDL, low-density lipoprotein; PPAR, peroxisome proliferator-activated receptor; SR-B1, scavenger receptor B1; VLDL, very-low-density lipoprotein.

► Lysosomal Storage Diseases

Disease	Deficiency and Accumulated Substrate	Features	
Tay-Sachs disease	↓ Hexosaminidase A ↑ GM_2 ganglioside *(whorled membranes in lysosomes)*	• Psychomotor retardation • Cherry red spots in macula • Onset in 1st 6 mos.; death <2 years	AR*
Niemann-Pick disease	↓ Sphingomyelinase ↑ Sphingomyelin *(zebra bodies in lysosomes)*	• Hepatosplenomegaly • Microcephaly • Mental retardation • Foamy macrophages • Neonatal onset	AR
Gaucher disease	↓ β-glucocerebrosidase ↑ Glucocerebroside	• Three clinical subtypes; type 1 is most common • Hepatosplenomegaly • Bone involvement, including fractures and bone pain • Neurologic defects (rare, types 2 and 3) • Mental retardation • Gaucher cells (enlarged macrophages with fibrillary cytoplasm)	AR*
Fabry disease	↓ α-galactosidase A ↑ Ceramide trihexoside	• Renal failure • Telangiectasias • Angiokeratomas • Peripheral neuropathy with pain in extremities	XR
Metachromatic leukodystrophy	↓ arylsulfatase A ↑ sulfatide	• Ataxia • Dementia • Seizures	AR
Hurler syndrome (MPSI)	↓ α-L-iduronidase ↑ dermatan sulfate ↑ heparan sulfate	• Coarse facial features • Corneal clouding • Hepatosplenomegaly • Skeletal deformities • Upper airway obstruction • Recurrent ear infections • Hearing loss • Hydrocephalus • Mental retardation • Death <10 years	AR
Hunter syndrome (MPSII)	↓ L-iduronate-2-sulfatase ↓ dermatan sulfate ↑ heparan sulfate	• Both mild and severe forms • Severe similar to Hurler but retinal degeneration instead of corneal clouding, aggressive behavior, and death <15 years • Mild form compatible with long life	XR
Krabbe disease	↓ galactocerebrosidase ↑ galactocerebroside	• Defective myelin sheaths • Peripheral neuropathy • Severe seizures	AR

► Eicosanoid Metabolism

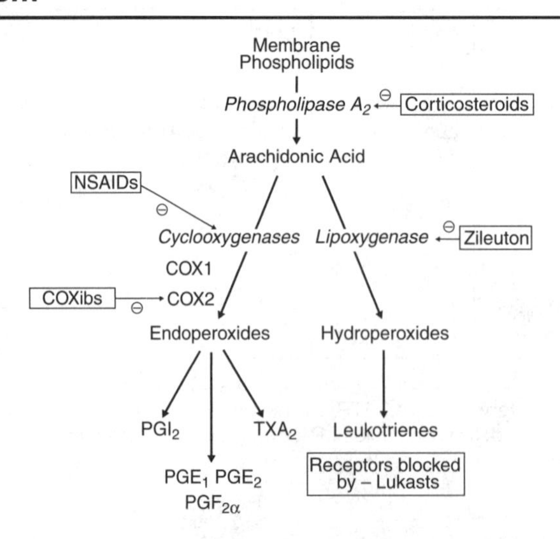

Definition of abbreviations: AR, autosomal recessive; COX, cyclooxygenase; NSAIDs, nonsteroidal anti-inflammatory drugs; XR, X-linked recessive.
*Common in Ashkenazi Jews

▶ Enzyme Kinetics

Whereas the thermodynamic equilibrium of a chemical reaction is determined by its **free energy** (ΔG), the rate at which the reaction reaches equilibrium is determined by its **activation energy** ($\Delta G^{\ddagger}$). Enzymes increase the rate of a reaction by reducing the energy of activation without affecting the equilibrium constant.

Michaelis-Menten Equation	Lineweaver-Burk Equation
$V = \frac{V_{max}[S]}{K_m + [S]}$ V = initial rate or velocity of reaction [S] = substrate concentration V_{max} = maximum rate of enzyme K_m = substrate concentration at $V_{max}/2$	$\frac{1}{V} = \frac{K_m}{V_{max}} \frac{1}{[S]} + \frac{1}{V_{max}}$ Reciprocal form of the Michaelis-Menten equation to achieve a straight line plot

In a typical enzyme-catalyzed reaction, the enzyme (E) is thought to bind reversibly to a substrate (S), forming a complex (ES), from which the product (P) dissociates as the reaction proceeds.

$$E + S \leftrightarrow E - S \rightarrow E + P,$$

where E is the enzyme, S is the substrate and P is the reaction product

The rate of a reaction as determined by both the concentration of enzyme (E) and substrate (S) is described by the **Michaelis-Menten equation**.

Michaelis-Menten Plot | **Lineweaver-Burk Plot**

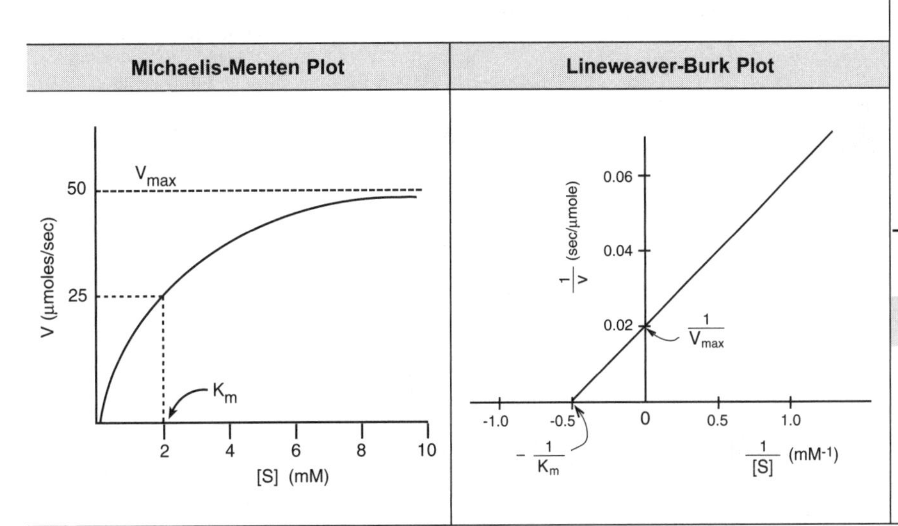

Classes of Inhibitors

Competitive, Reversible

(often substrate analogs that compete for the enzyme's binding site)

V_{max}: no effect
K_m: ↑

1/V

+ Competitive Inhibitor

No Inhibitor Present

0

1/[S]

Noncompetitive, Reversible

(bind outside active site but affects enzyme activity, possibly allosterically)

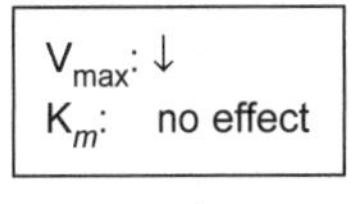

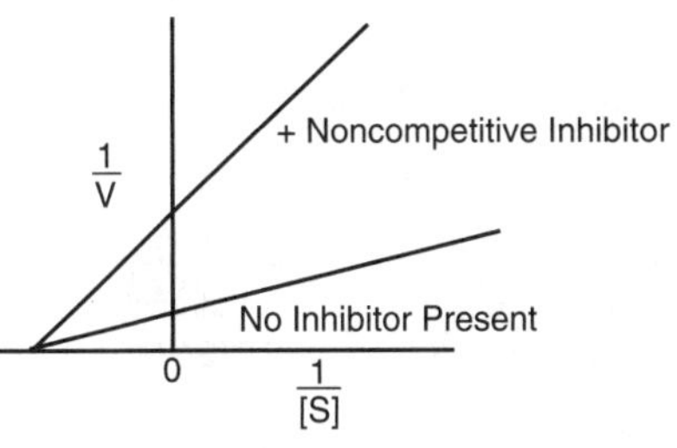

Irreversible (Inactivator)

(binds and inactivates enzyme permanently)

V_{max}: ↓
K_m: no effect

▶ Water-Soluble Vitamins

Vitamin or Coenzyme	Enzyme	Pathway	Deficiency
Biotin	Pyruvate carboxylase Acetyl-CoA carboxylase Propionyl-CoA carboxylase	Gluconeogenesis Fatty acid synthesis Odd-carbon fatty acids, Val, Met, Ile, Thr	Causes (rare): excessive consumption of raw eggs (contain avidin, a biotin-binding protein) Alopecia (hair loss), bowel inflammation, muscle pain
Thiamine (B_1)	Pyruvate dehydrogenase α-Ketoglutarate dehydrogenase Transketolase	PDH TCA cycle HMP shunt	Causes: alcoholism (alcohol interferes with absorption) Wernicke (ataxia, nystagmus, ophthalmoplegia) Korsakoff (confabulation, psychosis) High-output cardiac failure (wet beri-beri)
Niacin (B_3) NAD(H) NADP(H)	Dehydrogenases	Many	Pellagra may also be related to deficiency of tryptophan (corn major dietary staple), which supplies a portion of the niacin requirement Pellagra: diarrhea, dementia, dermatitis, and, if not treated, death
Folic acid THF	Thymidylate synthase Purine synthesis enzymes	Thymidine (pyrimidine) synthesis Purine synthesis	Causes: alcoholics and pregnancy (body stores depleted in 3 months) Homocystinemia with risk of deep vein thrombosis and atherosclerosis Megaloblastic (macrocytic) anemia Deficiency in early pregnancy causes neural tube defects in fetus
Cyanocobalamin (B_{12})	Homocysteine methyltransferase Methylmalonyl-CoA mutase	Methionine, SAM Odd-carbon fatty acids, Val, Met, Ile, Thr	Causes: pernicious anemia. Also in aging, especially with poor nutrition, bacterial overgrowth of terminal ileum, resection of the terminal ileum secondary to Crohn disease, chronic pancreatitis, and, rarely, vegans, or infection with *Diphyllobothrium latum* Megaloblastic (macrocytic) anemia Progressive peripheral neuropathy
Pyridoxine (B_6) PLP	Aminotransferases (transaminase): AST (SGOT), ALT (SGPT) δ-Aminolevulinate synthase	Protein catabolism Heme synthesis	Causes: isoniazid therapy Sideroblastic anemia Cheilosis or stomatitis (cracking or scaling of lip borders and corners of the mouth) Convulsions
Riboflavin (B_2) FAD(H_2)	Dehydrogenases	Many	Corneal neovascularization Cheilosis or stomatitis (cracking or scaling of lip borders and corners of the mouth) Magenta-colored tongue
Ascorbate (C)	Prolyl and lysyl hydroxylases Dopamine β-hydroxylase	Collagen synthesis Catecholamine synthesis Absorption of iron in GI tract	Causes: diet deficient in citrus fruits and green vegetables Scurvy: poor wound healing, easy bruising (perifollicular hemorrhage), bleeding gums, increased bleeding time, painful glossitis, anemia
Pantothenic acid CoA	Fatty acid synthase Fatty acyl CoA synthetase Pyruvate dehydrogenase α-Ketoglutarate dehydrogenase	Fatty acid metabolism PDH TCA cycle	Rare

Definition of abbreviations: ALT, alanine aminotransferase; AST, aspartate aminotransferase; CoA, coenzyme A; FAD(H_2), flavin adenine dinucleotide; HMP, hexose monophosphate shunt; NAD(H); nicotinamide adenine dinucleotide; NADP(H), nicotinamide adenine dinucleotide phosphate; PDH, pyruvate dehydrogenase; PLP, pyridoxal phosphate, SAM, S-adenosylmethionine; TCA, tricarboxylic acid cycle; THF, tetrahydrofolate.

► Lipid-Soluble Vitamins

Vitamin	Important Functions	Deficiency
D (cholecalciferol)	In response to hypocalcemia, helps normalize serum calcium levels	Rickets (in childhood): skeletal abnormalities (especially legs), muscle weakness After epiphysial fusion: osteomalacia
A (carotene)	Retinoic acid and retinol act as growth regulators, especially in epithelium Retinal is important in rod and cone cells for vision	Night blindness, metaplasia of corneal epithelium, dry eyes, bronchitis, pneumonia, follicular hyperkeratosis
K	Carboxylation of glutamic acid residues in many Ca^{2+}-binding proteins, importantly coagulation factors II, VII, IX, and X, as well as proteins C and S	Easy bruising, bleeding Increased prothrombin time, increased INR Associated with fat malabsorption, long-term antibiotic therapy, breast-fed newborns, infants of mothers who took anticonvulsants during pregnancy
E (α-tocopherol)	Antioxidant in the lipid phase; protects membrane lipids from peroxidation and helps prevent oxidation of LDL particles thought to be involved in atherosclerotic plaque formation	Hemolysis, neurologic problems, retinitis pigmentosa

Molecular Biology, Genetics, and Cell Biology

Chapter 3

► Nucleic Acid Structure

Nucleic acids, including **DNA** and **RNA**, are assembled from **nucleotides**, which contain a five-carbon sugar, a nitrogenous base, and phosphate. The sugar may be **ribose** (RNA) or **deoxyribose** (DNA). The base can be a **purine** (adenine or guanine) or **pyrimidine** (cytosine, uracil, thymidine). Phosphate groups link the 3′ carbon of one sugar to the 5′ carbon of the next, forming phosphodiester bonds. Base sequences are conventionally written in a **5′ → 3′** direction. Nucleotides lacking phosphate groups are called **nucleosides**. In prokaryotes and eukaryotes, RNA is generally single-stranded, while DNA is generally double-stranded in an **antiparallel** orientation, with two hydrogen bonds between base pairs **A and T** and three between **G and C**. Nuclear DNA forms a **double-helix**, which undergoes **supercoiling** via **topoisomerase** activity, and is generally associated with **histones** and other proteins to form **nucleosomes**, the basic packaging unit of **chromatin**.

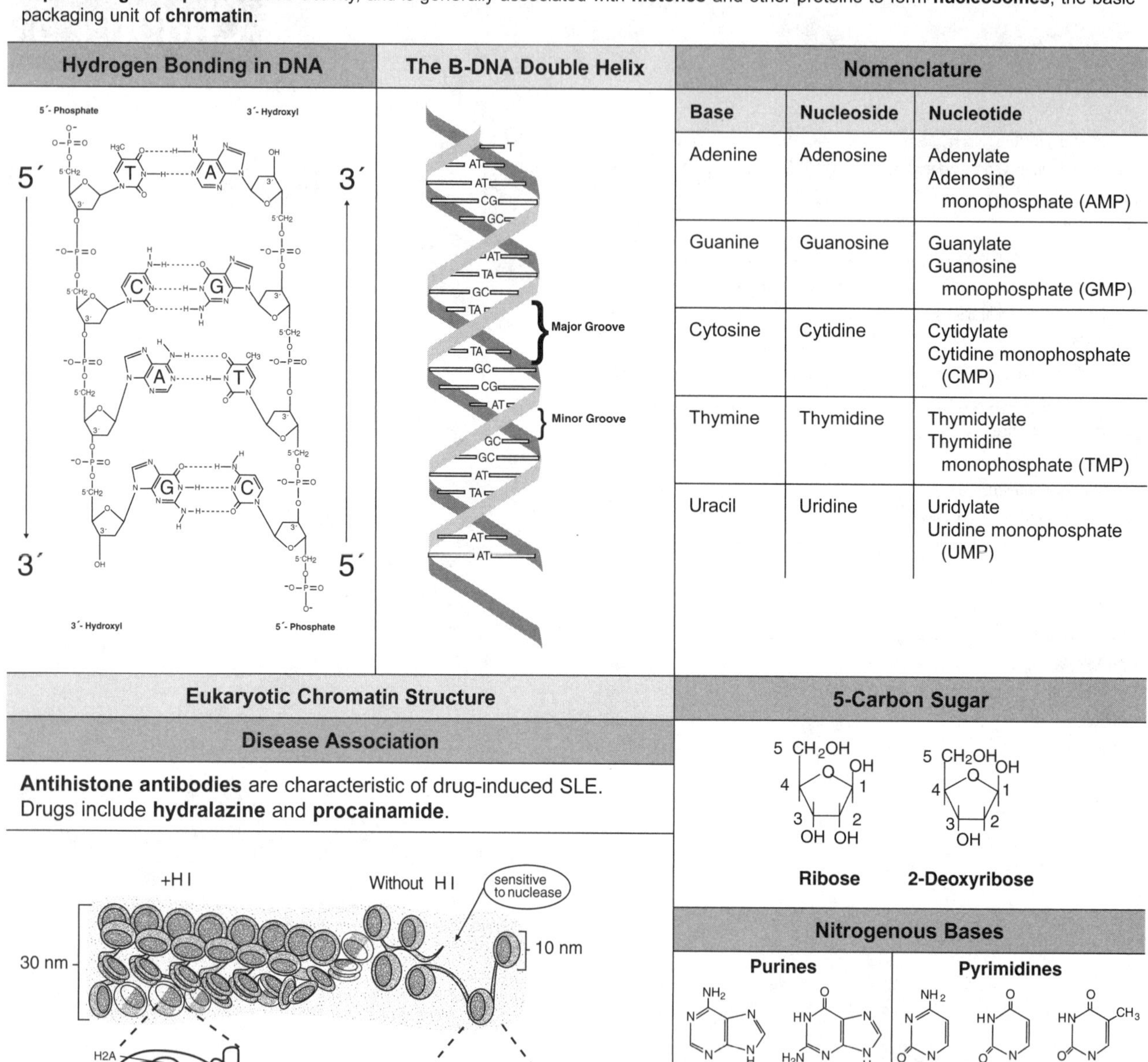

Nomenclature

Base	Nucleoside	Nucleotide
Adenine	Adenosine	Adenylate Adenosine monophosphate (AMP)
Guanine	Guanosine	Guanylate Guanosine monophosphate (GMP)
Cytosine	Cytidine	Cytidylate Cytidine monophosphate (CMP)
Thymine	Thymidine	Thymidylate Thymidine monophosphate (TMP)
Uracil	Uridine	Uridylate Uridine monophosphate (UMP)

Disease Association

Antihistone antibodies are characteristic of drug-induced SLE. Drugs include **hydralazine** and **procainamide**.

Chromatin

Euchromatin	Heterochromatin
Loosely packed	Tightly packed
Transcriptionally active	Transcriptionally inactive

► Nucleotide Synthesis and Salvage

Nucleotides for DNA and RNA synthesis can be generated by de novo synthesis or salvage pathways, both of which require **PRPP** generated from ribose-5-phosphate derived from the HMP shunt. **De novo synthesis** occurs mainly in the liver and generates new purine and pyrimidine bases from precursors. In contrast, **salvage pathways** reuse preformed bases derived from nucleotides during normal RNA turnover or released from dying cells or transported from the liver. Ribonucleotides are converted to deoxyribonucleotides for DNA synthesis by **ribonucleotide reductase**. Antineoplastic drugs that target ribonucleotide reductase (hydroxyurea), or an enzyme in the dTMP branch of pyrimidine synthesis (5-FU), or reduction of folate (methotrexate) preferentially inhibit DNA synthesis without compromising RNA synthesis and gene expression. Excretion of purine bases occurs in the form of **uric acid** from the kidneys.

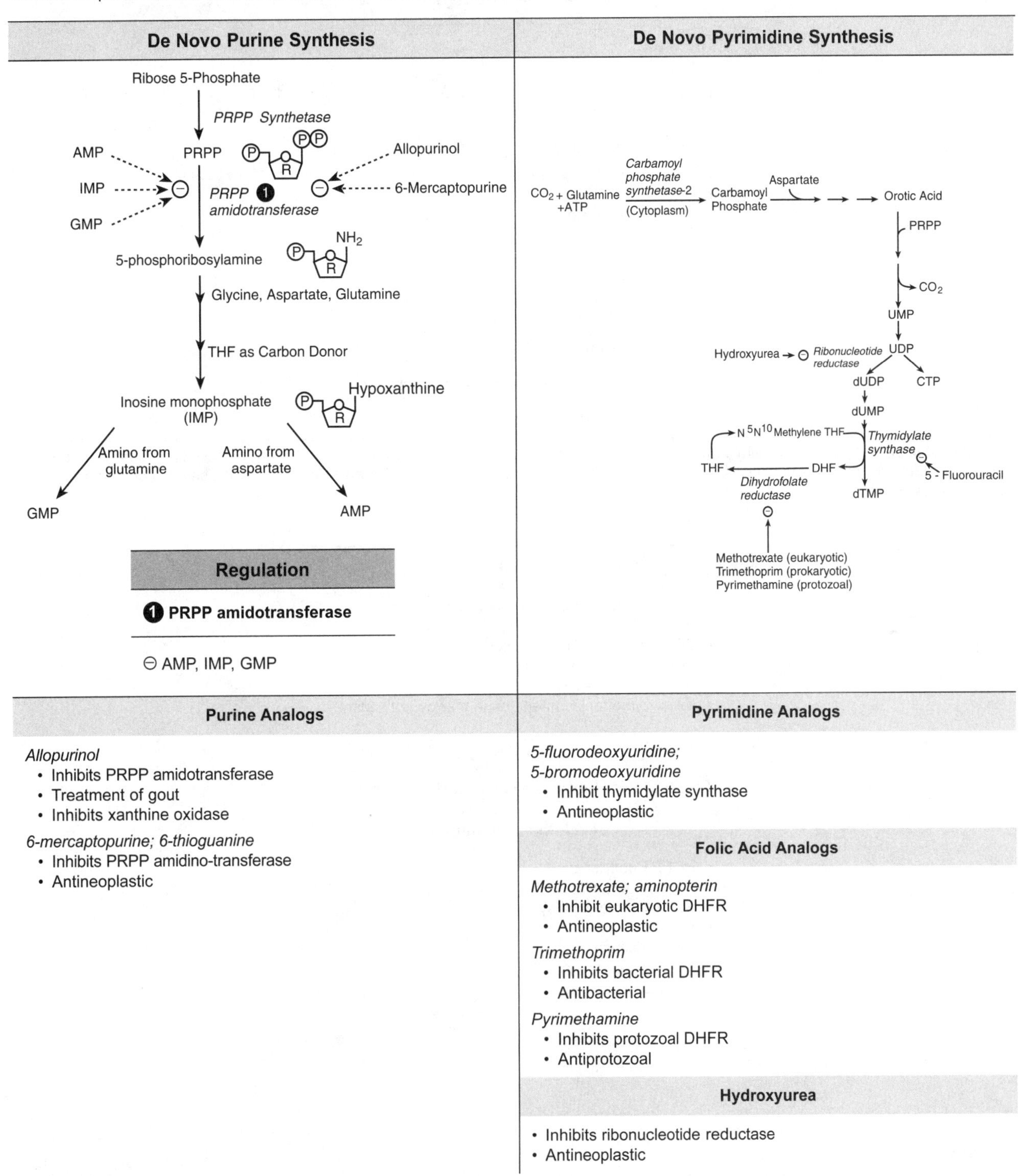

Purine Analogs

Allopurinol
- Inhibits PRPP amidotransferase
- Treatment of gout
- Inhibits xanthine oxidase

6-mercaptopurine; 6-thioguanine
- Inhibits PRPP amidino-transferase
- Antineoplastic

Pyrimidine Analogs

5-fluorodeoxyuridine;
5-bromodeoxyuridine
- Inhibit thymidylate synthase
- Antineoplastic

Folic Acid Analogs

Methotrexate; aminopterin
- Inhibit eukaryotic DHFR
- Antineoplastic

Trimethoprim
- Inhibits bacterial DHFR
- Antibacterial

Pyrimethamine
- Inhibits protozoal DHFR
- Antiprotozoal

Hydroxyurea

- Inhibits ribonucleotide reductase
- Antineoplastic

Definition of abbreviations: AMP, adenosine monophosphate; dTMP, deoxythymidine monophosphate; 5-FU, 5-flurouracil; IMP, inosine monophosphate; dUDP, deoxyuridine diphosphate; CTP, cytosine triphosphate; THF, tetrahydrofolate.

(*Continued*)

► Nucleotide Synthesis and Salvage (*Cont'd.*)

Purine Salvage Pathway and Excretion

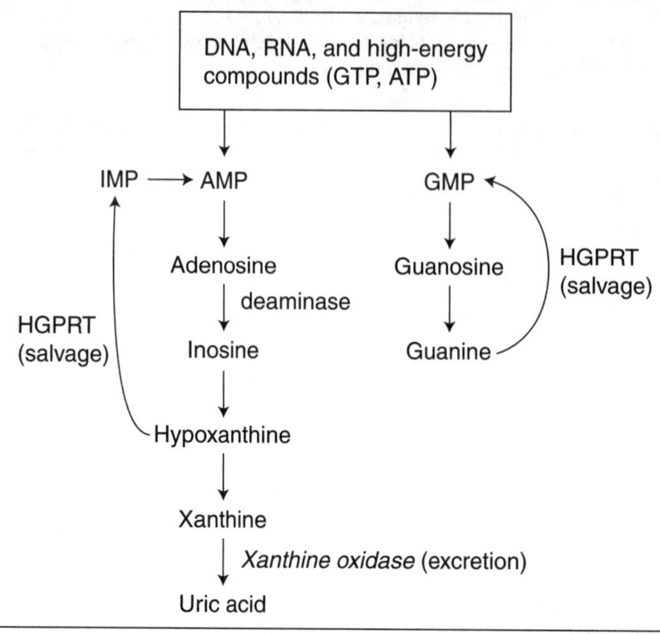

Definition of abbreviations: DHFR, dihydrofolate reductase; HGPRT, hypoxanthine-guanine phosphoribosyl pyrophosphate transferase; NSAID, nonsteroidal anti-inflammatory drug; PRPP, phosphoribosylpyrophosphate; THF, tetrahydrofolate; SCID, severe combined immunodeficiency disorder.

Disease Association
Adenosine Deaminase Deficiency
• SCID (no B- or T-cell function) • Multiple infections in children • Autosomal recessive • Tx: enzyme replacement, bone marrow transplant
Gout
↑ production or ↓ excretion of uric acid by kidneys
Lesch-Nyhan Syndrome
• HGPRT deficiency • Mental retardation (mild) • Spastic cerebral palsy • Self-mutilation • Hyperuricemia • X-linked recessive

▶ DNA Replication

DNA replication involves the synthesis of new DNA molecules in a 5′ → 3′ direction by **DNA polymerase** using the double-stranded DNA template. One strand (**leading strand**) is made continuously, while the other (**lagging strand**) is synthesized in segments. **Prokaryotic** chromosomes are closed, double-stranded circular DNA molecules with a single origin of replication that separates into two replication forks moving away in opposite directions. **Eukaryotic** chromosomes are double-stranded and linear with multiple origins of replication.

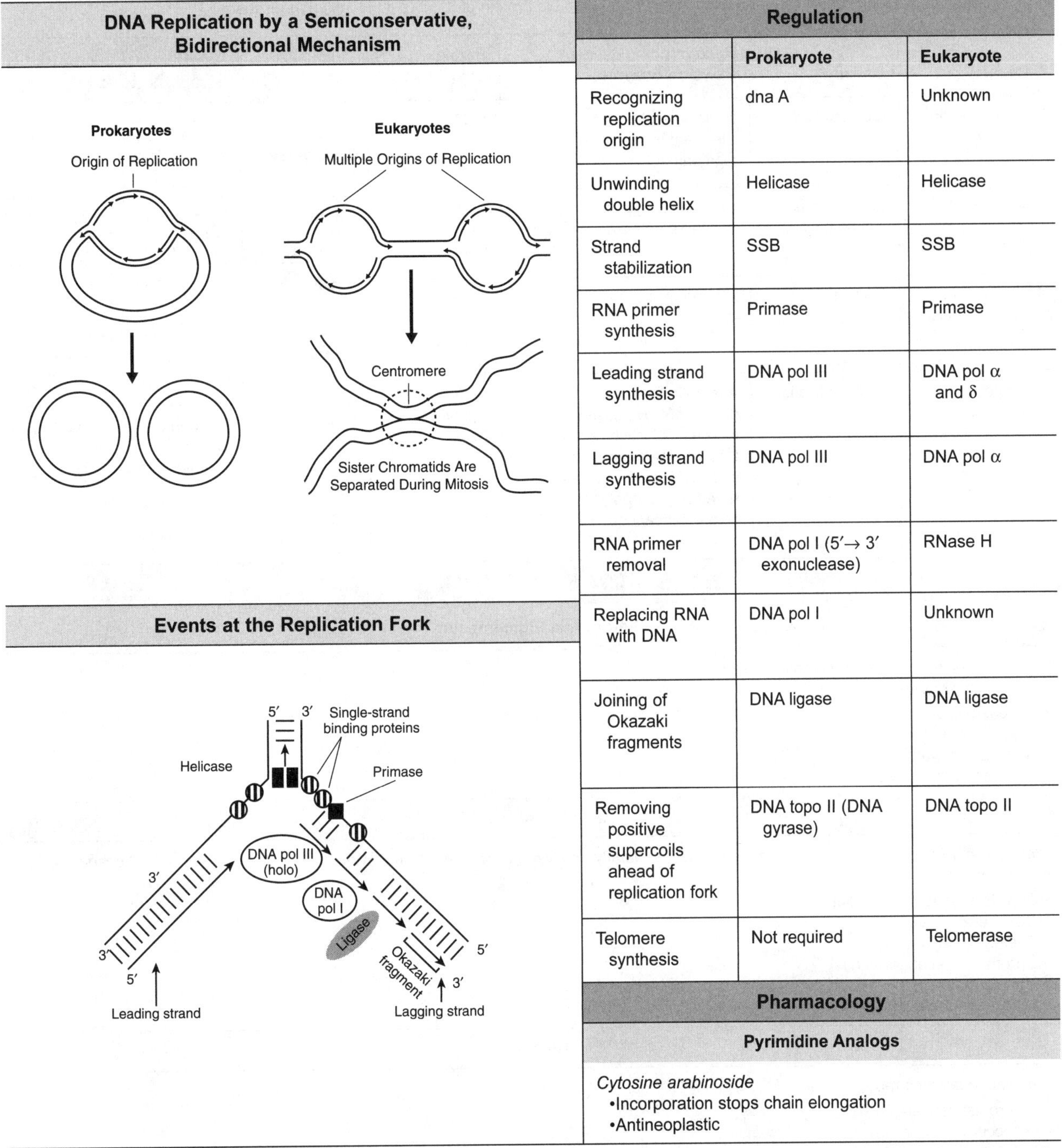

Regulation

	Prokaryote	Eukaryote
Recognizing replication origin	dna A	Unknown
Unwinding double helix	Helicase	Helicase
Strand stabilization	SSB	SSB
RNA primer synthesis	Primase	Primase
Leading strand synthesis	DNA pol III	DNA pol α and δ
Lagging strand synthesis	DNA pol III	DNA pol α
RNA primer removal	DNA pol I (5′→ 3′ exonuclease)	RNase H
Replacing RNA with DNA	DNA pol I	Unknown
Joining of Okazaki fragments	DNA ligase	DNA ligase
Removing positive supercoils ahead of replication fork	DNA topo II (DNA gyrase)	DNA topo II
Telomere synthesis	Not required	Telomerase

Pharmacology

Pyrimidine Analogs

Cytosine arabinoside
- Incorporation stops chain elongation
- Antineoplastic

Definition of abbreviations: DNA, deoxyribonucleic acid; DNA pol, DNA polymerase; DNA topo, DNA topoisomerase; RNA, ribonucleic acid; SSB, single-stranded DNA-binding protein.

► DNA Repair

DNA sequence and structure may be altered either during replication or by exposure to chemicals or radiation. Mutations include point mutations such as the **substitution** of one base with another. Substitution mutations in the third position of a codon **(wobble position)** are usually benign because several codons code for the same amino acid. Other types of mutations include (1) deletion or addition of one or two nucleotides (frameshift mutations), (2) large segment deletions (e.g., unequal crossover during meiosis), (3) mutations of 5′ or 3′ splice sites, or (4) triplet repeat expansion, which can lead to a longer, more unstable protein product (e.g., Huntington disease).

Damage	Cause	Recognition/ Excision Enzyme	Repair Enzymes
Thymine dimers (G_1)	UV radiation	Excision endonuclease (deficient in xeroderma pigmentosum)	DNA polymerase DNA ligase
Cytosine deamination (G_1)	Spontaneous/ chemicals	Uracil glycosylase AP endonuclease	DNA polymerase DNA ligase
Apurination or apyrimidination (G_1)	Spontaneous/ heat	AP endonuclease	DNA polymerase DNA ligase
Mismatched base (G_2)	DNA replication errors	A mutation on one of two genes, *hMSH2* or *hMLH1*, initiates defective repair of DNA mismatches, resulting in a condition known as hereditary nonpolyposis colorectal cancer—HNPCC.	DNA polymerase DNA ligase

Types of Mutations

Transition: A:T → G:C or G:C → A:T
Transversion: A:T → T:A or G:C → C:G

Silent	No change in AA	Sub
Missense	Change AA to another	Sub
Nonsense	Early stop codon	Sub or Ins/Del
Frameshift	Misreading of all codons downstream	Ins/Del

DNA Repair Defects

Xeroderma Pigmentosum

(defect in nucleotide excision-repair)

- Extreme UV sensitivity
- Excessive freckling
- Multiple skin cancers
- Corneal ulcerations
- Autosomal recessive

Ataxia Telangiectasia

(defect in ATM gene product, a member of PI-3 kinase family involved in mitogenic signal transduction, detection of DNA damage, and cell cycle control)

- Sensitivity to ionizing radiation
- Degenerative ataxia
- Dilated blood vessels
- Chromosomal aberrations
- Lymphomas
- Autosomal recessive

HNPCC

(defect in mismatch repair; usually hMSH2 or hMLH1 gene)

- Colorectal cancer
- $2/3$ occur in right colon
- Autosomal dominant
- Part of Lynch syndrome (a multi-cancer syndrome)

Definition of abbreviations: AA, amino acid; HNPCC, hereditary nonpolyposis colorectal cancer; Ins/Del, insert or deletion; Sub, substitution.

▶ Transcription and RNA Processing

Transcription involves the synthesis of an RNA in a 5′ → 3′ direction by an RNA polymerase using DNA as a template. An important class of RNA is messenger RNA (mRNA). Initiation of transcription occurs from a promoter region, which is the binding site of RNA polymerase, and stops at a termination signal. In **prokaryotes**, a single mRNA transcript can encode several genes **(polycistronic)**, and no RNA processing is required, allowing transcription and translation to proceed simultaneously. In **eukaryotes**, all mRNAs are **monocistronic**, but often include coding segments **(exons)** interrupted by noncoding regions **(introns)**. Eukaryotic mRNAs must therefore undergo extensive processing, including a 5′ cap, a 3′ tail, and removal of introns followed by exon splicing. Ribosomal RNA (rRNA) and transfer RNA (tRNA) are also produced by transcription.

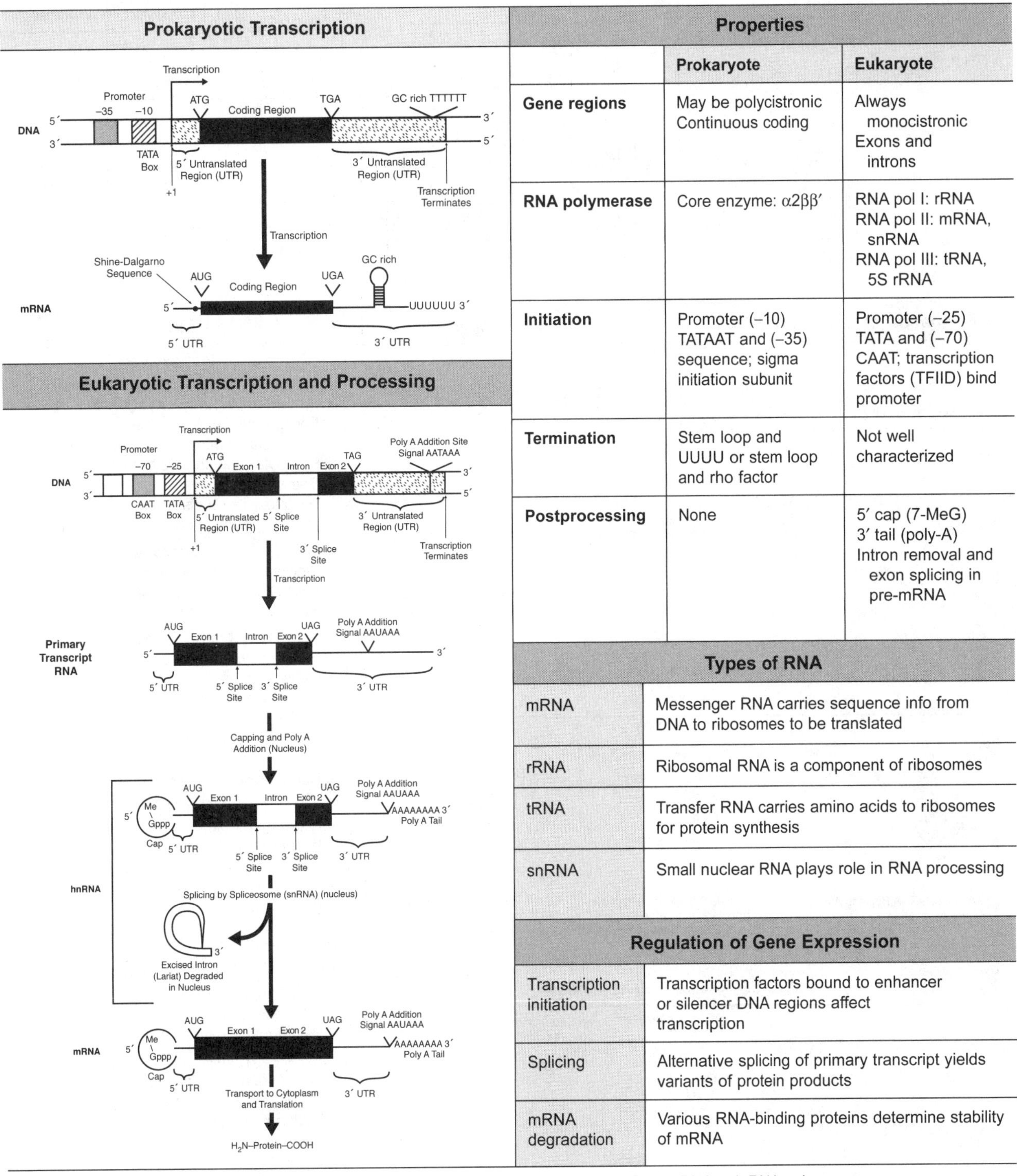

Properties		
	Prokaryote	**Eukaryote**
Gene regions	May be polycistronic Continuous coding	Always monocistronic Exons and introns
RNA polymerase	Core enzyme: $\alpha 2\beta\beta'$	RNA pol I: rRNA RNA pol II: mRNA, snRNA RNA pol III: tRNA, 5S rRNA
Initiation	Promoter (–10) TATAAT and (–35) sequence; sigma initiation subunit	Promoter (–25) TATA and (–70) CAAT; transcription factors (TFIID) bind promoter
Termination	Stem loop and UUUU or stem loop and rho factor	Not well characterized
Postprocessing	None	5′ cap (7-MeG) 3′ tail (poly-A) Intron removal and exon splicing in pre-mRNA

Types of RNA	
mRNA	Messenger RNA carries sequence info from DNA to ribosomes to be translated
rRNA	Ribosomal RNA is a component of ribosomes
tRNA	Transfer RNA carries amino acids to ribosomes for protein synthesis
snRNA	Small nuclear RNA plays role in RNA processing

Regulation of Gene Expression	
Transcription initiation	Transcription factors bound to enhancer or silencer DNA regions affect transcription
Splicing	Alternative splicing of primary transcript yields variants of protein products
mRNA degradation	Various RNA-binding proteins determine stability of mRNA

Definition of abbreviations: hnRNA, heterogeneous nuclear RNA; 7-MeG, 7-methylguanosine; RNA pol, RNA polymerase; UTR, untranslated region.

▶ Protein Translation

Translation involves the synthesis of protein from mRNA templates in **ribosomes** (complexes of proteins and ribosomal RNAs [**rRNA**]). Protein synthesis begins from an initiation codon (**AUG** = methionine) and ends at a stop codon (**UAA**, **UGA**, or **UAG**). Elongation involves transfer RNAs (**tRNA**), which have an anticodon region at one end to recognize the codon on the mRNA and an amino acid attached at the other end for covalent linkage to the growing polypeptide chain. Several ribosomes can simultaneously transcribe an mRNA, forming a polyribosome, or **polysome**.

tRNA

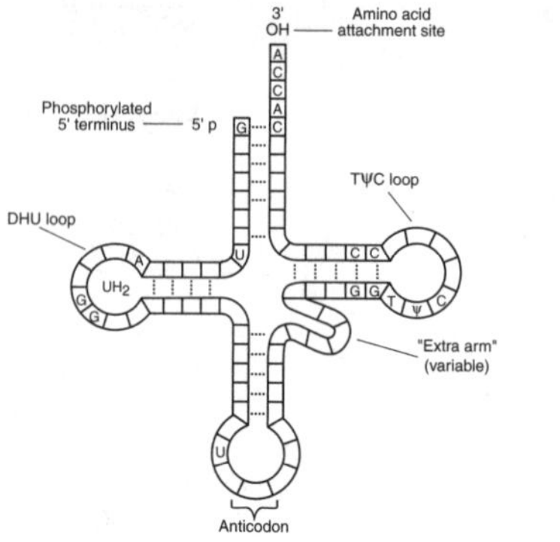

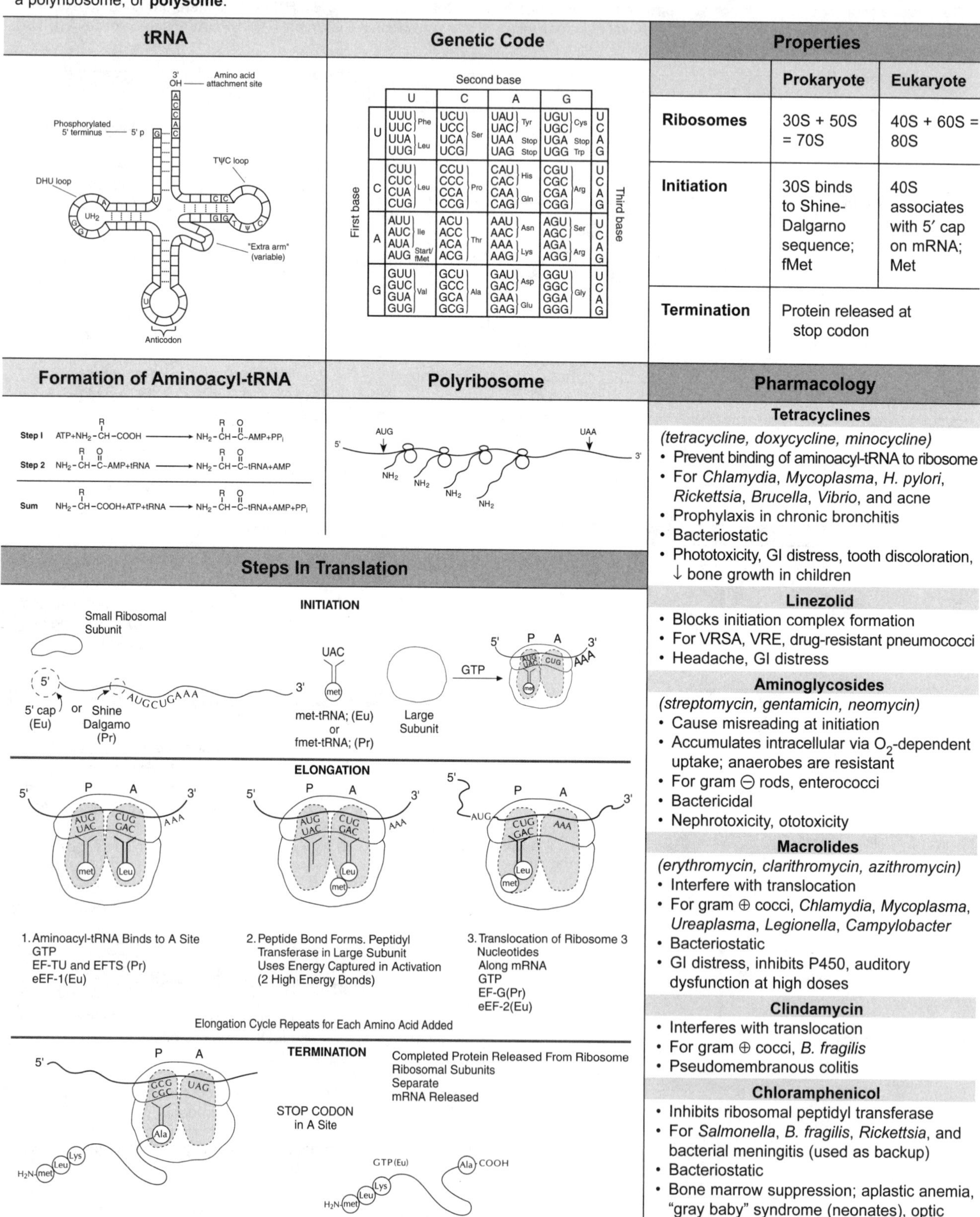

Genetic Code

Second base

First base	U	C	A	G	Third base
U	UUU Phe UUC Phe UUA Leu UUG Leu	UCU Ser UCC Ser UCA Ser UCG Ser	UAU Tyr UAC Tyr UAA Stop UAG Stop	UGU Cys UGC Cys UGA Stop UGG Trp	U C A G
C	CUU Leu CUC Leu CUA Leu CUG Leu	CCU Pro CCC Pro CCA Pro CCG Pro	CAU His CAC His CAA Gln CAG Gln	CGU Arg CGC Arg CGA Arg CGG Arg	U C A G
A	AUU Ile AUC Ile AUA Ile AUG Start/fMet	ACU Thr ACC Thr ACA Thr ACG Thr	AAU Asn AAC Asn AAA Lys AAG Lys	AGU Ser AGC Ser AGA Arg AGG Arg	U C A G
G	GUU Val GUC Val GUA Val GUG Val	GCU Ala GCC Ala GCA Ala GCG Ala	GAU Asp GAC Asp GAA Glu GAG Glu	GGU Gly GGC Gly GGA Gly GGG Gly	U C A G

Properties

	Prokaryote	Eukaryote
Ribosomes	30S + 50S = 70S	40S + 60S = 80S
Initiation	30S binds to Shine-Dalgarno sequence; fMet	40S associates with 5′ cap on mRNA; Met
Termination	Protein released at stop codon	

Formation of Aminoacyl-tRNA

Step 1 $ATP + NH_2\text{-}CH(R)\text{-}COOH \longrightarrow NH_2\text{-}CH(R)\text{-}C(=O)\text{-}AMP + PP_i$

Step 2 $NH_2\text{-}CH(R)\text{-}C(=O)\text{-}AMP + tRNA \longrightarrow NH_2\text{-}CH(R)\text{-}C(=O)\text{-}tRNA + AMP$

Sum $NH_2\text{-}CH(R)\text{-}COOH + ATP + tRNA \longrightarrow NH_2\text{-}CH(R)\text{-}C(=O)\text{-}tRNA + AMP + PP_i$

Polyribosome

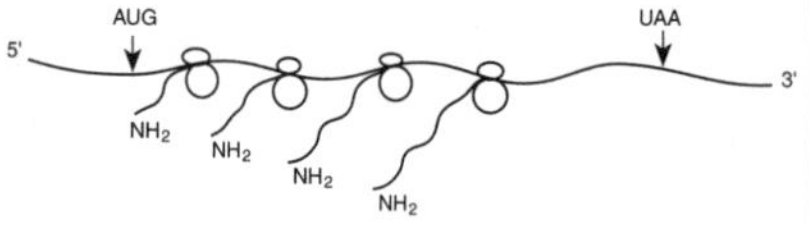

Steps In Translation

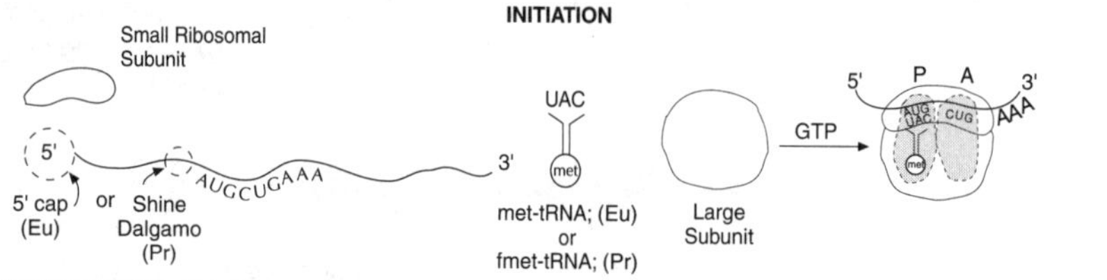

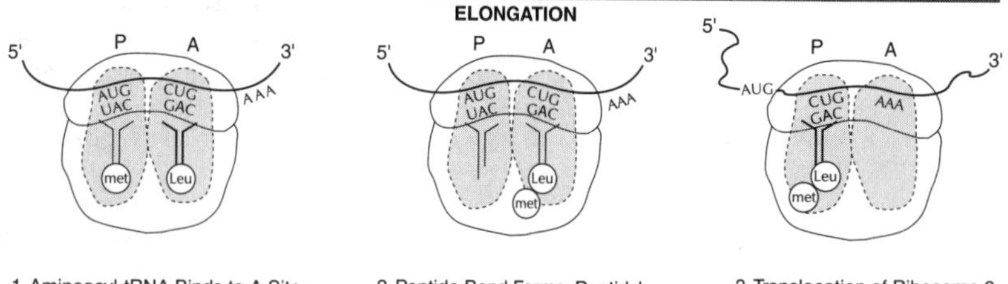

1. Aminoacyl-tRNA Binds to A Site
GTP
EF-TU and EFTS (Pr)
eEF-1(Eu)

2. Peptide Bond Forms. Peptidyl Transferase in Large Subunit Uses Energy Captured in Activation (2 High Energy Bonds)

3. Translocation of Ribosome 3 Nucleotides Along mRNA
GTP
EF-G(Pr)
eEF-2(Eu)

Elongation Cycle Repeats for Each Amino Acid Added

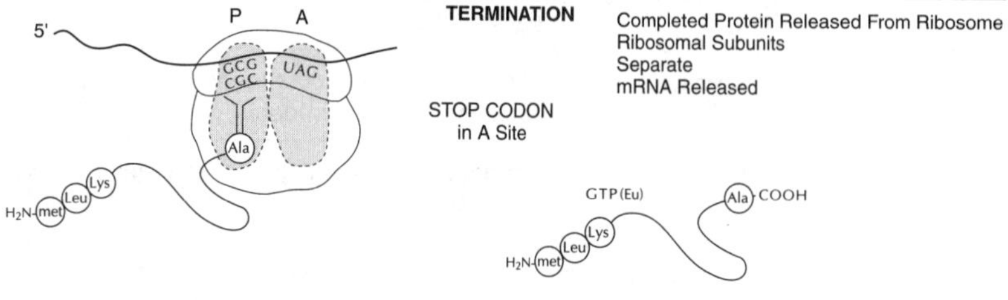

Pharmacology

Tetracyclines

(tetracycline, doxycycline, minocycline)

- Prevent binding of aminoacyl-tRNA to ribosome
- For *Chlamydia*, *Mycoplasma*, *H. pylori*, *Rickettsia*, *Brucella*, *Vibrio*, and acne
- Prophylaxis in chronic bronchitis
- Bacteriostatic
- Phototoxicity, GI distress, tooth discoloration, ↓ bone growth in children

Linezolid

- Blocks initiation complex formation
- For VRSA, VRE, drug-resistant pneumococci
- Headache, GI distress

Aminoglycosides

(streptomycin, gentamicin, neomycin)

- Cause misreading at initiation
- Accumulates intracellular via O_2-dependent uptake; anaerobes are resistant
- For gram ⊖ rods, enterococci
- Bactericidal
- Nephrotoxicity, ototoxicity

Macrolides

(erythromycin, clarithromycin, azithromycin)

- Interfere with translocation
- For gram ⊕ cocci, *Chlamydia*, *Mycoplasma*, *Ureaplasma*, *Legionella*, *Campylobacter*
- Bacteriostatic
- GI distress, inhibits P450, auditory dysfunction at high doses

Clindamycin

- Interferes with translocation
- For gram ⊕ cocci, *B. fragilis*
- Pseudomembranous colitis

Chloramphenicol

- Inhibits ribosomal peptidyl transferase
- For *Salmonella*, *B. fragilis*, *Rickettsia*, and bacterial meningitis (used as backup)
- Bacteriostatic
- Bone marrow suppression; aplastic anemia, "gray baby" syndrome (neonates), optic neuritis (children)

Definition of abbreviations: AA, amino acid; EF-2, elongation factor 2; fMet, formylmethionine; Met, methionine; VRE, vancomycin-resistant enterococci; VRSA, vancomycin-resistant *Staphylococcus aureus*.

▶ Post-Translational Modifications

Whereas cytoplasmic proteins are translated on free cytoplasmic ribosomes, secreted proteins, membrane proteins, and lysosomal enzymes have an ***N*-terminal hydrophobic signal sequence** and are translated on ribosomes associated with the rough endoplasmic reticulum (RER). After translation, proteins acquire more complex structures by being folded with the help of molecular **chaperones**. Misfolded proteins are targeted for destruction by **ubiquitin** and digested in cytoplasmic protein-digesting complexes called **proteasomes**.

Co- and Postranslational Covalent Modifications	
Glycosylation	Addition of oligosaccharides
Phosphorylation	Addition of phosphate groups by protein kinases
γ-carboxylation (vitamin K dependent)	Creation of Ca^{2+} binding sites
Prenylation	Addition of farnesyl/geranyl lipid groups to peripheral membrane proteins
Mannose phosphorylation	Addition of phosphates onto mannose residues to target protein to lysosomes

Protein Structure	
Primary	Amino acid sequence
Secondary	α-Helix or β-sheets
Tertiary	Higher order 3D structure
Quaternary	Multiple subunits

Synthesis of Secretory, Membrane, and Lysosomal Proteins

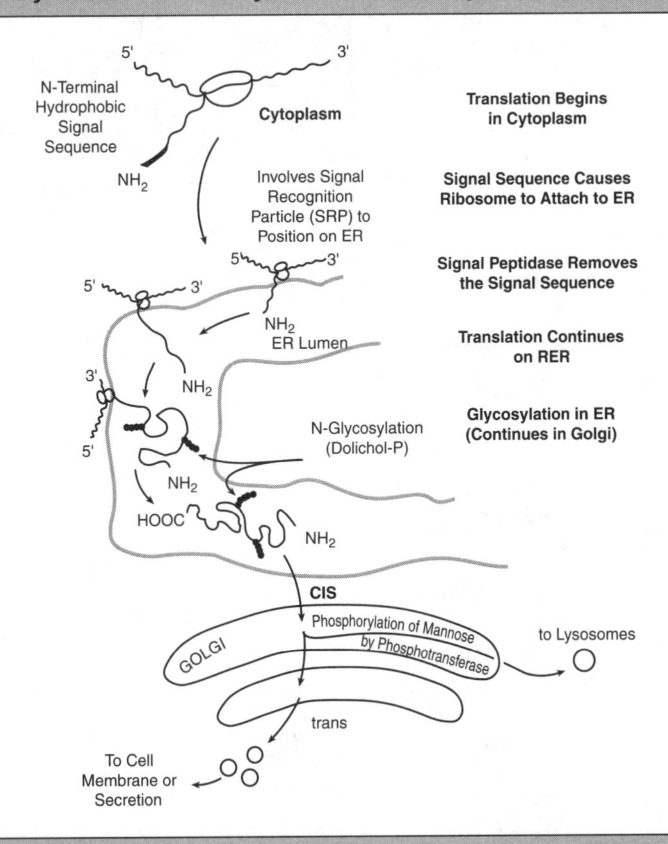

Disease Association

I-Cell Disease

(defect in mannose phosphorylation, causing lysosomal enzyme release into extracellular space and accumulation of undigested substrates in cell)

- Coarse facial features, gingival hyperplasia, macroglossia
- Craniofacial abnormalities, joint immobility, club-foot, claw-hand, scoliosis
- Psychomotor and growth retardation
- Cardiorespiratory failure
- Death in first decade
- 10–20-fold increase in lysosomal enzyme activity in serum

▶ Collagen Synthesis

Collagen is a structural protein composed of a triple helix of amino acid chains containing a repeating tripeptide Gly-X-Y-Gly-X-Y, where the unique amino acids **hydroxyproline** and **hydroxylysine** are frequently found in the X position. Hydroxylation of proline and lysine requires ascorbate (vitamin C), deficiency of which leads to scurvy.

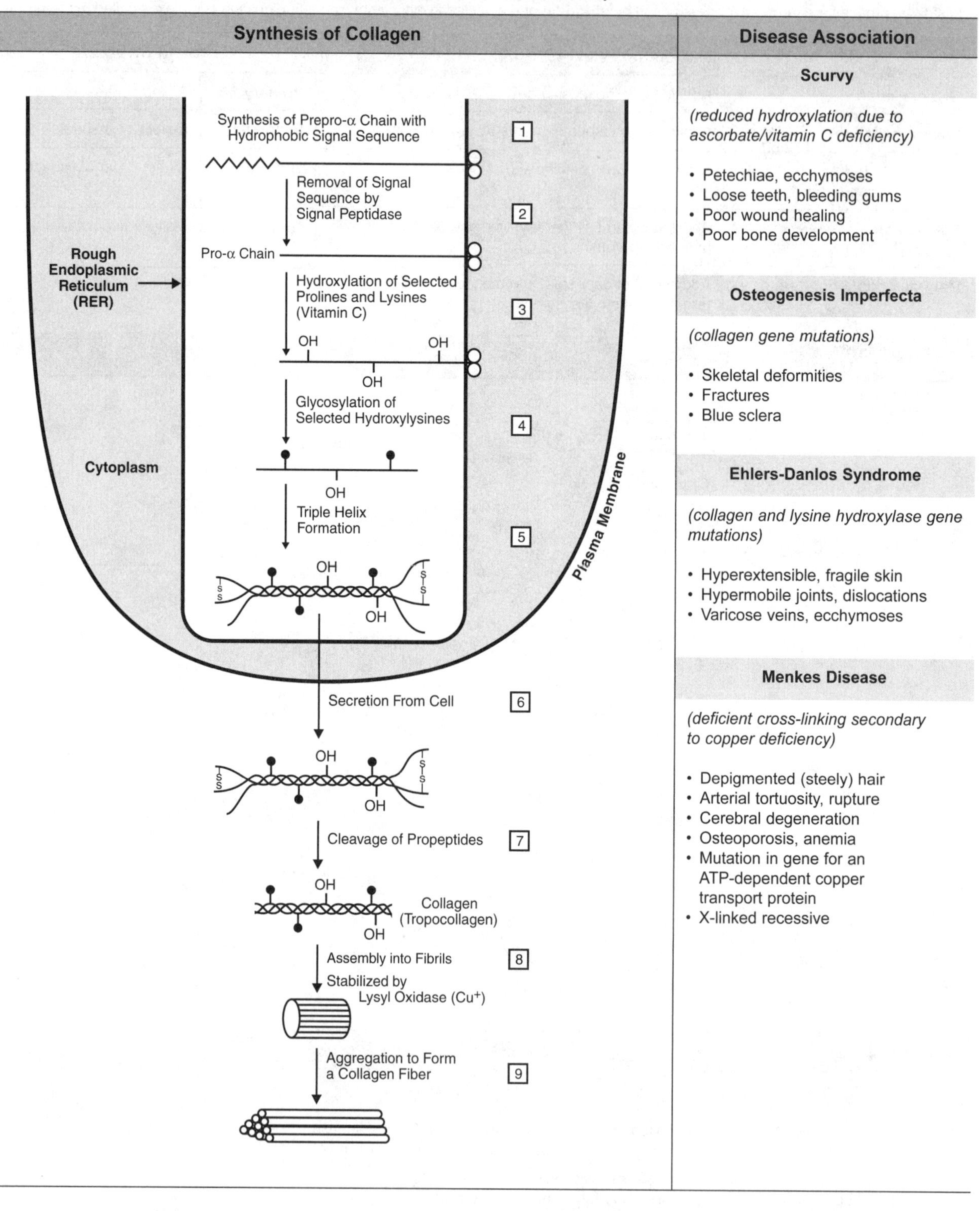

Disease Association

Scurvy

(reduced hydroxylation due to ascorbate/vitamin C deficiency)

- Petechiae, ecchymoses
- Loose teeth, bleeding gums
- Poor wound healing
- Poor bone development

Osteogenesis Imperfecta

(collagen gene mutations)

- Skeletal deformities
- Fractures
- Blue sclera

Ehlers-Danlos Syndrome

(collagen and lysine hydroxylase gene mutations)

- Hyperextensible, fragile skin
- Hypermobile joints, dislocations
- Varicose veins, ecchymoses

Menkes Disease

(deficient cross-linking secondary to copper deficiency)

- Depigmented (steely) hair
- Arterial tortuosity, rupture
- Cerebral degeneration
- Osteoporosis, anemia
- Mutation in gene for an ATP-dependent copper transport protein
- X-linked recessive

▶ Recombinant DNA

Recombinant DNA technology allows DNA fragments to be copied, manipulated, and analyzed in vitro. Eukaryotic DNA fragments may be **genomic DNA** containing both introns and exons, or **complementary DNA (cDNA)**, which is reverse-transcribed from mRNA and contains exons only. DNA fragments may be amplified by **polymerase chain reaction (PCR)**, cut with specific **restriction endonucleases**, and ligated into a **DNA vector**. These vectors can then be used for further manipulation or amplification of the DNA to produce genomic DNA or cDNA (expression) libraries, to generate recombinant proteins, or for incorporation into humans (**gene therapy**) or other animals (**transgenic animals**).

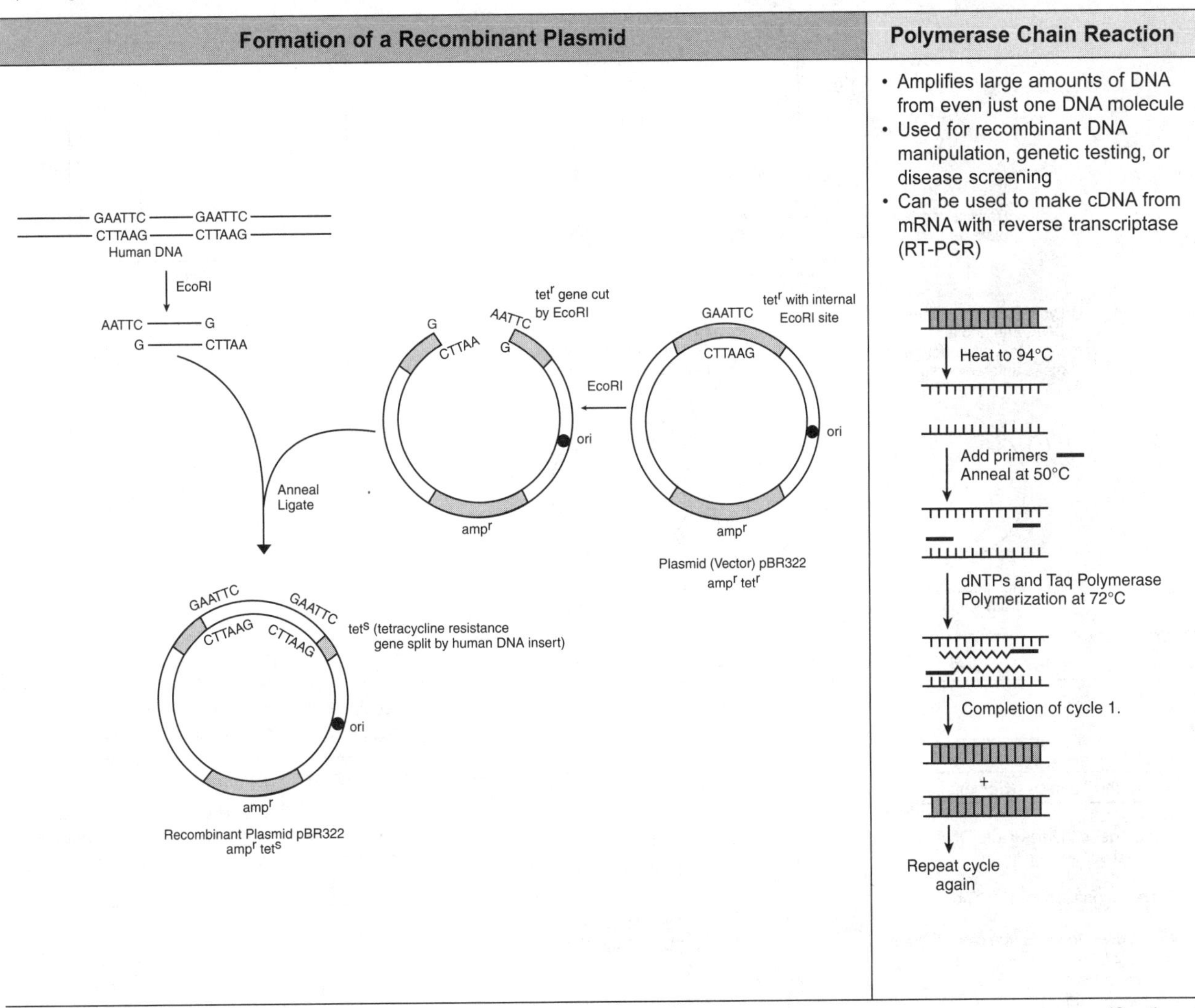

- Amplifies large amounts of DNA from even just one DNA molecule
- Used for recombinant DNA manipulation, genetic testing, or disease screening
- Can be used to make cDNA from mRNA with reverse transcriptase (RT-PCR)

(Continued)

► Recombinant DNA (*Cont'd.*)

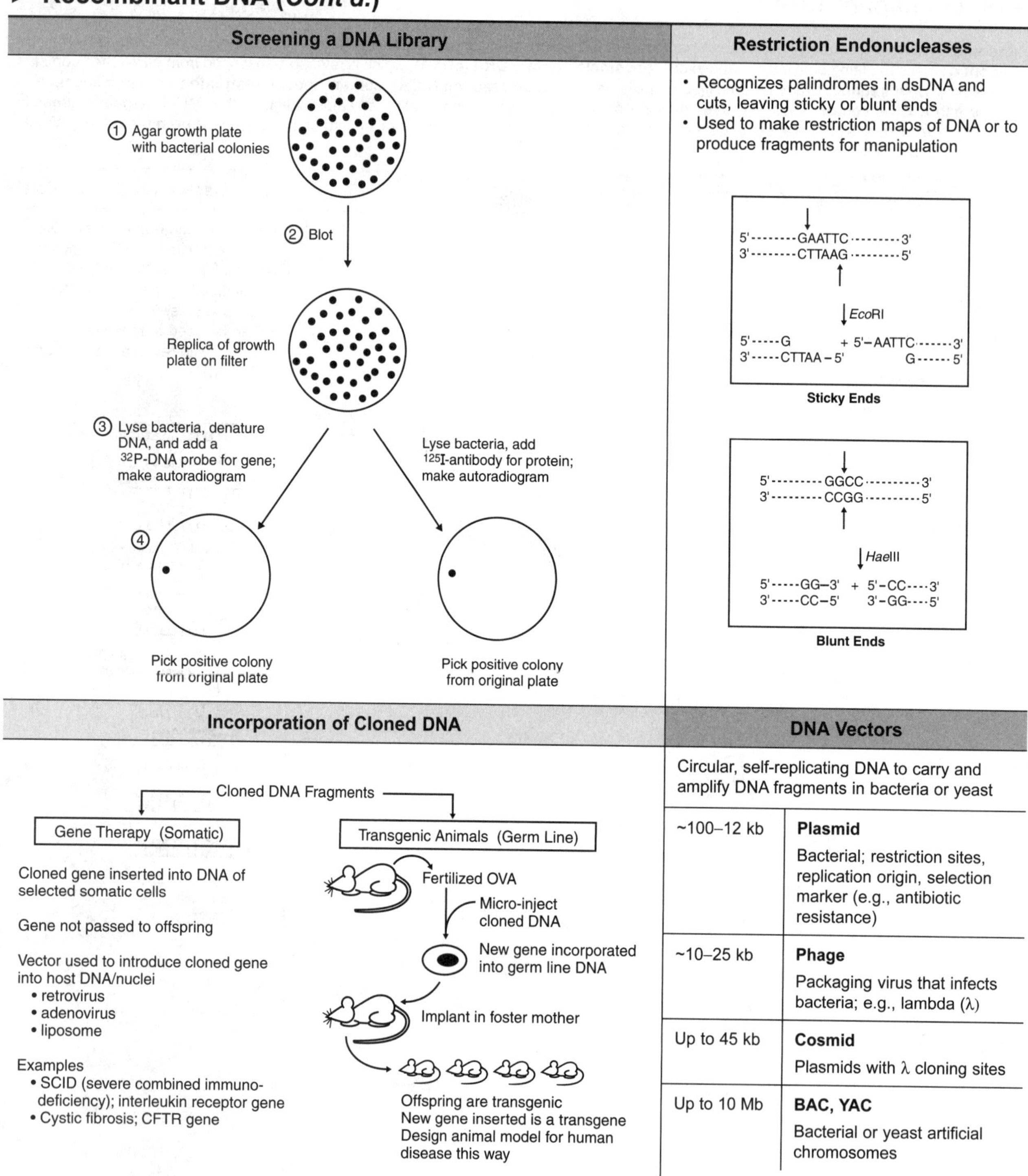

Definition of abbreviation: dsDNA, double-stranded DNA.

► Genetic Testing

The presence of specific DNA, RNA, and proteins can be identified by first separating these molecules by **gel electrophoresis**, transferring to a membrane by **blotting**, and finally detecting with radioactive nucleic acid probes (for DNA and RNA) or antibodies (for proteins). Direct detection in cells or tissues can also be performed using similar tools to identify mRNA (**in situ hybridization**) or proteins (**immunostaining**). In vitro detection of proteins can also be achieved by enzyme-linked immunosorbent assay (**ELISA**). Using these methods of detection, diversity between individuals or genetic mutations manifested by different restriction endonuclease sites (restriction fragment length polymorphisms [**RFLP**]) or expansion of highly repetitive sequences (e.g., **satellites**, **minisatellites**, and **microsatellites**) may be employed for genetic testing.

	DNA	RNA	Protein
Separation	Gel electrophoresis		
Blotting (probe)	Southern (^{32}P-DNA)	Northern (^{32}P-DNA)	Western (^{125}I or antibody)
Other detection	—	In situ hybridization	Immunostaining or ELISA

	Repeated Unit	Length of Repeat
Satellites	20–175 bp	0.1–1 Mb
Minisatellites	20–70 bp	Up to 20 kb
Microsatellites	2–4 bp	<150 bp

Sickle Cell Disease (Southern Blot; RFLP)

*Mst*II restriction digest of patient sample, followed by Southern blotting using a probe against β-globin gene, allows identification of either the normal or sickle allele.

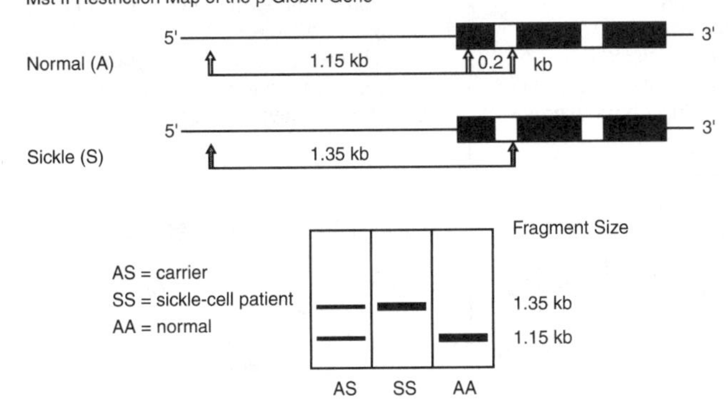

Paternity Testing (PCR; STRs or Microsatellites)

PCR amplification of STRs or microsatellite sequences can be used to match the banding pattern to each parent. The child should share one allele with each parent.

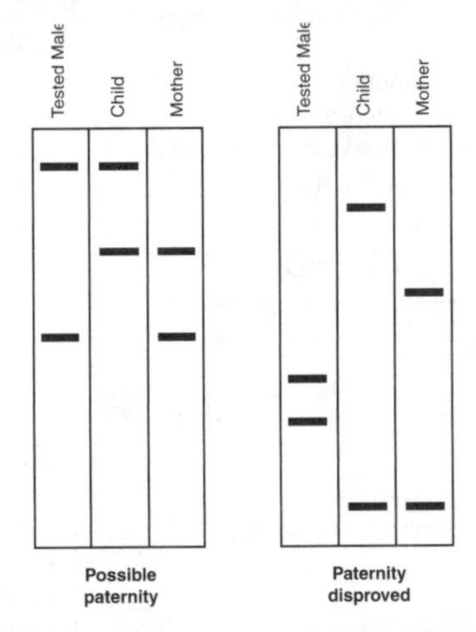

Cystic Fibrosis (PCR; ASO Dot Blot)

The most common CF mutation, ΔF508, can be detected by comparing PCR product sizes by gel electrophoresis or hybridization with allele-specific oligonucleotide (ASO) probes on a dot blot (a simplified form of Southern blot with no electrophoresis required).

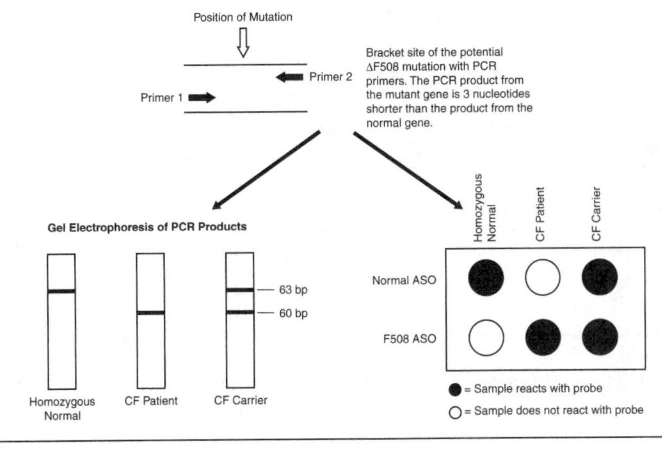

HIV Detection (ELISA and Western Blot)

Serum antibodies to HIV are first detected by ELISA and then confirmed by Western blot.

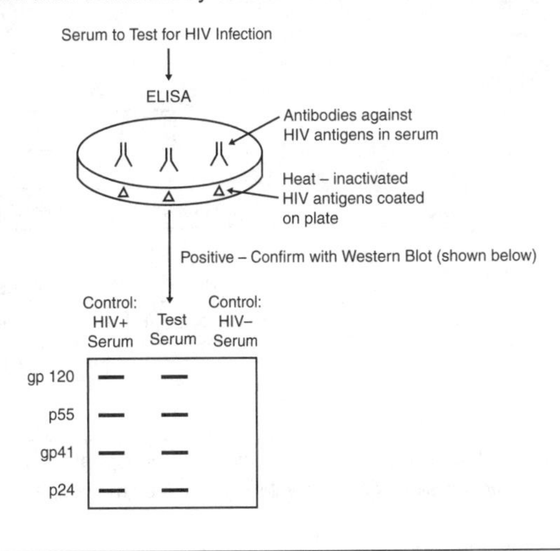

Definition of abbreviation: STR, short tandem repeats

▶ Patterns of Inheritance

Autosomal Dominant

- Affected individuals have an affected parent
- Either sex affected
- Variable to late onset (may be delayed to adulthood)
- Often encode structural proteins

- **Familial hypercholesterolemia**
- **Huntington disease**
- **Neurofibromatosis**
- **Marfan syndrome**
- **von Hippel-Lindau disease**

Autosomal Recessive

- Affected individuals usually have unaffected (carrier) parents
- Either sex affected
- Early uniform onset (infancy/childhood)
- Often encode catalytic proteins

- **Sickle cell anemia**
- **Cystic fibrosis**
- **Phenylketonuria (PKU)**
- **Kartagener syndrome**

Pedigree Analysis

Symbols

- Male
- Female
- Unknown Sex
- Affected
- Carrier of an Autosomal Recessive (Optional)
- Carrier of an X-linked Recessive (Optional)
- Stillborn or Abortion
- Dead
- Mating
- Consanguineous or Incestuous Mating
- Sibship
- Dizygous Twins
- Monozygous Twins

X-Linked Dominant

- Affected individuals have an affected parent
- Either sex affected
- No male-to-male transmission
- Females often have more mild and variable symptoms than males

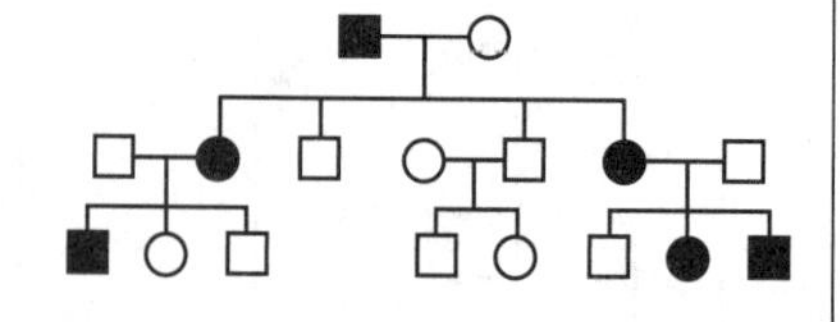

- **Fragile X syndrome**
- **Hypophosphatemic rickets**

X-Linked Recessive

- Affected individuals usually have unaffected (carrier) parents
- Usually affect males only
- No male-to-male transmission
- Female carriers sometimes show mild symptoms (manifesting heterozygote)

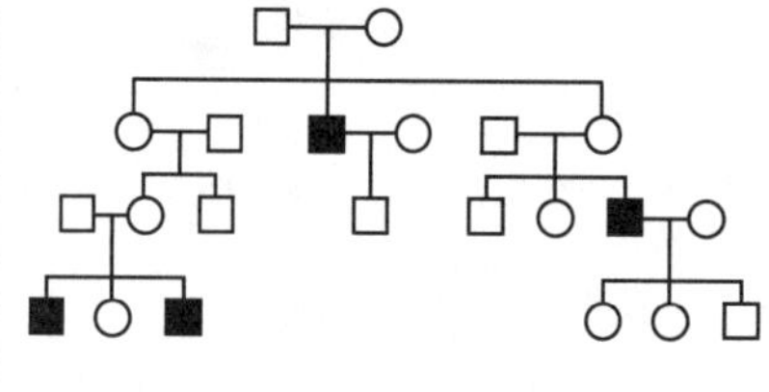

- **Duchenne muscular dystrophy**
- **Lesch-Nyhan syndrome**
- **G6PD deficiency**
- **Hemophilia A and B**

Mitochondrial Inheritance

- Inherited maternally because only mother contributes mitochondria during conception
- Either sex affected
- Usually neuropathies and myopathies because brain and muscle are highly dependent on oxidative phosphorylation

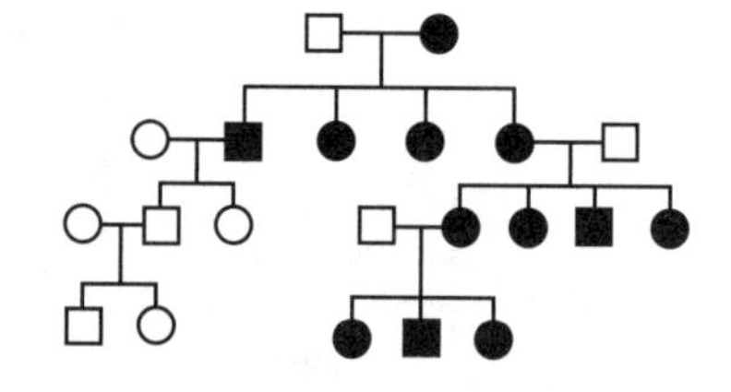

- **Leber hereditary optic neuropathy**
- **MELAS:** **m**itochondrial **e**ncephalomyopathy, **l**actic **a**cidosis, and **s**troke-like episodes
- **Myoclonic epilepsy** with ragged red muscle fiber

Decision Tree for Determining Mode of Inheritance*

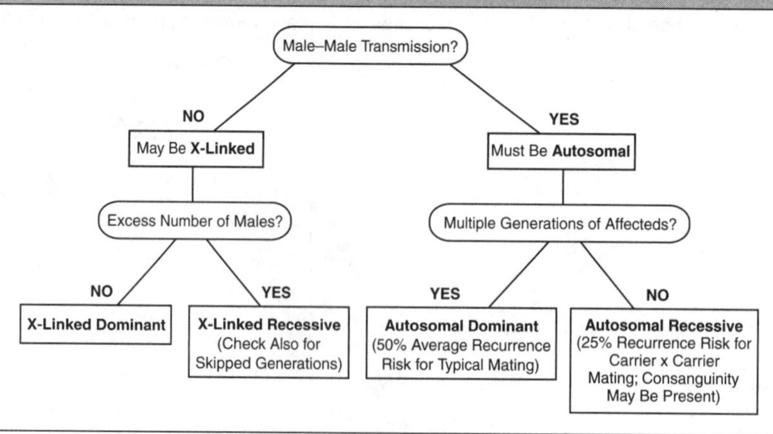

Definition of abbreviation: G6PD, glucose-6-phosphate dehydrogenase.

**Note:* If transmission occurs only through affected mothers and never through affected sons, the pedigree is likely to reflect mitochondrial inheritance.

► Single-Gene Disorders

Neurofibromatosis (NF) Type 1 (von Recklinghausen Disease)

(mutation in NF1 *tumor-suppressor gene on chromosome 17)*

- Multiple neurofibromas
- Café-au-lait spots (pigmented skin lesions)
- Lisch nodules (pigmented iris hamartomas)
- Increased risk of meningiomas and pheochromocytoma
- 90% of NF cases
- Autosomal dominant

Neurofibromatosis (NF) Type 2 (Bilateral Acoustic Neurofibromatosis)

(mutation in NF2 *tumor-suppressor gene on chromosome 22)*

- Bilateral acoustic neuromas
- Neurofibromas and café-au-lait spots
- Increased risk of meningiomas and pheochromocytoma
- 10% of NF cases
- Autosomal dominant

Marfan Syndrome

(mutation of fibrillin gene on chromosome 15)

- Skeletal abnormalities (tall build with hyperextensible joints)
- Subluxation of lens
- Cardiovascular defects (cystic medial necrosis, dissecting aortic aneurysm, valvular insufficiency)
- Autosomal dominant

von Hippel-Lindau Disease

(mutation in tumor-suppressor gene on chromosome 3)

- Hemangioblastomas in CNS and retina
- Renal cell carcinoma
- Cysts in internal organs
- Autosomal dominant

Cystic Fibrosis

(mutation in CFTR *chloride channel gene on chromosome 7q, leading to thick secretion of mucus plugs)*

- Recurrent pulmonary infections (*P. aeruginosa* and *S. aureus*)
- Pneumonia, bronchitis, bronchiectasis
- Pancreatic insufficiency; steatorrhea
- Fat-soluble vitamin deficiency
- Male infertility
- Biliary cirrhosis
- Meconium ileus
- Most common mutation: ΔF508
- Dx: ↑ NaCl in sweat; PCR and ASO probes
- Tx: *N*-acetylcysteine, respiratory therapy, enzyme replacement, vitamin supplement, inhaled bronchodilator, antibiotics
- Autosomal recessive

Features

Variable expression—
differences in severity of symptoms for same genotype; allelic heterogeneity can contribute to variable expression

Incomplete penetrance—
some individuals with disease genotype do not have disease phenotype

Delayed age of onset—
individuals do not manifest phenotype until later in life

Pleiotropy—
single disease mutation affects multiple organ systems

Locus heterogeneity—
same disease phenotype from mutations in different loci

Anticipation—
earlier age of onset and increased disease severity with each generation

Imprinting—
symptoms depend on whether mutant gene was inherited from father or mother; due to different DNA methylation patterns of parents (e.g., Prader-Willi versus Angelman syndrome)

Definition of abbreviations: ASO, allele-specific oligonucleotides; CFTR, cystic fibrosis transmembrane conductance regulator; CNS, central nervous system; Dx, diagnosis; PCR, polymerase chain reaction; Tx, treatment.

▶ Chromosomal Abnormalities

Aneuploidy refers to having a chromosome number that is not a multiple of the haploid number. It is the most common type of chromosomal disorder, and its incidence is related to increasing maternal age. Most arise from a **nondisjunction** event, when chromosomes fail to segregate during cell division. Nondisjunction during either phase of meiosis usually leads to spontaneous abortion, but sometimes results in live birth, often with severe physical deformities and mental retardation. Trisomies are the most common genetic cause of pregnancy loss. Nondisjunction during mitosis in the developing embryo can lead to cells in a single individual carrying different karyotypes, a condition known as **mosaicism**. Based on the **Lyon hypothesis**, females are naturally mosaics for genes on the X chromosome because one X chromosome in every cell is randomly inactivated to form a **Barr body**. Fluorescence in situ hybridization (FISH) can detect DNA sequences to identify deletions, translocations, and aneuploidies.

Nondisjunction During Meiosis I

Gametes
Metaphase of Meiosis 2
Metaphase of Meiosis 1
S, G2 Prophase
Nondisjunction During Meiosis 1
Disjunction During Meiosis 2

This figure shows the result of nondisjunction of one homologous pair (for example, chromosome 21) during meiosis 1. All other homologs segregate (disjoin) normally in the cell. Two of the gametes are diploid for chromosome 21. When fertilization occurs, the conception will be a trisomy 21 with Down syndrome. The other gametes with no copy of chromosome 21 will result in conceptions that are monosomy 21, a condition incompatible with a live birth.

Nondisjunction During Meiosis II

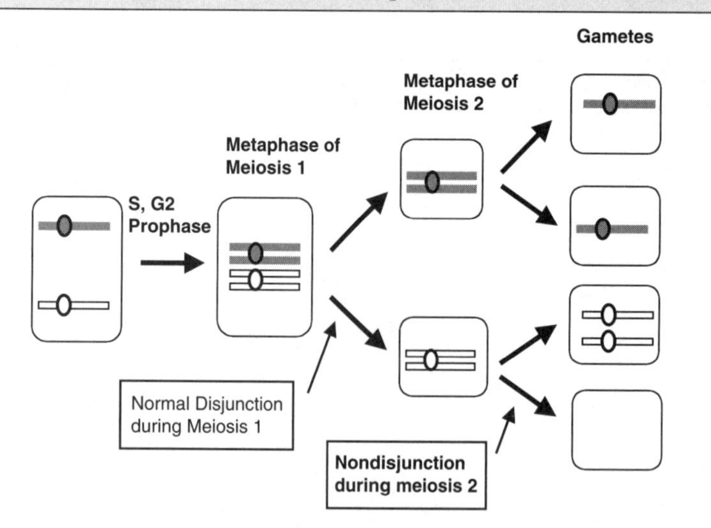

This figure shows the result of nondisjunction during meiosis 2. In this case, the sister chromatids of a chromosome (for example, chromosome 21) fail to segregate (disjoin). The sister chromatids of all other chromosomes segregate normally. One of the gametes is diploid for chromosome 21. When fertilization occurs, the conception will be a trisomy 21 with Down syndrome. One gamete has no copy of chromosome 21 and will result in a conception that is a monosomy 21. The remaining two gametes are normal haploid ones.

Autosomal Trisomies

Trisomy 21 (Down Syndrome)

- Most common chromosomal disorder
- Epicanthal folds, brachycephaly, flat nasal bridge, low-set ears, and short, broad hands with single transverse palmar crease
- Mental retardation
- Early-onset Alzheimer disease
- Congenital septal defects in heart
- ↑ risk of acute leukemia
- Incidence: 1/800 births (1/25 if age >45 years)
- 95% nondisjunction; 4% Robertsonian translocation

Trisomy 18 (Edwards Syndrome)

- Intrauterine growth retardation
- Mental retardation
- Failure to thrive
- Short sternum, small pelvis, rocker-bottom feet
- Cardiac, renal, and intestinal defects
- Usually death <1 year
- Incidence: 1/8,000 births

Trisomy 13 (Patau Syndrome)

- Microcephaly and abnormal brain development
- Cleft lip and palate, polydactyly
- Cardiac dextroposition and septal defects
- Incidence: 1/25,000 births

Sex Chromosome Aneuploidy

Turner Syndrome (45,XO)

- Short stature, webbed neck, shield chest
- Primary amenorrhea, infertility
- Coarctation of aorta
- Incidence: 1/6,000 female births

Klinefelter Syndrome (47,XXY)

- Eunuchoid body with lack of male secondary sex characteristics
- Hypogonadism, testicular atrophy
- Incidence: 1/2,000 male births

XYY Syndrome

- Excessively tall with severe acne
- ↑ risk of behavioral problems
- Incidence: 1/1,000 male births

▶ Other Chromosomal Abnormalities

In addition to aneuploidy, large segments of chromosomes may also undergo structural aberrations, including deletions, inversions, and translocations. **Deletions** occur when a chromosome loses a segment because of breakage. **Inversions** are rearrangements of the gene order within a single chromosome due to incorrect repair of two breaks. An inversion that includes a centromere is called a **pericentric** inversion, while one that does not involve the centromere is **paracentric**. Finally, **translocations** involve exchange of chromosomal material between nonhomologous chromosomes. **Reciprocal translocations** result when two nonhomologous chromosomes exchange pieces, while **Robertsonian translocations** involve any two acrocentric chromosomes that break near the centromeres and rejoin with a fusion of the q arms at the centromere and loss of the p arms.

Reciprocal Translocation

When one parent is a reciprocal translocation carrier:

- Adjacent segregation produces unbalanced genetic material and a likely loss of pregnancy
- Alternate segregation produces a normal haploid gamete (and diploid conception) or a liveborn who is a phenotypically normal translocation carrier

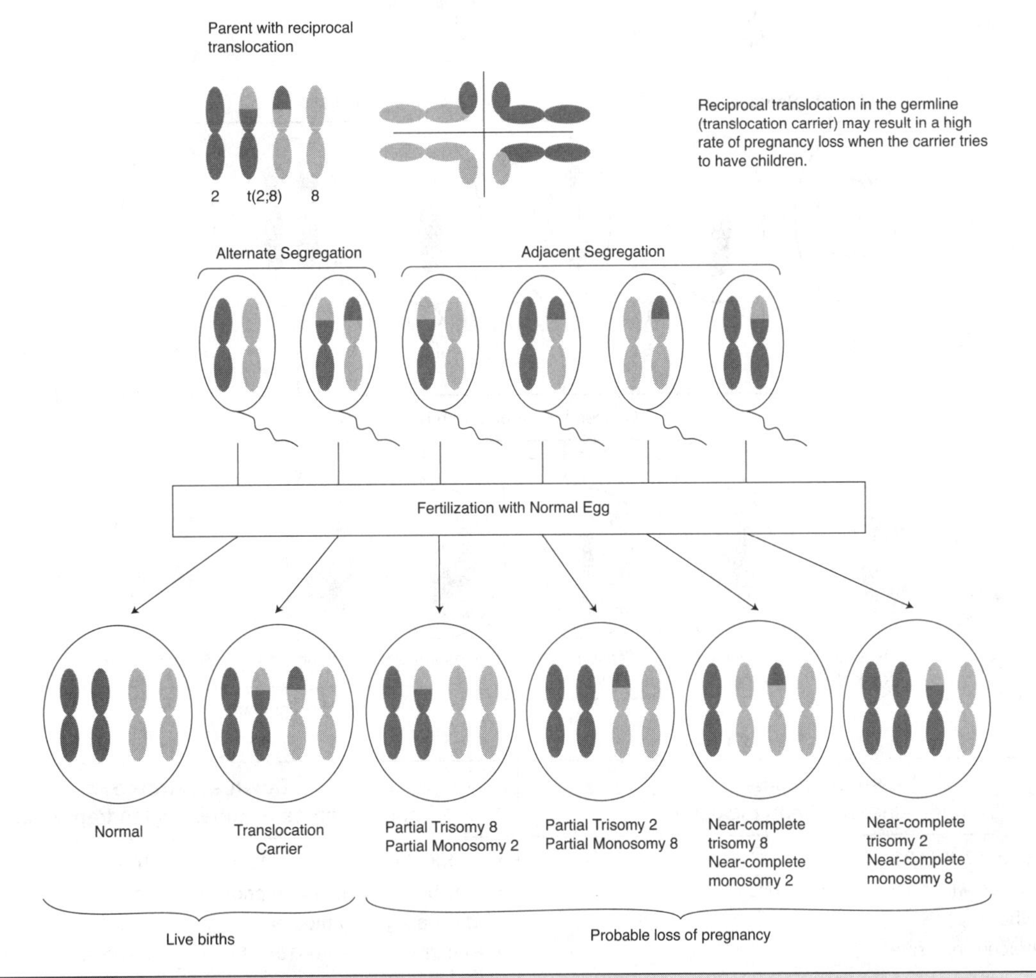

Reciprocal Translocations in Somatic Cells May Result in Cancer

t(9;22)	CML, ALL	*bcr-abl* fusion produces a fusion protein with tyrosine kinase activity.
t(8;14)	Burkitt lymphoma	*c-myc* oncogene (chromosome 8) placed near Ig heavy chain locus. Activating Ig heavy chain gene activates *c-myc.*
t(11;14)	Mantle cell lymphoma	*bcl-1* (chromosome 11) encodes cyclin D. Translocation places *bcl-1* near Ig heavy chain locus (chromosome 14).

Definition of abbreviations: ALL, acute lymphocytic leukemia; CML, chronic myelogenous leukemia.

(Continued)

▶ Other Chromosomal Abnormalities *(Cont'd.)*

Consequences of a Robertsonian Translocation in One Parent

Approximately 5% of Down syndrome cases result from a Robertsonian translocation affecting chromosomes 14 and 21. When a translocation carrier (in this case, a male) produces gametes, the translocation can segregate with the normal 14 or the normal 21. Although adjacent segregation usually results in pregnancy loss, it can result in a **trisomy 21**. Alternate segregation produces a normal haploid gamete (and diploid conception) or a liveborn who is a phenotypically normal translocation carrier.

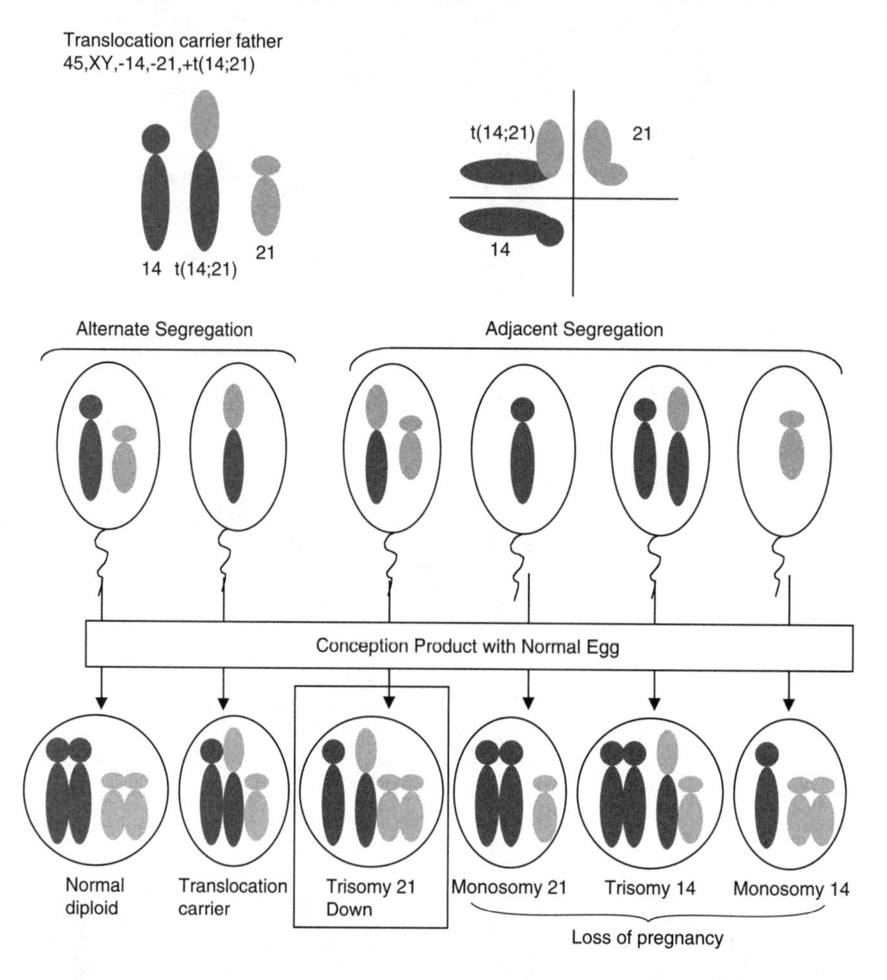

Down syndrome (nondisjunction during meiosis)	Down syndrome (parent carries a Robertsonian translocation)
• 47,XX,+21 or 47,XY,+21 • No association with prior pregnancy loss • Older mother • Very low recurrence rate	• 46,XX,-14,+t(14;21), or 46,XY,-14,+t(14;21) • *May* be associated with prior pregnancy losses • *May* be a young mother • Recurrence rate 10–15% if mom is translocation carrier; 1–2% if dad is translocation carrier

► Additional Diseases Resulting from Chromosomal Abnormalities

Disease	Abnormality	Characteristics
Cri-du-chat syndrome	Terminal or interstitial deletion of 5p	• Mental retardation • **Cat-like cry** • Microcephaly, low-set ears, micrognathia • Epicanthal folds
DiGeorge syndrome	Deletion of 22q11	• Hereditary absence of thymus and parathyroid glands due to **abnormal development of 3rd and 4th pharyngeal pouches** • T-cell deficiency • Cardiac outflow tract abnormalities • Abnormal facies • Hypoparathyroidism
Wilms tumor	Deletion of 11p13	• Malignant urinary tract tumors • □ diagnosed by age 4 • Tx: surgical removal
Angelman syndrome	Deletion of 15q11-q13 in **mother**	• **"Happy puppet" syndrome** • Always smiling but lacks speech • Hyperactive, hypotonic • Mental retardation, seizures • Dysmorphic facial features • Ataxic, puppet-like gait
Prader-Willi syndrome	Deletion of 15q11-q13 in **father**	• **Short stature** and **obese** with small hands and feet • **Hyperphagia** • Dysmorphic facial features • Mental retardation ***NOTE:*** **Angelman** and **Prader-Willi** are both examples of the effects of a deletion in an area affected by imprinting. A minority of cases are caused by uniparental disomy.

► Population Genetics

The **Hardy-Weinberg equilibrium** states that under certain conditions, if the population is large and randomly mating, the genotypic frequencies of the population will remain stable from generation to generation.

<table>
<tr><th>Hardy-Weinberg Conditions</th><th>Factors Affecting Equilibrium</th></tr>
<tr><td rowspan="8">1. No mutations
2. No selection against a genotype
3. No migration or immigration of the population
4. Random mating

If: frequency of A allele = p
frequency of a allele = q

Then: allele frequencies can be expressed as:

$$p + q = 1$$
genotypic frequencies at that locus can be expressed as:

$$p^2 + 2pq + q^2 = 1$$
where p^2 = frequency of genotype AA
$2pq$ = frequency of genotype Aa
q^2 = frequency of genotype aa</td><th>Natural Selection</th></tr>
<tr><td>Increases frequencies of genes that promote survival or fertility (e.g., malaria protection in sickle cell heterozygotes)</td></tr>
<tr><th>Genetic Drift</th></tr>
<tr><td>Gene frequency change due to finite population size</td></tr>
<tr><th>Gene Flow</th></tr>
<tr><td>Gene exchange between different populations</td></tr>
<tr><th>Linkage Disequilibrium</th></tr>
<tr><td>Preferential association of an allele at one locus with another allele at a nearby locus more frequently than by chance alone</td></tr>
</table>

▶ Subcellular Organelles

In contrast to simple prokaryotic cells that have a cell wall but no membrane-bound nucleus or organelles, eukaryotic cells are, in general, larger and lack a cell wall, but are composed of various subcellular membranous organelles with distinct functions.

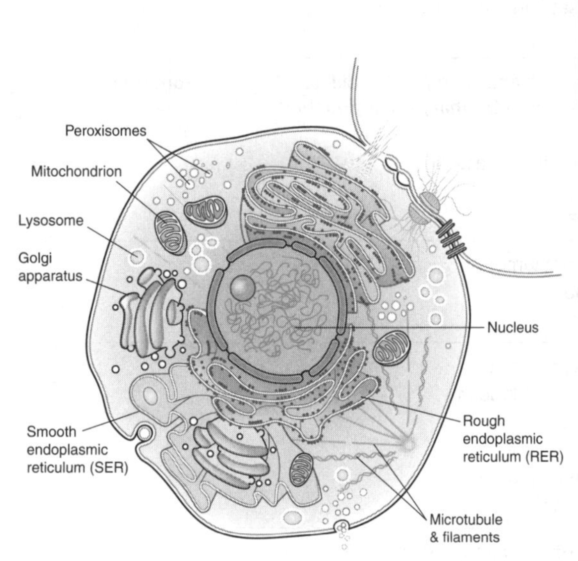

Prokaryotic	Eukaryotic
Small (1–10 μm)	Large (10–100 μm)
Thick, rigid cell wall	No cell wall
No membrane-bound organelles	Various subcellular membranous organelles
Non-membrane–bound nucleoid region	Nucleus with double-membrane envelope

Proteasome

- Small complexes of proteolytic enzymes in cytosol
- Digest (usually misfolded) proteins that are marked with ubiquitin
- Peptides produced are presented along with MHC I at the cell surface

Nucleus

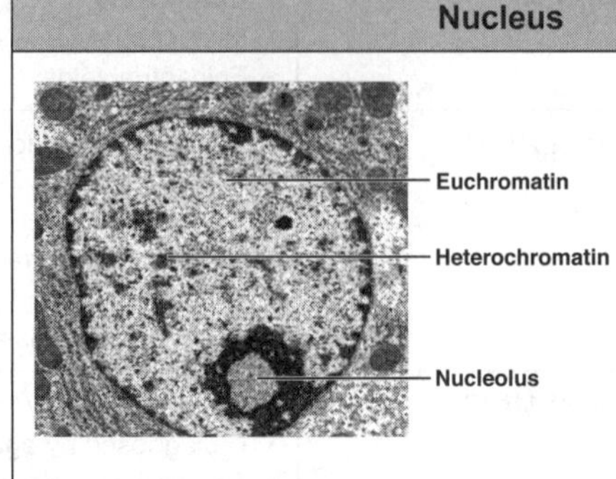

- Site of **DNA replication and transcription**
- Enclosed by nuclear envelope
- Contains **nucleolus** (site of ribosome synthesis); no membrane surrounds the nucleolus
- Contains DNA packaged with histones to form **chromatin**

Rough Endoplasmic Reticulum (RER)

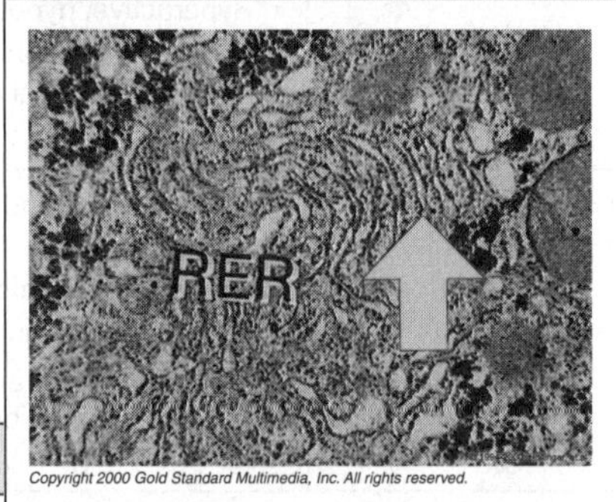

Copyright 2000 Gold Standard Multimedia, Inc. All rights reserved.

- Contains **ribosomes** for synthesizing proteins destined for RER, SER, Golgi, lysosomes, cell membrane, and secretion
- Cotranslational modifications, including ***N*-linked glycosylation** (proteins synthesized on free ribosomes are not usually glycosylated)

Ribosomes

- Site of **protein synthesis**
- Composed of ribosomal RNA (rRNA) and proteins forming large 60S + small 40S subunits
- Single mRNA simultaneously translated by several ribosomes is a **polysome**

Types of Ribosomes	
RER-Bound	**Free Cytosolic**
Proteins for RER, SER, Golgi apparatus, lysosomes, cell membrane, and secretion	Cytosolic, mitochondrial, nuclear, and peroxisomal proteins

(*Continued*)

► Subcellular Organelles (*Cont'd.*)

Golgi Apparatus

- Site of **post-translational modifications and protein sorting**
- Consists of disk-shaped cisternae in stacks
- ***Cis* (forming) face** associated with RER
- ***Trans* (maturing) face** oriented toward plasma membrane

Smooth Endoplasmic Reticulum (SER)

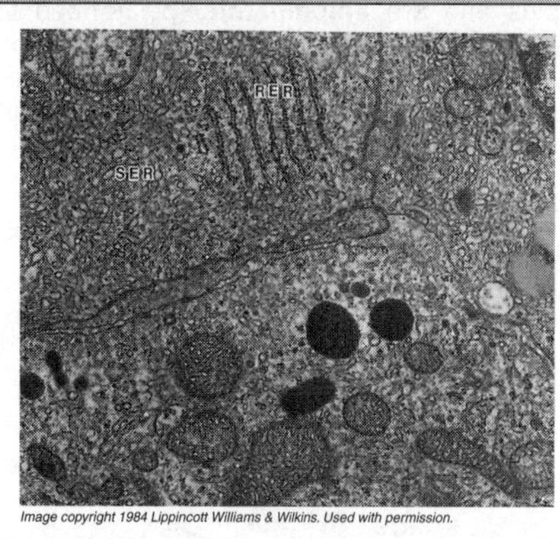

Image copyright 1984 Lippincott Williams & Wilkins. Used with permission.

- Involved in detoxification reactions, including **phase I hydroxylation** (via cytochrome P450) and **phase II conjugation** (addition of polar groups)
- Synthesis of phospholipids, lipoproteins, and sterols
- Sequesters Ca^{2+}; known as **sarcoplasmic reticulum** in striated and smooth muscle cells

Mitochondria

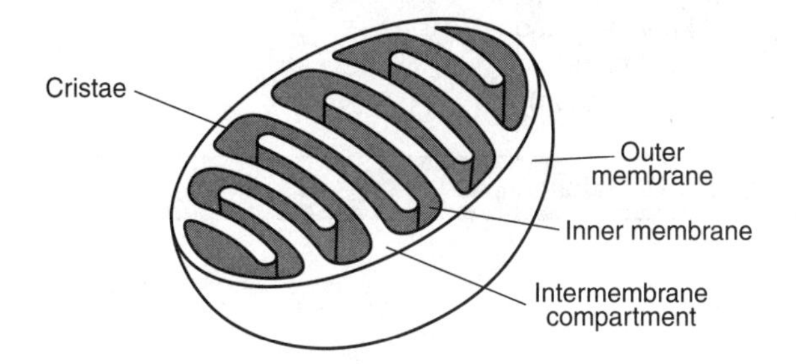

- Major function is ATP synthesis
- Similar to bacteria in size and shape; self-replicating
- Contain their own double-stranded circular DNA
- Smooth, permeable outer membrane; heavily infolded, impermeable inner membrane

Lysosomes

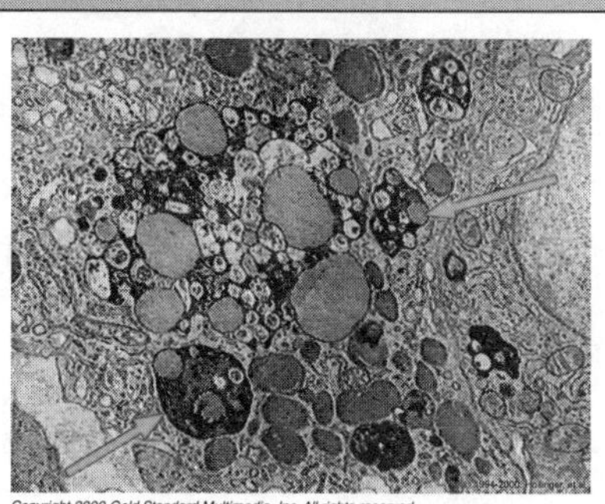

Copyright 2000 Gold Standard Multimedia, Inc. All rights reserved.

- Enzymatic degradation of extracellular or intracellular macromolecules
- Primary lysosome fuses with phagosomes or cellular organelles to form secondary lysosomes
- Acidic hydrolytic enzymes with optimal activity at pH 5
- Degradation of intracellular organelles known as **autophagy**

Endosomes

- Formed from endocytosed vesicles acquired by receptor-mediated endocytosis involving **clathrin-coated pits**
- Can fuse with primary lysosomes to form secondary lysosomes to degrade extracellular materials
- Exogenous peptides presented on membrane with MHC II on antigen presenting cells

Peroxisomes

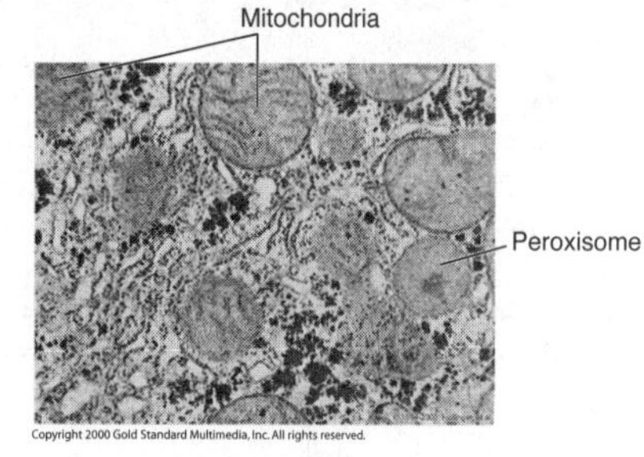

Copyright 2000 Gold Standard Multimedia, Inc. All rights reserved.

- Synthesis and degradation of hydrogen peroxide
- β-oxidation of very long chain fatty acids ($>C_{24}$)
- Phospholipid exchange reactions
- Bile acid synthesis

► Plasma Membrane

The **plasma membrane** of a cell is a bilayer of lipids and proteins. The lipids include phospholipids, unesterified cholesterol, and glycolipids and are **amphipathic** (polar head to interact with aqueous environment, and nonpolar tail to interact with the bilayer interior). Proteins may act as adhesion molecules, receptors, transporters, channels, or enzymes. Proteins embedded in the bilayer are **integral proteins**, whereas those loosely associated with the membrane are **peripheral proteins**. In general, ***N*-glycosylation** of proteins and lipids is associated with location on the external surface, whereas ***N*-myristoylation, prenylation**, and **palmitoylation** of proteins are associated with location on the cytoplasmic face of the plasma membrane.

Structure of Biologic Membranes

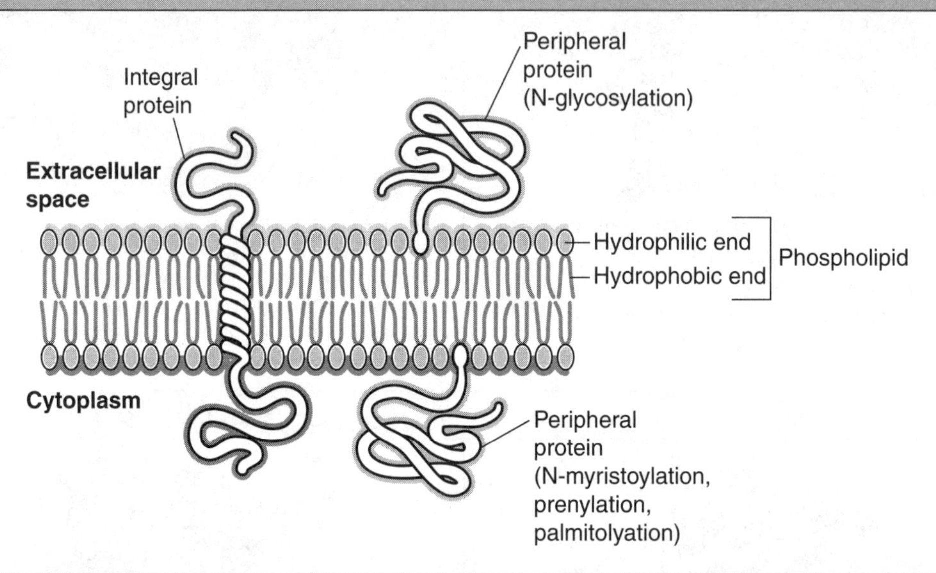

Types of Integral Membrane Proteins

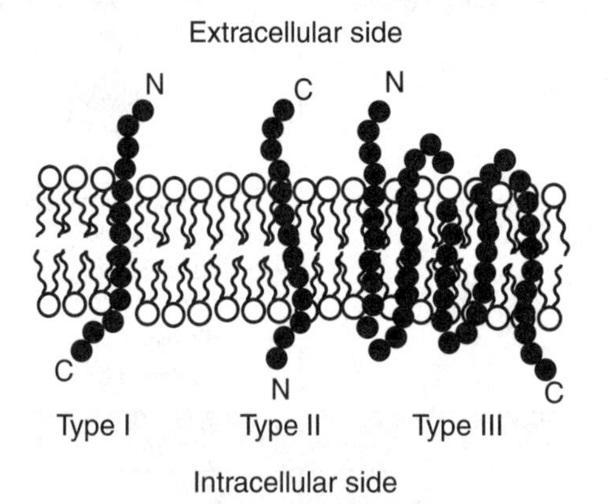

Types of Transport

Simple Diffusion	Facilitated Diffusion	Active Transport
• Movement of highly permeable molecules from a region of higher to lower concentration • e.g., O_2, CO_2, NO	• Passage of poorly permeable molecules from a region of higher to lower concentration via a carrier • e.g., glucose transporter	• Movement of molecules from a region of lower to higher concentration with *1)* energy expenditure via ATP hydrolysis, or *2)* cotransport using another molecule's chemical gradient • e.g., *1)* Na^+/K^+ pump, Ca^{2+}-ATPase • e.g., *2)* Na^+/glucose symporter

▶ Cytoskeleton

The cytoskeleton consists of a supportive network of tubules and filaments in the cytoplasm of eukaryotic cells. It is a dynamic structure responsible for cellular movement, changes in cell shape, and the contraction of muscle cells. It also provides the machinery for intracellular movement of organelles. The cytoskeleton is composed of three types of supportive structures: **microtubules*, intermediate filaments**, and **microfilaments**.

Microtubules*	Intermediate Filaments	Microfilaments
Tubulin (hollow cylindrical polymer of tubulin dimers)	• **Keratin** (epithelium) • **Vimentin** (nonepithelial) • **Neurofilament** (neurons)	**Actin** (double-stranded polymer twisted in helical pattern)
Function		
• Movement of chromosomes in mitosis or meiosis • Intracellular transport via motor proteins • Ciliary and flagellar motility	Structural	• Structural • Muscle contraction via interaction with myosin

Axoneme Structure

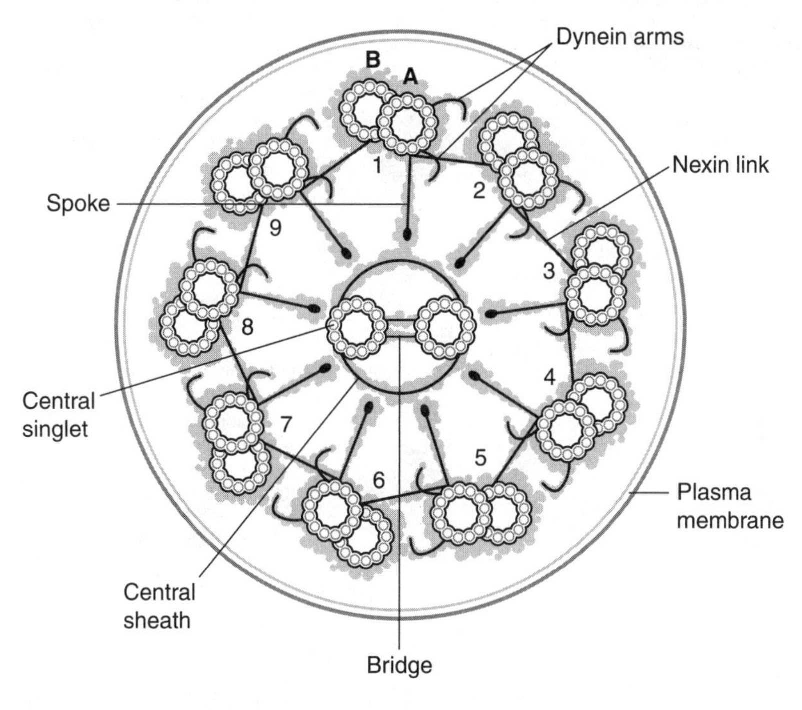

Motor Proteins

Kinesins for ⊖ → ⊕ anterograde direction

Dyneins for ⊕ → ⊖ retrograde direction

Disease Association

Chediak-Higashi Syndrome

(defect in microtubule polymerization in leukocytes)

- Recurrent pyogenic infections of respiratory tract and skin
- Partial albinism
- Photophobia, nystagmus, peripheral neuropathy, motor dysfunction, seizures
- Presents early in childhood

Kartagener Syndrome

(immotile cilia due to defect in axonemal proteins, such as dynein arms)

- Chronic cough, rhinitis, and sinusitis
- Situs inversus
- Fatigue and headaches
- Male infertility from immotile spermatozoa
- Autosomal recessive

Pharmacology

Colchicine

- Inhibits tubulin polymerization
- Used for gout

Vincristine/Vinblastine

- Inhibits tubulin polymerization
- Antineoplastic

Taxol

- Promotes tubulin polymerization
- Antineoplastic

*Microtubules are polarized structures with assembly/disassembly occurring at the ⊕ ends, which are oriented toward the cell's periphery.

▶ Cell Adhesion

A cell must physically interact via cell surface molecules with its external environment, whether it be the extracellular matrix or **basement membrane**. The basement membrane is a sheet-like structure underlying virtually all epithelia, which consists of **basal lamina** (made of type IV collagen, glycoproteins [e.g., laminin], and proteoglycans [e.g., heparin sulfate]), and **reticular lamina** (composed of reticular fibers). Cell junctions anchor cells to each other, seal boundaries between cells, and form channels for direct transport and communication between cells. The three types of junctional complexes include **anchoring, tight,** and **gap junctions**.

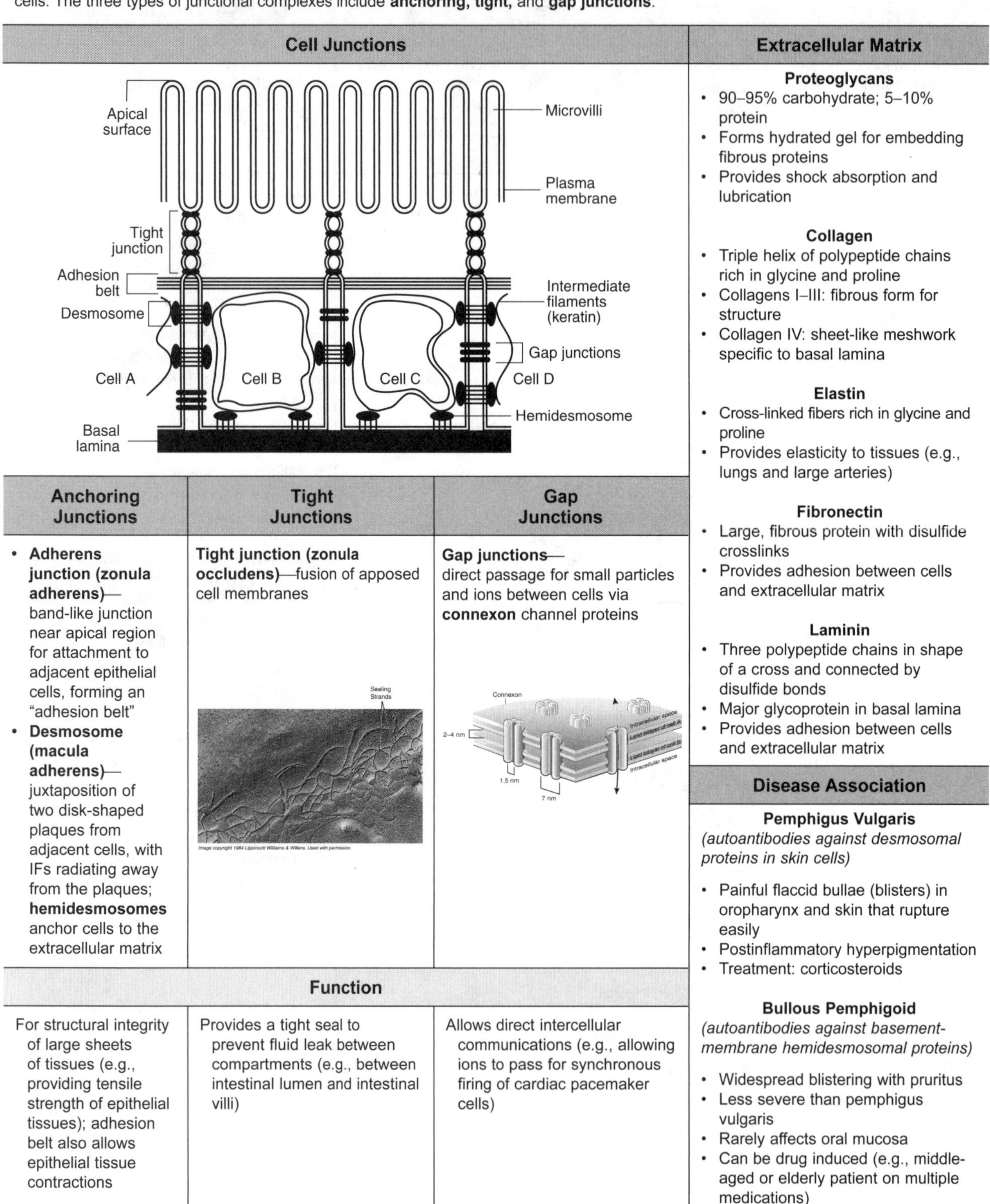

Cell Junctions

Anchoring Junctions	Tight Junctions	Gap Junctions
• **Adherens junction (zonula adherens)**—band-like junction near apical region for attachment to adjacent epithelial cells, forming an "adhesion belt" • **Desmosome (macula adherens)**—juxtaposition of two disk-shaped plaques from adjacent cells, with IFs radiating away from the plaques; **hemidesmosomes** anchor cells to the extracellular matrix	**Tight junction (zonula occludens)**—fusion of apposed cell membranes	**Gap junctions**—direct passage for small particles and ions between cells via **connexon** channel proteins
Function		
For structural integrity of large sheets of tissues (e.g., providing tensile strength of epithelial tissues); adhesion belt also allows epithelial tissue contractions	Provides a tight seal to prevent fluid leak between compartments (e.g., between intestinal lumen and intestinal villi)	Allows direct intercellular communications (e.g., allowing ions to pass for synchronous firing of cardiac pacemaker cells)

Image copyright 1984 Lippincott Williams & Wilkins. Used with permission.

Extracellular Matrix

Proteoglycans
- 90–95% carbohydrate; 5–10% protein
- Forms hydrated gel for embedding fibrous proteins
- Provides shock absorption and lubrication

Collagen
- Triple helix of polypeptide chains rich in glycine and proline
- Collagens I–III: fibrous form for structure
- Collagen IV: sheet-like meshwork specific to basal lamina

Elastin
- Cross-linked fibers rich in glycine and proline
- Provides elasticity to tissues (e.g., lungs and large arteries)

Fibronectin
- Large, fibrous protein with disulfide crosslinks
- Provides adhesion between cells and extracellular matrix

Laminin
- Three polypeptide chains in shape of a cross and connected by disulfide bonds
- Major glycoprotein in basal lamina
- Provides adhesion between cells and extracellular matrix

Disease Association

Pemphigus Vulgaris
(autoantibodies against desmosomal proteins in skin cells)

- Painful flaccid bullae (blisters) in oropharynx and skin that rupture easily
- Postinflammatory hyperpigmentation
- Treatment: corticosteroids

Bullous Pemphigoid
(autoantibodies against basement-membrane hemidesmosomal proteins)

- Widespread blistering with pruritus
- Less severe than pemphigus vulgaris
- Rarely affects oral mucosa
- Can be drug induced (e.g., middle-aged or elderly patient on multiple medications)
- Treatment: corticosteroids

Definition of abbreviations: IF, intermediate filament.

► Cell Cycle

The cell cycle consists of the mitosis phase (M), the presynthetic gap (G_1), the DNA synthesis phase (S), and the postsynthetic gap (G_2). Mitosis is the shortest phase, consisting of **prophase, metaphase, anaphase**, and **telophase**. Both G_1 and G_2 phases are variable in duration, with most cells spending much of their time in a stable, nondividing G_0 phase. Cells in G_2 have twice the amount of DNA as those in G_1.

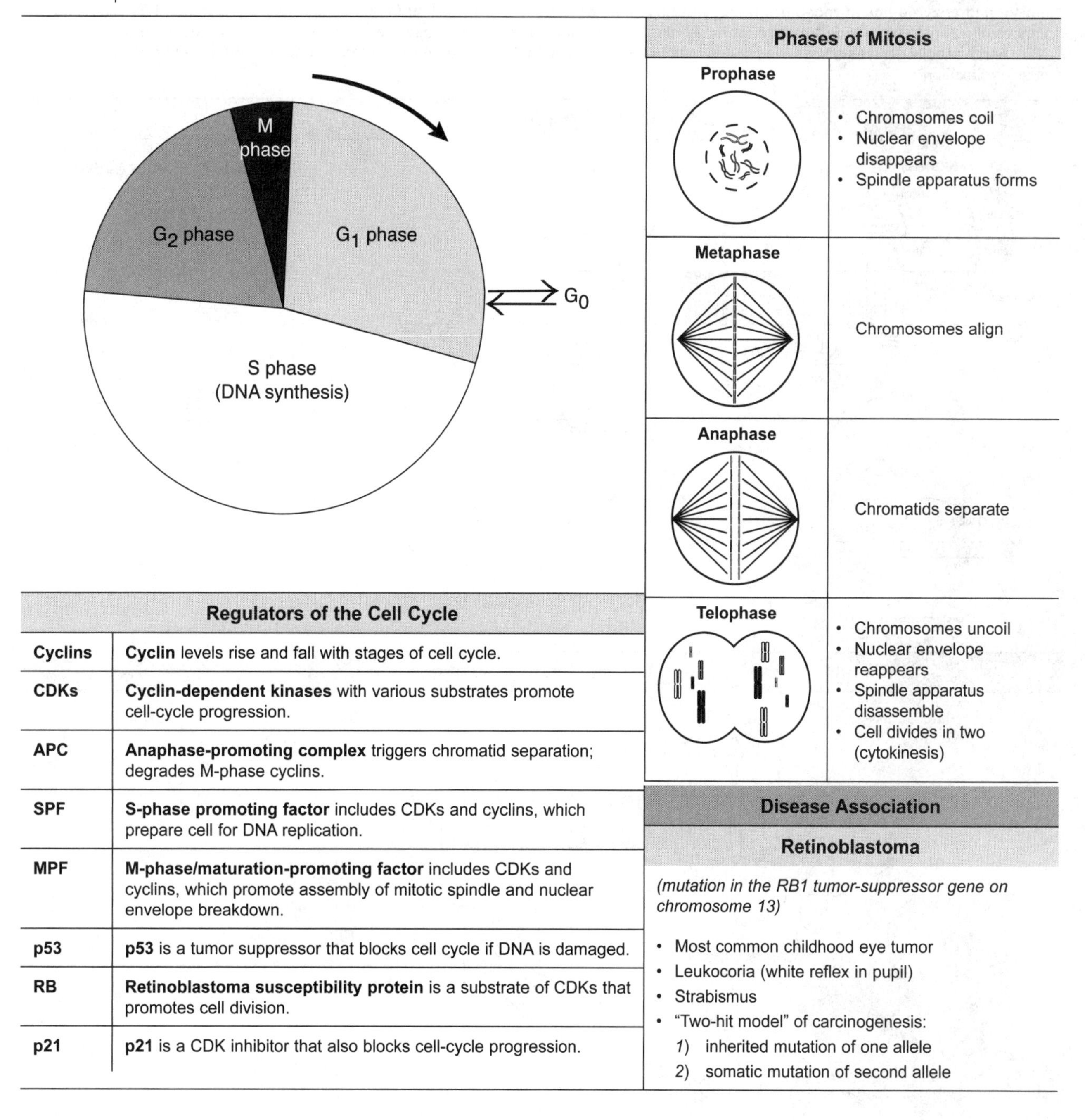

Phases of Mitosis	
Prophase	• Chromosomes coil • Nuclear envelope disappears • Spindle apparatus forms
Metaphase	Chromosomes align
Anaphase	Chromatids separate
Telophase	• Chromosomes uncoil • Nuclear envelope reappears • Spindle apparatus disassemble • Cell divides in two (cytokinesis)

Regulators of the Cell Cycle	
Cyclins	**Cyclin** levels rise and fall with stages of cell cycle.
CDKs	**Cyclin-dependent kinases** with various substrates promote cell-cycle progression.
APC	**Anaphase-promoting complex** triggers chromatid separation; degrades M-phase cyclins.
SPF	**S-phase promoting factor** includes CDKs and cyclins, which prepare cell for DNA replication.
MPF	**M-phase/maturation-promoting factor** includes CDKs and cyclins, which promote assembly of mitotic spindle and nuclear envelope breakdown.
p53	**p53** is a tumor suppressor that blocks cell cycle if DNA is damaged.
RB	**Retinoblastoma susceptibility protein** is a substrate of CDKs that promotes cell division.
p21	**p21** is a CDK inhibitor that also blocks cell-cycle progression.

Disease Association

Retinoblastoma

(mutation in the RB1 tumor-suppressor gene on chromosome 13)

- Most common childhood eye tumor
- Leukocoria (white reflex in pupil)
- Strabismus
- "Two-hit model" of carcinogenesis:
  1) inherited mutation of one allele
  2) somatic mutation of second allele

► Cell Signaling

In order to act on a cell, external molecules, such as hormones and neurotransmitters, must interact with a **receptor**. In general, small hydrophobic molecules (e.g., cortisol, sex hormones, thyroid hormone, and retinoids) can readily penetrate the plasma membrane to bind **intracellular receptors**, which often act as transcription factors to affect gene expression. Most other molecules bind to cell surface receptors, which include **ion-channel–linked receptors** (e.g., transmitter-gated channels), **G-protein–linked receptors** (the largest family), and **enzyme-linked receptors** (e.g., tyrosine kinase receptors). These cell surface receptors usually transmit their signal via a number of downstream **second messengers**, leading to a **signal transduction cascade**. One exception is the gaseous **nitric oxide** (NO), which readily diffuses across the plasma membrane to activate soluble **guanylate cyclase**, generate **cGMP**, and promote smooth muscle relaxation.

G-Protein–Coupled Receptor Systems

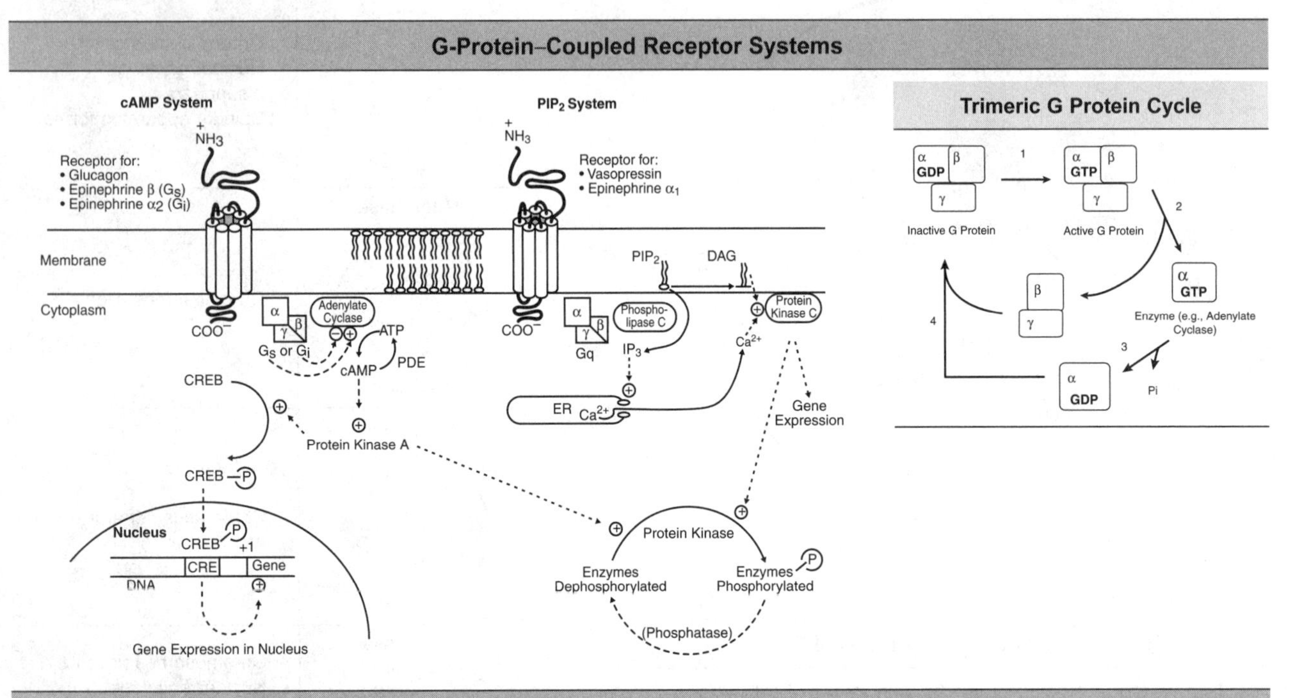

Tyrosine Kinase Receptor

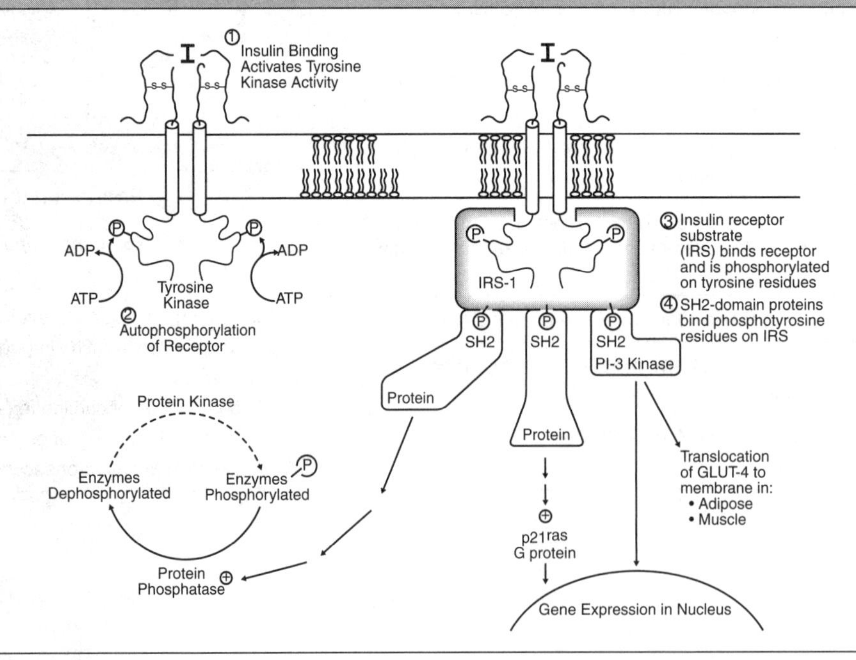

Definition of abbreviations: See next page. (*Continued*)

► Cell Signaling *(Cont'd.)*

Guanylate Cyclase

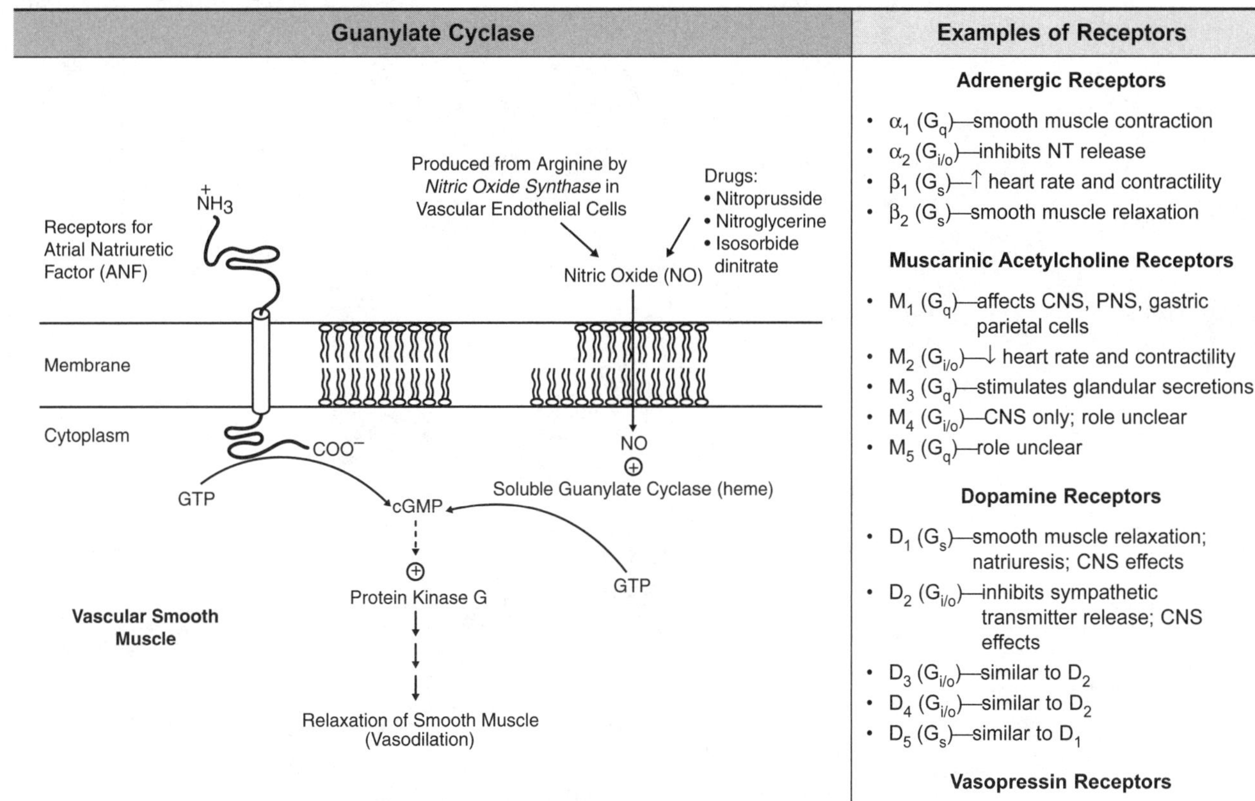

Examples of Receptors

Adrenergic Receptors

- α_1 (G_q)—smooth muscle contraction
- α_2 ($G_{i/o}$)—inhibits NT release
- β_1 (G_s)—↑ heart rate and contractility
- β_2 (G_s)—smooth muscle relaxation

Muscarinic Acetylcholine Receptors

- M_1 (G_q)—affects CNS, PNS, gastric parietal cells
- M_2 ($G_{i/o}$)—↓ heart rate and contractility
- M_3 (G_q)—stimulates glandular secretions
- M_4 ($G_{i/o}$)—CNS only; role unclear
- M_5 (G_q)—role unclear

Dopamine Receptors

- D_1 (G_s)—smooth muscle relaxation; natriuresis; CNS effects
- D_2 ($G_{i/o}$)—inhibits sympathetic transmitter release; CNS effects
- D_3 ($G_{i/o}$)—similar to D_2
- D_4 ($G_{i/o}$)—similar to D_2
- D_5 (G_s)—similar to D_1

Vasopressin Receptors

- V_1 (G_q)—smooth muscle contraction
- V_2 (G_s)—↑ H_2O reabsorption in kidney

Other Receptors

- Insulin (TK)—↑ glycogen synthesis; ↓ glycogenolysis
- Glucagon (G_s)—↑ glycogenolysis; ↓ glycogen synthesis
- IGF (TK)—↑ proliferation of various cell types
- PDGF (TK)—↑ proliferation of connective tissue, glial, and smooth muscle cells
- EGF (TK)—↑ proliferation of mesenchymal, glial, and epithelial cells
- ANF (GC)—smooth muscle relaxation; ↑ Na^+ and H_2O excretion in kidney
- NO (GC)—smooth muscle relaxation

Pharmacology

Nitrates	Sildenafil
(nitroglycerin, nitroprusside, isosorbide dinitrate) • Converted to NO, which activates guanylate cyclase, leading to ↑ in cGMP. cGMP causes smooth muscle relaxation of blood vessels. • For angina and pulmonary edema • Adverse effects: headache, hypotension	• Inhibits cGMP-dependent phosphodiesterase (PDE5), leading to cGMP buildup, causing smooth muscle relaxation and dilation of blood vessels leading to the corpus cavernosum • For erectile dysfunction and Raynaud phenomenon • Adverse effects: headache, hypotension

Definition of abbreviations: ANF, atrial natriuretic factor; ATP, adenosine triphosphate; cGMP, cyclic guanosine monophosphate; DAG, diacylglycerol; EGF, epidermal growth factor; ER, endoplasmic reticulum; GC, guanylate cyclase–coupled receptor; $G_{i/o}$, cAMP-inhibiting GPCR; G_q, PLC-activating GPCR; G_s, cAMP-activating GPCR; IGF, insulin-like growth factor; PDE, phosphodiesterase; PDGF, platelet-derived growth factor; PIP_2, phosphoinositol biphosphate; PLC, phospholipase C; NO, nitric oxide; TK, tyrosine kinase receptor.

Immunology

Chapter 4

Overview of the Immune System

▶ Characteristics of Innate versus Adaptive Immunity

The immune system can be divided into **two** complementary arms: the **innate** (native, natural) immune system and the **adaptive** (acquired, specific) immune system. These two arms work in concert with each other through soluble substances, such as antibodies, complement, and cytokines.

Characteristics	Innate	Adaptive
Specificity	For structures shared by groups of microbes	For specific antigens of microbial and nonmicrobial agents
Diversity	Limited	**High**
Memory	No	**Yes**
Self-reactivity	No	No
Components	**Innate**	**Adaptive**
Anatomic and chemical barriers	Skin, mucosa, chemicals (lysozyme, interferons α and β), temperature, pH	Lymph nodes, spleen, mucosal-associated lymphoid tissues
Blood proteins	**Complement**	**Antibodies**
Cells	**Phagocytes and NK cells**	**Lymphocytes** (other than NK cells)

▶ Overview of the Immune Response

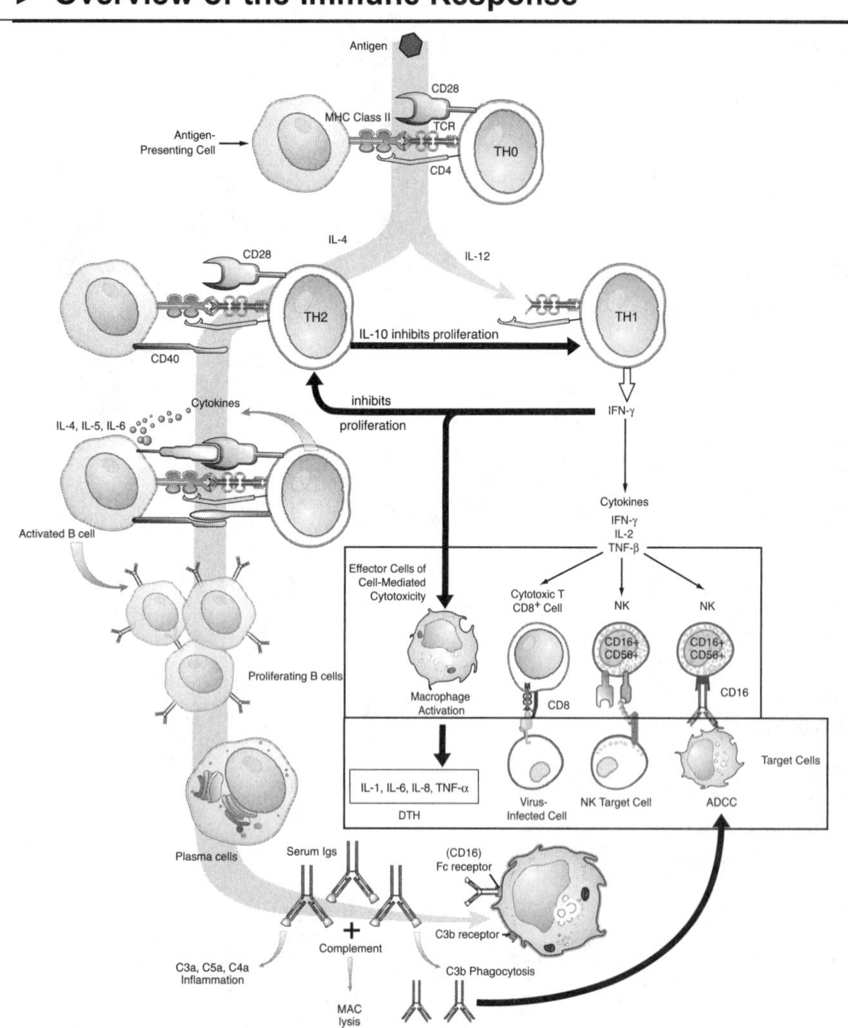

- Foreign materials are processed and presented to TH cells in the MHC class II groove.
- Extracellular pathogens stimulate production of TH2 cells, which provide the CD40L costimulatory signal and cytokines to induce B cells to differentiate into plasma cells, which produce antibody. Antibody may assist in phagocytosis (**opsonization**) or complement-mediated lysis.
- Intracellular pathogens stimulate production of TH1 cells, which stimulate the effector cells of cell-mediated immunity.
- Macrophages are induced to become more effective intracellular killers of the bacteria they ingest.
- Cytotoxic T cells kill virus-infected cells by recognition of peptides presented in the MHC class I molecule.
- Natural killer (NK) cells kill cells devoid of MHC class I that are infected with some viruses or have undergone malignant transformation.
- In antibody-dependent cell-mediated cytotoxicity (ADCC), abnormal surface molecules on infected or transformed cells are recognized by antibodies and targeted for extracellular lysis by NK cells, eosinophils, neutrophils, or macrophages.

Definition of abbreviations: MHC, major histocompatibility complex; TCR, T cell receptor; TH, T helper.

► Cells of the Immune System

Myeloid Cell	Location	Identifying Features	Function
Monocyte	Bloodstream, 0–900/µL	**Kidney bean-shaped nucleus, CD 14 (endotoxin receptor) positive**	**Phagocytic**, differentiate into tissue macrophages
Macrophage	Tissues	Ruffled membrane, cytoplasm with vacuoles and vesicles (CD14+)	**Phagocytosis**, secretion of **cytokines**
Dendritic cell	Epithelia, tissues	Long, cytoplasmic arms	**Antigen capture**, transport, and presentation (There are also plasmacytoid dendritic cells that look like plasma cells and produce interferon-α.)
Neutrophil	Bloodstream, 1,800–7,800/µL	**Multilobed** nucleus; **small pink** granules	**Phagocytosis** and activation of bactericidal mechanisms
Eosinophil	Bloodstream, 0–450/µL	**Bilobed** nucleus; **large pink** granules	**Killing of antibody-coated parasites**
Basophil	Bloodstream, 0–200/µL	**Bilobed** nucleus; **large blue** granules	**Nonphagocytic**, release pharmacologically active substances during **allergy**
Mast cell	Tissues, mucosa, and epithelia	**Small** nucleus; **cytoplasm** packed with **large blue** granules	Release of granules containing histamine, etc., during **allergy**

(Continued)

► Cells of the Immune System *(Cont'd.)*

Lymphoid Cell	Location	Identifying Features	Function
Lymphocyte	Bloodstream, 1,000–4,000/µL, lymph nodes, spleen, submucosa, and epithelia	**Large, dark** nucleus, small rim of cytoplasm	**B cells produce antibody** **TH (CD3+CD4+) cells regulate immune responses** **Cytotoxic T cells (CTLs: CD3+CD8+) kill altered or infected cells**
Natural killer (NK) lymphocyte	Bloodstream, ≤10% of lymphocytes	**Lymphocytes** with **large cytoplasmic** granules, **CD16+CD56+**	**Kill tumor/virus cell** targets or antibody-coated target cells
Plasma cell	Lymph nodes, spleen, mucosal-associated lymphoid tissues, and bone marrow	**Small dark** nucleus, **intensely staining Golgi** apparatus	End cell of B-cell differentiation, **produce antibody**

► Characteristics of Lymphoid Cells

Comparison of B- and T-lymphocyte Antigen Receptors

Property	B-Cell Antigen Receptor	T-Cell Antlgen Receptor
Idiotypes/lymphocyte	1	1
Isotypes/lymphocyte	2 (**IgM and IgD**)	1 (α/β)
Is secretion possible?	**Yes**	**No**
Number of combining sites/molecules	2	1
Mobility	**Flexible** (hinge region)	**Rigid**
Signal transduction molecules	Ig-α, Ig-β, **CD19, CD21**	**CD3**

► Generation of Receptor Diversity in B and T Lymphocytes

Mechanism	Cell in Which Expressed
Existence in genome of multiple V, D, J segments	B and T cells
VDJ recombination (gene segments are selected and recombined randomly to generate unique variable domains)	**B** and **T cells**
N-nucleotide addition (TdT adds nucleotides randomly where V, D, and J are joined)	**B cells (only heavy chain), T cells (both chains)**
Combinatorial association of heavy and light chains	B and T cells
Somatic hypermutation (mutations in variable domain coding occur during blastogenesis, and natural selection causes affinity maturation)	**B cells only,** after **Ag stimulation**

Definition of abbreviation: Tdt, terminal deoxyribonucleotidyl transferase.

► Human Major Histocompatibility Complex (MHC) Summary

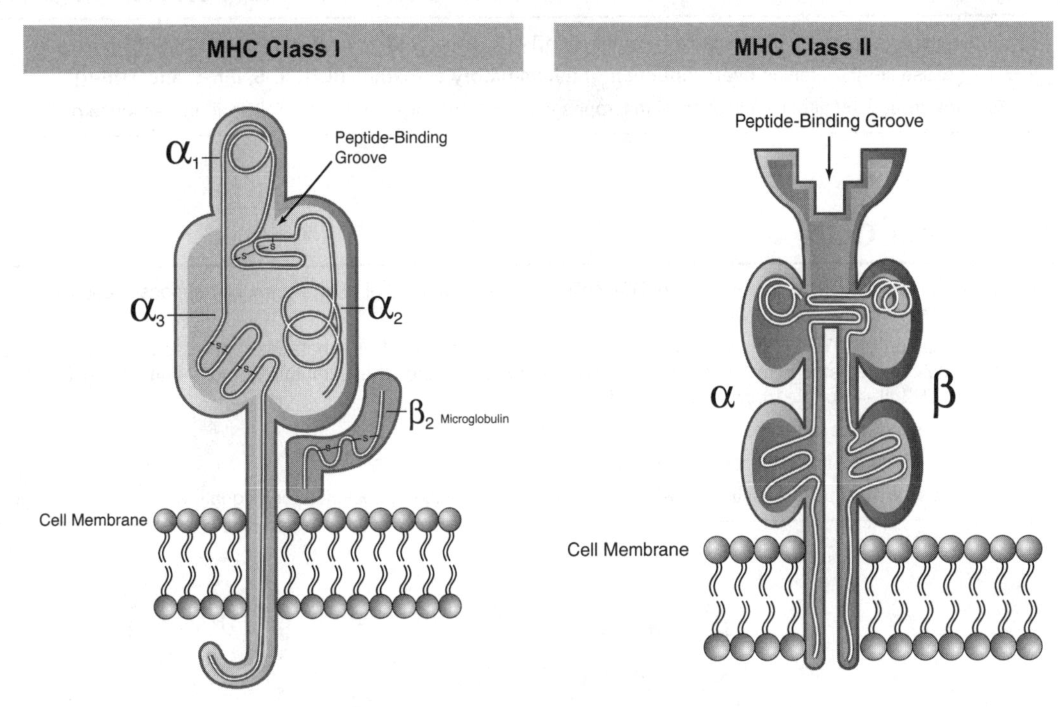

	MHC Class I	MHC Class II
Names	HLA-A, HLA-B, HLA-C	HLA-DP, HLA-DQ, HLA-DR
Tissue distribution	**All nucleated cells**, platelets	**B and T lymphocytes, antigen-presenting cells**
Recognized by	Cytotoxic T cells **(CD8+)**	Helper T cells **(CD4+)**
Peptides bound	Endogenously **synthesized**	Exogenously **processed**
Function	**Elimination of abnormal (infected) host cells** by cytotoxic T cells	**Presentation of foreign antigen** to helper T cells
Invariant chain	No	**Yes**
β_2-microglobulin	**Yes**	No

► Examples of HLA-Linked Immunologic Diseases

Disease	HLA Allele
Rheumatoid arthritis	DR4
Insulin-dependent diabetes mellitus	DR3/DR4
Multiple sclerosis, Goodpasture syndrome	DR2
Systemic lupus erythematosus	DR2/DR3
Ankylosing spondylitis, psoriasis, inflammatory bowel disease, Reiter syndrome	B27
Celiac disease	DQ2 or DQ8
Graves disease	B8

▶ Superantigens
(Staphylococcal Enterotoxins, Toxic-Shock Syndrome Toxin-1, and Streptococcal Pyrogenic Exotoxins)

- Superantigens cross-link the variable β domain of a T-cell receptor to an α chain of a class II MHC molecule.
- They cause life-threatening **overproduction of inflammatory cytokines (IL-1, IL-6, IFN-γ, and TNF-α)**.
- **Endotoxin** acts by direct stimulation of macrophages to produce IL-1, IL-6, and TNF-α in the absence of TH participation.

▶ T-Helper Cells

- Naive TH cells (TH0) differentiate into TH1 cells when a strong initial innate immune response leads to production of IL-12 from macrophages or IFN-γ from NK cells.
- Differentiation of a TH0 cell into a TH2 cell occurs in the absence of an innate immune response. TH1 cells secrete cytokines that stimulate cell-mediated immunity (CMI). TH2 cells produce cytokines that stimulate humoral immunity (HMI). IFN-γ, produced by TH1, inhibits TH2. IL-4 and IL-10, produced by TH2, inhibit TH1.
- TH17 cells increase inflammation and produce IL-17.
- T_{reg} cells decrease inflammation and produce IL-10.
- T cells that are exposed to antigen presenting cells in the absence of costimulatory signals will become anergic.

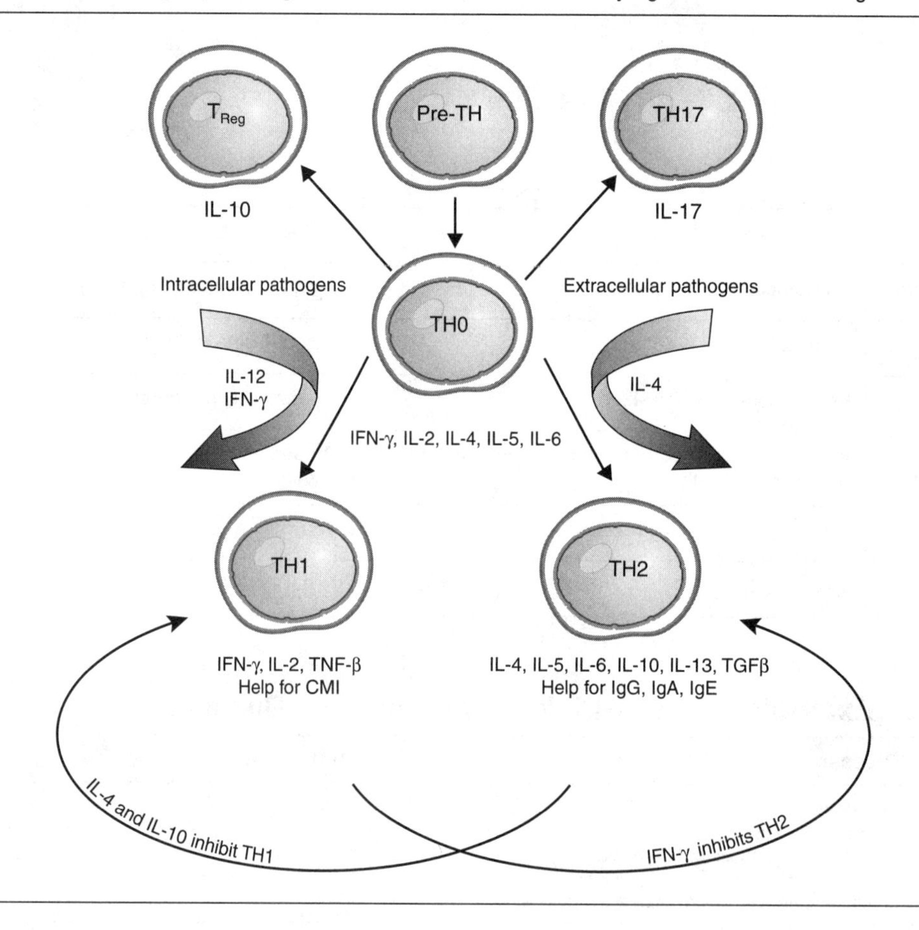

► Effector Cells in Cell-Mediated Immunity

Effector Cell	CD Markers	Antigen Recognition	MHC Recognition Required for Killing	Effector Molecules
CTL	TCR, CD3, **CD8**, CD2	**Specific**, TCR	Yes, **class I**	**Perforin, cytokines** (TNF-β, IFN-γ)
NK cell	**CD16, CD56**, CD2	ADCC: specific by IgG, otherwise recognizes lectins	No, **MHC I recognition inhibits**	**Perforin, cytokines** (TNF-β, IFN-γ)
Macrophage	**CD14**	**Nonspecific**	No	**Intracellular** mechanisms

► Key CD Markers

CD Designation	Cellular Expression	Known Functions
CD2 (LFA-2)	T cells, thymocytes, NK cells	Adhesion molecule
CD3	**T cells**, thymocytes	Signal transduction by the TCR
CD4	**TH cells**, thymocytes, monocytes, and macrophages	Coreceptor for **MHC class II** TH-cell activation, **receptor for HIV**
CD8	**CTLs**, some thymocytes	Coreceptor for **MHC class I**–restricted T cells
CD14 (LPS receptor)	Monocytes, macrophages, granulocytes	Binds LPS
CD16 (Fc receptor)	**NK cells**, macrophages, mast cells	Immune complex-induced cellular activation, ADCC
CD19 and 20	**B cells**	Coreceptor with CD21 for B-cell activation
CD21 (CR2, C3d receptor)	Mature B cells, follicular dendritic cells	Receptor for complement fragment C3d, forms coreceptor complex with CD19, **Epstein-Barr virus** receptor
CD25	Activated TH cells and T_{Reg}	Alpha chain of IL-2 receptor
CD28	T cells	T-cell receptor for costimulatory molecule B7
CD40	B cells, macrophages, dendritic cells, endothelial cells	Binds CD40L, **starts isotype switch**
CD56	**NK cells**	Not known

Definition of abbreviations: ADCC, antibody-dependent cell-mediated cytotoxicity; CTL, cytotoxic T lymphocytes; LPS, endotoxin (lipopolysaccharide); NK, natural killer; TCR, T-cell receptor.

► Cytokines

Cytokine	Source	Activity
Interleukin-1	Monocytes, macrophages	Stimulates cells, **endogenous pyrogen**
Interleukin-2	TH cells	**Induces proliferation**, enhances activity
Interleukin-3	TH cells, NK cells	Supports growth and differentiation of **myeloid cells**
Interleukin-4	TH2 cells	Stimulates activation, differentiation, class switch to IgG1 and **IgE**
Interleukin-5	TH2 cells	Stimulates proliferation and differentiation, class switch to **IgA**; in bone marrow, stimulates eosinophil production
Interleukin-6	Monocytes, macrophages, TH2 cells	Second endogenous pyrogen, promotes differentiation into plasma cells, **induces acute phase response**
Interleukin-7	Primary lymphoid organs	**Stimulates progenitor B- and T-cell production in bone marrow**
Interleukin-8	Macrophages, endothelial cells	**Chemokine** (chemotactic for neutrophils)
Interleukin-10	**TH2 cells**	Suppresses cytokine production of TH1 cells; diminishes inflammation
Interleukin-11	Bone marrow stroma	↑ platelet count
Interleukin-12	Macrophages	**Stimulates CMI**
Interleukin-17	TH17 cells	↑ inflammation and tissue damage associated with some autoimmune diseases
Interferon-α and -β	Leukocytes, fibroblasts	**Inhibits viral protein synthesis** by acting on uninfected cells
Interferon-γ	**TH1**, CTLs, NK cells	**Stimulates CMI, Inhibits TH2**, increases expression of class I and II MHC
Tumor necrosis factor-α and -β	CMI cells	Enhances CMI
Granulocyte and granulocyte-monocyte colony-stimulating factors (G-CSF and GM-CSF)	Macrophages and TH cells	Induce proliferation in bone marrow; counteract neutropenia following ablative chemotherapy

Definition of abbreviation: CMI, cell-mediated immunity.

► The Basic Structure of Immunoglobulin

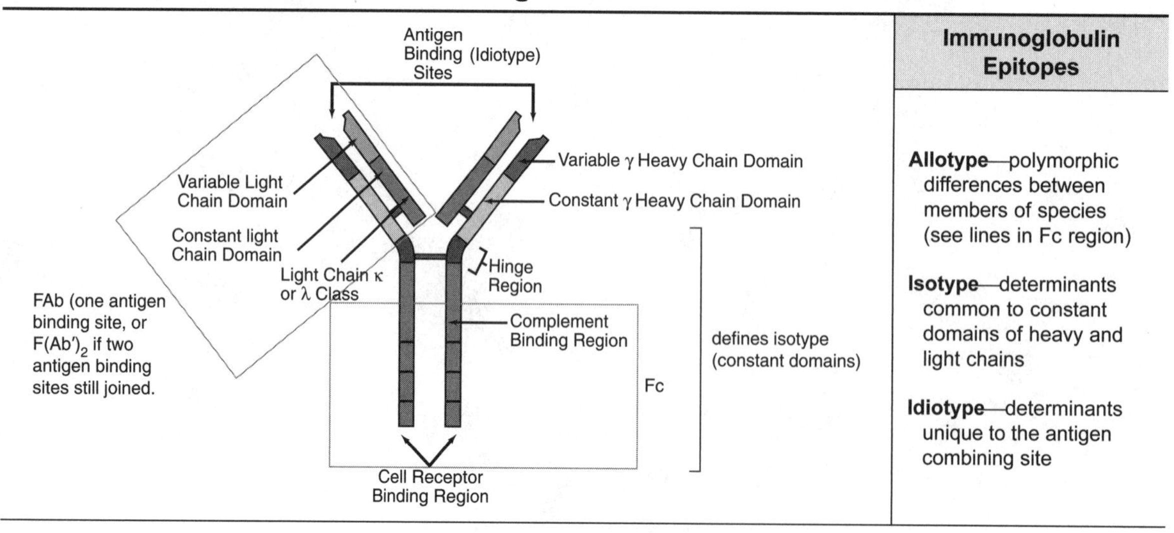

Immunoglobulin Epitopes
Allotype—polymorphic differences between members of species (see lines in Fc region)
Isotype—determinants common to constant domains of heavy and light chains
Idiotype—determinants unique to the antigen combining site

▶ Summary of the Biologic Functions of the Antibody Isotypes

	IgM	IgG	IgA	IgD	IgE
Heavy chain	μ	γ	α	δ	ε
Adult serum levels	40–345 mg/dL	650–1,500 mg/dL	75–390 mg/dL	Trace	Trace
Functions					
Complement activation, classic pathway	+	+	–	–	–
Opsonization	–	+	–	–	–
Antibody-dependent, cell-mediated cytotoxicity (ADCC)	–	+	–	–	–
Placental transport	–	+	–	–	–
Naive B-cell antigen receptor	+	–	–	+	–
Memory B-cell antigen receptor (one only)	–	+	+	–	+
Trigger mast cell granule release	–	–	–	–	+

▶ Precipitation of Ab-Ag Complexes

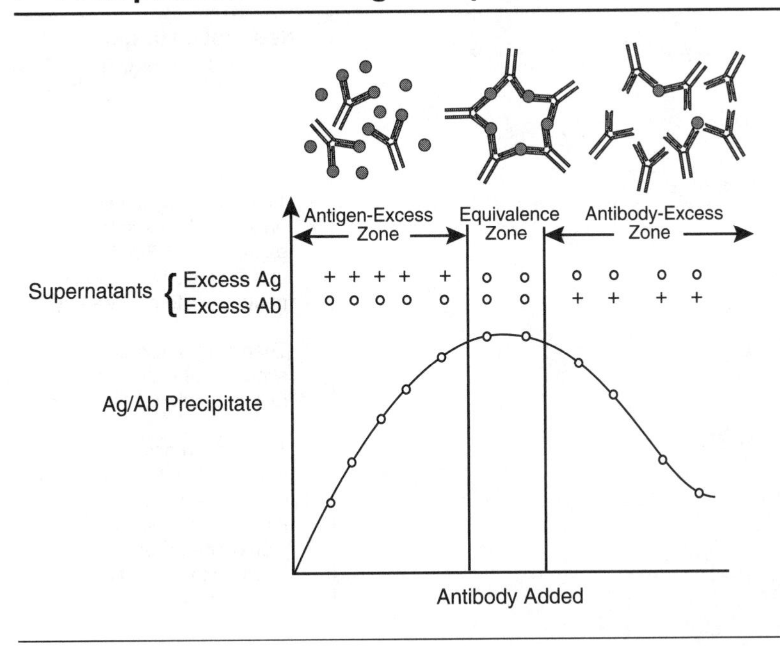

Prozone phenomenon—formation of suboptimal amount of precipitate in antibody excess (secondary syphilis)

Window period (hepatitis B)—the equivalence zone for HBsAg/HBsAb combination

Inflammation

► Acute Inflammation

Acute inflammation is an immediate response to injury, associated with redness, heat, swelling, pain, and loss of function. Understand the sequence of events of acute inflammation (extravasation, chemotaxis, phagocytosis, intracellular killing) and how these events set the stage for the subsequent adaptive immune response.

Hemodynamic changes	• Transient initial vasoconstriction, followed by massive dilation (mediated by histamine, bradykinin, and prostaglandins) • Increased vascular permeability (due to endothelial cell contraction and/or injury)—histamine, serotonin, bradykinin, leukotrienes (e.g., LTC_4, LTD_4, LTE_4) • Blood stasis due to increased viscosity allows neutrophils to marginate
Cellular response	• **Neutrophils:** (segmented) polymorphonuclear leukocytes (PMNs) are important mediators in acute inflammation • Neutrophils have **primary (azurophilic)** and **secondary (specific) granules:** – **Primary granules contain:** myeloperoxidase, phospholipase A2, lysozyme, acid hydrolases, elastase, defensins, and bactericidal permeability increasing protein (BPI) – **Secondary granules contain:** phospholipase A2, lysozyme, leukocyte alkaline phosphatase (LAP), collagenase, lactoferrin, vitamin B_{12}–binding proteins

► Neutrophil Margination and Extravasation

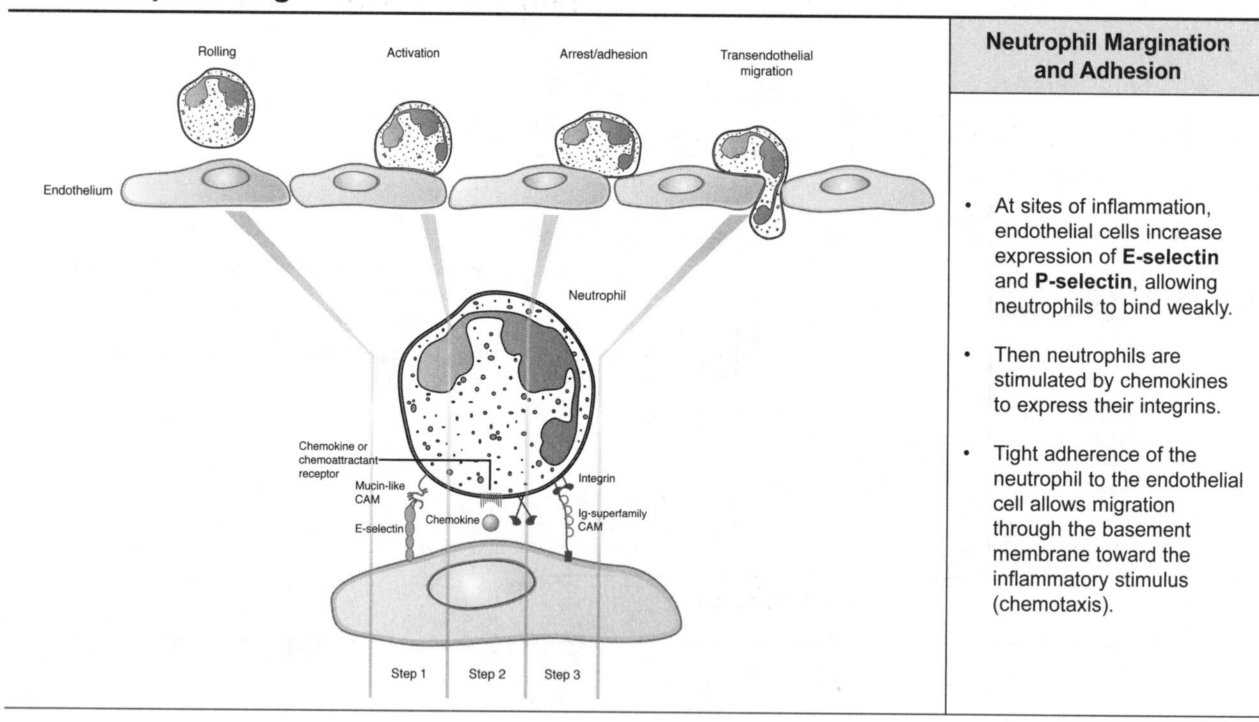

Neutrophil Margination and Adhesion

- At sites of inflammation, endothelial cells increase expression of **E-selectin** and **P-selectin**, allowing neutrophils to bind weakly.
- Then neutrophils are stimulated by chemokines to express their integrins.
- Tight adherence of the neutrophil to the endothelial cell allows migration through the basement membrane toward the inflammatory stimulus (chemotaxis).

▶ Neutrophil Migration/Chemotaxis

Chemoattractive Molecule	Origin
Chemokines (**IL-8**)	Tissue mast cells, platelets, neutrophils, monocytes, macrophages, eosinophils, basophils, lymphocytes
Complement split product **C5a**	Endothelial damage → activation Hageman factor → plasmin activation
Leukotriene B_4	Membrane phospholipids of macrophages, monocytes, neutrophils, mast cells → arachidonic acid cascade → lipoxygenase pathway
Formyl methionyl peptides	Released from **microorganisms**

▶ Other Chemical Mediators of Inflammation

Monoamine	Sources	Effects	Triggers for Release
Histamine	Basophils, platelets, and mast cells	Vasodilation and increased vascular permeability	• IgE-mediated mast cell reactions • Physical injury • Anaphylatoxins (C3a and C5a) • Cytokines (IL-1)
Serotonin	Platelets		Platelet aggregation
Enzyme	**Arachidonic Acid Product**	**Effects**	**Comments**
Cyclooxygenase	Thromboxane A_2	Vasoconstriction, platelet aggregation	Produced by platelets
	Prostacyclin (PGI_2)	**Vasodilation** and inhibits platelet aggregation	Produced by vascular endothelium
	PGE_2	Pain	—
	PGE_2, PGD_2, PGF_2	Vasodilation	—
Lipoxygenase	LTB_4	Neutrophil chemotaxis, **increased vascular permeability, vasoconstriction or vasodilation***	—
	LTC_4, LTD_4, LTE_4	**Bronchoconstriction, increased vascular permeability, vasoconstriction or vasodilation***	Slow-reacting substance of anaphylaxis

Kinin System
• Bradykinin—vasoactive peptide produced from kininogen by family of enzymes called kallikreins; degraded by different peptidases, including angiotensin-converting enzyme (ACE) • Activated **Hageman factor (factor XII)** converts prekallikrein → kallikrein • Kallikrein cleaves high molecular weight kininogen (HMWK) → **bradykinin** (produces increased vascular permeability, pain, vasodilation, bronchoconstriction)

***Can be tissue specific (e.g., vasoconstriction in kidneys and heart; vasodilation in skin and nasal mucosa)**

► Phagocytosis

- Several steps of phagocytosis: engulfment, fusion of the phagosome and lysosome, and digestion.
- Mechanisms of intracellular killing: NADPH oxidase-dependent, myeloperoxidase-dependent, and lysosome-dependent.
- **Opsonization** is the enhancement of phagocytosis with IgG and/or C3b.

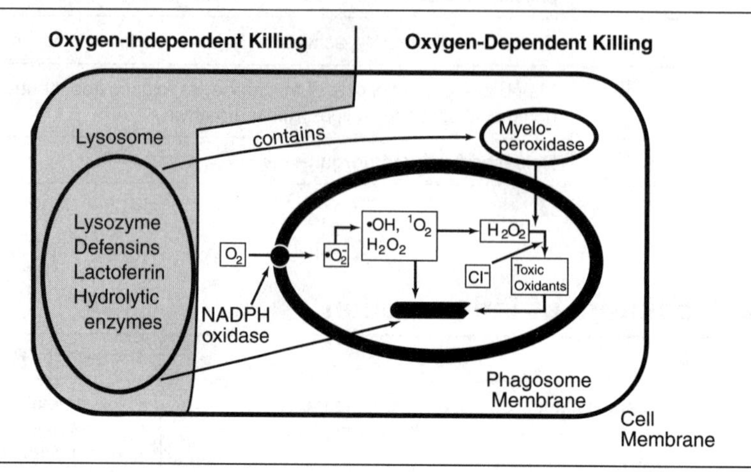

► Defects of Phagocytic Cells

Disease	Molecular Defect(s)	Symptoms
Chronic granulomatous disease (CGD)	Deficiency of **NADPH oxidase** (any one of four component proteins); failure to generate superoxide anion, other O_2 radicals	Recurrent infections with **catalase-positive** bacteria and fungi
Chédiak-Higashi syndrome	Granule structural defect	Recurrent infection with bacteria: chemotactic and degranulation defects; **absent NK** activity, **partial albinism**
Leukocyte adhesion deficiency	**Absence of CD18**—common β chain of the leukocyte integrins	Recurrent and chronic infections, failure to form pus, **does not reject umbilical cord** stump

► Complement Cascade

- The complement system is a set of interacting serum proteins that enhance inflammation (C3a, C4a, C5a) and opsonization (C3b) and cause lysis of particles (e.g., gram-negative bacteria) via C5b-9.
- The **alternative** pathway is initiated by **surfaces of pathogens**.
- The **classical** pathway is activated by **Ag/Ab complexes**.

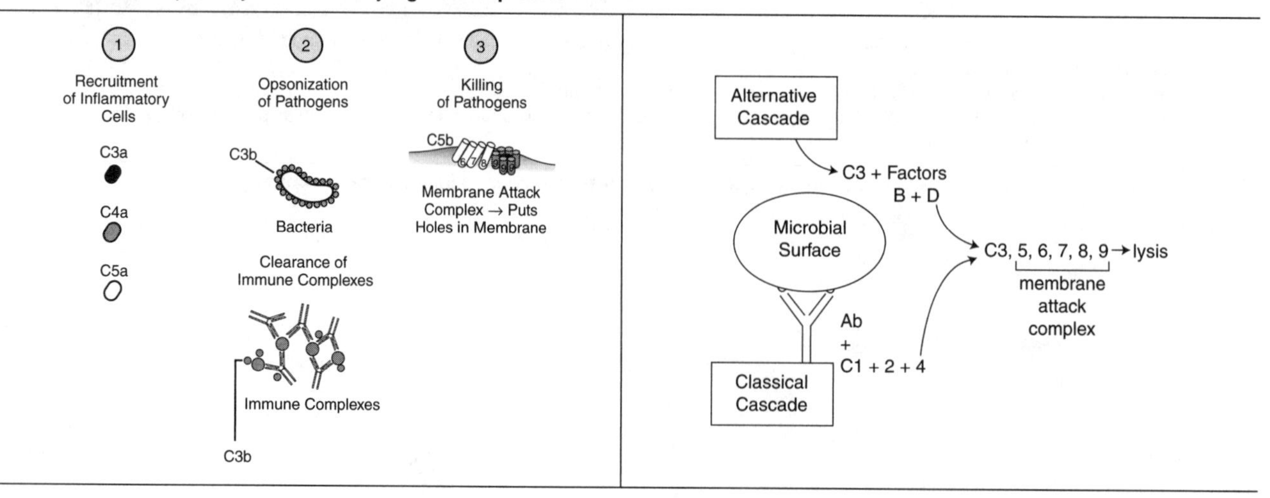

► Deficiencies of Complement or Its Regulation

Deficiencies in Complement Components	Deficiency	Signs/Diagnosis
Classical pathway	C1q, C1r, C1s, C2, C4	Marked increase in immune complex diseases, increased infections with pyogenic bacteria
Alternative pathway	Factor B, properdin	Increased neisserial infections
Both pathways	C3	Recurrent bacterial infections, immune complex disease
	C5, C6, C7, or **C8**	**Recurrent meningococcal and gonococcal infections**
Deficiencies in complement regulatory proteins	C1-INH (**hereditary angioedema**)	Serum depletion of C1, C2, C4 **Edema at mucosal surfaces**

► Summary of Acute Inflammation

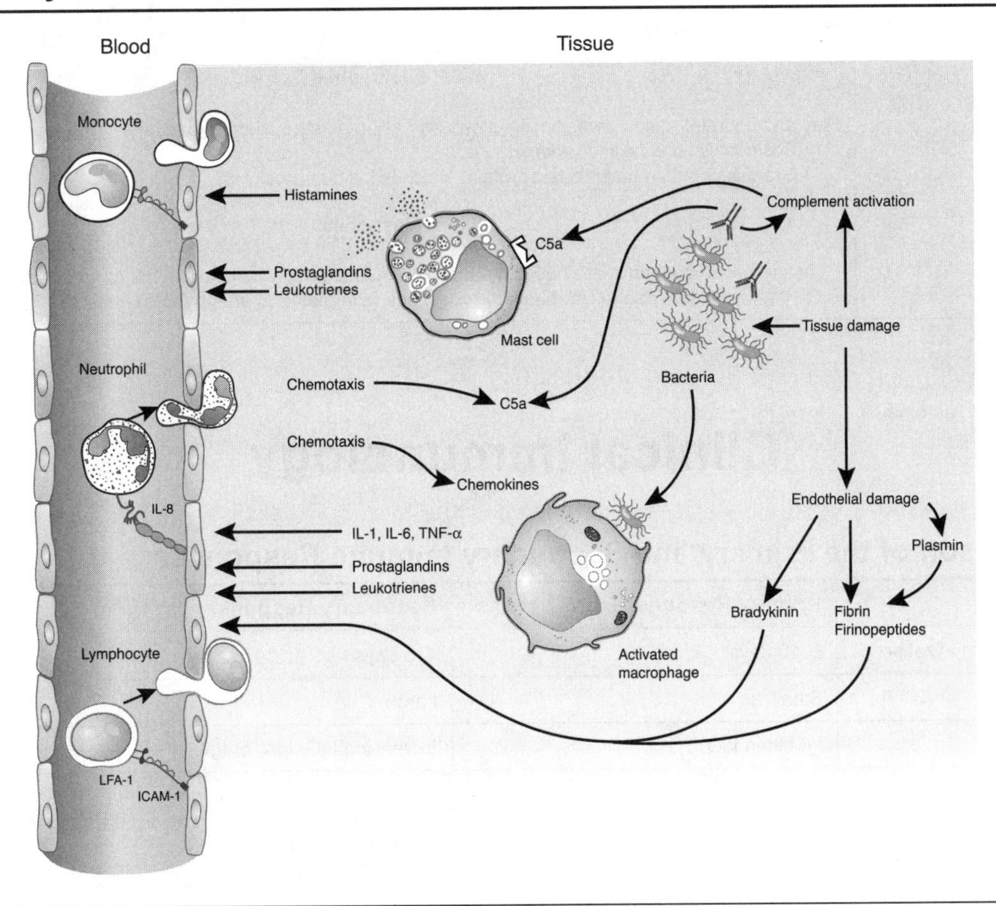

▶ Chronic Inflammation

<table>
<tr><th>Causes</th><th>Important Cell Types</th></tr>
<tr><td rowspan="8">• Following a bout of acute inflammation
• Persistent infections
• Infections with certain organisms (viruses, mycobacteria, parasites, fungi)
• Autoimmune diseases
• Response to foreign material</td><td>Macrophages</td></tr>
<tr><td>Derived from blood monocytes. During inflammation, macrophages are mainly recruited from the blood (circulating monocytes). Macrophages contain acid hydrolases, elastase, and collagenase, secrete monokines.
Chemotactic factors: C5a, MCP-1, MIP-1-α, PDGF, TGF-α
Tissue-based macrophages:
• Connective tissue (histiocyte)
• Lung (pulmonary alveolar macrophages)
• Liver (Kupffer cells)
• Bone (osteoclasts)
• Brain (microglia)
• Kidney (mesangial cells)</td></tr>
<tr><td>Lymphocytes (e.g., B cells, plasma cells)</td></tr>
<tr><td>• T cells
• Lymphocyte chemokine: lymphotaxin</td></tr>
<tr><td>Eosinophils</td></tr>
<tr><td>Play an important role in parasitic infections and IgE-mediated allergic reactions
• Eosinophilic chemokine: eotaxin
• Granules contain major basic protein, which is toxic to parasites</td></tr>
<tr><td>Basophils</td></tr>
<tr><td>• Tissue-based basophils are called mast cells, present in high numbers in the lung and skin
• Play an important role in IgE-mediated reactions (allergies and anaphylaxis), release histamine</td></tr>
</table>

Clinical Immunology

▶ Comparison of the Primary and Secondary Immune Responses

Feature	Primary Response	Secondary Response
Time lag after immunization	5–10 days	1–3 days
Peak response	Small	Large
Antibody isotype	IgM then IgG	Increasing IgG, IgA, or IgE
Antibody **affinity**	Variable to low	High (**affinity maturation**)
Inducing agent	All immunogens	**Protein antigens**
Immunization protocol	High dose of antigen (often with adjuvant)	Low dose of antigen (often without adjuvant)

▶ Types of Immunization Used in Medicine

Adjuvants increase immunogenicity nonspecifically. They are given with weak immunogens to enhance the response.

Type of Immunity	Acquired Through	Examples
Natural	Passive means	Placental IgG transport, colostrum
Natural	Active means	Recovery from infection
Artificial	Passive means	Immunoglobulins or immune cells given
Artificial	Active means	Vaccination

▶ Summary of Bacterial Vaccines

Organism	Vaccine	Vaccine Type
C. diphtheriae	DTaP	Toxoid
B. pertussis	DTaP	Toxoid
C. tetani	DTaP	Toxoid
H. influenzae	Hib	Capsular polysaccharide and protein
S. pneumoniae	PCV Pediatric	13 capsular serotypes and protein
	PPV Adult	23 capsular serotypes
N. meningitidis	MCV-4	4 capsular serotypes (Y, W-135, C, A) plus protein

▶ Summary of Viral Vaccines

Killed Vaccines	Live Viral Vaccines	Component Vaccines
Mnemonic: RIP-A (Rest In Peace Always—the killed viral vaccines): Rabies (killed human diploid cell vaccine) Influenza Polio (Salk) A Hepatitis	All but adenovirus are attenuated (**mnemonic: Mrr. V.Z. Mapsy**) Mumps Rotavirus Rubella Varicella – Zoster Measles Adenovirus (pathogenic [not attenuated] respiratory strains given in enteric coated capsules) Polio (Sabin) Small Pox Yellow Fever	Hepatitis B HPV (human papilloma virus)

► Defects of Humoral Immunity

Disease	Molecular Defect	Symptoms/Signs
Bruton X-linked hypogammaglobulinemia	Deficiency of a tyrosine kinase blocks B-cell maturation	Low immunoglobulin of all classes, **no circulating B cells**, **pre-B cells in bone marrow in normal numbers**, normal cell-mediated immunity
Selective IgA deficiency	Deficiency of IgA (most common)	Repeated **sinopulmonary and gastrointestinal infections**, ↑ atopic allergy
X-linked hyper-IgM syndrome	Deficiency of **CD40L** on activated T cells	**High serum titers of IgM without other isotypes** Normal B- and T-cell numbers, susceptibility to extracellular bacteria and opportunists
Common variable immunodeficiency	B-cell maturation defect and **hypogammaglobulinemia**	Both sexes affected, childhood onset, recurrent bacterial infections and increased susceptibility to *Giardia* Increased risk later in life to autoimmune disease, lymphoma, or gastric cancer

► Defects of T Cells and Severe Combined Immunodeficiencies

Category	Disease	Defect	Clinical Manifestation
Selective T-cell deficiency	**DiGeorge syndrome (velocardiofacial syndrome)**	Failure of formation of third and fourth pharyngeal pouches, **thymic aplasia**	Facial abnormalities, hypoparathyroidism, cardiac malformations, depression of T-cell numbers and absence of T-cell responses
	MHC class I deficiency	Failure of TAP 1 molecules to transport peptides to endoplasmic reticulum	**CD8+ T cells deficient**, CD4+ T cells normal, recurring viral infections, normal DTH, normal Ab production
	MHC class II deficiency	**Bare lymphocyte syndrome**	T cells present and responsive to nonspecific mitogens, no GVHD, **deficient in CD4+ T cells,** hypogammaglobulinemia; presents as SCID
Combined partial B- and T-cell deficiency	**Wiskott-Aldrich syndrome**	Defect in cytoskeletal glycoprotein, X-linked	Defective responses to bacterial polysaccharides and depressed IgM, gradual loss of humoral and cellular responses, **thrombocytopenia and eczema**
	Ataxia telangiectasia	Defect in kinase involved in the cell cycle	Ataxia (gait abnormalities), telangiectasia (capillary distortions in the eye), deficiency of IgA and IgE production
Complete functional B- and T-cell deficiency	Severe combined immunodeficiency (SCID)	Defects in common γ chain of IL-2 receptor (also present in receptors for IL-4, -7, -9, -15) X-linked	Chronic diarrhea; skin, mouth, and throat lesions; opportunistic (**fungal**) infections; low levels of circulating lymphocytes; cells unresponsive to mitogens
		Adenosine deaminase deficiency (results in toxic metabolic products in cells)	
		Defect in signal transduction from T-cell IL-2 receptors	

▶ Association Between Immunodeficiency Disease and Developmental Blocks

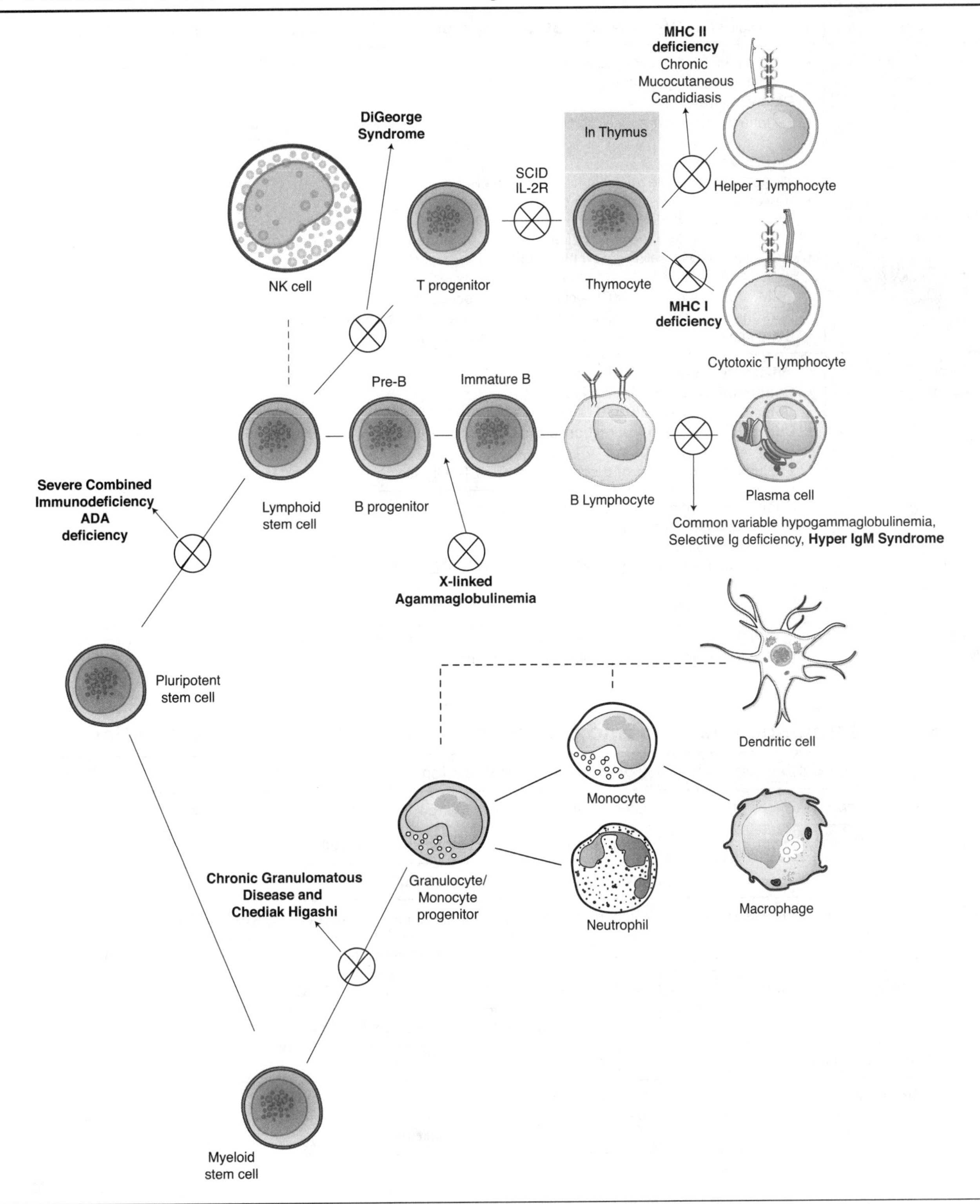

► Acquired Immunodeficiency Syndrome (AIDS)

Definition	When a patient is HIV-positive with CD4 count less than 200/mm^3 *or* HIV-positive with an AIDS-defining disease
Transmission	• Sexual contact: homosexual > heterosexual in U.S. – Cofactors: herpes and syphilis • Parenteral transmission: – Intravenous drug abuse – Hemophiliacs – Blood transfusions – Accidental needle sticks in hospital workers • Vertical transmission (mother to child)
Pathogenesis	• Human immunodeficiency virus (HIV; an enveloped retrovirus containing reverse transcriptase) • HIV infects CD4+ cells (gp120 binds to CD4) – CD4+ T cells (with CCR5 chemokine coreceptor) – Macrophages (with CXCR4 chemokine coreceptor) • **Persons who are homozygous CCR5 defective cannot be infected** • Entry into cell by fusion requires gp41 and coreceptors
Diagnosis and Monitoring	• **Initial screening:** Serologic, ELISA • **Confirmation:** Serologic, Western blot • **Detection of virus in blood (evaluate viral load):** RT-PCR • **Detect HIV infection in newborns of HIV+ mother (provirus):** PCR* • **Early marker of infection:** p24 antigen • **Evaluate progression of disease:** CD4:CD8 T-cell ratio
Treatment	• Combination antiretroviral treatment • Reverse transcriptase inhibitors • Protease inhibitors • Prophylaxis for opportunistic infections based on CD4 count • Fusion inhibitors • Integrase inhibitors

*RT-PCR tests for circulating viral RNA and is used to monitor the efficacy of treatment. PCR detects integrated virus (provirus DNA). Viral load has been demonstrated to be the best prognostic indicator during infection.

► Opportunistic Infection and Common Sites of Infection in AIDS Patients

Opportunistic Infection	Common Sites of Infection
Pneumocystis jiroveci (carinii)	Lung (pneumonia), bone marrow, start prophylaxis at **< 200 CD4**
Histoplasma capsulatum	Lung, disseminated, start prophylaxis at **< 100 CD4** in endemic area
Toxoplasma gondii	CNS, start prophylaxis at **< 100 CD4**
Mycobacterium avium-intracellulare	Lung, gastrointestinal tract, disseminated, start prophylaxis at **< 50 CD4**
Cytomegalovirus	Lung, retina, adrenals, and gastrointestinal tract, start prophylaxis at **< 50 CD4**
Cryptosporidium parvum	Gastrointestinal tract, start prophylaxis at **< 50 CD4**
Cryptococcus neoformans	CNS (meningitis), start prophylaxis at **< 50 CD4**
Coccidioides immitis	Lung, disseminated
Giardia lamblia	Gastrointestinal tract
Herpes simplex virus	Esophagus and CNS (encephalitis)
Candida albicans	Oropharynx and esophagus
Aspergillus spp.	CNS, lungs, blood vessels
JC virus	CNS (progressive multifocal leukoencephalopathy)
Mycobacterium tuberculosis	Lung, disseminated

▶ Other Complications of AIDS

Hairy leukoplakia	Associated with Epstein-Barr virus (EBV)
Kaposi sarcoma	Associated with human herpes virus 8 (HHV8) Common sites: skin, GI tract, lymph nodes, and lungs
Non-Hodgkin lymphoma	Tend to be high-grade B-cell lymphomas Extranodal CNS lymphomas common
Miscellaneous	Cervical cancer HIV wasting syndrome AIDS nephropathy AIDS dementia complex

▶ HIV Therapies

Anti-HIV therapy usually involves three or more anti-retroviral agents including antimetabolite inhibitors of retroviral reverse transcriptase and viral protease. These aggressive drug combinations (highly active antiretroviral therapy, **HAART**) are typically initiated early after HIV infection and will often reduce viral load, preserve CD-4 counts, and limit opportunistic infections. Drug combinations also slow the development of resistance to drugs as opposed to monotherapy.

Drug	Mechanism of Action	Side Effects and Comments
	Nucleoside reverse transcriptase inhibitors (NRTIs)	
Abacavir (ABC) Didanosine (ddI) Emtricitabine (FTC) Lamivudine (3TC) Stavudine (d4T) Tenofovir (TDF) Zalcitabine (ddC) Zidovudine (ZDV, AZT)	Converted to nucleoside triphosphates, which **inhibit reverse transcriptase**, leading the inhibition of viral replication by chain termination	Dose-limiting pancreatitis (ddI) Dose-limiting peripheral neuropathy (ddC, ddI, d4T) Hypersensitivity reactions (ABC) Lactic acidosis (ddI, d4T) Fanconi syndrome (TDF) Dose-limiting BMS (ZDV) Minimal toxicity (FTC, 3TC) Least toxic and also used for hepatitis B (3TC)
	Non-nucleoside reverse transcriptase inhibitors (NNRTIs)	
Dela**vir**dine (DLV) Efa**vir**enz (EFV) Etra**vir**ine (ETR) Ne**vir**apine (NVP)	**Inhibit reverse transcriptase** noncompetitively	Rash (occasional Steven-Johnson syndrome) Adverse CNS effects (EFV) P450 interactions Increased transaminase levels (DLV, EFV)
	Protease inhibitors	
Ataza**navir** (ATV) Daru**navir** (DRV) Fosampre**navir** (FPV) Indi**navir** (IDV) Nelfi**navir** (NFV) Rito**navir** (RTV) Saqui**navir** (SQV) Lopi**navir** + Rito**navir** (LPV/r)	**Inhibits aspartate protease** (HIV-1 protease encoded by ***pol* gene**), blocking structural protein formation of the mature virion core	Inhibit CYP3A4 (especially ritonavir) Crystalluria (indinavir, maintain hydration) Elevated transaminases Lopinavir is only available in combination with ritonavir
	Fusion inhibitor	
Enfuvirtide	**Binds gp41** and **inhibits fusion** of HIV-1 to CD4+ cells	Local injection reactions (100% patients) Hypersensitivity reactions
	CCR5 antagonist	
Maraviroc (MVC)	**Blocks CCR5 receptor**, preventing HIV entry into target cells	Hepatotoxicity, orthostatic hypotension
	Integrase inhibitor	
Raltegravir (RAL)	**Inhibits integrase**, thus blocking viral integration	CPK elevation

► Hypersensitivity Reactions

Type	Antibody	Complement	Effector Cells	Examples
I (immediate)	**IgE**	No	**Basophil, mast cell**	Hay fever, atopic dermatitis, **insect venom sensitivity**, **anaphylaxis** to drugs, some food allergies, allergy to animals and animal products, **asthma**
II (cytotoxic)	IgG, IgM	Yes	PMN, macrophages, NK cells	Autoimmune or drug-induced hemolytic anemia, transfusion reactions, **HDNB**, hyperacute graft rejection, **Goodpasture disease**, **rheumatic fever**
II (noncytotoxic)	IgG	**No**	None	Myasthenia gravis, Graves disease, type 2 diabetes mellitus
III (immune complex)	IgG, IgM	Yes	PMN, macrophages	**SLE,** polyarteritis nodosa, poststreptococcal glomerulonephritis, Arthus reaction, serum sickness
IV (delayed, DTH)	None	No	CTL, TH1, macrophages	**Tuberculin test**, tuberculosis, leprosy, **RA**, Hashimoto thyroiditis, poison ivy (**contact dermatitis**), **GVHD**, IDDM

Definition of abbreviations: GVHD, graft-versus-host disease; HDNB, hemolytic disease of the newborn; IDDM, insulin-dependent diabetes mellitus; RA, rheumatoid arthritis; SLE, systemic lupus erythematosus.

► Important Autoimmune Diseases

Autoantibodies	Clinical Features	Comments
Systemic lupus erythematosus: chronic systemic autoimmune disease characterized by a loss of self-tolerance and production of autoantibodies		
Antinuclear antibody (ANA) (>95%): **Anti-dsDNA** (40–60%) **Anti-Sm** (20–30%)	• Hemolytic anemia, thrombocytopenia, leukopenia • Arthritis • Skin rashes (including classic malar rash) • Renal disease • Libman-Sacks endocarditis • Serositis • Neurologic symptoms	• Females >> Males (M:F = 1:9), peak age 20–45 years, African American > Caucasian • Mechanism of injury: type II and III hypersensitivity reactions • Treatment: steroids and other immunosuppressants
Sjögren syndrome: an autoimmune disease characterized by destruction of the lacrimal and salivary glands, resulting in the inability to produce saliva or tears		
Antiribonucleoprotein antibodies: Anti-SS-A (Ro) Anti-SS-B (La)	• Keratoconjuctivitis sicca (dry eyes) and corneal ulcers • Xerostomia (dry mouth) • Mikulicz syndrome: enlargement of the salivary and lacrimal glands	• Females > males; age range: 30–50 years • Often associated with rheumatoid arthritis and other autoimmune diseases (e.g., SLE) • Increased risk of developing lymphoma

(Continued)

▶ Important Autoimmune Diseases *(Cont'd.)*

Autoantibodies	Clinical Features	Comments
Scleroderma (progressive systemic sclerosis): characterized by fibroblast stimulation and deposition of collagen in the skin and internal organs; females > males; age range: 20–55 years; activation of fibroblasts by growth factors/cytokines leads to fibrosis		
Diffuse Scleroderma		
Anti-DNA topoisomerase I antibodies (Scl-70) (70%)	Widespread skin involvement Early involvement of the visceral organs • Esophagus—dysphagia • GI tract—malabsorption • Pulmonary fibrosis—dyspnea on exertion • Cardiac fibrosis—arrhythmias • Kidney fibrosis—renal insufficiency	Raynaud phenomenon is seen in almost all patients and often preceeds other symptoms. **Treatment:** vasodilators, ACE inhibitors, NSAIDs, steroids, d-penicillamine
Limited scleroderma (e.g., CREST syndrome)		
Anticentromere antibodies	• Skin involvement of the face and hands • Late involvement of visceral organs (relatively benign clinical course)	(**C**alcinosis, **R**aynaud phenomenon, **E**sophageal dysmotility, **S**clerodactyly, **T**elangiectasia)

Transplantation Immunology

▶ Grafts Used in Medicine

Grafts	Definition
Autologous (**autografts**)	Tissue is moved from one location to another in the same individual
Isograft	Transplants between genetically identical individuals (monozygotic twins)
Allograft	Transplants between genetically different members of the same species
Xenograft	Transplants between members of different species

▶ Graft Rejection Reactions

Type of Rejection	Time Taken	Cause
Hyperacute	Minutes to hours	Preformed anti-donor antibodies and complement (type II hypersensitivity)
Accelerated	Days	Reactivation of sensitized T cells
Acute	Days to weeks	Primary activation of T cells
Chronic	Months to years	Causes are unclear: antibodies, immune complexes, slow cellular reaction, recurrence of disease (type IV hypersensitivity)
Graft versus host	Weeks to months	Grafted bone marrow T cells attack host (type IV hypersensitivity)

Immunology Techniques in Diagnosis

► Fluorescence Activated Cell Sorter

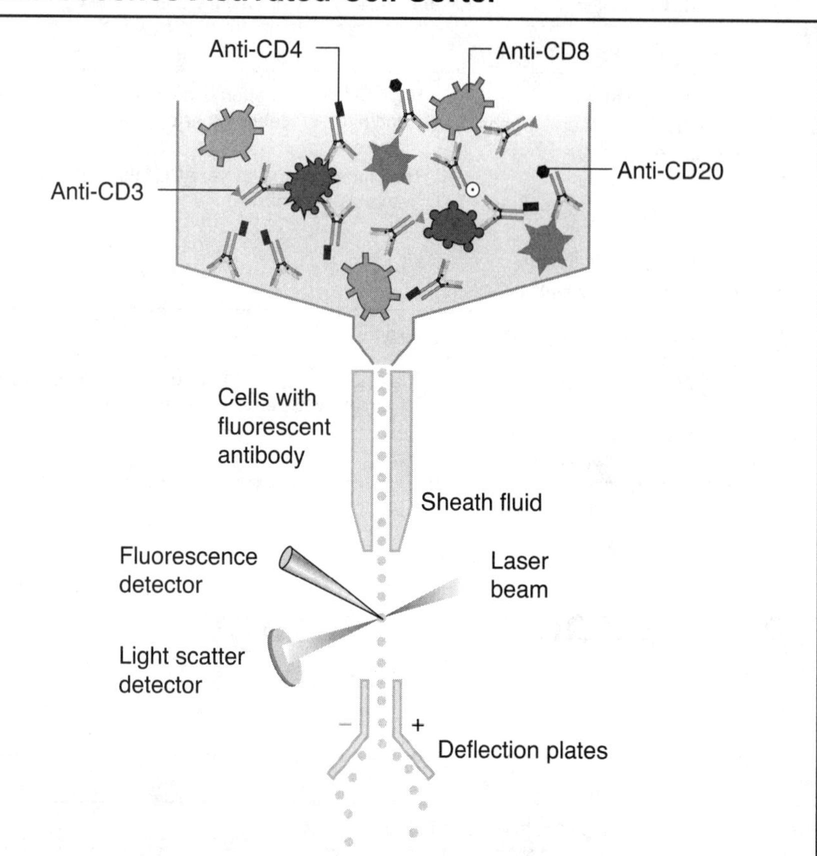

Computer-Generated Graphs

Double stained
3 2 1
Anti-CD8 (increasing intensity of fluorescence)
Anti-CD3 (increasing intensity of fluorescence)

Double stained
3 2 1
Anti-CD4
Anti-CD3

3 2 1
Anti-CD20
Anti-CD3

- Complex mixtures of cells are treated with fluorescent dye-labeled antibodies and run through the apparatus.
- The fluorescence activated cell sorter (FACS) separates the cells into populations based on their level of fluorescence with a particular dye.
- Each dot on the diagram represents a cell that has bound to a fluorescent-labeled antibody.
- Increasing fluorescence intensity with one dye is represented as a rise on the *y*-axis, and increasing fluorescence with the other dye occurs as you move right on the *x*-axis.
- Double-labeled cells are always found in the upper right quadrant.
- Cells that have only background fluorescence with either dye are found in the lower left quadrant.

The results of flow cytometry are often shown for question analysis. Be sure to know the key CD markers and the biologic functions of the cells that possess them.

Immunopharmacology

► Recombinant Cytokines and Clinical Uses

Cytokine	Clinical Uses
Aldesleukin (IL-2)	↑ Lymphocyte differentiation and ↑ NKs—used in renal cell cancer and metastatic melanoma
Oprelvekin (IL-11)	↑ Platelet formation—used in thrombocytopenia
Filgrastim (G-CSF)	↑ Granulocytes—used for bone marrow recovery
Sargramostim (GM-CSF)	↑ Granulocytes and macrophages—used for bone marrow recovery
Erythropoietin	Anemias, especially associated with renal failure
Thrombopoietin	Thrombocytopenia
Interferon-α	Hepatitis B and C, leukemias, malignant melanoma, Kaposi sarcoma
Interferon-β	Multiple sclerosis
Interferon-γ	Chronic granulomatous disease →↑ TNF

► Immunosuppressant Agents

Drug	Mechanism	Uses
Azathioprine	Converted to mercaptopurine, whose metabolites inhibit purine metabolism Cytotoxic to proliferating lymphocytes (especially T cells)	Autoimmune diseases (e.g., SLE, rheumatoid arthritis) and immunosuppression in renal allografts
Corticosteroids	↓ synthesis of prostaglandins, leukotrienes, cytokines; inhibit T-cell proliferation; at immunosuppressive doses, they are cytotoxic to some T cells	Cancer, organ transplants
Cyclophosphamide	Cytotoxic to proliferating lymphocytes (especially B cells)	• Autoimmune diseases, bone marrow transplants • Similar cytotoxic drugs: cytarabine, dactinomycin, methotrexate, vincristine
Cyclosporine	Antibiotic that binds to cyclophilin → inhibits calcineurin (cytoplasmic phosphatase) → ↓ activation of T-cell transcription factors → ↓ IL-2, IL-3, and interferon-γ	• DOC in organ or tissue transplantation (± mycophenolate ± steroids ± cytotoxic drugs) • Side effects: peripheral neuropathy, nephrotoxicity, hyperglycemia, hypertension, hyperlipidemia, hirsutism, gingival overgrowth, cholelithiasis
Tacrolimus	Antibiotic that binds to FK-binding protein (FKBP); also inhibits calcineurin (similar to cyclosporine)	• Used alternatively to cyclosporine in renal and liver transplants • Side effects similar to cyclosporine
Sirolimus (rapamycin)	Inhibits T-cell activation and proliferation in response to IL-2 by binding to mTOR	Immunosuppression after kidney transplantation in conjunction with cyclosporine and corticosteroids
Mycophenolate	Inhibits de novo purine synthesis	• Kidney, liver, heart transplants • Used with cyclosporine in renal transplants to ↓ cyclosporine dose
RhD immune globulin (RhoGAM™)	Antibody to red cell RhD antigens	Administer to RhD ⊖ mother within 72 h of Rh ⊕ delivery to prevent hemolytic disease of newborn in subsequent pregnancy

► Monoclonal Antibodies (MABs) and Clinical Uses

MAB	Clinical Uses
Abciximab	Antiplatelet (acute coronary symptoms, post-angioplasty)—antagonist of IIb/IIIa receptors
Daclizumab	Kidney transplants—blocks IL-2 receptors
Infliximab	Rheumatoid arthritis and Crohn disease—binds TNF-α
Muromonab	Allograft rejection block in renal transplants—binds CD3 on T cells
Palivizumab	Respiratory syncytial virus—blocks RSV fusion protein
Rituximab	Non-Hodgkin lymphoma—binds to CD20 antigen on B-cell surface protein
Trastuzumab	Breast cancer—antagonist to HER2/neu receptor

Microbiology

Chapter 5

General Principles of Microbiology

► Comparison of Medically Important Microbial Groups

Characteristic	Viruses	Bacteria	Fungi	Parasites
Diameter	Minute (0.02–0.3 μ)	Small (0.3–2 μ)	3–10 μ	15–25 μ (trophozoites)
Cell type	**Acellular**—no nucleus	**Prokaryotic cells**	**Eukaryotic cells**	
	• DNA or RNA • 1 nucleocapsid, except in segmented or diploid viruses	• DNA and RNA • 1 chromosome • **No histones**	• DNA and RNA • More than one chromosome	
	Replicates in host cells	DNA replicates continuously	G and S phases	
		Exons, **no introns**	Introns and exons	
	Some have polycistronic mRNA and post-translational cleavage	**Mono- and polycistronic mRNA**	**Monocistronic RNA**	
	Uses host organelles; obligate intracellular parasites	**No membrane-bound organelles**	Mitochondria and other membrane-bound organelles	
	No ribosomes	**70S** ribosomes (30S+50S)	**80S** ribosomes (40S+60S)	
Replication	Make and assemble viral components	**Binary fission (asexual)**	Cytokinesis with mitosis/meiosis	
Cellular membrane	Some are enveloped, but no membrane function	Membranes have **no sterols, except *Mycoplasma***, which have cholesterol	**Ergosterol** is major sterol	Sterols, such as **cholesterol**
Cell wall	No cell wall	**Peptidoglycan**	Complex carbohydrate **cell wall: chitin**, glucans, or mannans	No cell wall

Note: Prions are infectious proteins (contain no nucleic acids). They are the agents of kuru, mad cow disease, etc.

Bacteriology

► Bacterial Growth Curve

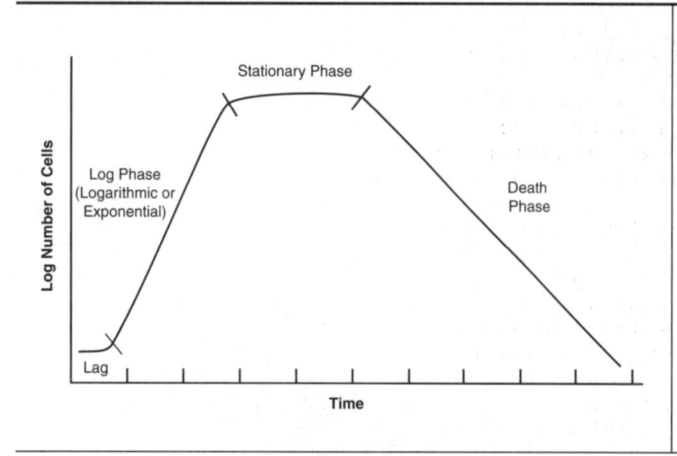

Lag Phase:
- Detoxifying medium
- Turning on enzymes to utilize medium

Log Phase:
- Rapid exponential growth
- Generation time—time it takes one cell to divide into two
- This is determined during log phase

Stationary Phase:
- Nutrients used up
- Toxic products begin to accumulate
- Number of new cells = the number of dying cells

Death Phase:
- Nutrients gone
- Toxic products kill cells
- Number of cells dying exceeds the number of cells dividing

► Features of Bacteria

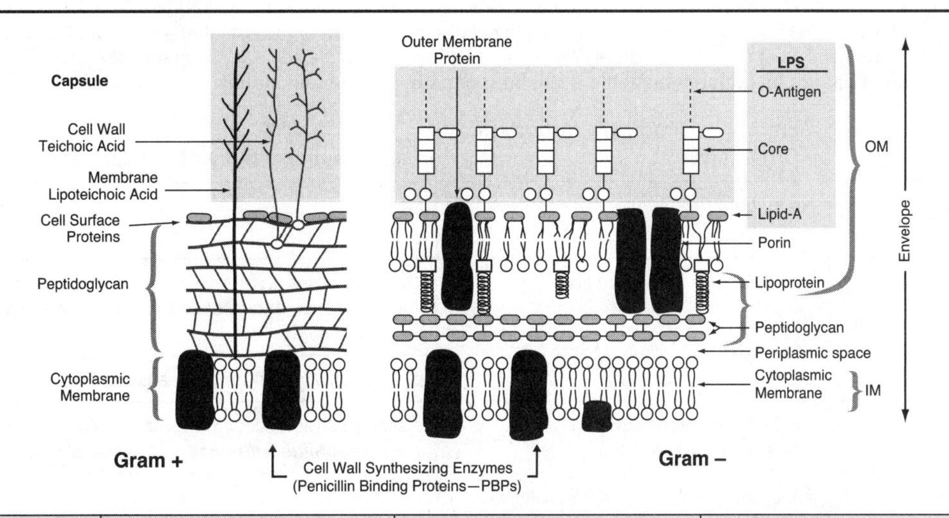

Envelope Structure	Gram ⊕ or ⊖	Chemical Composition	Function
Capsule (nonessential) = slime = glycocalyx	Both Gram ⊕ and Gram ⊖	**Polysaccharide gel** (except *B. anthracis*: poly-D-glutamate)	• **Antiphagocytic** • **Immunogenic** (except *S. pyogenes* and *N. meningitidis*, type B)
Outer membrane	**Gram** ⊖ only	Phospholipid/proteins LPS: • Lipid A • Polysaccharide	Hydrophobic membrane: • **LPS = endotoxin** • **Lipid A = toxic moiety** • PS = immunogenic portion
		Outer membrane proteins	Attachment, virulence, etc.
		Protein porins	Passive transport
Cell wall = peptidoglycan	Gram ⊕ (thick) Gram ⊖ (thin)	**Peptidoglycan**—open 3-D net of: • *N*-acetyl-glucosamine • *N*-acetyl-muramic acid • Amino acids (including DAP)	• Rigid support, cell shape, and protection from osmotic damage • Synthesis **inhibited by penicillins and** cephalosporins • **Confers Gram reaction**
	Gram ⊕ only	**Teichoic acids**	• Immunogenic, induces TNF-α, IL-1 • **Attachment**
	Acid-fast only	**Mycolic acids**	• Acid-fastness • **Resistance to drying and chemicals**
Periplasmic space	**Gram** ⊖ only	"Storage space" between the inner and outer membranes	• Enzymes to break down large molecules (β-lactamases) • Aids regulation of osmolarity
Cytoplasmic membrane = inner membrane = cell membrane = plasma membrane	Gram ⊕ Gram ⊖	Phospholipid bilayer with many embedded proteins	• Hydrophobic cell "sac" • Selective permeability and active transport • **Carrier for enzymes** for: – Oxidative metabolism – Phosphorylation – Phospholipid synthesis – DNA replication – Peptidoglycan cross linkage • **Penicillin binding proteins** (PBPs)

Definition of abbreviation: DAP, diaminopimelic acid; LPS, lipopolysaccharide.

▶ Normal Flora Organisms

Most infectious disease vignettes begin with the necessity to rule out the normal flora organisms that are cultured from the patient. Make sure you know those organisms that can confound a simple Gram-stain type of diagnosis, **e.g., notice in the oropharynx and vagina, normal flora organisms cannot be indistinguishable from pathogens by Gram stain alone!**

Site	Common or Medically Important Organisms	Less Common but Notable Organisms
Blood, internal organs	None, generally sterile	—
Cutaneous surfaces	*Staphylococcus epidermidis*	*Staphylococcus aureus, Corynebacteria* (diphtheroids), streptococci, anaerobes, e.g., peptostreptococci, *Candida* spp.
Nose	*Staphylococcus aureus*	*S. epidermidis,* diphtheroids, assorted streptococci
Oropharynx	**Viridans streptococci**, including *Streptococcus mutans*	Assorted streptococci, **nonpathogenic *Neisseria,* nontypeable *Haemophilus influenzae,*** *Candida albicans*
Gingival crevices	Anaerobes: *Bacteroides*, *Prevotella*, *Fusobacterium*, *Streptococcus*, *Actinomyces*	—
Stomach	None	—
Colon (microaerophilic/ anaerobic)	Adult: ***Bacteroides***/*Prevotella* (predominant organism), *Escherichia*, *Bifidobacterium*	*Eubacterium*, *Fusobacterium*, *Lactobacillus*, assorted gram-negative anaerobic rods *Enterococcus faecalis* and other streptococci
Vagina	**Lactobacillus**	Assorted streptococci, gram-negative rods, diphtheroids, yeasts, *Veillonella* (gram-negative diplococcus)

▶ Bacterial Toxins

Endotoxin
Endotoxin (lipopolysaccharide = LPS) is part of the gram-negative outer membrane. It is encoded on the chromosome. • The toxic portion is lipid A. LPS is heat stable and not strongly immunogenic, so it cannot be converted to a toxoid. • LPS activates macrophages, leading to release of TNF-α, IL-1, IL-6, and nitric oxide (NO). IL-1 is a major mediator of fever. Damage to the endothelium from bradykinin-induced vasodilation leads to shock. Coagulation (DIC) is mediated through the activation of Hageman factor. NO production causes hypotension, which contributes to shock.

Exotoxin
Exotoxins are protein toxins, generally quite toxic, and secreted by bacterial cells. They are encoded on plasmids or in lysogenic phage genomes. • Exotoxins can be modified by chemicals or heat to produce a toxoid that still is immunogenic, but no longer toxic, so it can be used as a vaccine. • Most are A-B (or two) component protein toxins. B component binds to specific cell receptors to facilitate the internalization of A (the active [toxic] component)

▶ Major Exotoxins

	Organism (Gram)	Toxin	Mode of Action	Role in Disease
Protein synthesis inhibitors	*Corynebacterium diphtheriae* (⊕)	**Diphtheria toxin**	• ADP ribosyl transferase **inactivates eEF-2** • *Targets:* **heart, nerves, epithelium**	Inhibits eukaryotic cell protein synthesis
	Pseudomonas aeruginosa (⊖)	**Exotoxin A**	• ADP ribosyl transferase **inactivates eEF-2** • *Target:* **liver**	Inhibits eukaryotic cell protein synthesis
	Shigella dysenteriae (⊖)	**Shiga toxin**	**Interferes with 60S ribosomal subunit**	• Inhibits protein synthesis in eukaryotic cells • Enterotoxic, cytotoxic, and neurotoxic
	Enterohemorrhagic *E. coli* (EHEC) (⊖)	**Verotoxin** (shiga-like)	**Interferes with 60S ribosomal subunit**	Inhibits protein synthesis in eukaryotic cells
Neurotoxins	*Clostridium tetani* (⊕)	**Tetanus toxin**	**Blocks release of glycine and GABA**	Inhibits neurotransmission in inhibitory synapses
	Clostridium botulinum (⊕)	**Botulinum toxin**	**Blocks release of acetylcholine**	Inhibits cholinergic synapses
Superantigens	***Staphylococcus aureus*** **(⊕)**	**TSST-1**	• Induces IL-1, IL-6, TNF-α, IFN-γ • Decreases liver clearance of LPS	Fever, increased susceptibility to LPS, rash, shock, capillary leakage
	Streptococcus pyogenes **(⊕)**	**Exotoxin A**, also called erythrogenic or pyrogenic toxin	**Similar to TSST-1**	Fever, increased susceptibility to LPS, rash, shock, capillary leakage, cardiotoxicity
cAMP inducers	Enterotoxigenic *Escherichia coli* (⊖)	Heat **labile toxin** (LT)	LT stimulates an adenylate cyclase by **ADP ribosylation of GTP-binding protein**	Both LT and ST promote secretion of fluid and electrolytes from intestinal epithelium
	Vibrio cholerae (⊖)	**Cholera toxin**	**Similar to *E. coli* LT**	Profuse, watery diarrhea
	Bacillus anthracis (⊕)	**Anthrax toxin (3 proteins** make 2 toxins)	• **EF = edema factor = adenylate cyclase** • **LF = lethal factor** • **PA = protective antigen (B component for both)**	• Decreases phagocytosis • Causes **edema, kills cells**
	Bordetella pertussis (⊖)	**Pertussis toxin**	**ADP ribosylates G_i,** the negative regulator of adenylate cyclase, leading to increased cAMP	• **Histamine sensitizing** • **Lymphocytosis promotion** (inhibits chemokine receptors) • **Islet activation**
Cytolysins	***Clostridium perfringens*** (⊕)	**Alpha toxin**	**Lecithinase**	• Damages cell membranes • **Myonecrosis**
	Staphylococcus aureus (⊕)	**Alpha toxin**	**Pore former**	Membrane becomes leaky

Definition of abbreviations: TSST-1, toxic shock syndrome toxin-1

► Important Pathogenic Factors and Diagnostic Enzymes

Factor	Function	Organisms
All **capsules**	**Antiphagocytic**	***S**treptococcus pneumoniae*, ***K**lebsiella pneumoniae*, ***H**aemophilus influenzae*, ***P**seudomonas aeruginosa*, ***N**eisseria meningitidis*, ***C**ryptococcus neoformans* (mnemonic: **s**ome **k**illers **h**ave **p**retty **n**ice **c**apsules) and many more
M protein	Antiphagocytic	Group A streptococci
A protein	Binds Fc of IgG to inhibit opsonization and phagocytosis	*Staphylococcus aureus*
Lipoteichoic acid	Attachment to host cells	All **gram-positive** bacteria
All **pili**	Attachment	Many gram-negatives
Pili of *N. gonorrhoeae*	**Antiphagocytic, antigenic variation**	*N. gonorrhoeae*
Hyaluronidase	Hydrolysis of ground substance	Group A streptococci
Collagenase	Hydrolysis of collagen	*Clostridium perfringens, Prevotella melaninogenica*
Urease	Increases pH of locale, contributes to kidney stones	***P**roteus, **U**reaplasma, **N**ocardia, **C**ryptococcus, **H**elicobacter* (mnemonic: **PUNCH**)
Kinases	Hydrolysis of fibrin	*Streptococcus, Staphylococcus*
Lecithinase	Destroys cell membranes	*Clostridium perfringens*
Heparinase	Thrombophlebitis	*Bacteroides*
Catalase	Destroys hydrogen peroxide **(major problem for CGD patients)**	• Most important: *Staphylococcus*, *Pseudomonas*, *Aspergillus*, *Candida*, Enterobacteriaceae • Most anaerobes lack catalase
IgA proteases	Destroy IgA, promote colonization of mucosal surfaces	*Neisseria, Haemophilus, Streptococcus pneumoniae*
Oxidase	Cytochrome c oxidase is the terminal electron acceptor	*Neisseria* and most gram-negatives, except the *Enterobacteriaceae*
Coagulase	Produces fibrin clot	***Staphylococcus aureus*** and *Yersinia pestis*

► Unusual Growth Requirements

Requirements in Culture	Organism
Factors **X and V**	***Haemophilus***
Cholesterol	*Mycoplasma*
High salt	*Staphylococcus aureus,* group D **enterococci** and ***Vibrio***
Cysteine	***Francisella, Legionella, Brucella,* and *Pasteurella*** (**mnemonic:** the 4 Sisters "ELLA" worship in the Cysteine Chapel)
High temperature (42° C)	***Campylobacter***
Lower than atmospheric oxygen pressure (microaerophilic); special CO_2 incubator	***Campylobacter*** and ***Helicobacter***

► Bacterial Genetics

Recombination	**Homologous**: The one-to-one exchange of linear extrachromosomal DNA for homologous alleles within the chromosome, using recombinase A **Site-specific**: The incorporation of extrachromosomal circles of DNA into another molecule of DNA using restriction endonucleases
Conjugation	The donation of chromosomal or plasmid genes from one bacterium to another through a conjugal bridge. • **F+ cells** have a fertility factor plasmid and serve as donors of plasmid DNA • **F– cells** do not have fertility factors and serve as the recipients of DNA in any cross. • **Hfr cells** have incorporated a fertility factor plasmid (now called episome) into their chromosome by site-specific recombination and serve as donors of chromosomal DNA • Conjugation is the most important means of transfer of drug resistance genes in gram-negative bacilli
Transduction	The delivery of bacterial genes from one bacterium to another via a virus vector. • **Generalized transduction**: Transfer of any genes from one bacterium to another via an accident in the assembly of a virus with a lytic life cycle. • **Specialized transduction**: The transfer of bacterial chromosomal genes located near the insertion site of a temperate phage from one bacterium to another via an accident of excision. • Transduction has been shown to be the means of transfer of drug resistance to methicillin (MRSA) and imipenem (*Pseudomonas*).
Transformation	The uptake and incorporation of free DNA from the environment by competent cells followed by homologous recombination. • Transformation is an important means of transfer of traits in bacteria that are naturally competent (*Streptococcus pneumoniae, Helicobacter, Neisseria,* and *Haemophilus influenzae*)
Lysogeny	The stable association of DNA molecules between a bacterium and a temperate phage. • It imparts the important traits: **C** = Cholera toxin, **O** = *Salmonella* O antigen, **B**= Botulinum toxin, **E** = Erythrogenic toxin of *Streptococcus pyogenes,* **D** = Diphtheria toxin, **S** = Shiga toxin. **Mnemonic: COBEDS:** when 2 people share a bed, someone gets a little bit pregnant (with phage). • Bacterial cells with stably integrated temperate phage DNA are said to have undergone **lysogenic conversion**.
Transposon	A mobile genetic element capable of movement within a cell. • They move by a variation of site-specific recombination and are responsible for the formation of multiple drug-resistance plasmids. • VRSA has arisen because *Enterococcus* donated a multi-drug resistance plasmid, produced by transposition, to MRSA.

Antibacterial Agents

► Mechanisms of Action of Antibacterial Agents

Mechanism of Action	Antibacterial Agents
Inhibition of bacterial cell-wall synthesis	Penicillins, cephalosporins, imipenem/meropenem, aztreonam, vancomycin
Inhibition of bacterial protein synthesis	Aminoglycosides, chloramphenicol, macrolides, tetracyclines, streptogramins, linezolid
Inhibition of DNA replication or transcription	Fluoroquinolones, rifampin
Inhibition of nucleic acid synthesis	Trimethoprim, flucytosine
Inhibition of folic acid synthesis	Sulfonamides, trimethoprim, pyrimethamine

► Cell Wall Synthesis Inhibitors

Class/Example	Mechanism of Action/Resistance	Spectrum	Toxicity/Notes
		Penicillins	
Narrow spectrum, β-lactamase sensitive: penicillin G, penicillin V	**Mechanism:** inhibit cross-linking of peptidoglycan component of cell wall by transpeptidases; action mediated by binding of penicillin-binding proteins (PBPs)	Gram-positives Clostridia Syphilis	—
Very narrow spectrum, β-lactamase resistant: methicillin, nafcillin, oxacillin		Gram-positives, especially *S. aureus*	Resistant staph emerging "MRSA"
Broad spectrum, aminopenicillins, β-lactamase sensitive: ampicillin, amoxicillin	**Resistance:** production of β-lactamases, which cleave the β-lactam ring structure; change in PBPs; change in porins	Gram-positives, enterococci, *H. influenzae* *L monocytogenes* *M. catarrhalis* *E. coli*	Activity may be augmented with penicillinase **β-lactamase inhibitors** (e.g., clavulanic acid, sulbactam, and tazobactam)
Extended spectrum, antipseudomonal, β-lactamase sensitive: mezlocillin, piperacillin, carbenicillin, ticarcillin, azlocillin		• Gram-negatives, including ***Pseudomonas*** • Gram-positives, including enterococci (mezlocillin and piperacillin)	—
		Cephalosporins	
First generation: cefazolin, cephalexin	**Mechanism:** inhibition of cell wall formation similar to penicillins **Resistance:** same as penicillins	Gram-positives *Proteus mirabilis* *E. coli* *Klebsiella pneumoniae*	• Cross-allergenicity with penicillins occurs in 5% • Anaphylaxis, but not rash, a contraindication in penicillin-sensitive pt. • Disulfiram-like effects (cefotetan)
Second generation: cefotetan, cefoxitin, cefuroxime, cefaclor		Less gram-positive activity and more gram-negative activity than first generation • *B. fragilis* • *H. influenzae* • *M. catarrhalis* • *P. mirabilis* • *E. coli* • *K. pneumoniae* • *Neisseria, Enterobacter*	
Third generation: ceftazidime, cefoperazone, cefotaxime, ceftriaxone		Less gram-positive activity and more gram-negative activity than second generation • *Serratia* sp. • *Borrelia burgdorferi* • *H. influenzae* • *Neisseria* • *Enterobacter* Some have anti-*Pseudomonas* activity	• Most penetrate **blood-brain** barrier (not cefoperazone) • Reserved for serious infections • Disulfiram-like effects (cefoperazone)
Fourth generation: cefepime		More gram-negative activity while retaining first-generation gram-positive activity	More resistant to β- lactamases

(Continued)

▶ Cell Wall Synthesis Inhibitors *(Cont'd.)*

Class/Example	Mechanism of Action/Resistance	Spectrum	Toxicity/Notes
Carbapenems and Monobactams			
Carbapenems (meropenem, imipenem)	Similar mechanism to penicillins and cephalosporins	Gram-positives, gram-negative rods, anaerobes	• **Nephrotoxic** • GI distress • Rash • CNS toxicity • **Cilastatin administered concurrently** with imipenem increases the drug's half-life and reduces nephrotoxicity • Beta-lactamase resistant
Monobactams (aztreonam)		Gram-negative rods	• GI distress with superinfection, vertigo, headache • Synergistic with aminoglycosides • Beta-lactamase resistant
Non-Beta Lactam Cell Wall Synthesis Inhibitors			
Vancomycin	Binds cell wall precursors (D-ala-D-ala muramyl pentapeptide), preventing polymerization, peptidoglycan elongation	Drug resistant gram-positives, e.g., MRSA sepsis (IV) or *C. difficile* (oral [not absorbed from lumen])	• Chills, fever, ototoxicity, nephrotoxicity • Flushing or **"red man syndrome"** upon rapid infusion • Resistant strains (VRSA, VRE) emerging

Definition of abbreviations: VRE, vancomycin-resistant enterococcus; VRSA, vancomycin-resistant *Staphylococcus aureus.*

▶ Summary of Mechanisms of Protein Synthesis Inhibition

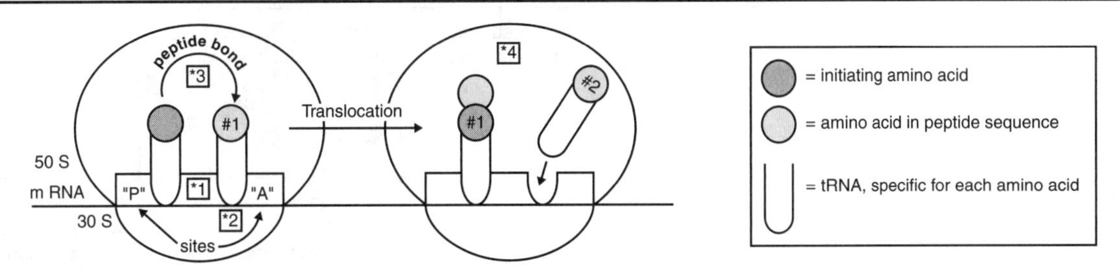

Event	Antibiotics and Binding Sites	Mechanism
1. Formation of initiation complex	Aminoglycosides (30S) Linezolid (50S)	Interfere with initiation codon functions—block association of 50S ribosomal subunit with mRNA-30S (bacteriostatic); misreading of code—incorporation of wrong amino acid (bactericidal)
2. Amino-acid incorporation	Tetracyclines (30S) Dalfopristin/quinupristin (50S)	Block the attachment of aminoacyl tRNA to acceptor site (bacteriostatic)
3. Formation of peptide bond	Chloramphenicol (50S)	Inhibit the activity of peptidyltransferase (bacteriostatic)
4. Translocation	Macrolides and clindamycin (50S)	Inhibit translocation of peptidyl tRNA from acceptor to donor site (bacteriostatic)

▶ Protein Synthesis Inhibitors

Drug/Class	Site of Inhibition	Spectrum	Mechanisms of Resistance	Toxicities	Notes
Chloramphenicol	50S	Wide spectrum, including: • *H. influenzae* • *N. meningitidis* • *Bacteroides* • *Rickettsia* • *Salmonella*	Plasmid-mediated acetyltransferases that inactivate the drug	• **"Gray baby" syndrome** (↓ glucuronyl transferase in neonates) • **Aplastic anemia**/bone marrow suppression • GI irritation	• Toxicity limits clinical use • Reserved for severe *Salmonella* and bacterial meningitis in β-lactam–sensitive patients
Tetracyclines: • tetracycline • doxycycline • minocycline	30S	Gram ⊕ and Gram ⊖: • *Rickettsia* • *Chlamydia* • *Mycoplasma* • *H. pylori* • *Brucella* • *Vibrio*	Plasmid-mediated efflux pumps and reduced uptake via transport systems	• GI irritation • **Tooth enamel dysplasia** • **Bone growth irregularities** • Hepatotoxicity • **Photosensitivity** • Vestibular toxicity	• **Fanconi syndrome** with expired tetracycline • Oral absorption limited by multivalent cations
Macrolides: • erythromycin, • azithromycin • clarithromycin	50S	Gram ⊕, some Gram ⊖: • *Chlamydia* • *Mycoplasma* • *Ureaplasma* • *Legionella* • *Campylobacter*	Methylation of binding site on 50S; increased efflux from multidrug exporters	• **GI irritation** • Cholestasis • Hepatitis • Skin rashes • ↓ CYP3A4 (except azithromycin)	Useful in atypical pneumonia
Ketolides • Telithromycin	50S	Similar spectrum to macrolides	Many macrolide-resistant strains are susceptible to ketolides	• Severe hepatotoxicity • Visual disturbances • Fainting	Its use is limited because of toxicity
Clindamycin	50S	Narrow spectrum: Gram ⊕, anaerobes	Methylation of binding site on 50S	• GI irritation • Skin rash • ***C. difficile*** **superinfection**	—
Aminoglycosides: • gentamicin • neomycin • tobramycin • streptomycin	30S	Gram ⊖ rods, aerobic only	Plasmid-mediated **group transferases**	Ototoxicity, nephrotoxicity	Neomycin for bowel prep (stays in bowel lumen)
Oxazolidinones: • linezolid	50S	Gram ⊕ cocci	Resistance rare	• Thrombocytopenia, neutropenia, esp. in immunocompromised • MAO inhibition (dietary and drug restrictions)	No cross-resistance with other protein synthesis inhibitors, so often reserved for resistant infections

▶ Folic Acid Synthesis Inhibitors

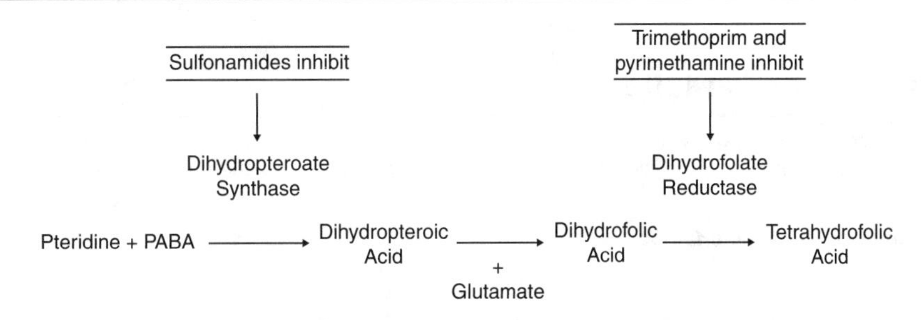

Class/Drug	Mechanism of Action	Spectrum	Mechanism of Resistance	Toxicity	Notes
Sulfonamides	PABA antimetabolite inhibits bacterial dihydropteroate synthase, thus curbing folate synthesis	Gram ⊖, gram ⊕, *Chlamydia*, *Nocardia*	Decreased accumulation of drugs, decreased affinity of drug for **dihydropteroate synthase**	• Hypersensitivity • Hemolytic anemia in G6PD-deficient • Nephrotoxicity • Kernicterus in newborns	Combined with trimethoprim for increased efficacy
Trimethoprim	Inhibits bacterial dihydrofolate reductase, thus inhibiting folate synthesis	*H. influenzae* *M. catarrhalis*	Production of bacterial **dihydrofolate reductase** with decreased affinity for drug	• Megaloblastic anemia • Leukopenia • Granulocytopenia	• Adverse effects may be reduced by concurrent **folinic acid** • Good for UTIs because it is excreted in urine unchanged

▶ DNA Replication Inhibitors

Class/Drug	Mechanism of Action	Spectrum	Mechanism of Resistance	Toxicity	Notes
Fluoroquinolones: • cipro**floxacin** • o**floxacin** • levo**floxacin** • moxi**floxacin**	Interfere with bacterial DNA topoisomerase II and IV (DNA gyrase), resulting in inhibition of DNA synthesis	Gram ⊖ rods, *Neisseria*, occasional gram ⊕	• Decreased intracellular drug concentrations through efflux pumps and altered porins • Alteration of drug's binding site	• **GI distress** • Skin rash • Superinfection • Tendonitis and tendon rupture	Contraindicated in pregnancy due to cartilage formation abnormalities in animal studies

▶ Miscellaneous

Class/Drug	Mechanism of Action	Spectrum	Mechanism of Resistance	Toxicity	Notes
Metronidazole	When reduced, interferes with nucleic acid synthesis (bactericidal)	Anaerobes (except *Actinomyces*)	Rare plasma-mediated resistance	• GI distress • Disulfiram-like reaction with alcohol • Peripheral neuropathy, ataxia	• Strong metallic taste • DOC in pseudomembranous colitis

Definition of abbreviations: DOC, drug of choice; G6PD, glucose-6-phosphate dehydrogenase; PABA, para-aminobenzoic acid; UTI, urinary tract Infection.

Parasites

► Obligate Intracellular Parasites

- Cannot be cultured on inert media. Intracellular organisms (both obligate and facultative) are protected from antibody and complement. Intracellular pathogens tend to elicit cell-mediated immune responses, so end pathologic lesion is frequently a granuloma
- **All rickettsiae, chlamydiae, *Mycobacterium leprae***
- **All viruses**
- ***Plasmodium, Toxoplasma gondii, Babesia***, *Leishmania, Trypanosoma cruzi* (amastigotes in cardiac muscle)

► Facultative Intracellular Parasites

Live inside phagocytic cells in the body, but can be cultured on inert media

- *Francisella tularensis*
- ***Listeria monocytogenes***
- ***Mycobacterium tuberculosis***
- *Brucella* species
- Nontuberculous mycobacteria
- *Salmonella typhi*
- *Legionella pneumophila*
- *Yersinia pestis*
- *Nocardia*
- *Borrelia burgdorferi*
- ***Histoplasma capsulatum***

► Protozoans

Common Name	Amebae	Flagellates	Apicomplexa (Intracellular)
Important genera	***Entamoeba*** *Naegleria* *Acanthamoeba*	LUMINAL (GUT, UG) ***Trichomonas*** ***Giardia*** HEMOFLAGELLATES *Leishmania* *Trypanosoma*	BLOOD/TISSUE ***Plasmodium*** ***Toxoplasma*** *Babesia* INTESTINAL ***Cryptosporidium*** *Isospora*

► Protozoan Parasites

Species	Disease/Organs Most Affected	Form/Transmission	Diagnosis
Entamoeba histolytica	• **Amebiasis:** dysentery • **Inverted, flask-shaped lesions** in large intestine with extension to peritoneum and liver, lungs, brain, and heart • Blood and pus in stool • Liver abscesses	• Cysts • Fecal-oral transmission: water, fresh fruits, and vegetables • Travel to tropics	• Trophozoites or cysts (with 4 nuclei) in stool • Nuclei have sharp central karyosome and fine chromatin "spokes" • Serology
Giardia lamblia	**Giardiasis:** Ventral sucking disk attaches to lining of duodenal wall, causing a **fatty**, foul-smelling diarrhea (diarrhea → ***malabsorption*** in duodenum, jejunum)	• Cysts • Fecal (human, beaver, muskrat, etc.), oral transmission: water, food, day care, oral-anal sex • Campers and hikers	• Trophozoites or cysts in stool or fecal antigen test (replaces "string" test) • "Falling leaf" motility
***Cryptosporidium* spp.**	Cryptosporidiosis: transient diarrhea in healthy; severe in immunocompromised hosts	• Cysts • Undercooked meat, water; not killed by chlorination	**Acid fast oocysts** in stool: biopsy shows dots (cysts) in intestinal glands
Trichomonas vaginalis (urogenital)	Trichomoniasis: frothy green, unpleasant smelling vaginal discharge, itching, burning vaginitis	• Trophozoites • **Sexual**	Motile trophozoites in methylene blue wet mount

► Free-Living Amebae That Occasionally Infect Humans

Species	Disease/Locale	Form/Transmission	Diagnosis
Naegleria	**Primary amebic meningoencephalitis** (PAM): severe prefrontal headache, nausea, high fever, often an altered sense of smell; often fatal	• Free-living amebae picked up while swimming or **diving in very warm fresh water** • Penetrates cribriform plate	• Motile trophozoites in CSF • Culture on plates seeded with gram ⊖ bacteria; amebae will leave trails
Acanthamoeba	**Keratitis; GAE** in immunocompromised patients; insidious onset but progressive to death	• Free-living amebae in contaminated **contact lens solution (airborne cysts)** • Not certain for GAE; inhalation or contact with contaminated soil or water	• Star-shaped cysts on biopsy; rarely seen in CSF • Culture as above

Definition of abbreviations: CSF, cerebrospinal fluid; GAE, granulomatous amaebic encephalitis.

▶ *Plasmodium* Life Cycle

Each *Plasmodium* has two distinct hosts:

- A vertebrate, such as the human (intermediate host), where asexual phase (schizogony) takes place in the liver and red blood cells
- An arthropod (definitive) host (*Anopheles* mosquito), where gametogony (sexual phase) and sporogony take place

Disease is caused by a variety of mechanisms, including metabolism of hemoglobin and lysis of infected cells, leading to anemia and to agglutination of the infected RBCs. Paroxysms (chills, fever spike, and malarial rigors) occur when the infected RBCs are lysed, liberating a new crop of merozoites.

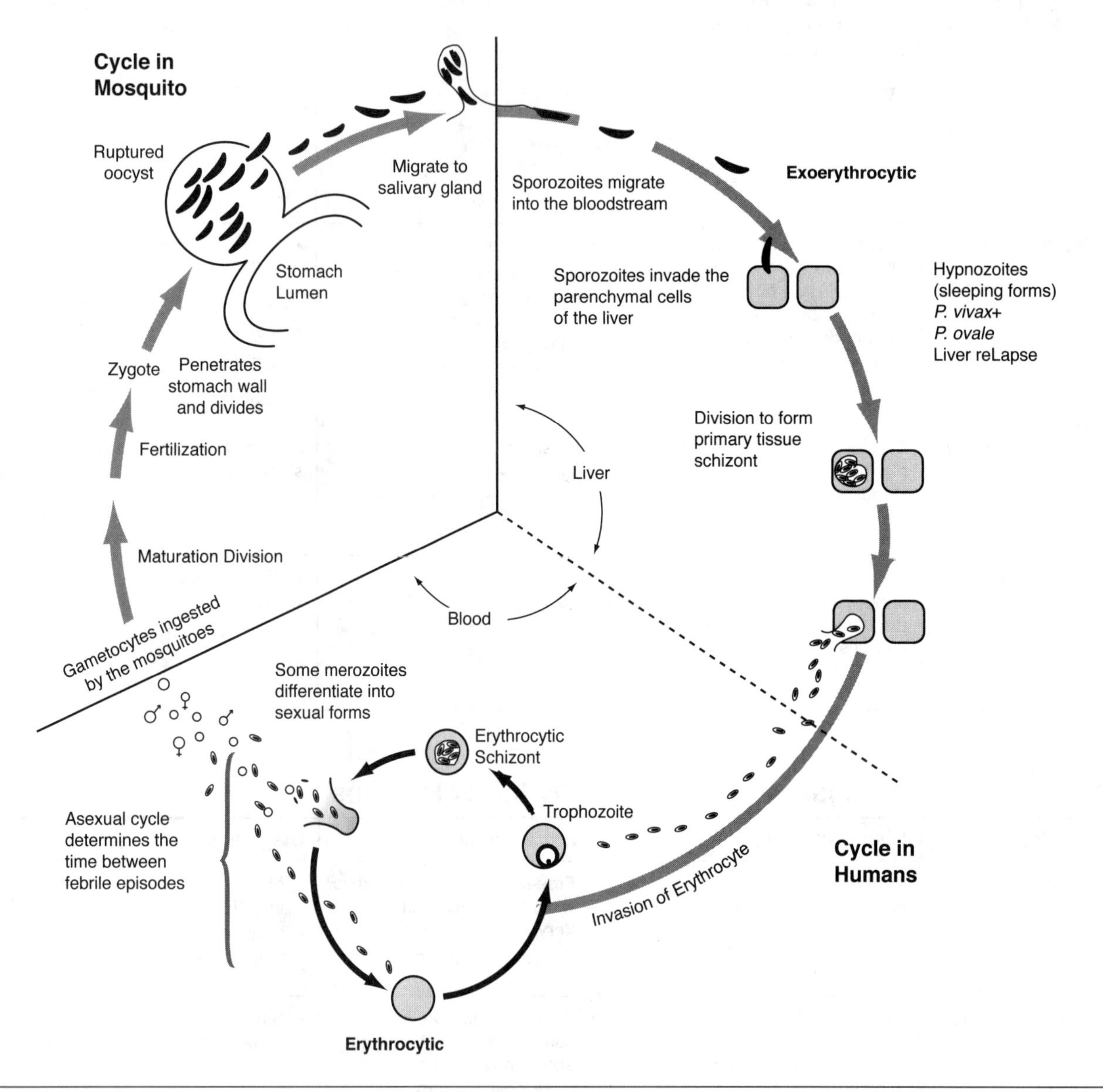

Definition of abbreviation: RBCs, red blood cells.

▶ *Plasmodium* Species

Species	Disease	Important Features	Blood Smears	Liver Stages
Plasmodium vivax	Benign tertian	48-hour fever spikes	Enlarged host cells; ameboid trophozoites	Persistent hypnozoites Relapse*
Plasmodium ovale	Benign tertian	48-hour fever spikes	Oval, jagged, infected RBCs	Persistent hypnozoites Relapse
Plasmodium malariae	Quartan or malarial	72-hour fever spikes; recrudescence*	Bar and band forms; rosette schizonts	No persistent stage*
Plasmodium falciparum	Malignant tertian	Irregular fever spikes; causes cerebral malaria; most dangerous	Multiple ring forms crescent-shaped gametes	No persistent stage*

Definition of abbreviations: PO, by mouth; RBCs, red blood cells.

***Re**cru**d**escence is a recurrence of symptoms from low levels of organisms remaining in **red** cells. Re**l**apse is an exacerbation from **l**iver stages (hypnozoites).

†Use quinine sulfate plus pyrimethamine-sulfadoxine.

▶ Antimalarial Drugs

1. Suppressive (to avoid infection)
2. Therapeutic (eliminate erythrocytic)
3. Radical cure (eliminate exoerythrocytic)
4. Gametocidal (destruction of gametocytes)

Successful treatment is accomplished with chloroquine followed by primaquine. Chloroquine therapy is suppressive, therapeutic, and gametocidal, whereas primaquine eliminates the exoerythrocytic form.

Chloroquine-Sensitive Malaria	
P. falciparum	Chloroquine
P. malariae	Chloroquine
P. vivax	Chloroquine plus primaquine
P. ovale	Chloroquine plus primaquine
Chloroquine-Resistant Malaria	

Prophylaxis: mefloquine; backup drugs: doxycycline, atovaquone-proguanil
Treatment: quinine ± either doxycycline, clindamycin, or pyrimethamine

▶ Adverse Effects of Antimalarial Drugs

Drug	Side Effects	Contraindications and Cautions
Chloroquine, hydroxychloroquine	GI distress, pruritus, headache, dizziness, hemolysis, retinopathy	Avoid in psoriasis
Mefloquine	NVD, dizziness, syncope, extrasystoles, CNS effects (rare)	Avoid in seizures, psychiatric disorders, and in cardiac conduction defects
Primaquine	GI distress, headache, dizziness, neutropenia, hemolysis	Avoid in pregnancy, G6PD deficiency, and autoimmune disorders
Quinine	GI distress, cinchonism, CNS effects, hemolysis, hematotoxicity	Avoid in pregnancy

Definition of abbreviations: CNS, central nervous system; GI, gastrointestinal; G6PD, glucose-6-phosphate dehydrogenase; NVD, nausea, vomiting, diarrhea.

► Hemoflagellates

Hemoflagellates (Trypanosomes and *Leishmania*) infect blood and tissues. They are found in:

- Human blood as **trypomastigotes** with flagellum and an undulating membrane.
- Intracellular as **amastigotes (oval cells having neither the flagellum nor undulating membrane)**. *Leishmania* spp. have only amastigotes in the human.

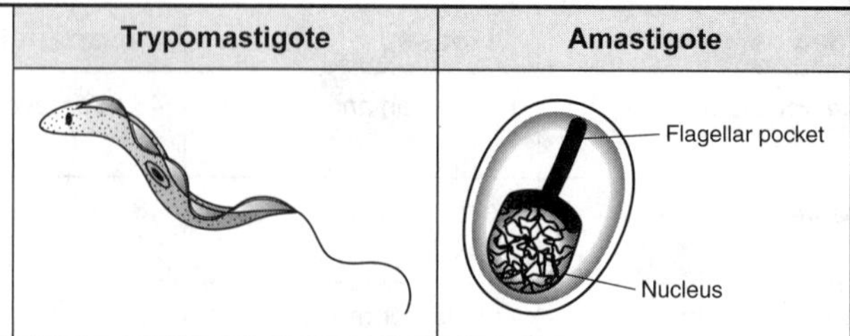

Species	Disease	Vector/Form/ Transmission	Diagnosis
Trypanosoma cruzi*	• **Chagas disease** (American trypanosomiasis) • Latin America • Swelling around eye (**Romaña's sign**): common early sign • Cardiac muscle (dilated cardiomyopathy), megaesophagus, megacolon	**Reduviid bug (kissing or cone bug)** passes trypomastigote **in feces**	Blood films, **trypomastigotes**
Trypanosoma brucei gambiense ***Trypanosoma b. rhodesiense***	• **African sleeping sickness** (African trypanosomiasis) • **Antigenic variation**	Trypomastigote in saliva of **tsetse fly**	• **Trypomastigotes in blood films**, CSF • High immunoglobulin levels in CSF
Leishmania donovani† complex	• **Visceral leishmaniasis** • Kala-azar	**Sandfly** bite	**Amastigotes in macrophages** in bone marrow, liver, spleen
Leishmania (about 15 different species)	Cutaneous leishmaniasis (Oriental sore, etc.)	**Sandfly** bite	**Amastigotes in macrophages** in cutaneous lesions
Leishmania braziliensis complex	Mucocutaneous leishmaniasis	**Sandfly** bite	Same

Definition of abbreviations: CSF, cerebrospinal fluid.

**T. cruzi*: An estimated 0.5 million Americans are infected, creating some risk of transfusion transmission in the United States. In babies, acute infections are often serious and involve the CNS. In older children and adults, mild acute infections may become chronic with the risk of development of cardiomyopathy and heart failure.

† All *Leishmania*: intracellular, sandfly vector, stibogluconate

▶ Miscellaneous Apicomplexa Infecting Blood or Tissues

Species	Disease/Locale of Origin	Transmission	Diagnosis
Babesia (primarily a disease of cattle) Humans: *Babesia microti*, WA1 and MO1 strains	Babesiosis (hemolytic, **malaria-like**) Same range as Lyme: N.E., N. Central, California, and N.W. United States	• ***Ixodes* tick** • Coinfections with ***Borrelia*** and *Ehrlichia*	Giemsa stain of blood smear
Toxoplasma gondii	• Most common parasitic disease • Infections **after birth are most commonly asymptomatic** or mild; may mimic mononucleosis • Produces **severe disease in AIDS** or other immunocompromised **(ring-enhancing lesions in brain)** • **Primary maternal infection during pregnancy may infect fetus:** – **Severe** congenital infections (intracerebral calcifications, chorioretinitis, hydro- or microcephaly, seizures) **if *Toxoplasma* crosses the placenta early** – **Later term congenital infection** may produce **progressive blindness**	**Cat** is **essential definitive** host; many other animals are intermediate hosts Mode: • Raw meat in U.S.; pork is #1 • Contact with cat feces	Serology**:** High IgM or rising IgM acute infection

▶ Major Protozoal Infections and Drugs of Choice

Infection	Drug of Choice	Comments
Amebiasis	Metronidazole	Diloxanide for noninvasive intestinal amebiasis
Giardiasis	Metronidazole or furazolidone	—
Trichomoniasis	Metronidazole	Treat both partners
Toxoplasmosis	Pyrimethamine and sulfadiazine	TMP-SMX is also prophylactic against *Pneumocystis jiroveci* in AIDS
Leishmaniasis	Stibogluconate	—
Trypanosomiasis	• Nifurtimox (Chagas disease) • Arsenicals, pentamidine, suramin (African sleeping sickness)	—

▶ Metazoans: Worms*

Phylum	Flatworms (Platyhelminthes)		Roundworms (Nemathelminthes)
Class (common name)	**Trematodes** (flukes)	**Cestodes** (tapeworms)	**Nematodes**† (roundworms)
Genera	*Fasciola* *Fasciolopsis* *Paragonimus* *Clonorchis* ***Schistosoma***	***Diphyllobothrium*** *Hymenolepis* *Taenia* *Echinococcus*	***Necator*** ***Enterobius*** (*W*)*uchereria/Brugia* ***Ascaris*** and ***Ancylostoma*** *Toxocara, Trichuris,* and *Trichinella* *Onchocerca* *Dracunculus* Eyeworm (*Loa loa*) *Strongyloides*

*Metazoans also include the Arthropoda, which serve mainly as intermediate hosts (the crustaceans) or as vectors of disease (the Arachnida and Insecta).

†Nematodes **mnemonic** (turn the "W" upside down)

► Trematode (Fluke) Diseases

Trematodes:

- Are commonly called flukes, which are generally flat and fleshy, leaf-shaped worms
- Are hermaphroditic, except for *Schistosoma*, which has separate males and females
- Have complicated life cycles occurring in two or more hosts
- Have operculated eggs (except for *Schistosoma*)
- **The first intermediate hosts are snails**

Organism	Common Name	Acquisition	Progression in Humans	Important Ova
Schistosoma mansoni ***S. japonicum***	**Intestinal schistosomiasis**	Contact with water; skin penetration	Skin penetration (itching) → mature in veins of mesentery → eggs cause granulomas in liver (**portal hypertension** and liver fibrosis in chronic cases)	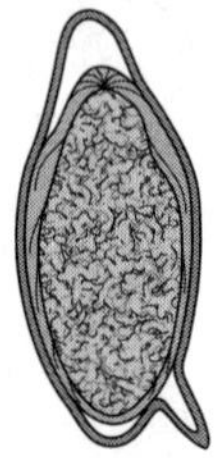 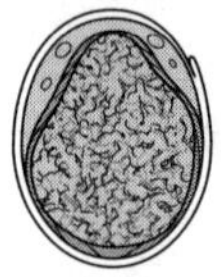
Schistosoma haematobium	**Vesicular schistosomiasis**	Contact with water; skin penetration	Skin penetration (itching) → mature in bladder veins; chronic infection has high association with **bladder carcinoma in Egypt and Africa**	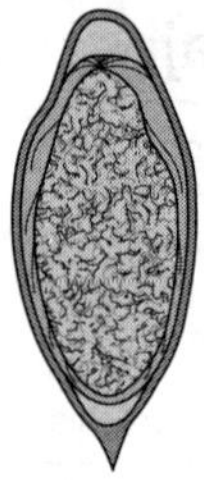
Nonhuman schistosomes	**Swimmer's itch**	Contact with water; skin penetration (Great Lakes)	Penetrate skin, producing **dermatitis** without further development in humans; itching is most intense at 2 to 3 days	—
Clonorchis sinensis	**Chinese liver fluke**	Raw fish ingestion	Inflammation of **biliary tract**, pigmented gallstones, **cholangiocarcinoma**	Operculated eggs
Paragonimus westermani	**Lung fluke**	Raw crabs, crayfish	**Hemoptysis**, secondary bacterial infection of lung	Operculated eggs

► Gastrointestinal Cestodes (Tapeworms)

- Consist of three basic portions: the head or scolex; a “neck” section, which produces the proglottids; and the segments or proglottids, which mature as they move away from the scolex
- Are diagnosed by finding eggs or proglottids in the feces
- Have complex life cycles involving extraintestinal larval forms in intermediate hosts; when humans are intermediate host, these infections are generally more serious than intestinal infections with adult tapeworms

Cestode (Common Name)	Form/ Transmission	Human Host Type	Disease/Organ Involvement/ Symptoms (Sx)	Diagnosis
***Taenia solium* (pork tapeworm)** IH: swine; rare: humans DH: humans,developing and Slavic countries	Water, vegetation, food contaminated with **eggs** Autoinfection	IH	**Cysticercosis**/eggs → larva develop in brain, eye, heart, lung, etc. **Epilepsy** with onset after age 20	Biopsy
	Rare/raw pork containing the **cysticerci** ingested by humans	DH	• **Intestinal tapeworm** • Sx: same as for *Taenia saginata*	**Proglottids** or **eggs** in feces
***Diphyllobothrium latum* (fish tapeworm)** 2 IHs: crustaceans → fish; rare: humans DH: humans/mammals; cool lake regions	Drinking pond water containing copepods (crustaceans) carrying the **larval** forms or frog/snake poultices	IH	**Sparganosis**/larvae penetrate intestinal wall and encyst	Biopsy
	Rare, raw pickled fish containing a **sparganum**	DH	**Intestinal tapeworm** (up to 10 meters)/small intestine, **megaloblastic anemia**	Proglottids or eggs in feces
Echinococcus granulosus IH: herbivores; rare: humans DH: carnivores in sheep-raising areas	Ingestion of eggs	IH	**Hydatid cyst disease**; liver and lung, where cysts containing brood capsules develop	Imaging, serology

Definition of abbreviations: IH, intermediate host; DH, definitive host.

► Roundworms (Nematodes)

Roundworms are transmitted by:

- Ingestion of eggs (*Enterobius*, *Ascaris*, or *Trichuris*)
- Direct invasion of skin by larval forms (*Necator*, *Ancylostoma*, or *Strongyloides*)
- Ingestion of meat containing larvae (*Trichinella*)
- Infection involving insects transmitting the larvae with bites (*Wuchereria*, *Loa loa*, *Mansonella*, *Onchocerca*, and *Dracunculus*)

► Roundworms (Nematodes) Transmitted By Eggs

Species	Disease/Organs Most Affected	Form/Transmission	Diagnosis
Enterobius vermicularis (Most frequent helminth parasite in U.S.)	**Pinworms**, large intestine, perianal itching	• **Eggs**/person to person • **Autoinfection**	• **Scotch tape swab** of perianal area • Ova have flattened side with larvae inside
Trichuris trichiura	**Whipworm** cecum, appendicitis, and rectal prolapse	**Eggs** ingested	**Barrel-shaped eggs with bipolar plugs** in stools
Ascaris lumbricoides (Most common helminth worldwide; largest roundworm)	**Ascariasis** Ingest egg → larvae migrate through lungs (cough) and mature in small intestine; may obstruct intestine or bile duct	**Eggs** ingested	**Bile stained, knobby eggs** Adult 35 to 40 cm
Toxocara canis or cati (Dog/cat ascarids)	**Visceral larva migrans** Larvae wander aimlessly until they die, cause inflammation	**Eggs** ingested/from handling puppies or from eating dirt in yard (pica)	Clinical findings and serology

▶ Roundworms (Nematodes) Transmitted by Larvae

Species	Disease/Organs	Form/Transmission	Diagnosis
Necator americanus (New World hookworm)	**Hookworm** infection Lung migration → pneumonitis Bloodsucking → anemia	Filariform **larva penetrates intact skin of bare feet**	Fecal larvae (up to 13 mm) and ova: oval, transparent with 2–8 cell-stage visible inside Fecal occult blood may be present
Ancylostoma braziliense ***Ancylostoma caninum*** (dog and cat hookworms)	**Cutaneous larva migrans**/ intense skin itching, snake-like tracks	Filariform larva penetrates intact skin but cannot mature in humans	Usually a presumptive diagnosis; exposure
Strongyloides stercoralis	**Threadworm** strongyloidiasis: *Early:* pneumonitis, abdominal pain, diarrhea *Later:* malabsorption, ulcers, bloody stools	Filariform **larva penetrates intact skin; autoinfection** leads to indefinite infections unless treated	Larvae in stool, serology
Trichinella spiralis	Trichinosis: larvae encyst in muscle → pain	**Viable encysted larvae in meat** are consumed: wildgame meat	Muscle biopsy; clinical findings: **fever, myalgia, splinter hemorrhages, eosinophilia**

▶ Filarial Nematodes

Species	Disease	Transmission/Vector	Diagnosis
Wuchereria bancrofti; Brugia malayi	Elephantiasis	Mosquito	Microfilariae in blood, eosinophilia
Loa loa (African eye worm)	Pruritus, calabar swellings	Chrysops, mango flies	Microfilariae in blood, eosinophilia
Onchocerca volvulus	River blindness, itchy "leopard" rash	Blackflies	Skin snips from calabar swellings
Dracunculus medinensis (Guinea worm, fiery serpent)	Creeping eruptions, ulcerations, rash	Drinking water with infected copepods	Increased IgE; worm eruption from skin

▶ Drugs for Helminthic Infections

Most intestinal nematodes (worms)	• Mebendazole (↓ glucose uptake and ↓ microtubular structure) • Pyrantel pamoate (NM agonist → spastic paralysis)
Most cestodes (tapeworms) and trematodes (flukes)	• Praziquantel (↑ Ca^{2+} influx, ↑ vacuolization)

Virology

► Viral Structure and Morphology

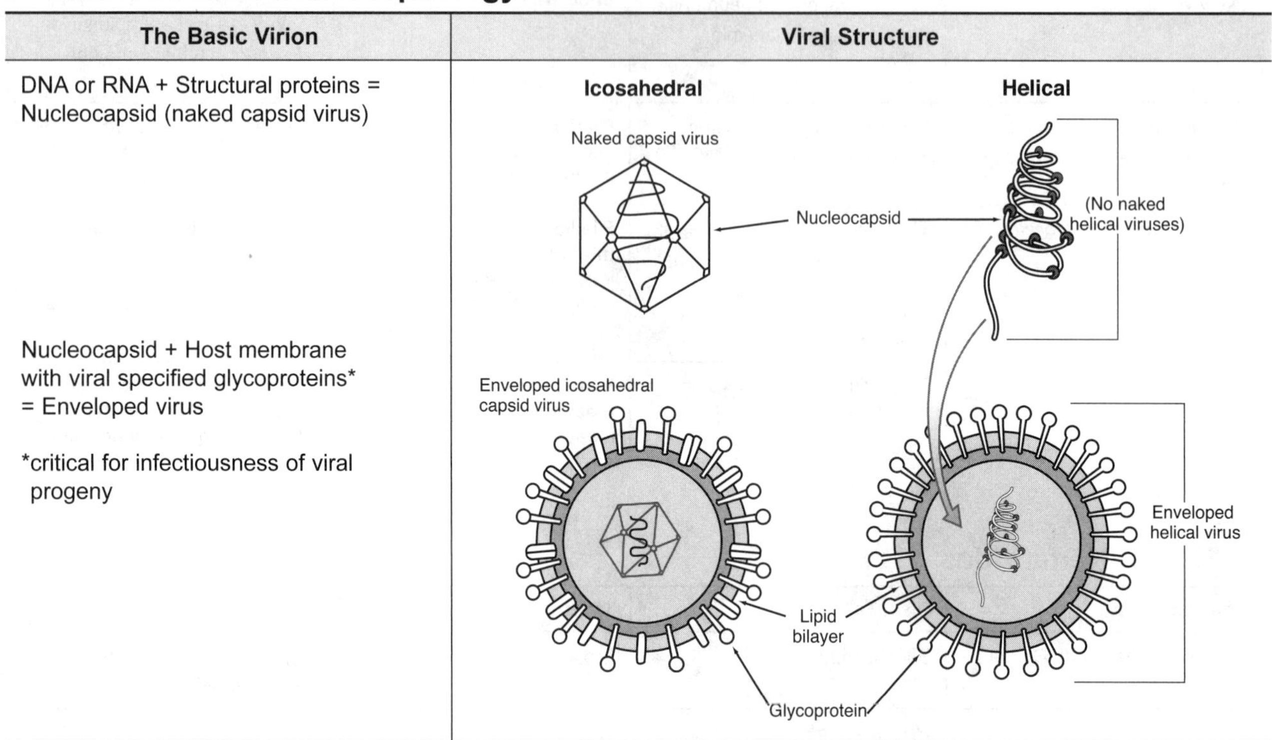

The Basic Virion	Viral Structure
DNA or RNA + Structural proteins = Nucleocapsid (naked capsid virus)	
Nucleocapsid + Host membrane with viral specified glycoproteins* = Enveloped virus *critical for infectiousness of viral progeny	

► DNA Viruses

General Comments About DNA Viruses:

- All are double-stranded except Parvo
- All are icosahedral except Pox, which are brick-shaped complex
- All replicate in the nucleus except Pox

Virus Family	DNA Type	Polymerase	Envelope	Area of Replication	Major Viruses
Parvovirus	**ssDNA**	No	**Naked**	Nucleus	**B19**
Papilloma	dsDNA Circular	No	**Naked**	Nucleus	Papilloma
Polyomavirus	dsDNA Circular	No	**Naked**	Nucleus	Polyoma
Adenovirus	dsDNA Linear	No	**Naked**	Nucleus	Adenovirus
Hepadnavirus	Partially dsDNA Circular	**Yes**	Enveloped	Nucleus via RNA intermediate	HBV
Herpesvirus	dsDNA Linear	No	Enveloped **Nuclear**	Nucleus, assembled in nucleus	**HSV, VZV, EBV, CMV** HHV-6, HHV-8
Poxvirus	dsDNA Linear	**Yes**	Enveloped	**Cytoplasm**	Variola, vaccinia Molluscum contagiosum

Definition of abbreviations: CMV, cytomegalovirus; ds, double-stranded; EBV, Epstein-Barr virus; HBV, hepatitis B virus; HSV, herpes simplex virus; ss, single-stranded; VZV, varicella-zoster virus.

Note: Viruses are listed from top to bottom in order of increasing size. If you know them in this order, then you can remember that the smallest 4 are naked (wearing more clothing makes you larger).

Mnemonic: **P**ardon **P**a**P**a **A**s **He** **H**as **P**ox

Mnemonic: **Naked viruses PAPP** (you need to be naked for a Pap smear)

▶ RNA Viruses

General Comments About RNA Viruses:

- All are single-stranded except Reo
- Most are enveloped. Only naked ones are **H**epe, **C**alici, **P**ico, and **R**eo (**Mnemonic: H**elp! **CPR!**)
- Some are segmented; **Mnemonic: ROBA: R**eo, **O**rthomyxo, **B**unya, and **A**rena – ROBA sounds like ROBOT
- Positive sense RNA = (+)RNA (can be used itself as mRNA)
- Negative sense RNA = (–)RNA
  - Complementary to mRNA
  - Cannot be used as mRNA
  - Requires virion-associated, RNA-dependent RNA polymerase (as part of mature virus)
- Others that carry a polymerase are Reo, Arena, and Retro

▶ Double-Stranded RNA Viruses: Reoviridae

	RNA Structure	Polymerase	Envelope	Shape	Major viruses
Reovirus	• Linear dsRNA • **10–11 segments**	Yes	**Naked**	• Icosahedral • **Double shelled**	• Reovirus • **Rotavirus**

▶ Positive-Sense RNA Viruses

Family	RNA Structure	Polymerase	Envelope	Shape	Area of Replication	Major Viruses
Calicivirus	• ss⊕RNA, linear • Nonsegmented	No	**Naked**	Icosahedral	Cytoplasm	• Norwalk • Norolike
Hepevirus	• ss⊕RNA, linear • Nonsegmented	No	**Naked**	Icosahedral	Cytoplasm	Hepatitis E
Picornavirus	• ss⊕RNA, linear • Nonsegmented	No	**Naked**	Icosahedral	Cytoplasm	• Polio, ECHO, Entero • Rhino, coxsackie • Hepatitis A
Flavivirus	• ss⊕RNA, linear • Nonsegmented	No	Enveloped	Icosahedral	Cytoplasm	Yellow fever, dengue, SLE, hepatitis C, **West Nile virus**
Togavirus	• ss⊕RNA, linear • Nonsegmented	No	Enveloped	Icosahedral	Cytoplasm	Rubella WEE, EEE, VEE
Coronavirus	• ss⊕RNA, linear • Nonsegmented	No	Enveloped	**Helical**	Cytoplasm	Coronaviruses **SARS agent**
Retrovirus	• **Diploid** • ss⊕RNA, linear • Nonsegmented	**RNA-dependent DNA polymerase**	Enveloped	Icosahedral or truncated conical	**Nucleus**	HIV HTLV Sarcoma

Definition of abbreviations: EEE, eastern equine encephalitis; HIV, human immunodeficiency virus; HTLV, human T-cell lymphocytotropic virus; VEE, Venezuelan equine encephalitis; WEE, Western equine encephalitis.

Mnemonic: Again from smallest (top) to largest (bottom): **C**all **H**enry, **P**ico, and **Flo** **To** **Co**me **R**ightaway
Mnemonic: Picornaviruses: **PEE Co Rn A** viruses (**P**olio, **E**ntero, **E**cho, **Co**xsackie, **R**hino, Hep **A**)

▶ Negative-Sense RNA Viruses

Virus	RNA	Polymerase	Envelope	Shape	Area Where Multiplies	Major Viruses
Paramyxovirus	• ss⊖RNA, linear • Nonsegmented	Yes	Yes	Helical	Cytoplasm	Mumps, measles RSV, parainfluenza
Rhabdovirus	• ss⊖RNA, linear • Nonsegmented	Yes	Yes	**Bullet-shaped**, helical	Cytoplasm	Rabies, VSV
Filovirus	• ss⊖RNA, linear • Nonsegmented	Yes	Yes	Helical	Cytoplasm	Marburg, Ebola
Orthomyxovirus	• ss⊖RNA, linear • **8 segments**	Yes	Yes	Helical	**Cytoplasm and nucleus**	Influenza
Bunyavirus	• ss⊖RNA, linear to circular • 3 segments, ambisense	Yes	Yes	Helical	Cytoplasm	California and LaCrosse encephalitis, Hantavirus
Arenavirus	• ss⊖RNA, circular, • 2 segments: 1 ⊖ sense, 1 ambisense	Yes	Yes	Helical	Cytoplasm	Lymphocytic choriomeningitis virus, Lassa fever

Definition of abbreviations: ds, double-stranded; RSV, respiratory syncytial virus; ss, single-stranded; VSV, vesicular stomatitis virus.

Mnemonic (in order of increasing size): **P**ain **R**esults **F**rom **O**ur **B**unions **A**lways (Pain is a negative thing!)
Or, to remind you of life-cycle: **B**ring **A** **P**olymerase **Or** **F**ail **R**eplication

▶ Viral Genetics

Phenotypic mixing	• Related viruses coinfect cell (virus A and virus B) • Resulting proteins on the surface are a mixture capsid of AB around nucleic acid of either A or B
Phenotypic masking	• Related viruses coinfect cell (virus A and virus B) • Capsid of proteins of virus A form around nucleic acid of B
Complementation	• Two related defective viruses infect the same cell; if they are defective in different genes, viral progeny (still with mutated DNA) will be formed • If they are defective in the same gene, no progeny will be formed • Coinfection of hepatitis B and D is a clinical example of complementation where HBV supplies the needed surface antigen for hepatitis D
Genetic reassortment (genetic shift)	• Two different strains of a segmented RNA virus infect the same cell. • Major new genetic combinations are produced through "shuffling," resulting in stable and dramatic changes • This results in **pandemics** of disease
Genetic drift	• Minor antigenic changes from mutation • Occurs in many viruses, particularly RNA types • Most noted in HIV and influenza
Viral vectors	• Recombinant viruses are produced that have combinations of human replacement genes with the defective viral nucleic acid

► Antiviral Agents

As viruses rely on host machinery to produce viral products, selectivity must be achieved by targeting minute differences in viral enzymes. This may be accomplished at any stage in the viral "life cycle" including adsorption, penetration, nucleic acid synthesis, late protein synthesis, protein processing, viral product packaging, and viral release.

Class/Agent	Mechanism of Action	Spectrum/Clinical Applications	Mechanism(s) of Resistance	Toxicity/Notes
ANTIHERPETICS				
Acyclovir Famciclovir Valacyclovir	**Inhibit viral DNA polymerases** Activated by viral thymidine kinase (TK); • Famciclovir is oral prodrug converted to penciclovir • Valacyclovir is oral prodrug of acyclovir • Mechanism of penciclovir same as acyclovir	HSV, VZV (esp. famciclovir, valacyclovir)	Decreased activity or loss of thymidine kinase/DNA polymerase	• Fairly well-tolerated (esp. oral), some nausea/vomiting • IV use associated with seizure, delirium, crystalluria (maintain hydration) • Famciclovir and valacyclovir have much greater oral bioavailability and longer $t_{1/2}$ than acyclovir
Ganciclovir	Similar to acyclovir	CMV (e.g., CMV retinitis), HSV, VZV	Similar to acyclovir	Dose-limiting leukopenia, thrombocytopenia; crystalluria (maintain hydration)
Foscarnet	**Inhibits DNA and RNA polymerases**; does not require activation by kinases (may be effective in acyclovir-, ganciclovir-resistant strains)	• CMV retinitis in AIDS patients • Acyclovir-resistant mucocutaneous HSV in immuno-compromised patients		• Dose limiting nephrotoxicity • Electrolyte imbalance (can lead to seizures)
ANTI-INFLUENZA DRUGS				
Amantadine Rimantadine	Block viral penetration/ uncoating of influenza A virus via interaction with viral M2 protein	• Influenza A (prophylaxis) • Amantadine also used in Parkinson disease to stimulate dopamine release	Current isolates are resistant Resistance due to mutations in M2 protein (no cross resistance to neuraminidase inhibitors)	• Ataxia • Increased seizure activity • Dizziness & hypotension • Rimantadine better tolerated in elderly
Oseltamivir Zanamivir	**Inhibit neuraminidases** made by influenza A and B (enzymes that promote virion release and prevent clumping of these virions), decreasing viral spread	Prophylaxis, but may ↓ duration of flu symptoms by 2-3 days	Mutations to viral neuraminidase	Oseltamivir: oral prodrug; GI discomfort Zanamivir: inhalational drug; cough, bronchospasm in asthmatics
Ribavirin	Inhibits viral RNA synthesis by altering the nucleotide pools and normal messenger RNA formation	Influenza A & B, Parainfluenza, RSV, paramyxoviruses HCV (combined with α–interferon), HIV	Unknown	Dose-dependent hemolytic anemia

(Continued)

► Antiviral Agents *(Cont'd.)*

Class/Agent	Mechanism of Action	Spectrum/Clinical Applications	Mechanism(s) of Resistance	Toxicity/Notes
HIV THERAPY: For details on HIV therapy, see Ch. 4, General Principles of Immunology				
Interferons†	Interferons are a class of related proteins with antiviral, antiproliferative, and immune regulating activity. They induce the synthesis of a number of antiviral proteins (e.g., RNAse and a protein kinase) that protect the cell against subsequent challenges by a variety of viruses.	• Hepatitis B & C • Kaposi sarcoma • Leukemias • Malignant melanoma	Anti-interferon antibodies are seen with prolonged use	• Interferons can cause influenza-like symptoms, especially in the first week of therapy • Bone marrow suppression • Profound fatigue, myalgia, weight loss, and increased susceptibility to bacterial infections • Depression is seen in up to 20% of patients

Definition of abbreviations: CMV, cytomegalovirus; HCV, hepatitis C virus; HSV, herpes simplex virus; RSV, respiratory syncytial virus; VZV, varicella zoster virus.

†For more information on interferons and other immunosuppressants, see Chapter 4, General Principles of Immunology

Mycology

► Mycology: Overview

Fungi are eukaryotic organisms with complex carbohydrate cell walls (the reason they frequently calcify in chronic infections) and ergosterol as their major membrane sterol (which is targeted with nystatin and the imidazoles). Morphologic and geographic clues are very important in determining the identity of the organism.

Fungi come in two basic forms	• **Hyphae**—filamentous forms may either have cross walls (septate) or lack them (aseptate) • **Yeasts**—single-celled oval/round forms • **Dimorphic fungi**—may convert from hyphal to yeast forms (key examples: *Histoplasma*, *Blastomyces*, *Coccidioides*, and *Sporothrix*). **Mnemonic: H**eat **C**hanges **B**ody **S**hape: yeast in the heat, mold in the cold
Pseudohyphae	• Hyphae formed by budding off yeasts; formed by ***Candida albicans***; the basis of the germ tube test for diagnosis of invasive *C. albicans*
Spores are used for reproduction and dissemination	• **Conidia**—asexual spores form off hyphae • **Blastoconidia**—asexual spores like buds on yeasts • **Arthroconidia**—asexual spores formed with joints between • **Spherules** with **endospores**—sexual spores in tissues (*Coccidioides*)

▶ Nonsystemic Fungal Infections

Organism	Disease	Notes
Malassezia furfur	**Pityriasis** or **tinea versicolor**	• Superficial infection of keratinized cells • **Hypopigmented spots on the chest/back** (blotchy suntan) • KOH mount of skin scales: "spaghetti and meatballs," yeast clusters and short, curved septate hyphae • Treatment is topical selenium sulfide; recurs.
	Fungemia	• In **premature infants**
Candida albicans, *Candida* spp.	**Cutaneous or mucocutaneous candidiasis**	• Causes oral thrush and vulvovaginitis in immunocompetent individuals • Source of opportunistic infections in hospitalized and immunocompromised (*see* Opportunistic Mycoses) **Pseudohyphae** **Budding Yeasts** **Germ Tubes** **True Hyphae**
Trichophyton, Microsporum, Epidermophyton	**Tinea** (capitis, barbae, corporis, cruris, pedis)	• Infects **skin, hair, and nails** • **Monomorphic** filamentous fungi • KOH mount shows **arthroconidia, hyphae** • **Pruritic lesions with serpiginous borders and central clearing**
Sporothrix schenckii Hyphae with sleeves or rosettes of conidia	• **Sporotrichosis (rose gardener's disease)** • **Pulmonary sporotrichosis** (in alcoholics/homeless)	Dimorphic fungus: • Environmental form: hyphae with rosettes and sleeves of conidia • Tissue form: cigar-shaped yeast

► Systemic Fungal Infections

Organism	Disease	Notes
	General Comments	
Histoplasma ***Coccidioides*** ***Blastomyces***	• Acute pulmonary (asymptomatic or self-resolving in about 95% of the cases) • Chronic pulmonary • Disseminated infections	**Diagnosis:** • Sputum cytology (calcofluor white staining helpful) • **Sputum cultures** on blood agar and **special fungal media** (inhibitory mold agar, Sabouraud's agar) • **Peripheral blood cultures are useful for *Histoplasma* because it circulates in RES cells**
	Specific Organisms	
Histoplasma capsulatum	Fungus flu (a pneumonia) • Asymptomatic or flu-like • **Hepatosplenomegaly** may be present • May **disseminate in AIDS patient**	**Dimorphic fungus:** • **Environmental form: hyphae** with **microconidia** and **tuberculate macroconidia** – Endemic region: **Eastern Great Lakes, Ohio, Mississippi,** and **Missouri River beds** – Found in **soil (dust) enriched with bird or bat feces** (caves, chicken coops) • **Tissue form: small intracellular yeasts** with narrow neck on bud; **no capsule** • **Facultative intracellular parasite** found in **RES cells** (tiny; can get 30 or so in a human cell)
Coccidioides immitis (endospores/spherules) (arthroconidia)	Coccidioidomycosis (San Joaquin Valley fever)	• Dimorphic fungus • Asymptomatic to **self-resolving pneumonia** • Desert bumps (erythema nodosum) • **Pulmonary lesions may calcify** • **May disseminate in AIDS and immunocompromised** (meningitis, mucocutaneous lesions) • Has a tendency to **disseminate in third trimester of pregnancy**
Blastomyces dermatitidis	Blastomycosis	• Dimorphic fungus • Environmental form: hyphae with conidia • Tissue form: broad-based budding yeast • Pulmonary disease • Disseminated disease

Definition of abbreviations: RES, reticuloendothelial.

▶ Opportunistic Fungi

Aspergillus fumigatus	• **Allergic bronchopulmonary aspergillosis**/asthma, cystic fibrosis • **Fungus ball:** free in preformed lung cavities • **Invasive aspergillosis**/severe neutropenia, CGD, CF, burns – Invades tissues, causing infarcts and hemorrhage – Nasal colonization →**pneumonia or meningitis** – **Cellulitis**/in burn patients; may also disseminate	• **Dichotomously branching** • **Generally acute angles** • **Septate** • Compost pits, moldy marijuana • May cause disease in immunocompromised patients **Treatment:** itraconazole, amphotericin B; depends on severity of infection
Candida albicans (and other spp. of *Candida*)	• Involvement of the oral cavity and digestive tract • Septicemia, endocarditis in IV drug abusers • Mucocutaneous candidiasis	**Diagnosis:** • KOH: pseudohyphae, true hyphae, budding yeasts • Septicemia: culture lab identification: biochemical tests/formation of germ tubes **Treatment:** • Topical imidazoles or oral imidazoles; nystatin • Disseminated: amphotericin B or fluconazole
Cryptococcus neoformans	• **Meningitis/Hodgkin, AIDS (the dominant meningitis)** • **Acute pulmonary** (usually asymptomatic)/**pigeon breeders**	• **Encapsulated yeast (monomorphic)** • **Environmental source:** ~~Soil enriched~~ with pigeon droppings • **Diagnosis of meningitis: CSF** – Detect capsular antigen in CSF (by latex particle agglutination or counter immunoelectrophoresis) – **India ink mount** (misses 50%) of CSF sediment to find budding yeasts with capsular "halos" – Cultures (urease ⊕ yeast) • **Treatment:** amphotericin B plus flucytosine until afebrile and culture ⊖, then fluconazole
Mucor, Rhizopus, Absidia (Zygomycophyta family)	**Rhinocerebral infection** (mucormycosis) caused by *Mucor* (or other Zygomycophyta)	• Nonseptate, filamentous fungi • Characterized by paranasal swelling, necrotic tissues, hemorrhagic exudates from nose and eyes, mental lethargy • Occurs in **ketoacidotic diabetic patients** and **leukemic patients** • These fungi penetrate without respect to anatomic barriers, progressing rapidly from sinuses into brain tissue • **Diagnosis:** KOH of tissue; broad, ribbon-like nonseptate hyphae with about 90° angles on branches • **Treatment:** débride necrotic tissue and start amphotericin B fast; high fatality rate because of rapid growth and invasion
Pneumocystis jiroveci (formerly *carinii*)	Pneumonia in AIDS patients, malnourished babies, premature neonates, other immunocompromised	• An exudate with foamy or **honeycomb appearance on H & E stain** • **Patchy infiltrative (ground-glass appearance) on x-ray** • **Diagnosis: silver-staining cysts** in bronchial alveolar lavage fluids or biopsy • Treatment: trimethoprim/sulfamethoxazole, pentamidine

Definition of abbreviations: CF, cystic fibrosis; CSF, cerebrospinal fluid; CGD, chronic granulomatous disease; KOH, potassium hydroxide.

► Antifungal Agents

Because fungi are eukaryotic, finding selectively toxic antifungal agents is difficult. Consequently, treating fungal infections poses a clinical challenge, especially in immunocompromised patients. Fungal cell membranes contain **ergosterol**, a sterol not found in mammalian tissue. Thus, this difference provides the basis for most systemically administered antifungal agents.

Class/Agent	Mechanism of Action	Spectrum/Clinical Use	Mechanism(s) of Resistance	Toxicity/Notes
Agents for Systemic Infections				
Amphotericin B	Binds ergosterol, causing formation of artificial pores, thus altering membrane permeability, killing the cell	Widest antifungal spectrum: *Aspergillus* *Coccidioides* *Blastomyces* *Candida albicans* *Cryptococcus* *Histoplasma* *Mucor* *Sporothrix schenckii*	Very uncommon; ↓ or structurally altered ergosterol	• Fever and chills ("cytokine storm") • **Nephrotoxicity** limits dosing (cumulative over lifetime) • Reversible anemia (secondary to ↓ erythropoietin) • Arrhythmias • IV only
Flucytosine	Permease allows entry, deaminated to 5-FU, then converted to 5-FdUMP (thymidylate synthase inhibitor)	Narrow spectrum: *Cryptococcus* *Candida albicans* (systemic)	Rapid if used as a single agent; ↓ activity of fungal permeases and deaminases	• Reversible **bone marrow suppression** • Alopecia • Typically **combined with amphotericin B** or fluconazole
Azoles: fluconazole, itraconazole, voriconazole, ketoconazole	Inhibit synthesis of ergosterol, leading to altered membrane permeability	Varies: *Candida* *Coccidioides* *Cryptococcus* *Aspergillus* *Histoplasma*	↓ sensitivity of target enzymes	• Vomiting and diarrhea • Skin rash • Hepatotoxicity (rare) • **Gynecomastia** • ↓ P450
Echinocandin/ caspofungin	Inhibits synthesis of β-1,2 glycan, a component of fungal cell walls	*Candida* *Aspergillus*	—	• Not very toxic • Headache • Infusion-related reactions
Systemic Agents for Superficial Infections				
Griseofulvin	• Uptake by energy-dependent transport • Interferes with microtubule formation in dermatophytes • May inhibit polymerization of nucleic acids	• Dermatophytes of the hair and scalp • Accumulates in keratin	↓ in transport/uptake	• Confusion and vertigo • Headache • Blurred vision • Nausea/vomiting • ↑ P450 • GI irritation • **Disulfiram-like reaction** with ethanol
Terbinafine	Inhibits squalene epoxidase (for sterol biosynthesis)	Accumulates in keratin, used in onychomycosis		• GI irritation • Rash • Headache • Taste disturbance
Azoles (*see* above)	—	—	—	—
Topical Antifungals†				
Nystatin	Disrupts membrane by binding ergosterol	*Candida*, especially in oral candidiasis (thrush)	Same as amphotericin B	• Contact dermatitis • Stevens-Johnson syndrome

†Topical azoles, such as miconazole and clotrimazole are also widely used.

Embryology

Chapter 6

General Principles of Embryology

General Principles of Embryology

► Early Embryology

Week 1

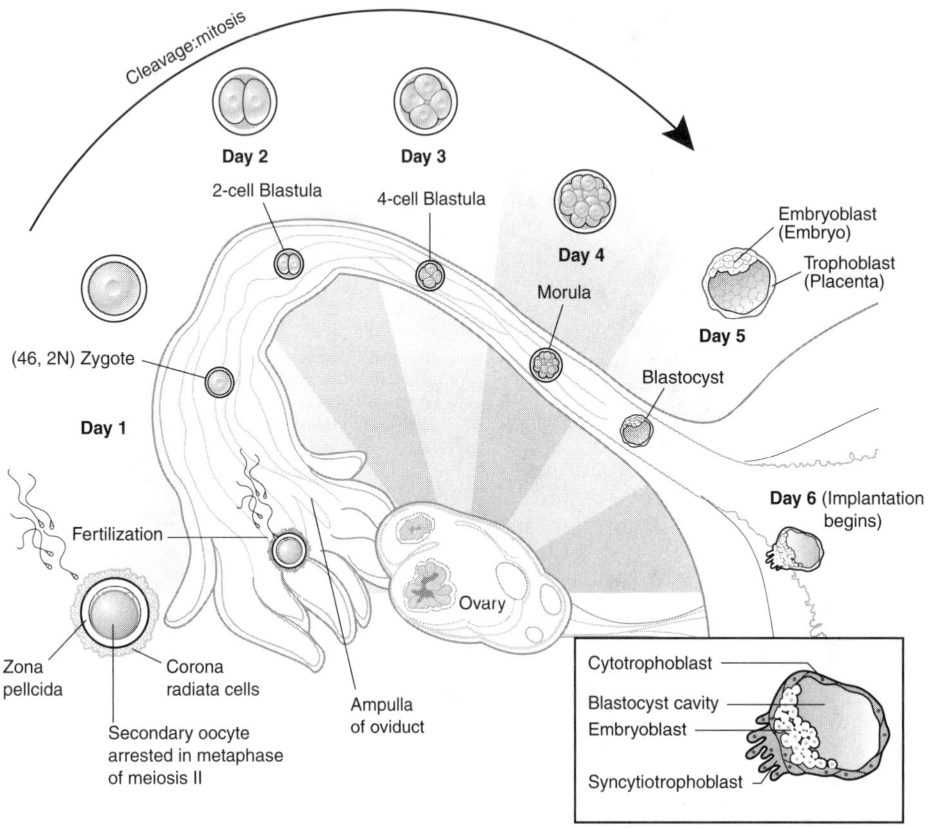

Fertilization occurs in the **ampulla of the uterine tube** when the male and female pronuclei fuse to form a **zygote**. At fertilization, the secondary oocyte rapidly completes meiosis II.

During the first 4 to 5 days of the first week, the zygote undergoes rapid mitotic division (**cleavage**) in the oviduct to form a **blastula**, consisting of increasingly smaller **blastomeres**. This becomes the **morula** (32-cell stage).

A **blastocyst** forms as fluid develops in the morula. The blastocyst consists of an inner cell mass known as the **embryoblast**, and the outer cell mass known as the **trophoblast** becomes the placenta.

At the end of the first week, the trophoblast differentiates into the **cytotrophoblast** and **syncytiotrophoblast** and then implantation begins. Implantation usually occurs in the **posterior superior wall** of the uterus.

Clinical Correlation: Ectopic Pregnancy	
Tubal	The **most common form** of ectopic pregnancy Usually occurs when the blastocyst **implants within the ampulla** of the uterine tube because of delayed transport **Risk factors:** endometriosis, pelvic inflammatory disease (PID), tubular pelvic surgery, or exposure to diethylstilbestrol (DES) **Clinical signs:** abnormal or brisk uterine bleeding, sudden onset of abdominal pain that may be confused with appendicitis, missed menstrual period (e.g., LMP 60 days ago), positive human chorionic gonadotropin (hCG) test, culdocentesis showing intraperitoneal blood, positive sonogram
Abdominal	Most commonly occurs in the **rectouterine pouch** (pouch of Douglas)

(Continued)

► Early Embryology *(Cont'd.)*

Week 2

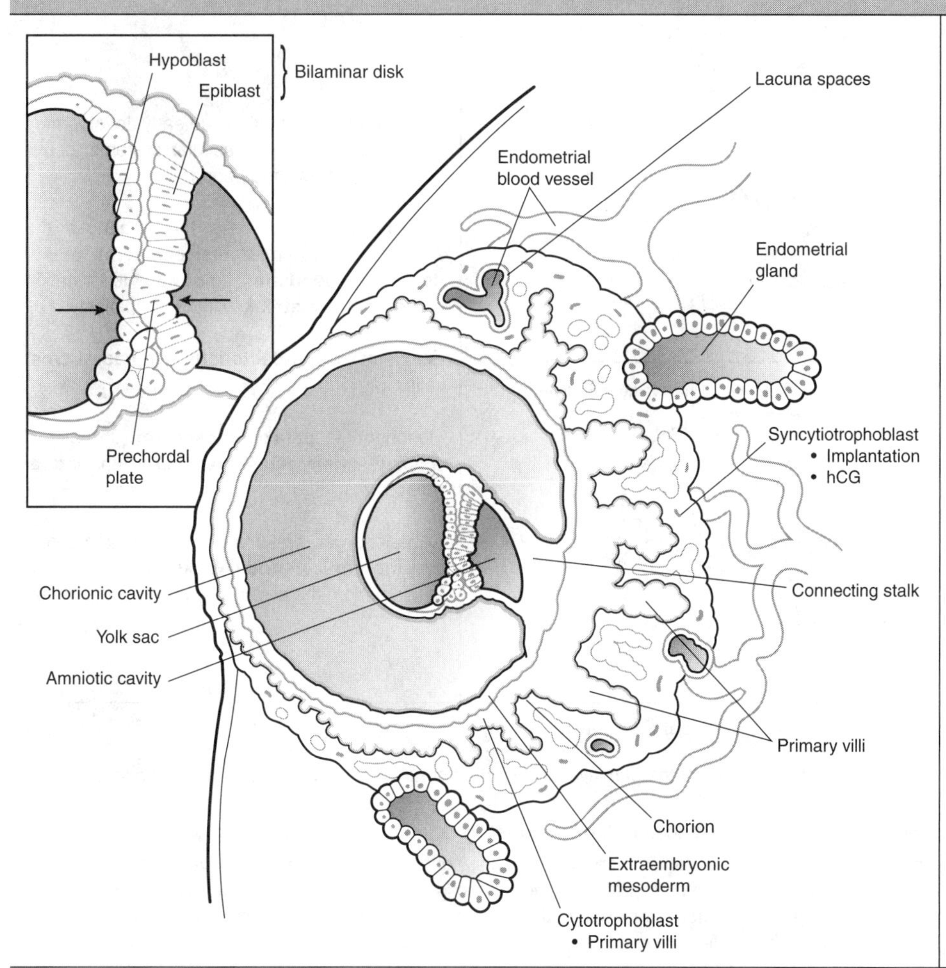

The embryoblast differentiates into the **epiblast** and **hypoblast**, forming a **bilaminar embryonic disk**.

The epiblast forms the **amniotic cavity**, and hypoblast cells migrate from the **primary yolk sac**.

The **prechordal plate**, formed from fusion of epiblast and hypoblast cells, is the site of the future **mouth**.

Extraembryonic mesoderm is derived from the epiblast. **Extraembryonic somatic mesoderm** lines the cytotrophoblast, forms the connecting stalk, and covers the amnion. **Extraembryonic visceral mesoderm** covers the yolk sac.

The connecting stalk suspends the conceptus within the chorionic cavity. The wall of the chorionic cavity is called the **chorion**, consisting of extraembryonic somatic mesoderm, the cytotrophoblast, and the syncytiotrophoblast.

Clinical Correlation

Human chorionic gonadotropin (hCG) is a glycoprotein produced by the syncytiotrophoblast. It stimulates progesterone production by the corpus luteum. hCG can be assayed in maternal blood or urine and is the basis for early pregnancy testing. hCG is detectable throughout pregnancy. **Low hCG** levels may predict a spontaneous abortion or ectopic pregnancy. **High hCG** levels may predict a multiple pregnancy, hydatidiform mole, or gestational trophoblastic disease.

(Continued)

► Early Embryology *(Cont'd.)*

Weeks 3 Through 8

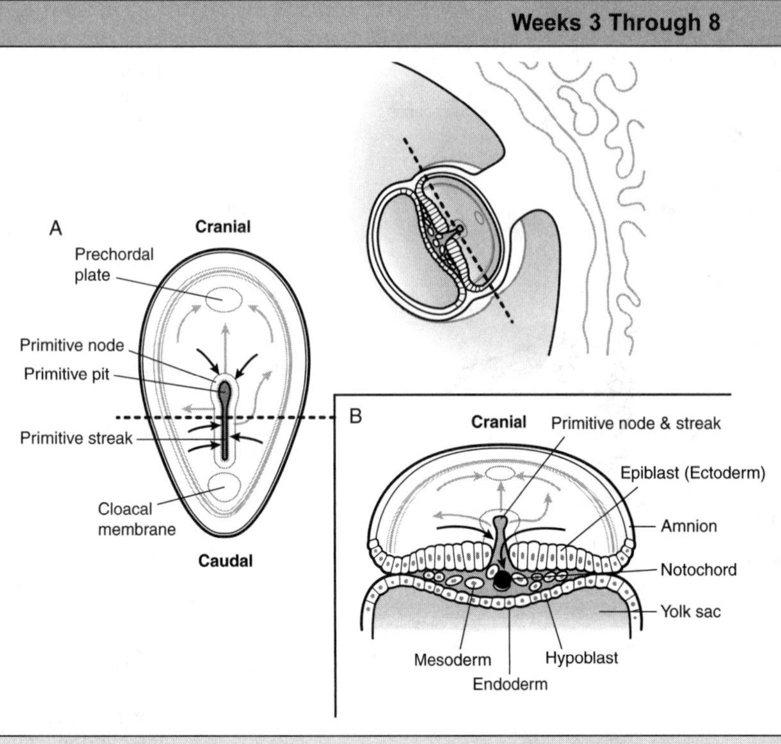

Third week: gastrulation and early development of nervous and cardiovascular systems; corresponds to first missed period

Gastrulation—a process that produces the three primary germ layers: **ectoderm, mesoderm**, and **endoderm**; begins with the formation of the **primitive streak** within the epiblast

Ectoderm → **neuroectoderm** and **neural crest** cells

Mesoderm → **paraxial mesoderm** (35 pairs of somites), **intermediate mesoderm**, and **lateral mesoderm**

All major organ systems begin to develop during the **embryonic period (Weeks 3–8)**. By the end of this period, the embryo begins to look human.

Clinical Correlation

Sacrococcygeal teratoma: a tumor that arises from remnants of the primitive streak; often contains various types of tissue (bone, nerve, hair, etc.).

Chordoma: a tumor that arises from remnants of the notochord, found either intracranially or in the sacral region

Caudal dysplasia (sirenomelia): a constellation of syndromes ranging from minor lesions of the lower vertebrae to complete fusion of lower limbs. Occurs as a result of abnormal gastrulation, in which migration of mesoderm is disturbed. Associated with VATER (vertebral defects, anal atresia, tracheoesophageal fistula, and renal defects) or VACTERL (vertebral defects, anal atresia, cardiovascular defects, tracheoesophageal fistula, renal defects, and upper limb defects)

Hydatidiform mole: results from the partial or complete replacement of the trophoblast by dilated villi

- **In a complete mole, there is no embryo;** a haploid sperm fertilizes a blighted ovum and reduplicates so that the karyotype is 46,XX, with all chromosomes of paternal origin. In a **partial mole**, there is a haploid set of maternal chromosomes and usually two sets of paternal chromosomes so that the typical karyotype is 69,XXY.
- **Molar pregnancies have high levels of hCG, and 20% develop into a malignant trophoblastic disease, including choriocarcinoma.**

► Germ-Layer Derivatives

Ectoderm	
Surface ectoderm	Epidermis, hair, nails, inner and external ear, tooth enamel, lens of eye, anterior pituitary (from Rathke pouch), major salivary, sweat, and mammary glands; epithelial lining of nasal and oral cavities and ear
Neuroectoderm	CNS (brain and spinal cord), retina and optic nerve, pineal gland, neurohypophysis, astrocytes, oligodendrocytes, ependymal cells
Neural crest	Adrenal medulla, ganglia (sensory, autonomic), melanocytes, Schwann cells, meninges (pia, arachnoid), pharyngeal arch cartilage, bones of the skull, odontoblasts, parafollicular (C) cells, laryngeal cartilage, aorticopulmonary septum, endocardial cushions (abnormal development can lead to many congenital defects)
Mesoderm	Muscle (smooth, cardiac, skeletal), connective tissue, serous membranes, bone and cartilage, blood and blood vessels, lymphatics, cardiovascular organs, adrenal cortex, gonads and internal reproductive organs, spleen, kidney and ureter, dura mater
Endoderm	Epithelial parts: GI tract, tonsils, thymus, pharynx, larynx, trachea, bronchi, lungs, urinary bladder, urethra, tympanic cavity, auditory tube and other pharyngeal pouches Parenchyma: liver, pancreas, tonsils, thyroid, parathyroids, glands of GI tract, submandibular and sublingual glands

► Placenta

The placenta permits exchange of nutrients and waste products between maternal and fetal circulations.

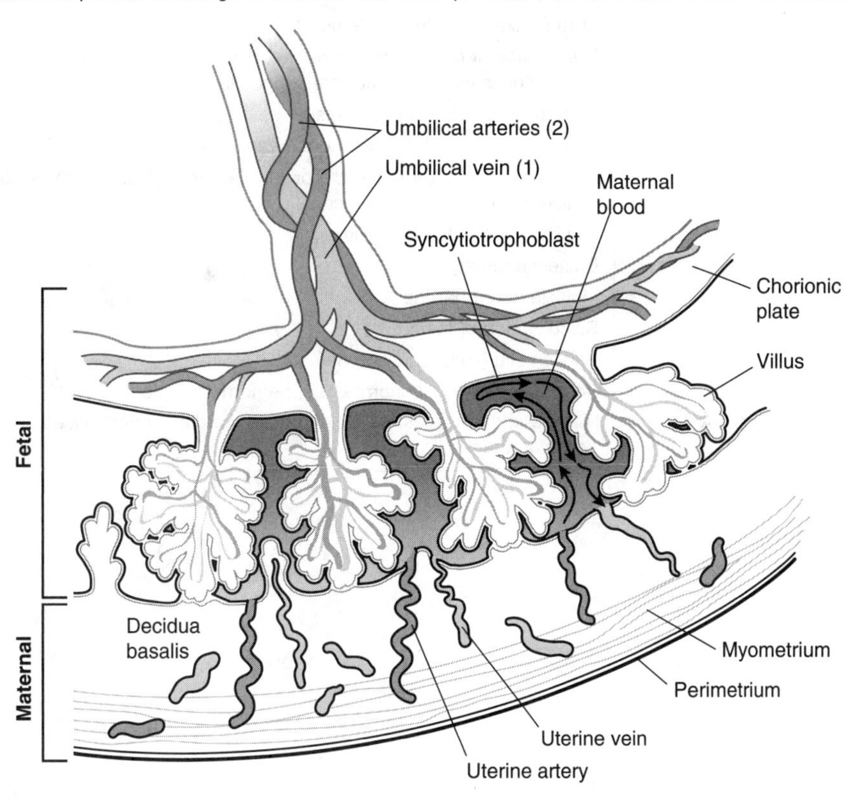

Fetal component	Chorionic plate and villi **Syncytiotrophoblast:** outer layer of chorionic villi; secretes hCG **Cytotrophoblast:** inner layer of chorionic villi
Maternal component	**Decidua basalis:** Maternal blood vessels from the decidua conduct blood into the intervillous spaces of the placenta, where floating villi are present
Placental barrier	The syncytiotrophoblast, cytotrophoblast, basement membrane, fetal capillary endothelium separate the maternal and fetal blood
Umbilical cord	1 **umbilical vein** supplies **oxygenated** blood from the placenta to the fetus 2 **umbilical arteries** carries **deoxygenated** blood back from the fetus to the placenta **Urachus**: Removes nitrogenous waste from the fetal bladder

► Twinning

The type of twinning depends on the point in time in which cleavage occurs. If the twinning occurs very early on (before the chorion forms), two separate chorions will form (dichorionic). All other possible types of twins are monochorionic because the chorion has already formed. Fraternal twins are **dizygotic** and identical twins are **monozygotic**.

Placental morphology	Time of cleavage	Type of twin
Dichorionic diamniotic	Very early (Days 1–3)	Fraternal or identical
Monochorionic diamniotic	Early (Days 4–8)	Identical
Monochorionic monoamniotic	Later (Days 8–13): identical Even later (Days 13–15): conjoined	Identical, conjoined

► Teratogens

Teratogen	Effect
ACE inhibitors; ARBS	Renal damage
Aminoglycosides	CN VIII toxicity
Androgens	Masculinization of female fetus
Anticonvulsants	Neural tube defects (carbamazepine, valproic acid), phenytoin (fetal hydantoin syndrome); multiple congenital defects
Antithyroid drugs	Congenital goiter, hypothyroidism
Diethylstilbestrol (DES)	Vaginal clear cell adenocarcinoma; vaginal adenosis
Ethanol	Fetal alcohol syndrome (e.g., growth retardation, facial abnormalities, microcephaly, cardiac defects)
Folate antagonists	Multiple congenital anomalies
Lithium	Ebstein anomaly
Radiation	Multiple abnormalities
Tetracyclines	Discoloration of teeth
Thalidomide	Phocomelia (limb reduction defects)
Warfarin	Bone and cartilage abnormalities, hemorrhage, etc.
Vitamin A excess; vitamin A derivatives	Multiple abnormalities, e.g., cleft palate, cardiac abnormalities, mental retardation

Physiology

Chapter 7

▶ Physiologic Terminology

Equilibrium	Equilibrium occurs when the balance of opposing forces has reached the **lowest free energy state**, and as a result, a given variable has reached a constant value.
Steady state	Steady state is a condition in which a variable is maintained within narrow limits by regulating an opposing activity. This process requires energy.
Negative feedback	This is a common system that acts to oppose changes in the internal environment. Negative feedback systems promote stability and act to restore steady-state function after a perturbation.
Positive feedback	This is a less common system (also called a vicious cycle) that acts to magnify a change in the internal environment; the initial change in a system is increased as a result of feedback activity. In a viable organism, any positive feedback system is ultimately overridden by one or more negative feedback systems.

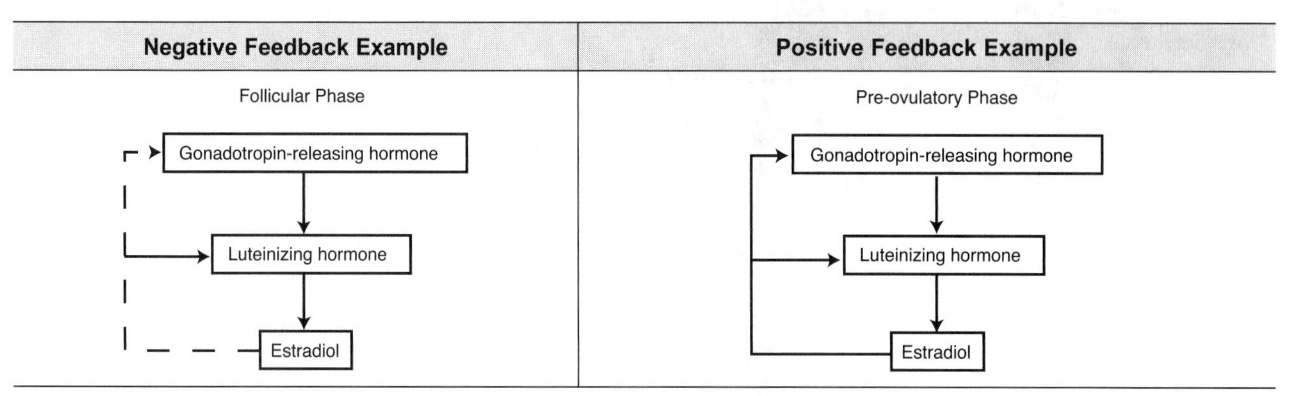

The figures above show the negative feedback relationship between estrogens and the gonadotropins that dominates during the follicular phase, which transforms into a positive feedback relationship, producing the LH surge prior to ovulation. Solid arrows show direct relationships (stimulation) and dashed arrows show inverse relationships (inhibition).

▶ Transport

Type of Transport	Energy Source	Example	Other Characteristics	Clinical Correlation
Simple diffusion	Passive	Pulmonary gases	—	Pulmonary edema decreases diffusion
Carrier-mediated or facilitated diffusion	Passive	Glucose uptake by muscle	• Insulin controls carrier population • Chemical specificity	Insulin-dependent glucose uptake impaired in diabetes mellitus
Primary active transport	Direct use of ATP	Na^+/K^+-ATPase Ca^{2+}-ATPase	Antiport (countertransport) Regulates cellular Ca^{2+}	Inhibition by cardiac glycosides Deficiency causes neuromuscular disorders
Secondary active transport	Electrochemical gradient for sodium is most common driving force	Na^+-glucose in kidney Na^+-H^+ exchange	Symport (cotransport) Antiport (countertransport)	Osmotic diuresis results when transporters saturated Renal tubular acidosis

Key Points

- Passive processes are directly related to concentration gradients.
- Active processes create or increase a concentration gradient and thus depend upon metabolic energy.
- Both carrier-mediated facilitated diffusion and active transport can be saturated; maximum transport rate depends on population and activity of transport molecules.
- Primary active transport proteins have an ATPase as part of their structure.
- Most secondary active transport depends upon the electrochemical gradient of sodium ions, which in turn depends on the activity of the primary active transporter, sodium-potassium ATPase.

► Diffusion Kinetics

Simple diffusion	The rate is estimated by **Fick's law of diffusion**: $J = -DA(\Delta C/\Delta X)$ J = net flux, D = diffusion coefficient, A = surface area, ΔC = concentration or pressure gradient, ΔX = diffusion distance Changes in surface area or diffusion distance are most important in disease states (e.g., the decrease in surface area caused by destruction of alveoli in emphysema or the decreased diffusion of oxygen during pulmonary edema related to increased diffusion distance).
Carrier-mediated transport	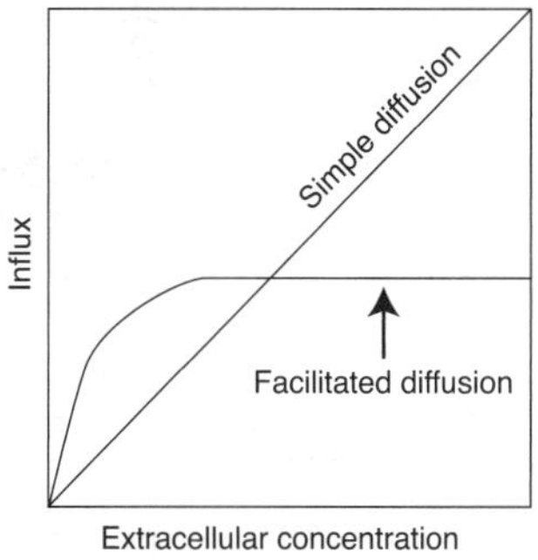 Influx; Simple diffusion; Facilitated diffusion; Extracellular concentration Assumption: intracellular concentration is negligible 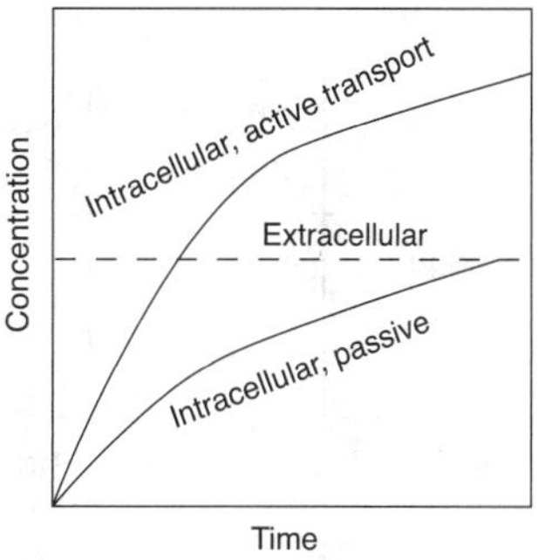 Concentration; Intracellular, active transport; Extracellular; Intracellular, passive; Time Assumption: extracellular concentration does not change **Facilitated diffusion** increases transport rate above that capable with **simple diffusion**, but has saturation kinetics (*left*). **Active transport** can produce a concentration gradient, and **passive processes** will lead to an equilibrium state (*right*).

► Fluid Volume Compartments and Distribution

	% of Body Weight	Fraction of TBW	Markers Used to Measure Volume	Primary Cations	Primary Anions
Total body water (TBW)	60	1.0	Tritiated H_2O D_2O, antipyrine	—	—
Intracellular fluid volume (ICF)	40	⅔	TBW – ECF*	K^+	Organic phosphates; protein
Extracellular fluid volume (ECF)	20	⅓	Inulin Mannitol	Na^+	Cl^- HCO_3^-
Plasma	5	1/12 (¼ of ECF)	RISA Evans blue	Na^+	Cl^- HCO_3^- Plasma proteins
Interstitial fluid	15	¼ (¾ of ECF)	ECF-plasma volume*	Na^+	Cl^- HCO_3^-

Principles of Fluid Distribution

1. Osmolarity of the ICF and ECF are equal.
2. Intracellular volume changes only when extracellular osmolarity changes.
3. All substances enter or leave the body by passing through the extracellular compartment.

Measurement of Fluid Volumes

$\text{Volume} = \dfrac{\text{Mass}}{\text{Concentration}}$	*Example:* 100 mg of inulin is infused. After equilibration, its concentration = 0.01 mg/mL. What is patient's ECF volume? *Answer:* ECF = 100 mg/0.01 mg/mL = 10,000 mL, or 10 L

Osmolarity and Mass

1. Plasma osmolarity in mOsm/L can be quickly estimated as twice the plasma sodium concentration in mmol/L. More rigorously, plasma osmolarity (mOsm/L) = (2 × serum sodium [mEq/L]) + (BUN [mg/dL]/2.8) + (glucose [mg/dL]/18).
2. Mass of solutes in the TBW, ICF, or ECF in mOsm is calculated by the relevant volume multiplied by the osmolarity.

Definition of abbreviation: RISA, radio-iodinated serum albumin.
*Indirect measurement

► Summary of Volume and Osmolarity Changes of Body Fluids

Body osmolarity and intracellular and extracellular fluid volumes change in clinically relevant situations. The **Darrow-Yannet diagram** *(right)* represents this information. The *y*-axis is solute concentration or osmolarity. The *x*-axis is the volume of ICF and ECF. The *solid line* represents the control state, and the *dashed line* represents changes in volume or osmolarity. In this example, osmolarity ↓, ICF volume ↑, and ECF volume ↓.

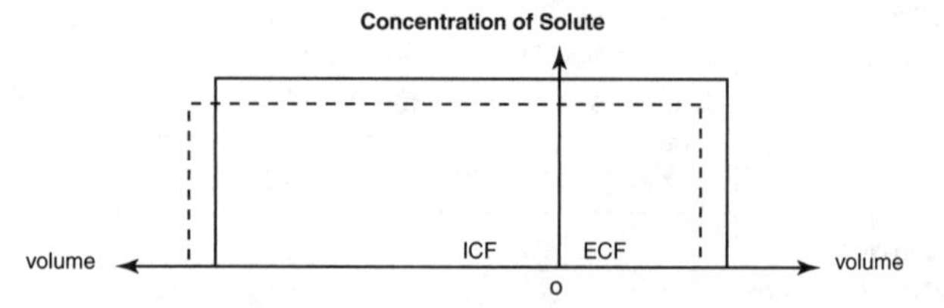

Type	Examples	ECF Volume	Body Osmolarity	ICF Volume	D-Y Diagram
Isosmotic volume contraction (loss of isotonic fluid)	Diarrhea, hemorrhage, vomiting	↓	No change	No change	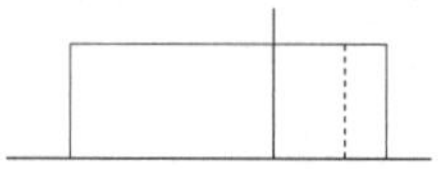
Isosmotic volume expansion (gain of isotonic fluid)	Isotonic saline infusion	↑	No change	No change	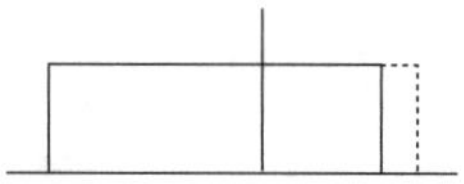
Hyperosmotic volume contraction (loss of water)	Dehydration, diabetes insipidus	↓	↑	↓	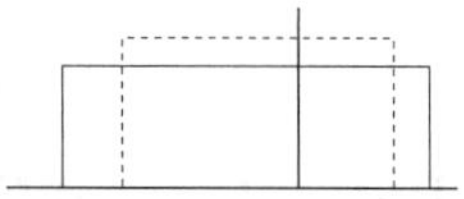
Hyperosmotic volume expansion (gain of NaCl)	Excessive NaCl intake, hypertonic mannitol, chronic aldosterone excess	↑	↑	↓	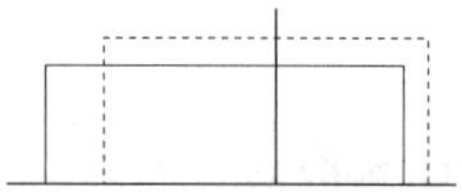
Hyposmotic volume contraction (loss of NaCl)	Adrenal insufficiency	↓	↓	↑	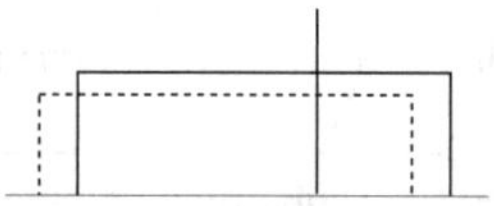
Hyposmotic volume expansion (gain of water)	SIADH, water intoxication	↑	↓	↑	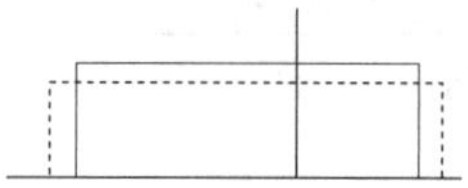

Definiton of abbreviation: SIADH, syndrome of inappropriate antidiuretic hormone.

► Membrane Potentials

Equilibrium potential

Amount of voltage needed to balance the chemical force due to its concentration gradient
The **Nernst equation** can determine this. For a monovalent cation:

$$E_x = \frac{60\ \text{mV}}{Z} \log_{10} \frac{[X]_o}{[X]_i}$$; E_x = equilibrium potential, [X] = ion concentration (out and in), Z = charge

(Na^+ = 1, Cl^- = –1, Ca^{2+} = 2)

Resting membrane (RM) potential

- RM is the potential difference across a cell membrane in millivolts (mV); –70 mV is typical.
- RM occurs because of an **unequal distribution of ions** between the ICF and ECF and the **selective permeability** of the membrane to ions. **Proteins (anions)** in cells that do not diffuse help establish the electrical potential across the membrane.
- The relative effect of an ion on the membrane potential is in proportion to the conductance or permeability of that ion. The **greater the conductance**, the closer the membrane will approach the **equilibrium potential** of that ion.
- The resting potential of cells is negative inside. **Hyperpolarization** occurs when the membrane potential becomes more negative. **Depolarization** occurs when the membrane potential becomes less negative or even positive.

Chord conductance equation

$E_m = (g_K/\Sigma g \times E_K) + (g_{Na}/\Sigma g \times E_{Na}) + (g_{Cl}/\Sigma g \times E_{Cl})$

E_m = membrane potential in mV; g = conductance of individual ion; Σg = total conductance of cell membrane; E = equilibrium potential of individual ion from Nernst equation

- Used to calculate membrane potential; useful to evaluate effects of ion conductance and concentrations; explains action potential, synaptic potentials, and electrolyte disorders

Properties of ions in a typical neuron

Ion	Extracellular (mM)	Intracellular (mM)	Equilibrium Potential (mV)	Conductance
Na^+	150.0	15.0	+60	Very low
K^+	5.5	150.0	–90	High
Cl^-	125.0	9.0	–70	High

- Because potassium and chloride have a high conductance, their equilibrium potentials dominate the membrane potential, so the inside of the cell is negative.
- Changes in EC K^+ can produce large changes in the membrane potential:

$\uparrow$ EC $K^+ \rightarrow$ depolarization; $\downarrow$ EC $K^+ \rightarrow$ hyperpolarization.

- Changes in EC Na^+ have little effect on membrane potential, but an increase in Na^+ conductance $\rightarrow$ depolarization.

Definition of abbreviation: EC, extracellular.

▶ Action Potential

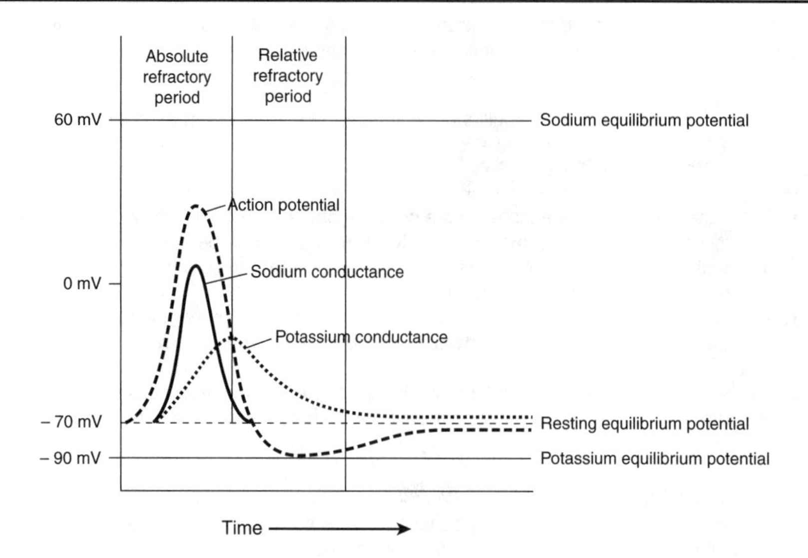

The action potentials (AP) of excitable cells involve the opening and closing of voltage-gated channels for sodium, potassium, and in some cells, calcium. The figure above shows a neuronal AP. The opening of a channel increases conductance.
Steps are shown below.

- Inward currents depolarize the membrane potential to **threshold**.
- **Voltage-gated Na^+ channels open**, causing an **inward Na^+ current**. The membrane potential approaches the Na^+ equilibrium potential. These channels can be blocked by **tetrodotoxin**.
- These Na^+ channels close rapidly **(inactivation)**, even though the membrane is still depolarized.
- Depolarization slowly opens K^+ channels, **increasing K^+ conductance (outward current)**, leading to **repolarization**.
- **Absolute refractory period:** An AP cannot be elicited because Na^+ channels are closed.
- **Relative refractory period:** Only a greater than normal stimulus can produce an AP because the K^+ conductance is still higher than at rest.
- The **Na^+-K^+ pump** restores ion concentrations. It is **electrogenic** (3 Na^+ pumped out for every 2 K^+ pumped in).

Clinical Correlations

Calcium has a low resting conductance and does not contribute to the resting potential. It has a very positive equilibrium potential, so when conductance increases (e.g., cardiac and smooth muscle cells), **depolarization** occurs. Calcium concentration affects the action potentials and force of contraction of **cardiac and smooth muscle**.

Hypercalcemia stabilizes excitable membranes, leading to flaccid paralysis of skeletal muscle. **Hypocalcemia** destabilizes membranes, leading to spontaneous action potentials and spasms.

Abnormal increases and decreases of **extracellular potassium** have severe consequences for cardiac conduction and rhythm.

Renal and gastrointestinal disorders are likely to cause abnormalities of electrolytes and alteration of resting potentials and action potentials.

Pathology

Chapter 8

Cellular Injury and Adaptation

Changing conditions in the cell's environment can produce changes from adaptation to injury or even cell death. The cellular response to injury depends on the *type, duration*, and *severity* of injury, the *type of cell* injured, *metabolic state*, and *ability to adapt*.

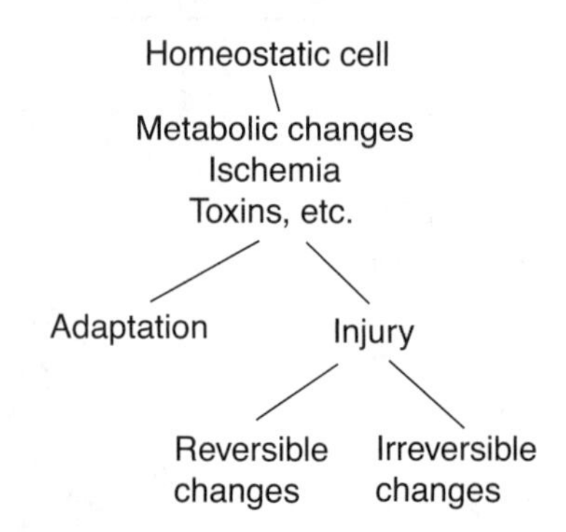

CAUSES	EXAMPLES
Hypoxia (most common)	• Ischemia (e.g., arteriosclerosis, thromboembolus) • Cardiopulmonary failure • Severe anemia
Infections	Viruses, bacteria, parasites, rickettsiae, fungi, prions
Immunologic reactions	• Hypersensitivity reactions • Autoimmune
Congenital/metabolic disorders	• Inborn errors of metabolism (e.g., phenylketonuria, galactosemia, glycogen storage diseases)
Chemical injury	• Drugs (e.g., therapeutic, drugs of abuse) • Poisons • Pollution • Occupational exposure (e.g., CCl_4, CO, asbestos)
Physical injury	Trauma, burns, frostbite, radiation
Nutritional or vitamin imbalance	• Inadequate calorie/protein intake (e.g., marasmus, kwashiorkor, anorexia nervosa) • Excess caloric intake (e.g., obesity, atherosclerosis) • Vitamin deficiencies • Hypervitaminosis

Important Mechanisms of Cell Injury

- Oxygen-free radicals (superoxide [$O_2^{\cdot -}$], hydroxyl radical [OH·], hydrogen peroxide [H_2O_2]) damage DNA, proteins, lipid membranes, and circulating lipids (LDL) by peroxidation
- Decreased oxidative phosphorylation
- ATP depletion
- Increased cell-membrane permeability
- Influx of calcium—activates a wide spectrum of enzymes: proteases, ATPases, phospholipases, endonucleases
- Mitochondrial dysfunction, formation of mitochondrial permeability transition (MPT) channels
- Release of cytochrome c is a trigger for apoptosis

Protective Factors Against Free Radicals

1. *Antioxidants*—vitamins A, E, and C
2. *Superoxide dismutase*—superoxide → hydrogen peroxide
3. *Glutathione peroxidase*—hydroxyl ions or hydrogen peroxide → water
4. *Catalase*—hydrogen peroxide → oxygen and water

► Direct and Indirect Results of Reversible Cell Injury

Direct Result	Consequences	Pathophysiologic Correlates
Decreased synthesis of ATP by oxidative phosphorylation	Decreased function of Na^+/K^+ ATPase → influx of Na^+ and water, efflux of K^+, and swelling of the ER	Cellular swelling (hydropic swelling), swelling of endoplasmic reticulum, membrane blebs, myelin figures
Increased glycolysis → glycogen depletion	Increased lactic acid production → decreased intracellular pH	Tissue acidosis
Ribosomes detach from rough ER	Decreased protein synthesis	Lipid deposition (fatty change)

► As the Degree of Cellular Injury Worsens . . .

Severe plasma membrane damage	Massive influx of calcium, efflux of intracellular enzymes and proteins into the circulation	Markers of cellular damage detectable in serum (LDH, CK, ALT, AST, troponin, etc.)
Calcium influx into mitochondria	Irreparable damage to oxidative phosphorylation	Mitochondrial densities
Lysosomal contents leak out	Lysosomal hydrolases are activated intracellularly	Autolysis, heterolysis, nuclear changes (pyknosis, karyorrhexis, karyolysis)

► Irreversible Injury and Cell Death

Morphologic Pattern	Characteristics
Coagulative necrosis	**Most common** (e.g., heart, liver, kidney) Proteins denatured, nucleus is lost, but cellular shape is maintained
Liquefactive necrosis	**Abscesses**, **brain infarcts**, pancreatic necrosis Cellular destruction by hydrolytic enzymes
Caseous necrosis	Seen in **tuberculosis** Combination of coagulation and liquefaction necrosis → soft, friable, and "cottage-cheese–like" appearance
Fat necrosis	Caused by the action of lipases on fatty tissue (e.g., with **pancreatic damage**) Chalky white appearance
Fibrinoid necrosis	Eosinophilic homogeneous appearance—resembles fibrin
Gangrenous necrosis	*Common sites:* lower limbs, gallbladder, GI tract, and testes Dry gangrene—coagulative necrosis Wet gangrene—liquefactive necrosis
Apoptosis	A specialized form of **programmed cell death**, an active process under genetic control Often affects only single cells or small groups of cells Mediated by a cascade of **caspases** (digest nuclear and cytoskeletal proteins and activate endonucleases)

Myocardial ischemia is a good example of cellular injury and death.

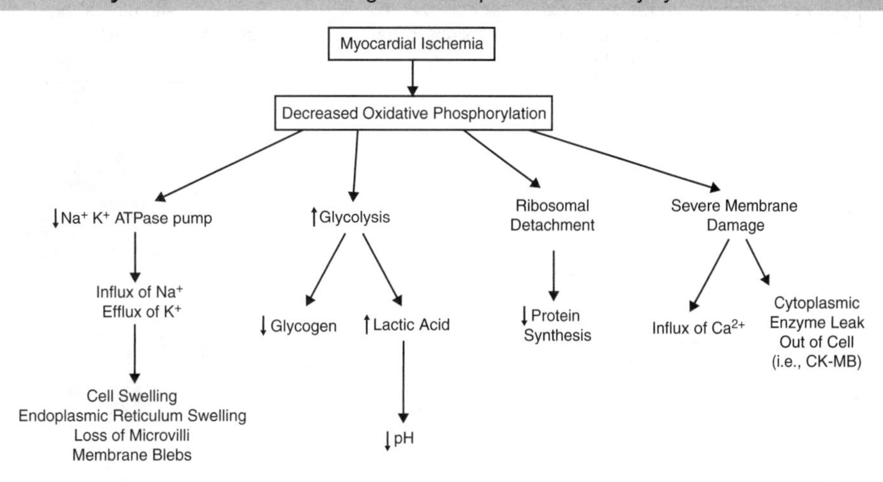

► Apoptosis

Morphology	Stimuli for Apoptosis	Genetic Regulation	Physiologic Examples	Pathologic Examples
• Cells shrink, cytoplasm is dense and eosinophilic • Nuclear chromatin condenses, then fragments • Cell membrane blebs • Cell fragments (apoptotic bodies) are phagocytized by adjacent cells or macrophages • **Lack of inflammatory response**	• Cell injury and DNA damage • Lack of hormones, cytokines, or growth factors • Receptor-ligand signals: – **Fas** binding to Fas ligand – Tumor necrosis factor (**TNF**) binding to TNF receptor 1 (TNFR1)	• ***bcl-2*** (inhibits apoptosis) • ***p-53*** (stimulates apoptosis)	• Embryogenesis—organogenesis and development • Hormone-dependent apoptosis (menstrual cycle) • Selective death of lymphocytes in thymus	• Viral hepatitis (Councilman body) • Graft versus host disease • Cystic fibrosis (CF)—duct obstruction and pancreatic atrophy

► Adaptive Cellular Responses to Injury (Potentially Reversible)

Types	Definitions	Causes
Atrophy	**Decrease in cell size and functional ability.** Cells shrink; lipofuscin granules can be seen microscopically. EM—autophagosomes	• Deceased workload/disuse • Ischemia • Lack of hormonal or neural stimulation • Malnutrition • Aging
Hypertrophy	**An increase in cell size and functional ability** mediated by growth factors, cytokines, and other trophic stimuli, leading to increased expression of genes and increased protein synthesis. May coexist with hyperplasia.	• Increased mechanical demand, e.g., striated muscle of weight lifters, cardiac muscle in hypertension • Increased endocrine stimulation, e.g., puberty, pregnancy, lactation
Hyperplasia	**An increase in the number of cells** in a tissue or organ, mediated by growth factors, cytokines, and other trophic stimuli. Increased expression of growth-promoting genes (proto-oncogenes), increased DNA synthesis, and cell division	**Physiologic causes:** • Compensatory (e.g., after partial hepatectomy) • Hormonal stimulation (e.g., breast development at puberty) • Antigenic stimulation (e.g., lymphoid hyperplasia) **Pathologic causes:** • Endometrial hyperplasia • Prostatic hyperplasia of aging
Metaplasia	**A reversible change of one cell type to another**	Irritation
Dysplasia	**An abnormal proliferation of cells** characterized by changes in cell size, shape, and loss of cellular organization; premalignant, e.g., cervical dysplasia	Similar to stimuli that produce cancer (e.g., HPV, esophageal reflux)

Regeneration and Repair

Wound healing involves **regeneration** of cells in a damaged tissue, along with repair of the connective tissue matrix.

▶ Capacity of Cells to Regenerate

Cell Type	Properties	Examples
Labile	Regenerate throughout life	Skin and mucosal lining cells Hematopoietic cells Stem cells
Stable	Replicate at a low level throughout life Have the capacity to divide if stimulated by some initiating event	Hepatocytes Proximal tubule cells Endothelium
Permanent	Cannot replicate	Neurons and cardiac muscle

Growth Factors and Cytokines Involved in Growth Repair

- Transforming growth factor α (TGF-α)
- Platelet-derived growth factor (PDGF)
- Fibroblast growth factor (FGF)
- Vascular endothelial growth factor (VEGF)
- Epidermal growth factor (EGF)
- Tumor necrosis factor α (TNF-α) and IL-1

▶ Connective Tissue Components

Collagen production requires vitamin C (wound healing impaired in scurvy) and copper. Different types of collagen are found in different body sites.

- Type I collagen is the most common form.
- Type II collagen is found in cartilage.
- Type III collagen is an immature form found in granulation tissue.
- Type IV collage is found in basement membranes.

Other extracellular matrix components include:

- adhesion molecules
- proteoglycans
- glycosaminoglycans

▶ Wound Healing

Primary Union (healing by first intention)

Occurs with clean wounds when there has been little tissue damage and wound edges are closely approximated, e.g., a surgical incision

Secondary Union (healing by second intention)

Occurs in wounds that have large tissue defects (the two skin edges are not in contact):

- Granulation tissue fills in the defects
- Often accompanied by significant wound contraction; may produce large residual scar

Keloids

Occur as result of excessive scar collagen deposition/ hypertrophy, especially in dark-skinned people

Connective Tissue Diseases

Scurvy

Vitamin C deficiency first affecting collagen with highest hydroxyproline content, such as that found in blood vessels. Thus, an early symptom is bleeding gums.

Ehlers-Danlos (ED) Syndrome

Defect in collagen synthesis or structure. There are many types. ED type IV is a defect in type III collagen.

Osteogenesis Imperfecta

Defect in collagen type I

Staining Methods

Stain	Cell Type/Component
Hematoxylin (stains blue to purple)	Nuclei, nucleoli, bacteria, calcium
Eosin (stains pink to red)	Cytoplasm, collagen, fibrin, RBCs, thyroid colloid
Prussian blue	Iron
Congo red	Amyloid
Periodic acid-Schiff (PAS)	Glycogen, mucin, mucoprotein, glycoprotein, as well as fungi
Gram stain	Microorganisms
Trichrome	Collagen
Reticulin	Reticular fibers in loose connective tissue
Immunohistochemical (antibody) stains:	
• Cytokeratin	• Epithelial cells
• Vimentin	• Connective tissue
• Desmin	• Muscle
• Prostate-specific antigen (PSA)	• Prostate
• S-100, neuron specific enolase, neurofilament	• Neurons, neuronally derived, neural crest–derived growths
• GFAP	• Glial cells (including astrocytes)
• Tartrate-resistant acid phosphatase (TRAP)	• Hairy cell leukemia
• Bombesin	• Small cell lung cancer, carotid body tumor

Definition of abbreviations: GFAP, glial fibrillary acidic protein.

Amyloidosis

An accumulation of various insoluble fibrillar in various tissues. It stains with **Congo red** and shows **apple-green birefringence** with polarized light.

Disease	Major Fibril Protein	Precursor Protein
Primary amyloidosis (e.g., plasma cell disorders)	AL	Kappa or lambda chains
Secondary amyloidosis (e.g., neoplasia, rheumatoid arthritis, SLE, TB, Crohn disease, osteomyelitis, familial Mediterranean fever)	AA	Serum amyloid associated (SAA)
Renal hemodialysis	$A\beta_2M$	β_2 microglobin
Senile cerebral amyloidosis	$A\beta$	β-amyloid precursor protein
Cardiac amyloidosis	ATTR	Transthyretin
Medullary carcinoma of thyroid	A Cal	Calcitonin
Type 2 diabetes, pancreatic islet-cell tumors	A Amylin	Amylin

Definition of abbreviations: ATTR, amyloid transthyretin; SLE, systemic lupus erythematosus; TB, tuberculosis.

Neoplasia

Carcinogenesis is a multistep process involving multiple genetic changes from inherited germ-line mutations or acquired mutations, leading to monoclonal expansion of a mutated cell.

Progression of epithelial cancer: normal epithelium → atypical hyperplasia → dysplasia → carcinoma in situ → invasion, metastasis (collagenases, hydrolases aid in penetration of barriers, such as basement membrane)

▶ Basic Terms

Anaplasia	Loss of cell differentiation and tissue organization
Atypical hyperplasia	Increased cell number with morphologic abnormalities
Carcinoma	Malignant tumor of epithelium
Carcinoma in situ	Malignant tumor of epithelium that does not penetrate basement membrane to underlying tissue
Desmoplasia	Excessive fibrous tissue formation in tumor stroma
Dysplasia	Abnormal atypical cellular proliferation
Metaplasia	Replacement of one type of adult cell or tissue by another (within the same germ cell line) that is not normally present in that site
Metastasis	Secondary, discontinuous malignant growth (spread), such as a lung metastasis of a colon carcinoma
Sarcoma	A mesenchymal (nonepithelial) malignant tumor

▶ Selected Risk Factors for Cancer

Geographic/Racial	Stomach cancer (Japan); hepatocellular carcinoma (Asia)	
Occupational/Environmental exposures	**Aflatoxin:** hepatocellular carcinoma **Alkylating agents:** leukemia, lymphoma, other cancers **Aromatic "–amines" and "azo–" dyes:** hepatocellular carcinoma **Arsenic:** squamous cell carcinomas of skin and lung, angiosarcoma of liver **Asbestos:** bronchogenic carcinoma, mesothelioma **Benzene:** leukemia	**Chromium and nickel:** bronchogenic carcinoma **Cigarette smoke:** multiple malignancies **CCl_4:** fatty change and centrilobular necrosis of the liver **Ionizing radiation:** thyroid cancer, leukemia **Naphthylamine:** bladder cancer **Nitrosamines:** gastric cancer **Polycyclic aromatic hydrocarbons:** bronchogenic carcinoma **Ultraviolet exposure:** skin cancers **Vinyl chloride:** angiosarcoma of liver
Age	Increases risk of most cancers (exceptions: Wilms, etc.)	
Hereditary predisposition	Familial retinoblastoma; multiple endocrine neoplasia, familial polyposis coli	
Acquired risk factors	Cervical dysplasia, endometrial hyperplasia, cirrhosis, ulcerative colitis, chronic atrophic gastritis	

▶ Mechanisms of Carcinogenesis

There are many proposed mechanisms of carcinogenesis. The most important mutations involve growth-promoting genes (proto-oncogenes), growth-inhibiting tumor suppressor genes, and genes regulating apoptosis.

Clinically Important Oncogenes

Proto-oncogenes are normal cellular genes involved with growth and cellular differentiation. **Oncogenes** are derived from proto-oncogenes by changing the gene sequence (resulting in a new gene product, oncoprotein) or a loss of gene regulation → overexpression of the normal gene product. Oncogenes lack regulatory control and are overexpressed → unregulated cellular proliferation.

Oncogene	Tumor	Gene Product	Mechanism of Activation
		Growth factor	
hst-1/int-2	Cancer of the stomach, breast, bladder, and melanoma	Fibroblast growth factor	Overexpression
sis	Astrocytoma	Platelet-derived growth factor	Overexpression
		Growth factor receptor	
erb-B1	SCC of lung	Epidermal growth factor receptor	Overexpression
erb-B2	Breast, ovary, lung	Epidermal growth factor receptor	Amplification
erb-B3	Breast	Epidermal growth factor receptor	Overexpression
ret	MEN II and III, familial thyroid (medullary) cancer	Glial neurotrophic factor receptor	Point mutation
		Signal transduction proteins	
abl	CML, ALL	*bcr-abl* fusion protein with tyrosine kinase activity	Translocation t(9;22)
Ki-ras	Lung, pancreas, and colon	GTP-binding protein	Point mutation
c-myc	Burkitt lymphoma	Nuclear regulatory protein	Translocation t(8;14)
L-myc	Small cell lung carcinoma	Nuclear regulatory protein	Amplification
N-myc	Neuroblastoma	Nuclear regulatory protein	Amplification
		Cell-cycle regulatory proteins	
bcl-1	Mantle cell lymphoma	Cyclin D_1 protein	Translocation t(11;14)
CDK4	Melanoma, glioblastoma multiforme	Cyclin-dependent kinase	Amplification
c-kit	Gastrointestinal stromal tumor (GIST)	KIT (also called CD117) is a stem cell factor receptor, a receptor tyrosine kinase	Overexpression or mutation

Definition of abbreviations: ALL, acute lymphocytic leukemia; CML, chronic myelogenous leukemia; MEN, multiple endocrine neoplasia; SCC, squamous cell carcinoma.

(Continued)

▶ Mechanisms of Carcinogenesis *(Cont'd.)*

Inactivation of Tumor Suppressor Genes

Tumor suppressor genes encode proteins that regulate and suppress cell proliferation by inhibiting progression through the cell cycle. Inactivation of these genes → uncontrolled cellular proliferation.

Gene	Chromosome	Tumors
VHL	3p25	Von Hippel-Lindau disease, renal cell carcinoma
WT-1	11p13	Wilms tumor
WT-2	11p15	Wilms tumor
Rb	13q14	Retinoblastoma, osteosarcoma
p53	17p13.1	Lung, breast, colon, etc.
BRCA-1	17q12-21	Hereditary breast and ovary cancers
BRCA-2	13q12-13	Hereditary breast cancer
APC	5q21	Adenomatous polyps and colon cancer
DCC	18q21	Colon cancer
NF-1	17q11.2	Neurofibromas
NF-2	22q12	Acoustic neuromas, meningiomas
p16	9p	Melanoma
DPC4	18q21	Pancreatic cancer

Failure of Apoptosis Is Another Cause of Cancer

bcl-2

- **Prevents apoptosis**
- Overexpressed in follicular lymphomas t(14:18) (chromosome 14 [immunoglobulin heavy chain gene]; chromosome 18 [*bcl-2*])

bax, bad, bcl-xS, bid

- **Promote apoptosis**
- ***p53*** promotes apoptosis in mutated cells by stimulating *bax* synthesis; inactivation → failure of apoptosis

c-myc

- Promotes cellular proliferation
- When associated with *p53* → promotes apoptosis; when associated with *bcl-2* → inhibits apoptosis

▶ Oncogenic Viruses

Specific virus	Human T-cell leukemia virus (HTLV-1)*	Hepatitis B (HBV) Hepatitis C (HCV)*	Epstein-Barr	Human papilloma virus (types 16, 18 in genital sites)	Human herpesvirus 8 (HHV-8)
Associated disease	Adult T-cell leukemia/ lymphoma	Hepatocellular carcinoma	Burkitt lymphoma, B-cell lymphoma, nasopharyngeal carcinoma	Cervical, vulvar, vaginal, penile, and anal carcinoma; some head and neck cancers	Kaposi sarcoma

*RNA oncogenic viruses (all other are DNA).

▶ Serum Tumor Markers

These are usually normal cellular components that are increased in neoplasms but may also be elevated in non-neoplastic conditions. Can be used for screening, monitoring of treatment efficacy, and detecting recurrence.

Marker	Associated Cancers
α-fetoprotein (AFP)	Hepatocellular carcinoma, nonseminomatous testicular germ-cell tumors
β-human chorionic gonadotropin (hCG)	Trophoblastic tumors, choriocarcinoma
Calcitonin	Medullary carcinoma of the thyroid
Carcinoembryonic antigen (CEA)	Carcinomas of the lung, pancreas, stomach, breast, colon
CA-125	Ovarian epithelial carcinoma
CA19-9	Pancreatic adenocarcinoma
Placental alkaline phosphatase	Seminoma
Prostatic acid phosphatase	Prostate cancer
Prostate-specific antigen (PSA)	Prostate cancer
S-100	Melanoma, neural-derived tumors, astrocytoma

▶ Paraneoplastic Syndromes

Syndrome	Neoplasm	Mechanism
Carcinoid syndrome	Carcinoid tumor (metastatic, bronchial, ovarian)	Serotonin, bradykinin
Cushing syndrome	Small cell carcinoma of the lung, neural tumors	ACTH, ACTH-like peptide
Hypercalcemia	Squamous cell carcinoma of the lung; breast, renal, and ovarian carcinomas	PTH-related peptide, TGF-α, TNF, IL-1
Lambert-Eaton myasthenic syndrome	Small cell carcinoma of the lung	Antibodies against presynaptic voltage-gated Ca^{2+} channels at the neuromuscular junction
Polycythemia	Renal cell carcinoma, hepatocellular carcinoma, cerebellar hemangioblastoma	Erythropoietin
SIADH	Small cell carcinoma of the lung; intracranial neoplasms	ADH

Definition of abbreviations: ACTH, adrenocorticotropic hormone; IL, interleukin; TGF, transforming growth factor; TNF, tumor necrosis factor; PTH, parathyroid hormone; SIADH, syndrome of inappropriate antidiuretic hormone

► Grading and Staging

Grade	• An estimate of the cytologic malignancy of a tumor, including the degree of anaplasia and number of mitoses. • Nuclear size, chromatin content, nucleoli, and nuclear-to-cytoplasmic ratio are all used.
Stage	• The clinical estimate of the extent of spread of a malignant tumor. Low stage means a localized tumor. Stage rises as tumors spread locally then metastasize. • **TNM** is typically used; **T** = size of **t**umor; **N** = **n**ode involvement; **M** = **m**etastases

► Cancer Incidence and Mortality

Incidence	**Males**	**Females**
	Prostate: 29%	Breast: 30%
	Lung and bronchus: 14%	Lung and bronchus: 14%
	Colon and rectum: 9%	Colon and rectum: 9%
Mortality	**Males**	**Females**
	Lung and bronchus: 28%	Lung and bronchus: 26%
	Prostate: 11%	Breast: 15%
	Colon and rectum: 8%	Colon and rectum: 9%

Pharmacology/Therapeutics

Chapter 9

Pharmacokinetics

► Key Concepts

Volume of distribution (apparent)	$V_d = \frac{Dose}{C^0}$ C^0= plasma concentration at time zero	• V_d estimates the fluid volume into which the drug has distributed (one needs to extrapolate plasma concentration at time zero). • The lower the C^0, the higher the V_d, and vice versa (i.e., inversely related). • Drugs stored in nonfluid compartments like fat may have a V_d greater than TBW (e.g., lipid-soluble drugs, thiopental). • Drugs that bind strongly to plasma proteins have a V_d that approaches plasma volume. • Approximate V_d values (weight 70 kg)—plasma volume (3 L), blood (5 L), extracellular fluid (12–14 L), TBW (40–42 L)
Clearance	$Cl = \frac{\text{Rate of drug elimination}}{\text{Plasma drug concentration}}$	Clearance is the theoretical volume of blood totally cleared of drug/unit time. It represents the ratio of drug elimination to its plasma concentration. For a drug with first-order elimination, clearance is constant.
Elimination	Elimination is synonymous with termination of drug action. The two primary mechanisms are hepatic metabolism to inactive metabolites or renal excretion.	

First-Order Elimination		Zero-Order Elimination	
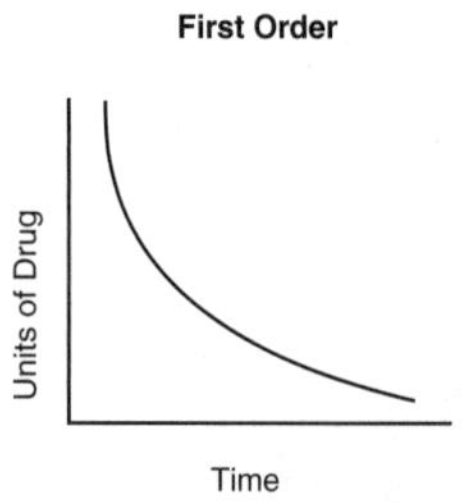 First Order Units of Drug Time	A constant **fraction** of drug is eliminated with time. Most drugs have first-order elimination.	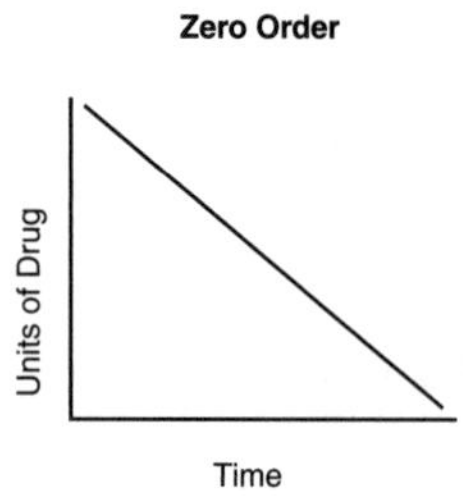 Zero Order Units of Drug Time	A constant **amount** of drug is eliminated with time. Half-life is not applicable. *Examples:* ethanol (except at low blood levels), phenytoin, and aspirin (high doses)

Half-life	$t_{1/2} = \frac{0.7 \times V_d}{Cl}$	Half-life is the time it takes for the amount or concentration of a drug to fall to 50% of a previous estimate. It is constant for drugs eliminated by first-order kinetics.
Steady state	Plasma drug concentrations remain relatively constant over time. As a rule, **it takes 4 to 5 half-lives to achieve steady state** (rise in concentration: 50% at 1 half-life; 75% at 2 half-lives; 87.5% at 3 half-lives; 93.75% at 4 half-lives. For elimination, decline in concentration is similar: 50% remains at 1 half-life; 25% at 2 half-lives; 12.5% at 3 half-lives; 6.25% at 4 half-lives.)	
Bioavailability (F)	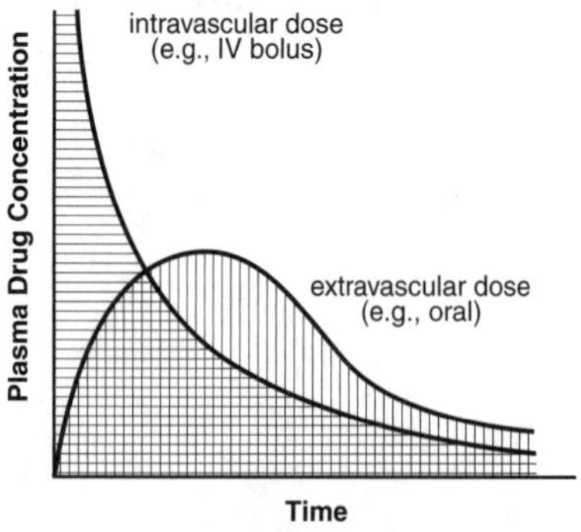 Plasma Drug Concentration intravascular dose (e.g., IV bolus) extravascular dose (e.g., oral) Time	The fraction of the administered drug that reaches the systemic circulation. $F = \frac{AUC_{PO}}{AUC_{IV}}$ (AUC = area under the curve, PO = oral, IV = intravenous) By definition, intravenous drug administration has an F of 1 (100%).

Definition of abbreviations: AUC, area under the curve; F, bioavailability; IV, intravenously; PO, by mouth; TBW, total body weight.

(Continued)

► Key Concepts *(Cont'd.)*

First-pass effect	With oral administration, drugs are absorbed into the portal circulation and initially distributed to the liver. For some drugs, their rapid hepatic metabolism decreases bioavailability. This can be avoided by giving the drug by an alternate route (e.g., sublingual, transdermal).	
Maintenance dose	$MD = \frac{Cl \times C_P}{F}$	A maintenance dose is given to maintain a relatively constant plasma concentration. It is equal to the rate of elimination.
Loading dose	$LD = \frac{V_d \times C_P}{F}$	If therapeutic plasma concentrations are needed quickly and the V_d is large, a loading dose may be given to produce the desired drug levels (fill up the V_d) without the typical delay of 4 to 5 half-lives.

Definition of abbreviation: C_P, desired plasma concentration

Drug Metabolism

Biotransformation is the conversion of drugs to more water-soluble drugs that can be more readily excreted. There are two main types of metabolic reactions: **phase I** and **phase II**.

Phase I	The parent drug becomes more water-soluble by oxidation, reduction, or hydrolysis by **cytochrome P450** isozymes (also called **mixed function oxidases**) located in **smooth endoplasmic reticulum** in liver, and in the GI, lungs, and kidney to a lesser extent. Often the drug is converted to compounds with little or no pharmacologic activity; in other instances, the metabolites retain pharmacologic activity.			
CYP450	**Substrate Example**	**Inducers**	**Inhibitors**	**Genetic Polymorphisms**
1A2	Theophylline Acetaminophen	Aromatic hydrocarbons (smoke) Cruciferous vegetables	Quinolones Macrolides	No
2C9	Phenytoin Warfarin	General inducers*	—	Yes
2D6	Many cardiovascular and CNS drugs	None known	Haloperidol Quinidine	Yes
3A4	60% of drugs in PDR	General inducers*	General inhibitors† Grapefruit juice	No
Phase II	**Conjugation reactions** may follow phase I or can occur on the parent drug. Examples include **glucuronidation**, **acetylation**, **sulfation**, **glutathione conjugation**, and **methylation**. Phase II reactions lead to inactive molecules, which are polar and usually renally excreted. All phase II reactions use enzymes called transferases.			

Definition of abbreviations: PDR; *Physician's Desk Reference.*

* General inducers: anticonvulsants (barbiturates, phenytoin, carbamazepine), antibiotics (rifampin), chronic alcohol, glucocorticoids.

† General inhibitors: antiulcer medications (cimetidine, omeprazole), antibiotics (erythromycin, clarithromycin, macrolides) protease inhibitors, azole antifungals, acute alcohol.

Pharmacodynamics

► Key Concepts

Agonist	An agonist is a drug that binds to a receptor and activates it.
Partial agonist	A partial agonist is a drug that binds to a receptor but does not elicit a 100% response. It will elicit a partial response when administered alone. When administered with a full agonist, it acts as an **antagonist** because it displaces the full agonist from the receptor.
Graded dose–response curve	A graded dose–response curve depicts increasing responses to increasing drug doses. **Potency:** the measure of how much drug is required to produce a given effect. It is typically expressed as the concentration that can elicit a 50% response (EC_{50}). **Efficacy:** the maximal effect a drug can produce. It is also known as maximal efficacy. A partial agonist has a lower efficacy than a full agonist. It can be less potent (*C*) or more potent (*A*) than a full agonist (*B*). *(See figure to left.)*
Competitive antagonist	A competitive antagonist binds to the receptor without activating the effector system. It **can** be overcome by increasing the agonist dose. This is seen as a **parallel right shift** in the dose–response curve.
Noncompetitive antagonist	A noncompetitive antagonist binds to the receptor without activating the effector system. It **cannot** be overcome by increasing the agonist dose. This is seen as a **downward shift** in the dose–response curve.
K_d	The concentration of drug that binds to 50% of the receptors
Physiologic antagonist	Substances that produce opposing physiologic effects, but do not exert their mechanism of action at the same receptor
Quantal dose–response curve	A quantal dose–response curve depicts the dose of drug needed to produce a predetermined response in a population. It is the percent of population responding versus log (dose).
Therapeutic index (TI)	The therapeutic index is the ratio of the drug dose required to produce a toxic or lethal effect to the dose needed for a therapeutic effect. $TI = \frac{TD_{50}}{ED_{50}}$ or $\frac{LD_{50}}{ED_{50}}$ ED_{50}, TD_{50}, and LD_{50} are the median effective, toxic, and lethal doses in 50% of the studied population, respectively. **↑ TI, safe drug; ↓ TI, unsafe drug**

► Signal Transduction

G-protein–coupled receptors*	These receptors consist of one polypeptide with seven-transmembrane–spanning regions. When bound by an agonist, the trimeric (α, β, γ) GTP-binding protein (G protein) is activated. The α component usually interacts with the effector molecules. The most common G proteins and their receptors are as follows:

G protein	Receptors	Effector	Second messenger response
G_s	β_1, β_2,D_1, H_2, V_2	Adenylyl cyclase	↑ cAMP
G_i	α_2, M_2, D_2	Adenylyl cyclase	↓ cAMP
G_q	α_1, M_1, M_3, H_1, V_1	Phospholipase C	↑ IP_3 (↑ Ca_i^{2+}), DAG

Ligand-gated channels	Activation of receptors within ion channels may directly open the channel, e.g., **nicotinic ACh** (Na^+/K^+), **$GABA_A$** (Cl^-), **NMDA** (Ca^{2+}/Na^+) receptors, or may regulate the ion channel's response to an agonist, e.g., benzodiazepine or barbiturate sites on the $GABA_A$ receptor.
Intracellular receptors	Lipid-soluble agents diffuse across the plasma membrane to bind intracellular receptors (e.g., **steroid** receptors, **thyroid** receptors). This permits receptor binding to nuclear DNA sequences that modify gene expression.
Ligand-regulated transmembrane enzymes*	These receptors have extracellular ligand binding sites and intracellular catalytic sites. Ligand binding causes dimerization and activates the enzyme activity (often a **tyrosine kinase**). *Examples:* **insulin** and **growth factor** receptors.
Transmembrane receptors that activate a separate tyrosine kinase	These also form dimers when activated, then activate a separate cytoplasmic tyrosine kinase (Janus kinases; JAKs). The kinase phosphorylates STAT factors (signal transducers and activators of transcription). STAT dimers then regulate transcription. *Examples:* **cytokine** and **growth hormone** receptors.

*Figures of these pathways can be found on pages 84–85, Cell Signaling.

Toxicology

► Management of the Poisoned Patient

There are many facets of managing the poisoned patient, including supportive care, poison identification, decontamination, enhancing elimination, and administration of antidotes.

Decontamination	**Syrup of ipecac** (induces vomiting), **gastric lavage** ("stomach pumping"), **activated charcoal** (absorbs drug)
Enhancing elimination	• Many drugs are weak acids or bases and can be in nonionized or ionized forms. **Nonionized** drugs are **lipid soluble** (cross membranes); **ionized drugs** are **water soluble** (renally excreted) • The urine pH can be manipulated to increase drug elimination **Weak acid:** R-COOH ↔ R-COO^- + H^+ (e.g., aspirin) (Increase excretion by giving **bicarbonate** to trap drug in basic environment) **Weak base:** R-NH^+_3 ↔ R-NH_2 + H^+ (e.g., amphetamines) (Increase excretion with **NH_4Cl** to trap drug in acidic environment)

► Signs, Symptoms, and Treatments for Common Toxic Syndromes

Compounds	Signs and Symptoms of Toxicity	Treatment
AChE inhibitors	Miosis, salivation, sweating, GI cramps, diarrhea, seizures, anxiety/agitation, muscle fasciculations followed by muscle paralysis (including diaphragm), respiratory failure, coma	Respiratory support; atropine plus **pralidoxime** (2-PAM, AChE-reactivating agent for organophosphate inhibitors)
Atropine and muscarinic blockers	↑ HR, ↑ BP, hyperthermia (hot, dry skin), ↓ GI motility, urinary retention, mydriasis, delirium, hallucinations, seizures, coma	Control CV symptoms and hyperthermia plus **physostigmine** (crosses BBB)
Carbon monoxide (>10% carboxyHb)	Headache, dizziness, nausea/vomiting, shortness of breath, chest pain, ↑ HR, ↓ BP, arrhythmias, confusion, coma	Hyperbaric O_2 and decontamination
CNS stimulants	Anxiety/agitation, hyperthermia (warm, sweaty skin), mydriasis, ↑ HR, ↑ BP, psychosis, seizures	Control CV symptoms, hyperthermia, and seizures; BZs or antipsychotics may be beneficial
Opioid analgesics	Lethargy, sedation, coma, ↓ HR, ↓ BP, miosis, hypoventilation, respiratory failure, ↓ GI motility	Ventilatory support; **naloxone** at frequent intervals
Salicylates	Confusion, lethargy, hyperventilation, ototoxicity, hyperthermia, dehydration, hypokalemia, acid-base disturbances, seizures, coma	Correct acidosis and electrolytes, urinary alkalinization, possible hemodialysis
Sedative-hypnotics and ethanol	Disinhibition, lethargy, stupor, coma, ataxia, nystagmus, hypothermia, respiratory failure	Ventilatory support—**flumazenil** if BZs implicated
SSRIs	Agitation, confusion, coma, muscle rigidity, hyperthermia, seizures, autonomic instability	Control hyperthermia and seizures—possible use of cyproheptadine and BZs
Tricyclic antidepressants	Mydriasis, hyperthermia (hot, dry skin), **3 Cs** (convulsions, coma, and cardiotoxicity)	Control seizures and hyperthermia, correct acidosis plus possible antiarrhythmics

Definition of abbreviations: AChE, acetylcholinesterase; BBB, blood–brain barrier; BP, blood pressure; BZs, benzodiazepines; CNS, central nervous system; CV, cardiovascular; GI, gastrointestinal; HR, heart rate; SSRIs, selective serotonin reuptake inhibitors.

▶ Signs, Symptoms, and Treatments for Heavy Metal Poisoning

Metals	Source	Signs and Symptoms	Treatment*
Arsenic	Wood preservatives, insecticides, occupational, environmental	**Acute:** GI distress, garlic breath, "rice water" stools, hypotension **Chronic:** paresthesias, stocking-glove neuropathy, pallor from anemia, skin **Arsine gas:** headache, N/V, abdominal pain, dyspnea, jaundice	Dimercaprol, penicillamine, succimer Supportive care
Iron	Iron supplements, multivitamin supplements	Occurs mainly in children; severe GI distress, GI bleeding, hepatocellular injury, seizures, shock, coma	Deferoxamine
Lead	Tap water, leaded paint chips, glazed kitchenware, etc.	**Acute:** abdominal (colic, N/V), CNS (headaches, ataxia, seizures, coma, encephalopathy) **Chronic:** anemia (inhibits heme synthesis), neuropathy (wristdrop), GI symptoms, nephropathy, developmental delays, growth retardation, decreased fertility, stillbirths	EDTA, dimercaprol, succimer
Mercury	Dental amalgams, electroplating, batteries, wood preservatives, occupational, contaminated foods, old thermometers	**Acute:** vapor inhalation (elemental)—chest pain, dyspnea, pneumonitis, confusion **Acute:** tremors, gingivitis, CNS disturbances, GI distress, renal failure **Chronic:** renal failure, dementia, acrodynia **Organic:** CNS (paresthesias, auditory and visual loss, movement disorders)	Dimercaprol, penicillamine, succimer

Definition of abbreviations: CNS, central nervous system; GI, gastrointestinal; EDTA, ethylenediaminetetraacetic acid; N/V, nausea and vomiting.

*Need to remove patient from source; decontamination is also important part of management.

► Summary of Antidotes

Antidote	Type of Poisoning
Acetylcysteine	Acetaminophen
Atropine	AChE inhibitors
Deferoxamine	Iron
Digoxin antibodies	Digoxin
Dimercaprol (BAL)	Arsenic, mercury (inorganic, elemental), lead (with EDTA if severe poisoning); succimer and unithiol now used more frequently
EDTA	Primarily for lead poisoning
Esmolol	Theophylline, caffeine, β agonists
Ethanol	Methanol, ethylene glycol
Flumazenil	Benzodiazepines, zolpidem, suggested for zaleplon
Fomepizole	Methanol, ethylene glycol
Glucagon	β blockers
Naloxone	Opioid analgesics
Oxygen	Carbon monoxide
Penicillamine	Copper, Wilson disease, adjunctive in iron and arsenic intoxication
Physostigmine	Anticholinergics: atropine, antihistamine, antiparkinsonian—*not* tricyclic antidepressants
Pralidoxime (2-PAM)	Organophosphate cholinesterase inhibitors
Protamine	Heparins
Succimer	Lead, arsenic, mercury
Vitamin K	Warfarin and coumarin anticoagulants
Activated charcoal	Nonspecific: all oral poisonings except Fe, CN, Li, solvents, mineral acids, or corrosives

Definition of abbreviations: AChE, acetylcholinesterase; EDTA, edetate calcium disodium, ethylenediaminetetraacetic acid.

Autacoids

Autacoids are endogenously produced substances that do not fit well in other classifications such as hormones or neurotransmitters. The autacoids include histamine, serotonin, vasoactive peptides, and prostaglandins (see page 54).

▶ Histamine

Synthesis	Histidine —(Histidine decarboxylase)→ Histamine
Location	Circulating **basophils** and tissue **mast cells**, GI tract, skin, lung
Degranulation	• Liberation of histamine from mast cells via IgE-mediated hypersensitivity reactions, trauma, drugs, and venoms • ↓ cAMP favors release; ↑ cAMP (via β-adrenergic and glucocorticoid stimulation) ↓ release • Other substances released include kallikrein, kinins, prostaglandins, SRS-A

Histamine Receptors			
Receptor	**Second Messenger**	**Distribution**	**Action**
H_1	G_q; ↑ IP_3, DAG	Smooth muscle	Vasodilation (via NO), ↑ bronchoconstriction, activates nociceptive receptors
H_2	G_s; ↑ cAMP	Stomach, smooth muscle	↑ gastric acid secretion

Pharmacologic Agents			
Agent	**Mechanism of Action**	**Clinical Uses**	**Notes/Toxicity**
H_1 antagonists: Diphenhydramine Promethazine Chlorpheniramine Hydroxyzine Fexofenadine* Loratadine* Cetirizine*	Competitively inhibit H_1 receptors	• Allergic reactions • Motion sickness • OTC: sleep aids and cold medications	• Muscarinic block (sedation, dry mouth)
H_2 antagonists: Cimetidine Ranitidine Famotidine Nizatidine	Competitively inhibit H_2 receptors → reduce gastric acid secretion†	• Peptic ulcer disease† • GERD • Zollinger Ellison syndrome	*Cimetidine:* P450 inhibition, antiandrogen effect

Definition of abbreviations: GERD, gastroesophageal reflux disease; GI, gastrointestinal; IgE, immunoglobulin E; SRS-A, slow-reacting substance of anaphylaxis.

*No CNS entry (less sedating) and no muscarinic block; loratadine (Claritin®), fexofenadine (Allegra®), cetirizine (Zyrtec®)

†Not as efficacious as proton pump inhibitors

▶ Serotonin

Synthesis and degradation	Tryptophan —(Tryptophan hydroxylase)→ 5HT —(MAO_A)→ 5HIAA
Location	Enterochromaffin cells in the gut, CNS neurons, platelets (primarily just storage)

Serotonin Receptors

Receptor	Second Messenger	Action
$5HT_{1(A, B, D, E, F)}$	G_i; ↓ cAMP	• CNS • Behavioral effects (sleep, feeding, thermoregulation, anxiety) • Vasoconstriction
$5HT_{2(A, B, C)}$	G_q; ↑ IP_3, DAG	• CNS • Behavioral effects • Smooth muscle contraction • Platelet aggregation
$5HT_3$	Ion channel	• CNS (area postrema), PNS • Emesis • Anxiety
$5HT_4$	G_s; ↑ cAMP	• CNS: neuronal excitation • GI motility

Pharmacologic Agents

Agent	Mechanism of Action	Clinical Uses	Notes/Toxicity
Sumatriptan Naratriptan Other "-triptans"	$5HT_{1D}$ agonist	Migraine headaches	—
Buspirone	$5HT_{1A}$ partial agonist	Anxiety disorders	Lower addiction potential than other drugs like benzodiazepines
Ondansetron Granisetron Other "-setrons"	$5HT_3$ antagonist	Emesis	Mainly for postoperative or chemotherapy-induced nausea and vomiting
SSRIs: Citalopram Fluoxetine Fluvoxamine Paroxetine Sertraline	Selectively block 5HT reuptake	• Anxiety disorders • Depression	• Sexual dysfunction • Drug interactions: **serotonin syndrome** with MAO inhibitors, TCAs, meperidine, and St. John's wort
Ergot alkaloids: Ergonovine Ergotamine Methysergide Bromocriptine* LSD	Agonists, partial agonists, and antagonists at 5HT and α-adrenergic receptors; some are agonists at DA receptors*	• Postpartum hemorrhage (ergonovine, ergotamine) • Migraine headaches (ergotamine [for acute attacks], methysergide [prophylaxis]) • Parkinson disease, hyperprolactinemia (bromocriptine, pergolide) • Abuse (LSD)	**Ergotism** ("St. Anthony's Fire"): – Mental disorientation – Hallucination – Convulsions – Muscle cramps – Dry gangrene of extremities
MAO inhibitors: Phenelzine Tranylcypromine	Inhibit metabolism of 5HT, NE, and DA by MAO	Depression	• Non-selective MAO inhibitors • Tyramine (red wine, cheese) ingestion → hypertensive crisis

Definition of abbreviations: DA, dopamine; 5-HT, 5-hydroxytrypamine; 5-HIAA, 5-hydroxyindoleacetic acid; LSD, D-lysergic acid diethylamide; MAO, monoamine oxidase; SSRI, selective serotonin reuptake inhibitor; TCA, tricyclic antidepressants.

► Angiotensin II

Synthesis and Actions

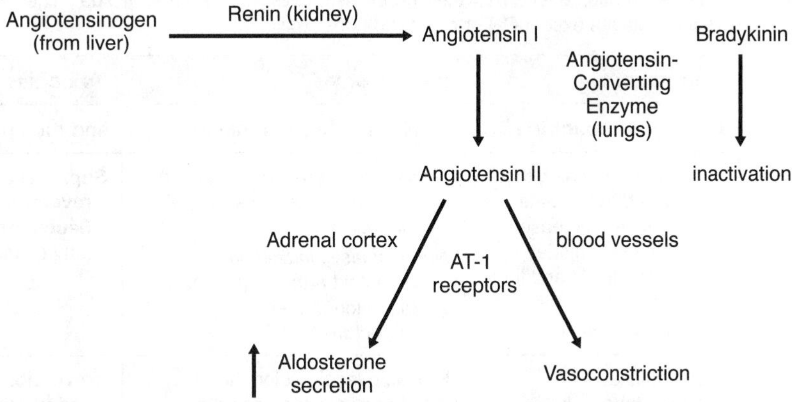

Angiotensinogen (from liver) → Renin (kidney) → Angiotensin I

Angiotensin I → Angiotensin-Converting Enzyme (lungs) → Angiotensin II

Bradykinin → inactivation

Angiotensin II → AT-1 receptors

Adrenal cortex → ↑ Aldosterone secretion

blood vessels → Vasoconstriction

Pharmacologic Agents

Agent	Mechanism of Action	Clinical Uses	Notes/Toxicity
ACE Inhibitors Captopril Enalapril* Benazepril*	Inhibits conversion of AT-I to AT-II by ACE Also inhibits the inactivation of bradykinins by kininase II, potentiating their vasodilatory effect	• Hypertension • CHF • Myocardial infarction • Diabetic nephropathy	• Dry cough • Angioedema • Hypotension • Acute renal failure • ↓ Aldosterone • Hyperkalemia • Fetal renal toxicity; contraindicated in pregnancy
AT_1-receptor antagonists: Losartan Valsartan	Competitively inhibit receptor for AT-II	• Hypertension • CHF • Diabetic nephropathy	• Common alternative to ACE-inhibitors if patient cannot tolerate adverse reactions (e.g., cough) • Does not block BK degradation • Similar BP effects and toxicity profile as ACE inhibitors (but no cough)
Renin inhibitor: Aliskiren	Blocks formation of AT-I (and therefore AT-II) via renin inhibition	Hypertension	• Similar BP effects as ACE inhibitors • Does not block BK degradation • Headache, diarrhea

Definition of abbreviations: ACE, angiotensin-converting enzyme; AT-I, angiotensin I; AT-II, angiotensin II; BK, bradykinin; CHF, congestive heart failure.

*Ester prodrugs of ACE inhibitor converted to active form by liver

Antineoplastic Agents

These agents are used to treat various neoplasms. Although the mechanism of action varies, each agent hinders cell replication in some way. Specificity relies on differential effect between neoplastic cells and normal tissue. The mechanism of action may be cell-cycle specific (affecting cells in all stages except G_0) or cell-cycle nonspecific.

Class	Mechanism	Indications	Toxicities
Antimetabolites	**Cell-cycle specific (CCS).** Inhibit synthesis of nucleic acids and thus protein synthesis.		
Methotrexate	• A folic acid analog that inhibits dihydrofolate reductase; decreased dTMP levels hinder DNA synthesis and thus protein synthesis • **S-phase** specific	*Neoplastic indications:* leukemia, lymphomas, breast cancer, choriocarcinoma *Nonneoplastic indications:* rheumatoid arthritis, psoriasis, termination of pregnancy (e.g., ectopic)	**Suppresses bone marrow reversibly**; folinic acid (**leucovorin**) is used to **"rescue"**; fatty change in liver
5-Fluorouracil (5-FU)	• Pyrimidine antimetabolite is converted to **5-F-dUMP**, which when bound to folic acid, **inhibits thymidylate synthase**. This prevents dTMP synthesis, thus inhibiting DNA and protein synthesis. • **S-phase** specific	Breast, ovarian, colon, head and neck cancers, basal cell carcinomas and keratoses (use topically)	**Irreversible myelosuppression** and **photosensitivity**, GI irritation, alopecia
Cytarabine (Ara-C)	• Pyrimidine antimetabolite • Inhibits DNA polymerases • **S-phase** specific	Acute leukemias	BMS, GI irritation, ↑ doses → neurotoxicity
6-Mercaptopurine (6-MP)	• Activated by hypoxanthine-guanine phosphoribosyltransferase (**HGPRT**) • **Inhibits purine synthesis**, inhibiting nucleic acid synthesis • **S-phase** specific	Aacute leukemias, CML, non-Hodgkin lymphoma	• BMS, hepatotoxicity—coadministration with allopurinol increases toxicity (6-MP metabolized by xanthine oxidase) • Azathioprine forms 6-MP
Alkylating Agents	**Cell cycle-nonspecific (CCNS).** This class of agents causes alkylation of DNA, leading to cross-linking, abnormal base pairing, or DNA strand breakage.		
Busulfan	Alkylates DNA	CML	**Pulmonary fibrosis, hyperpigmentation,** and adrenal insufficiency
Cyclophosphamide	Alkylates DNA—attacks guanine N7, induces cross-linking	Non-Hodgkin lymphoma; ovarian and breast cancers; neuroblastoma	BMS and **hemorrhagic cystitis** (can be ↓ by **mesna**, which traps **acrolein**, a toxic metabolite)
Nitrosoureas (lomustine, carmustine)	• Alkylates DNA • Crosses blood–brain barrier	Brain tumors	**Neurologic**

Definition of abbreviations: BMS, bone marrow suppression; CML, chronic myelogenous leukemia.

(Continued)

► Antineoplastic Agents (*Cont'd.*)

Class	Mechanism	Indications	Toxicities
Cisplatin, carboplatin	Alkylates DNA	Testicular, bladder, lung, and ovarian carcinomas	**Nephrotoxic**, neurotoxicity (**deafness, tinnitus**)
Procarbazine	Alkylates DNA	Hodgkin disease (MOPP*)	BMS, pulmonary toxicity, neurotoxic, leukemogenic
Antibiotics	Structurally dissimilar subclass of drugs. Mechanisms of action vary.		
Doxorubicin	Intercalates DNA, creating breaks. Hinders DNA replication and transcription.	Hodgkin lymphoma (ABVD†), breast, endometrial, lung, ovarian CAs, myeloma, sarcomas	**Cardiotoxic—dexrazoxane** (treatment to inhibit free radical formation; may protect against cardiotoxicity), BMS, alopecia, GI distress
Bleomycin	• Generates free radicals → DNA strand scission • **G_2 phase**	Lymphomas, testicular, skin CA	**Pulmonary fibrosis**, mucocutaneous reactions (blisters, alopecia), hypersensitivity reactions
Hormones/ Hormone Antagonists	May inhibit hormone-dependent tumor growth.		
Prednisone	Induces apoptosis of lymphoid cells	Chronic lymphocytic leukemia (CLL), Hodgkin lymphoma (MOPP*), autoimmune disease	Typical symptoms of glucocorticoid excess, including **Cushing syndrome**
Tamoxifen	**Selective estrogen receptor modulator (SERM)**. Prevents estrogen from binding estrogen receptor–positive breast CA cells, leading to involution of estrogen-dependent tumors.	Breast cancer	**Hot flashes**, increased risk of **endometrial carcinoma**
Plant Alkaloids	Cell-cycle specific drugs. Most prevent the assembly of microtubules and the formation of the mitotic spindle.		
Vinblastine	• Inhibits microtubule/ spindle formation • **M-phase** specific	Lymphoma, Wilms tumor, choriocarcinoma	BMS
Vincristine	• Inhibits microtubule/ spindle formation • **M-phase** specific	Same as vinblastine, MOPP* (is Oncovin)	Neurotoxic, GI distress
Paclitaxel	• Stabilizes microtubules so that spindle cannot break down • **M-phase** specific	Ovarian and breast carcinomas	BMS
Etoposide	• Inhibits topoisomerase II, ↑ DNA degradation • **Late S/early G_2** phase	Small cell carcinoma, prostate cancer, testicular carcinoma	BMS, GI irritation, alopecia

†ABVD: Adriamycin® (doxoribicin), bleomycin, vinblastine, decarbazine
*MOPP: mechlorethamine, vincristine (Oncovin®), procarbazine, prednisone

Natural Medicinals and Nutritional Supplements

Please note that these substances are **not FDA-approved** for any conditions and there are variable amounts of evidence regarding efficacy. However, many of your patients may be taking herbal medicines such as these, and familiarity with these agents is useful.

► Natural Medicinals

Name	Medicinal Use(s)	Possible Mechanism(s)	Side Effects/Interactions
Echinacea	↓ Cold symptoms	↑ ILs and TNF	GI distress, dizziness, headache
Ephedra (Ma Huang)	Bronchodilator, cold symptoms, mild CNS stimulant	Indirect acting sympathomimetic	Insomnia, palpitations, tachycardia; higher doses: hypertension, cardiac arrhythmias, toxic psychosis
Garlic	↓ Cholesterol, atherosclerosis	Inhibits HMG-CoA reductase and ACE	Allergies, hypotension, antiplatelet actions; caution advised when used with anticoagulants
Gingko	Intermittent claudication; cognitive improvement	Antioxidant, free radical scavenger, ↑ NO	Anxiety, GI distress, insomnia, antiplatelet actions; **caution advised when used with anticoagulants**
Ginseng	Possible ↑ in mental and physical performance	Unknown	Insomnia, nervousness, hypertension, mastalgia, vaginal bleeding
Milk thistle	Limits hepatic injury from *Amanita* mushroom poisoning, viral hepatitis, alcohol, acetaminophen	Inhibits P450	None reported
Saw palmetto	BPH treatment	5α-reductase inhibitor and androgen receptor antagonist	GI pain, ↓ libido, headache, hypertension
St. John's wort	Depression	May enhance brain 5HT functions	Major drug interactions: **serotonin syndrome** with SSRIs, MAOIs; induces P450, leading to ↓ effects of multiple drugs

► Purified Nutritional Supplements

Name	Medicinal Use(s)	Side Effects
Coenzyme Q10 (ubiquinone)	Hypertension, coronary artery disease, chronic stable angina, CHF	Well-tolerated, some GI disturbance
Dehydroepiandrosterone (DHEA)	Androgen precursor advocated for treatment of AIDS (↑ CD4 in females), Alzheimer disease and "aging," diabetes, hypercholesterolemia, and SLE (↓ in symptoms and "flare-ups" in females)	*Females:* androgenization and concern regarding CV disease and breast cancer *Males:* feminization in young and concern in elderly regarding BPH and cancer
Glucosamine	Osteoarthritis	Well tolerated, some GI disturbances
Melatonin	Serotonin derivative used for "jet-lag" and sleep disorders	Drowsiness, sedation, headache. Contraindicated in pregnancy, in women trying to conceive (↓ LH), and in nursing mothers (↓ prolactin)

Drug Development and Testing

Preclinical	IND	Phase 1	Phase 2	Phase 3	NDA	Phase 4
In vitro testing; animal testing (at least 2 different species)	Proposal to the FDA for human testing; includes preclinical data and proposals for the clinical trials	25-50 normal volunteers[†]	100-200 patients	1000-5000 patients; double-blind studies used	Request to market new agent	Post-marketing surveillance (if FDA approved NDA)
Biologic activity and safety		Safety and pharmacokinetics	Efficacy	Efficacy, side effects		Physicians report toxicities that may be now apparent with a larger sample size.

**Definition of abbreviations:* IND: Investigational New Drug application; NDA: New Drug Application

[†] If the agent is known to be toxic (e.g., antineoplastic, AIDS drug), volunteer patients will be used instead

Section II

Organ Systems

The Nervous System

Chapter 1

Development of the Nervous System

Neurulation

- **Neurulation** begins in the third week of fetal development.
- The **notochord** induces the overlying ectoderm to form the neural plate.
- By end of the third week, **neural folds** grow over midline and fuse to form **neural tube**.
- During closure, **neural crest** cells form from neuroectoderm.

- **Neural tube** → brain and spinal cord (plus lower motoneurons, preganglionic neurons)
- Brain stem and spinal cord have an **alar** plate (**sensory**) and a **basal** plate (**motor**); plates are separated by the **sulcus limitans**.
- **Neural tube** → 3 primary vesicles → 5 primary vesicles

- **Neural crest** → sensory and postganglionic neurons

- **Peripheral NS (PNS):** cranial nerves (12 pairs) and spinal nerves (31 pairs)

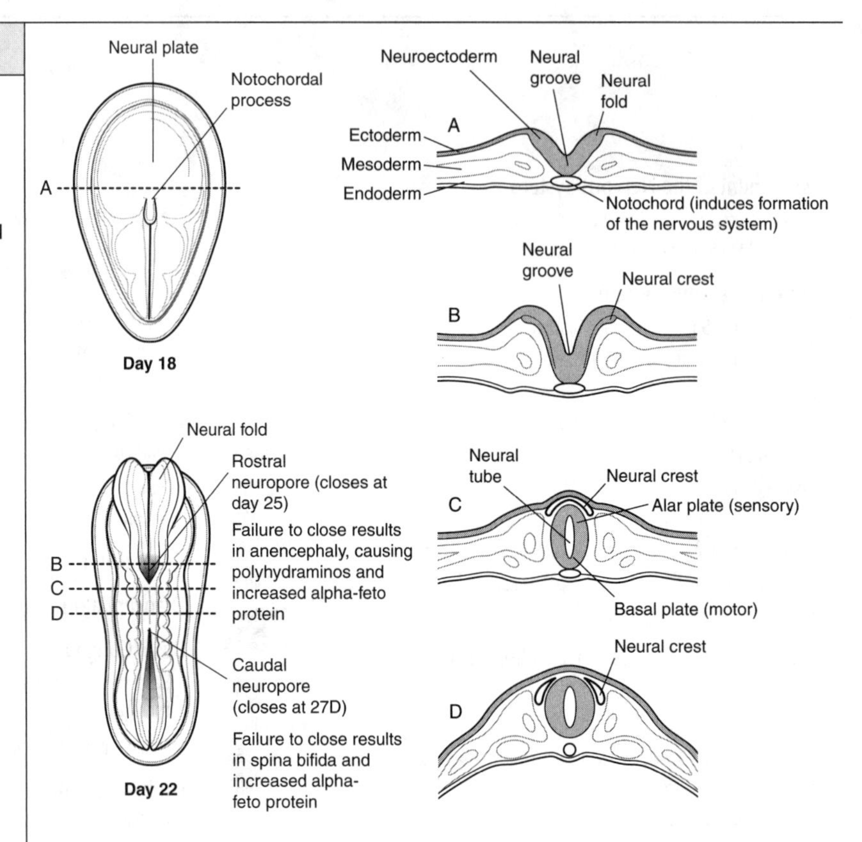

▶ Central Nervous System Development

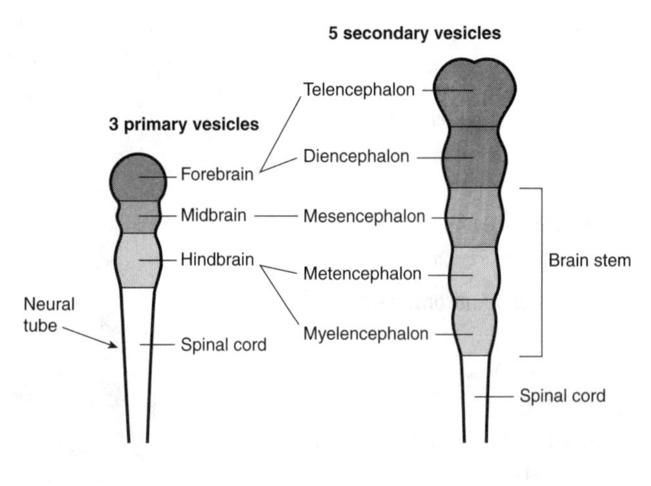

Adult Derivatives		
Structures		**Ventricles**
Telencephalon	Cerebral hemispheres, most of basal ganglia	Lateral ventricle
Diencephalon	Thalamus, hypothalamus, subthalamus, epithalamus (pineal gland), retina	Third ventricle
Mesencephalon	Midbrain	Cerebral aqueduct
Metencephalon	Pons, cerebellum	Fourth ventricle
Myelencephalon	Medulla	

► Congenital Malformations of the Nervous System

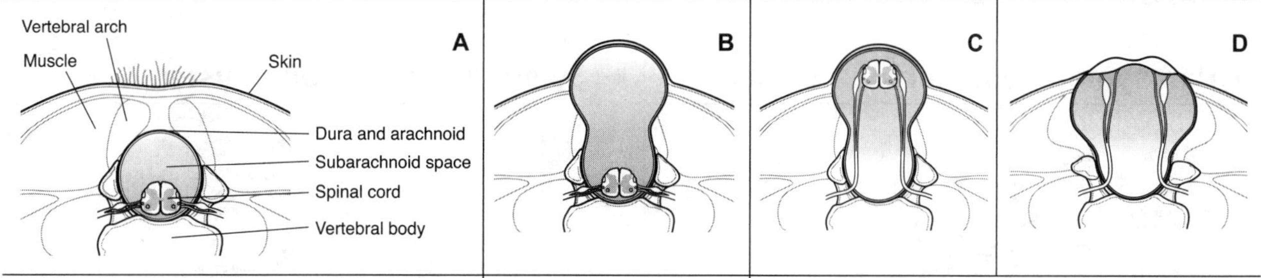

Condition	Types	Description
Anencephaly	—	Failure of **anterior neuropore** to close Brain does not develop Incompatible with life Increased AFP
Spina bifida	Failure of **posterior neuropore** to close	
	Spina bifida occulta **(Figure A)**	Mildest form **Vertebrae** fail to form around spinal cord **No** increase in **AFP** Asymptomatic
	Spina bifida with meningocele **(Figure B)**	**Meninges** protrude through vertebral defect Increase in AFP
	Spina bifida with meningomyelocele **(Figure C)**	**Meninges** and **spinal cord** protrude through vertebral defect Increase in AFP
	Spina bifida with myeloschisis **(Figure D)**	Most severe **Spinal cord** can be seen **externally** Increase in AFP
A Vertebral arch Muscle Skin Dura and arachnoid Subarachnoid space Spinal cord Vertebral body	B	C D
Arnold-Chiari malformation	Type I	Most common Mostly **asymptomatic** Downward displacement of cerebellar tonsils through foramen magnum
	Type II	More often **symptomatic** Downward displacement of **cerebellar vermis** and **medulla** through foramen magnum Compression of IV ventricle → obstructive **hydrocephaly** Frequent lumbar **meningomyelocele** Frequent association with **syringomyelia**
Dandy-Walker malformation		Failure of foramina of Luschka and Magendie to open → **dilation of IV ventricle** Agenesis of cerebellar vermis and splenium of the corpus callosum
Hydrocephalus		Most often caused by stenosis of cerebral aqueduct CSF accumulates in ventricles and subarachnoid space Increased head circumference
Holoprosencephaly		Incomplete separation of cerebral hemispheres One ventricle in telencephalon Seen in trisomy 13 (Patau)

Definition of abbreviation: AFP, α-fetoprotein.

Peripheral Nervous System

► Autonomic and Somatic Nervous Systems

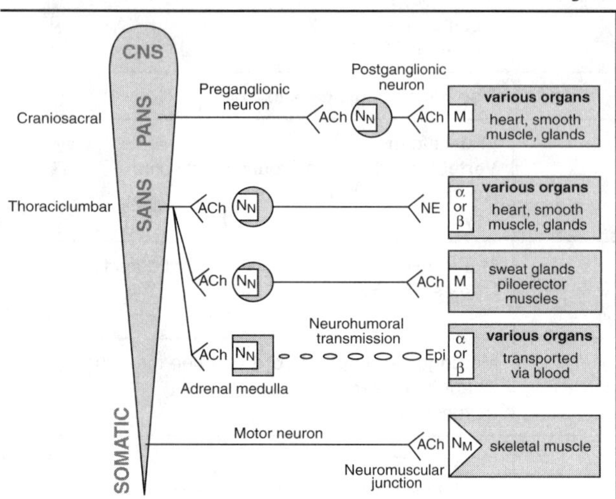

- **Somatic NS:** 1 neuron (from spinal cord → effector organ)
- **Autonomic NS:** 2 neurons (from spinal cord → effector organ)
  - **Preganglionic** neuron: cell body in CNS
  - **Postganglionic** neuron: cell body in ganglia in PNS
  - **Parasympathetic:** long preganglionic, short postganglionic
  - **Sympathetic:** short preganglionic, long postganglionic (except adrenal medulla)

Definition of abbreviations: N_N, neuronal nicotinic receptor; N_M, muscle nicotinic receptor; NE, norepinephrine; M, muscarinic receptor; ACh, acetylcholine.

Note: Arrows indicate lesion sites that result in Horner syndrome.

► Parasympathetic Nervous System

1. Ciliary ganglion
Pupillary sphincter
Ciliary m.
2. Submandibular ganglion
Submandibular gland
Sublingual gland
3. Pterygopalatine ganglion
Lacrimal gland
Nasal mucosa
Oral mucosa
Parotid gland
4. Otic ganglion
Viscera of the thorax and abdomen (foregut and midgut)
Terminal ganglia
III VI IV V VII a VIII IX X XI
Midbrain
Pons
Medulla
C1
T1
L1
S2
S3
S4
Terminal ganglia
Hindgut and pelvic viscera (including the bladder, erectile tissue, and rectum)
Pelvic splanchnics
Preganglionic
Postganglionic

► Sympathetic Nervous System

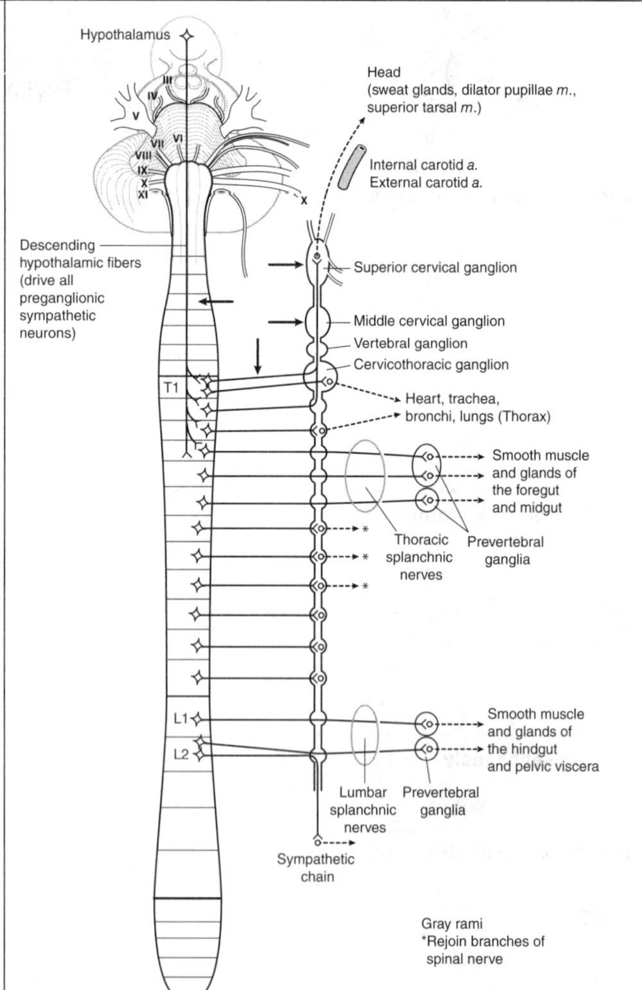

Note: Arrows indicate lesion sites that result in Horner syndrome.

▶ Parasympathetic = Craniosacral Outflow

Origin	Site of Synapse	Innervation
Cranial nerves III, VII, IX	4 cranial ganglia (ciliary, submandibular, pterygopalatine, otic)	Glands and smooth muscle of the **head**
Cranial nerve X	Terminal ganglia (in or near the walls of viscera)	Viscera of the neck, **thorax**, **foregut**, and **midgut**
Pelvic splanchnic nerves (S2, S3, S4)	Terminal ganglia (in or near the walls of viscera)	**Hindgut** and **pelvic viscera** (including bladder, rectum, and erectile tissue)

▶ Sympathetic = Thoracolumbar Outflow

Origin	Site of Synapse	Innervation
Spinal cord levels T1–L2	Sympathetic chain ganglia (paravertebral ganglia)	Smooth and cardiac muscle and glands of **body wall** and **limbs**; head and **thoracic viscera**
Thoracic splanchnic nerves T5–T12	Prevertebral ganglia (collateral; e.g., celiac, aorticorenal superior mesenteric ganglia)	Smooth muscle and glands of the **foregut** and **midgut**
Lumbar splanchnic nerves L1, L2	Prevertebral ganglia (collateral; e.g., inferior mesenteric and pelvic ganglia)	Smooth muscle and glands of the **pelvic viscera** and **hindgut**

▶ Autonomic Effects on Organ Systems

As a general rule, the **sympathetic autonomic nervous system (SANS)** mediates **"fight or flight"** responses, such as increasing heart rate and contractility, dilating airways and pupils, inhibiting GI and GU functions, and directing blood flow away from skin and GI tract and toward skeletal muscles. In contrast, the **parasympathetic autonomic nervous system (PANS)** causes the body to **"rest and digest,"** reducing heart rate and contractility, contracting airways and pupils, inducing secretion from lacrimal and salivary glands, and promoting GI and GU motility. Blood vessels are solely innervated by SANS nerve fibers.

	Sympathetic "Fight or Flight"		Parasympathetic "Rest and Digest"	
Organ	**Action**	**Receptor**	**Action**	**Receptor**
	Cardiovascular			
Heart				
SA node	↑ heart rate	β_1, (β_2)	↓ heart rate	M_2
Atria	↑ contractility		↓ contractility	
AV node	↑ conduction velocity and automaticity[1]		↓ conduction velocity and automaticity[1]	
Ventricles	↑ contractility		↓ contractility (slight)	
Arterioles[2]	Contract: ↑ resistance	α_1	—	$(M_3)^3$
Veins	Contract: ↑ venous pressure	α_1	—	—
Kidney	Renin release	β_1	—	—
	Respiratory			
Bronchiolar smooth muscle	Relax: ↓ resistance	β_2	Contract: ↑ resistance	M_3

(Continued)

[1]When acting as a pacemaker; otherwise, the SA node suppresses automaticity in these cells

[2]β_2 receptors that mediate relaxation and decrease resistance are also present on coronary arteries and arterioles. Low doses of epinephrine (or β_2 agonists) act selectively on β_2 receptors and can decrease systemic vascular resistance, but increased sympathetic tone increases systemic vascular resistance because vasoconstriction dominates.

[3]M_3 receptors are on vascular endothelium (not smooth muscle, like the adrenergic receptors) and cause vasodilation via nitric oxide (NO) generation; this has little physiologic significance because vasculature is not innervated by PANS, but is more important with muscarinic agonist administration.

► Autonomic Effects on Organ Systems (*Cont'd.*)

	Sympathetic "Fight or Flight"		Parasympathetic "Rest and Digest"	
Organ	**Action**	**Receptor**	**Action**	**Receptor**
Gastrointestinal				
GI smooth muscle				
Walls	↓ GI motility	α_2,[4] β_2	↑ GI motility	M_3
Sphincters	Contracts	α_1	Relaxes	M_3
Glandular secretion	—	—	Increases	M_3
Liver	Gluconeogenesis	β_2, α	—	—
	Glycogenolysis	β_2, α	—	—
Fat cells	Lipolysis	β_3	—	—
Genitourinary				
Bladder				
Walls	Relaxes	β_2	Contracts	M_3
Sphincters	Contracts	α_1	Relaxes	M_3
Uterus, pregnant	Relaxes	β_2	—	—
	Contracts	α	Contracts	M_3
Penis, seminal vesicles	Ejaculation	α	Erection	M
Skin				
Sweat glands	Secretion	M,[5] α[6]	—	—
Eye				
Radial dilator muscle	Contracts (dilates pupil)	α_1	—	—
Pupillary sphincter muscle	—	—	Contracts (constricts pupil)	M_3
Ciliary muscle	Relaxes—far vision	β	Contracts—near vision	M_3

[4]Probably via inhibition of cholinergic nerve terminals
[5]Generalized
[6]Localized (e.g., palms of hands)

▶ Cholinergic Transmission

Acetylcholine (ACh) is synthesized from acetate and choline in synaptic nerve terminals via **choline acetyltransferase (ChAT)** and stored in synaptic vesicles and released by Ca^{2+} influx upon depolarization. The **uptake of choline** into the nerve terminal is the **rate-limiting step** of ACh synthesis and can be blocked by **hemicholinium**. ACh then binds to postsynaptic receptors to elicit somatic (N_M) or autonomic (N_N and M) effects. Signal termination occurs by degradation of ACh by **acetylcholinesterase (AChE)**.

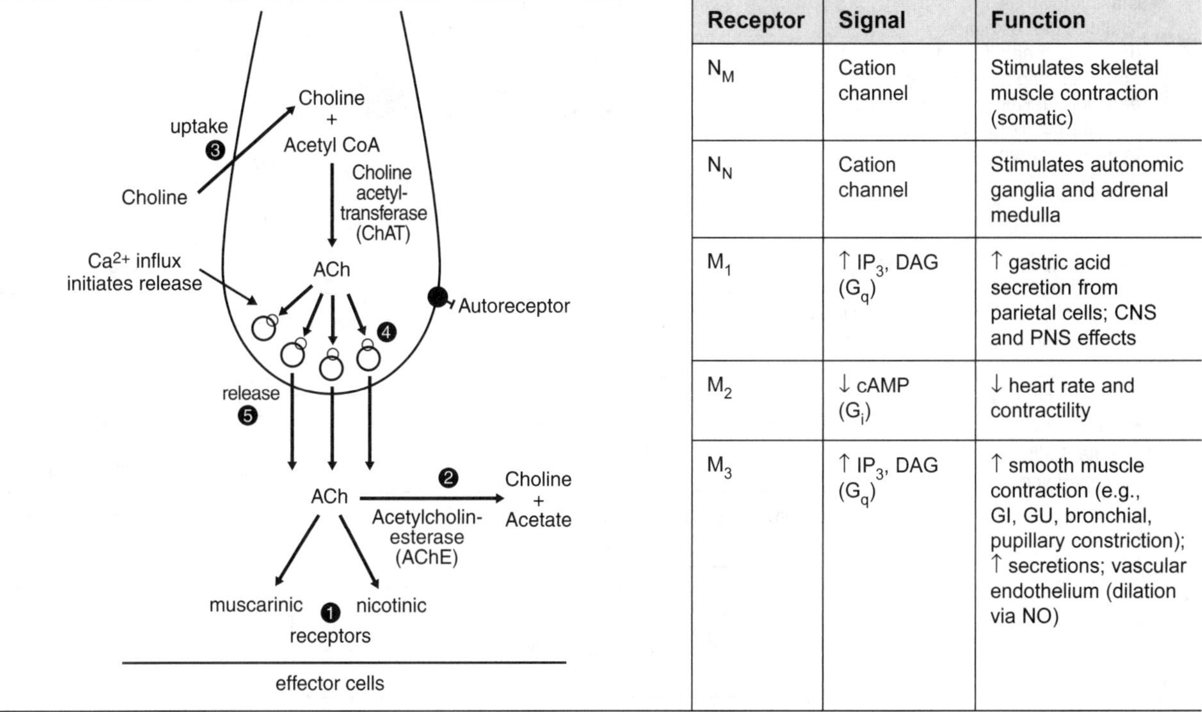

Receptor	Signal	Function
N_M	Cation channel	Stimulates skeletal muscle contraction (somatic)
N_N	Cation channel	Stimulates autonomic ganglia and adrenal medulla
M_1	↑ IP_3, DAG (G_q)	↑ gastric acid secretion from parietal cells; CNS and PNS effects
M_2	↓ cAMP (G_i)	↓ heart rate and contractility
M_3	↑ IP_3, DAG (G_q)	↑ smooth muscle contraction (e.g., GI, GU, bronchial, pupillary constriction); ↑ secretions; vascular endothelium (dilation via NO)

Definition of abbreviations: DAG, diacylglycerol; GI, gastrointestinal; GU, genitourinary; IP_3, inositol triphosphate.

Note: The numbers in the figure correspond to the numbers in the table below.

▶ Cholinergic Pharmacology

Mechanism of Action	Agent	Clinical Uses	Other Notes and Toxicity
❶ ACh receptor			
Nicotinic (N_M) agonist/antagonists (skeletal muscle)	(*See* neuromuscular blockers, page 427.)		
Nicotinic (N_N) antagonists (ganglion blockers)	Hexamethonium Mecamylamine	Rarely used due to toxicities	Blocks both SANS and PANS; effectively reduces predominant tone (*see table below*)
Cholinergic agonists M = muscarinic N = nicotinic B = both	Bethanechol (M)	Ileus (postop/neurogenic) Urinary retention	Heart block, cardiac arrest Syncope
	Methacholine (M)	Diagnosis of bronchial hyperreactivity in asthma	Partially sensitive to cholinesterase (others listed here are resistant)
	Pilocarpine (M)	Glaucoma (topical) Xerostomia	—
	Carbachol (B)	Glaucoma (topical)	—
	Nicotine (N)	Smoking deterrence	—

(Continued)

► Cholinergic Pharmacology (*Cont'd.*)

Mechanism of Action	Agent	Clinical Uses	Other Notes and Toxicity
Muscarinic antagonists Class known as "Belladonna" alkaloids (meaning "beautiful lady") from their origin from *Atropa belladonna*, which was used to dilate pupils (believed to make women look more attractive). Other drugs with anti-muscarinic effects: • Antihistamines • Tricyclics • Antipsychotics • Quinidine • Amantadine • Meperidine	Atropine	Counteracts cholinergic toxicity Antidiarrheal Mydriatic agent for eye exams Reversal of sinus bradycardia and heart block	Mydriasis and cycloplegia ***(blind as a bat)*** Decreased secretions ***(dry as a bone)*** Vasodilation ***(red as a beet)*** Delirium and hallucinations ***(mad as a hatter)*** Hyperthermia Tachycardia Urinary retention and constipation Sedation, amnesia
	Homatropine Cyclopentolate Tropicamide	Ophthalmology (topical), for mydriasis	—
	Ipratropium	Asthma and COPD	Localized effect because is a quaternary amine; few antimuscarinic side effects
	Scopolamine	Motion-sickness Antiemetic	(*See side effects for atropine.*)
	Benztropine Trihexyphenidyl Biperiden	Parkinsonism Acute extrapyramidal symptoms from neuroleptics	
	Glycopyrrolate Dicyclomine	Reduces hypermotility in GI and GU tracts	
❷ Metabolism of ACh (AChE inhibitors)			
Short-acting:	Edrophonium	Diagnosis of myasthenia gravis	Seizures
Tertiary amines:	Physostigmine	Glaucoma Reversal of anticholinergic toxicity	Carbamylating inhibitor Seizures
Quaternary amines:	Neostigmine Pyridostigmine Ambenonium	Ileus, urinary retention Myasthenia gravis Reversal of nondepolarizing neuromuscular blockers	Carbamylating inhibitors
Lipid-soluble:	Donepezil Tacrine Rivastigmine Galantamine	Alzheimer disease	Hepatotoxicity GI bleeding

(Continued)

► Cholinergic Pharmacology (*Cont'd.*)

Mechanism of Action	Agent	Clinical Uses	Other Notes and Toxicity
Organophosphates:	Echothiophate	Glaucoma	
• Long-acting and irreversible	Malathion Parathion	Insecticide	**Mnemonic: DUMBBELSS** **D**iarrhea **U**rination **M**iosis **B**radycardia **B**ronchoconstriction **E**xcitation (CNS and muscle) **L**acrimation **S**alivation **S**weating Acute treatment with **atropine** and **pralidoxime (2-PAM)** to regenerate AChE before aging occurs
• Causes time-dependent **aging**, which permanently inactivates AChE	Sarin VX	Nerve agents (chemical warfare)	
❸ Reuptake inhibitors	Hemicholinium	Only used in research settings	Blocks choline reuptake, slowing ACh synthesis
❹ Vesicular transport inhibitors	Vesamicol	Only used in research settings	Blocks ACh uptake into vesicles, preventing storage and leading to ACh depletion
❺ Vesicle release inhibitors	Botulinum toxin	Blepharospasm Strabismus/Hyperhydrosis Cervical dystonia Cosmetic (removes wrinkles)	Prevents fusion of cholinergic vesicles to membrane, thereby inhibiting ACh release

Definition of abbreviations: ACh, acetylcholine; AChE, acetylcholinesterase; ChAT, choline acetyltransferase; COPD, chronic obstructive pulmonary disorder.

► Predominant Tone and the Effect of Ganglionic Blockers

The effects of ganglionic blockers can be easily predicted if you know the predominant autonomic tone to a particular effector organ. The effect of the blockade will be the opposite of what the predominant tone causes. In general, the predominant tone to blood vessels and sweat glands is sympathetic, and most everything else is parasympathetic.

Site	Predominant Tone	Effect of Ganglionic Blockade
Arterioles	Sympathetic	Dilation (↓ blood pressure)
Veins	Sympathetic	Dilation (↓ venous return)
Heart	Parasympathetic	↑ heart rate
GI tract	Parasympathetic	↓ motility and secretions
Eye	Parasympathetic	Pupillary dilation, focus to far vision
Urinary bladder	Parasympathetic	Urinary retention
Salivary glands	Parasympathetic	Dry mouth
Sweat glands	Sympathetic (cholinergic)	Anhidrosis

► Adrenergic Transmission

Norepinephrine (NE), **epinephrine (EPI)**, and dopamine are part of the **catecholamine** family, which are synthesized from **tyrosine**. The first step of the synthetic pathway is carried out by **tyrosine hydroxylase**; it is also the **rate-limiting** step. NE levels inside the presynaptic terminal may also be regulated by metabolism by **monoamine oxidase (MAO)**. Once released, NE binds to various adrenergic receptors to transmit its signal. NE primarily binds α_1, α_2, and β_1 receptors, whereas EPI (released by the adrenal medulla) binds α_1, α_2, β_1, and β_2 receptors. **Reuptake** (especially uptake-1) and diffusion are most important in the termination of action of NE (and DA). Metabolism occurs via **catechol-*O*-methyltransferase ([COMT]** extracellular) and **MAO** (intracellular). Metabolites such as **metanephrine**, **normetanephrine**, **vanillylmandelic acid (VMA)** can be measured in the urine and are used in the diagnosis of diseases such as pheochromocytoma.

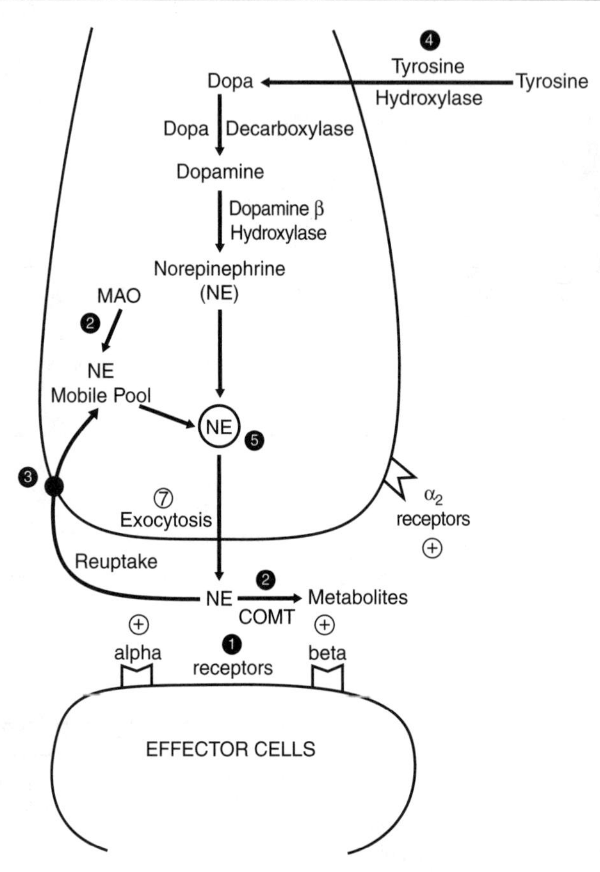

Receptor	Signal	Function
α_1	↑ IP_3, DAG (G_q)	↑ smooth muscle contraction (in vascular walls, radial dilator muscle [eye], and GI and bladder sphincters); ↑ glycogenolysis in liver
α_2	↓ cAMP (G_i)	Inhibits neurotransmitter release; inhibits insulin release and lipolysis
β_1	↑ cAMP (G_s)	↑ heart rate and contractility; ↑ AV conduction velocity; ↑ renin secretion
β_2		↑ smooth muscle relaxation (in vascular, bronchial, GI, and bladder walls); ↑ glycogenolysis in liver, ↑ insulin release
β_3		↑ lipolysis from adipose tissues
		Mnemonic: You can get an adrenaline rush from a **QISSS** ($G_qG_iG_sG_sG_s$).

Definition of abbreviations: DAG, diacylglycerol; GI, gastrointestinal; GU, genitourinary; IP_3, inositol triphosphate.

Note: The numbers in the figure correspond to the numbers in the following table.

► Adrenergic Pharmacology

Mechanism of Action	Agent	Clinical Uses	Other Notes and Toxicity
❶ Adrenergic receptor			
α-adrenergic agonists	α_1: Phenylephrine	Nasal congestion Vasoconstriction Mydriasis (topical)	—
	α_1: Methoxamine	Paroxysmal supraventricular tachycardia	Bradycardia (vagal reflex)
	α_2: Clonidine Methyldopa	Hypertension	Decreases sympathetic outflow Methyldopa: prodrug converted to methylnorepinephrine
α-adrenergic antagonists	**Nonselective:** Phentolamine	Pheochromocytoma	**Reversible**, competitive inhibitor Postural hypotension, reflex tachycardia
	Nonselective: Phenoxybenzamine	Pheochromocytoma	**Irreversible** inhibitor Postural hypotension, reflex tachycardia
	α_1: Prazosin Doxazosin Terazosin Tamsulosin	Hypertension Benign prostatic hypertrophy	Postural hypotension on first dose
	α_2: Yohimbine	Impotence Postural hypotension	Clinical use limited
	α_2: Mirtazapine	Depression	May also block 5HT receptors
β-adrenergic agonists	$\beta_1 = \beta_2$: Isoproterenol	Bronchospasm Heart block and bradyarrhythmias	Clinical use limited ↓ BP, ↑ HR Arrhythmias
	$\beta_1 > \beta_2$: Dobutamine	Acute heart failure	—
	β_2: Albuterol Terbutaline Metaproterenol Salmeterol	Asthma	Tachycardia, skeletal muscle tremor Pulmonary delivery to minimize side effects Salmeterol is long-acting Suppress premature labor (terbutaline)
β-adrenergic antagonists	β_1: Acebutolol Atenolol Esmolol Metoprolol	Hypertension Angina Chronic heart failure (carvedilol, metoprolol) Arrhythmia Glaucoma (timolol) Migraine, tremor, thyrotoxicosis (propanolol)	Sedation Decreases libido Bradycardia β_1 selectivity safer in asthma, diabetes, and vascular diseases Pindolol and acebutolol have **intrinsic sympathomimetic activity** (useful in asthmatics); should not be used in patients with recent MI
	Nonselective: Pindolol Propranolol Nadolol Timolol		
	α and β: Labetalol Carvedilol	CHF (carvedilol) Hypertension (labetalol)	Similar to above Liver damage (labetalol)

Definition of abbreviation: CHF, congestive heart failure.

(Continued)

► Adrenergic Pharmacology (*Cont'd.*)

Mechanism of Action	Agent	Clinical Uses	Other Notes and Toxicity
❷ Metabolism			
MAO inhibitors MAO-A: mainly liver but Anywhere MAO-B: mainly in Brain	Phenelzine Tranylcypromine	Depression	Nonselective MAO-A/B inhibitors Tyramine (red wine, cheese) ingestion → hypertensive crisis Insomnia Postural hypotension MAO-AIs: moclobemide (reversible), clorgyline available in Europe
	Selegiline Rasagiline	Parkinson disease	Selective MAO-B inhibitor
COMT inhibitors	Entacapone Tolcapone	Parkinson disease	—
❸ Reuptake inhibitors	Cocaine	Local anesthesia Abuse	Inhibits monoamine reuptake Addiction
Tricyclic antidepressants	Amitriptyline Imipramine Desipramine Nortriptyline	Depression	Inhibits monoamine reuptake Sedation Postural hypotension Tachycardia Atropine-like effects
❹ Synthesis inhibitors	α-Methyltyrosine (metyrosine)	Hypertension	Inhibits tyrosine hydroxylase Only for hypertension associated with pheochromocytoma
❺ Drugs affecting release			
Adrenergic neuron blockers	Reserpine	Hypertension	Inhibits monoamine vesicular uptake, leading to neurotransmitter depletion Sedation, depression
	Guanethidine	Hypertension	Inhibits NE release from sympathetic nerve endings Requires neuronal uptake to work, so interferes with other drugs that require uptake carrier (e.g., cocaine, amphetamine, cyclic antidepressants)
Indirect-acting sympathomimetics	Amphetamine Methylphenidate	Narcolepsy and ADHD	Displaces NE from mobile pool Addiction Restlessness and rebound fatigue
	Ephedrine Pseudoephedrine	For vasoconstriction	Displaces NE from mobile pool Pseudoephedrine: OTC for nasal congestion

Definition of abbreviations: ADHD, attention deficit and hyperactivity disorder; COMT, catechol-O-methyltransferase; MAO, monoamine oxidase; NE, norepinephrine; OTC, over the counter.

Meninges, Ventricular System, and Venous Drainage

► Meninges and Meningeal Spaces

The meninges consist of three connective tissue membranes that surround the brain and spinal cord. Meningeal spaces are spaces or potential spaces adjacent to the meninges.

Meninges	Meningeal Space	Anatomic Description	Clinical Correlate
	Epidural space	Tight potential space between dura and skull Contains **middle meningeal artery**	**Epidural hematoma** Temporal bone fracture → rupture of **middle meningeal artery** **Lens-shaped biconvex** hematoma
Dura		Tough outer layer; dense connective tissue	
	Subdural space	Contains **bridging veins**	**Subdural hematoma** Rupture of bridging veins **Crescent-shaped** hematoma
Arachnoid		Delicate, nonvascular connective tissue	
	Subarachnoid space	Contains **CSF** Ends at S2 vertebra	**Subarachnoid hemorrhage** **"Worst headache of my life"** Often caused by **berry aneurysms** **Lumbar puncture** between L4, L5 discs
Pia		Thin, highly vascular connective tissue Adheres to brain and spinal cord	

► Meningitis (infection of the meninges, especially the pia and arachnoid)

Acute purulent (bacterial) meningitis	• **Headache, fever, nuchal rigidity**, obtundation; coma may occur • Meninges opaque; neutrophilic exudate present • Sequelae: hydrocephalus, herniation, cranial nerve impairment
Acute aseptic (viral) meningitis	• Leptomeningeal inflammation (lymphomonocytic infiltrates) due to viruses (Enterovirus most frequent) • Fever, signs of meningeal irritation, depressed consciousness, but low mortality
Mycobacterial meningoencephalitis	• Can be caused by *Mycobacterium tuberculosis* or atypical mycobacteria • Usually involves the **basal surface** of the brain with tuberculomas within the brain and dura mater • Frequent in AIDS patients, particularly by *Mycobacterium avium-intracellulare* (MAI)
Fungal meningoencephalitis	• *Candida, Aspergillus, Cryptococcus*, and *Mucor* species most frequent agents • *Aspergillus* and *Mucor* attack blood vessels → vasculitis, rupture of blood vessels, and hemorrhage • *Cryptococcus* causes diffuse meningoencephalitis

► Organisms Causing Bacterial Meningitis by Age Group

Neonates	Infants/Children	Adolescents/Young Adults	Elderly
Group B streptococci ***Escherichia coli*** *Listeria monocytogenes*	***Haemophilus influenzae* B** (if not vaccinated) *Streptococcus pneumoniae* *Neisseria meningitidis*	***Neisseria meningitidis***	***Streptococcus pneumoniae*** *Listeria monocytogenes*

▶ CSF Parameters in Meningitis

Condition	Cells/µl	Glucose (mg/dL)	Proteins (mg/dL)	Pressure (mm H_2O)
Normal values	<5 lymphocytes	45–85 (50–70% of blood glucose)	15–45	70–180
Purulent (bacterial)	Up to 90,000 neutrophils	Decreased (<45)	Increased (>50)	Markedly elevated
Aseptic (viral)	100–1,000 most lymphocytes	Normal	Increased (>50)	Slightly elevated
Granulomatous (mycobacterial/fungal)	100–1,000 most lymphocytes	Decreased (<45)	Increased (>50)	Moderately elevated

▶ Viral Encephalitis

Pathology: perivascular cuffs, microglial nodules, neuron loss, and neuronophagia
Clinical: fever, headache, mental status changes, often progressing to coma

Arthropod-Borne	Herpes Simplex	Rabies	HIV
St. Louis, California, Eastern equine, Western equine, Venezuelan	Characteristic **hemorrhagic necrosis of temporal lobes**	**Negri bodies** in hippocampal and Purkinje neurons of the cerebellum	**Microglial nodules, multinucleated giant cells**

▶ Ventricular System and Venous Drainage

The brain and spinal cord float within a protective bath of cerebrospinal fluid (CSF), which is produced by the lining of the ventricles, the choroid plexus. CSF circulation begins in the ventricles and then enters the subarachnoid space to surround the brain and spinal cord.

Ventricles and CSF Circulation

Third ventricle
Foramen of Monro
Body of lateral ventricle
Anterior horn of the lateral ventricle
Posterior horn of lateral ventricle (occipital)
Inferior horn of lateral ventricle (temporal)
Cerebral aqueduct (of Sylvius)
Fourth ventricle
Central canal

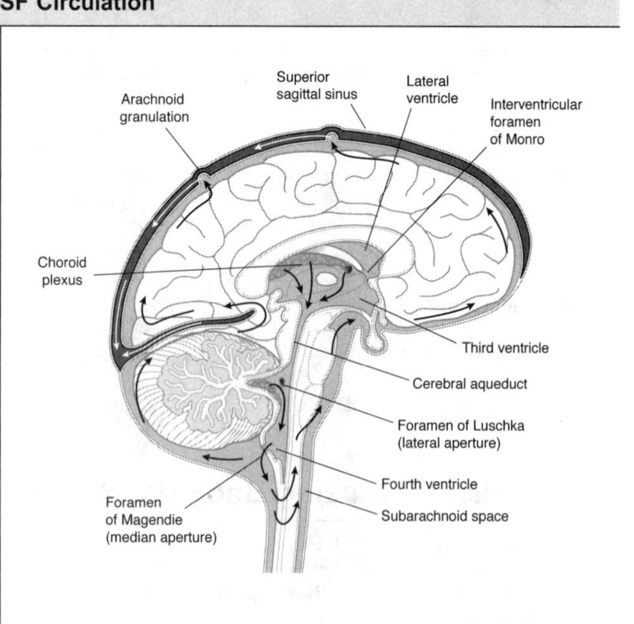

Lateral ventricles $\xrightarrow{A}$ third ventricle $\xrightarrow{B}$ fourth ventricle → subarachnoid space (via foramina of Luschka and foramen of Magendie)
(**A,** interventricular foramen of Monro; **B,** cerebral aqueduct.)

(Continued)

► Ventricular System and Venous Drainage (*Cont'd.*)

CSF Production and Barriers

Choroid plexus—contains **ependymal cells** and is found in the lateral, third, and fourth ventricles. **Secretes CSF.** Tight junctions form **blood-CSF barrier**.

Blood-brain barrier—formed by capillary endothelium with tight junctions; astrocyte foot processes contribute.

Once CSF is in the subarachnoid space, it goes up over convexity of the brain and enters the venous circulation by passing through **arachnoid granulations** into the **superior sagittal sinus**.

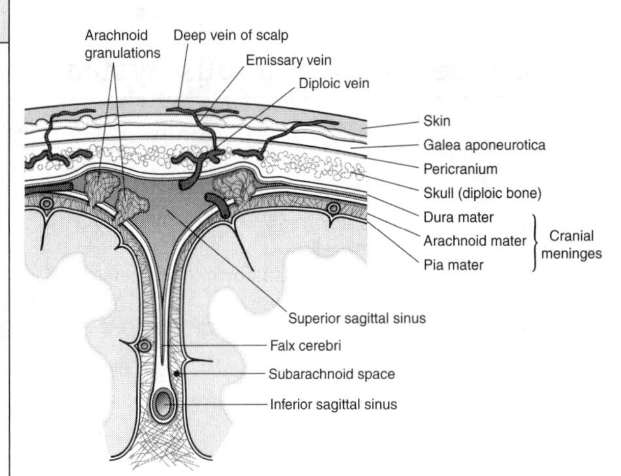

Sinuses

Superior sagittal sinus (in superior margin of falx cerebri)—drains into two **transverse sinuses**. Each of these drains blood from the **confluence of sinuses** into **sigmoid sinuses**. Each sigmoid sinus exits the skull (via **jugular foramen**) as the **internal jugular veins**.

Inferior sagittal sinus (in inferior margin of falx cerebri)— terminates by joining with the great cerebral vein of Galen to form the **straight sinus** at the falx cerebri and tentorium cerebelli junction. This drains into the confluence of sinuses.

Cavernous sinus—a plexus of veins on either side of the **sella turcica**. Surrounds internal carotid artery and cranial nerves III, IV, V, and VI. It drains into the transverse sinus (via the **superior petrosal sinus**) and the internal jugular vein (via the **inferior petrosal sinus**).

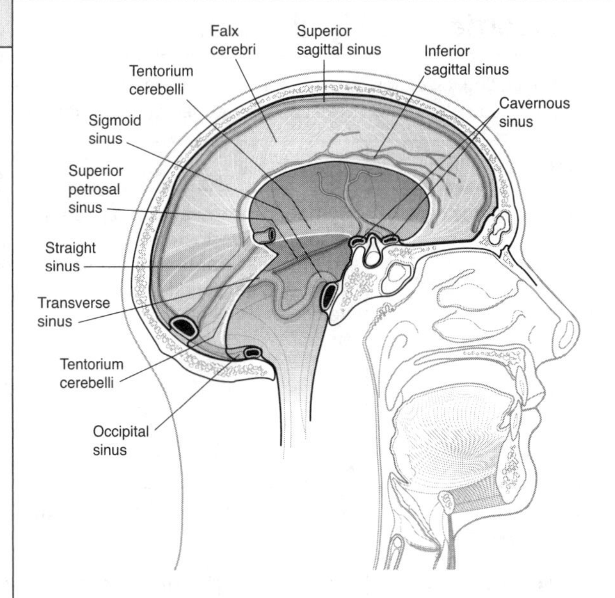

► Hydrocephalus

Excess Volume or Pressure of CSF, Leading to Dilated Ventricles

Noncommunicating	**Obstruction of flow within ventricles**; most commonly occurs at narrow points, e.g., foramen of Monro, cerebral aqueduct, fourth ventricle
Communicating	**Impaired CSF reabsorption** in arachnoid granulations or obstruction of flow in subarachnoid space
Normal pressure	CSF is not absorbed by arachnoid villi (a form of communicating hydrocephalus). CSF pressure is usually normal. Ventricles chronically dilated. Produces **triad** of **dementia, ataxic gait,** and **urinary incontinence**.
Hydrocephalus ex vacuo	Descriptive term referring to excess CSF in regions where brain tissue is lost due to atrophy, stroke, surgery, trauma, etc.

Neurohistology and Pathologic Correlates

► Cell Types of the Nervous System

Neurons	Glia (non-neuronal cells of the nervous system)
Composed of: • **Dendrites:** receive info and transmit to the cell body • **Cell body:** contains organelles • **Axon:** transmits electrical impulse down to the nerve terminals • **Nerve terminals:** contain synaptic vesicles and release neurotransmitter into the synapse	• **Oligodendrocytes:** form **myelin** in the **CNS** (one cell myelinates many axons) • **Astrocytes:** control microenvironment of neurons and help maintain the blood/brain barrier with foot processes • **Schwann cells:** form **myelin** in the **PNS** (one cell myelinates one internode) • **Ependymal cells:** ciliated neurons that line the ventricles and central canal of the spinal cord • **Microglia:** are **phagocytic** and are part of the **mononuclear phagocyte system**

► Disorders of Myelination

Demyelinating diseases are acquired conditions involving selective damage to myelin. Other diseases (e.g., infectious, metabolic, inherited) can also affect myelin and are generally called **leukodystrophies**.

Disease	Symptoms	Notes
Multiple sclerosis (MS)	**Symptoms separated in space and time** Vision loss (**optic neuritis**) **Internuclear ophthalmoplegia** (MLF degeneration) Motor and sensory deficits Vertigo Neuropsychiatric	Occurs twice as often in **women** Onset often in **third or fourth decade** Higher prevalence in **temperate zones** **Relapsing–remitting course** is most common Well-circumscribed **demyelinated plaques** often in periventricular areas Chronic inflammation; axons initially preserved Type IV hypersensitivity Increased IgG **(oligoclonal bands)** in CSF **Treatment:** high-dose steroids, interferon-beta, glatiramer (Copaxone®), natalizumab
Metachromatic leukodystrophy	Varied neurologic and psychiatric symptoms	**Arylsulfatase A deficiency**
Progressive multifocal leukoencephalopathy (PML)	Varied neurologic symptoms, dementia	Caused by **JC virus** Affects **immunocompromised, especially AIDS** Demyelination, astrogliosis, lymphohistiocytosis
Central pontine myelinolysis (CPM)	**Pseudobulbar palsy** **Spastic quadriparesis** Mental changes May produce the "locked-in" syndrome Often fatal	Focal demyelination of central area of basis pontis (affects corticospinal, corticobulbar tracts) Seen in **severely malnourished, alcoholics, liver disease** Probably **caused by overly aggressive correction of hyponatremia**
Guillain-Barré syndrome	Acute symmetric ascending inflammatory neuropathy **Weakness begins in lower limbs and ascends; respiratory failure can occur** in severe cases Autonomic dysfunction may be prominent Cranial nerve involvement is common Sensory loss, pain, and paresthesias occur **Reflexes invariably decreased or absent**	Two-thirds of patients have **history of respiratory or GI illness 1–3 weeks prior to onset**, 30% post *Campylobacter* Elevated CSF protein with normal cell count **(albuminocytologic dissociation)**
Charcot-Marie-Tooth disease	Slowly progressing weakness in distal limbs; usually lower extremities before upper extremities Abnormal proprioception/vibration sensation	Most common inherited neurologic disorder Onset usually in 1st two decades of life Different constellation of symptoms can occur, demyelination or axonal disease

(Continued)

▶ Disorders of Myelination (*Cont'd.*)

Acute disseminated (postinfectious) encephalomyelitis	Symptoms can be similar to MS, but is associated with constitutional symptoms, mental status changes, seizures, and fewer dorsal column abnormalities	Nonvasculitic inflammatory demyelinating disease History of preceding infectious illness or immunization Most often in prepubertal children

Definition of abbreviation: MLF, medial longitudinal fasciculus.

▶ Tumors of the CNS and PNS

One half of brain and spinal cord tumors are metastatic. Some differences between primary and metastatic tumors are listed below:

Primary	Metastatic
Poorly circumscribed	Well circumscribed
Usually single	Often multiple
Location varies by specific type	Usually located at the junction between gray and white matter

▶ Primary Tumors

Tumor	Features	Pathology
Glioblastoma multiforme (grade IV astrocytoma)	• **Most common primary brain tumor** • **Highly malignant** • Usually lethal in 8–12 months	• Can cross the midline via the corpus callosum ("**butterfly glioma**") • Areas of necrosis surrounded by rows of neoplastic cells (**pseudopalisading necrosis**)
Pilocytic astrocytoma (grade 1)	• Tumor of children and young adults • Usually in **posterior fossa in children**	• **Rosenthal fibers** • Immunostaining with **GFAP**
Oligodendroglioma	• Slow growing • Long survival (average 5–10 years)	**"Fried-egg" appearance**—perinuclear halo
Ependymoma	• Ependymal origin • Can arise in IV ventricle and lead to **hydrocephalus**	Rosettes and pseudorosettes
Medulloblastoma	• Highly malignant cerebellar tumor • A type of primitive neuroectodermal tumors (PNET)	Blue, small, round cells with Homer-Wright pseudorosettes
Meningioma	• Second most common primary brain tumor • Dural convexities; parasagittal region	• Attaches to the dura, compresses underlying brain without invasion • Microscopic—**psammoma bodies**
Schwannoma	• Third most common primary brain tumor • Most frequent location: **CN VIII at cerebellopontine angle** • **Hearing loss, tinnitus** • Good prognosis after surgical resection	• Antoni A (hypercellular) and B (hypocellular) areas • **Bilateral acoustic schwannomas—pathognomonic for neurofibromatosis type 2**
Retinoblastoma	• Sporadic—unilateral • **Familial—bilateral; associated with osteosarcoma**	Small, round, blue cells; may have rosettes
Craniopharyngioma	• Derived from odontogenic epithelium (remnants of **Rathke pouch**) • Usually children and young adults • Often calcified • Symptoms due to encroachment on pituitary stalk or optic chiasm • Benign but may recur	Histology resembles **adamantinoma** (most common tumor of tooth)

Definition of abbreviation: GFAP, glial fibrillary acidic protein.

Spinal Cord

The spinal cord is divided internally into 31 segments that give rise to **31 pairs of spinal nerves** (from rostral to caudal): **8 cervical**, **12 thoracic**, **5 lumbar**, **5 sacral**, and **1 coccygeal**. Each segment is divided into an inner butterfly-like gray matter containing neuronal cell bodies and a surrounding area of white matter. The ventral horn contains alpha and gamma motoneurons; the intermediate horn contains preganglionic neurons and Clarke's nucleus; and the dorsal horn contains sensory neurons. The outer covering of the spinal cord is the white matter containing ascending and descending axons that form tracts located within funiculi.

► General Spinal Cord Features

Conus medullaris	Caudal end of the spinal cord (S3–S5). In adult, ends at the L2 vertebra
Cauda equina	Nerve roots of the lumbar, sacral, and coccygeal spinal nerves
Filum terminale	Slender pial extension that tethers the spinal cord to the bottom of the vertebral column
Doral root ganglia	Cell bodies of primary sensory neurons
Dorsal and ventral roots	Each segment has a pair
Dorsal	**In (sensory)**
Ventral	**Out (motor)**
Spinal nerve	Formed from dorsal and ventral roots (mixed nerve)
Cervical enlargement	(C4–T1) → branchial plexus → upper limbs
Lumbar enlargement	(L2–S3) → lumbar and sacral plexuses → lower limbs

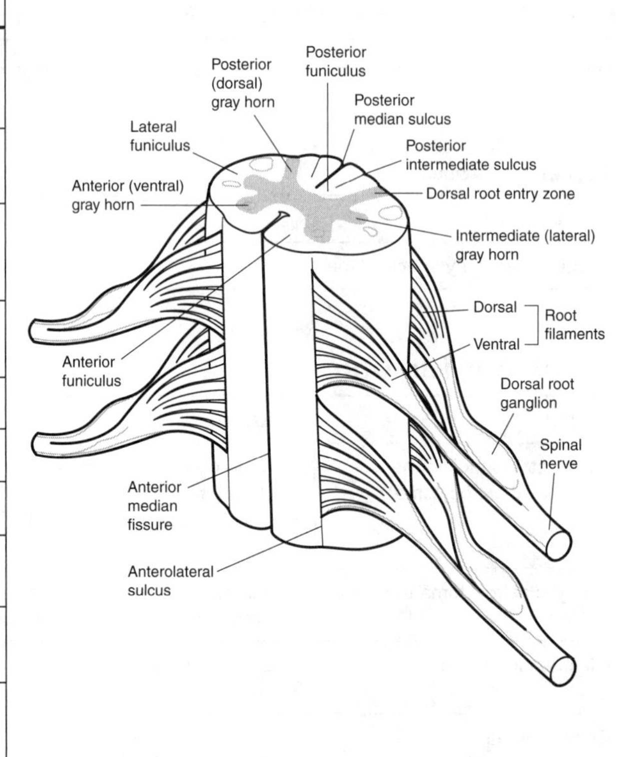

The following figure shows additional features of the spinal cord, such as the gray and white communicating rami, which are part of the autonomic nervous system.

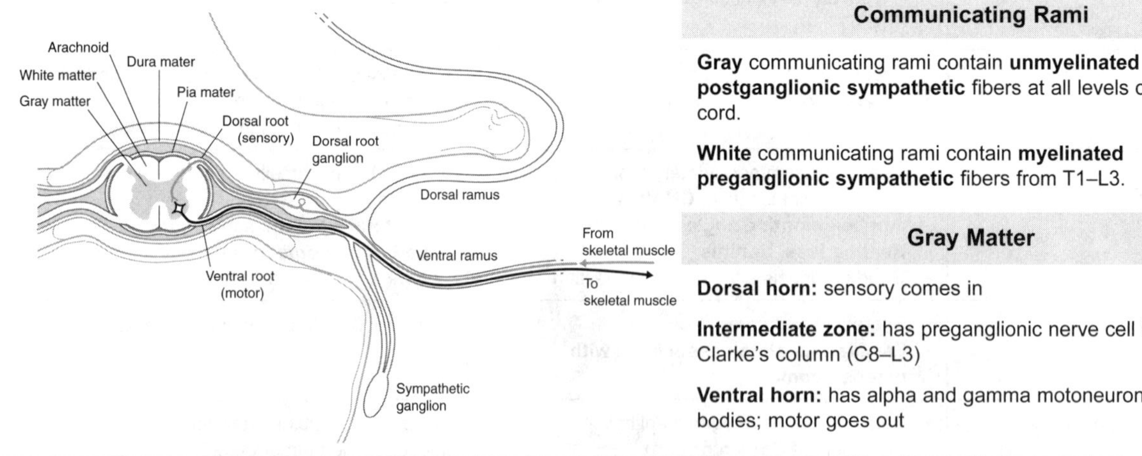

Communicating Rami

Gray communicating rami contain **unmyelinated postganglionic sympathetic** fibers at all levels of spinal cord.

White communicating rami contain **myelinated preganglionic sympathetic** fibers from T1–L3.

Gray Matter

Dorsal horn: sensory comes in

Intermediate zone: has preganglionic nerve cell bodies; Clarke's column (C8–L3)

Ventral horn: has alpha and gamma motoneuron cell bodies; motor goes out

White Matter

Tract: a collection of axons with the same origin, function, and termination; are named by stating site of origin followed by site of termination, e.g., corticospinal (cortex → spinal cord)
Fasciculi: bundle of axons
Funiculi: a region of white matter (dorsal, lateral, ventral) that may have functionally different fasciculi

► Descending Pathways

The most essential descending pathway is that which mediates voluntary skilled motor activity. This is formed by the **upper motor neuron (UMN; corticospinal tract)** and the **lower motor neuron (LMN)**.

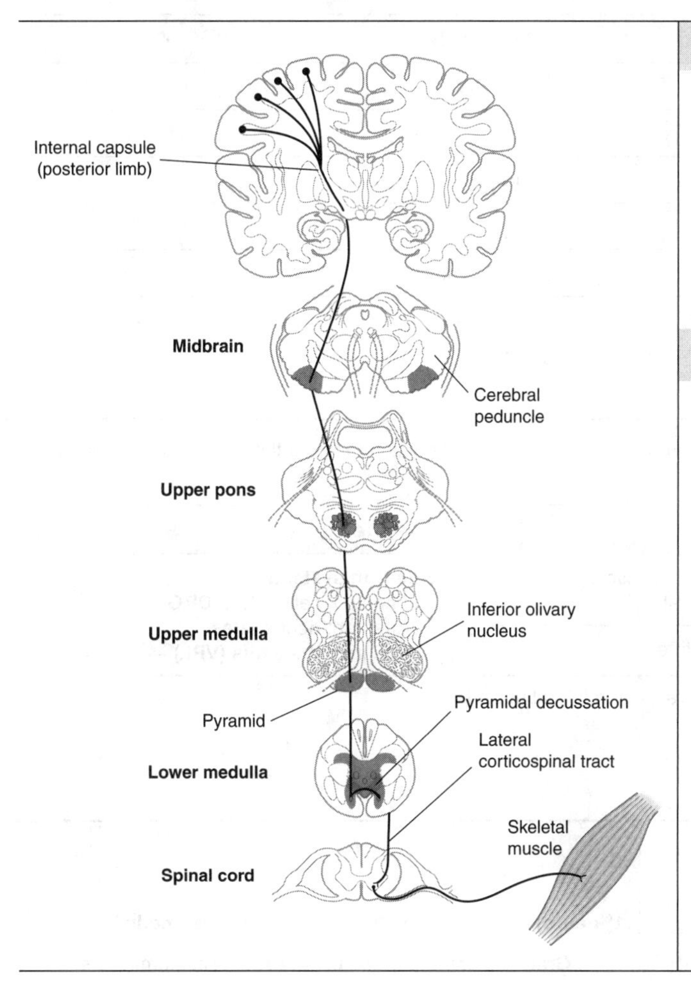

Voluntary Motor System

Function: voluntary refined movements of distal extremities

Components:

1. **Upper motor neuron (UMN):** cortex → ventral horn
2. **Lower motor neuron (LMN):** ventral horn → skeletal muscle

Lateral Corticospinal Tract

Cortex (primary motor, premotor, primary sensory)
↓
Posterior limb of **internal capsule**
↓
Crus cerebri (midbrain)
↓
Base of **pons**
↓
Pyramids (medulla)
↓
Pyramidal decussation (80–90% cross)
↓
Lateral corticospinal tract (spinal cord)
↓
Ventral horn (spinal cord): **synapse**

Note: The 10–20% of fibers that do not cross descend as the anterior corticospinal tract.

► Upper versus Lower Motor Neuron Lesions

Clinical Signs	Upper Motor Neuron	Lower Motor Neuron
Paralysis	Spastic	Flaccid
Muscle tone	Hypertonia	Hypotonia
Muscle bulk	Disuse atrophy	Atrophy, fasciculations
Deep tendon reflexes	Hyperreflexia	Hyporeflexia
Pathologic reflexes	Babinski	None
Area of body involved:		
Size:	Large	Small
Side:	Contralateral if above decussation; ipsilateral if below decussation	Ipsilateral

► Commonly Tested Muscle Stretch Reflexes

The **deep tendon (stretch, myotatic) reflex** is monosynaptic and ipsilateral. The **afferent limb** consists of a **muscle spindle receptor, Ia sensory neuron,** and **efferent limb (lower motor neuron)**. These reflexes are useful in the clinical exam.

Reflex	Cord Segment Involved	Muscle Tested
Knee (patellar)	L2–L4	Quadriceps
Ankle	S1	Gastrocnemius
Biceps	C5–C6	Biceps
Triceps	C7–C8	Triceps
Forearm	C5–C6	Brachioradialis

► Ascending Pathways

The two most important ascending pathways use a three-neuron system to convey sensory information to the cortex. Key general features are listed below.

Pathway	Function	Overview
Dorsal column–medial lemniscus	Discriminative touch, conscious proprioception, vibration, pressure	**3 neuron system:** 1° neuron: cell body in **DRG** 2° neuron: **decussates** 3° neuron: **thalamus (VPL) → cortex**
Anterolateral (spinothalamic)	Pain and temperature	

Definition of abbreviations: DRG, dorsal root ganglia; VPL, ventral posterolateral nucleus.

► Dorsal Column–Medial Lemniscus

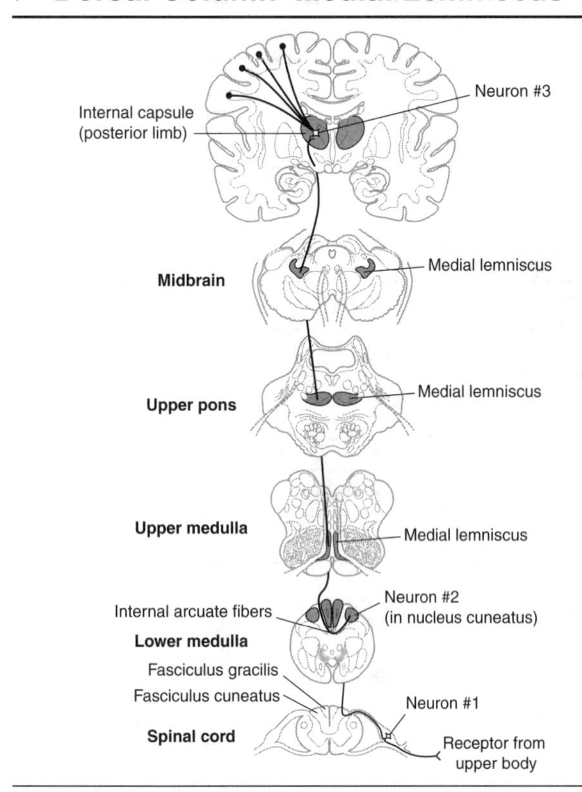

Pathway

1° neuron: cell body in **DRG**, synapses in **lower medulla**

- **Gracile fasciculus** from **lower** extremities; terminates in **gracile nucleus**
- **Cuneate fasciculus** from **upper** extremities; terminates in **cuneate nucleus**

2° neuron: decussates as **internal arcuate fibers**; ascends as **medial lemniscus**; synapses in **VPL** of the thalamus

3° neuron: VPL → somatosensory cortex

Clinical Correlation

If lesion is **above decussation** → **contralateral** loss of function

If lesion is in **spinal cord** → **ipsilateral** loss of function

Lesion can → **Romberg sign** (sways when standing, feet together, and eyes closed; swaying with eyes open indicates cerebellar dysfunction)

► Anterolateral (Spinothalamic)

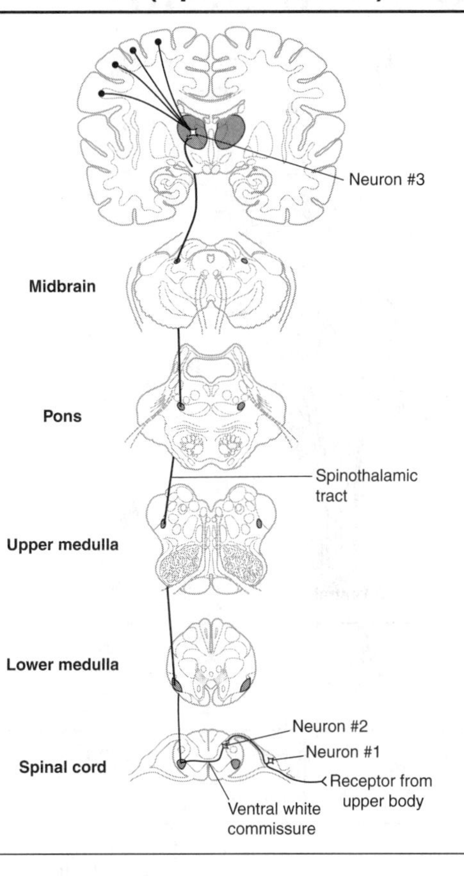

Pathway

1° neuron: cell body in **DRG**, enters cord via dorsolateral tract of Lissauer (ascends or descends one or two segments), **synapses in dorsal horn** of spinal cord. Fibers are **A-δ** (fast) or **C** (slow).

2° neuron: decussates as **ventral white commissure**; ascends in **lateral spinothalamic tract**; synapses in **VPL**

3° neuron: VPL → somatosensory cortex

Clinical Correlation

Lesion of lateral spinothalamic tract → loss of pain and temperature sensation on **contralateral** body, starting one or two segments below lesion

► Classic Spinal Cord Lesions

There are several classic and very testable spinal cord syndromes. An understanding of basic spinal cord anatomy makes the symptoms easy to predict.

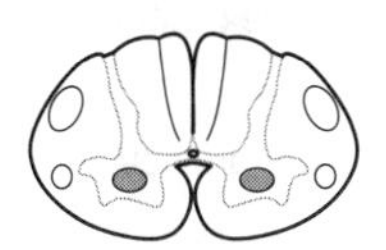

Polio
a. Flaccid paralysis
b. Muscle atrophy
c. Fasciculations
d. Areflexive

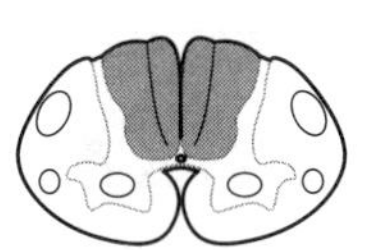

Tabes dorsalis
a. Bilateral dorsal column signs below lesions
b. Associated with late stage syphilis, plus Romberg sign: sways with eyes closed; Argyll Robertson pupils

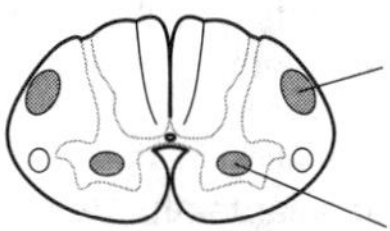

Amyotrophic lateral sclerosis (ALS)
a. Primary lateral sclerosis (corticospinal tract)
- Spastic paralysis in lower limbs
- Increased tone and reflexes
- Flaccid paralysis in upper limbs

b. Progressive spinal muscular atrophy (ventral horn)

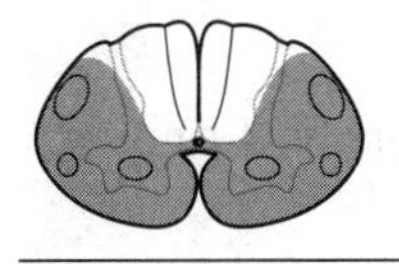

Anterior spinal artery (ASA) occlusion
a. DC spared
b. All else bilateral signs

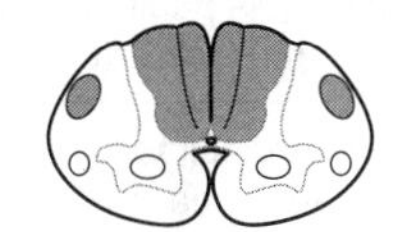

Subacute combined degeneration
a. Vitamin B_{12}, pernicious anemia; AIDS
b. Demyelination of the:
- Dorsal columns
- Corticospinal tracts (CST)

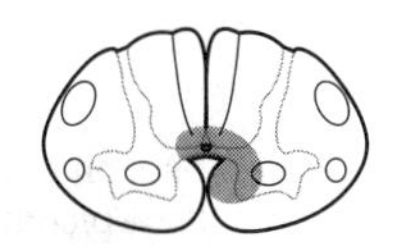

Syringomyelia
a. Cavitation of the cord (usually cervical)
b. Bilateral loss of pain and temperature at the level of the lesion
c. As the disease progresses, there is muscle weakness; eventually flaccid paralysis and atrophy of the upper limb muscles due to destruction of ventral horn cells

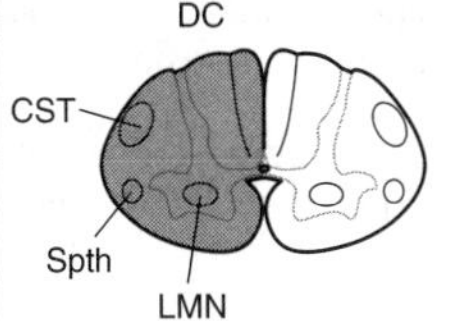

Hemisection: Brown-Séquard syndrome
a. DC: Ipsilateral loss of position and vibratory senses at and below level of the lesion
b. Spinothalamic tract: Contralateral loss of P&T below lesion and bilateral loss at the level of the lesion
c. CST: Ipsilateral paresis below the level of the lesion
d. LMN: Flaccid paralysis at the level of the lesion
e. Descending hypothalamics: Ipsilateral Horner syndrome (if cord lesion is above T_1)
- Facial hemianhydrosis
- Ptosis (slight)
- Miosis

Definition of abbreviations: DC, dorsal column; LMN, lower motor neuron.

Cranial Nerves and Brain Stem

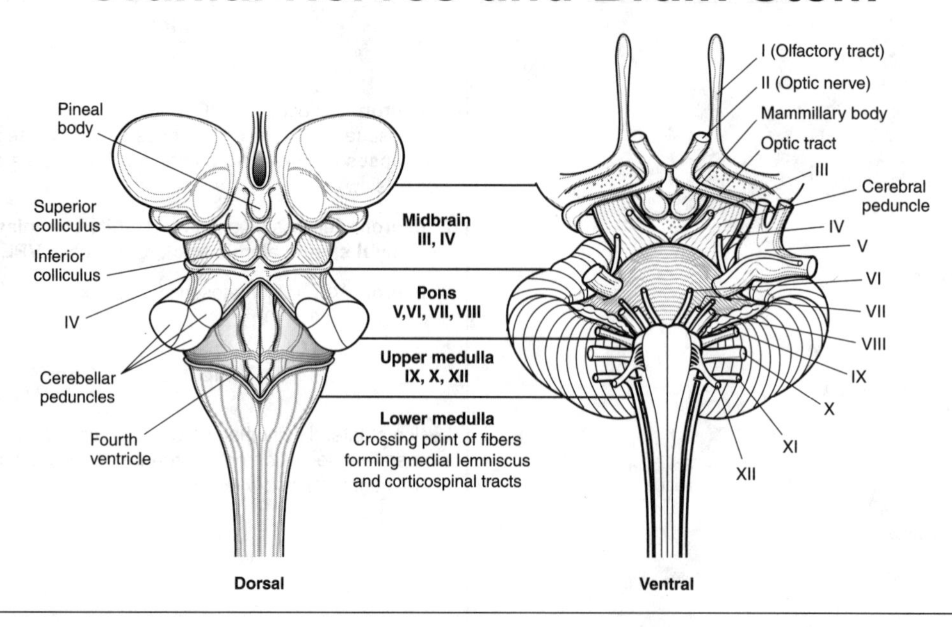

▸ Cranial Nerves: Functional Features

CN	Name	Type	Function	Results of Lesions
I	Olfactory	Sensory	Smells	Anosmia
II	Optic	Sensory	Sees	Visual field deficits (anopsia) Loss of light reflex with III Only nerve to be affected by MS
III	Oculomotor	Motor	Innervates SR, IR, MR, IO extraocular muscles: adduction (MR) most important action Raises eyelid (levator palpebrae superioris)	Diplopia, external strabismus Loss of parallel gaze Ptosis
			Constricts pupil (sphincter pupillae) Accommodates (ciliary muscle)	Dilated pupil, loss of light reflex with II Loss of near response
IV	Trochlear	Motor	Superior oblique—depresses and abducts eyeball (makes eyeball look down and out) Intorts	Weakness looking down with adducted eye Trouble going down stairs Head tilts away from lesioned side
V	Trigeminal	Mixed		
	Ophthalmic (V1)		General sensation (touch, pain, temperature) of forehead/scalp/cornea	V1—loss of general sensation in skin of forehead/scalp Loss of blink reflex with VII
	Maxillary (V2)		General sensation of palate, nasal cavity, maxillary face, maxillary teeth	V2—loss of general sensation in skin over maxilla, maxillary teeth
	Mandibular (V3)		General sensation of anterior two thirds of tongue, mandibular face, mandibular teeth	V3—loss of general sensation in skin over mandible, mandibular teeth, tongue, weakness in chewing
			Motor to muscles of mastication (temporalis, masseter, medial and lateral pterygoids) and anterior belly of digastric, mylohyoid, tensor tympani, tensor palati	Jaw deviation toward weak side Trigeminal neuralgia—intractable pain in V2 or V3 territory

Definition of abbreviations: IO, inferior oblique; MR, medial rectus; IR, inferior rectus; MS, multiple sclerosis; SR, superior rectus.

(*Continued*)

► Cranial Nerves: Functional Features (*Cont'd.*)

CN	Name	Type	Function	Results of Lesions
VI	Abducens	Motor	Lateral rectus—abducts eyeball	Diplopia, internal strabismus Loss of parallel gaze, "pseudoptosis"
VII	Facial	Mixed	To muscles of facial expression, posterior belly of digastric, stylohyoid, stapedius Salivation (submandibular, sublingual glands) Taste in anterior two thirds of tongue/palate Tears (lacrimal gland)	Corner of mouth droops, cannot close eye, cannot wrinkle forehead, loss of blink reflex, hyperacusis; Bell palsy—lesion of nerve in facial canal Alteration or loss of taste (ageusia) Eye dry and red
VIII	Vestibulocochlear	Sensory	Hearing Angular acceleration (head turning) Linear acceleration (gravity)	Sensorineural hearing loss Loss of balance, nystagmus
IX	Glossopharyngeal	Mixed	Sense of pharynx, carotid sinus/body Salivation (parotid gland) Taste and somatosensation of posterior one third of tongue Motor to one muscle—stylopharyngeus	Loss of gag reflex with X
X	Vagus	Mixed	To muscles of palate and pharynx for swallowing except tensor palati (V) and stylopharyngeus (IX) To all muscles of larynx (phonates) Sensory of larynx and laryngopharynx Sensory of GI tract To GI tract smooth muscle and glands in foregut and midgut	Nasal speech, nasal regurgitation Dysphagia, palate droop Uvula pointing away from affected side Hoarseness/fixed vocal cord Loss of gag reflex with IX Loss of cough reflex
XI	Accessory	Motor	Head rotation to opposite side (sternocleidomastoid) Elevates and rotates scapula (trapezius)	Weakness turning head to opposite side Shoulder droop
XII	Hypoglossal	Motor	Tongue movement (styloglossus, hyoglossus, genioglossus, and intrinsic tongue muscles—palatoglossus is by X)	Tongue pointing toward same (affected) side on protrusion

► Skull Base Anatomy

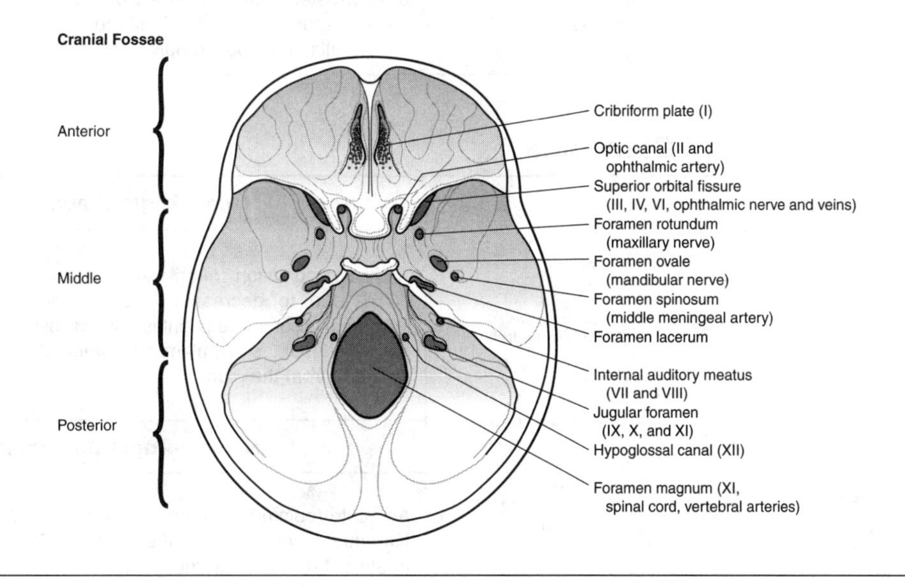

Visual System

► Visual Field Defects

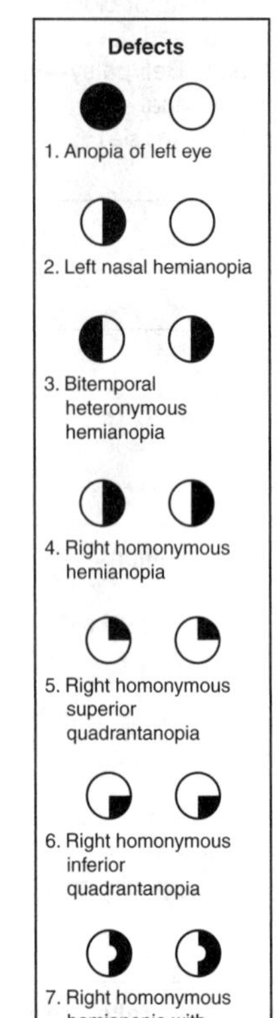

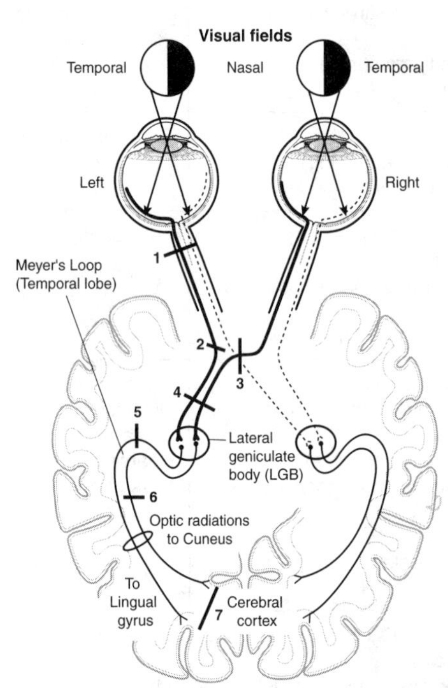

Notes

Like a camera, the lens inverts the image of the visual field, so the nasal retina receives information from the temporal visual field, and the temporal retina receives information from the nasal visual field.

At the **optic chiasm**, optic nerve fibers from the nasal half of each retina cross and project to the contralateral optic tract.

Most fibers from the **optic tract** project to the **lateral geniculate body (LGB)**; some also project to the pretectal area (light reflex), the superior colliculi (reflex gaze), and the suprachiasmatic nuclei (circadian rhythm). The LGB projects to the **primary visual cortex** (striate cortex, Brodmann area 17) of the occipital lobe via the optic radiations.

- Visual information from the lower retina (upper contralateral visual field) → temporal lobe (**Meyer loop**) → **lingual gyrus**
- Visual information from the upper retina (lower contralateral visual field) → parietal lobe → **cuneus gyrus**

Clinical Correlate (Some Causes of Lesions)

1. Optic neuritis, central retinal artery occlusion
2. Internal carotid artery aneurysm
3. Pituitary adenoma, craniopharyngioma

5. Middle cerebral artery (MCA) occlusion

6, 7. Posterior cerebral artery occlusion
Macula is spared in 7 due to collateral blood supply from MCA.

► Anatomy of the Eye and Glaucoma

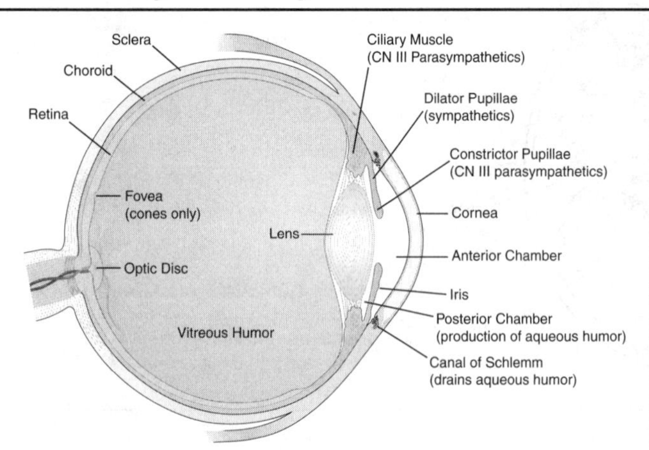

Open-Angle Glaucoma

A chronic condition (often with increased intraocular pressure [IOP]) due to decreased reabsorption of aqueous humor, leading to progressive (painless) visual loss and, if left untreated, blindness. IOP is a balance between fluid formation and its drainage from the globe.

Narrow-Angle Glaucoma

An acute (painful) or chronic (genetic) condition with increased IOP due to blockade of the canal of Schlemm. Emergency treatment prior to surgery often involves cholinomimetics, carbonic anhydrase inhibitors, and/or mannitol.

► Pharmacology of the Eye

The eye is predominantly innervated by the parasympathetic nervous system. Therefore, application of muscarinic antagonists or ganglionic blockers has a large effect by blocking the parasympathetic nervous system.

Structure	Predominant Receptor	Receptor Stimulation	Receptor Blockade
Pupillary sphincter ms. (iris)	M_3 receptor (PANS)	Contraction → miosis	Relaxation → mydriasis
Radial dilator ms. (iris)	α receptor (SANS)	Contraction → mydriasis	Relaxation → miosis
Ciliary ms.	M_3 receptor (PANS)	Contraction → accommodation for near vision	Relaxation → focus for far vision
Ciliary body epithelium	β receptor (SANS)	Secretion of aqueous humor	Decreased aqueous humor production

Definition of abbreviations: ms., muscle; PANS, parasympathetic nervous system; SANS, sympathetic nervous system.

► Drugs Used to Treat Glaucoma

Drug Class	Drug	Mechanism
Cholinomimetics (miotics)	Pilocarpine (M agonist) Carbachol (M agonist) Physostigmine (AChEI) Echothiophate (AChEI)	Contracts ciliary muscle and opens trabecular meshwork, increasing the outflow of aqueous humor through the canal of Schlemm
Beta blockers	Timolol (nonselective) Betaxolol (β_1)	Blocks actions of NE at ciliary epithelium to ↓ aqueous humor secretion
Prostaglandins	Latanoprost ($PGF_{2\alpha}$ analog)	↑ aqueous humor outflow; can darken the iris
Alpha agonists	Epinephrine Dipivefrin	↑ aqueous humor outflow
Alpha-2 agonists	Apraclonidine Brimonidine	↓ aqueous humor secretion
Diuretics	Acetazolamide (oral; CAI) Dorzolamide, brinzolamide (topical; CAI) Mannitol (for narrow-angle; osmotic)	CAI: ↓ HCO_3^- → ↓ aqueous humor secretion

Definition of abbreviations: AChEI, acetylcholinesterase inhibitor; CAI, carbonic anhydrase inhibitor; M, muscarinic cholinergic receptor.

► Pupillary Light Reflex Pathway

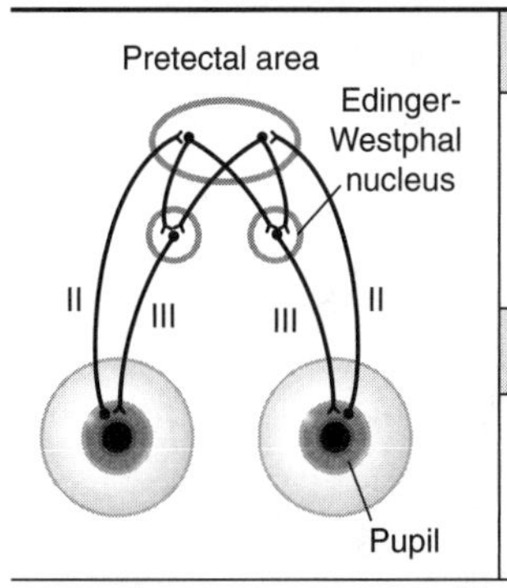

Afferent Limb: CN II

Light stimulates ganglion retinal cells → impulses travel up **CNII**, which projects **bilaterally** to the **pretectal nuclei** (midbrain)

The pretectal nucleus projects **bilaterally** → **Edinger-Westphal nuclei (CN III)**

Efferent Limb: CN III

Edinger-Westphal nucleus (preganglionic parasympathetic) → **ciliary ganglion** (postganglionic parasympathetic) → **pupillary sphincter muscle** → **miosis**

Note: This is a simplified diagram; the ciliary ganglion is not shown.

Because cells in the pretectal area supply the Edinger-Westphal nuclei bilaterally, shining light in one eye → constriction in the ipsilateral pupil (direct light reflex) and the contralateral pupil (consensual light reflex).

Because this reflex does not involve the visual cortex, a person who is cortically blind can still have this reflex.

► Accommodation–Convergence Reaction

When an individual focuses on a nearby object after looking at a distant object, three events occur:

1. **Accommodation**
2. **Convergence**
3. **Pupillary constriction (miosis)**

In general, stimuli from light → visual cortex → superior colliculus and pretectal nucleus → Edinger-Westphal nucleus (1, 3) and oculomotor nucleus (2).

Accommodation: Parasympathetic fibers contract the ciliary muscle, which relaxes suspensory ligaments, allowing the lens to increase its convexity (become more round). This increases the refractive index of the lens, thereby focusing a nearby object on the retina.

Convergence: Both medial rectus muscles contract, adducting both eyes.

Pupillary constriction: Parasympathetic fibers contract the pupillary sphincter muscle → miosis.

► Clinical Correlations

Pupillary Abnormalities	
Argyll Robertson pupil (pupillary light-near dissociation)	No direct or consensual light reflex; accommodation-convergence intact Seen in **neurosyphilis**, diabetes
Relative afferent (Marcus Gunn) pupil	Lesion of afferent limb of pupillary light reflex; diagnosis made with swinging flashlight Shine light in Marcus Gunn pupil → pupils do not constrict fully Shine light in normal eye → pupils constrict fully Shine light immediately again in affected eye → apparent dilation of both pupils because stimulus carried through that CN II is weaker; seen in multiple sclerosis
Horner syndrome	Caused by a lesion of the oculosympathetic pathway; syndrome consists of miosis, ptosis, apparent enophthalmos, and hemianhidrosis
Adie pupil	Dilated pupil that reacts sluggishly to light, but better to accommodation; often seen in women and often associated with loss of knee jerks
Transentorial (uncal) herniation	Increased intracranial pressure → leads to uncal herniation → CN III compression → fixed and dilated pupil, "down-and-out" eye, ptosis

▶ Eye Movement Control Systems

Extraocular Muscles: Function and Innervation			
CN III	**Medial rectus:** adducts eye **Superior rectus:** elevates, intorts, adducts eye **Inferior rectus:** depresses, extorts, adducts eye **Inferior oblique:** elevates, extorts, abducts eye	**CN IV**	**Superior oblique:** depresses, intorts, abducts eye
		CN VI	**Lateral rectus:** Abducts eye

Two important eye movements are **abduction** (away from nose, **CN VI**) and **adduction** (toward nose, **CN III**).

For the eyes to move together **(conjugate gaze)**, the oculomotor nuclei and abducens nuclei are interconnected by the **medial longitudinal fasciculus (MLF)**.

Horizontal gaze is controlled by two gaze centers:

1. **Frontal eye field** (contralateral gaze)
2. **PPRF** (paramedial pontine reticular formation, ipsilateral gaze)

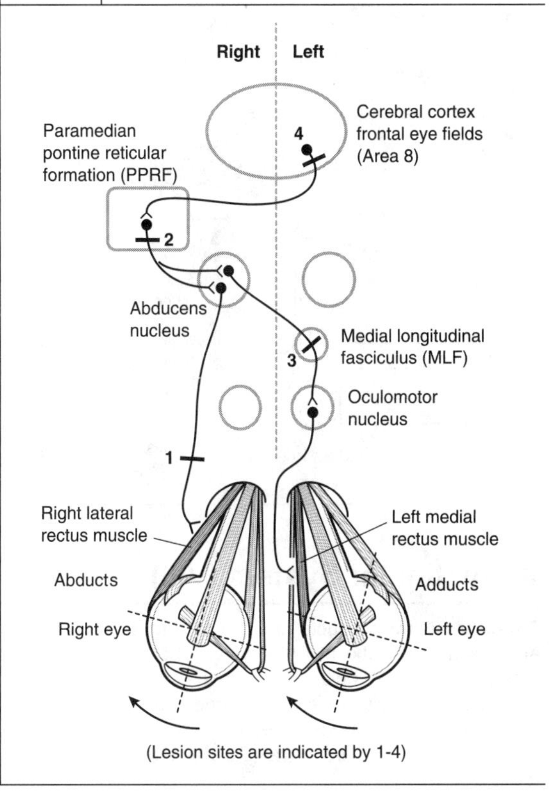

(Lesion sites are indicated by 1-4)

Clinical Correlation

Lesion Examples	Symptoms
1. Right CN VI	Right eye cannot look right
2. Right PPRF	Neither eye can look right
3. Left MLF	**Internuclear ophthalmoplegia (INO)** Left eye cannot look right; convergence is intact (this is how to distinguish an INO from an oculomotor lesion); right eye has nystagmus; seen in multiple sclerosis
4. Left frontal eye field	Neither eye can look right; but slow drift to left

► Trigeminal Nerve (V)

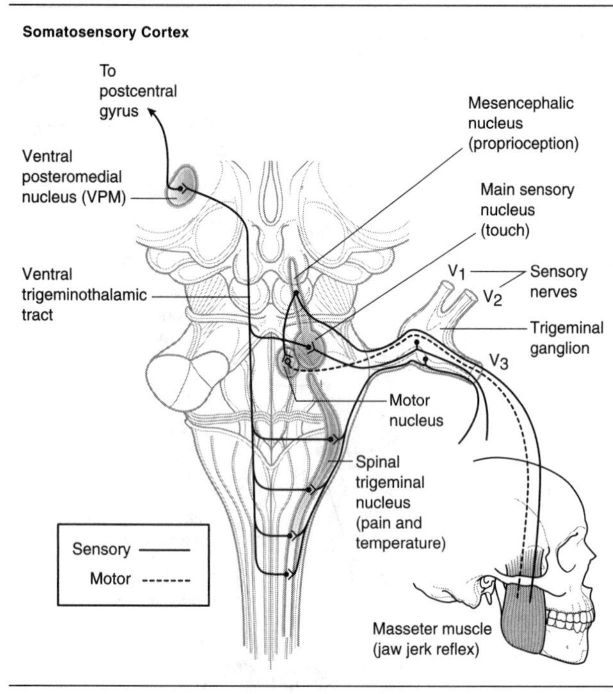

► Facial Nerve (VII)

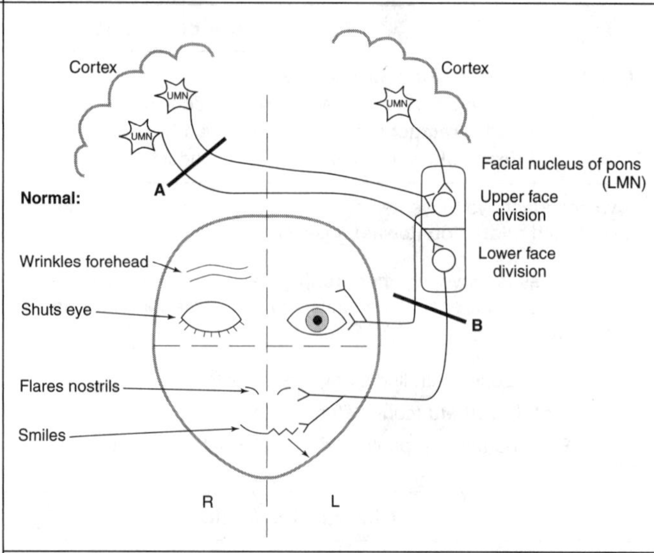

A. **Upper motor neuron lesion** → weakness of **contralateral lower face**

B. **Lower motor neuron lesion** → paralysis of **ipsilateral upper and lower face (Bell palsy)**

► Vestibular System (VIII)

Three semicircular ducts respond to **angular acceleration and deceleration** of the head. The **utricle** and **saccule** respond to **linear acceleration** and the pull of **gravity**. There are four **vestibular nuclei** in the medulla and pons, which receive information from CN VIII. Fibers from the vestibular nuclei join the MLF and supply the motor nuclei of CNs III, IV, and VI, thereby regulating conjugate eye movements. Vestibular nuclei also receive and send information to the **flocculonodular lobe** of the cerebellum.

Vestibulo-Ocular Reflex

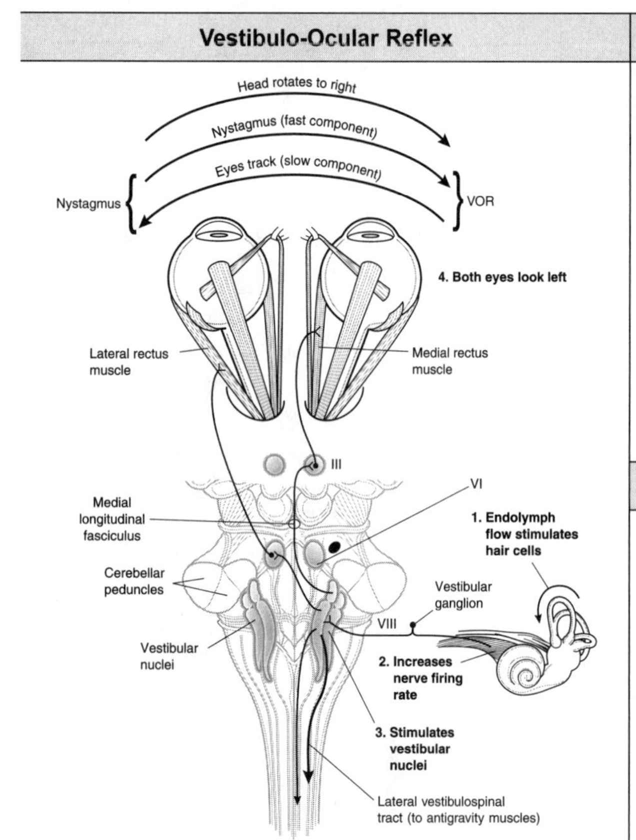

Caloric Test

This stimulates the horizontal semicircular ducts; can be used as a test of brain stem function in unconscious patients.

Normal results:

- **Cold water** irrigation of ear → nystagmus to opposite side
- **Warm water** irrigation of ear → nystagmus to same side
- **COWS:** cold opposite, warm same

Clinical Correlation

Vertigo: the perception of rotation. Usually severe in peripheral disease and mild in brain stem disease. Chronic vertigo suggests a central lesion.

Ménière disease: characterized by abrupt, recurrent attacks of vertigo lasting minutes to hours; accompanied by deafness or tinnitus and is usually in one ear. Nausea and vomiting may occur. Due to distention of fluid spaces in the cochlear and vestibular parts of the labyrinth.

▶ Auditory System (VIII)

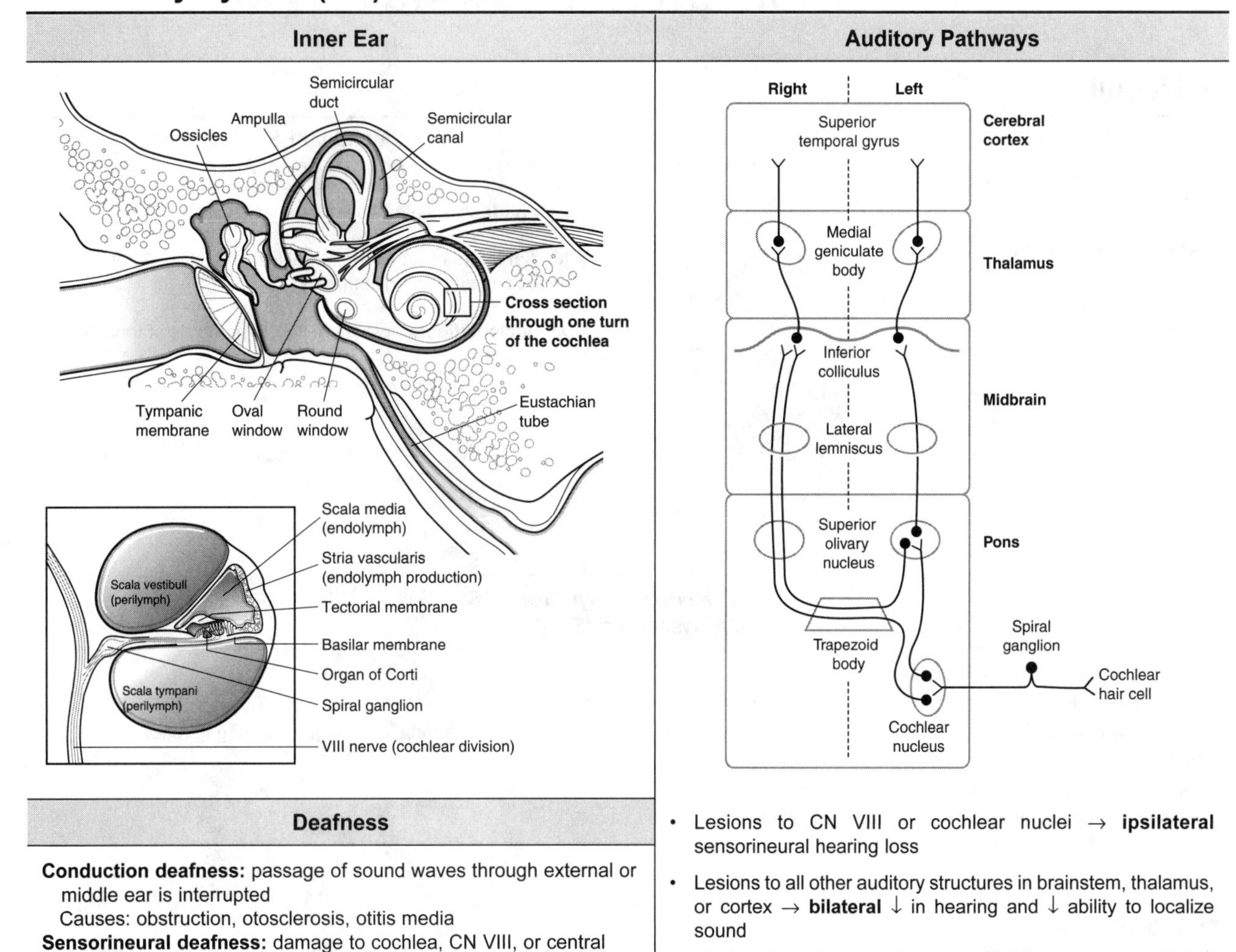

Deafness

Conduction deafness: passage of sound waves through external or middle ear is interrupted
Causes: obstruction, otosclerosis, otitis media

Sensorineural deafness: damage to cochlea, CN VIII, or central auditory connections

- Lesions to CN VIII or cochlear nuclei → **ipsilateral** sensorineural hearing loss
- Lesions to all other auditory structures in brainstem, thalamus, or cortex → **bilateral** ↓ in hearing and ↓ ability to localize sound

Auditory Tests

Weber test: place tuning fork on vertex of skull. If **unilateral conduction deafness** → vibration is louder in affected ear; if **unilateral sensorineural deafness** → vibration is louder in normal ear

Rinne test: place tuning fork on mastoid process (bone conduction) until vibration is not heard, then place fork in front of ear (air conduction). If **unilateral conduction deafness** → no air conduction after bone conduction is gone; if **unilateral sensorineural deafness** → air conduction present after bone conduction is gone

Brain Stem Lesions

▶ Medulla

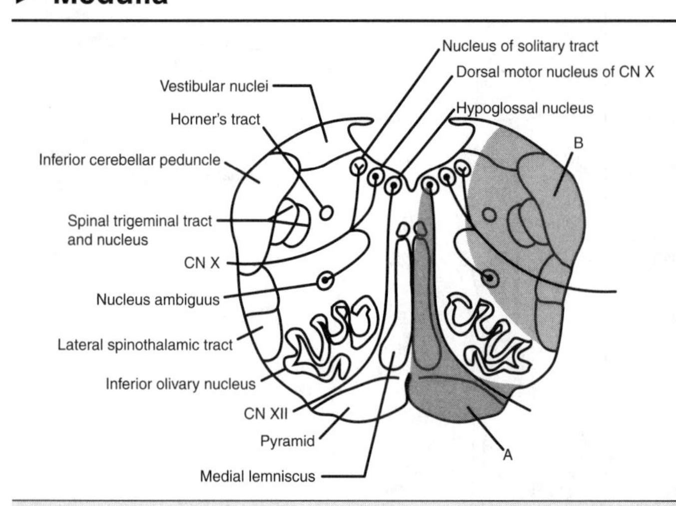

Medial Medullary Syndrome (A)
Anterior Spinal Artery

Pyramid: contralateral spastic paresis (body)

Medial lemniscus: contralateral loss of tactile, vibration, conscious proprioception (body)

XII nucleus/fibers: ipsilateral flaccid paralysis of tongue

Lateral Medullary Syndrome (B)
PICA, Wallenberg Syndrome

Inferior cerebellar peduncle: ipsilateral limb ataxia
Vestibular nuclei: vertigo, nausea/vomiting, nystagmus (away from lesion)
Nucleus ambiguus (CN IX, X, XI): ipsilateral paralysis of larynx, pharynx, palate → dysarthria, dysphagia, loss of gag reflex
Spinal V: ipsilateral pain/temperature loss (face)
Spinothalamic tract: Contralateral pain/temperature loss (body)
Descending hypothalamics: ipsilateral Horner syndrome

▶ Pons

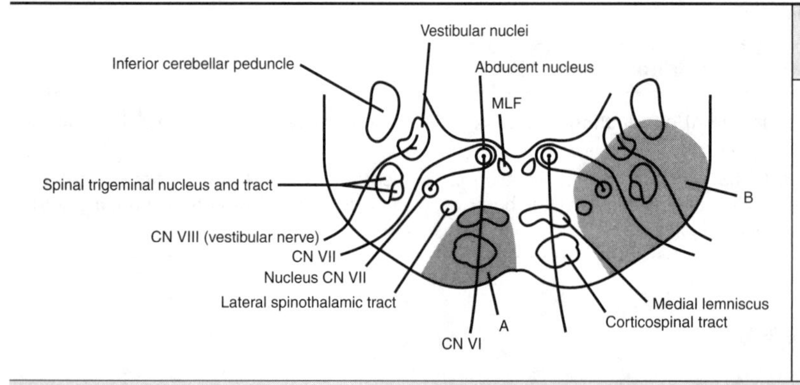

Medial Pontine Syndrome (A)
Paramedian Branches of Basilar Artery

Corticospinal tract: contralateral spastic hemiparesis

Medial lemniscus: contralateral loss of tactile/position/vibration sensation on body

Fibers of VI: medial strabismus

Lateral Pontine Syndrome (B)
AICA

Middle cerebellar peduncle: ipsilateral ataxia
Vestibular nuclei: vertigo, nausea and vomiting, nystagmus
Facial nucleus and fibers: ipsilateral facial paralysis; ipsilateral loss of taste (anterior 2/3 tongue), lacrimation, salivation, and corneal reflex; hyperacusis
Spinal trigeminal nucleus/tract: ipsilateral pain/temperature loss (face)
Spinothalamic tract: contralateral pain/temperature loss (body)
Cochlear nucleus/VIII fibers: ipsilateral hearing loss
Descending sympathetics: ipsilateral Horner syndrome

▶ Midbrain

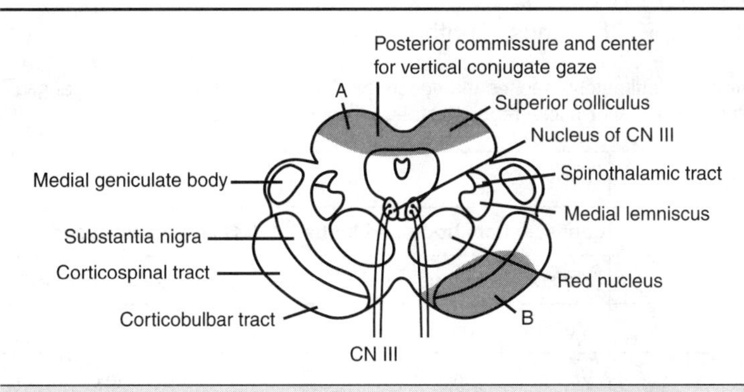

Dorsal Midbrain (Parinaud) Syndrome (A) Tumor in Pineal Region
Superior colliculus/pretectal area: paralysis of upward gaze, various pupillary abnormalities **Cerebral aqueduct:** noncommunicating hydrocephalus

Medial Midbrain (Weber) Syndrome (B) Branches of PCA
Fibers of III: ipsilateral oculomotor palsy (lateral strabismus, dilated pupil, ptosis) **Corticospinal tract:** contralateral spastic hemiparesis **Corticobulbar tract:** contralateral spastic hemiparesis of lower face

Cerebellum

The cerebellum controls posture, muscle tone, learning of repeated motor functions, and coordinates voluntary motor activity. Diseases of the cerebellum result in disturbances of gait, balance, and coordinated motor actions, but there is no paralysis or inability to start or stop movement.

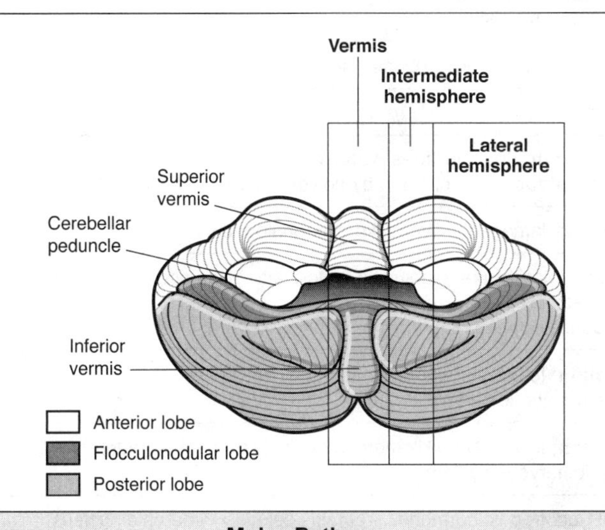

Input
Climbing fibers (from inferior olivary nucleus of medulla), **mossy fibers** (from vestibular nucleus, spinal cord, pons); most input via **ICP** and **MCP**

Output
From deep cerebellar nucleus (fastigial, interpositus, dentate); most output via **SCP**

Three Layers
Molecular, Purkinje, granule cell

Major Pathway	Dysfunction
Purkinje cells → deep cerebellar nucleus; dentate nucleus → contralateral VL → 1° motor cortex → pontine nuclei → contralateral cerebellar cortex	• **Hemisphere lesions** → ipsilateral symptoms; **intention** tremor, dysmetria, dysdiadochokinesia, scanning dysarthria, nystagmus, hypotonia • **Vermal lesions** → truncal ataxia

Definition of abbreviations: ICP, inferior cerebellar peduncle; MCP, middle cerebellar peduncle; SCP, superior cerebellar peduncle; VL, ventral lateral nucleus.

Diencephalon

Thalamus, Hypothalamus, Epithalamus, Subthalamus

Thalamus—serves as a major sensory relay for information that ultimately reaches the neocortex. Motor control areas (basal ganglia, cerebellum) also synapse in the thalamus before reaching the cortex. Other nuclei regulate states of consciousness.

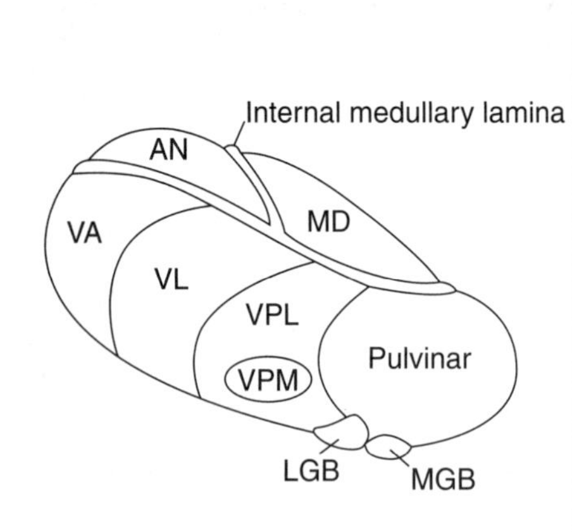

THALAMIC NUCLEI	INPUT	OUTPUT
VPL	Sensory from **body and limbs**	Somatosensory cortex
VPM	Sensory from **face**	Somatosensory cortex
VA/VL	**Motor** info from BG, cerebellum	Motor cortices
LGB	**Visual** from optic tract	1° visual cortex
MGB	**Auditory** from inferior colliculus	1° auditory cortex
AN	Mammillary nucleus (via mamillothalamic tract)	Cingulate gyrus (part of **Papez** circuit)
MD	(Dorsomedial nucleus). Involved in **memory** Damaged in **Wernicke-Korsakoff** syndrome	
Pulvinar	Helps integrate somesthetic, visual, and auditory input	
Midline/intralaminar	Involved in **arousal**	

Hypothalamus—helps maintain homeostasis; has roles in the autonomic, endocrine, and limbic systems

HYPOTHALAMIC NUCLEI	FUNCTIONS AND LESIONS
Lateral hypothalamic	**Feeding center;** lesion → starvation
Ventromedial	**Satiety center;** lesion → hyperphagia, obesity, savage behavior
Suprachiasmatic	Regulates circadian rhythms
Supraoptic and paraventricular	Synthesizes **ADH** and **oxytocin; regulates water balance** Lesion → **diabetes insipidus**, characterized by polydipsia and polyuria
Mammillary body	Input from hippocampus; damaged in Wernicke encephalopathy
Arcuate	Produces hypothalamic releasing and inhibiting factors and gives rise to tuberohypophysial tract Has neurons that produce dopamine (prolactin-inhibiting factor)
Anterior	**Temperature regulation;** lesion → hyperthermia Stimulates the parasympathetic nervous system
Posterior	**Temperature regulation;** lesion → poikilothermia (inability to thermoregulate) Stimulates sympathetic nervous system
Preoptic area	Regulates release of gonotrophic hormones; contains sexually dimorphic nucleus Lesion before puberty → arrested sexual development; lesion after puberty → amenorrhea or impotence
Dorsomedial	Stimulation → savage behavior

Epithalamus—Consists of pineal body and habenular nuclei. The **pineal body** aids in the regulation of **circadian rhythms**.

Subthalamus—The **subthalamic nucleus** is involved in **basal ganglia** circuitry. Lesion → **hemiballismus** (contralateral flinging movements of one or both extremities)

Definition of abbreviations: ADH, antidiuretic hormone; AN, anterior nuclear group; BG, basal ganglia; LBG, lateral geniculate body; MD, mediodorsal nucleus; MGB, medial geniculate body; VA, ventral anterior nucleus; VL, ventral lateral nucleus; VLP, ventroposterolateral nucleus; VPM, ventroposteromedial nucleus.

Basal Ganglia

The basal ganglia initiate and provide gross control over skeletal muscle movements. The basal ganglia are sometimes called the **extrapyramidal** nervous system because they modulate the pyramidal (corticospinal) nervous system.

► Basal Ganglia Components

Striatum (caudate and putamen)	**Subthalamic nucleus** (diencephalon)
Globus pallidus (external and internal segments)	**Lentiform nucleus** (globus pallidus and putamen)
Substantia nigra (midbrain)	**Corpus striatum** (lentiform nucleus and caudate)

Together with the cerebral cortex and **VA/VL** thalamic nuclei, these structures form two parallel but antagonistic circuits known as the **direct** and **indirect pathways**. The direct pathway increases cortical excitation and promotes movement, and the indirect pathway decreases cortical excitation and inhibits movement. Hypokinetic movement disorders (e.g., Parkinson disease) result in a lesion of the direct pathway, and hyperkinetic movement disorders (e.g., Huntington disease, hemiballismus) result from indirect pathway lesions.

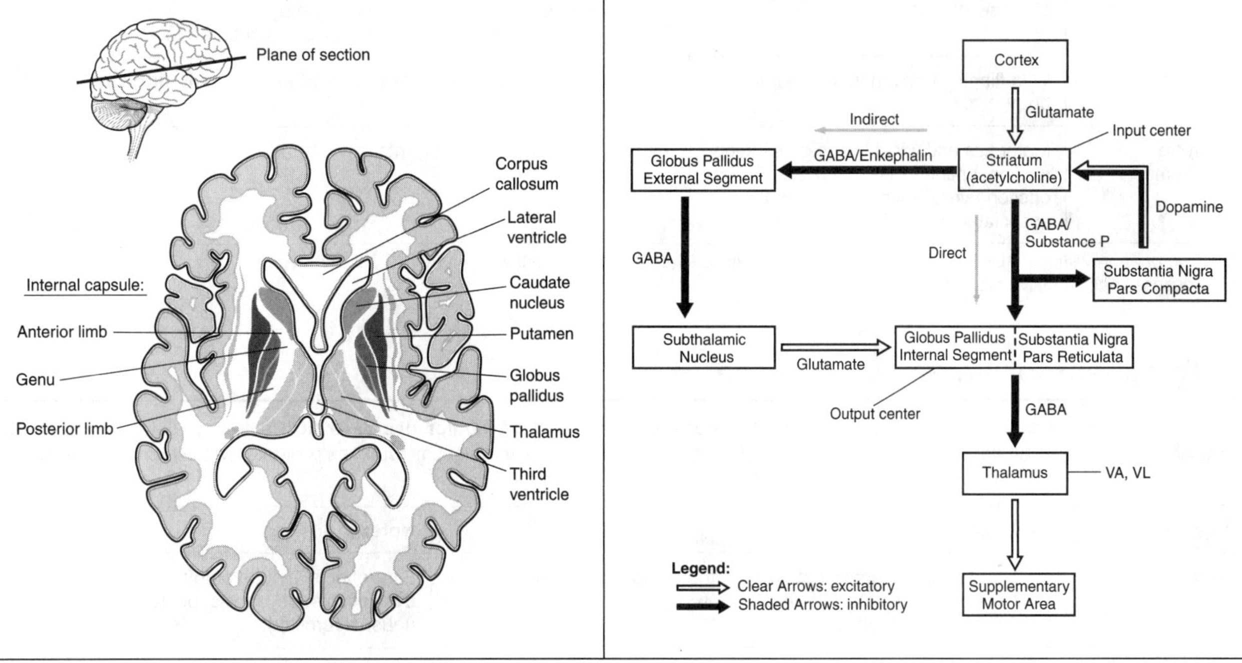

Definition of abbreviation: VA/VL, ventral anterior/ventral lateral thalamic nuclei.

► Diseases of the Basal Ganglia

Disease	Clinical Manifestations	Notes
Parkinson disease	Bradykinesia, cogwheel rigidity, pill rolling (resting) tremor, shuffling gate, stooped posture, masked facies, depression, dementia	**Loss of pigmented dopaminergic neurons from substantia nigra** **Lewy bodies:** intracytoplasmic eosinophilic inclusions, contain α-synuclein Known causes of parkinsonism: infections, vascular, and toxic insults (e.g., **MPTP**)
Huntington disease	Chorea (multiple, rapid, random movements), athetosis (slow writhing, movements), personality changes, dementia Onset: 20–40 years	**Degeneration** of GABAergic neurons in **neostriatum**, causing atrophy of neostriatum (and ventricular dilatation) **Autosomal dominant** **Unstable nucleotide repeat** on gene in chromosome 4, which codes for huntingtin protein Disease shows **anticipation** and **genomic imprinting** **Treatment:** antipsychotic agents, benzodiazepines, anticonvulsants
Wilson disease (hepatolenticular degeneration)	Tremor, asterixis, parkinsonian symptoms, chorea, neuropsychiatric symptoms; fatty change, hepatitis, or cirrhosis of liver	**Autosomal recessive defect in copper transport** Accumulation of copper in liver, brain, and eye (Descemet membrane, producing **Kayser-Fleischer ring**) Lesions in basal ganglia (especially putamen) **Treatment: penicillamine** (a chelator), zinc acetate (blocks absorption)
Hemiballism	Wild flinging movements of half the body	Hemorrhagic destruction of **contralateral subthalamic nucleus** Hypertensive patients
Tourette syndrome	Motor tics and vocal tics (e.g., snorting, sniffing, uncontrolled and often obscene vocalizations), commonly associated with OCD and ADHD	**Treatment:** Antipsychotic agents

Definition of abbreviations: ADHD, attention deficit hyperactivity disorder; MPTP, 1-methyl-4-phenyl-1,2,3,6-tetrahydropyridine; OCD, obsessive-compulsive disorder.

► Treatment for Parkinson Disease

The pharmacologic goal in the treatment of Parkinson disease is to **increase DA and/or decrease ACh** activity in the striatum, thereby correcting the DA/ACh imbalance. Additional treatment strategies include surgical intervention, such as pallidotomy, thalamotomy, deep brain stimulation, and transplantation.

Agents	Mechanism	Notes
Dopamine precursor: L-dopa	Dopamine precursor that crosses the blood–brain barrier (BBB) Converted to DA by DOPA decarboxylase (L-aromatic amino acid decarboxylase)	Side effects include on/off phenomena, dyskinesias, psychosis, postural hypotension, nausea/vomiting
DOPA decarboxylase inhibitor: carbidopa	Inhibits DOPA decarboxylase in the periphery, preventing L-dopa from being converted to dopamine in the periphery; instead, L-dopa crosses the BBB and is converted to dopamine in the brain	Often given in combination with L-dopa (Sinemet®)
DA agonists: bromocriptine pramipexole ropinirole	Stimulates D_2 receptors in the striatum; pramipexole and ropinirole are also D_3 agonists	Pramipexole and ropinirole now considered first-line drugs in the initial management of PD; pramipexole and ropinirole are also used in **restless legs syndrome**
MAO B inhibitor: selegiline rasagiline	Inhibits MAO type B, which preferentially metabolizes dopamine	Not to be taken with SSRIs (serotonin syndrome) or meperidine
Antimuscarinics: benztropine trihexyphenidyl biperiden	Blocks muscarinic receptors in the striatum	↓ tremor and rigidity, have little effect on bradykinesia Antimuscarinic side effects
COMT inhibitors: entacapone tolcapone	COMT inhibitors increase the efficacy of L-dopa. COMT metabolizes L-dopa to 3-*O*-methyldopa (3OMD), which competes with L-dopa for active transport into the CNS	Used as an adjunct to L-dopa/carbidopa, increases the "on" time
Amantadine	Increases dopaminergic neurotransmission, antimuscarinic; also an antiviral	Antimuscarinic effects, livedo reticularis

Limbic System

The limbic system is involved in emotion, memory, attention, feeding, and mating behaviors. It consists of a core of cortical and diencephalic structures found on the medial aspect of the hemisphere. The limbic system modulates feelings, such as fear, anxiety, sadness, happiness, sexual pleasure, and familiarity.

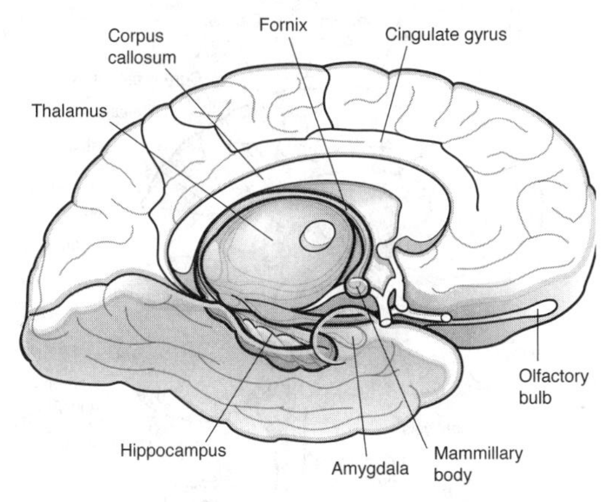

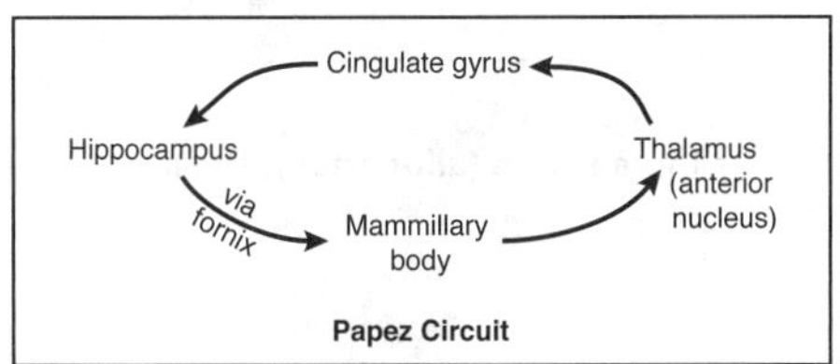

Papez Circuit

Limbic Structures and Function

- Hippocampal formation (hippocampus, dentate gyrus, the subiculum, and entorhinal cortex)
- Amygdala
- Septal nuclei
- The hippocampus is important in learning and memory. The amygdala attaches an emotional significance to a stimulus and helps imprint the emotional response in memory.

Limbic Connections

- The limbic system is interconnected with anterior and dorsomedial nuclei of the thalamus and the mammillary bodies.
- The cingulate gyrus is the main limbic cortical area.
- Limbic-related structures also project to wide areas of the prefrontal cortex.
- Central projections of olfactory structures reach parts of the temporal lobe and the amygdala.

Papez Circuit

Axons of hippocampal pyramidal cells converge to form the fimbria and, finally, the fornix. The fornix projects mainly to the mammillary bodies in the hypothalamus. The mammillary bodies project to the anterior nucleus of the thalamus (mammillothalamic tract). The anterior nuclei project to the cingulate gyrus, and the cingulate gyrus projects to the entorhinal cortex (via the cingulum). The entorhinal cortex projects to the hippocampus (via the perforant pathway).

▶ Clinical Correlations

Anterograde Amnesia

Bilateral damage to the medial temporal lobes, including the **hippocampus**, results in a profound loss of the ability to acquire new information.

Wernicke Encephalopathy and Korsakoff Syndrome

Wernicke encephalopathy typically occurs in alcoholics who have a **thiamine deficiency**. Patients present with ocular palsies, confusion, and gait ataxia. If the thiamine deficiency is not corrected in time, patients can develop **Korsakoff syndrome**, characterized by **anterograde amnesia**, retrograde amnesia, and confabulation. Lesions are found in the **mammillary bodies** and the **dorsomedial nuclei of the thalamus**. Wernicke encephalopathy is reversible; Korsakoff syndrome is not.

Klüver-Bucy Syndrome

Klüver-Bucy syndrome results from bilateral lesions of the anterior temporal lobes, including the **amygdala**. Symptoms include placidity (decrease in aggressive behavior), psychic blindness (visual agnosia), increased oral exploratory behavior, hypersexuality and loss of sexual preference, hypermetamorphosis (visual stimuli are repeatedly approached as if they were new), and anterograde amnesia.

Cerebral Cortex

The cerebral cortex is highly convoluted with bulges (**gyri**) separated by spaces (**sulci**). Several prominent sulci separate the cortex into four lobes: **frontal**, **parietal**, **temporal**, and **occipital**. The figures below show the lateral and medial views of the right cerebral hemispheres.

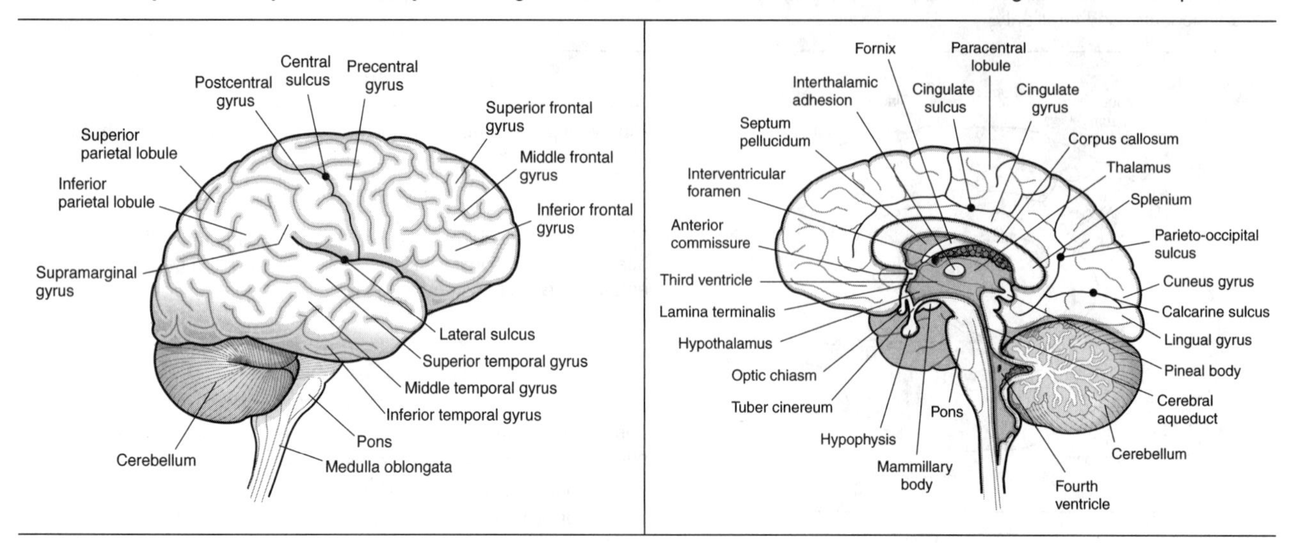

Most of the cortex has **six layers** (**neocortex**). The olfactory cortex and hippocampus have **three layers** (**allocortex**). The figure below shows the six-layered cortex:

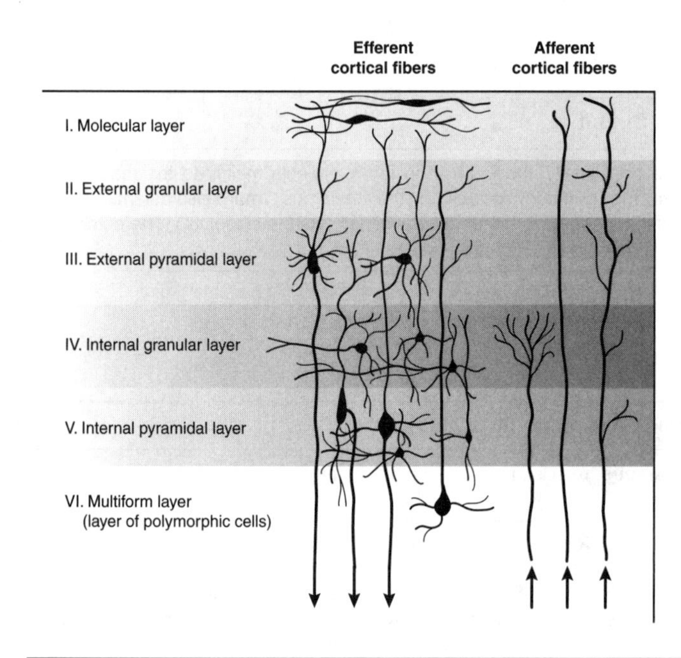

Key Afferents/Efferents

Layer IV receives thalamocortical inputs.

Layer V gives rise to corticospinal and corticobulbar tracts.

► Key Features of Lobes

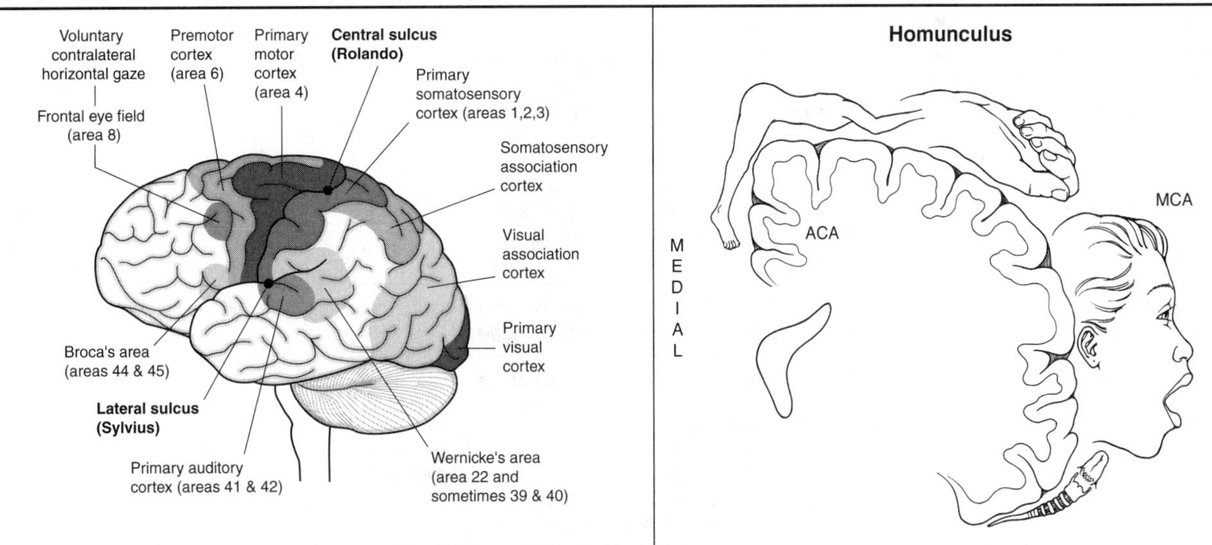

Lobes	Important Regions	Deficit After Lesion
Frontal	Primary motor and premotor cortex	Contralateral spastic paresis (region depends on area of homunculus affected; see figure above)
	Frontal eye fields	Eyes deviate to ipsilateral side
	Broca speech area*	**Broca aphasia (expressive, nonfluent aphasia):** patient can understand written and spoken language, but speech and writing are slow and effortful; patients are aware of their problem; often associated with contralateral facial and arm weakness
	Prefrontal cortex	**Frontal lobe syndrome:** symptoms can include poor judgment, difficulty concentrating and problem solving, apathy, inappropriate social behavior
Parietal	Primary somatosensory cortex	Contralateral hemihypesthesia (region depends on area of homunculus affected)
	Superior parietal lobule	Contralateral astereognosis and sensory neglect, apraxia
	Inferior parietal lobule	**Gerstmann syndrome** (if dominant hemisphere): right/left confusion, dyscalculia and dysgraphia, finger agnosia, contralateral hemianopia or lower quadrantanopia
Temporal	Primary auditory cortex	Bilateral damage → deafness Unilateral leads to slight hearing loss
	Wernicke area*	**Wernicke aphasia (receptive, fluent aphasia):** patient cannot understand any form of language; speech is fast and fluent, but not comprehensible
	Hippocampus	Bilateral lesions lead to inability to consolidate short-term to long-term memory
	Amygdala	**Klüver-Bucy syndrome:** hyperphagia, hypersexuality, visual agnosia
	Olfactory bulb, tract, primary cortex	Ipsilateral anosmia
	Meyer loop (visual radiations)	Contralateral upper quadrantanopia ("pie in the sky")
Occipital	Primary visual cortex	Blindness

*In the dominant hemisphere. Eighty percent of people are left-hemisphere dominant.

Alzheimer Disease

- Alzheimer disease accounts for 60% of all cases of dementia. The incidence increases with age.
- **Clinical:** insidious onset, progressive memory impairment, mood alterations, disorientation, aphasia, apraxia, and progression to a bedridden state with eventual death
- Five to 10% of AD cases are hereditary, early onset, and transmitted as an autosomal dominant trait.

Genetics of Alzheimer Disease

Gene	Location	Notes
Amyloid precursor protein (*APP*) gene	Chromosome 21	Virtually all Down syndrome patients are destined to develop AD in their forties. Down patients have triple copies of the *APP* gene.
Presenilin-1 gene	Chromosome 14	Majority of hereditary AD cases—early onset
Presenilin-2 gene	Chromosome 1	Early onset
Apolipoprotein E gene	Chromosome 19	Three allelic forms of this gene: epsilon 2, epsilon 3, and epsilon 4 The allele epsilon 4 of apolipoprotein E (*ApoE*) increases the risk for AD, epsilon 2 confers relative protection

Pathology of Alzheimer Disease

Lesions involve the neocortex, hippocampus, and subcortical nuclei, including forebrain cholinergic nuclei (i.e., basal nucleus of Meynert). These areas show atrophy, as well as characteristic microscopic changes. The earliest and most severely affected areas are the hippocampus and temporal lobe, which are involved in learning and memory.

Intra- and extracellular accumulation of abnormal proteins	**Aβ amyloid:** 42-residue peptide from a normal transmembrane protein, the amyloid precursor protein (APP) **Abnormal *tau*** (a microtubule-associated protein)
Senile plaques	Core of Aβ amyloid surrounded by dystrophic neuritic processes associated with microglia and astrocytes
Neurofibrillary tangles (NFT)	Intraneuronal aggregates of insoluble cytoskeletal elements, mainly composed of abnormally phosphorylated tau forming **paired helical filaments** (PHF)
Cerebral amyloid angiopathy (CAA)	Accumulation of Aβ amyloid within the media of small and medium-sized intracortical and leptomeningeal arteries; associated with intracerebral hemorrhage
Granulovacuolar degeneration (GVD) and Hirano bodies (HBs)	GVD and HBs develop in the hippocampus and are less significant diagnostically

Treatment of Alzheimer Disease

AChE Inhibitors
(rivastigmine, donepezil, galantamine, tacrine)

These agents prevent the metabolism of ACh to counteract the depletion in ACh in the cerebral cortex and hippocampus. Rivastigmine and tacrine also inhibit BuChE. Indicated for mild to moderate AD.

NMDA Antagonist
(memantine)

This is the newest class of agents used for the treatment of AD. It is hypothesized that overstimulation of NMDA receptors contributes to the symptomatology of AD. Indicated for moderate to severe AD. Often used in combination with the AChEIs.

Definition of abbreviations: AChE, acetylcholinesterase; BuChE, butyrylcholinesterase; NMDA, *N*-methyl-D-aspartate.

► Creutzfeldt-Jakob Disease (CJD)

Mechanism of Disease

Caused by a **prion protein** (PrP = 30-kD protein normally present in neurons encoded by gene on chromosome 20); PrP^{c} = normal conformation = alpha-helix; PrP^{sc} = abnormal conformation = beta-pleated sheet. PrP^{sc} facilitates conformational change of other PrP^{c} molecules into PrP^{sc}.

Spontaneous change from one form to another → sporadic cases (85% of total) of CJD

Mutations of PrP → hereditary cases (15% of total) of CJD

Pathology

Spongiform change: vacuolization of the neuropil in gray matter (especially cortex) due to large membrane-bound vacuoles within neuronal processes

- Associated with neuronal loss and astrogliosis
- **Kuru plaques** are deposits of amyloid of altered PrP protein

Clinical Manifestations

Rapidly progressive dementia, memory loss, startle myoclonus or other involuntary movements. EEG changes. Death within 6–12 months.

Variant CJD (vCJD)

Affects young adults. May be related to bovine spongiform encephalopathy (BSE; mad cow disease). Pathologically similar to CJD.

► Pick Disease (Lobar Atrophy)

- Rare cause of dementia
- Striking **atrophy of frontal and temporal lobes** with sparing of posterior structures
- Microscopic: swollen neurons (Pick cells) or neurons containing **Pick bodies** (round to oval inclusions that stain with silver stains)

► CNS Trauma

Concussion	Occurs with a change in momentum of the head (impact against a rigid surface) Loss of consciousness and reflexes, temporary respiratory arrest, and amnesia for the event
Contusion	Bruising to the brain resulting from impact of the brain against inner calvarial surfaces, especially along crests of orbital gyri (frontal lobe) and temporal poles Coup (site of injury) and contrecoup (site diametrically opposite) develop when the head is **mobile** at the time of impact.
Diffuse axonal injury	Injury to white matter due to acceleration/deceleration produces damage to axons at nodes of Ranvier with impairment of axoplasmic flow Poor prognosis, related to duration of coma

► Cerebral Herniations

Subfalcine (cingulate)	Cingulate gyrus displaced underneath the falx to the opposite side with compression of anterior cerebral artery
Transtentorial (uncal)	The uncus of the temporal lobe displaced over the free edge of the tentorium Compression of the third nerve, with pupillary dilatation on the same side Infarct in dependent territory Advanced stages: **Duret hemorrhage** within the central pons and midbrain
Cerebellar tonsillar	Displacement of cerebellar tonsils through the foramen magnum Compression of medulla leads to cardiorespiratory arrest

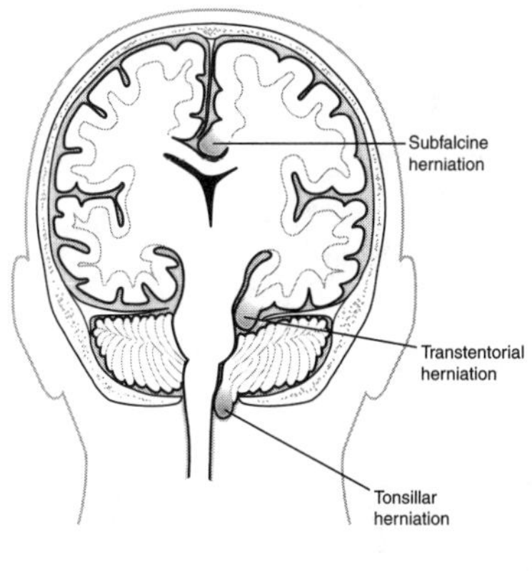

Blood Supply

The cortex is supplied by branches of the **two internal carotid arteries** and **two vertebral arteries**.

- On the ventral surface of the brain, the anterior cerebral and middle cerebral branches of the internal carotid arteries connect with the posterior cerebral artery (derived from the basilar artery) and form the **circle of Willis**. This circle of vessels is completed by the anterior and posterior communicating arteries.
- The **middle cerebral artery** mainly supplies the lateral surface of the frontal, parietal, and upper aspect of the temporal lobe. Deep branches also supply part of the basal ganglia and internal capsule.
- The **anterior cerebral artery** supplies the medial aspect of the frontal and parietal lobes.
- The entire occipital lobe, lower aspect of temporal lobe, and the midbrain are supplied by the **posterior cerebral artery**.

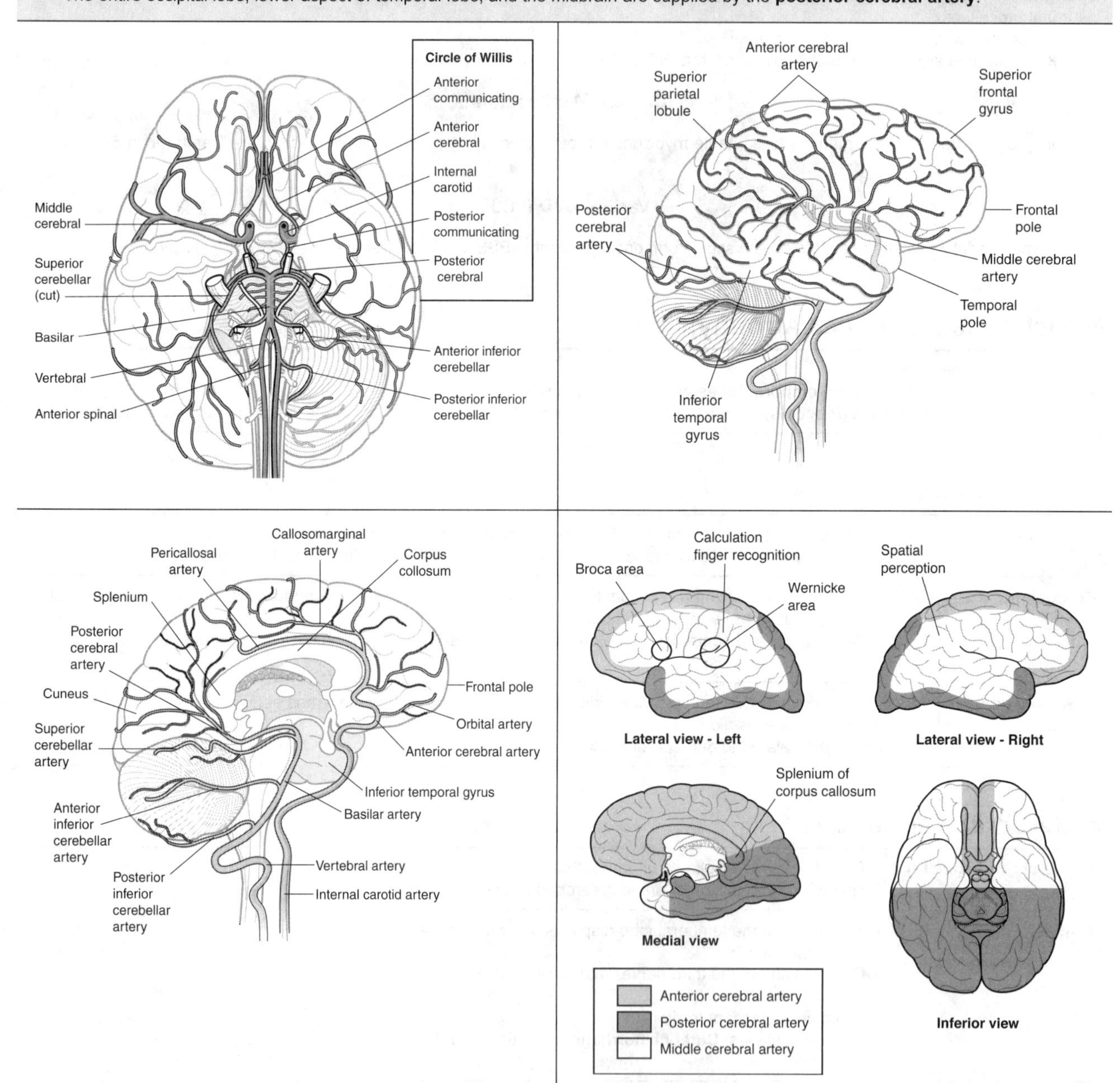

(*Continued*)

► Blood Supply (*Cont'd.*)

System	Primary Arteries	Branches	Supplies	Deficits after Stroke
Vertebrobasilar (posterior circulation)	**Vertebral arteries**	**Anterior spinal artery**	Anterior 2/3 of spinal cord	Dorsal columns spared; all else bilateral
		Posterior cerebellar (PICA)	Dorsolateral medulla	See brain stem lesions on pages 212–213.
	Basilar artery	Pontine arteries	Base of pons	
		Anterior inferior cerebellar (AICA)	Inferior cerebellum, cerebellar nuclei	
		Superior cerebellar artery	Dorsal cerebellar hemispheres; superior cerebellar peduncle	
		Labyrinthine artery (sometimes arises from AICA)	Inner ear	
	Posterior cerebral arteries	—	Midbrain, thalamus, occipital lobe	**Contralateral hemianopia with macular sparing** Alexia without agraphia*
Internal carotid (anterior circulation)	Ophthalmic artery	Central artery of retina	Retina	Blindness
	Posterior communicating artery	—	—	Second most common **aneurysm** site (often with CN III palsy)
	Anterior cerebral artery	—	Primary motor and sensory cortex (leg/foot)	Contralateral spastic paralysis and anesthesia of **lower limb** Frontal lobe abnormalities
	Anterior communicating artery	—	—	Most common site of **aneurysm**
	Middle cerebral artery	Outer cortical	Lateral convexity of hemispheres	Contralateral spastic paralysis and anesthesia of **upper limb/face** **Gaze palsy** **Aphasia*** Gerstmann syndrome* Hemi inattention and neglect of contralateral body†
		Lenticulostriate	Internal capsule, caudate, putamen, globus pallidus	

*If dominant hemisphere is affected (usually the left).
†Right parietal lobe lesion

▶ Cerebrovascular Disorders

Disorder	Types	Key Concepts
Cerebral infarcts	Thrombotic	**Anemic/pale** infarct; usually atherosclerotic complication
	Embolic	**Hemorrhagic/red** infarct; from heart or atherosclerotic plaques; **middle cerebral artery** most vulnerable to emboli
	Hypotension	**"Watershed"** areas and **deep cortical layers** most affected
	Hypertension	**Lacunar** infarcts; **basal ganglia** most affected
Hemorrhages	Epidural hematoma	Almost always traumatic Rupture of **middle meningeal artery** after skull fracture Lucid interval before loss of consciousness ("talk and die" syndrome)
	Subdural hematoma	Usually caused by trauma Rupture of **bridging veins** (connect brain and sagittal sinus)
	Subarachnoid hemorrhage	**Ruptured berry aneurysm** is most frequent cause Predisposing factors: Marfan syndrome, Ehlers-Danlos type 4, adult polycystic kidney disease, hypertension, smoking
	Intracerebral hemorrhage	Common causes: hypertension, trauma, infarction

Seizures and Anticonvulsants

▶ Seizures

Partial	Occur in localized region of brain; can become secondarily generalized **Drugs of choice:** carbamazepine, phenytoin, valproic acid **Backup and adjuvants:** most newer drugs are also effective
Simple	• **Consciousness unaffected** • Can be motor, somatosensory or special sensory, autonomic, psychic
Complex	• Consciousness is impaired • The four "A"s: aura, alteration of consciousness, automatisms, amnesia • Often called "psychomotor" or "temporal lobe seizures"
Generalized	Affects **entire brain**
Absence	• Impaired consciousness (usually abrupt onset and brief); automatisms sometimes occur • Begin in childhood, often end by age 20 • Also called **petit mal** • **Drugs of choice:** ethosuximide (1st line if only absence is present), valproic acid • **Backup and adjuvants:** clonazepam, lamotrigine, topiramate
Tonic-clonic	• Alternating tonic (stiffening) and clonic (movements); loss of consciousness • Also called grand mal • **Drugs of choice:** valproic acid, carbamazepine, phenytoin, phenobarbital • **Backup and adjuvants:** lamotrigine, topiramate, levetiracetam, phenobarbital, others
Myoclonic	Single or multiple myoclonic jerks • **Drug of choice:** valproic acid • **Backup and adjuvants:** clonazepam, topiramate, lamotrigine, levetiracetam
Status epilepticus	• Seizure activity (often tonic-clonic), continuous or intermittent (without recovery of consciousness) for at least 30 minutes; life-threatening • **Drugs of choice:** diazepam, lorazepam, phenytoin, fosphenytoin • **Backup and adjuvants:** phenobarbital, general anesthesia

► Anticonvulsants

Drug	Mechanism	Notes
Benzodiazepines	↑ frequency of $GABA_A$ (Cl^-) receptor opening	Sedation, dependence, tolerance
Carbamazepine	Blocks Na^+ channels	Diplopia, ataxia, blood dyscrasias (agranulocytosis, aplastic anemia), P450 induction, teratogenic
Ethosuximide	Blocks T-type Ca^{2+} channels (thalamus)	GI distress, headache, lethargy, hematotoxicity, Stevens-Johnson syndrome
Phenobarbital	↑ duration of $GABA_A$ (Cl^-) receptor opening	Induction of cytochrome P450, sedation, dependence, tolerance
Phenytoin	Blocks Na^+ channels	Gingival hyperplasia, hirsutism, sedation, anemia, nystagmus, diplopia, ataxia, teratogenic (fetal hydantoin syndrome), P450 induction, zero-order kinetics
Valproic acid	Blocks Na^+ channels, inhibits GABA transaminase	GI distress, hepatotoxic (rare but can be fatal), inhibits drug metabolism, neural tube defects

► Newer Agents

Drug	Side Effects
Felbamate	Aplastic anemia, hepatoxicity
Gabapentin	Sedation, dizziness
Lamotrigine	Life-threatening rash, Stevens-Johnson syndrome
Levetiracetam	Neuropsychiatric effects, sedation
Tiagabine	Sedation, dizziness
Topiramate	Sedation, dizziness, ataxia, anomia, renal stones, weight loss
Vigabatrin	Sedation, dizziness, visual field defects, psychosis

Opioid Analgesics and Related Drugs

Opioid analgesics act by stimulating receptors for endogenous opioid peptides (e.g., enkephalins, β-endorphin, dynorphins). Opioid receptors are G-protein coupled, and the three major classes are μ **(mu)**, κ **(kappa)**, and δ **(delta)**. β-Endorphin has the greatest affinity for the μ receptor, dynorphins for the κ receptor, and enkephalins for the δ receptor. The effects of specific drugs depend on the receptor subtype with which they interact, and whether they act as full agonists, partial agonists, or antagonists. The μ receptor is primarily responsible for analgesia, respiratory depression, euphoria, and physical dependence.

► Individual Agents

	Strong Agonists	
Morphine	Full μ agonist	Prototype of this class; poor oral bioavailability; histamine release
Methadone		Orally active; long duration; useful in maintenance
Meperidine		Muscarinic antagonist (no miosis or smooth muscle contraction); forms normeperidine → possible seizures; do not combine with MAO inhibitors or SSRIs
	Additional strong agonists: fentanyl, heroin (schedule I), levorphanol	
	Moderate Agonist	
Codeine	Partial μ agonist	Antitussive, often given in combination with NSAIDs
	Additional moderate agonists: oxycodone, hydrocodone	
	Partial Agonist	
Buprenorphine	Partial μ agonist	Binds tightly to receptor, so is more resistant to naloxone reversal
	Mixed Agonist–Antagonists	
Pentazocine **Nalbuphine** **Butorphanol**	κ agonist/ μ antagonist	In general, less analgesia than morphine Can cause hallucinations and nightmares Has less respiratory depression and abuse liability than the pure agonists
	Antitussives	
Codeine **Dextromethorphan**	Codeine is by prescription; dextromethorphan is available OTC	
	Antidiarrheals	
Diphenoxylate **Loperamide**	Diphenoxylate used in combination with atropine to prevent abuse Loperamide available OTC	
	Antagonists	
Naloxone (IV) **Naltrexone (PO)** **Nalmefene (IV)**	All are used in the management of acute opioid overdose. Naloxone has a short half-life and may require multiple doses. Naltrexone ↓ craving for ethanol and is used in alcohol dependency programs.	

▶ Characteristics of Opioid Analgesics

Effects and Side Effects	
• Analgesia • Sedation • Respiratory depression • Constipation • Smooth muscle: (except meperidine) ↑ tone: biliary tract (biliary colic), bladder, ureter ↓ tone: uterus (prolongs labor), vascular	• Euphoria • Cough suppression • Nausea and vomiting • Pupillary miosis (except meperidine) • Cardiovascular: Cerebrovascular dilation (esp. with ↑ P_{CO_2}) leads to ↑ intracranial pressure ↓ BP may occur; bradycardia

Clinical Uses	Chronic Effects
Analgesia, cough suppression, treatment of diarrhea, preoperative medications and adjunct to anesthesia, management of pulmonary edema	• Pharmacodynamic tolerance (tolerance does not develop to constipation or miosis) • **Dependence:** Psychological and physical • **Abstinence syndrome (withdrawal):** anxiety, hostility, GI distress (cramps and diarrhea) gooseflesh ("cold turkey"), muscle cramps and spasms ("kicking the habit"), rhinorrhea, lacrimation, sweating, yawning • Abstinence syndrome can be precipitated in tolerant individuals by administering an opioid antagonist
Overdose	
• **Classic triad:** Respiratory depression, miosis (pinpoint pupils), coma • Diagnosis confirmed with naloxone (short duration, may need repeat dosing); give supportive care	

Contraindications and Cautions	
• Use of full agonists with weak partial agonists. Weak agonists can precipitate withdrawal from the full agonist. • Use in patients with pulmonary dysfunction (acute respiratory failure). Exception: pulmonary edema • Use in patients with head injuries (possible increased intracranial pressure)	• Use in patients with hepatic/renal dysfunction (drug accumulation) • Use in patients with adrenal or thyroid deficiencies (prolonged and exaggerated responses) • Use in pregnant patients (possible neonatal depression or dependence)

Local Anesthetics

Local anesthetics **block voltage-gated sodium channels**, preventing sensory information from being transmitted from a local area to the brain. These agents are initially uncharged and diffuse across the axonal membrane to enter the cytoplasm. Once inside, they become ionized and block the Na^+ channels from the **inside**. These agents bind best to channels that are open or recently inactivated, rather than resting, and therefore work better in rapidly firing fibers (**use-dependence**). Infection leads to a more acidic environment, making the basic local anesthetic more likely to be ionized, therefore higher doses may be required. **Order of blockade:** small fibers > larger fibers; myelinated fibers > unmyelinated fibers. **Modality blocked:** autonomic and pain > touch/pressure > motor. There are two main classes: **amides** and **esters**.

Amides (metabolized in **liver**)	**Esters** (metabolized by **plasma cholinesterases**)	**Side Effects**
Bupivacaine (L) Etidocaine (L) Ropivacaine (L) Lidocaine (M)* Mepivacaine (M) Prilocaine (M)	Tetracaine (L) Cocaine (M)† Procaine (S) Benzocaine‡	1. **Neurotoxicity:** lightheadedness, nystagmus, restlessness, convulsions 2. **Cardiovascular toxicity**: ↓ CV parameters (except cocaine which ↑ HR and BP); bupivacaine especially notable for CV toxicity 3. **Allergic reaction:** esters via PABA formation; switch to amides if allergic to esters
Notes		
Duration of action can be increased by coadministration of a vasoconstrictor (e.g., epinephrine) to limit blood flow **Hint:** Amide drugs have two "i"s, and ester drugs have one "i" in their names.		

Definition of abbreviations: L, long acting; M, medium acting; S, short acting.
*Also a IB antiarrhythmic
†Primarily used topically; sympathomimetic; drug of abuse (schedule II)
‡Topical only

General Anesthetics

The ideal general anesthetic produces unconsciousness, analgesia, skeletal muscle relaxation, loss of reflexes, and amnesia. There are two broad classes of general anesthetics: **inhalational** and **intravenous**.

► Inhalational Anesthetics

Definitions

- **Solubility:** blood:gas partition coefficient
- **Minimum alveolar anesthetic concentration (MAC):** the minimal alveolar concentration at which 50% of patients do not respond to a standardized painful stimulus. **MAC ∝ 1/potency**

General Principles

- Drugs with a ↑ **solubility** have **slow induction and recovery times** (if it is soluble in blood, then it takes longer to achieve the partial pressure required for anesthesia).
- Drugs with a ↓ **solubility** have **rapid induction and recovery times** (if it is not very soluble in blood, it quickly achieves the partial pressure required for anesthesia).
- Anesthesia is **terminated by redistribution** of the agent from the brain to the blood.
- Anesthetics that undergo **hepatic metabolism** tend be **more toxic**.

General Side Effects

- Sometimes when administered with muscle relaxants (especially succinylcholine) → **malignant hyperthermia**. Treat with dantrolene.
- Most ↓ BP moderately.

Anesthetic	Solubility	MAC (%)	Metabolism (%)	Unique Side Effects/Properties
Nitrous oxide	0.5	>100	0	Low potency; because the **MAC is >100**, it cannot provide complete anesthesia; good for induction; **good analgesia and amnesia**
Desflurane	0.4	6.5	<0.1	**Pulmonary irritant**
Sevoflurane	0.7	2.0	3	—
Isoflurane	1.4	1.3	0.2	—
Enflurane	1.8	1.7	8	**Proconvulsant**
Halothane	2.3	0.8	20	**Halothane hepatitis; sensitizes heart to arrhythmogenic effects of catecholamines**
Methoxyflurane	12	0.2	> 70	**Nephrotoxicity**

► Intravenous Anesthetics

Drug Class	Agents	Unique Properties
Barbiturates	Thiopental, methohexital	**Redistribution** from brain terminates effects, but hepatic metabolism is required for elimination; used mainly for **induction** or short procedures; hyperalgesic; ↓ respiration, cardiac function, and cerebral blood flow
Benzodiazepines	Midazolam	Good amnesic; respiratory depression (can be reversed with **flumazenil**)
Dissociative	Ketamine	Patient remains conscious, but has amnesia, catatonia, and analgesia; related to phencyclidine (PCP)—causes **emergence reactions** (hallucinations, excitation, disorientation); **CV stimulant**
Opioids	Fentanyl, alfentanil, remifentanil, morphine	Chest wall rigidity, respiratory depression (can reverse with naloxone); **neuroleptanesthesia**: fentanyl + droperidol + nitrous oxide
Miscellaneous	Propofol, etomidate	Propofol—**rapid induction, antiemetic** Etomidate—**rapid induction; minimal CV or respiratory effects;** pain and myoclonus on injection; nausea

The Cardiovascular System

Chapter 2

Embryology

Cardiovascular Anatomy

Cardiovascular Physiology

Cardiovascular Pathology

Cardiovascular Pharmacology

Cardiovascular Embryology

► Development of the Heart Tube

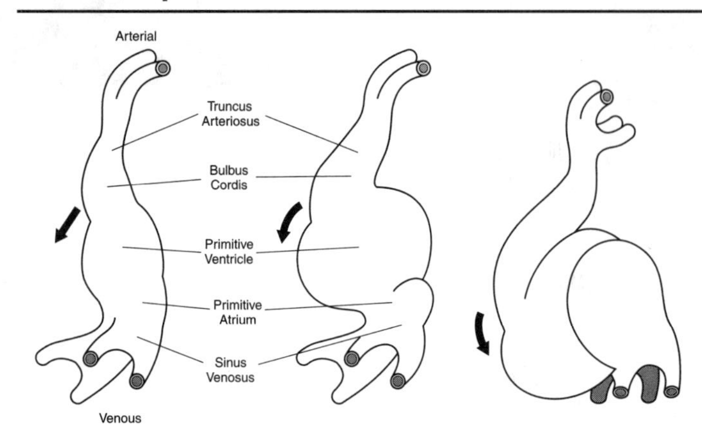

The primitive heart tube is formed from lateral plate mesoderm. The primitive heart tube undergoes dextral looping (bends to the right) and forms five dilatations. Four of the five dilatations become subdivided by a septum. Most of the common congenital cardiac anomalies result from defects in the formation of these septa.

► Adult Structures Derived from the Dilatations of the Primitive Heart

Embryonic Dilatation	Adult Structure
Truncus arteriosus (neural crest)	Aorta Pulmonary trunk
Bulbus cordis	Smooth part of right ventricle (**conus arteriosus**) Smooth part of left ventricle (**aortic vestibule**)
Primitive ventricle	Trabeculated part of right ventricle Trabeculated part of left ventricle
Primitive atrium	Trabeculated part of right atrium Trabeculated part of left atrium
Sinus venosus (the only dilatation that does not become subdivided by a septum)	Right—smooth part of right atrium (**sinus venarum**) Left—coronary sinus Oblique vein of left atrium

► Atrial Septum

Formation of the Atrial Septum

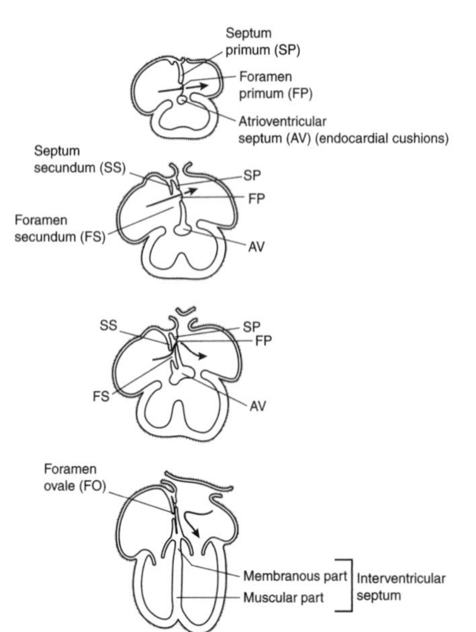

Atrial septal defects are called **ASDs**.

Secundum-type ASDs are caused by excessive resorption of the SP or reduced size of the SS or both. This results in an opening between the right and left atria. If the ASD is small, clinical symptoms may be delayed as late as age 30. This is the most clinically significant ASD.

The **foramen ovale** (FO) is the fetal communication between the right and left atria. It remains patent in up to 25% of normal individuals throughout life, although paradoxical emboli may pass through a large patent FO. **Premature closure of the FO** is the closure of the FO during prenatal life. This results in hypertrophy of the right side of the heart and underdevelopment of the left side.

► Ventricular Septum

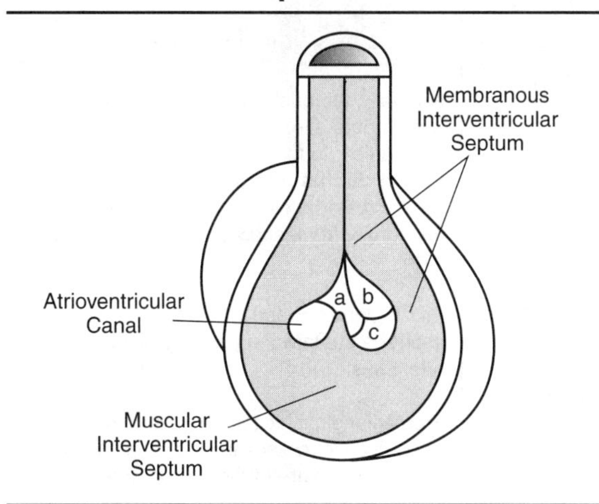

A **membranous ventricular septal defect (VSD)** is caused by the failure of the membranous interventricular septum to develop, and it results in **left-to-right shunting** of blood through the interventricular foramen. Patients with left-to-right shunting complain of **excessive fatigue upon exertion**.

Left-to-right shunting of blood is not cyanotic but causes increased blood flow and pressure to the lungs (pulmonary hypertension). Pulmonary hypertension causes marked proliferation of the tunica intima and media of pulmonary muscular arteries and arterioles. Ultimately, the pulmonary resistance becomes higher than systemic resistance and causes **right-to-left shunting** of blood and "late" **cyanosis**. At this stage, the condition is called **Eisenmenger complex**. VSD is the most common congenital cardiac anomaly.

Figure legend: a, right bulbar ridge; b, left bulbar ridge; c, AV cushions.

► Aorticopulmonary Septum

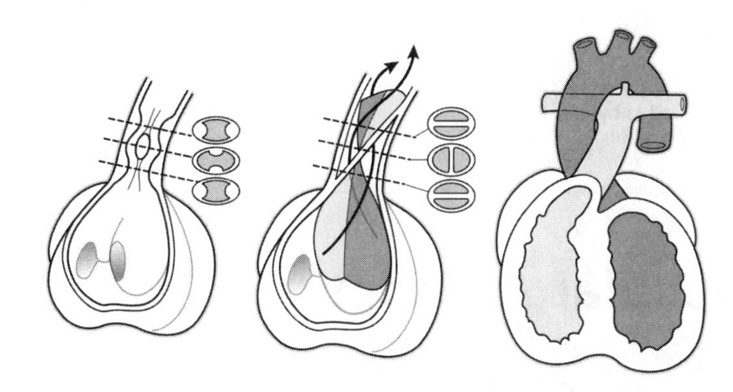

Neural crest cells migrate into the truncal and bulbar ridges of the truncus arteriosus, which grow in a spiral fashion and fuse to form the aorticopulmonary (AP) septum. The AP septum divides the truncus arteriosus into the **aorta** (dark gray) and **pulmonary trunk** (light gray).

Transposition of the Great Vessels
Occurs when the AP septum fails to develop in a spiral fashion and results in the aorta opening into the right ventricle and the pulmonary trunk opening into the left ventricle. This causes **right-to-left** shunting of blood with resultant **cyanosis**. Infants born alive with this defect must have other defects (like a PDA or VSD) that allow mixing of oxygenated and deoxygenated blood.
Tetralogy of Fallot
Occurs when the AP septum fails to align properly and results in (1) pulmonary stenosis, (2) overriding aorta, (3) interventricular septal defect, and (4) right ventricular hypertrophy. This causes **right-to-left** shunting of blood with resultant "early" **cyanosis**, which is usually present at birth. Tetralogy of Fallot is the most common congenital cyanotic cardiac anomaly.
Persistent Truncus Arteriosus
Occurs when there is only partial development of the AP septum. This results in a condition in which only one large vessel leaves the heart and receives blood from both the right and left ventricles. This causes **right-to-left** shunting of blood with resultant **cyanosis**. This defect is always accompanied by membranous ventricular septal defect.

Definition of abbreviation: PDA, patent ductus arteriosus.

► Fetal Circulation

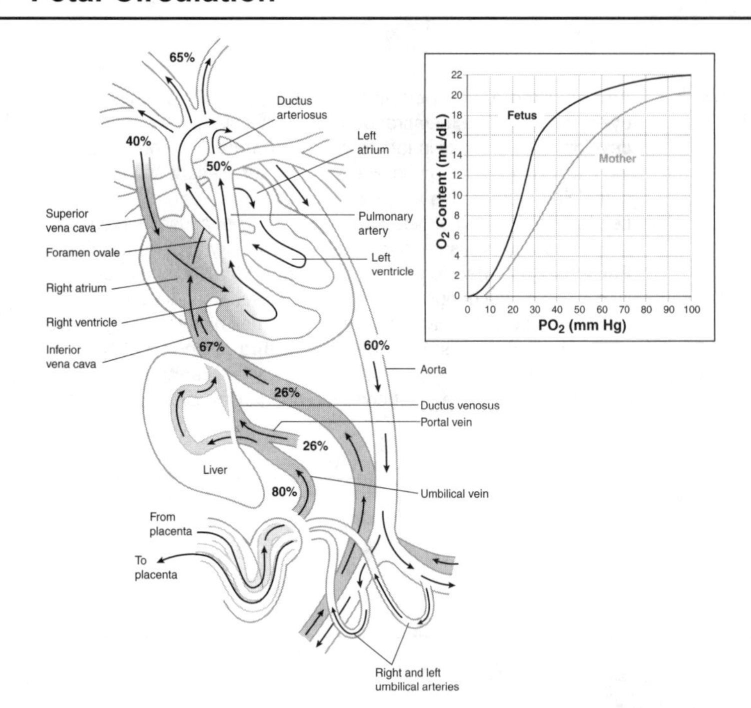

In the fetal circulation, the ductus venosus allows fetal blood to bypass the liver, and the foramen ovale and the ductus arteriosus allow fetal blood to bypass the lungs. Note the sites where the oxygen saturation level of fetal blood is the highest (umbilical vein) and the lowest (ductus arteriosus).

Clinical Correlation

Normally, the ductus arteriosus closes within a few hours after birth, via smooth muscle contraction, to form the ligamentum arteriosum. **Patent ductus arteriosus (PDA)** occurs when the ductus arteriosus (connection between the pulmonary trunk and aorta) fails to close after birth.

Prostaglandin E and intrauterine or neonatal asphyxia **sustain the patency** of the ductus arteriosus.

Prostaglandin inhibitors (e.g., indomethacin), acetylcholine, histamine, and catecholamines **promote closure** of the ductus arteriosus.

PDA is common in premature infants and cases of maternal rubella infection. It causes a left-to-right shunting of blood. (Note: During fetal development, the ductus arteriosus is a right-to-left shunt).

Cardiovascular Anatomy

► Structures of the Mediastinum

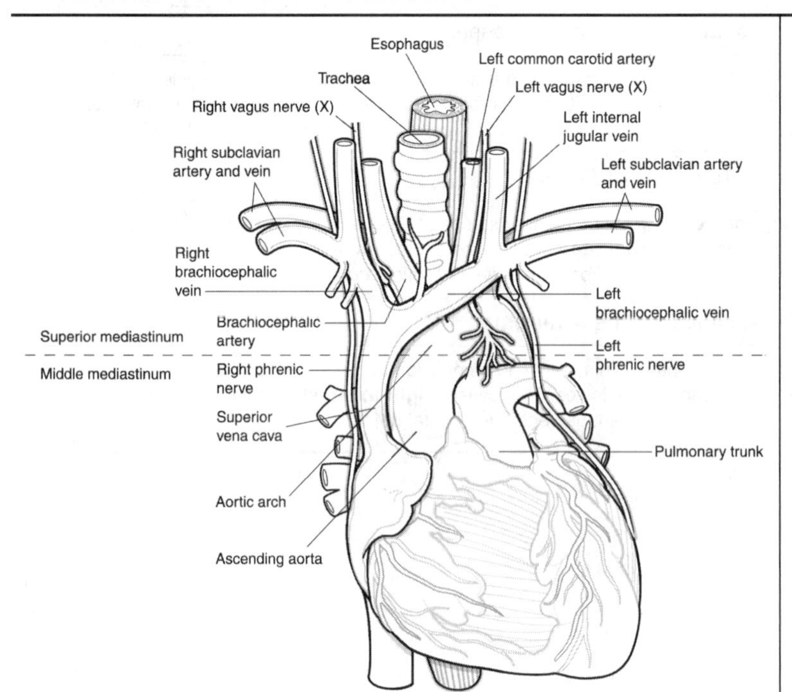

The thoracic cavity is divided into the **superior mediastinum** above the plane of the sternal angle and the **inferior mediastinum** (anterior, middle, and posterior mediastina) below that sternal plane. The superior mediastinum contains the thymic remnants, superior vena cava and its brachiocephalic tributaries, aortic arch and its branches, trachea, esophagus, thoracic duct, and the vagus and phrenic nerves.

The **anterior mediastinum** is anterior to the heart and contains remnants of the thymus. The **middle mediastinum** contains the heart and great vessels, and the **posterior mediastinum** contains the thoracic aorta, esophagus, thoracic duct, azygos veins, and the vagus nerves. The inferior vena cava passes through the diaphragm at the caval hiatus at the level of the eighth thoracic vertebra; the esophagus through the esophageal hiatus at the tenth thoracic vertebra; and the aorta courses through the aortic hiatus at the level of the twelfth thoracic vertebra.

▶ Arterial Supply to the Heart

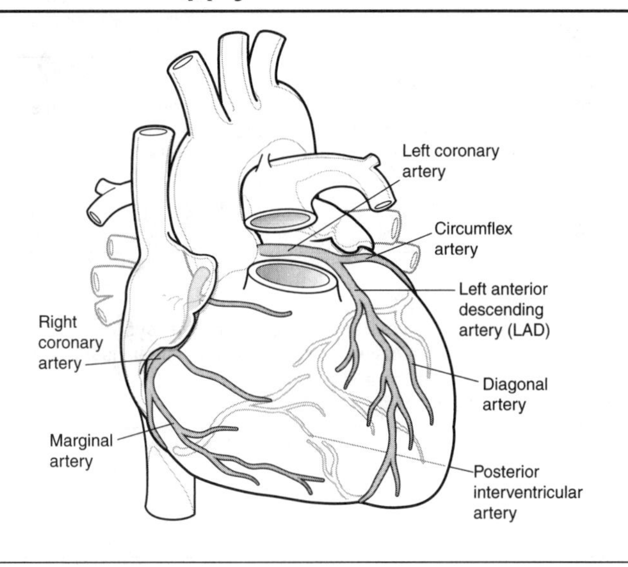

Arterial supply to the heart muscle is provided by the **right and left coronary arteries**, which are branches of the ascending aorta. The **right coronary artery** supplies the right atrium, the right ventricle, the sinoatrial and atrioventricular nodes, and parts of the left atrium and left ventricle. The distal branch of the right coronary artery (in 70% of subjects, "right dominant") is the **posterior interventricular artery** that supplies, in part, the posterior aspect of the interventricular septum. The **left coronary artery** supplies most of the left ventricle, the left atrium, and the anterior part of the interventricular septum. The two main branches of the left coronary artery are the anterior interventricular artery (LAD) and the circumflex artery.

In a myocardial infarction, the LAD is obstructed in 50% of cases, the right coronary in 30%, and the circumflex artery in 20% of cases.

▶ Chambers and Valves of the Heart

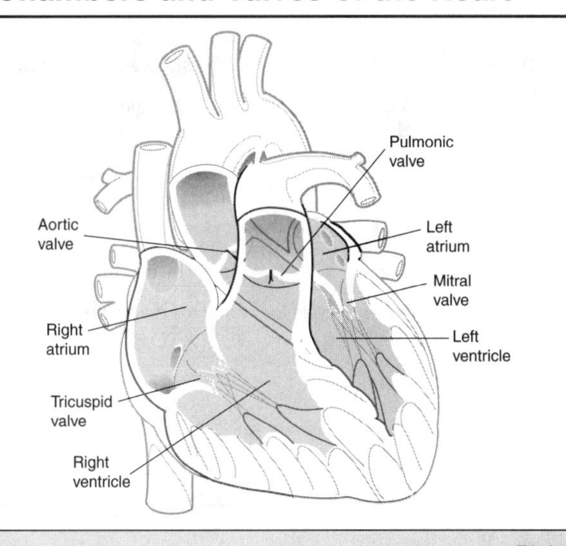

Right Atrium

The right atrium receives venous blood from the entire body (except for blood from the pulmonary veins).

The **auricle** is derived from the fetal atrium; it has rough myocardium known as pectinate muscles.

The **sinus venarum** is the smooth-walled portion of the atrium, which receives blood from the superior and inferior venae cavae.

The **crista terminalis** is the vertical ridge that separates the smooth from the rough portion of the right atrium; it extends longitudinally from the superior vena cava to the inferior vena cava. The SA node is in the upper part of the crista terminalis.

The right AV **(tricuspid valve)** communicates with the right ventricle.

Right Ventricle

The right ventricle receives blood from the right atrium via the tricuspid valve; outflow is to the pulmonary trunk via the pulmonary semilunar valve.

The **trabeculae carneae** are the ridges of myocardium in the ventricular wall.

The **papillary muscles** project into the cavity of the ventricle and attach to cusps of the AV valve by the strands of the chordae tendineae. Papillary muscles contract during ventricular contraction to keep the cusps of the AV valves closed.

The **chordae tendineae** control closure of the valve during contraction of the ventricle.

The **infundibulum** is the smooth area of the right ventricle leading to the pulmonary valve.

Left Atrium

The left atrium receives oxygenated blood from the lungs via the pulmonary veins. There are four openings: the upper right and left and the lower right and left pulmonary veins.

The left AV orifice is guarded by the **mitral (bicuspid) valve**; it allows oxygenated blood to pass from the left atrium into the left ventricle.

(Continued)

► Chambers and Valves of the Heart (*Cont'd.*)

Left Ventricle
Blood enters from the left atrium through the mitral valve and is pumped out to the aorta through the aortic valve. **Trabeculae carneae**, the ridges of myocardium in the ventricular wall, are normally three times thicker than those of the right ventricle. **Papillary muscles** (usually two large ones) are attached by the chordae tendineae to the cusps of the bicuspid valve. The **aortic vestibule** leads to the aortic semilunar valve and ascending aorta; the right and left coronary arteries originate from the right and left aortic sinuses at the root of the ascending aorta.

Clinical Correlation
Murmurs
Murmurs in valvular heart disease result when there is valvular insufficiency or a stenotic valve. For most of **ventricular systole**, the mitral valve should be closed and the aortic valve should be open, so that "common systolic valvular defects" include mitral insufficiency and aortic stenosis. For most of **ventricular diastole**, the mitral valve should be open and the aortic valve should be closed, so that "common diastolic valvular defects" include mitral stenosis and aortic insufficiency.

► Borders of the Heart

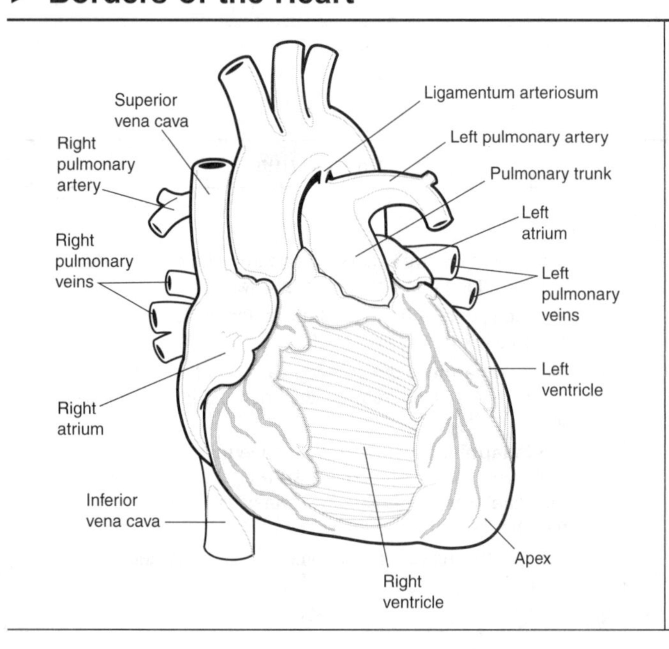

The **external** surface of the heart consists of several borders:

- the right border is formed by the right atrium
- the left border is formed by the left ventricle
- the base formed by the two atria
- the apex at the tip of the left ventricle

The **anterior** surface is formed by the right ventricle.

The **posterior** surface is formed mainly by the left atrium.

A **diaphragmatic** surface is formed primarily by the left ventricle.

▶ Cross-Sectional Anatomy of the Thorax

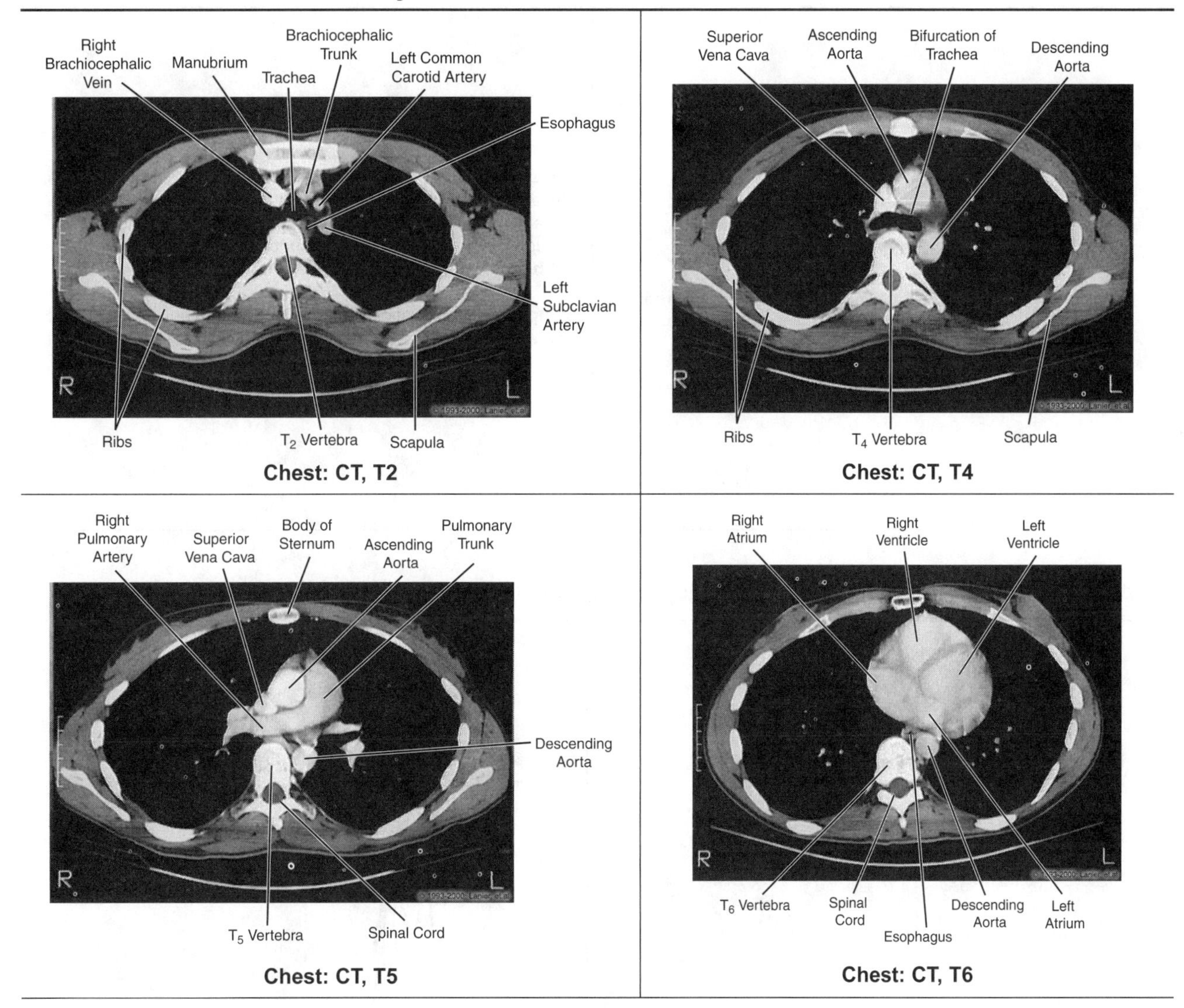

Chest: CT, T2

Chest: CT, T4

Chest: CT, T5

Chest: CT, T6

Images copyright 2005 DxR Development Group Inc. All rights reserved.

► Conducting System of the Heart

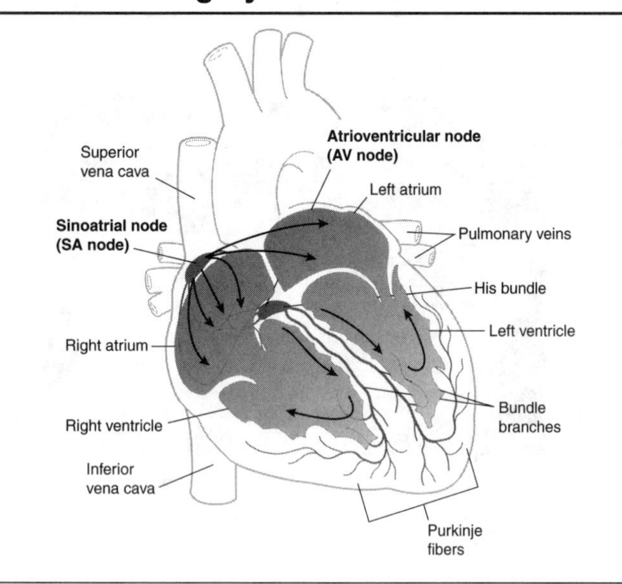

The **sinoatrial node** initiates the impulse for cardiac contraction. The **atrioventricular node** receives the impulse from the sinoatrial node and transmits that impulse to the ventricles through the **bundle of His**. The bundle divides into the **right and left bundle branches** and **Purkinje fibers** to the two ventricles.

Sympathetic innervation from the T1 to T5 spinal cord segments increases the heart rate, while the **parasympathetics** by way of the vagus nerves slow the heart rate.

► Associations of Common Traumatic Injuries with Vessel and Nerve Damage

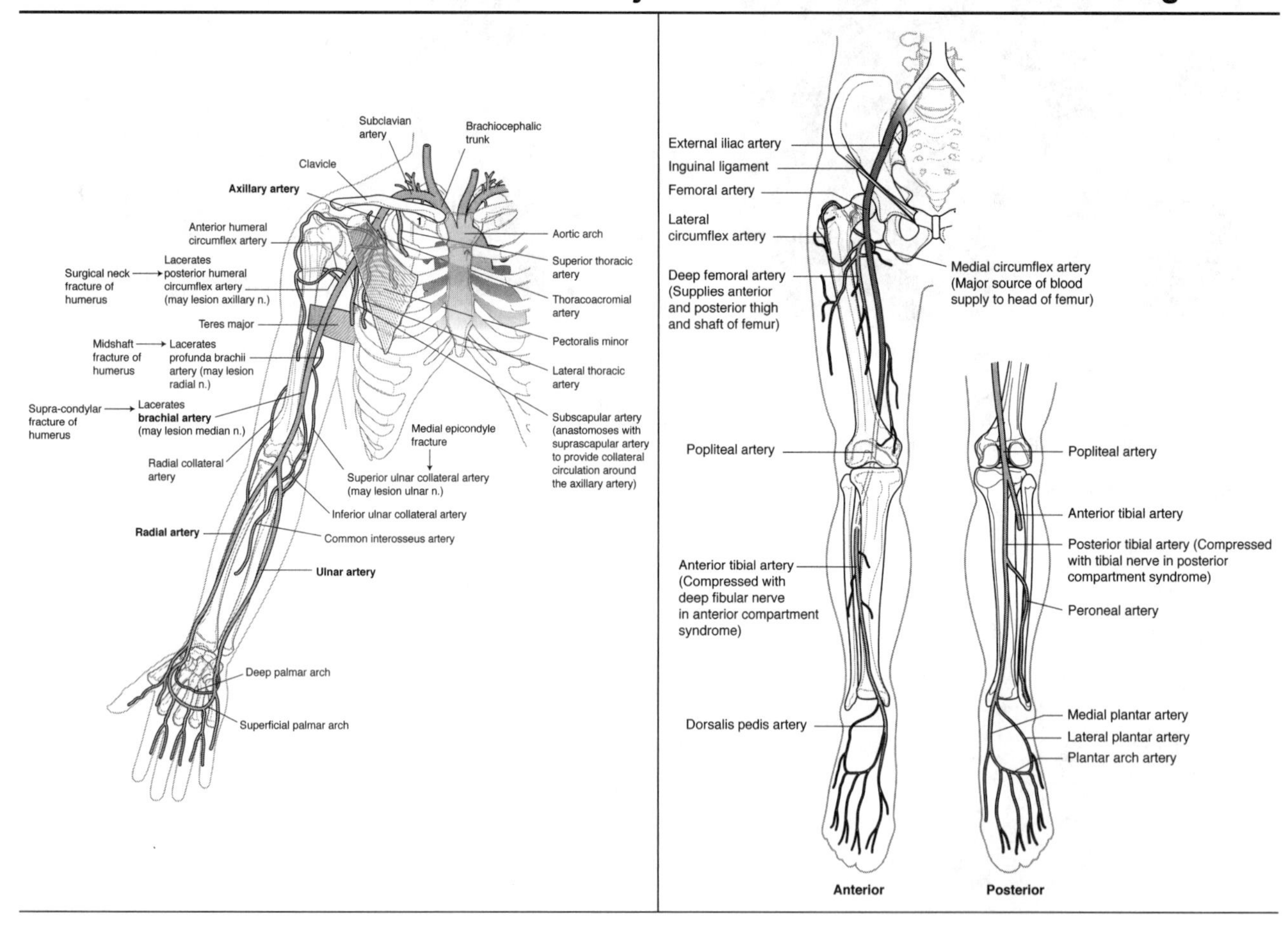

Cardiovascular Physiology

▶ Comparison of Cardiac Action Potentials

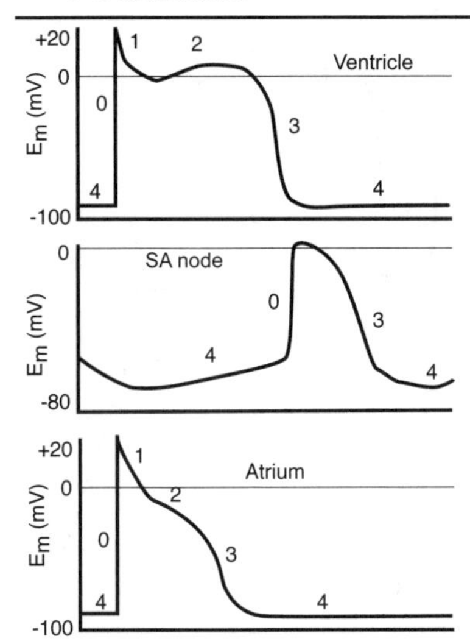

▶ Important Implications of Ion Currents

Force development
Ca^{2+} current of plateau (phase 2) has major influence

Timing/heart rate
i_f is increased by sympathetics → increased heart rate; parasympathetics increase gK^+ → decreased heart rate

Premature beats
Action potential amplitude and shape not all-or-none; early beats abnormal with low force

Susceptible period
Arrhythmia risk high during relative refractory period

▶ Cardiac Action Potentials—Ionic Mechanisms

Unique Cardiac Ion Channels	
iK	Delayed rectifier; slow to open/close; depolarization opens
iK_1	Inward rectifier; open at rest; depolarization closes it
L-type	Ca^{2+}, slow channel, long acting; depolarization opens
i_f	Na^+; "funny channel"; repolarization opens it; causes the pacemaker spontaneous depolarization

Ionic Basis of Ventricular AP

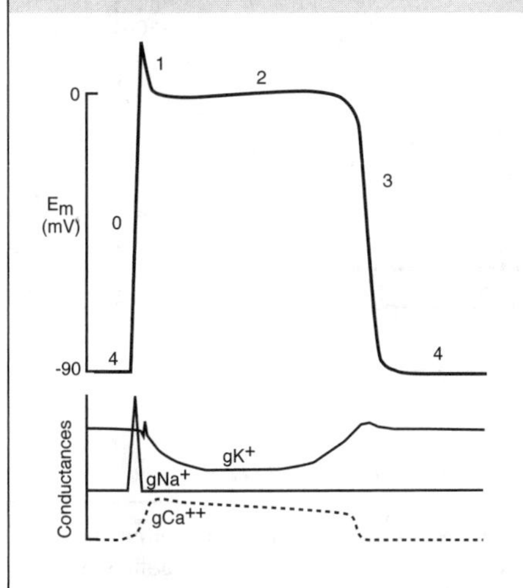

Conductances show changes only and do not reflect absolute values of different ions.

Ionic Basis of SA Node AP

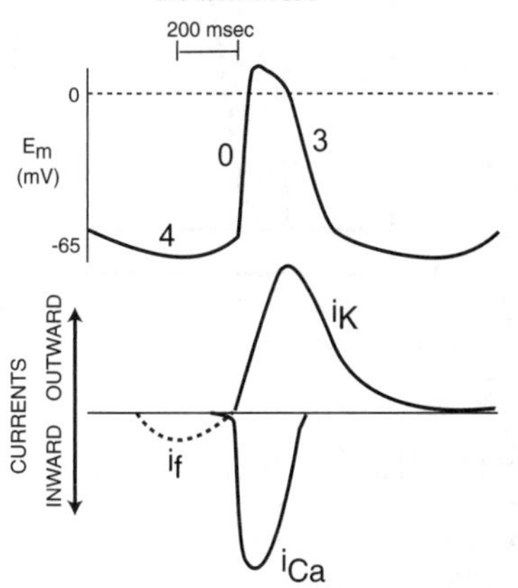

Ventricles and Atria	SA Node and AV Node
Phase 4—resting potential ↑ gK^+ occurs via iK_1 channels iK channels are closing or closed	**Phase 4**—pacemaker ↑ gNa^+ via i_f "funny channel" High gK^+ but ↓ as iK channels close
Phase 0—upstroke ↑ gNa^+ via typical fast Na^+ channels ↓ gK^+ as iK_1 channels close	**Phase 0**—upstroke ↑ gCa^{2+} via T-type (fast, transient) channels, then L-type (slow) open
Phase 1—rapid partial repolarization ↓ gNa^+ as fast channels close ↑ gK^+ transiently via iK_{to}	No **phase 1** because no fast sodium channels
Phase 2—plateau ↑ gCa^{2+}: slow (L-type) channels at end of phase 2, ↑ gK^+ via iK channels	**Phase 2** usually absent
Phase 3—repolarization ↓ gCa^{2+} as L-type channels close ↑ gK^+ via iK; then iK_1 opens	**Phase 3**—repolarization ↓ gCa^{2+} as slow channels close ↑ gK^+ via iK channels

► Refractory Periods

Summation is difficult to achieve in cardiac muscle and tetany does not occur. In fact, the abnormal shape of action potentials initiated during the relative refractory period reduces calcium influx and thus contractile force, as shown.

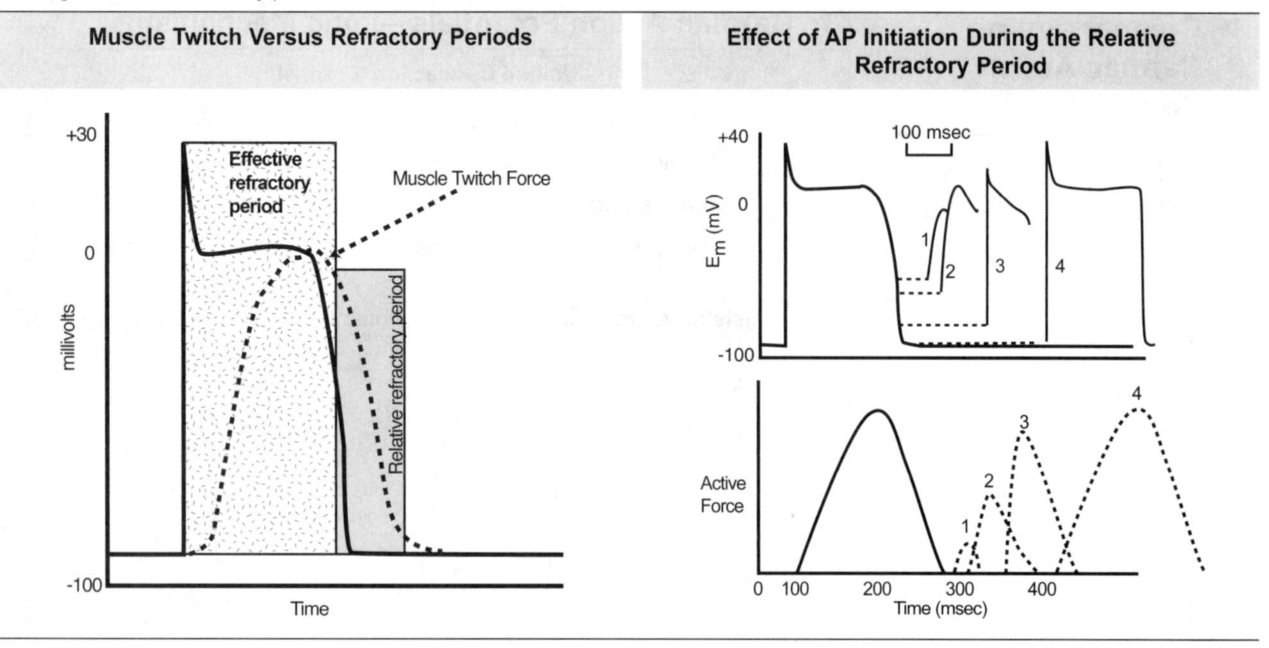

► Basic Principles of the Electrocardiogram

A moving wave of depolarization in the heart produces a positive deflection as it moves toward the positive terminals of the ECG electrodes. A depolarizing wave moving away from the positive (toward the negative) terminals produces a negative deflection. A wave of depolarization moving at right angles to the axis of the electrode terminals produces no deflection. Upon repolarization, the reverse occurs.

► Sequence of Myocardial Excitation and Conduction

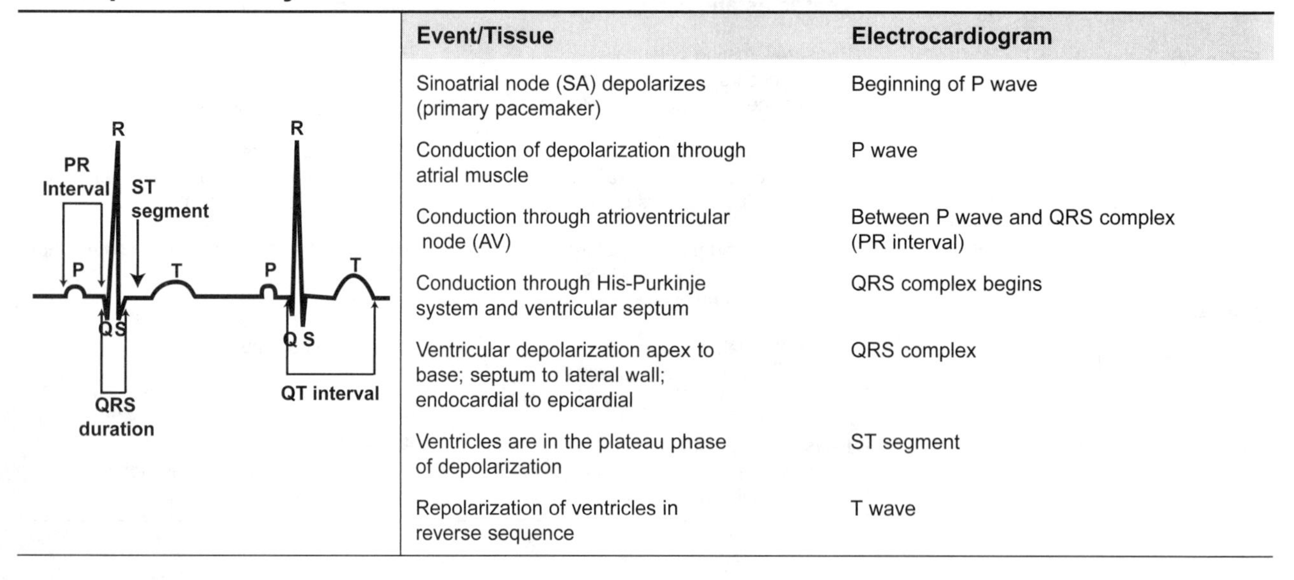

Event/Tissue	Electrocardiogram
Sinoatrial node (SA) depolarizes (primary pacemaker)	Beginning of P wave
Conduction of depolarization through atrial muscle	P wave
Conduction through atrioventricular node (AV)	Between P wave and QRS complex (PR interval)
Conduction through His-Purkinje system and ventricular septum	QRS complex begins
Ventricular depolarization apex to base; septum to lateral wall; endocardial to epicardial	QRS complex
Ventricles are in the plateau phase of depolarization	ST segment
Repolarization of ventricles in reverse sequence	T wave

Important ECG Values		
PR interval	0.12–0.20 sec	Length measures AV conduction time
QRS duration	<0.12 sec	Measures ventricular conduction time
QT interval	0.35–0.45 sec	Total time of ventricular depolarization and repolarization Varies with heart rate, age
Heart rate, normal resting	60–100 beats/min	<60/min = bradycardia; >100/min = tachycardia

▶ Principles of the Electrocardiogram (EKG or ECG)

Moving Electrical Charge		Creates Electrical Field Movement That Causes Ion Currents in Skin
Standard limb leads (frontal plane)	I	⊖ right arm; ⊕ left arm; positive lead at 0°
	II	⊖ right arm; ⊕ left leg; positive lead at +60°
	III	⊖ left arm; ⊕ left leg; positive lead at +120°
Augmented limb leads (frontal plane)	aVR; aVL, aVF	aVR positive lead at –150°; aVL positive lead at –30°; aVF positive lead at +90°
Precordial	V_1–V_6 (chest)	Horizontal plane; positive leads front; negative leads back of chest
Chart speed	25 mm/sec	Each horizontal mm = 0.04 sec (40 msec)
Voltage	1 mV/10 mm	Measure ± voltages of QRS, add to get net voltage of each lead
Mean electrical axis	Vector sum of two leads	Measure of overall wave of ventricular depolarization: normal axis, left or right axis deviation

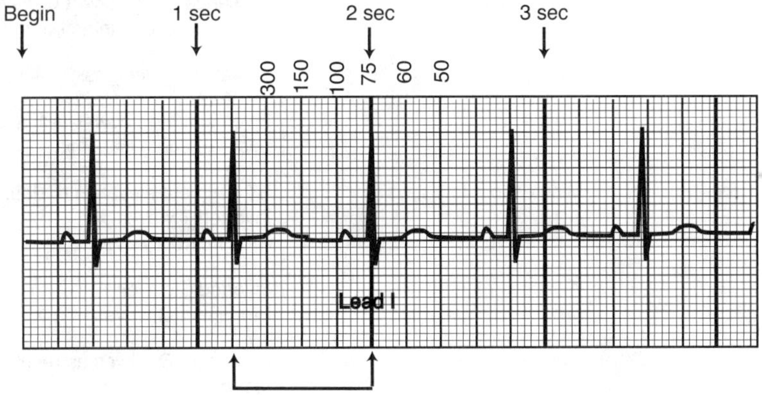

Begin 1 sec 2 sec 3 sec

300 150 100 75 60 50

Lead I

Four intervals = 75 beats/min
OR
Four beats in 3 sec = 4 X 20 = 80 beats/min

Estimation of Heart Rate	
Triplet Method	**Interval**
How many dark lines are between R waves (5 mm apart)? 1 = 300/min; 2 = 150; 3 = 100; 4 = 75; 5 = 60; 6 = 50	Measure time interval (longer interval is better) Count R waves Multiply count to convert to 1 minute e.g., four R waves in 3 sec = 4 × 20 = 80 beats/min

► Important Rhythms to Recognize

Rhythm	Characteristics
Sinus rhythm Normal rate = 60–100 Bradycardia rate <60 Tachycardia rate >100	Each beat originates in the SA node; therefore, the P wave precedes each QRS complex; PR interval is normal
AV conduction block	Abnormal conduction through AV node
First degree	PR interval (> 0.20 sec); 1:1 correspondence; P wave:R wave
Second degree Mobitz I (Wenckebach)	Progressively increased PR interval; then dropped (missing) QRS and then repeat of sequence
Second degree Mobitz II	Regular but prolonged PR interval; unexpected dropped QRS; may be a regular pattern, such as 2:1 = 2 P waves:1 QRS complex or 3:1, etc.
Third degree (complete)	No correlation of P waves and QRS complexes; usually high atrial rate and lower ventricular rate
Premature ventricular contraction (PVC)	Large, wide QRS complex originates in ectopic focus of irritability in ventricle; may indicate hypoxia
Ventricular tachycardia	Repeated large, wide QRS complexes like PVCs; Rate 150–250/min; acts like prolonged sequence of PVCs
Ventricular fibrillation	Total loss of rhythmic contraction; totally erratic shape

► Evolution of an Infarction: Signs on the EKG

Features to observe	QRS complex	Presence of prominent Q waves in leads where normally absent: infarct damage
	ST segment	Elevation or depression: acute injury
	T wave	Inversion; e.g., downward in lead where usually positive: acute ischemia
Acute myocardial infarction (MI)	Minutes to a few days	ST segment elevation or depression Inverted T waves Prominent Q waves
Resolving infarction (healing)	Weeks to months	Inverted T waves Prominent Q waves
Stable (old) MI	Months to years	Prominent Q waves as result of MI persist for the rest of life

Caution: Not all infarctions produce Q waves. Inverted T waves and/or ST abnormalities should always be investigated, even in absence of significant Q waves.

► Identifying Location of an Infarction

Location	Principal Feature of ECG	Vessel Involvement
Posterior	Large R with ST depression in V1 and V2: Mirror test or reversed transillumination	Right coronary artery
Lateral	Q waves in lateral leads I and AVL	Circumflex coronary artery
Inferior	Q waves in inferior leads II, III & AVF	Right or left coronary artery
Anterior	Q waves in V1, V2, V3 & V4	Anterior descending coronary artery

► Mean Electrical Axis (MEA)

Definition	• Overall direction and force (vector) of the events of ventricular depolarization: obtained by vector sum of net voltage of two leads or by quadrant method using leads I and aVF • MEA tends to shift toward large mass and away from an MI
Normal axis	• Expected in the absence of cardiac disease • R wave: lead I, +; lead II, +; lead III, +
Left axis deviation	• May indicate left heart enlargement, as in hypertrophy or left dilated failure • Abnormally prolonged (slow) left ventricular conduction • Right heart MI, expiration, obesity, lying down • R wave: lead I, +; lead II, +; lead III, –
Right axis deviation	• Right ventricular hypertrophy or dilation • Prolonged right conduction • Left heart MI, inspiration, tall lanky people, standing up • R wave: lead I, –; lead II, +; lead III, +
Extreme right axis deviation	Difficult interpretation; one example: depolarization proceeding from abnormal focus in LV apex

Einthoven's Triangle: Leads I, II, and III*

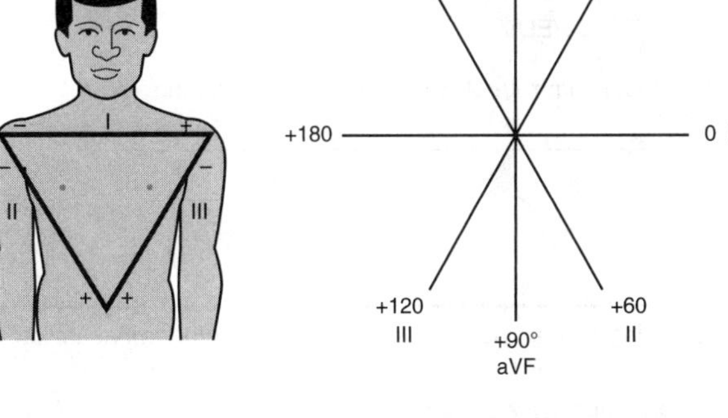
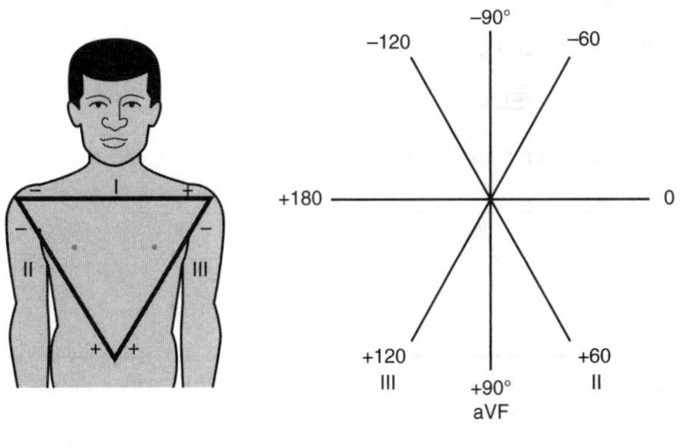

Vector Cardiogram

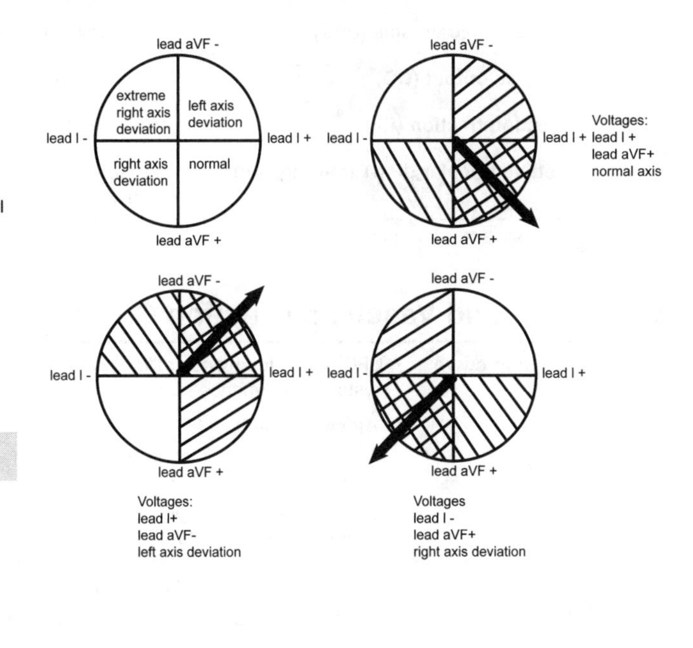

Essentials of the EKG

Heart rate

Rhythm

Axis deviation

Hypertrophy

Infarction

*The figure adds aVF because the quadrant method of determining axis uses leads I and aVF.

▶ Cardiac Mechanical Performance

Factor	Definition	Effects
Preload	Cardiac muscle cell length (sarcomere length) **before contraction begins**	↑ preload causes ↑ active force development up to a limit
Afterload	Load on the heart during ejection of blood from the ventricle	↑ afterload ↓ the volume of blood ejected during a beat
Contractility	Capacity of the heart to produce active force at a specified preload	High contractility ↑ ability to work Low contractility ↓ ability to work
Rate	Heart rate (HR): number of cardiac cycles per minute	↑ output of blood per minute, but ↓ output per beat; very high rate (>≈150/min) ↓ output

▶ Cardiac Performance: Definitions

Stroke volume (SV)	Blood ejected from ventricle per beat = EDV – ESV
End diastolic volume (EDV)	Volume of blood in ventricle at end of diastole; the preload
End systolic volume (ESV)	Volume of blood remaining in ventricle at end of systole
Cardiac output (CO)	Volume of blood per minute pumped by the heart; CO = SV × HR
Ejection fraction (EF)	Measure of contractility: EF = SV/EDV
Left ventricular dP/dT (mm Hg/sec)	Measure of contractility: maximum rate of change of pressure during isovolumic contraction

▶ Cardiac and Vascular Function Curves

Cardiac function curve (CFC)	• CFC generated by controlling preload and measuring cardiac output, stroke volume or other measure of systolic performance • ↑ preload improves actin-myosin interdigitation and thus ↑ SV, CO, etc. • CFC shifts **up** with ↑ **contractility; down** with ↓ **contractility**; so a new curve is produced when contractility changes • Moving to a **different point on the same CFC** is a change only of **preload:** moving to a **different CFC** is change of **contractility**
Vascular function curve (VFC)	• VFC relates venous return to right atrial pressure • ↑ **blood volume shifts VFC up,** ↓ **volume shifts VFC down**
Equilibrium point	Cardiac output is determined by both CFC and VFC. Intersection of the CFC and VFC is the stable operating point; if contractility or blood volume changes, the system will operate at the intersection of the two new curves.

▶ Cardiovascular Responses in Exercise

	Heart Rate	TPR	MAP	Cardiac Output
Aerobic, Dynamic	Increased	Decreased	Minimal change	Increased
Anaerobic, Static	Increased	Increased	Increased	Increased, unchanged, or decreased*

*Cardiac output during static or anaerobic exercise is highly dependent on the type and intensity of exercise.

Definition of abbreviations: MAP, mean arterial pressure; TPR, total peripheral resistance.

► Cardiac and Vascular Function Curves: Examples

Stability of Typical CFC and VFC

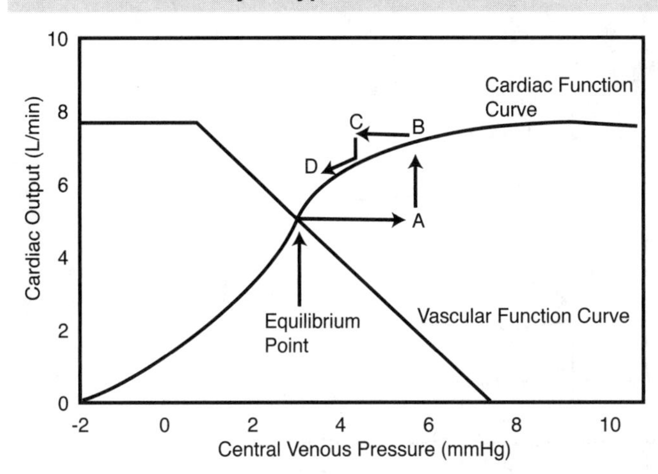

Diagram shows that cardiac output (CO, 5 L/min) changes only transiently when CFC and VFC are not changed. **Point A:** venous pressure is increased from 3 to 6 mm Hg because of sudden removal of blood from arterial system and injection into venous system. This causes **CO** to increase to **point B**. CO then returns to **equilibrium point** in steps (B → C, C → D) as blood is pumped from venous system back to arterial system.

Changes in Blood Volume

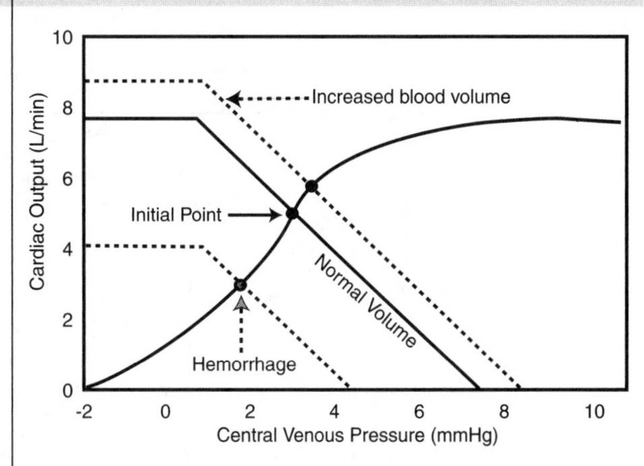

Increased blood volume (e.g., transfusion) shifts the VFC upward, which increases preload. Increased CO follows. Decreased blood volume (e.g., hemorrhage) shifts the VFC downward, which decreases preload. Decreased CO follows. Increases and decreases in preload produce increases and decreases in CO by the **Frank-Starling** mechanism.

Sympathetic Stimulation of Heart

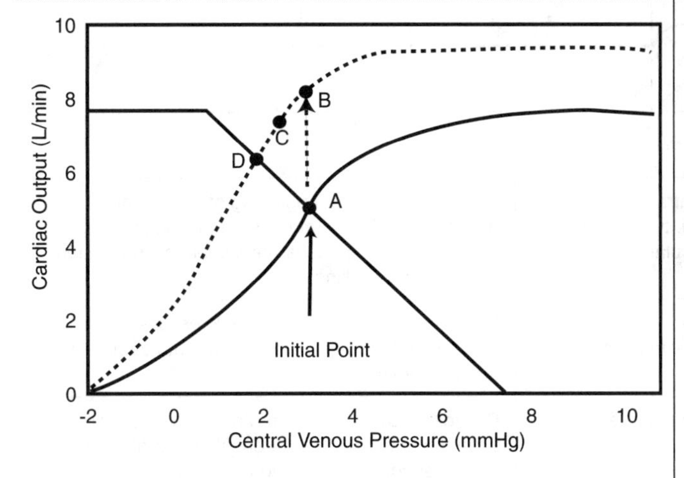

Increased contractility by cardiac sympathetic nerve stimulation shifts the CFC upward *(dashed line)*; however, this does not change the VFC. The initial large increase in CO *(point B)* returns to **point D** on the VFC as blood is transferred from the venous system to the arterial system.

Changes in CO After Heart Failure

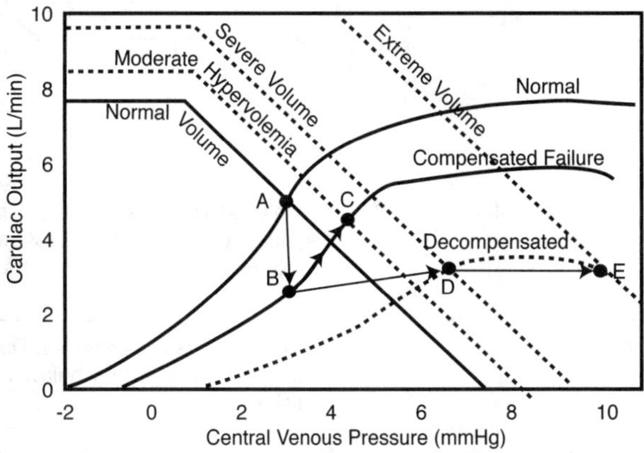

Reduced contractility shifts CO down **(point B)**, but preload immediately increases to intersect with the normal volume curve as shown. Within hours to days, blood volume increases, shifting VFC upward, and **point C** becomes the equilibrium point. With progressive failure, blood volume cannot increase enough to maintain CO at a normal level. **(point D)**. Blood volume continues to increase, which overstretches the heart **(point E)**.

► The Cardiac Cycle: the Wigger's Diagram

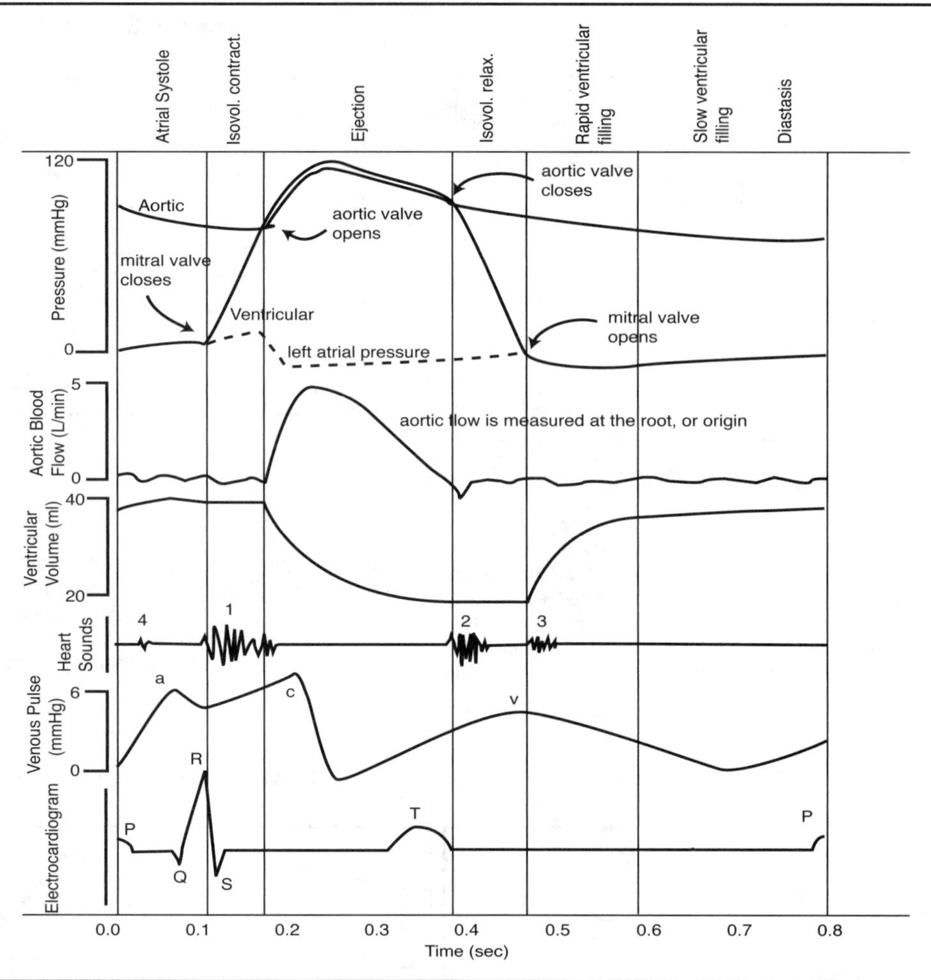

Left ventricular pressure	**Systole:** begins at isovolumic contraction, ends at beginning of isovolumic relaxation: two phases are isovolumic contraction and ventricular ejection	**Diastole:** begins at beginning of isovolumic relaxation and ends at onset of isovolumic contraction: two phases are isovolumic relaxation and ventricular filling
Aortic pressure	Maximum is systolic pressure. During ejection, aortic pressure is slightly below ventricular pressure.	Minimum is diastolic pressure. Pressure falls during diastole as blood flows from aorta into capillaries and then veins.
Left atrial pressure	Systole; isolated from ventricular pressure because mitral valve is closed	Diastole, blood flows from atrium into ventricle because mitral valve is open. Note mitral closed during isovolumic relaxation.
Aortic flow (measured at root)	Systolic ejection begins when ventricular pressure exceeds aortic diastolic and aortic valve opens.	Ejection ends when rapidly falling ventricular pressure causes aortic valve to close.
Ventricular volume	Maximum at end of diastole; does not change during isovolumic contraction because mitral and aortic valves are closed.	Minimum at end of ejection phase; does not change during isovolumic relaxation (both valves closed).
Heart sounds	Systole: S_1 caused by sound of mitral closure	S_2 caused by sound of aortic valve closure
Venous pulse	Rises with atrial systole	Drops as atrium fills
EKG	QRS begins before isovolumic contraction	T wave begins during late ejection phase

▶ Cardiac Pressure–Volume Loops (PV Loops)

Phase	Pressure	Volume
Filling	Slightly ↑	Large ↑; point C = EDV
Isovolumic contraction	Rapid ↑; maximum dp/dt	No change, valves closed
Ejection	Continues to rise	↓ as ejection proceeds
Isovolumic relaxation	Rapid ↓	No change, valves closed
Applications		
Area within loop = stroke work output	Increase work by ↑ stroke volume (volume work) or by ↑ LVP (pressure work)	
Decreased blood volume (hemorrhage, dehydration, urination)	Line C–D shift left (↓ preload); ↓ stroke volume ↓ stroke work	
Increase in contractility (sympathetics, or β-adrenergic drugs, digitalis)	Line F–A shifts left (↓ ESV) a major effect; slight ↓ EDV; overall ↑ stroke volume, ↑ stroke work	
Decreased contractility, as in heart failure	Loop shifts to right and systolic pressure is lower: ↑↑ ESV, ↑ EDP, ↓ SV, ↓ stroke work	
Volume expansion (normal heart)	Line C–D shifts right (↑ EDV); ↑ SV; ↑ stroke work	

Normal PV Curve

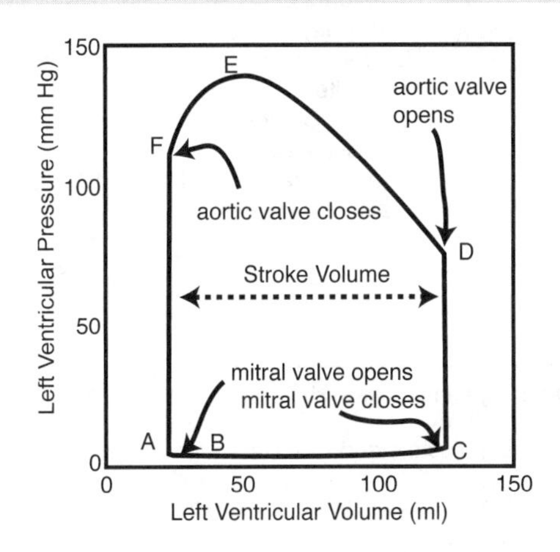

A–B: rapid filling
B–C: reduced or slower filling
C: end diastolic volume (EDV)
C–D: isovolumic contraction
D–F: ejection phase
F: end systolic volume (ESV)
F–A: isovolumic relaxation

- Area within loop = **stroke work**
- Increase work by ↑ stroke volume (volume work) or by ↑ LVP (pressure work)

Blood Volume Changes

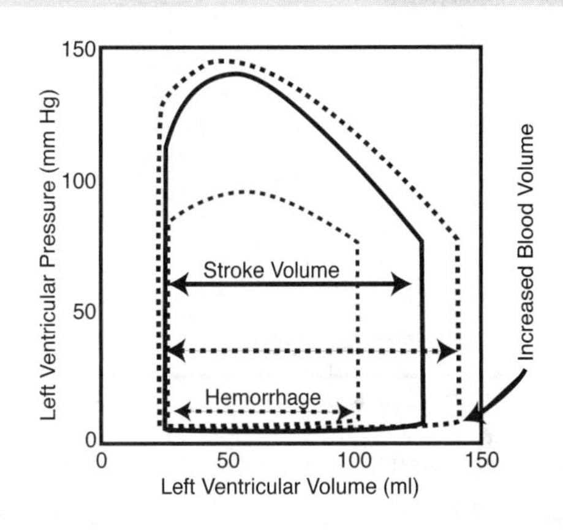

- **Decreased blood volume** (hemorrhage, dehydration): Line C–D shifts **left** (↓ preload); ↓ stroke volume, ↓ stroke work
- **Volume expansion** (normal heart): Line C–D shifts **right** (↑ EDV), ↑ SV, ↑ stroke work
- Diastolic dysfunction tends to ↓ EDV despite ↑ EDP
- Systolic dysfunction tends to ↑ ESV and ↓ systolic LVP

(Continued)

▶ Cardiac Pressure–Volume Loops (PV Loops; *Cont'd.*)

Increased Afterload

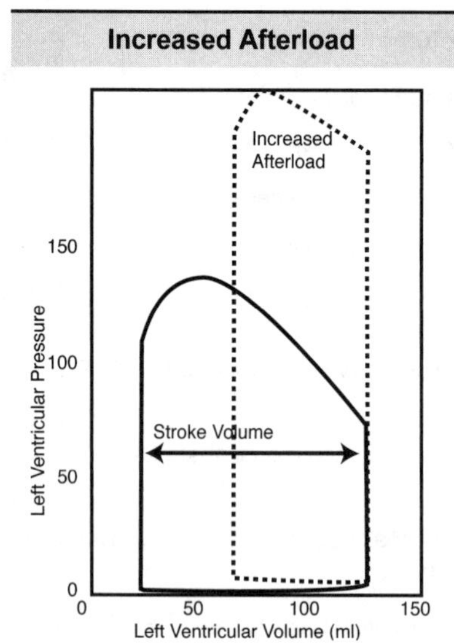

With increased afterload (e.g., ↑ aortic pressure), the velocity of shortening and the distance shortened are both decreased. Thus, ESV increases, causing SV to decrease.

Decreased Afterload

Decreased afterload produces the opposite changes as increased afterload. Thus, ESV decreases and SV increases.

Progressive Heart Failure

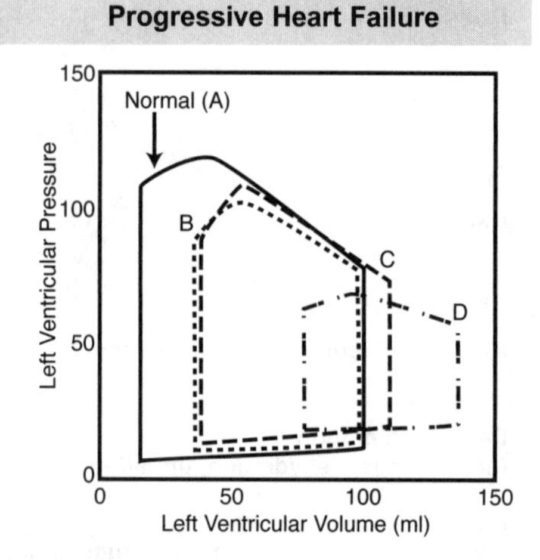

A: Normal
B: Acute loss of contractility without compensation
C: Compensated LV failure (SV partially restored because of moderate increase in preload)
D: Decompensated failure (SV remains low despite ↑↑↑ in preload)
Overall: Curves shift to the right and systolic pressures ↓.
Heart failure: ↑↑↑ ESV, ↑ EDV, ↓ SV, ↓ stroke work

▶ The Cardiac Valves

Mitral	Between LA and LV	Open during filling	Closed during ventricular systole and isovolumic relaxation
Aortic	Between LV and aorta	Open during ejection	Closed during diastole and isovolumic contraction
Tricuspid	Between RA and RV	Open during filling	Closed during ventricular systole and isovolumic relaxation
Pulmonic	Between RV and pulmonary artery	Open during ejection	Closed during diastole and isovolumic contraction

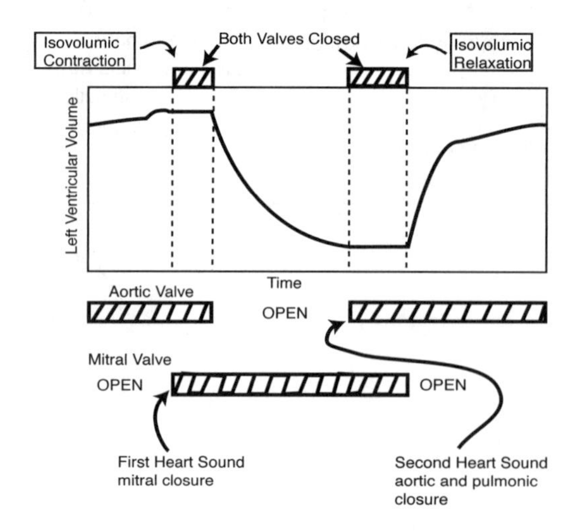

► Valvular Disorders

Aortic Stenosis

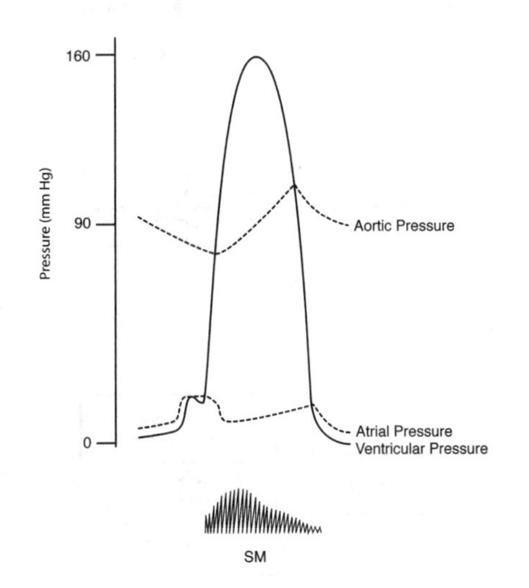

- Discrepancy of systolic LV and systolic aortic pressures
- Causes crescendo-decrescendo systolic murmur

Aortic Regurgitation

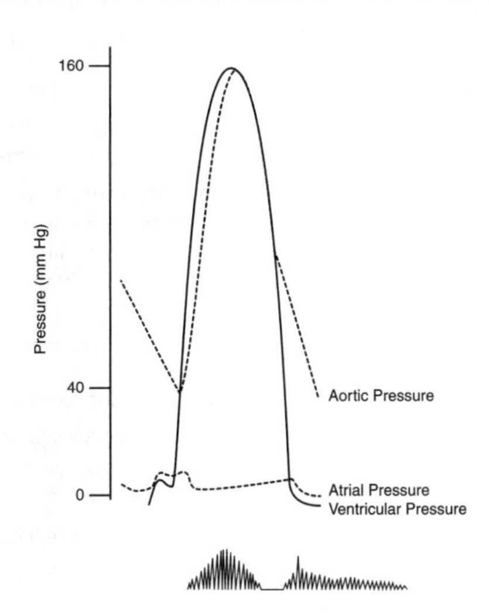

- Diastolic aortic P decreases rapidly as blood flows back into ventricle; ventricular diastolic P is elevated.
- Causes diastolic murmur

Mitral Valve Stenosis

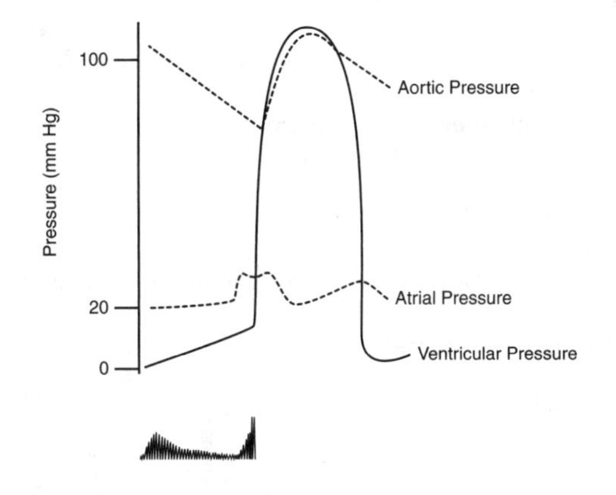

- Discrepancy of diastolic LVP and left atrial P during filling
- Causes diastolic murmur

Mitral Regurgitation

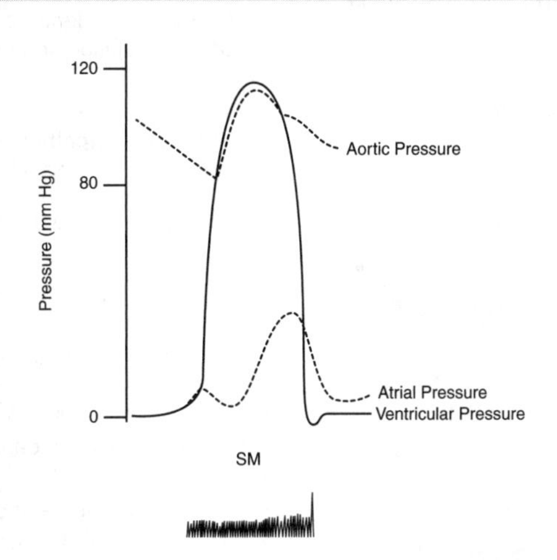

- Incompetent valve allows backflow into left atrium during ventricular systole
- Causes systolic murmur

Definition of abbreviation: SM, systolic murmur.

► Hemodynamics

Poiseuille's equation:

$Q = (P_1 - P_2)/R$

- Flow (Q); P_1 (input pressure); P_2 (output pressure); R (resistance); $(P_1 - P_2)$ = pressure gradient
- ↑ pressure gradient → ↑ flow
- ↑ resistance → ↓ flow

Series circuits:

$R_T = R_1 + R_2 + R_3 \ldots R_n$

- R_T = total resistance
- Flow is equal at all points in series circuit; pressure drops across each resistor
- **Adding more resistors in series increases R_T.** Pressure drop increases along circuit with constant flow, and flow decreases with constant input pressure (P_1).
- Various types of blood vessels lie in series.

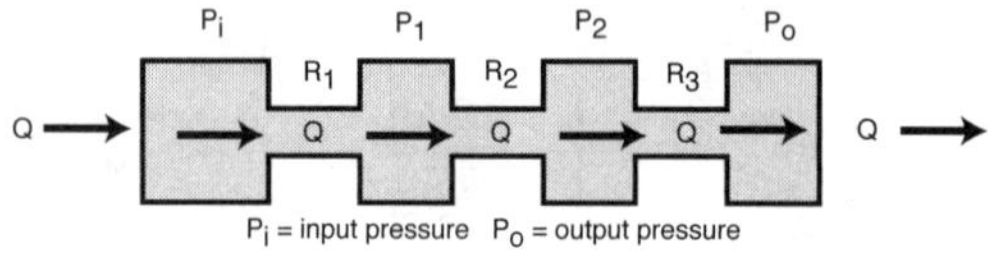

P_i = input pressure P_o = output pressure

Parallel circuits:

$1/R_T = 1/R_1 + 1/R_2 + 1/R_3 \ldots 1/R_n$

- Flow divided between parallel resistors
- R_T is always lower than the lowest resistor
- **Adding more resistors in parallel decreases R_T.**
- Produces low resistance circuit
- Organs lie in parallel

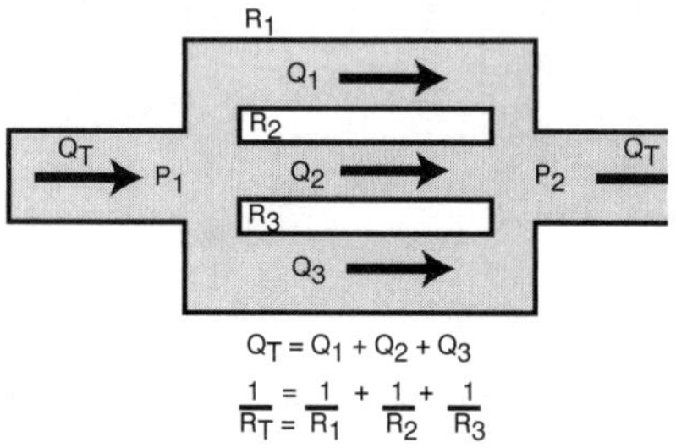

$$Q_T = Q_1 + Q_2 + Q_3$$

$$\frac{1}{R_T} = \frac{1}{R_1} + \frac{1}{R_2} + \frac{1}{R_3}$$

Hydraulic Resistance Equation:

$R = (P_1 - P_2)/Q = 8\eta l/\pi r^4$

- η = viscosity; l = length; r = radius
- **Viscosity ↑ by ↑ hematocrit**
- **Viscosity ↓ in anemia**
- l is usually constant; r changes greatly for normal regulation and in disease.
- 2× radius = 1/16 R → 16 × flow
- ½ radius = 16 × R → 1/16 × flow
- **Control of radius is the dominant mechanism to control resistance.**

Total peripheral resistance (TPR)

- Resistance of peripheral circuit: aorta → right atrium
- **TPR ↑ by sympathetics, angiotensin II, and other vasoconstrictors**
- Highest TPRs in **arterioles**; also main site of blood flow regulation

Total peripheral resistance equation

TPR = (MAP – RAP)/CO

- Mean arterial pressure (MAP); right atrial pressure (RAP)
- Pressure gradient is between aorta and right atrium.
- TPR is calculated from MAP and cardiac output (CO). RAP is assumed to be 0 mm Hg, unless specified.
- TPR is also known as SVR (systemic vascular resistance)

Compliance (C):

$C = \Delta V / \Delta P$

- ΔV = volume change; ΔP = pressure change
- High **compliance** means vessels easily distended by blood.
- **Elasticity** is inverse of compliance; vessels are stiff when elasticity is high

Pulse pressure (PP):

PP = SP – DP

- SP = systolic pressure; DP = diastolic pressure

↓ compliance (e.g., arteriosclerosis) → ↑ SP and ↓ DP, so PP ↑

- **Compliance:** systemic veins > pulmonary circuit > systemic arteries (volume of blood is in same order)

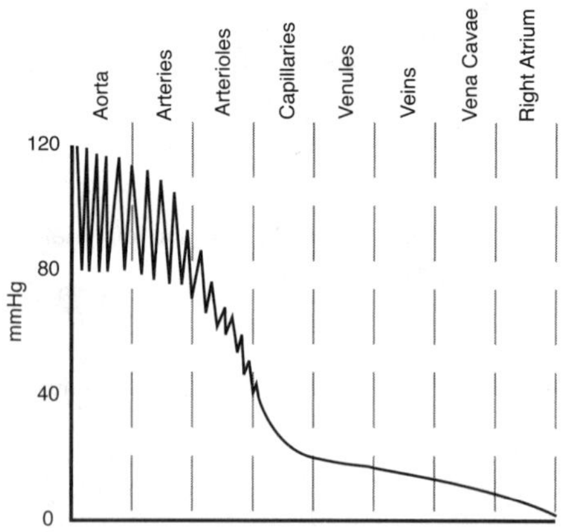

- **MAP = diastolic + 1/3 (pulse pressure)**
- MAP = 80 + 1/3(120 – 80) = 80 = 13 = 93 mm Hg

(Continued)

▶ Hemodynamics (Cont'd.)

Cardiac output (Fick method)	• $CO = \dot{V}O_2/(Ca - Cv)$ • $\dot{V}O_2$ = oxygen consumption, Ca = arterial oxygen content, Cv = venous oxygen content	• Used to measure cardiac output; most accurate if Ca is pulmonary venous and Cv is pulmonary arterial

▶ Area-Velocity Relationship

- **V∝ 1/cross sectional area**
- **$V \propto 1/r^2$,**
- V = velocity; r = radius

- **Assuming total flow is equal in all vessel types, velocity increases as radius decreases.**

- **The aorta is a single large vessel, but its total area is small compared with numerous capillaries in parallel.**

- If low capillary velocity allows adequate time for diffusion, exchange is **perfusion limited**.

- If velocity is high, metabolic exchange may become **diffusion limited**.

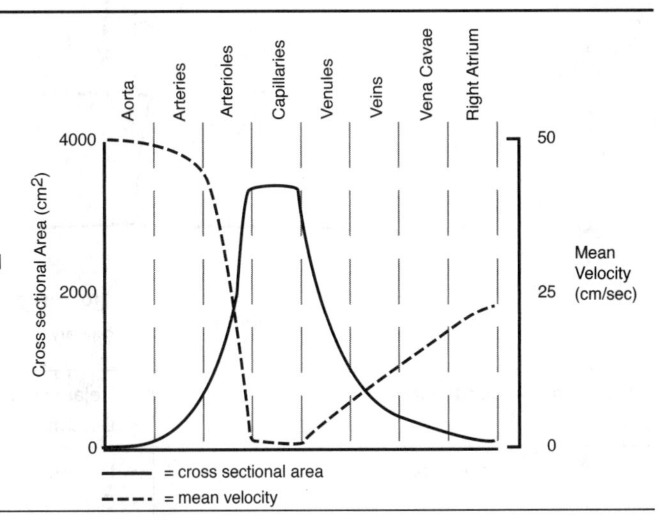

▶ Pressures of the Cardiovascular System

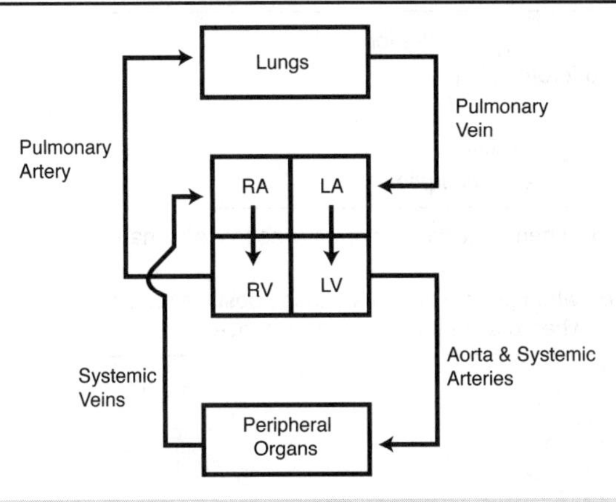

- The cardiac output and stroke volume of the left and right heart are nearly equal.

- The mean pressures are different because the systemic resistance is about 6× higher.

- The pulse pressure in the pulmonary circulation is lower because its compliance is higher.

Pressures in the Pulmonary Circulation (mm Hg)	
Right ventricle	25/0
Pulmonary artery	25/8
Mean pulmonary artery	14
Capillary	7–9
Pulmonary vein	5
Left atrium	<5
Pressure gradient	15 – 5 = 10

Pressures in the Systemic Circulation (mm Hg)	
Left ventricle	120/0
Aorta	120/80
MAP	93
Capillary: skeletal Renal glomerular	<30 45–50
Peripheral veins	<15
Right atrium	0
Pressure gradient	93 – 0 = 93

▶ Factors That Control Filtration and Reabsorption in Capillaries

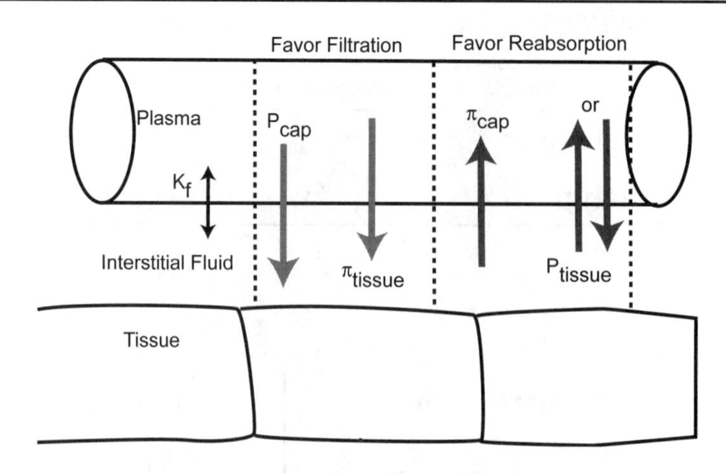

Filtration = K_f (forces favor – forces opposed)
Filtration = K_f [($P_{capillary}$ + π_{tissue}) – (P_{tissue} + $\pi_{capillary}$)]

Ultrafiltration coefficient (K_f)	• Related to surface area, capillary porosity; different in each tissue/organ • Determines amount of ultrafiltration to given filtration pressure
Capillary hydrostatic pressure (P_{cap})	• **Favors filtration;** ↑ P_{cap} → ↑ filtration • Controlled by input pressure, arteriolar diameter, venous pressure
Tissue (interstitial) oncotic pressure (π_{tissue})	• **Favors filtration;** ↑ π_{tissue} →↑ filtration • Directly related to [protein] in interstitial fluid • **Example:** ↑ permeability (e.g., sepsis) →↑ π_{tissue} →↑ filtration
Capillary (plasma) oncotic pressure (π_{cap})	• **Opposes filtration;** ↑π_{cap} →↓ filtration • Directly related to [protein] in plasma • **Examples:** – Liver failure ↓π_{cap} → edema – Dehydration ↑π_{cap} → reabsorption
Tissue (interstitial) hydrostatic pressure (P_{tissue})	• Increases filtration when negative (is normally negative in many but not all tissues) • Opposes filtration when positive; edema causes positive pressure in interstitium, even when pressure is normally negative.

▶ Factors That Alter Capillary Flow and Pressure

	Resistance	Capillary Flow	Capillary Pressure	Example
Arteriole dilation	↓	↑	↑	β-adrenergic agonist, α-adrenergic blocker, decreased sympathetic nervous system activity, metabolic dilation, ACE inhibitors
Arteriole constriction	↑	↓	↓	α-adrenergic agonist, β-adrenergic blocker, increased sympathetic nervous system activity, angiotensin II
Venous dilation	↓	↑	↓	Increased metabolism of tissue
Venous constriction	↑	↓	↑	Physical compression, increased sympathetic activity
Increased arterial pressure	N	↑	↑	Increased cardiac output, volume expansion
Decreased arterial pressure	N	↓	↓	Decreased cardiac output, hemorrhage, dehydration
Increased venous pressure	N	↓	↑	Congestive heart failure, physical compression
Decreased venous pressure	N	↑	↓	Hemorrhage, dehydration

▶ Wall Tension: Law of Laplace

T = P × r

- Tension (T) in wall
- **↑ pressure and ↑ radius → ↑ tension**

Applications

Arterial aneurysm:

- Weak wall balloons
- Vessel radius ↑, causing ↑ wall tension.
- ↑ tension causes ↑ radius (vicious cycle), increasing risk of rupture

Dilated heart failure:

- ↑ ventricular volume → ↑ ventricular radius, which in turn causes ↑ wall tension.
- Thus, dilated ventricle must work harder than normal heart

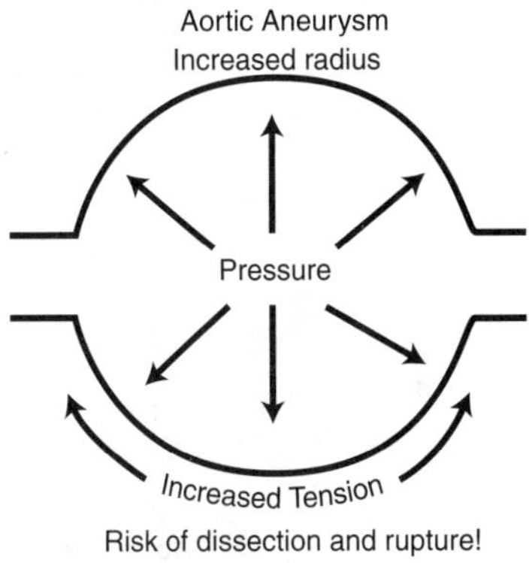

► Autonomic Control of Heart and Circulation

	Sympathetic	Parasympathetic
Heart		
Transmitter-receptor	Norepinephrine: β_1-adrenergic	Acetylcholine: muscarinic
Heart rate	↑ rate (i_f and Ca^{2+} currents): ⊕ chronotropic	↓ rate (↑ gK, ↓ i_f): ⊖ chronotropic
Contractility	↑ force (dp/dt, EF): ⊕ inotropic ↑ gCa^{2+}, ↑ cAMP, ↑ Ca^{2+} release from SR, ↓ duration: fast, strong, short duration	Modest effects: ⊖ inotropic ↓cAMP; ↓gCa^{2+} mainly at very high levels of activity only
Conduction	↑ atrial and ventricular conduction ↑ conduction AV node ↓ PR interval of EKG	↓ atrial and ventricular conduction ↓ conduction AV node ↑ PR interval of EKG
Arteries/arterioles	Norepinephrine: α_1 (mainly) constriction Epinephrine (adrenal): β_2 dilation High levels: α_1 constriction	No direct innervation of vascular smooth muscle
Veins	Norepinephrine: α_1 (mainly) constriction but not usually much ↑ resistance, rather, ↓ capacitance shifts blood toward heart	No direct innervation

► Control of Organ Blood Flow

Organ/Tissue	Neural	Metabolic/Other
Skeletal muscle	**Resting:** α-adrenergic constriction **Exercising:** β-adrenergic dilation (epinephrine, adrenal medulla)	• **Metabolic vasodilation dominates in exercise** • Compression during static exercise blocks flow
Skin	• Thermoregulatory center • α-adrenergic constriction only	• Heat dilates, cold constricts, a direct effect
Heart	• α-adrenergic constriction • β-adrenergic dilation • Overridden by metabolism	• **Metabolic dilation is dominant** • ↑ Cardiac work → ↑ O_2 consumption →↑ coronary flow • Compression during systole, so most coronary flow is during diastole
Brain	Not generally under neural control: autoregulation	• **Metabolism dominates:** ↑ CO_2 → dilation

► Autoregulation

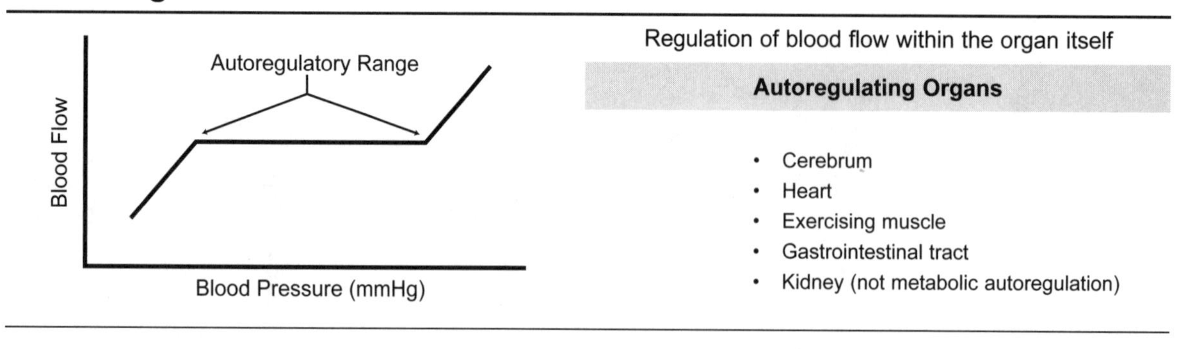

Regulation of blood flow within the organ itself

Autoregulating Organs

- Cerebrum
- Heart
- Exercising muscle
- Gastrointestinal tract
- Kidney (not metabolic autoregulation)

▶ Basics of Integrated Control of Arterial Pressure and Cardiac Output

	Control	Major Actions
Baroreceptors: Carotid sinus (primary) Aortic arch (secondary)	↑ pressure →↑ activity ↓ pressure →↓ activity	↑ activity → ↑ PNS and ↓ SNS ↓ activity → ↓ PNS and ↑ SNS
Sympathetic (SNS)	↑ pressure → ↓ SNS ↓ pressure → ↑ SNS	• Vasoconstriction (↑ α-adrenergic) → ↑ TPR • ↑ Contractility heart (↑ β-adrenergic) → ↓ ESV → ↑ SV • ↑ Heart rate (β-adrenergic) • ↑ Cardiac output (CO)
Parasympathetic (PNS)	↑ pressure → ↑ PNS ↓ pressure → ↓ PNS	• ↓ heart rate (major); ↓ contractility (minor) • At rest, PNS dominant control of heart rate • ↓ cardiac output
Renal	↑ MAP → ↓ renin, angiotensin II (AII) ↑ MAP →↑ urination →↓ volume →↓ preload	• AII vasoconstricts • **Renal control of blood volume and TPR dominant long-term control of blood pressure and CO**

▶ Selected Applications of the Diagram

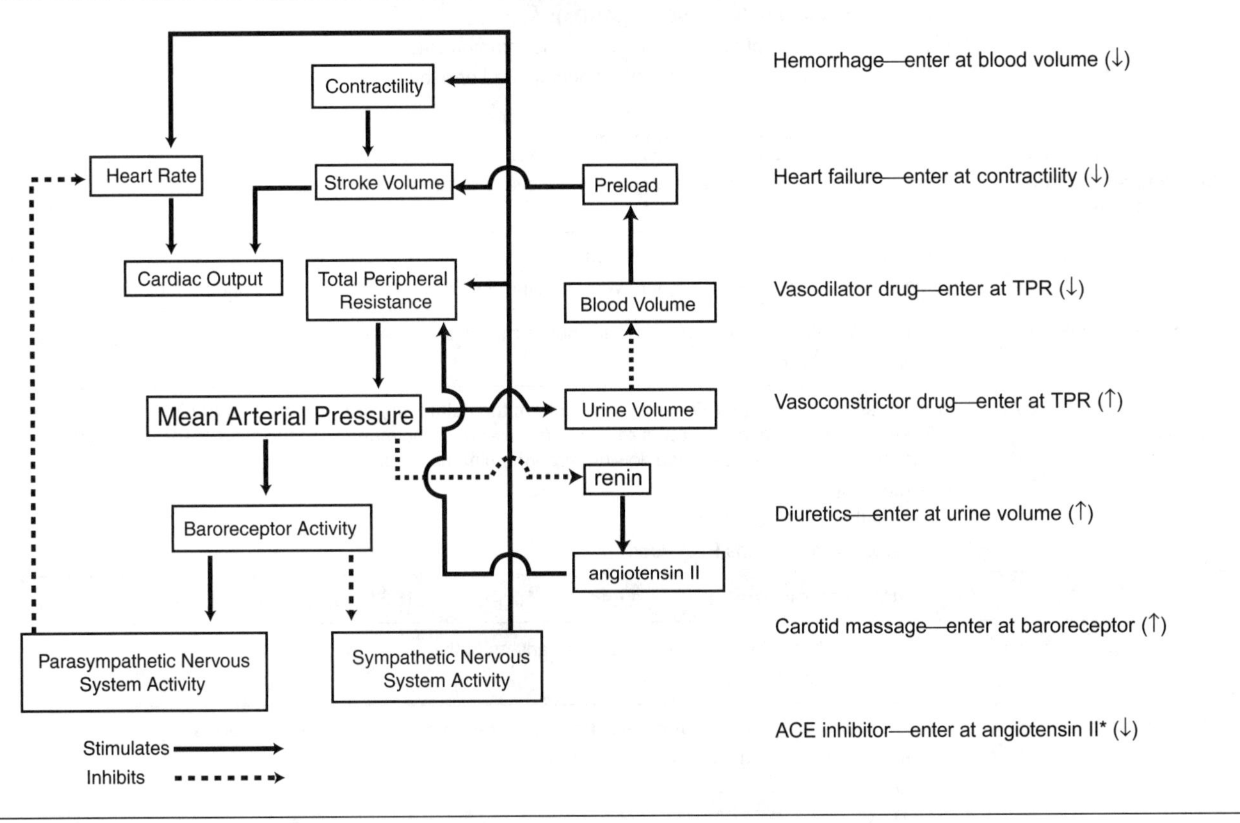

**Note:* also would increase urine flow through reduction of aldosterone (not shown in figure)

Cardiovascular Pathology

► Congenital Abnormalities of the Heart

- Congenital abnormalities occur before the end of week 16 (completion of heart development).
- Clinical significance depends on degree of shunt.
- Up to 90% of congenital heart disease is of unknown etiology.
- **Maternal rubella:** exposure at fifth to tenth week can lead to PDA, ASDs, and VSDs.
- **Fetal alcohol syndrome:** cardiovascular defects, including VSD

Acyanotic (Late Cyanosis) Congenital Heart Disease (Left-to-Right Shunts)

Initially a left-to-right shunt; causes chronic right heart failure and secondary pulmonary hypertension. Increased pressure causes **reversal of shunt flow** with late onset cyanosis: **Eisenmenger syndrome**

Ventricular septal defect (VSD)	• Usually of membranous interventricular septum • Often associated with other defects, including **trisomy 21**
Atrial septal defect (ASD)	**Ostium primum defect (5% of ASDs):** • Defect in lower atrial septum above the atrioventricular valves • Associated with anomaly of AV valves • Requires antibiotic prophylaxis for invasive procedures **Ostium secundum defects (90% of ASDs):** • Defect is in center of the atrial septum at the foramen ovale • Results from abnormalities of septum primum and/or septum secundum • AV valves normal • Antibiotic prophylaxis not needed
Complete endocardial cushion defect	Combination of ASD, VSD, and a common atrioventricular valve
Sinus venosus	• Defect in the upper part of the atrial septum • May cause anomalous pulmonary venous return into superior vena cava or right atrium
Patent foramen ovale	• Remnant of the foramen ovale, usually not of clinical significance • Risk of paradoxical emboli
Patent ductus arteriosus (PDA)	In the fetal circulation, the PDA shunts blood from the pulmonary artery into aorta. At birth, the pressure differential changes, and the flow is reversed (from aorta to pulmonary artery). This leads to pulmonary hypertension due to excess blood flowing through pulmonary artery. **Pharmacology:** • Indomethacin: closes PDA • Prostaglandin E: keeps PDA open

Cyanotic Congenital Heart Disease (Right-to-Left Shunts)

- Right-to-left shunt bypasses the lungs and hence produces cyanosis as early as birth.
- Paradoxical embolism (DVT causes systemic infarct) may occur.

Tetralogy of Fallot	• **Most common cyanotic congenital heart disease in older children and adults** • Associated with trisomy 21 • Four lesions: *1)* VSD *2)* An overriding aorta that receives blood from both ventricles *3)* Right ventricular hypertrophy *4)* Pulmonic stenosis (right ventricular outflow obstruction)

(Continued)

Cyanotic Congenital Heart Disease (Right-To-Left Shunts; *Cont'd.*)	
Transposition of the great vessels	• Failure of the **truncoconal septum** to spiral • The aorta arises from the right ventricle, and the pulmonary artery arises from the left ventricle, producing two closed loops. This is fatal if a shunt (e.g., PDA, VSD, ASD, patent foramen ovale) is not present to mix the venous and systemic blood.
Persistent truncus arteriosus	• Failure of the truncus arteriosus to separate into the aorta and pulmonary arteries • Usually a membranous VSD • Truncus arteriosus receives blood from both ventricles, so cyanosis results.

Definition of abbreviation: DVT, deep venous thrombosis.

► Obstructive Congenital Heart Disease

Does not usually cause cyanosis	
Coarctation of the aorta	**Preductal (infantile) type:** • Narrowing of the aorta proximal to the opening of the ductus arteriosus • May be associated with cyanosis of the lower half of the body • Often requires surgery **Postductal (adult) type:** • Narrowing of the aorta distal to the opening of the ductus arteriosus • **Most common type**; allows survival into adulthood • **Disparity in pressure between the upper and lower extremities** • Collateral circulation leads to rib notching
Pulmonic valve stenosis or atresia	• Unequal division of the truncus arteriosus so that the pulmonary trunk has no lumen or opening at the level of the pulmonary valve • May cause cyanosis if severe
Aortic valve stenosis or atresia	**Complete atresia:** incompatible with life **Bicuspid aortic valves:** asymptomatic, can lead to infective endocarditis, left ventricular overload, and sudden death • Calcify fifth to sixth decade (tricuspid aortic valves usually calcify 10 years later) • Most common cause of aortic stenosis (more than rheumatic fever)

Diseases Associated with Congenital Heart Defects
Marfan syndrome: 1/3 patients have aortic dilatation and incompetence, aortic dissection, and ASD
Down syndrome: 20% of patients may have congenital cardiovascular disease
Turner syndrome: Coarctation of the aorta
22q11 syndromes (DiGeorge syndrome and velocardial facial syndrome): Truncus arteriosus and tetralogy of Fallot
Congenital rubella: Septal defects, patent ductus arteriosus, pulmonary artery stenosis
Maternal diabetes: Transposition of great vessels

▶ Ischemic Heart Disease

- **Leading cause of death**
- Most angina pectoris caused by severe atherosclerotic narrowing of coronary arteries
- Result of decreased supply (anemia, carbon monoxide, pulmonary disease) and/or increased demand (exertion, hypertrophy)
- Sudden cardiac death the presenting symptom in 25% of patients with IHD

Angina pectoris	• Paroxysmal substernal or precordial chest pain • Transient **myocardial ischemia** without myocardial infarction
Stable angina pectoris	• Paroxysms are associated with a **fixed** amount of exertion, e.g., after walking three blocks • Typical attacks last less than 10 minutes and are **relieved with rest** or sublingual nitroglycerin • ECG may show **ST segment depression** (ischemia limited to subendocardium)
Prinzmetal angina	• Vasospasm causes decreased blood flow through atherosclerotic vessels • This form of attack frequently occurs at rest with **ST-segment elevation** on ECG • Treat with calcium channel blockers
Unstable angina	• Chest pain with progressively less exertion, then occurring at rest, **often precedes myocardial infarction** • May be unresponsive to nitroglycerin

▶ Myocardial Infarction

- Ischemic necrosis of myocardium, most commonly transmural, but can be subendocardial
- Highest incidence of fatal MI: 55 to 64 years old
- Risk factors: male sex, hypertension, hypercholesterolemia, cigarette smoking, family history, diabetes mellitus, oral contraceptive use, sedentary lifestyle, type A personality, family history of MI in men under 45, women under 55 years of age

Clinical features	• Acute, severe, crushing chest pain, often radiating to the jaw or left arm; diaphoresis; little or no chest pain may be present in diabetic and elderly patients • ECG: **ST elevation and T-wave inversion** with or without Q-waves • **Elevated cardiac enzymes**
Prognosis	• Sudden cardiac death: secondary to a fatal arrhythmia, occurs in 25% • Mortality after myocardial infarction: 35% in the first year, 45% in second year, and 55% in third year • Complications: arrhythmias, CHF, cardiogenic shock, systemic emboli from mural thrombi, aneurysm • Wall/papillary muscle rupture (3–7 days after infarct), postinfarction pericarditis (Dressler syndrome; 2–10 weeks postinfarction)
Treatment and management	• **Coronary artery bypass:** saphenous vein or internal mammary artery grafts restore circulation; grafts last approximately 10 years before restenosis typically occurs • **Angioplasty** (balloon dilatation) also restores circulation, half re-stenose in 1 year • **Fibrinolytic therapy** (e.g., tissue plasminogen activators): highly effective in reducing mortality if administered early in the course of the MI.

▶ Appearance of Infarcted Myocardium

Time	Gross	Histologic
1 hour	No gross changes evident	Intracellular edema
6–12 hours		Wavy myocardial fibers, vacuolar degeneration, contraction band necrosis, beginning neutrophilic infiltrate
12–24 hours	Pale, cyanotic, edematous	
24–48 hours	Well-demarcated, soft, pale	Neutrophilic infiltrate, increased cytoplasmic eosinophilia, and coagulation necrosis become evident
3–10 days	Infarct becomes soft, yellow, surrounded by hyperemic rim	Monocytic infiltrate predominates at 72 hours
2 weeks	Infarct area is surrounded by granulation tissue that is gradually replaced by scar tissue.	

Cardiac Enzymes
Troponin I peaks first (4 h): remains elevated 7–10 days
CK-MB peaks within 24 h: remains elevated 2–3 days
LDH peaks later (about 2 days): remains elevated 8–14 days
AST also rises and falls predictably in myocardial infarction, but may indicate liver damage instead

Definition of abbreviations: AST, aspartate aminotransferase; CK-MB, creatine kinase MB fraction; LDH, lactate dehydrogenase.

▶ Rheumatic Fever and Rheumatic Heart Disease

Acute rheumatic fever	• Onset is typically 1–3 weeks after group A β-hemolytic streptococcal pharyngitis, otitis media • Children: 5–15 years old • Declining incidence secondary to penicillin use • Antistreptococcal antibodies cross-react with host connective tissue • Diagnosed using Jones criteria (two major or one major and two minor)	
Jones Criteria	**Major**	**Minor**
Mnemonic: **J**ones ♥ carditis **N**odules **E**rythema marginatum **S**ydenham chorea	• **Migratory polyarthritis**—large joints that become red, swollen, and painful • **Erythema marginatum**—macular skin rash, often in "bathing suit" distribution • **Sydenham chorea**—involuntary, choreiform movements of the extremities • **Subcutaneous nodules** • **Carditis**—may affect the endocardium, myocardium, or pericardium; myocarditis causes most deaths during the acute stage	• Previous rheumatic fever • Fever • Arthralgias • Prolonged PR interval • Elevated ESR • Leukocytosis • Elevated C-reactive protein
Rheumatic heart disease	• Repeated bouts of endocarditis and inflammatory insult lead to scarring and thickening of the valve leaflets with nodules along lines of closure • **Mitral valve** most commonly (75–80%) affected; fibrosis and deformity lead to "fish mouth" or "buttonhole" stenosis. Next in frequency are the aortic and mitral valves together. Tricuspid and pulmonic valves are rarely affected. • **Aschoff bodies** are pathognomonic lesions; focal collections of perivascular fibrinoid necrosis surrounded by inflammatory cells including large histiocytes (**Anitschkow cells**) • Left atrial dilatation, mural thrombi, and right ventricular hypertrophy • Predisposes to infective endocarditis	

► Congestive Heart Failure (CHF)

Types	Etiology	Comments
Left-sided heart failure	• Ischemic heart disease • Aortic stenosis • Aortic insufficiency • Hypertension • Cardiomyopathies	• Increased back pressure produces pulmonary congestion and edema • Dyspnea, orthopnea, paroxysmal nocturnal dyspnea, and cough • Renal hypoperfusion stimulates renin-angiotensin-aldosterone axis • Retention of salt and water compounds the pulmonary edema
Right-sided heart failure	• Left-sided heart failure • Cor pulmonale • Pulmonary stenosis • Pulmonary insufficiency	• Chronic passive congestion of the liver (**nutmeg liver**), peripheral edema, ascites, jugular venous distension • Renal hypoperfusion with salt and water retention
Cor pulmonale	• Parenchymal disease (e.g., COPD, causing increased pulmonary vascular resistance) • Vascular disease (e.g., vasculitis, shunts, multiple emboli)	• Cor pulmonale is right ventricular failure, resulting specifically from pulmonary hypertension • May be acute (massive pulmonary embolus) or chronic.

► Shock

Decreased Effective Circulatory Volume

Causes	• Decreased cardiac output (myocardial infarction, arrhythmia, tamponade) • Reduction of blood volume (hemorrhage, adrenal insufficiency, fluid loss) • Pooling in periphery: massive vasodilation caused by bacterial toxins and vasoactive substances
Complications	• Cellular hypoxia, lactic acidosis • Encephalopathy • Myocardial necrosis and infarcts • Pulmonary edema, adult respiratory distress syndrome • Acute tubular necrosis
Stages	**Compensated:** reflex tachycardia, peripheral vasoconstriction **Decompensated:** ↓ blood pressure, ↑ tachycardia, metabolic acidosis, respiratory distress, and ↓ renal output **Irreversible:** irreversible cellular damage, coma, and death

▶ Endocarditis

Classic Signs	
Janeway lesions—erythematous, nontender lesions on palms and soles **Roth spots**—retinal hemorrhages **Osler nodes**—erythematous, tender lesions on fingers and toes (Also see anemia, **splinter hemorrhages)**	
Acute	• Organism—high virulence; *Staphylococcus aureus* (50%) and streptococci (35%) • Affects **previously normal valves** • Often involves the **tricuspid valve in intravenous drug users** • Vegetations may form myocardial abscesses, septic emboli, or destroy the HACEK valve, causing insufficiency. • High fever with chills
Subacute bacterial endocarditis	• Organism—low virulence. ***Streptococcus viridans***, gram-negative bacilli, *Staphylococcus epidermidis*, gram-negative bacilli (both are normal oral flora), *Staphylococcus epidermidis* (IV drug use) • *Candida* is a rare cause (associated with indwelling vascular catheters) • Affects **previously abnormal valves** • More insidious onset, with positive blood cultures, fatigue, low-grade fever without chills, splinter hemorrhages
Nonbacterial thrombotic (marantic) endocarditis	• Associated with chronic illness • Mitral valve most commonly affected • Sterile, small vegetations, loosely adhering along lines of closure • May embolize and provide a nidus for infective endocarditis
Nonbacterial verrucous (Libman-Sacks) endocarditis	• Mitral and tricuspid valvulitis in patients with systemic lupus erythematosus (SLE) • Small, warty vegetations on **both sides** of their valve leaflets • Does not embolize and rarely provides a nidus for infection

▶ Myocarditis

- May have dilatation and hypertrophy of all four chambers, diffuse or patchy pallor with foci of hemorrhage, peripheral edema
- Inflammatory lesions with characteristic cellular infiltrate:
  - Neutrophilic—bacterial myocarditis
  - Mononuclear—viral myocarditis
  - Eosinophilic—Fiedler myocarditis

Noninfectious myocarditis	Collagen vascular diseases, rheumatic fever, SLE, and drug allergies
Viral myocarditis	• **Most common form of myocarditis** • **Coxsackie B** (positive-sense RNA viruses, picornavirus family); also, polio, rubella, and influenza • Self-limited, but may be recurrent and lead to cardiomyopathy and death. • 1/3 of AIDS patients show focal myocarditis on autopsy
Bacterial myocarditis	Diphtheria (toxin-mediated), meningococci
Protozoal	• ***Trypanosoma cruzi***: Chagas disease; myocardial pseudocysts can lead to CHF • Toxoplasmosis also causes pseudocysts • Myocardial involvement appears days to weeks after the primary infection • May be asymptomatic versus acute onset of dyspnea, tachycardia, weakness, or severe CHF • Most recover fully

▶ Valvular Heart Disease

Mitral valve prolapse	• Mitral leaflets (usually posterior) project into left atrium during systole, leading to insufficiency • 7% of the United States population, most commonly in **young women** • Seen in most patients with **Marfan syndrome** • Characteristic **midsystolic click** and high-pitched murmur • Usually asymptomatic, but may have associated dyspnea, tachycardia, chest pain • *Complications:* atrial thrombosis, calcification, infective endocarditis, systemic embolization
Mitral stenosis	• Stenosis may be combined with mitral valve prolapse • Increased left atrial pressure and enlarged left atrium • Early diastolic **opening snap** • *Complications:* pulmonary edema, left atrial enlargement, chronic atrial fibrillation, atrial thrombosis and systemic emboli
Aortic valve insufficiency	• **Acute** (infective endocarditis) • Sudden left ventricular failure, increased left ventricular filling pressure, inadequate stroke volume • **Chronic** (aortic root dilation): – Volume overload, eccentric hypertrophy – Wide pulse pressure (**bounding pulse**) – Etiologies include congenitally bicuspid aortic valve, rheumatic heart disease, or syphilis
Aortic valve stenosis	• Rheumatic heart disease, bicuspid aortic valve • Thickening and fibrosis of valve cusps without fusion of valve commissures (fusion present in rheumatic heart disease) • Asymptomatic until late, presents with angina, syncope, and CHF • Systolic ejection click • *Complications:* sudden death, secondary to an arrhythmia or CHF

▶ Cardiomyopathies

Diseases Not Related to Ischemic Injury

Dilated (congestive) cardiomyopathy	• Gradual dilatation of all four chambers, producing cardiomegaly, ↓ contractility, stasis, formation of mural thrombi. • Death from progressive CHF, thromboembolism, or arrhythmia • *Etiologies:* idiopathic, alcohol (reversible), doxorubicin (irreversible), thiamine deficiency, pregnancy, postviral
Hypertrophic cardiomyopathy (idiopathic hypertrophic subaortic stenosis)	• Marked **asymmetric hypertrophy** of the ventricular septum, **left ventricular outflow obstruction** • Decreased cardiac output may cause dyspnea, angina, atrial fibrillation, syncope, sudden death • Classic case: **young adult athletes** who die during strenuous activity • *Etiologies:* genetic (50%, autosomal dominant pattern)
Restrictive (infiltrative) cardiomyopathy	• Diastolic dysfunction (impaired filling) • Infiltration of extracellular material within myocardium • *Etiologies:* elderly—cardiac amyloidosis (may induce arrhythmias); young (<25 years old)—sarcoidosis associated with systemic sarcoidosis • Secondary cardiomyopathy: metabolic disorders, nutritional deficiencies

► Pericardial Disease

- Usually secondary; local spread from adjacent mediastinal structures
- Primary pericarditis is usually due to systemic viral infection, uremia, and autoimmune diseases

Fibrinous pericarditis	• Exudate of fibrin • *Etiologies:* post myocardial infarction, trauma, rheumatic fever, radiation, SLE • Loud pericardial friction rub with chest pain, fever
Serous pericarditis	• Small exudative effusion with few inflammatory cells • *Etiologies:* nonbacterial, immunologic reaction (rheumatic fever, SLE), uremia, or viral • Usually asymptomatic
Suppurative pericarditis	• Purulent exudate; leads to constrictive pericarditis and cardiac insufficiency • *Etiologies:* bacterial, fungal, or parasitic infection • May have systemic signs of infection and a soft friction rub
Hemorrhagic pericarditis	• Exudate of blood with suppurative or fibrinous component • *Etiologies:* tuberculosis or a malignant neoplasm; organization • May lead to constrictive pericarditis
Caseous pericarditis	• Caseous exudate with fibrocalcific constrictive pericarditis • *Etiologies:* tuberculosis

► Pericardial Effusion

Pericardial effusion is leakage of fluid (transudate or exudate) into the limited pericardial space. It may result in cardiac tamponade. Generally, the rate of filling rather than the absolute volume determines the degree of tamponade.

Serous effusion	• *Etiology:* hypoproteinemia or CHF • Develops slowly, rarely causing cardiac compromise
Serosanguineous effusion	• *Etiology:* history of trauma (e.g., cardiopulmonary resuscitation), tumor, or TB • Develops slowly, rarely causing cardiac compromise
Hemopericardium	• *Etiology:* penetrating trauma, ventricular rupture (after myocardial infarction), or aortic rupture • Develops quickly; **can cause cardiac tamponade** and death

► Cardiac Neoplasms

Primary tumors (rare, majority benign)	**Myxoma: most common primary cardiac tumor in adults** • Most occur in the left atrium • May be any size; sessile or pedunculated • Complications include ball-valve obstructions of the mitral valve, embolization of tumor fragments **Rhabdomyoma: most common primary cardiac tumor in children** (especially those with **tuberous sclerosis**)
Metastases	Lung and lymphoma, predominantly involving the pericardium

► Vasculitides

Disease	Involvement	Clinical	Comments	Treatment
Buerger disease (thromboangiitis obliterans)	• Involves **small and medium-sized arteries** and veins in the **extremities** • **Microabscesses** and segmental **thrombosis** lead to vascular insufficiency, ulceration, **gangrene**	Causes severe pain (claudication) and Raynaud phenomenon in affected extremity	• Neutrophilic vasculitis that tends to involve the extremities of young men (usually under 40) who smoke heavily • Common in Israel, India, Japan, and South America	Smoking cessation
Churg-Strauss syndrome (allergic granulomatosis and angiitis)	Lung, spleen, kidney	Associated with bronchial **asthma**, **granulomas**, and **eosinophilia**	• Variant of polyarteritis nodosa • **P-ANCA ⊕**	Corticosteroids, occasionally immunosuppressants
Henoch-Schönlein purpura	• Affects small vessels, most commonly in the skin, joints, and gastrointestinal system	• Typically develops after a URI • "Palpable purpura" skin rash on buttocks and legs • Arthralgias • GI symptoms: abdominal pain, intestinal hemorrhage, melena • Nephritis in cases with IgA nephropathy	• Most common form of childhood systemic vasculitis (peak at age 5) • IgA-mediated leukocytoclastic vasculitis with circulating IgA immune complexes • Linked to HLA-B35 and human parvovirus	• Most patients treated with supportive therapy • Usually self-limited • Cyclophosphamide sometimes used in cases with severe IgA nephropathy
Kawasaki disease (mucocutaneous lymph node syndrome)	• Segmental necrotizing vasculitis involves large, medium-sized, and small arteries • **Coronary arteries** commonly affected (70%)	• Fever • Conjunctivitis • Erythema and erosions of the oral mucosa • Generalized maculopapular skin rash • Lymphadenopathy • Mortality rate is 1–2% due to rupture of a **coronary aneurysm** or coronary thrombosis	Commonly affects infants and **young children** (age <4) in **Japan**, Hawaii, and U.S. mainland	IV immunoglobulin (IVIG), aspirin, sometimes anticoagulants
Microscopic polyangiitis	• Involves small vessels in a pattern resembling Wegener granulomatosis, but without granulomas • Resembles polyarteritis nodosa in some vessels • Segmental fibrinoid necrosis of the media may be present • Some vessels show infiltration with fragmented neutrophils = **leukocytoclastic angiitis**	• Fever, malaise, myalgia, weight loss • Skin rash in 50%, including ulcerations and gangrene • Can affect vessels in many organ systems • Roughly ¾ of patients survive 5 years	• 80% are ANCA positive, with 60% P-ANCA positive and 40% C-ANCA positive • Immune complexes are rare • Etiology unclear	Prednisone and cyclophosphamide to induce remission or treat relapse; methotrexate or azathioprine to maintain remission

(Continued)

► Vasculitides (*Cont'd.*)

Disease	Involvement	Clinical	Comments	Treatment
Polyarteritis nodosa	• Small and medium-sized arteries in skin, joints, peripheral nerves, kidney, heart, and GI tract • Lesions will be at different stages (acute, healing, healed)	• Affects young adults (male > female) • Low-grade fever, weight loss, malaise • Hematuria, renal failure, hypertension • Abdominal pain, diarrhea, GI bleeding • Myalgia and arthralgia	• Hepatitis B antigen (HBsAg) ⊕ in 30% of cases	Corticosteroids and cyclophosphamide (often fatal without treatment)
Sturge Weber syndrome	• Congenital vascular disorder affecting capillary-type vessels • Angiomas of the leptomeninges and skin • Cutaneous angiomas (**port-wine stains**) involving the skin of the face in the distribution of ophthalmic and maxillary divisions of trigeminal nerve • Leptomeningeal angiomatosis	• Seizures and other neurologic manifestations due to altered blood flow ("vascular steal") in brain adjacent to leptomeningeal angiomas • Glaucoma, blindness • Mental deficiency may be present • May have hemiparesis contralateral to leptomeningeal angiomatosis	• Thought to be due to a failure of regression of embryonal vessels that normally occurs around the ninth week of gestation • Microscopically, the port-wine stain, which is a type of nevus flammeus, shows dilated and ectatic capillaries	• Laser therapy can ameliorate the port-wine stain • Medical treatment of secondary conditions such as seizures, glaucoma, and headaches • Some patients require surgery for intractable seizures or other major neurologic problems
Takayasu arteritis (pulseless disease)	• Granulomatous vasculitis with massive intimal fibrosis that tends to involve **medium-sized to large arteries**, including the aortic arch and major branches • Produces characteristic narrowing of arterial orifices	• Fever, night sweats, muscle and joint aches, loss of pulse in upper extremities • May lead to visual loss and other neurologic abnormalities	Most common in **Asia, especially in young and middle-aged women** (ages 15–45)	Corticosteroids
Temporal arteritis (giant cell arteritis)	• Usually segmental granulomatous involvement of small and medium-sized arteries, esp. the **cranial arteries** (temporal, facial, and ophthalmic arteries) • Multinucleated giant cells and **fragmentation of the internal elastic lamina** seen in affected segments.	• Headache, facial pain, tenderness over arteries, and visual disturbances (**can progress to blindness**) • Fever, malaise, weight loss, muscle aches, anemia • Patient usually middle-aged to elderly female • Elevated ESR	• Most common form of vasculitis • Associated with HLA-DR4 • **Polymyalgia rheumatica:** systemic flu-like symptoms; joint involvement also present	Corticosteroids (important to avoid blindness)
Wegener granulomatosis	• Necrotizing granulomatous vasculitis that affects small arteries and veins • Classically involves **nose**, **sinuses**, **lungs**, and **kidneys**	• Middle-aged adults, males > females • Bilateral pneumonitis with nodular and cavitary pulmonary infiltrates • Chronic sinusitis • Nasopharyngeal ulcerations • Renal disease (focal necrotizing glomerulonephritis)	Associated with **C-ANCA** (autoantibody against **proteinase 3**)	Corticosteroids, immunosuppressants

► Additional Vascular Diseases

Arteriolosclerosis

- Refers to small artery and arteriolar changes, leading to luminal narrowing that are most often seen in patients with diabetes, hypertension, and aging
- Hyaline and hyperplastic (onion-skinning) types

Atherosclerosis

Characterized by lipid deposition and intimal thickening of large and medium-sized arteries. Abdominal aorta more likely involved than the thoracic aorta. Within the abdominal aorta, lesions tend to be more prominent around the ostia. After the abdominal aorta, others commonly affected are the coronary, popliteal, and internal carotid arteries.

Key process: intimal thickening and lipid accumulation produces atheromatous plaques

The earliest lesion is the **fatty streak**, which is seen almost universally in children and may represent reversible precursor. The progression of the disease is thought in part due to a response to injury from such agents as hypertension, hyperlipidemia, and tobacco smoke. This leads to inflammatory cell adherence, migration, and proliferation of smooth muscle cells from the media into the intima.

The **mature plaque** has a **fibrous cap**, a cellular zone composed of **smooth muscle cells**, **macrophages**, and lymphocytes and a **central core** composed of necrotic cells, **cholesterol clefts, and lipid-filled foam cells** (macrophages). **Complicated plaques** are seen in advanced disease. These plaques may rupture, form fissures or ulcerate, leading to myocardial infarcts, strokes, and mesenteric artery occlusion. Damage to the cell wall predisposes to aneurysm formation.

Major Risk Factors	Minor Risk Factors
Hyperlipidemia, hypertension, smoking, diabetes	Male sex, obesity, sedentary lifestyle, stress, elevated homocysteine, oral contraceptive use, increasing age, familial/ genetic factors

Mönckeberg Medial Calcific Sclerosis

Asymptomatic medial calcification of medium-sized arteries

Raynaud Disease

An idiopathic small artery vasospasm that causes **blanching and cyanosis of the fingers and toes**; the term Raynaud phenomenon is used when similar changes are observed secondary to a systemic disease, such as scleroderma or systemic lupus erythematosus.

▶ Hypertension

Most cases (90%) are idiopathic and termed essential hypertension. The majority of the remainder is secondary to intrinsic renal disease; less commonly, narrowing of the renal artery. Infrequent secondary causes include primary aldosteronism, Cushing disease, and pheochromocytoma.

Renal causes of hypertension can usually be attributed to increased renin release. This converts angiotensinogen to angiotensin I, which is converted to angiotensin II in the lung. Angiotensin II causes arteriolar constriction and stimulates aldosterone secretion and therefore sodium retention, which leads to an increased intravascular volume.

Classification of Blood Pressure for Adults*		
BP Classification	**Systolic BP (mm Hg)**	**Diastolic BP (mmHg)**
Normal	< 120	and < 80
Prehypertension	120-139	or 80-89
Stage 1 Hypertension	140-159	or 90-99
Stage 2 Hypertension	≥ 160	or ≥ 100

*Guidelines from 7th Report of the Joint National Committee on Prevention, Detection, Evaluation, and Treatment of High Blood Pressure (JNC 7)

▶ Aneurysm

- A congenital or acquired weakness of the vessel wall media, resulting in a localized dilation or outpouching
- *Complications:* thrombus formation, compression of adjacent structures, and rupture with risk of sudden death

Type of Aneurysm	Associations	Anatomic Location	Comments
Atherosclerotic	Atherosclerosis, hypertension	Usually involve **abdominal aorta**, often below renal arteries	Half of aortic aneurysms >6 cm in diameter will rupture within 10 years
Syphilitic	Syphilitic obliterative endarteritis of vasa vasorum	**Ascending aorta** (aortic root)	May dilate the aortic valve ring, causing aortic insufficiency
Marfan syndrome	Lack of **fibrillin** leads to poor elastin function	Ascending aorta (aortic root)	May dilate the aortic valve ring, causing aortic insufficiency
Dissecting aneurysm (aortic dissection)	**Hypertension, cystic medial necrosis** (e.g., Marfan syndrome)	Blood enters intimal tear in aortic wall and spreads through media	Presents with severe tearing pain
Berry aneurysm	Congenital; some associated with **adult polycystic kidney disease**	Classic location: **Circle of Willis**	Rupture leads to **subarachnoid hemorrhage**

▶ Venous Disease

Deep vein thrombosis	Involves deep leg veins	Major complication: **pulmonary embolus**
Varicose veins	Dilated, tortuous veins caused by increased intraluminal pressure	• Superficial veins of legs • Hemorrhoids • **Esophageal varices**

► Vascular Tumors

Angiosarcoma	• Malignant vascular tumor with a high mortality • Occurs most commonly in skin, breast, liver, soft tissues
Glomus tumor	• Small, painful tumors most often found under fingernails
Hemangioma	• Common, benign tumors that may involve skin, mucous membranes, or internal organs
Hemangioblastoma	• Associated with **von Hippel-Lindau disease** • Tends to involve the central nervous system and retina
Kaposi sarcoma	• Low-grade malignancy of endothelial cells • Viral etiology: **human herpesvirus 8** (HHV8) • Most often seen in AIDS patients in the U.S.

► Edema and Shock

Edema	• Fluid is maintained with vessels via balance between hydrostatic pressure ("pushing fluid out") and oncotic pressure ("pulling fluid in"). • Most causes of edema can be related to either increased hydrostatic pressure or reduced plasma osmotic pressure. Other causes included lymphatic obstruction, sodium retention. • Clinically, may see pitting edema in extremities (dependent) or massive generalized edema (**anasarca**).

Increased Hydrostatic Pressure	Reduced Plasma Osmotic Pressure
Local: deep vein thrombosis *Generalized:* congestive heart failure	Cirrhosis, nephrotic syndrome, protein losing enteropathy

Shock	Three major variants: cardiogenic, septic, and hypovolemic			
Type	**Comments**	**Heart Rate**	**Systemic Vascular Resistance**	**Cardiac Output**
Cardiogenic	Intrinsic pump failure. As the heart fails, stroke volume decreases, with compensatory increases in heart rate and systemic vascular resistance.	↑	↑	↓
Septic	Endotoxin mediated. Massive peripheral vasodilation with a decrease in systemic vascular resistance. There is peripheral pooling of blood (decreased effective circulatory volume). The heart compensates with an increase in heart rate.	↑	↓	↑
Hypovolemic	Blood loss. The effective circulatory volume decreases through actual loss. The heart is able to attempt to compensate with an increase in heart rate.	↑	↑	Unchanged

Cardiovascular Pharmacology

► Antiarrhythmic Drugs

Drugs	Mechanism of Action	Effect	Indications	Toxicities	Notes
Class I: Na^+ Channel Blockers (Local Anesthetics) These agents block the open or inactivated channel preferentially and therefore block frequently depolarized (e.g., abnormal) tissue better **(use dependence; state dependence)**. Class I drugs are subdivided into three groups based on their effect on AP duration.					
Class IA quinidine procainamide disopyramide	• ↓ **Na^+ influx** • **Slows phase 0 depolarization** in His-Purkinje fibers and cardiac muscle	• ↓ K^+ efflux (↑ **AP duration**, ↑ ERP, slows conduction)	Atrial and ventricular arrhythmias	• *Quinidine:* **cinchonism** (headache, tinnitus, vertigo), ↑ QT interval, **torsades de pointes**, autoimmune reactions (e.g., thrombocytopenia) • *Procainamide:* reversible **SLE-like syndrome**	• Hyperkalemia enhances cardiotoxic effects • Quinidine enhances digoxin toxicity
Class IB lidocaine mexiletine tocainide	• ↓ **Na^+ influx** in ischemic or depolarized Purkinje and ventricular tissue (little effect on atrial or normal tissue) • **Shortens phase 3 repolarization**	• ↓ **AP duration** • Prolongs diastole	Ventricular arrhythmias (e.g., post MI, digitalis toxicity)	CNS toxicity	Hyperkalemia enhances cardiotoxic effects
Class IC flecainide propafenone encainide	• ↓ **Na^+ influx** • **Markedly slows phase 0 depolarization** in His-Purkinje fibers and cardiac muscle	• **No effect on AP duration** • Slows conduction velocity • Increase QRS duration	Refractory ventricular arrhythmias (used as last resort)	**Proarrhythmic**	Can precipitate cardiac arrest and sudden death in patients with preexisting cardiac abnormalities
Class II: Beta Blockers These drugs slow AV conduction.					
propranolol metoprolol esmolol	• **β-adrenoreceptor blockade** – ↓ cAMP – ↓ Ca^{2+} current • ↓ **phase 0 depolarization in AV node** • ↓ **phase 4 depolarization in SA node**	• ↓ AV node conduction • ↑ PR interval	• SVT • Post-MI arrhythmia prophylaxis	Impotence, bradycardia, depression, worsens asthma	• Used post MI; has a protective effect • May mask premonitory signs of hypoglycemia • **Esmolol**-very short acting

(Continued)

► Antiarrhythmic Drugs (*Cont'd.*)

Drugs	Mechanism of Action	Effect	Indications	Toxicities	Notes
Class III: K^+ Channel Blockers These agents prolong the AP and increase the ERP.					
sotalol ibutilide dofetilide amiodarone	↓ **K^+ current** (delayed rectifier current), **prolonging phase 3 repolarization** of AP	• ↑ **AP duration** • ↑ERP	Atrial fibrillation/ flutter, ventricular arrhythmias, refractory arrhythmias	• *General:* **torsade de pointes**, sinus bradycardia • *Amiodarone:* **pulmonary fibrosis**, **hepatotoxicity**, **cutaneous**, **photosensitivity**, corneal deposits, thyroid dysfunction • *Bretylium:* new arrhythmias, ↓ BP • *Sotalol:* excessive β blockade	• *Sotalol:* also class II
Class IV: Ca^{2+} Channel Blockers By blocking L-type Ca^{2+} channels, these agents slow AV node conduction.					
verapamil diltiazem	**Block L-type Ca^{2+} channels**	Decreased conductivity SA/AV nodes	• Atrial fibrillation/ flutter • Atrial automaticities • AV nodal reentry	Constipation, dizziness, flushing, AV block, strong negative inotropic effect, hypotension	—
Unclassified					

Adenosine: used for AV nodal arrhythmias; extremely short acting
Mg^{2+}: used in digitalis-induced arrhythmias, torsade de pointes
K^+: used in digitalis-induced arrhythmias; ↓ other ectopic pacemakers
Digoxin: used in rapid atrial flutter/fibrillation, AV nodal reentrant arrhythmias

► Antihypertensives

Class	Drugs	Mechanism	Side Effects/Notes
SYMPATHOPLEGICS			
α_1 antagonists	Prazosin Doxazosin Terazosin	Block α_1 receptors on arterioles and venules	Orthostatic hypotension and syncope, especially with the first dose
α_2 agonists	Clonidine	Decrease sympathetic outflow by stimulating α_2 receptors in the CNS	Rebound hypertension, dry mouth, sedation, bradyarrhythmias
	Methyldopa		Sedation, hemolytic anemia; it is a prodrug that is converted to α-methyl norepinephrine
β blockers	Propranolol Atenolol Metoprolol, others	Block postsynaptic β receptors	CV disturbances, impotence, sleep disturbances, sedation, asthma
Postganglionic sympathetic terminal blockers	Reserpine	Destroys adrenergic synaptic vesicles, decreasing NE release	Rarely used; depression, sedation, dry mouth, edema, bradycardia, night terrors
	Guanethidine	Depletes NE and blocks NE release	Rarely used; orthostatic hypotension, sexual dysfunction; uses uptake site to enter nerve terminal
Ganglionic blockers	Hexamethonium Mecamylamine	Ganglionic nicotinic antagonists that inhibit postganglionic sympathetic neurons	Rarely used; side effects result from blocking both sympathetic and parasympathetic tone
VASODILATORS			
Ca^{2+} channel blockers	Amlodipine Diltiazem Nifedipine Verapamil	Block L-type Ca^{2+} channels in cardiac and smooth muscle	Constipation, edema, headache, bradycardia, GI disturbances, dizziness, AV block, CHF, tachycardia (nifedipine)
Drugs acting through nitric oxide (NO)	Hydralazine	Release endothelial NO → stimulation of smooth muscle guanylate cyclase → ↑ cGMP	Reversible lupus erythematosus-like syndrome, edema; arteriolar dilation
	Nitroprusside		For hypertensive emergencies, arteriolar and venous dilation; cyanide poisoning
Drugs acting by opening K^+ channels	Minoxidil	Open K^+ channels → hyperpolarization of vascular smooth muscle	For severe hypertension; hirsutism, pericardial effusion, edema
	Diazoxide		For hypertensive emergencies; hypoglycemia
D_1 agonist	Fenoldopam	Vasodilate renal vessels	For hypertensive emergencies
INHIBITORS OF ANGIOTENSIN			
Angiotensin-converting enzyme inhibitors (ACEIs)	Captopril Enalapril Fosinopril Ramipril Lisinopril	Block formation of angiotensin II, leading also to a ↓ in aldosterone	Dry cough, hyperkalemia, angioedema, renal damage in preexisting renal disease; contraindicated in pregnancy (fetal renal damage)
Angiotensin II receptor blockers (ARBs)	Losartan Valsartan Candesartan	Block angiotensin II at AT1 receptor; ↓ in aldosterone	Renal damage in preexisting renal disease, hyperkalemia; contraindicated in pregnancy (fetal renal damage)
Renin inhibitor	Aliskiren	↓ angiotensin I (and therefore ATII and aldosterone)	Hyperkalemia

(Continued)

▶ Antihypertensives (*Cont'd.*)

Class	Drugs	Mechanism	Side Effects/Notes
		DIURETICS	
Thiazides	Hydrochlorothiazide Metolazone	Inhibit Na^+/Cl^- transporter	Useful in mild hypertension; ↓K^+, ↓Mg^{2+}, ↑Ca^{2+}, ↓Na^+, ↑uric acid, ↑glucose, ↑LDL cholesterol, ↑triglycerides
Loop diuretics	Furosemide	Inhibit $Na^+/K^+/2Cl^-$ transporter	Used in moderate to severe hypertension; ↓K^+, ↓Mg^{2+}, ↓Ca^{2+}, ↓Na+, ↑uric acid, ↑glucose, ↑LDL cholesterol, ↑triglycerides
Aldosterone antagonists	Spironolactone	Aldosterone antagonist in the distal convoluted tubule	Hyperkalemia, metabolic acidosis, gynecomastia; can be safely used in pregnancy

Concomitant Disease States

Disease State	Agents Initially Indicated	Agents Contraindicated	Other Notes
Pregnancy	Methyldopa Hydralazine	ACE; ARB	
Diabetes	ACEI, ARB	BB (high dose)	Additional treatment with lower-dose BB is acceptable. Diuretics are also good secondary agents
Heart Failure	ACEI	BB (high dose), verapamil; diltiazem	Additional treatment with BB (low dose), ARB, diuretics (all classes); can use select CBs (amlodipine, felodipine)
COPD / Asthma	CB	BB (Non-selective high dose)	ACE not recommended due to chronic cough side effect
Chronic Kidney Disease	ACEI, ARB		Loop diuretics can be used; do not use other diuretics in renal insufficiency
Benign Prostatic Hypertrophy	Alpha blockers		
Severe Depression		BB, reserpine, methyldopa	
Post-MI	BB, spironolactone, verapamil, diltiazem		ACEI also acceptable
Recurrent Stroke Prevention	Diuretics, ACEI		

Definition of Abbreviations: ACEI, Angiotensin-Converting Enzyme Inhibitor; ARB, Angiotensin Receptor Blocker; BB, Beta Blocker; CB, Calcium Channel Blocker

► Antianginal Drugs

Angina pectoris, the primary symptom of ischemic heart disease, occurs in periods of inadequate oxygen delivery to the myocardium. Classically, this symptom is described as a crushing, pressure-like pain that occurs during periods of exertion. Strategies used to treat this condition include:

- increasing oxygen delivery through increased perfusion
- decreasing myocardial oxygen demands

	Drug Class		
	Nitrates (nitroglycerin, isosorbide dinitrate)	**Calcium Channel Blockers** (nifedipine, verapamil, diltiazem)	**Beta blockers** (propranolol, atenolol, metoprolol)
Molecular mechanism	Generation of endothelial NO activates GC → ↑ cGMP → dephosphorylates MLCK → relaxation of vascular smooth muscle	Inhibits voltage-gated "L-type" Ca^{2+} channels and ↓ Ca^{2+} influx in cardiac and vascular smooth muscle → ↓ muscle contractility	β-adrenergic antagonism
Physiologic mechanism	• **Venodilation → ↓ preload →** ↓ afterload • **↓ myocardial O_2 demand**	• Arteriolar vasodilation → ↓ **afterload** • ↓ myocardial O_2 demand • ↓ AV node conduction velocity	↓ contractility, ↓ HR, ↓ BP (mild), ↓ myocardial O_2 demand, ↓ AV node conduction velocity
Indications	Acute angina (nitroglycerin), pulmonary edema	Angina, HTN, SVT (except nifedipine)	Angina, HTN, arrhythmia
Adverse effects	Reflex tachycardia, orthostatic hypotension, headache, tachyphylaxis	Cardiac depression, peripheral edema, constipation	Impotence, depression, bradycardia
Notes	Contraindicated in patients taking sildenafil → hypotension and sudden death	Selectivity for vascular Ca^{2+} channels: **Nifedipine > diltiazem > verapamil** Verapamil primarily affects myocardium	Non-CV indications include migraine, familial tremor, stage fright, thyrotoxicosis, glaucoma; beta blockers with ISA are contraindicated in angina/MI patients

Definition of abbreviations: AV, atrioventricular; cGMP, cyclic guanosine monophosphate; GC, guanylate cyclase; HTN, hypertension; ISA, intrinsic sympathomimetic activity; SVT, supraventricular tachycardia.

► Drugs Used in Heart Failure

Heart failure results when tissue demands for circulation cannot be met by an ailing myocardium. Inadequate cardiac output secondary to decreased contractility leads to decreased exercise tolerance and muscle fatigue. Neurohumoral responses to this physiologic shortcoming play an integral role in the pathogenesis of heart failure; thus, drugs used to treat this condition may be aimed at these responses. Physiologically, these drugs may reduce afterload, reduce preload, or increase contractility.

Drug Class	Mechanism of Action	Effects	Indications	Toxicities	Notes
ACE inhibitors captopril enalapril lisinopril	Inhibits angiotensin-converting enzyme (ACE) → ↓ angiotensin II and ↑ bradykinin	• **Decreased aldosterone** → ↓ fluid retention • **Vasodilation** → ↓ preload and afterload	• CHF • Post-MI to prevent pathologic remodeling • Hypertension • Chronic renal disease	**Dry cough**, hypotension, proteinuria, **fetal renal toxicity**, angioedema	• Cornerstone of CHF therapy • Prophylactic in post-MI because they oppose "remodeling" that leads to heart failure
Cardiac glycosides digoxin	Inhibits Na^+/K^+ ATPase → ↑ intracellular Na^+ → ↓ Na^+ gradient → ↓ Na^+-Ca^{2+} exchange → ↑ intracellular Ca^{2+}	• Increased myocardial Ca^{2+} → **increased contractility** • Delayed conduction at AV node. (parasympathomimetic effect)	• CHF (because ↑ contractility) • Atrial fibrillation (because ↓ AV conduction)	**Yellow vision,** nausea, vomiting, diarrhea, anorexia, hallucination, **life-threatening arrhythmias**	• **Hypokalemia** enhances toxicity • **Quinidine** → ↑ dig toxicity (↓ dig clearance) • **Digoxin antibodies** (FAb fragments) used in overdose • Digoxin does not improve survival following MI

(Continued)

► Drugs Used in Heart Failure (*Cont'd.*)

Drug Class	Mechanism of Action	Effects	Indications	Toxicities	Notes
Angiotensin II–receptor blockers losartan candesartan	Block angiotensin II receptors	Same as ACE inhibitors	Same as ACE inhibitors	**Fetal renal toxicity**, no cough	Not as well studied as ACEIs, but seem to have same efficacy
Vasodilators nitroglycerin nitroprusside isosorbide dinitrate hydralazine	↑ nitric oxide → cGMP → vasodilation	*Nitroglycerin, isosorbide dinitrate:* predominantly venodilators *Nitroprusside:* dilation of arteries = veins	CHF, HTN, angina, pulmonary edema	Tachycardia, headache hypotension	• *Nitroprusside, nitroglycerin* (extended release): used in acute HF • *Hydralazine, isosorbide dinitrate:* used in chronic HF
Beta-receptor antagonists carvedilol labetalol metoprolol	These agents were once contraindicated in heart failure; now they are used to reduce the progression of mild to moderate heart failure. • *Carvedilol, labetalol:* nonselective β antagonist, α_1 antagonist • *Metoprolol:* β_1 antagonist • Contraindicated in patients with asthma or severe bradycardia				
Beta-1 agonists dobutamine dopamine	Used in **acute** heart failure *Dopamine:* ↓ dose → improves renal blood flow; moderate dose: stimulates myocardial contractility; ↑ doses → vasoconstrictor (alpha$_1$ receptors); used for cardiogenic shock *Dobutamine:* β_1 selective				
Diuretics	Used to reduce symptoms of fluid retention (pulmonary congestion, edema); loop diuretics most effective, thiazides can be effective in mild cases				

► Antihyperlipidemics

Drug Class/Agents	Mechanism	Side Effects/Comments
HMG-CoA Reductase Inhibitors ("-statins": lovastatin, atorvastatin, fluvastatin pravastatin, simvastatin, rosuvastatin,)	Inhibit rate-limiting step in cholesterol synthesis ↓ Liver cholesterol ↑ LDL-receptor expression ↓ LDL ↑ HDL ↓ VLDL synthesis ↓ Triglycerides	• Myalgia, myopathy (check creatine kinase) • Rhabdomyolysis • ↑ serum aminotransferases • Teratogenic • P450 inhibitors can ↑ risk of hepatotoxicity, myopathy
Bile Acid Sequestrants (cholestyramine, colestipol, colesevelam)	↓ Enterohepatic recirculation of bile salts, leading to: ↑ Synthesis of new bile salts by liver ↓ Liver cholesterol ↑ LDL-receptor expression ↓ LDL	• GI disturbances • Malabsorption of lipid-soluble vitamins • ↓ Absorption of drugs (e.g., warfarin, thiazides, digoxin, pravastatin, fluvastatin)
Niacin	Liver: ↓ VLDL synthesis Adipose tissue: ↓ lipolysis ↓ VLDL ↓ LDL ↑ HDL ↓ Triglycerides	Flushing (↓ by aspirin and over time), pruritus, hepatoxicity
Fibrates (gemfibrozil, fenofibrate)	Ligands for PPAR-α → activation of lipoprotein lipases ↓ Triglycerides ↓ VLDL and IDL Modest ↓ LDL	• Gallstones • Myopathy (especially when combined with reductase inhibitors) • Can ↑ LDL in some patients, so often combined with other cholesterol-lowering agents
Ezetimibe	Blocks intestinal absorption of cholesterol ↓ LDL	Possible ↑ of hepatotoxicity with reductase inhibitors

The Respiratory System

Chapter 3

Respiratory Embryology and Histology

► Development of the Respiratory System

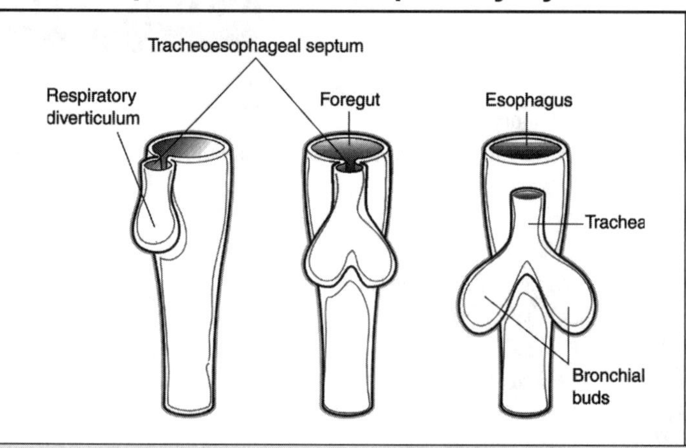

The **respiratory (laryngotracheal) diverticulum** forms in the ventral wall of the foregut. The lung bud forms at the distal end of the diverticulum and divides into two **bronchial buds**. These branch into the **main bronchi**, **lobar bronchi**, and **segmental bronchi**. The right bud divides into three main bronchi, and the left divides into two.

The **tracheoesophageal septum** divides the foregut into the esophagus and trachea.

Clinical Correlate

A **tracheoesophageal fistula** is an abnormal communication between the trachea and esophagus caused by a malformation of the tracheoesophageal septum. 90% occur between the esophagus and **distal third of the trachea**. It is generally associated with **esophageal atresia** and **polyhydramnios**. Symptoms include gagging and cyanosis after feeding and the reflux of gastric contents into the lungs, causing pneumonitis.

► The Alveoli and Blood-Gas Barrier

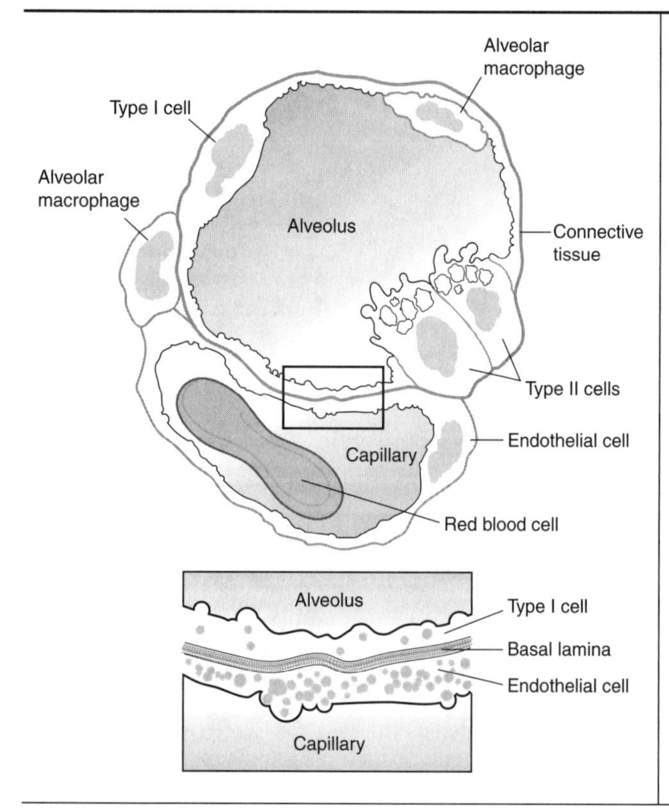

The **conducting zone** of the lungs does not participate in gas exchange and is **anatomic dead space**. It is composed of the trachea, bronchi, bronchioles, and terminal bronchioles. The trachea and bronchi contain pseudostratified ciliated columnar cells and goblet cells (secrete mucous). Bronchioles and terminal bronchioles contain ciliated epithelial cells and Clara cells (which secrete a surfactant-like substance, aid in detoxification, and are stem cells for the ciliated cells).

The **respiratory zone** carries out gas exchange and consists of respiratory bronchioles, alveolar ducts, and alveoli.

Terminal bronchioles divide into **respiratory bronchioles**, which contain alveoli and branch to form alveolar ducts. The ducts terminate in alveolar sacs and are lined by squamous alveolar epithelium.

Alveoli are thin-walled sacs responsible for gas exchange. They contain:

- **Type I epithelial cells**, which provide a thin surface for gas exchange.
- **Type II epithelial cells**, which produce **surfactant**.
- **Alveolar macrophages**, which are derived from monocytes and remove particles and other irritants via phagocytosis.

There are approximately 300 million alveoli in **each** lung.

Gross Anatomy

▶ Pharynx and Related Areas

The **pharynx** is a passageway shared by the digestive and respiratory systems. It has lateral, posterior, and medial walls throughout but is open anteriorly in its upper regions (**nasopharynx**, **oropharynx**), communicating with the nasal cavity and the oral cavity.

The **nasopharynx** is the region of the pharynx located directly posterior to the nasal cavity. It communicates with the nasal cavity through the **choanae** (i.e., posterior nasal apertures).

The **oropharynx** is the region of the pharynx located directly posterior to the oral cavity. It communicates with the oral cavity through a space called the **fauces**. The fauces are bounded by two folds, consisting of mucosa and muscle, known as the **anterior and posterior pillars**.

- The **anterior pillar of the fauces**, also known as the **palatoglossal fold**, contains the **palatoglossus muscle**.
- The **posterior pillar of the fauces**, also known as the **palatopharyngeal fold**, contains the **palatopharyngeus muscle**.
- The **tonsillar bed** is the space between the pillars that houses the **palatine tonsil**.

The **laryngopharynx** is the region of the pharynx that surrounds the larynx. It extends from the tip of the epiglottis to the cricoid cartilage. Its lateral extensions are known as the **piriform recesses**.

▶ The Larynx

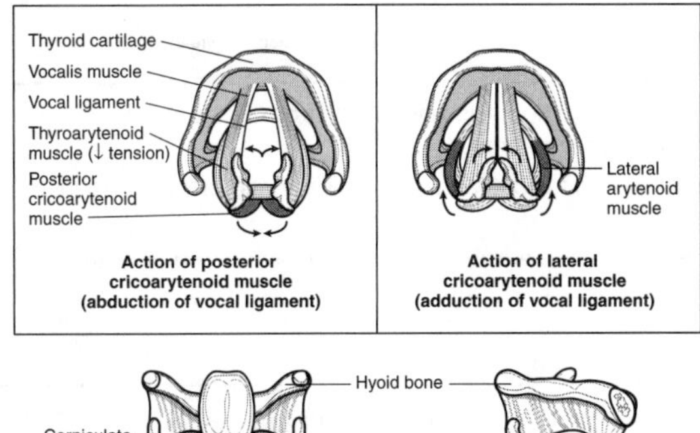

Action of posterior cricoarytenoid muscle (abduction of vocal ligament)

Action of lateral cricoarytenoid muscle (adduction of vocal ligament)

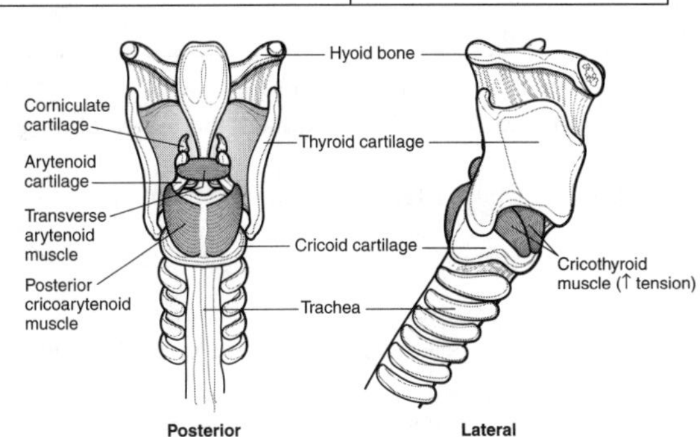

The larynx is the **voice box**. It also maintains a patent airway and acts as a sphincter during lifting and pushing.

Skeleton of the larynx:

- Three unpaired laryngeal cartilages (i.e., thyroid, cricoid, epiglottis) and three paired cartilages (i.e., arytenoid, cuneiform, corniculate)
- The fibroelastic membranes include the thyrohyoid membrane and the cricothyroid membrane (conus elasticus).The free, upper border of the latter is specialized to form the vocal ligament on either side.

▶ Intrinsic Muscles of the Larynx*

Muscle	Function
Posterior cricoarytenoid	Abducts vocal fold
Lateral cricoarytenoid	Adducts vocal fold
Cricothyroid	Tenses vocal fold
Thyroarytenoid (including vocalis)	Relaxes vocal fold
Thyroepiglotticus	Opens laryngeal inlet
Aryepiglotticus	Closes laryngeal inlet
Oblique and transverse arytenoids	Close laryngeal inlet

*Note that the cricothyroid is innervated by the external laryngeal nerve, a branch of the superior laryngeal branch of the vagus nerve. All other intrinsic laryngeal muscles are supplied by the recurrent laryngeal branch of the vagus nerve.

▶ Pleura and Pleural Cavities

Parietal pleura lines the inner surface of the thoracic cavity; **visceral pleura** follows the contours of the lung itself. Inflammation of the central part of the diaphragmatic pleura may produce pain referred to the shoulder (phrenic nerve; C3, C4, and C5).

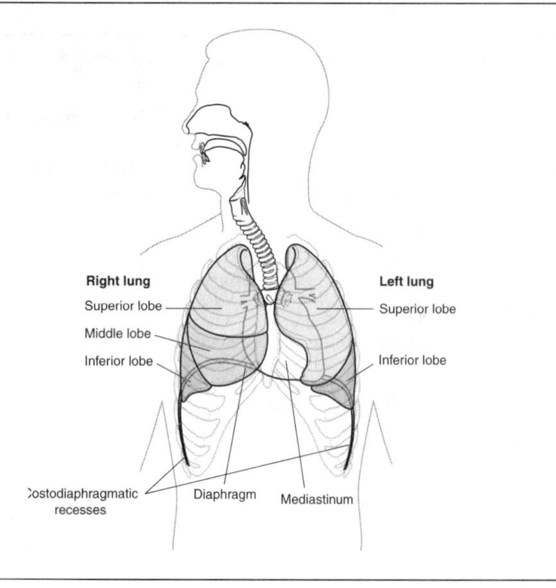

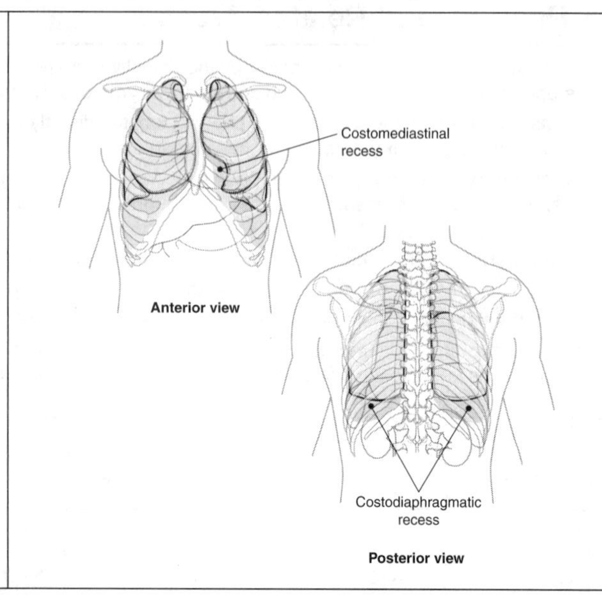

- The **costal line of reflection** is where the costal pleura becomes continuous with the diaphragmatic pleura from Rib 8 in the midclavicular line, to Rib 10 in the midaxillary line, and to Rib 12 lateral to the vertebral column.
- **Costodiaphragmatic recesses** are spaces below the inferior borders of the lungs where costal and diaphragmatic pleurae are in contact.
- The **costomediastinal recess** is a space where the left costal and mediastinal parietal pleurae meet, leaving a space due to the cardiac notch of the left lung. This space is occupied by the lingula of the left lung during inspiration.

Structure of the Lungs

- The **right lung** is divided by the oblique and horizontal fissures into three lobes: superior, middle, and inferior.
- The **left lung** has only one fissure, the oblique, which divides the lung into upper and lower lobes. The **lingula** of the upper lobe corresponds to the middle lobe of the right lung.
- **Bronchopulmonary segments** of the lung are supplied by the segmental (tertiary) bronchus, artery, and vein. There are 10 on the right and eight on the left.

Arterial Supply

- **Right and left pulmonary arteries** arise from the pulmonary trunk. The pulmonary arteries deliver deoxygenated blood to the lungs from the right side of the heart.
- **Bronchial arteries** supply the bronchi and nonrespiratory portions of the lung. They are usually branches of the thoracic aorta.

Venous Drainage

- There are **four pulmonary veins**: superior right and left and inferior right and left.
- Pulmonary veins carry oxygenated blood to the left atrium of the heart.
- The **bronchial veins** drain to the azygos system. They share drainage from the bronchi with the pulmonary veins.

Lymphatic Drainage

- Superficial drainage is to the bronchopulmonary nodes; from there, drainage is to the tracheobronchial nodes.
- Deep drainage is to the pulmonary nodes; from there, drainage is to the bronchopulmonary nodes.
- Bronchomediastinal lymph trunks drain to the right lymphatic and the thoracic ducts.

Innervation of Lungs

- Anterior and posterior pulmonary plexuses are formed by vagal (parasympathetic) and sympathetic fibers.
- Parasympathetic stimulation has a bronchoconstrictive effect.
- Sympathetic stimulation has a bronchodilator effect.

Respiratory Physiology

▶ Lung Volumes and Capacities

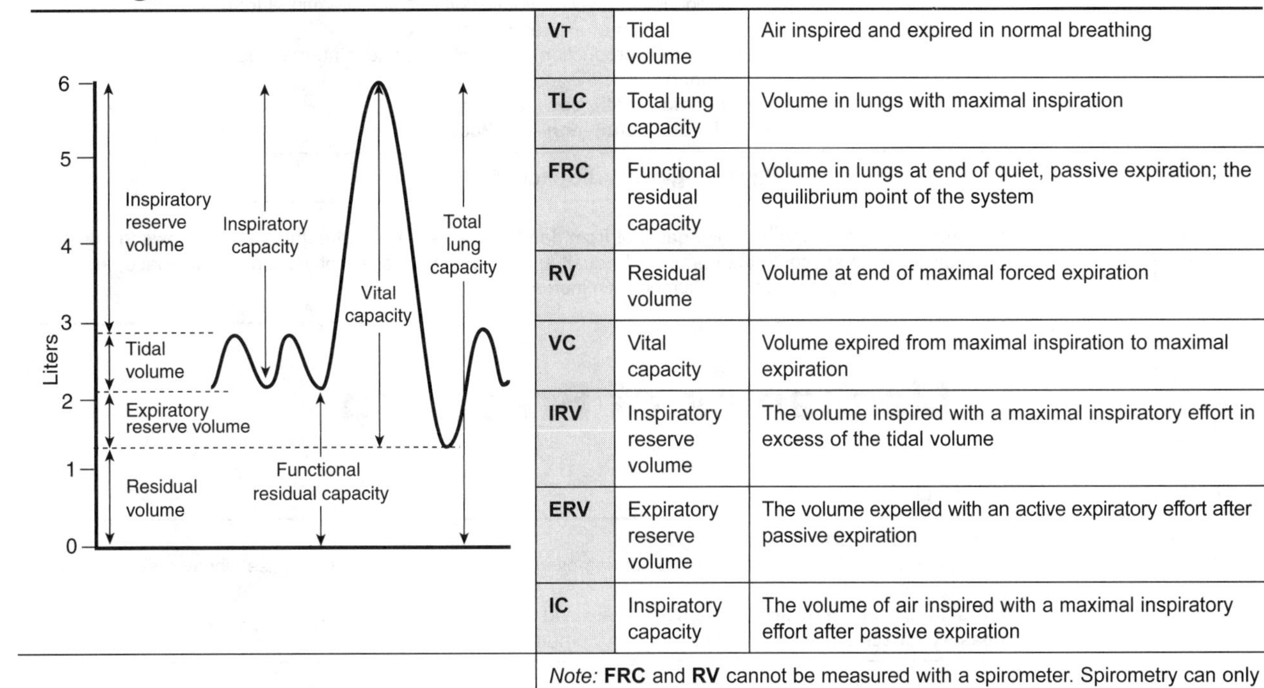

VT	Tidal volume	Air inspired and expired in normal breathing
TLC	Total lung capacity	Volume in lungs with maximal inspiration
FRC	Functional residual capacity	Volume in lungs at end of quiet, passive expiration; the equilibrium point of the system
RV	Residual volume	Volume at end of maximal forced expiration
VC	Vital capacity	Volume expired from maximal inspiration to maximal expiration
IRV	Inspiratory reserve volume	The volume inspired with a maximal inspiratory effort in excess of the tidal volume
ERV	Expiratory reserve volume	The volume expelled with an active expiratory effort after passive expiration
IC	Inspiratory capacity	The volume of air inspired with a maximal inspiratory effort after passive expiration

Note: **FRC** and **RV** cannot be measured with a spirometer. Spirometry can only measure changes in volume.

▶ Dead Space and Ventilation

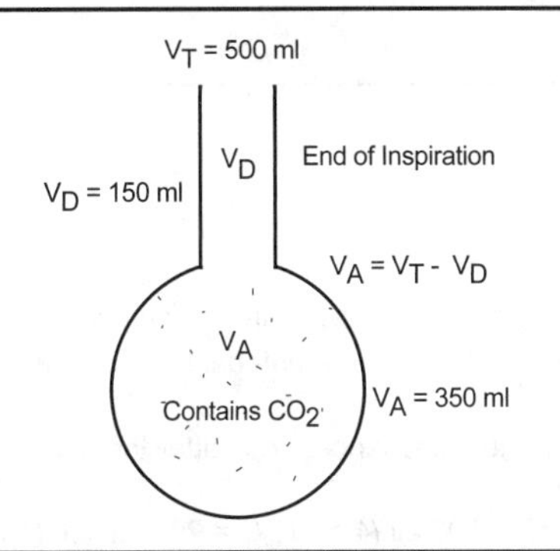

VD = dead space (no gas exchange)

Anatomic VD = conducting airways

Alveolar VD = alveoli with poor blood flow (ventilated but not perfused)

Physiologic VD = anatomic + alveolar dead space

Standard Symbols

A = alveolar
a = arterial
V = volume
V̇ = minute ventilation
P = pressure
P_{ACO_2} = alveolar pressure of CO_2
P_{aCO_2} = arterial pressure of CO_2
P_{ECO_2} = P_{CO_2} in expired air

Abbreviation	Name	Definition	Normal Values
VD	Dead space	Volume that does not exchange gas with blood	150 mL
VA	Alveolar volume	Portion of tidal volume that reaches alveoli during inspiration	350 mL
VT	Tidal volume: VT = VA + VD	Amount of gas inhaled and exhaled during normal breathing – the sum of dead space volume and alveolar volume	500 mL
n	Respiratory frequency	Breaths/minute	15/min
V̇E	Total ventilation: VTn = VAn + VDn; V̇E = V̇A + V̇D	Total ventilation per minute; VTn = (350 mL × 15/min) + (150 mL × 15/min) = 7,500 mL/min; V̇E = 5,250 mL/min + 2,250 mL/min = 7,500 mL/min	7,500 mL/min

(Continued)

► Dead Space and Ventilation *(Cont'd.)*

$\dot{V}_A$	Alveolar ventilation $\dot{V}_A = (V_T - V_D) \times n$	Amount of inspired air that reaches the alveoli each minute. It is the effective part of ventilation.	5,250 mL/min
	$\dot{V}_A = \frac{\dot{V}_{CO_2}}{P_{CO_2}} \times K$	The adequacy of alveolar ventilation can be determined from the concentration of expired carbon dioxide. • $\dot{V}_{CO_2}$ = CO_2 production (generally assume is normal and constant) • **↑ alveolar ventilation → ↓ Pa_{CO_2}** • **↓ alveolar ventilation → ↑ Pa_{CO_2}**	
		Physiologic Dead Space	
$\frac{V_D}{V_T} = \frac{Pa_{CO_2} - PE_{CO_2}}{Pa_{CO_2}}$	All expired CO_2 comes from alveolar gas, not from dead space gas. Therefore, the fraction shows the dilution of CO_2 by the dead space. In the normal individual, anatomic dead space = physiologic dead space, and V_D/V_T = 0.2–0.35. In lung disease, this number can increase.		

Mechanics of Breathing

► Muscles of Breathing

Muscles of inspiration	• **Diaphragm—most important** • Other muscles of inspiration are used primarily during exercise or in diseases that increase airway resistance (e.g., asthma): – **External intercostal muscles** (move ribs upward and outward) – **Accessory muscles** (elevate first two ribs and sternum)
Muscles of expiration	• Expiration is **passive** during quiet breathing. • Muscles of expiration are used during exercise or increased airway resistance (e.g., asthma): – **Abdominal muscles** (help push diaphragm up during exercise or increased airway resistance) – **Internal intercostal muscles** (pull ribs downward and inward)

► Elastic Properties of the Lung

Pressure-Volume Curve

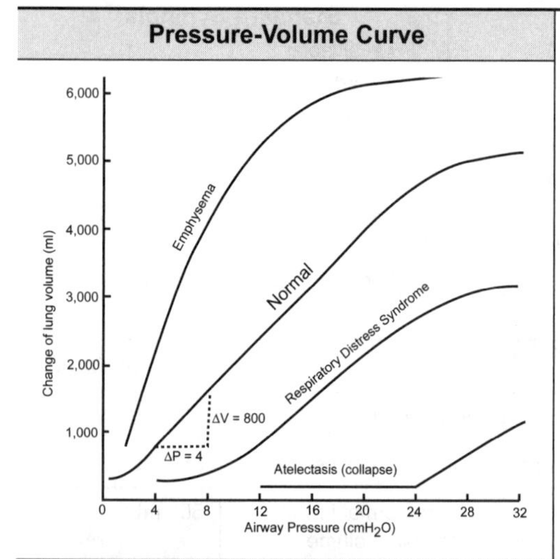

- **Compliance (ΔV/ΔP)** is used to estimate the distensibility of the lungs. It is inversely related to **elasticity** (tendency of a material to recoil when stretched).
- The steeper the slope, the higher the compliance. The flatter the slope, the lower the compliance (stiffer).
- Normal curve: Compliance = ΔV/ΔP = 800 mL/4 cm H_2O = 200 mL/cm H_2O
- Atelectasis requires an extreme effort to open collapsed alveoli.
- Compliance of lungs also ↑ with age.

Clinical Correlation: Changes in Lung Compliance		
↑ Compliance	**Emphysema**	• Less elastic recoil of lungs, so FRC ↑ • Chest wall expands and becomes **barrel-shaped** • Also, ↑ RV, ↑ TLC, ↓ FVC, ↑ R_{aw}
↓ Compliance*	**Fibrosis, respiratory distress syndrome**	• Tendency of lungs to collapse ↑, so FRC ↓ • Also, ↓↓ TLC, ↓ RV, ↓↓ FVC

Definition of abbreviation: R_{aw}, airway resistance.
***Restrictive lung disease:** a condition that reduces the ability to inflate the lungs (e.g., ↓ compliance).

▶ Elastic Properties of the Lung and Chest Wall

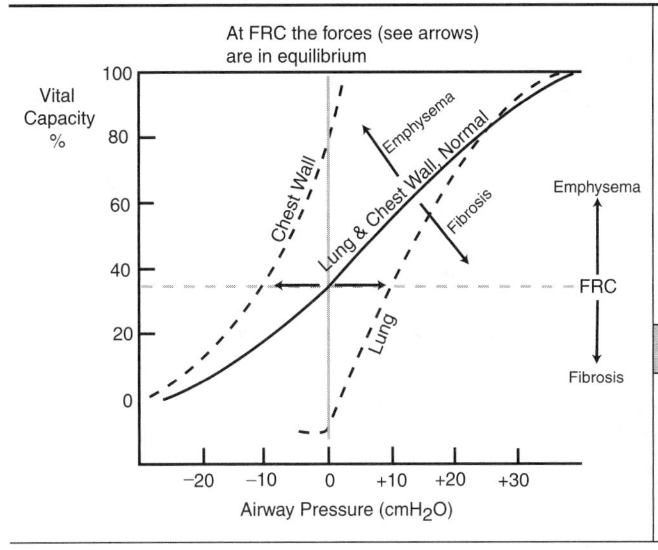

The figure to the left shows the pressure–volume relationships of the lung, the chest wall, and the lung and chest wall together.

- At FRC, the system is at equilibrium and the airway pressure = 0 cm H_2O. At FRC, the elastic recoil of the lungs tends to collapse the lungs. The tendency of the lungs to collapse is balanced exactly by the tendency of the chest wall to spring outward.
- The result of the opposing forces of the lungs and chest wall cause the intrapleural pressure (PIP) to be negative (a vacuum). The PIP is the pressure in the intrapleural space, which lies between the lungs and chest wall.

Clinical Correlation

If sufficient air is introduced into the intrapleural space, the PIP becomes atmospheric (0 mm Hg), and the lungs and chest wall follow their normal tendencies: the lungs collapse and the chest wall expands. This is a **pneumothorax**.

▶ Surface Tension

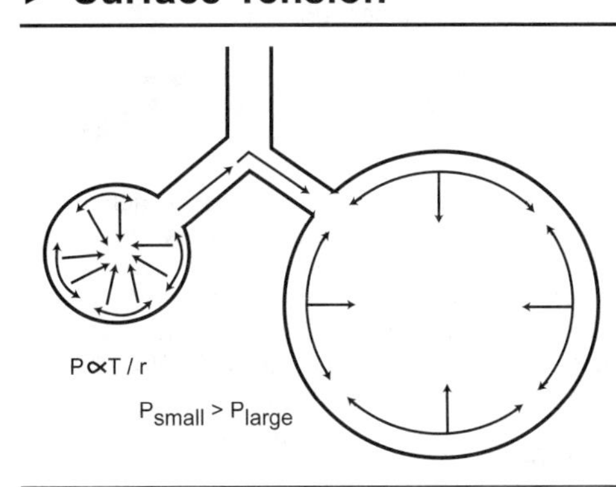

- The attractive forces between adjacent molecules of liquid are stronger than those between liquid and gas, creating a collapsing pressure.
- **Laplace's Law:** $P = \frac{2T}{r}$, where P = collapsing pressure
  T = surface tension
  r = radius of alveoli
- Large alveoli (↑r) have low collapsing pressures (easy to keep open).
- Small alveoli (↓r) have high collapsing pressures (difficult to keep open).
- **Surfactant** reduces surface tension (T). With ↓ surfactant (e.g., premature infants), smaller alveoli tend to collapse (**atelectasis**).
- **Surfactant**, produced by **type II alveolar cells**, ↑ **compliance.**

▶ Airway Resistance

Airflow	$Q = \frac{\Delta P}{R}$	where Q = airflow ΔP = pressure gradient R = airway resistance
Airway resistance	$R = \frac{8\eta l}{p r^4}$	where R = resistance η = viscosity of inspired gas l = airway length r = airway radius **Medium-sized bronchi** are the major sites of airway resistance (not the smaller airways because there are so many of them).
Changes in airway resistance	• **Bronchial smooth muscle:** – Parasympathetic nervous system → bronchoconstriction via M_3 muscarinic receptors (↑ resistance) – Sympathetic nervous system → bronchodilation via β_2 receptors (↓ resistance) • **Lung volume:** ↑ lung volume → ↓ resistance (greater radial traction on airways) ↓ lung volume → ↑ resistance • **Viscosity or density of inspired gas:** ↑ density → ↑ resistance (deep sea diving) ↓ density → ↓ resistance (breathing helium)	

► The Breathing Cycle

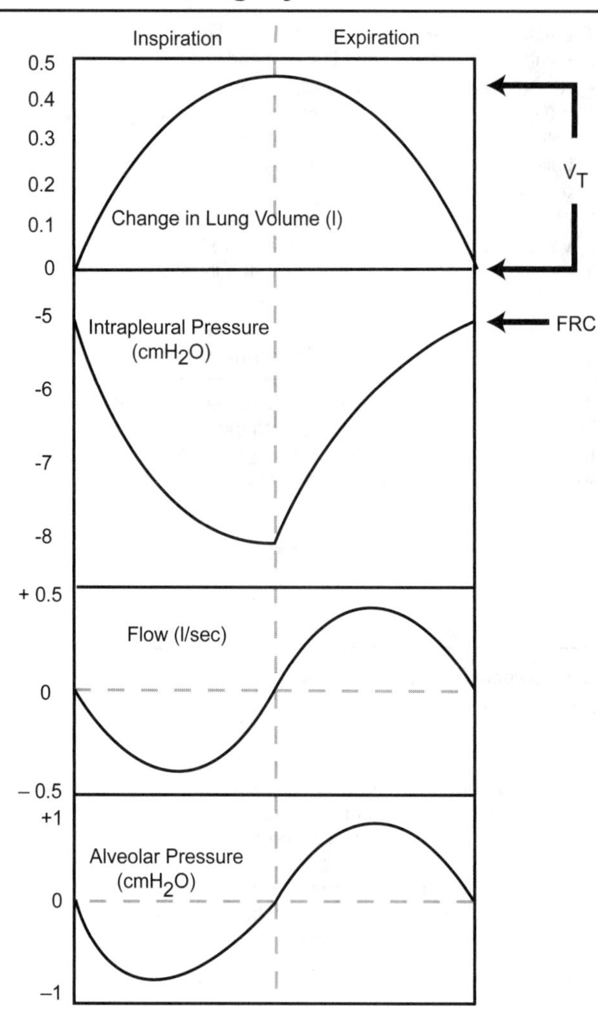

At rest (FRC): $P_A = P_{ATM} = 0$ mm Hg

Inspiration

1. Inspiratory muscles contract.
2. Thoracic volume ↑.
3. P_{IP} becomes more negative.
4. Lungs expand (also causes P_{IP} to be more negative because of ↑ elastic recoil).
5. P_A becomes negative.
6. Air flows in down pressure gradient ($P_{ATM} - P_A$).

Expiration

1. Muscles relax.
2. Thoracic volume ↓.
3. P_{IP} is less negative.
4. Lungs recoil inward (also causes P_{IP} to be less negative).
5. P_A becomes positive.
6. Air flows out down pressure gradient ($P_A - P_{ATM}$).

Clinical Correlation

Obstructive lung disease: a condition that causes an abnormal increase in R_{aw}. **Chronic obstructive pulmonary disease (COPD)** such as emphysema → destruction of elastic tissue → ↑ lung compliance → collapse of airways on expiration (**dynamic compression**). This occurs in normal individuals during a forced expiration but can occur during normal expiration in COPD. COPD patients learn to expire slowly and with pursed lips.

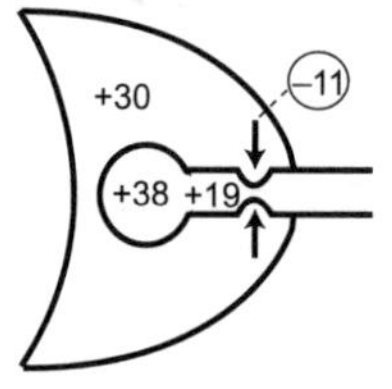

Definition of abbreviations: FRC, functional residual capacity; P_A, alveolar pressure; P_{ATM}, atmospheric pressure.

► Pulmonary Disease

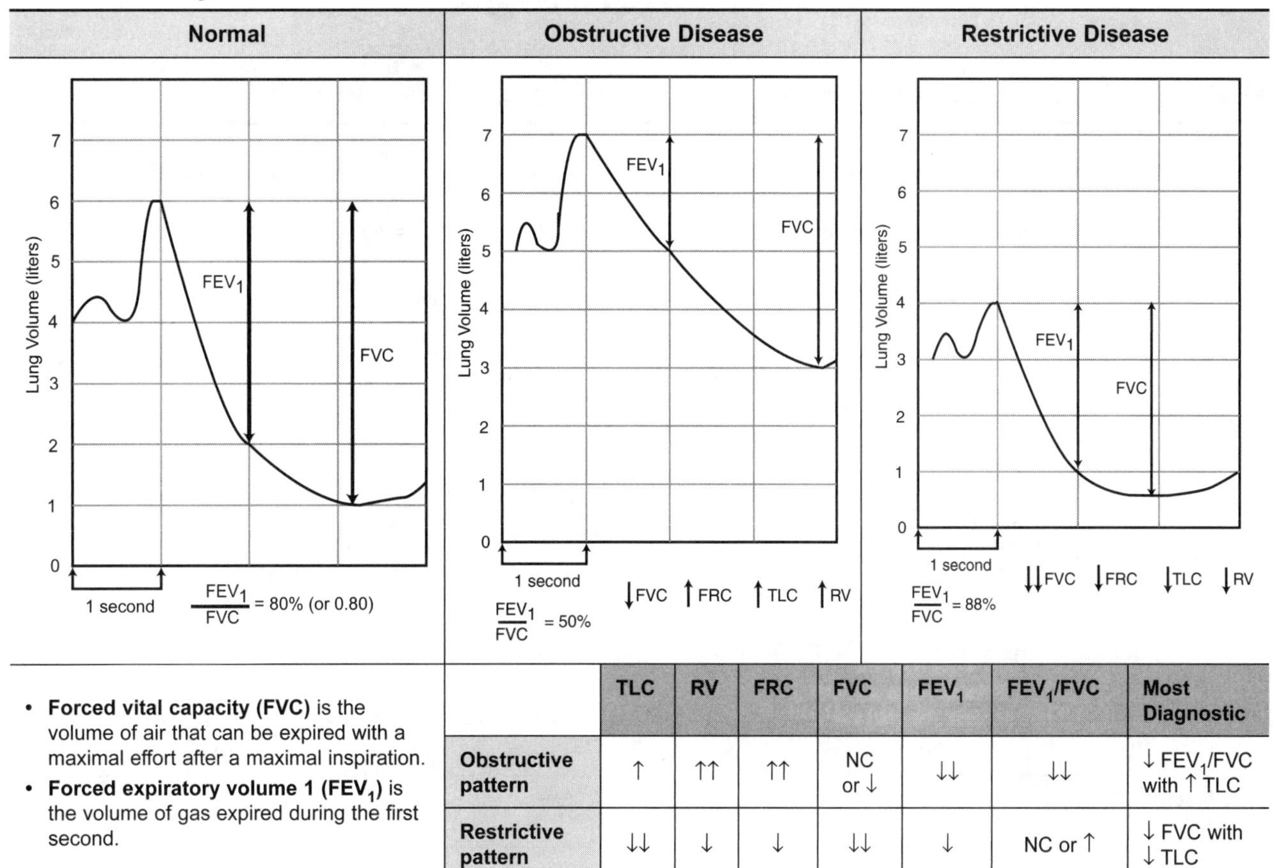

- **Forced vital capacity (FVC)** is the volume of air that can be expired with a maximal effort after a maximal inspiration.
- **Forced expiratory volume 1 (FEV_1)** is the volume of gas expired during the first second.

	TLC	RV	FRC	FVC	FEV_1	FEV_1/FVC	Most Diagnostic
Obstructive pattern	↑	↑↑	↑↑	NC or ↓	↓↓	↓↓	↓ FEV_1/FVC with ↑ TLC
Restrictive pattern	↓↓	↓	↓	↓↓	↓	NC or ↑	↓ FVC with ↓ TLC

Definition of abbreviation: NC, no change.

► Summary of Classic Lung Diseases

Disease	Pattern	Characteristics
Asthma	Obstructive	R_{aw} is ↑ and expiration is impaired. All measures of expiration are ↓ (FVC, FEV_1, FEV_1/FVC). Air is trapped →↑ FRC.
COPD	Obstructive	• Combination of **chronic bronchitis** and **emphysema** • There is ↑ **compliance**, and expiration is impaired. Air is trapped →↑ FRC. – **"Blue bloaters"** (mainly bronchitis): impaired alveolar ventilation → severe hypoxemia with cyanosis and ↑ $Paco_2$. They are blue and edematous from right heart failure. – **"Pink puffers"** (mainly emphysema): alveolar ventilation is maintained, so they have normal $Paco_2$ and only mild hypoxemia. They have a reddish complexion and breathe with pursed lips at an ↑ respiratory rate.
Fibrosis	Restrictive	• There is ↓ **compliance**, and inspiration is impaired. • **All lung volumes are decreased**, but because FEV_1 decreases less than FVC, FEV_1/FVC may be increased or normal.

Gas Exchange

► Partial Pressures of O_2 and CO_2

Dalton's Law of Partial Pressures: Partial pressure (p_{gas}) = total pressure (P_T) × fractional gas concentration (F_{gas})

Alveolar gas equation: $P_{AO_2} = P_{IO_2} - P_{CO_2}/R$

Alveolar ventilation equation: $\dot{V}_A = \dot{V}_{CO_2} \times K/P_{aCO_2}$; ($K = P_B - P_{H_2O} = 760 - 47 = 713$)

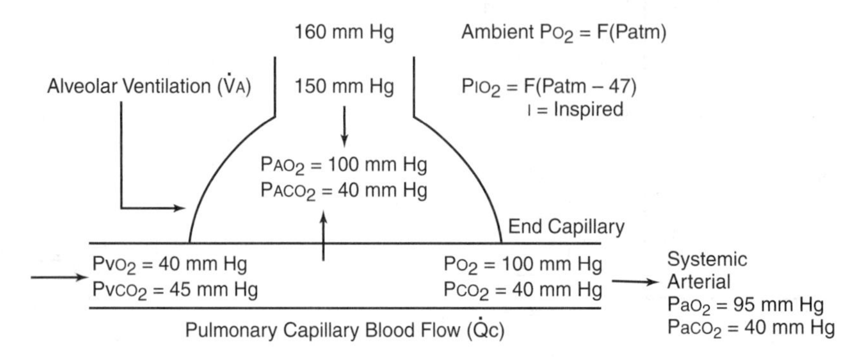

A = alveolar, a = systemic arterial

	Equation	O_2	CO_2
Dry inspired air (any altitude)	F_{gas}	0.21	0
Dry air at sea level	$P_{gas} = F_{gas} \times P_B$	0.21 (760) = 160	0
Inspired, humidified tracheal air (P_{IO_2})	$P_{gas} = F_{gas} \times (P_B - P_{H_2O})$	0.21 (760 – 47) = **150**	0
Alveolar air ($P_{A_{gas}}$)	**O_2:** $P_{AO_2} = P_{IO_2} - P_{ACO_2}/R$ **CO_2:** $\dot{V}_{CO_2}K/R$	150 – 40/0.8 = **100**	(280 mL/min × 713)/5,000 mL/min = **40**
Systemic arterial blood (Pa_{gas})	—	**100** (completely equilibrates with alveolar O_2 if no lung disease)	**40** (CO_2 is from pulmonary capillaries and equilibrates with alveolar gas)
Mixed venous blood ($P\bar{v}_{CO_2}$)	—	**40** (O_2 has diffused from arterial blood into tissues)	**45** (CO_2 has diffused from tissues to venous blood)

All pressures are expressed in mm Hg.

Definition of abbreviations: K, constant; P_{ACO_2}, partial pressure of alveolar carbon dioxide; P_{AO_2}, partial pressure of alveolar oxygen; P_B, barometric pressure; P_{H_2O}, water vapor pressure; P_{IO_2}, partial pressure of inspired oxygen; $\dot{V}_{CO_2}$, CO_2 production; R, respiratory exchange ratio.

► Diffusion

Fick's Law of Diffusion	• $V_{gas} \propto D(P_1 - P_2) \times A/T$ where V_{gas} = diffusion of gas, D = diffusion coefficient of a specific gas, A = surface area, T = thickness. • A and T are physical factors that change mainly in disease. • D of CO_2 >>> O_2
Time course in pulmonary capillary	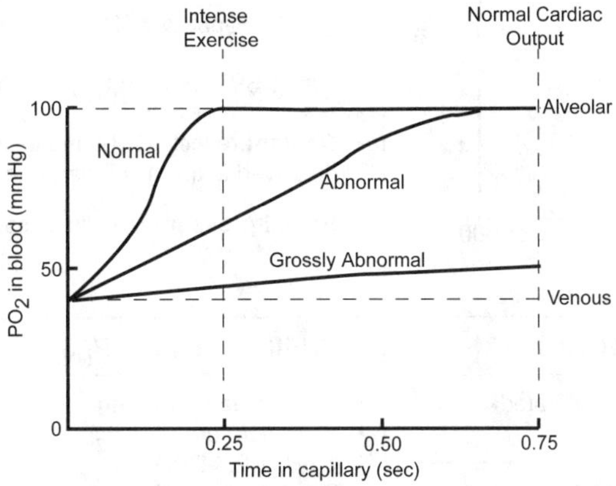 Intense Exercise; Normal Cardiac Output; Alveolar; Normal; Abnormal; Grossly Abnormal; Venous; PO_2 in blood (mmHg): 0, 50, 100; Time in capillary (sec): 0.25, 0.50, 0.75 • A red blood cell remains in capillary for 0.75 seconds (s) • Equilibrium is reached in 0.25 s in normal lung at resting state. • Exercise reduces equilibration time, but there is still enough reserve for full equilibration of oxygen in a healthy individual.

Perfusion-Limited Gases	Diffusion-Limited Gases
Gases that **equilibrate** between the alveolar gas and pulmonary capillaries are **perfusion-limited**. The amount of gas transferred is **not dependent** on the properties of the blood-gas barrier. • **O_2** (under normal conditions) • **N_2O** (nitrous oxide) • **CO_2**	Gases that **do not equilibrate** between the alveolar gas and the pulmonary capillaries are **diffusion-limited**. The amount of gas transferred **is dependent** on the properties of the blood-gas barrier. • **O_2:** – Blood-gas barrier is thickened in **fibrosis**. – Surface area is ↓ in **emphysema**. – **Intense exercise** ↓ time for equilibration in pulmonary capillaries (can occur in normal lungs). – **Low O_2 gas mixture** (less partial pressure gradient, can occur in normal lungs) • **CO:** Binds so avidly to Hb, Paco does not ↑ much. Used to measure the **pulmonary diffusing capacity**.

▶ Oxygen Transport and the Hemoglobin–O_2 Dissociation Curve

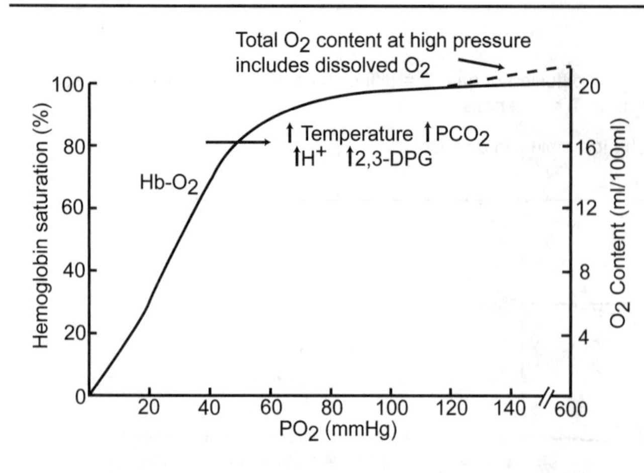

- Each hemoglobin (Hb) molecule has four subunits.
- Each subunit has a heme moiety with an iron in the ferrous state (Fe^{2+}), and two α and two β polypeptide chains.
- **O_2 capacity:** maximal amount of O_2 that can bind to Hb
- **O_2 content*:** Total O_2 in blood (bound + dissolved)
  = (O_2 capacity × % saturation) + dissolved O_2
  $= (1.39 \times Hb \times \frac{Sat}{100}) + 0.003\ P_{O2}$
- **Content** reflects O_2 bound to Hb (the amount of O_2 that is dissolved is trivial compared to bound).
- **Partial pressure** reflects dissolved O_2.

Key Pressures	Key Saturation %	Shift to Right (↑ P_{50})	Shift to Left (↓ P_{50})
Pao_2 = 100 mm Hg	Almost 100% saturated	• **Facilitates unloading** • ↑ temperature, ↑ Pco_2, ↓ pH, ↑ 2,3-DPG • Exercising muscle is hot, acidic, and hypercarbic	• **Facilitates loading** • ↓ temperature, ↓ Pco_2, ↑ pH, ↓ 2,3-DPG • CO poisoning
$P\bar{v}o_2$ = 40 mm Hg	75% saturated		
P_{50} = 27 mm Hg	50% saturated		

Definition of abbreviations: Hb, hemoglobin concentration; Sat, saturation; P_{50}, Po_2 at 50% saturation.
*1.39 mL of O_2 binds 1 g of Hb (some texts use 1.34 or 1.36).

▶ Additional Changes in the Hemoglobin–O_2 Dissociation Curve

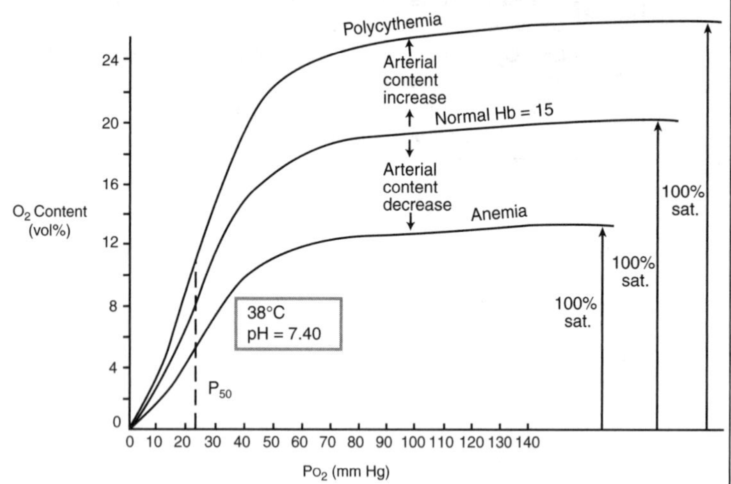

- **Polycythemia** and **anemia** change arterial **O_2 content**.
- Pao_2 and P_{50} remain the same.

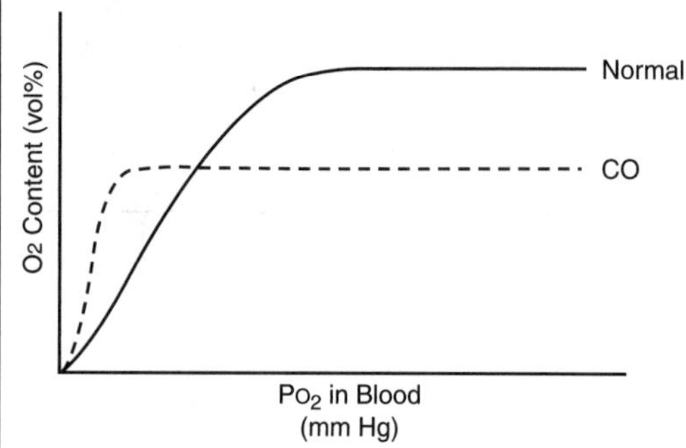

CO poisoning is dangerous for three reasons:

1. CO **left-shifts** the curve (↓ P_{50}), causing ↓ O_2 unloading in tissues.
2. CO has 240 times greater affinity for Hb as O_2, thus ↓ **the O_2 content** of blood.
3. CO inhibits cytochrome oxidase

► CO_2 Transport

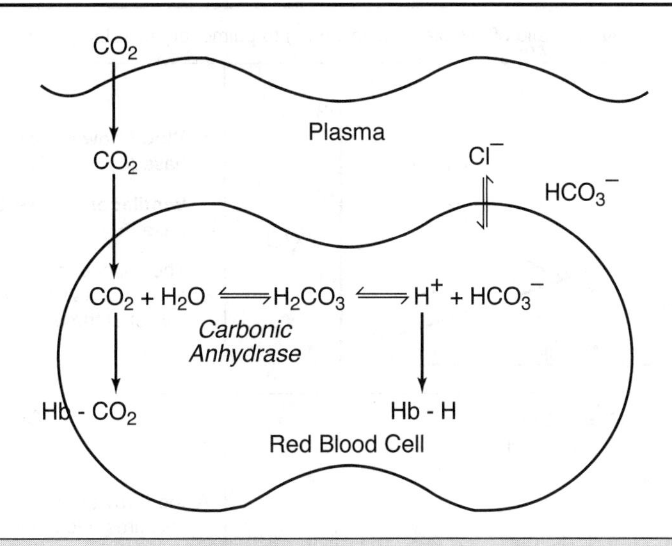

Forms of CO_2
Percentages reflect contribution in arterial blood.
1. **HCO_3^-** = 90%
2. Carbamino compounds (combination of CO_2 with proteins, especially Hb) = 5%
3. Dissolved CO_2 = 5%

► Pulmonary Blood Flow

Resistance (R)	Very low
Compliance	Very high
Pressures	Very low compared with systemic circulation
Effect of P_{AO_2}	• **Alveolar hypoxia → vasoconstriction.** • This is a local effect and the opposite of other organs, where hypoxia → vasodilation. • This directs blood away from hypoxic alveoli to better ventilated areas • This is also why fetal pulmonary vascular resistance is so high. Pulmonary resistance ↓ when the first breath oxygenates the alveoli, causing pulmonary blood flow to rise.
Gravity	Upright posture: greatest flow in base; lowest in apex
Filter	Removes small clots from circulation
Vasoactive substances	Converts angiotensin I → AII; inactivates bradykinin; removes prostaglandin E_2 and $F_{2\alpha}$ and leukotrienes

▶ Ventilation–Perfusion Relationships

$\dot{V}/\dot{Q}$: the ratio of alveolar ventilation ($\dot{V}$) to pulmonary blood flow ($\dot{Q}$).

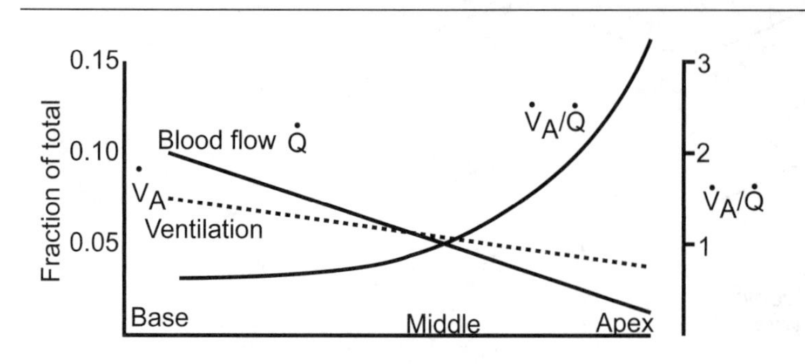

- **Blood flow** is lowest at the apex and highest at the base (gravity effect).
- **Ventilation** is lowest at the apex and highest at the base.
- The change in ventilation is not as great as blood flow, so the **$\dot{V}/\dot{Q}$ ratio** is highest at the apex and lowest at the base.

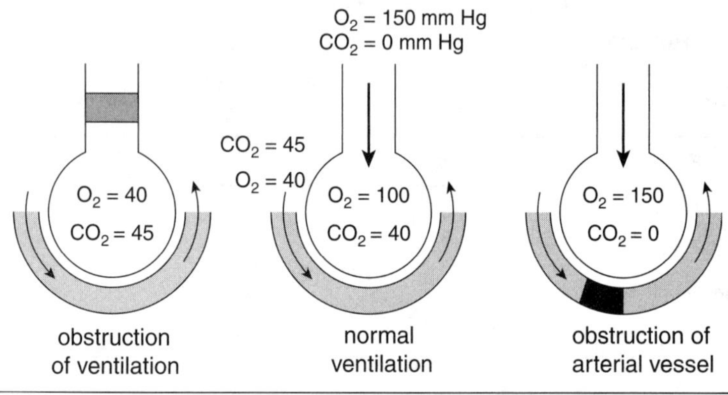

Changes in $\dot{V}/\dot{Q}$

A good way to remember the changes in partial pressures due to alterations in $\dot{V}/\dot{Q}$ is to think of the most extreme cases.

- If ventilation is 0 (airways blocked), $\dot{V}/\dot{Q}$ = 0 (a **shunt**). No gas exchange occurs and the PA_{O_2} and PA_{CO_2} are the same as mixed venous blood.
- If perfusion is 0 (embolism), $\dot{V}/\dot{Q}$ = ∞ (**dead space**). No gas exchange occurs and PA_{O_2} and PA_{CO_2} are the same as inspired air.

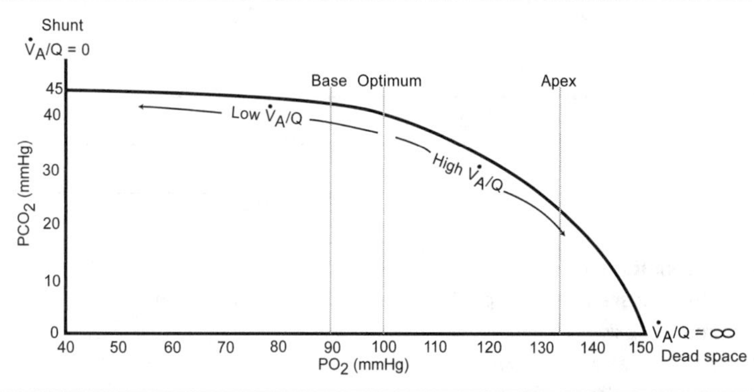

Summary of Changes

	Apex	Base
$\dot{V}A$	↓↓	↑
$\dot{Q}$	↓↓↓	↑↑
$\dot{V}A/\dot{Q}$	↑↑	↓
P_{O_2}	↑	↓
P_{CO_2}	↓	↑

► Disorders That Affect Arterial Oxygen Pressure or Content

	Notes	Pa_{O_2}	Pa_{CO_2}	Ca_{O_2}	A-a*	Response to Supplemental O_2 ☠
Hypoventilation[1]	Drugs (e.g., opiates, barbiturates), head trauma, chest wall dysfunction	↓	↑	↓	NC	↑ Pa_{O_2}; ↑ Ca_{O_2}
↓ inspired pO_2	↑ altitude	↓	↓[2]	↓	NC	↑ Pa_{O_2}; ↑ Ca_{O_2}
Diffusion limitation[3]	P_{AO_2} and Pa_{O_2} do not fully equilibrate	↓	NC[2]	↓	↑	↑ Pa_{O_2}; ↑ Ca_{O_2}
Shunt[4]	Venous blood mixes with arterial system, bypassing ventilated areas of lung	↓	NC[2]	↓	↑	Poor
$\dot{V}$/Q mismatch	Ventilation and perfusion are mismatched in the lung	↓	NC[2]	↓	↑	↑ Pa_{O_2}; ↑ Ca_{O_2}
CO poisoning	Exhaust fumes	NC	NC	↓	NC	↑ Pa_{O_2}; NC in Ca_{O_2}[5]
↓ [Hb]	Anemia	NC	NC	↓	NC	↑ Pa_{O_2}; NC in Ca_{O_2}[6]

Definition of abbreviation: NC, no change or minimal response.

*An increase in the A-a gradient indicates a problem with gas exchange.

☠ *Warning:* Supplemental O_2 in CNS depression and lung disease can shut off the hypoxic drive to ventilation and cause cessation of spontaneous breathing.

[1]Hypoventilation: ↓ $\dot{V}_A$, so ↑ P_{ACO_2} → ↓ P_{AO_2} → ↓ Pa_{O_2} (As ↑ CO_2 diffuses into alveoli from blood, it displaces O_2.)

[2]Result of hypoxia-induced increase of ventilation

[3]Diffusion limitation causes disease (blood-gas barrier has less surface area or is thickened), ↓ transit time during intense exercise, exercise at high altitude

[4]Abnormal shunts (often congenital, e.g., Tetralogy of Fallot)

[5]Slow response of Ca_{O_2} due to extremely high affinity of Hb for carbon monoxide

[6]Can only increase dissolved O_2 because Hb is saturated at normal Pa_{O_2}.

► Control of Ventilation

Brain Stem Respiratory Centers		
Medulla	Rhythm generator	**Inspiratory center:** generates breathing rhythm **Expiratory center:** not active during normal, passive expiration; involved in active expiration (e.g., exercise)
Pons	Regulates medulla	**Apneustic center:** stimulates prolonged inspiration **Pneumotaxic center:** terminates inspiration
Cortex	Conscious and emotional response	Lesions above the pons eliminate voluntary control, but basic breathing pattern remains intact.

► Chemoreceptors

	Central (Medulla) Chemoreceptors (Respond to Changes in pH of CSF)	Peripheral Chemoreceptors (Carotid and Aortic Bodies)
O_2	No response	• ↓ PaO_2 (<60 mm Hg) → stimulates chemoreceptors → increased ventilation. • *Note:* stimulated by changes in pressure, not O_2 content; thus, not stimulated by anemia.
CO_2	↑ Pco_2 → stimulate chemoreceptors → ↑ **ventilation** (**via** H^+)	• ↑ $Paco_2$ → stimulation →↑ ventilation. • Central chemoreceptor response is most important during normal breathing.
H^+	↑ H^+ → stimulation ↑ ventilation detects H^+ in CSF; 80–95% of response to hypercapnia	• ↑ H^+ → stimulation → ↑ ventilation. • *Note:* Most of the response to metabolic acidosis is peripheral because fixed acids penetrate blood/brain barrier poorly.

Clinical Correlation

Patients with severe lung disease can have chronic CO_2 retention, and the pH of their CSF can return to normal despite their hypercapnia. Having lost their CO_2 stimulus to ventilate, their hypoxic ventilatory drive becomes very important. If these patients are given enriched O_2 to breathe to correct their hypoxemia, their primary ventilatory drive will be removed, which can cause severe depression of ventilation.

Note: Other receptors such as pulmonary stretch receptors, irritant receptors, and joint and muscle receptors also have roles in the regulation of ventilation.

► Response to High Altitude

Parameter	Response
P_AO_2	↓ (because P_B is ↓)
PaO_2	↓ (hypoxemia because P_AO_2 is ↓)
Respiratory rate	↑ (hypoxic stimulation of peripheral chemoreceptors)
$Paco_2$ and P_ACO_2	↓ (hyperventilation due to hypoxemia)
Arterial pH	↑ (because of respiratory alkalosis); later becomes normal (renal compensation)
[Hb]	↑ (polycythemia)
Hb % saturation	↓ (because ↓ Po_2)
Pulmonary vascular resistance	↑ (hypoxic vasoconstriction); this plus polycythemia lead to ↑ work and hypertrophy of right heart
[2,3-DPG]	↑
Hemoglobin-O_2 curve	Right-shift (because of ↑ 2,3-DPG)
Acute mountain sickness	Hypoxemia and alkalosis cause headache, fatigue, nausea, dizziness, palpitations, and insomnia. Treatment with acetazolamide can be therapeutic.
Chronic mountain sickness	Reduced exercise tolerance, fatigue, hypoxemia, polycythemia

Respiratory Pathology

► Ear, Nose, Throat, and Upper Respiratory System Infections

Type of Infection	Case Vignette/Key Clues	Diagnosis	Common Causative Agents	Pathogenesis	Treatment
Sinusitis	Sinus pain; low-grade fever	Gram ⊕ coccus, catalase ⊖	*Streptococcus pneumoniae*	Capsule, IgA protease	Penicillin
		Gram ⊖ rod, chocolate agar	*Haemophilus influenzae*	Capsule, IgA protease, endotoxin	Amoxicillin
		Gram ⊖ coccus	*Moraxella catarrhalis*	β-lactamase producer	Ceftriaxone
Oral cavitary disease	Sore mouth with thick, white coating that can be scraped off easily to reveal painful red base	Gram ⊕ yeast, germ tube test	*Candida albicans*	Overgrowth of normal flora, immunocompromised, overuse of antibiotics	Nystatin, miconazole
Sore throat	Inflamed tonsils/pharynx, abscesses; cervical lymphadenopathy, fever, ± stomach upset; ± sandpaper rash	Rapid antigen test; gram ⊕, catalase ⊖ coccus; β-hemolytic, bacitracin sensitive	*Streptococcus pyogenes*	Exotoxins A–C (superantigens)	Penicillin
	White papules with red base on posterior palate and pharynx, fever	Virus culture or PCR	Coxsackie A	Unknown	None
	Throat looks like *Streptococcus* with severe fatigue, lymphadenopathy, fever, ± rash	Heterophile ⊕ (Monospot test); mononucleosis; 70% lymphocytosis (Downey type II cells = CTLs)	Epstein-Barr virus	Infects B lymphocytes by attachment to CD21, causes ↑ CTLs	Supportive
	Low-grade fever with a 1–2 day gradual onset of membranous nasopharyngitis and/or obstructive laryngotracheitis; bull neck from lymphadenopathy; elevated BUN; abnormal ECG; little change in WBC; unvaccinated, dislodged membrane bleeds profusely	Gram ⊕ nonmotile rod, Loeffler medium, ELEK test	*Corynebacterium diphtheriae*	Diphtheria toxin inactivates eEF-2 in heart, nerves, epithelium; pseudomembrane → airway obstruction when dislodged	Penicillin, antitoxin
Common cold	Rhinitis, sneezing, coughing; seasonal peaks	Clinical	Rhinoviruses (summer–fall) Coronaviruses (winter–spring)	Virus attaches to ICAM-1 on respiratory epithelium	Supportive
Acute otitis media	Red, bulging tympanic membrane, fever 102–103°F; pain goes away if drum ruptures or if ear tubes are patent	Gram ⊕ coccus, catalase ⊖	*Streptococcus pneumoniae*	Capsule, IgA protease	Penicillin
		Gram ⊖ rod, chocolate agar	*H. influenzae* (nontypeable)	Capsule, IgA protease, endotoxin	Amoxicillin
		Gram ⊖ diplococcus	*Moraxella catarrhalis*	β-lactamase producer	Ceftriaxone

(Continued)

► Ear, Nose, Throat, and Upper Respiratory System Infections *(Cont'd.)*

Type of Infection	Case Vignette/Key Clues	Diagnosis	Common Causative Agents	Pathogenesis	Treatment
Otitis externa	Ear pain	Gram ⊕, catalase ⊕, coagulase ⊕	*Staphylococcus aureus*	Normal flora enter abrasions	β-lactamase-resistant penicillin
		Gram ⊕ yeast, germ tube test	*Candida albicans*	Normal flora enter abrasions	Nystatin, miconazole
		Gram ⊖ rod, urease ⊕, oxidase ⊖, swarming motility	*Proteus*	From water source	Susceptibility testing*
		Gram ⊖ rod, oxidase ⊕, blue-green pigments	*Pseudomonas aeruginosa*	From water source	Susceptibility testing*
Malignant otitis externa	Severe ear pain in diabetic; life-threatening	Gram ⊖ rod, oxidase ⊕, blue-green pigments, fruity odor	*Pseudomonas aeruginosa*	Capsule	Susceptibility testing*

Definition of abbreviations: CTLs, cytotoxic T lymphocytes.

*Because there is so much drug resistance in these genera, susceptibility testing is necessary.

► Middle Respiratory Tract Infections

Disease	Case Vignette/ Key Clues	Diagnosis	Common Causative Agents	Pathogenesis	Treatment
Epiglottitis	Inflamed epiglottis; patient often 2–3 years old and **unvaccinated**; **thumbprint** sign on lateral x-ray	Gram ⊖ rod, chocolate agar (requires hemin and NAD)	*Haemophilus influenzae*	Capsule (polyribitol phosphate) inhibits phagocytosis; IgA protease	Ceftriaxone
Croup	**Infant** with fever, sharp barking cough, inspiratory stridor, hoarse phonation; **steeple** sign on AP x-ray	Detect virus in respiratory washings	Parainfluenza virus (croup)	Viral cytolysis; multinucleated giant cells formed	Ribavirin
Laryngotracheitis, laryngotracheobronchitis	Hoarseness, burning retrosternal pain	Detect virus in respiratory washings	Parainfluenza virus	Viral cytolysis; multinucleated giant cells formed	Ribavirin
Bronchitis, bronchiolitis	Wheezy; **infant** or child ≤5 years old	Direct immunofluorescence for viral Ags	RSV	Fusion protein creates syncytia	Ribavirin
	>5 years old	Slow growth on Eaton medium, cold agglutinins	*Mycoplasma pneumoniae*, viruses	Release of O_2 radicals causes necrosis of epithelium	Symptomatic

Definition of abbreviations: Ag, antigens; NAD, nicotinamide adenine dinucleotide; RSV, respiratory syncytial virus.

▶ Pneumonia

Lobar pneumonia and bronchopneumonia—acute inflammation and consolidation (solidification) of the lung due to an extracellular bacterial agent. Lobar affects entire lobe (opacification = consolidation on x-ray); bronchopneumonia (patchy consolidation around bronchioles on x-ray). Associated with high fever and productive cough.

Interstitial (atypical) pneumonia causes interstitial pneumonitis without consolidation and can be due to viral agents (influenza virus; parainfluenza; RSV, especially in young children; adenovirus; CMV, especially in immunocompromised; varicella), *Mycoplasma pneumoniae, and Pneumocytis jiroveci.*

Clinical:

- Fever and chills
- Cough (may be productive)
- Tachypnea
- Pleuritic chest pain
- Decreased breath sounds, rales, and dullness to percussion
- Elevated WBC count with a left shift

Type of Infection	Case Vignette/ Key Clues	Diagnosis	Most Common Causative Agents	Pathogenesis	Treatment
Pneumonia—typical	Adults (including alcoholics), **rust-colored sputum** Lobar pneumonia or less commonly, bronchopneumonia	Gram ⊕ diplococcus, α hemolytic, catalase ⊖ lysed by bile, inhibited by Optochin	*Streptococcus pneumoniae, Haemophilus influenzae* (much less common)	Capsule antiphagocytic IgA protease	Third-generation cephalosporin, azithromycin
	Neutropenic patients, burn patients, CGD, CF	Gram ⊖ rod, oxidase ⊕, blue-green pigments	*Pseudomonas aeruginosa*	Opportunist	Sensitivity testing required
	Foul-smelling sputum, aspiration possible	Culture of sputum	Anaerobes, mixed infection *(Bacteroides, Fusobacterium, Peptococcus)*	Aspiration of vomitus → enzyme damage → anaerobic foci	Empiric antibiotic therapy (amoxicillin/ clavulanate, gentamicin)
	Alcoholic with aspiration, facultative anaerobic, gram ⊖ bacterium with huge capsule, **currant jelly sputum**	Gram ⊖ rod, lactose fermenting, oxidase ⊖	*Klebsiella pneumoniae*	Capsule protects against phagocytosis	Susceptibility testing necessary
Pneumonia—atypical	Poorly nourished, unvaccinated baby/ child; giant cell pneumonia with hemorrhagic rash, Koplik spots	Serology	Measles: malnourishment ↑ risk of pneumonia and blindness	Cytolysis in lymph nodes, skin, mucosa Syncytia → giant cell pneumonia	Supportive
	Pneumonia teens/ young adults; bad hacking, dry cough "walking pneumonia"	Serology, cold agglutinins	*Mycoplasma pneumoniae* (most common cause of pneumonia in school-age children)	Adhesin causes adhesion to mucus; oxygen radicals cause necrosis of epithelium	Doxycycline, azithromycin

(Continued)

▶ Pneumonia *(Cont'd.)*

Type of Infection	Case Vignette/ Key Clues	Diagnosis	Most Common Causative Agents	Pathogenesis	Treatment
Pneumonia—atypical (*cont'd.*)	Air-conditioning exposure, common showers, especially >50 years, heavy smoker, drinker	Direct fluorescent antibody	*Legionella* spp.	Intracellular in macrophages	Macrolide
	Bird exposure ± hepatitis	Direct fluorescent antibody, intracytoplasmic inclusions	*Chlamydophilia psittaci*	Obligate intracellular	Tetracycline, erythromycin
	AIDS patients with staccato cough; **"ground glass"** x-ray; biopsy: honeycomb exudate with silver staining cysts; premature infants	Silver-staining cysts in alveolar lavage	*Pneumocystis jiroveci (carinii)*	Attaches to type I pneumocytes, causes excess replication of type II pneumocytes	Trimethoprim sulfamethoxazole
	Primary influenza pneumonia Secondary (bacterial)	Virus culture	Influenza virus	Cytolysis in respiratory tract; cytokines contribute; secondary infections common	Oseltamivar, zanamivar
Acute pneumonia or chronic cough with weight loss, night sweats	Over 55 years, HIV ⊕, or immigrant from developing country	Auramine-rhodamine stain of sputum acid-fast bacilli	*Mycobacterium tuberculosis*	Facultative intracellular parasite → cell-mediated immunity and DTH	Multidrug therapy
	Dusty environment with bird or bat fecal contamination (Missouri chicken farmers, Ohio river)	Intracellular yeast cells in sputum	*Histoplasma capsulatum*	Facultative intracellular	Amphotericin B
	Desert sand S.W. United States	Endospores in spherules in tissues	*Coccidioides immitis*	Acute, chronic lung infection, dissemination	Amphotericin B
	Rotting, contaminated wood, same endemic focus as *Histoplasma* and east coast states	Broad-based budding yeast cells in sputum or skin	*Blastomyces dermatitidis*	Acute, chronic lung infection, dissemination	Ketoconazole
Sudden acute respiratory syndromes	Travel to Far East, Toronto, winter, early spring	Serology, virus, isolation	SARS agent	Replication in cells of upper respiratory tree	None
	"Four Corners" region (CO, UT, NM, AZ), spring, inhalation rodent urine	Serology, virus, isolation	Hantavirus (Sin Nombre)	Virus disseminates to CNS, liver, kidneys, endothelium	Ribavirin

Definition of abbreviations: CGD, chronic granulomatous disease; CMV, cytomegalovirus; CF, cystic fibrosis; DTH, delayed-type hypersensitivity; RSV, respiratory syncytial virus.

► Granulomatous Diseases

Tuberculosis

Causes **caseating granulomas** containing acid-fast mycobacteria; transmission is by inhalation of aerosolized bacilli; increasing incidence in the U.S., secondary to AIDS

- Primary tuberculosis (initial exposure) can produce a **Ghon complex**, characterized by a subpleural caseous granuloma above or below the lobar fissure, accompanied by hilar lymph node granulomas.
- Secondary tuberculosis (reactivation or reinfection) tends to involve the **lung apex**.
- Progressive pulmonary tuberculosis can take the forms of cavitary tuberculosis, miliary pulmonary tuberculosis, and tuberculous bronchopneumonia. Miliary tuberculosis can also spread to involve other body sites.

Clinical: fevers and night sweats, weight loss, cough, hemoptysis, positive skin test (PPD)

Sarcoidosis

Sarcoidosis is a granulomatous disease of unknown etiology; affects females > males, ages 20–60; most common in African American women. **Noncaseating granulomas** occur in any organ of the body; hilar and mediastinal adenopathy are typical.

Clinical: Cough, shortness of breath, fatigue, malaise, skin lesions, eye irritation or pain, fever/night sweats

Labs: ↑ serum **angiotensin-converting enzyme** (ACE)
Schaumann bodies: laminated calcifications
Asteroid bodies: stellate giant-cell cytoplasmic inclusions

► Obstructive Lung Disease

Increased resistance to airflow secondary to obstruction of airways

Chronic obstructive pulmonary disease (COPD) includes chronic bronchitis, emphysema, asthma, and bronchiectasis.

Disease	Characteristics	Clinical Findings
Chronic bronchitis	• Persistent cough and copious **sputum production for at least 3 months** each year in 2 consecutive years • Highly **associated with smoking** (90%)	• Cough, sputum production, dyspnea, frequent infections • Hypoxia, cyanosis, weight gain
Emphysema	• Associated with destruction of alveolar septa, resulting in enlarged air spaces and a loss of elastic recoil, and producing overinflated, enlarged lungs • Thought to be due to protease/antiprotease imbalance *Gross:* • **Overinflated, enlarged lungs** • Enlarged, grossly visible air spaces • Formation of apical blebs and bullae (centriacinar type)	• Progressive dyspnea • Pursing of lips and **use of accessory muscles** to breathe • **Barrel chest** • Weight loss
Centriacinar (centrilobular) emphysema	• **Proximal respiratory bronchioles** involved • Most common type (95%) • **Associated with smoking** • Worst in apical segments of upper lobes	
Panacinar (panlobular) emphysema	• **Entire acinus involved**; distal alveoli spared • Less common • **Alpha-1-antitrypsin deficiency** • *Distribution:* entire lung; worse in bases of lower lobes	
Asthma	• Due to hyperreactive airways, resulting in episodic **bronchospasm**, producing **wheezing**, severe **dyspnea**, and coughing. • Inflammation, edema, hypertrophy of mucous glands with **goblet cell hyperplasia** and **mucus plugs** are characteristic findings. • Hypertrophy of bronchial wall smooth muscle, thickened basement membranes	

(Continued)

▶ Obstructive Lung Disease *(Cont'd.)*

Extrinsic asthma	• **Type I hypersensitivit**y reaction • **Allergic** (**atopic**)—most common type • Childhood and young adults; ⊕ **family history** • Allergens: pollen, dust, food, molds, animal dander, etc. • Occupational exposure: fumes, gases, and chemicals
Intrinsic asthma	• Unknown mechanism • **Respiratory infections** (usually viral) • **Stress** • **Exercise** • **Cold** temperatures • Drug induced (**aspirin**)
Bronchiectasis	• An abnormal permanent airway dilatation due to **chronic necrotizing infection** • Most patients have underlying lung disease, such as bronchial obstruction, necrotizing pneumonias, **cystic fibrosis**, or **Kartagener syndrome**.

▶ Drugs for Asthma

Class	Agents	Mechanism	Comments
		Bronchodilators	
β_2 agonists	Albuterol Terbutaline Metaproterenol Salmeterol Formoterol	Stimulate β_2 receptors → ↑ cAMP → smooth muscle relaxation	• Generally have the advantage of **minimal cardiac side effects** • Most often used as **inhalants** • Salmeterol and formoterol are **long-acting** agents, so are useful for prophylaxis • Cause skeletal muscle **tremors**; some **CV** side effects (tachycardia, arrhythmias) can still occur
Non-selective β agonists	Epinephrine (α_1,α_2,β_1,β_2) Isoproterenol (β_1, β_2)	Stimulate β_2 receptors → ↑ cAMP → smooth muscle relaxation	• Not used as much as the β_2 agonists • Epinephrine is used for **acute** asthma attacks and for **anaphylaxis**
Muscarinic antagonists	Ipratropium Tiotropium	Block muscarinic receptors, inhibiting vagally induced bronchoconstriction	• Used as an inhalant; there are minimal systemic side effects • Used in asthma and COPD • β_2 agonists are generally preferred for acute bronchospasm • These are useful in **COPD** because it decreases bronchial secretions and has fewer CV side effects • Tiotropium is longer-acting
Methylxanthines	Theophylline	Inhibit PDE; block adenosine receptors	• Available orally • Major use is for asthma (although β_2 agonists are first-line)
		Leukotriene Antagonists	
Leukotriene antagonists	Zafirlukast Montelukast	Block LTD_4 (and LTE_4) leukotriene receptors	• Orally active; not used for acute asthma episodes • Prevents exercise-, antigen-, and aspirin-induced asthma
5-lipoxygenase inhibitors	Zileuton	Block leukotriene synthesis	
		Anti-inflammatory Agents	
Corticosteroids	Beclomethasone Prednisone, prednisolone Others	Inhibit phospholipase A_2 → ↓ arachidonic acid synthesis	• ↓ inflammation and edema • Used orally and inhaled • IV use in status asthmaticus
Release inhibitors	Cromolyn Nedocromil	Inhibit mast cell degranulation	• Can prevent allergy-induced bronchoconstriction • Available as nasal spray, oral, eye drops

► Restrictive Lung Disease

(Decreased Lung Volumes and Capacities)

Examples
Chest wall disorders: obesity, kyphoscoliosis, polio, etc.
Intrinsic lung disease:
Adult respiratory distress syndrome (ARDS)
Neonatal respiratory distress syndrome (NRDS)
Pneumoconioses (silicosis, asbestosis, "black lung" disease from coal dust)
Sarcoidosis
Idiopathic pulmonary fibrosis (Hamman-Rich syndrome)
Goodpasture syndrome
Wegener granulomatosis
Eosinophilic granuloma
Collagen-vascular diseases
Hypersensitivity pneumonitis
Drug exposure

► Respiratory Distress Syndromes

Adult Respiratory Distress Syndrome (ARDS)

- Diffuse damage to the alveolar epithelium and capillaries, resulting in progressive respiratory failure unresponsive to oxygen therapy
- **Causes:** shock, sepsis, trauma, gastric aspiration, radiation, oxygen toxicity, drugs, pulmonary infections, and many others
- **Clinical presentation:** dyspnea, tachypnea, hypoxemia, cyanosis, and use of accessory respiratory muscles
  - X-ray: **bilateral lung opacity** ("white out")
  - Gross: heavy, **stiff, noncompliant lungs**
  - Micro:
    - Interstitial and intra-alveolar edema
    - Interstitial inflammation
    - Loss of type I pneumocytes
    - **Hyaline membrane formation**
  - Overall mortality 50%

Neonatal Respiratory Distress Syndrome (NRDS)

Also known as **hyaline membrane disease of newborns**. Causes respiratory distress within hours of birth and is seen in infants with **deficiency of surfactant** secondary to prematurity (gestational age of <28 weeks has a 60% incidence), maternal diabetes, multiple births, or C-section delivery

- Clinical presentation: often normal at birth, but within a few hours develop increasing respiratory effort, tachypnea, nasal flaring, use of accessory muscle of respiration, an expiratory grunt, cyanosis
- X-ray: "**ground-glass**" reticulogranular densities
- Labs: lecithin: sphingomyelin ratio <2
- Micro: atelectasis and **hyaline membrane formation**
- Treatment: surfactant replacement and oxygen
- Complications of oxygen treatment in newborns:
  - **Bronchopulmonary dysplasia**
  - **Retrolental fibroplasia** (retinopathy of prematurity)
- **Prevention:** delay labor and corticosteroids to mature the lung

► Tumors of the Lung and Pleura

Bronchogenic carcinoma is the leading cause of cancer deaths among both men and women.

Major Risk Factors

- **Cigarette smoking**
- Occupational exposures (asbestosis, uranium mining, radiation)
- Air pollution

Histologic Types

Adenocarcinoma, bronchioloalveolar carcinoma, squamous cell carcinoma, small cell carcinoma, and large cell carcinoma

- Oncogenes
  - **L-*myc*:** small cell carcinomas
  - **K-*ras*:** adenocarcinomas
- Tumor suppressor genes
  - **p53** and the retinoblastoma gene

Complications

- **Spread to hilar, bronchial, tracheal, or mediastinal nodes** in 50% of cases
- **Superior vena cava syndrome** (obstruction of SVC by tumor)
- Esophageal obstruction
- **Recurrent laryngeal nerve** involvement (hoarseness)
- **Phrenic nerve** damage, causing diaphragmatic paralysis
- Extrathoracic metastasis to adrenal, liver, brain, and bone
- **Pancoast tumor:** compression of cervical sympathetic plexus, leading to ipsilateral **Horner syndrome** (miosis, ptosis, anhidrosis)

Type of Cancer/ (Percentage of Total)	Association with Smoking; Sex Preference	Location	Pathology
Adenocarcinoma/35% **Bronchioloalveolar carcinoma** (5%)—subset of adenocarcinoma	**Less strongly related;** female > male; most common lung cancer in nonsmokers	**Peripheral**, may occur in scars; can have associated pleural involvement	Forms glands and may produce mucin
Squamous cell/30%	**Strongly related;** male > female	**Central**	• Invasive nests of squamous cells, intercellular bridges (desmosomes); keratin production ("**squamous pearls**") • Hyperparathyroidism secondary to increased secretion of parathyroid related peptide
Small cell (oat cell) carcinoma/20%	**Strongly related;** male > female	**Central**	• Very aggressive; micro: small round cells • Frequently associated with paraneoplastic syndromes, including production of ACTH (Cushing syndrome), ADH, and parathyroid related peptide
Large cell/10%	—	—	Large anaplastic cells without evidence of differentiation
Carcinoid/<5%	—	Bronchial	May produce carcinoid syndrome

► Diseases of the Pleura

Pleural effusion	• Accumulation of **fluid in the pleural cavity**; may be a transudate or exudate • Chylous fluid in pleural space secondary to obstruction of thoracic duct (usually by tumor) = chylothorax
Pneumothorax	• **Air in the pleural cavity**, often due to penetrating chest wall injuries • *Spontaneous pneumothorax:* young adults with rupture of emphysematous blebs • *Tension pneumothorax:* life-threatening shift of thoracic organs across midline • Clinical: ↓ breath sounds, hyperresonance, tracheal shift to opposite side
Mesothelioma	• Highly malignant tumor of pleura (and peritoneum) • Pleural mesothelioma is associated with **asbestos exposure** in 90% of cases (**bronchogenic carcinoma** also strongly associated with asbestos exposure)

► Pulmonary Vascular Disorders

Pulmonary edema	Fluid accumulation within the lungs that can be due to many causes, including left-sided heart failure, mitral valve stenosis, fluid overload, nephrotic syndrome, liver disease, infections, drugs, shock, and radiation
Pulmonary emboli	Mostly arise from **deep vein thrombosis** in the leg (also arise from pelvic veins) and may be asymptomatic, cause pulmonary infarction, or cause sudden death; severity related to size of embolus and other comorbid conditions
Pulmonary hypertension	Pulmonary hypertension is increased artery pressure, usually due to increased vascular resistance or blood flow; can be idiopathic or related to underlying COPD, interstitial disease, pulmonary emboli, mitral stenosis, left heart failure, and congenital heart disease with left-to-right shunt

The Renal and Urinary System

Chapter 4

Renal Embryology and Anatomy

▶ Development of the Kidney and Ureter (Mesoderm)

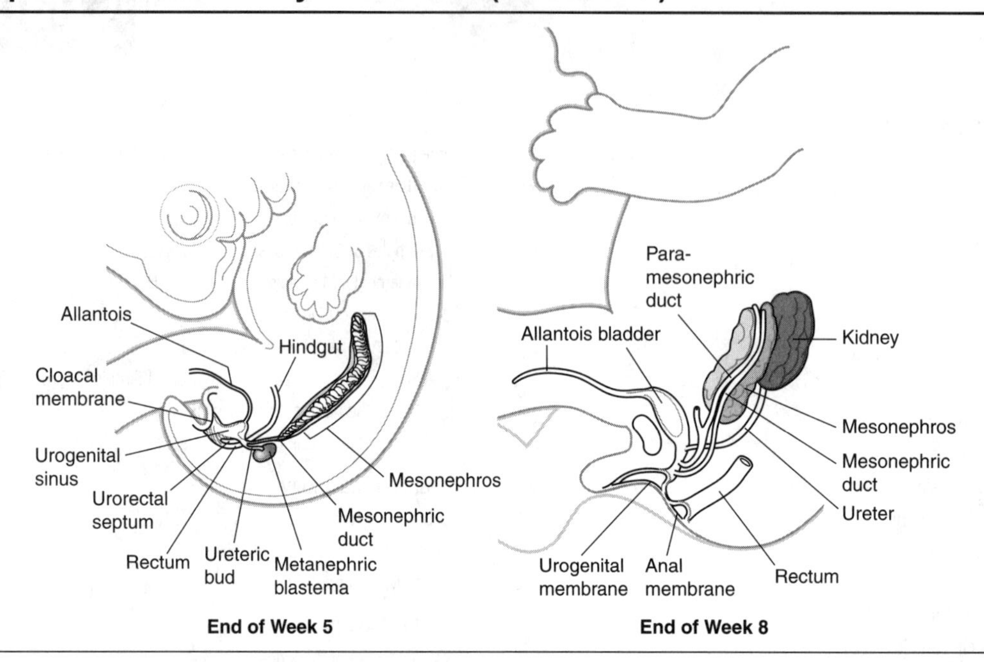

The **ureteric bud** penetrates the **metanephric mass**, which condenses around the diverticulum to form the metanephrogenic blastema. The **bud** dilates to form the **renal pelvis**, which subsequently splits into the cranial and caudal **major calyces**, which form the **minor calyces**. One to 3 million collecting tubules develop from the minor calyces, forming the renal pyramids.

Penetration of collecting tubules into the metanephric mass induces cells of the tissue cap to form nephrons, or excretory units.

- The **proximal nephron forms Bowman capsule**, whereas the **distal nephron connects to a collecting tubule**.
- Lengthening of the excretory tubule gives rise to the **proximal convoluted tubule**, the **loop of Henle**, and the **distal convoluted tubule**.

The kidneys develop in the pelvis but appear to ascend into the abdomen as a result of fetal growth of the lumbar and sacral regions. With their ascent, the ureters elongate, and the kidneys become vascularized by lateral splanchnic arteries, which arise from the abdominal aorta.

▶ Development of the Bladder and Urethra (Endoderm)

Bladder
• The **urorectal septum** divides the cloaca into the **anorectal canal** and the **urogenital sinus** by Week 7. The upper and largest part of the urogenital sinus becomes the **urinary bladder**, which is initially continuous with the **allantois**. As the lumen of the allantois becomes obliterated, a fibrous cord, the **urachus**, connects the apex of the bladder to the umbilicus. In the adult, this structure becomes the **median umbilical ligament**. • The **mucosa** of the trigone of the bladder is initially formed from mesodermal tissue, which is replaced by **endodermal epithelium**. The smooth muscle of the bladder is derived from splanchnic mesoderm.
Urethra
• The **male urethra** is anatomically divided into three portions: **prostatic**, **membranous**, and **spongy** (penile). The **prostatic urethra, membranous urethra, and proximal penile urethra** develop from the narrow portion of the **urogenital sinus** below the urinary bladder. The **distal spongy urethra** is derived from the **ectodermal cells** of the glans penis. • The **female urethra** is derived entirely from the **urogenital sinus** (endoderm).

► Congenital Abnormalities

Renal agenesis	Failure of one or both kidneys to develop because of early degeneration of the ureteric bud. Unilateral genesis is fairly common; bilateral agenesis is fatal (associated with oligohydramnios, and the fetus may have **Potter sequence**: clubbed feet, pulmonary hypoplasia, and craniofacial anomalies).
Pelvic and horseshoe kidney	Pelvic kidney is caused by a failure of one kidney to ascend. **Horseshoe kidney** (usually **normal renal function**, predisposition to calculi) is a fusion of both kidneys at their ends and failure of the fused kidney to ascend.
Double ureter	Caused by the early splitting of the ureteric bud or the development of two separate buds.
Patent urachus	**Failure of the allantois to be obliterated**. It causes urachal fistulas or sinuses. In male children with congenital valvular obstruction of the prostatic urethra or in older men with enlarged prostates, a patent urachus **may cause drainage of urine through the umbilicus**.

► Gross Anatomy of the Kidney

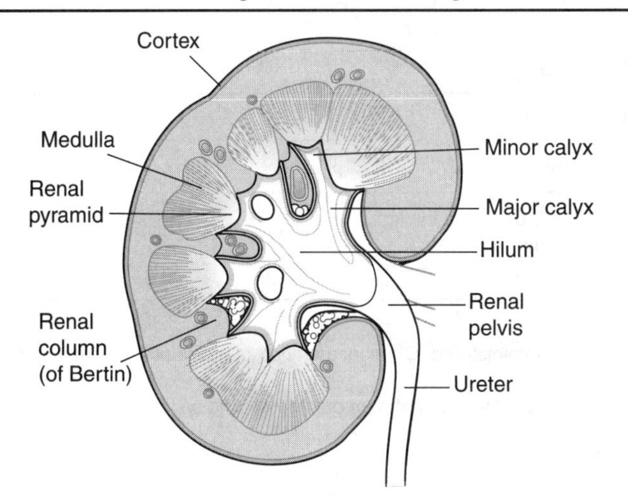

The kidney is divided into three major regions:

Hilum

Located medially and is the point of entrance and exit for the renal artery and vein, and ureter

Cortex

Forms the outer zone of the kidney, as well as several renal columns, which penetrate the entire depth of the kidney

Medulla

Appears as a series of medullary pyramids. The apex of each pyramid directs the urinary stream into a minor calyx, which then travels to the major calyx, through the renal pelvis, and into the ureter

Relation of the Kidneys to the Posterior Abdominal Wall

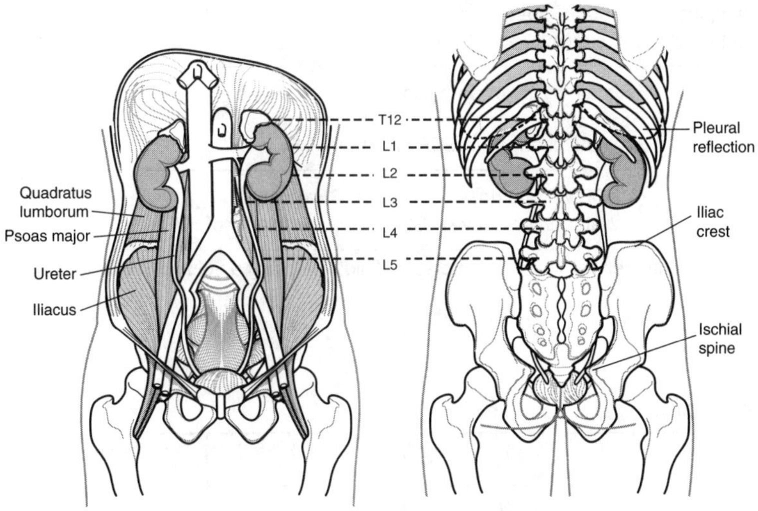

Clinical Correlation

Blockage by Renal Calculi

The most common sites of ureteral constriction susceptible to blockage by renal calculi are:

- where the renal pelvis joins the ureter
- where the ureter crosses the pelvic inlet
- where the ureter enters the wall of the urinary bladder

Renal Physiology

▶ Basic Functions of the Kidneys

Fluid balance	Maintain normal extracellular (ECF) and intracellular (ICF) fluid volumes
Electrolytes	Balance excretion with intake to maintain normal plasma concentrations
Wastes	Excrete metabolic wastes (nitrogenous products, acids, toxins, etc.)
Fuels	Reabsorb metabolic fuels (glucose, lactate, amino acids, etc.)
Blood pressure	Regulate ECF volume for the long-term control of blood pressure
Acid–base	Regulate absorption and excretion of H^+ and HCO_3^- to control acid–base balance

▶ Fluid Balance: Estimation of Fluid Volumes

Basic Concepts

Body fluid compartments: TBW = ICF + ECF   ECF = plasma + ISF   TBW = ICF + (plasma + ISF)

Estimating body fluid volumes:
(must assume normal hydration)

TBW in Liters = 0.6 × weight in kg
ICF = 0.4 × weight
ECF = 0.2 × weight

Measuring body fluid volumes (indicator dilution principle):

V = Q/C, where V = body fluid volume, Q = quantity of indicator administered, C = concentration of indicator after dilution in body fluid compartment

Indicators: must disperse evenly in compartment, must disperse only in compartment of interest, and cannot be metabolized or excreted (no indicator is perfect)

TBW indicators: D_2O, ^3H_2O, antipyrine ($C_{11}H_{12}N_2O$)
ECF indicators: ^{22}Na, inulin, mannitol
PV indicators: ^{125}I-albumin, Evans blue dye, ^{51}Cr-red blood cells

Osmolarity vs. molarity: Ionic substances dissociate, covalently bonded substances do not, therefore:

- 100 mM glucose = 100 mOsm/L
- 100 mM NaCl = 200 mOsm/L

Osmolarity of ECF always = osmolarity of ICF: Estimate by 2× plasma [Na^+]

All water and solutes pass through ECF: Evaluate changes in ECF first, then ICF

ICF volume is controlled by ECF osmolarity:

- ECF volume does not control ICF volume
- Water enters/leaves ICF to keep osmolarity of ICF = ECF

Definition of abbreviations: ECF, extracellular fluid; ICF, intracellular fluid; ISF, interstitial fluid; PV, plasma volume; TBW, total body water.

▶ Normal Values
(Body weight = 100 kg, 300 mOsm/L)

	ECF	ICF	TBW
Volume (liters)	20	40	60
Osmolarity (mOsm/L)	300	300	300
Solute mass in compartment (mOsm)	6,000	12,000	18,000

► Fluid and Electrolyte Abnormalities

Diarrhea

- Weight loss = 4 kg (assume all weight loss is fluid loss)
- Solutes lost from ECF
- Isoosmotic loss: osmolarity remains at 300 mOsm/L; calculate TBW, ECF, ICF

	ECF	ICF	TBW
Volume (liters)	V = 4,800/300 = 16	ICF = TBW – ECF 56 – 16 = 40	60 – 4 = 56
Osmolarity (mOsm/L)	300	300	300
Solute mass in compartment (mOsm)	6,000 – (300 × 4) = 4,800	12,000	56 × 300 = 16,800

Conclusion: Isoosmotic volume gain or loss only changes ECF volume; ICF volume is unchanged.

Sweating

- Weight loss = 8 kg (assume all weight loss is fluid loss)
- Solutes lost from ECF
- Loss of hypoosmotic fluid, osmolarity = 330 mOsm/L; calculate TBW, ECF, ICF

	ECF	ICF	TBW
Volume (liters)	V = 5,160/330 = 15.6	ICF = TBW – ECF 52 – 15.6 = 36.4	60 – 8 = 52
Osmolarity (mOsm/L)	330	330	330
Solute mass in compartment (mOsm)	6,000 – 840 = 5,160	12,000	52 × 330 = 17,160 = loss of 840 from ECF

Conclusion: ECF loss of water and solutes, but ↑ osmolarity → ↓ ICF volume due to water shift.

Drink 3 L Pure Water

- Weight gain = 3 kg
- No solutes lost or gained
- Gain of hypoosmotic fluid; calculate new osmolarity, TBW, ECF, ICF

	ECF	ICF	TBW
Volume (liters)	V = 6,000/285.7 = 21	V = 12,000/285.7 = 42	60 + 3 = 63
Osmolarity (mOsm/L)	285.7	285.7	18,000/63 = 285.7
Solute mass in compartment (mOsm)	21 × 285.7 = 6,000	42 × 285.7 = 12,000	63 × 285.7 = 18,000

Conclusion: Addition of pure water ↓ osmolarity, causing proportionate increases in ICF and ECF volumes.

► Processes in the Formation of Urine

Filtration	• Glomerular capillaries—same forces as described in cardiovascular section • Glomerular filtration rate **(GFR)**
Secretion (S)	Active transport of solutes from plasma (via interstitial fluid) into tubular fluid
Reabsorption (R)	Passive or active movement of solutes and water from tubular fluid back into capillaries
Excretion of X (E_x)	• Result of the balance of filtration, secretion, and reabsorption • $\mathbf{E_x = \dot{V} \times [X]_{urine}}$; $\dot{V}$ = urine flow; $[X]_{urine}$ = urinary concentration of X
Filtered load (F)	• $\mathbf{F = GFR \times [solute]_{plasma}}$ • Solute fraction bound to plasma protein does not filter
Mass balance	**E = F + S – R**
Transport	$F_x > E_x$ = net reabsorption of X; $F_x < E_x$ = net secretion of X

► Examples Using Mass Balance to Evaluate Renal Processing of a Solute

Data	Calculations	Conclusions
• GFR = 100 mL/min • $[Subst.]_{plasma}$ = 1 µg/mL • $\dot{V}$ = 2 mL/min • $[Subst.]_{urine}$ = 75 µg/mL	F = 100 × 1 = 100 µg/min E = 2 × 75 = 150 µg/min	Subst. must be secreted because the amount excreted is > the amount filtered. S ≈ 50 µg/min*
• GFR = 100 mL/min • $[Subst.]_{plasma}$ = 3 µg/mL • $\dot{V}$ = 1 mL/min • $[Subst.]_{urine}$ = 100 µg/mL	F = 100 × 3 = 300 µg/min E = 1 × 100 = 100 µg/min	Subst. must be reabsorbed because the amount filtered is > the amount excreted. R ≈ 200 µg/min†
• GFR = 100 mL/min • $[Subst.]_{plasma}$ = 2 µg/mL • $\dot{V}$ = 4 mL/min • $[Subst.]_{urine}$ = 50 µg/mL	F = 100 × 2 = 200 µg/min E = 4 × 50 = 200 µg/min	Subst. is neither secreted nor reabsorbed because the amount excreted = the amount filtered.¶

Definition of abbreviation: Subst., substance.

*Assumes no reabsorption

†Assumes no secretion

¶Still possible that S = R, but this is unlikely.

▶ Clearance and Its Applications

Definition	Volume of plasma from which a substance is removed (cleared) per unit time
Concept	Relates the excretion of a substance to its concentration in plasma
Calculation	$C_s = (U_s \times \dot{V})/P_s$ where, C_s = clearance of substance, U_s = urine concentration of substance, $\dot{V}$ = urine flow, P_s = plasma concentration of substance
Application: GFR	• **Inulin clearance** can be used to calculate **GFR**. • Rationale: inulin is filtered, but is neither secreted nor reabsorbed. Therefore, Clearance of inulin = $(U_{[inulin]} \times \dot{V})/P_{[inulin]}$ = GFR • **Creatinine clearance** is the best clinical measure of GFR because it is produced continually by the body and is freely filtered but not reabsorbed. Creatinine is partially secreted, but its clearance is still a reasonable clinical estimate of GFR.
Application: **Renal plasma flow (RPF)** **Renal blood flow (RBF)**	**Para-aminohippuric acid (PAH)** is filtered and secreted; at low doses it is almost completely cleared from blood flowing through the kidneys during a single pass. Therefore, PAH clearance = RPF. However, PAH clearance is a measure of *plasma* flow rather than *blood* flow. Renal blood flow (RBF) can be calculated as follows: **RBF = RPF / (1 – hematocrit)**, using hematocrit as decimal proportion, e.g., 0.40
Application: **Free water clearance (C_{H_2O})**	• C_{H_2O} is not a true clearance. It is the volume of water that would have to be added to or removed from urine to make the urine isoosmotic to plasma. • C_{H_2O} is the difference between water excretion (urine flow) and osmolar clearance. Thus: $C_{H_2O} = \dot{V} - C_{osm} = \dot{V} - (U_{osm} \times \dot{V})/P_{osm}$ • **Positive C_{H_2O}:** Urine is hypotonic, ADH is low, "free water" has been removed from body, and plasma osmolarity is being increased. • **Negative C_{H_2O}:** Urine is hypertonic, ADH is high, water is being conserved, and plasma osmolarity is being decreased. • Use C_{H_2O} to evaluate whether urine osmolarity is appropriate. If P_{osm} is elevated, C_{H_2O} should be negative to compensate; if P_{osm} is low, C_{H_2O} should be positive.

Definition of abbreviation: ADH, antidiuretic hormone.

▶ Structure of the Nephron

(Approximately 1 million nephrons/kidney)

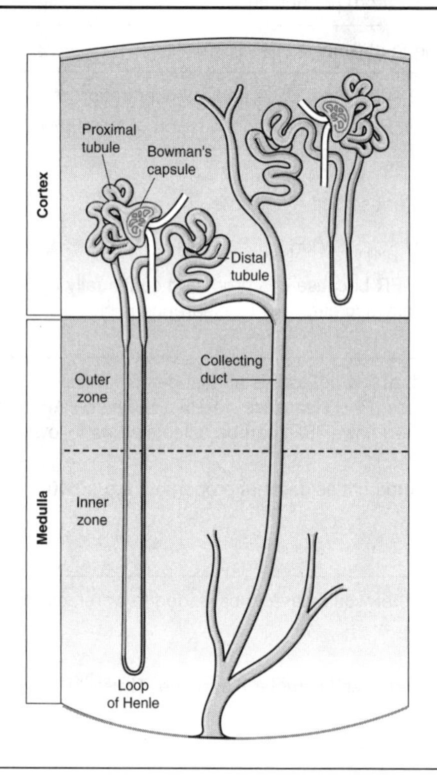

Sequence of Fluid Flow in a Nephron	
Structure	**Basic Function**
Bowman capsule	Formation of filtrate
Proximal tubule (PT)	Reabsorption of H_2O, solutes; some secretion
Descending thin loop of Henle (DTL)	Reabsorption of H_2O, no solute transport
Ascending thin loop of Henle (ATL)	Reabsorption of solutes, not water
Thick ascending loop of Henle (TAL)	Reabsorption of solutes, not water
Early distal tubule (EDT)	Reabsorption of solutes; special sensory region
Late distal tubule (LDT); cortical collecting duct (CCD)	Reabsorption of solutes and water; regulation of acid–base status
Medullary collecting duct (MCD)	Reabsorption of H_2O; final control of urine volume and osmolarity

▶ Association of Blood Vessels With the Nephron

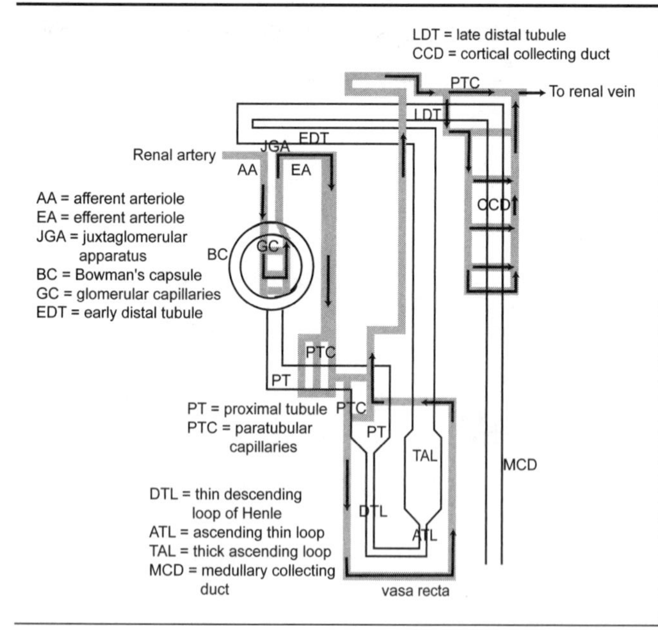

Each tubular segment has a blood supply that allows exchange of water and/or solute.

Sequence of Blood Flow in Nephrons	
Blood Vessel	**Tubular Segment**
Afferent arteriole	No exchange
Glomerular capillaries	Bowman capsule
Efferent arteriole	No exchange
Peritubular capillaries	Sequence: proximal tubule → loop of Henle → distal tubule → collecting duct
Vasa recta: specialized portion of peritubular capillaries that perfuse medulla 1–2% of blood flow	Loop of Henle, collecting duct

▶ The Juxtaglomerular Apparatus (JGA)

The **juxtaglomerular apparatus** includes the site of **filtration** (Bowman capsule [BC] and glomerular capillaries), as well as the arterioles and macula densa tubule cells found at the end of the thick ascending limb. Macula densa cells sense tubular fluid NaCl and contribute to **autoregulation** and regulation of **renin secretion**.

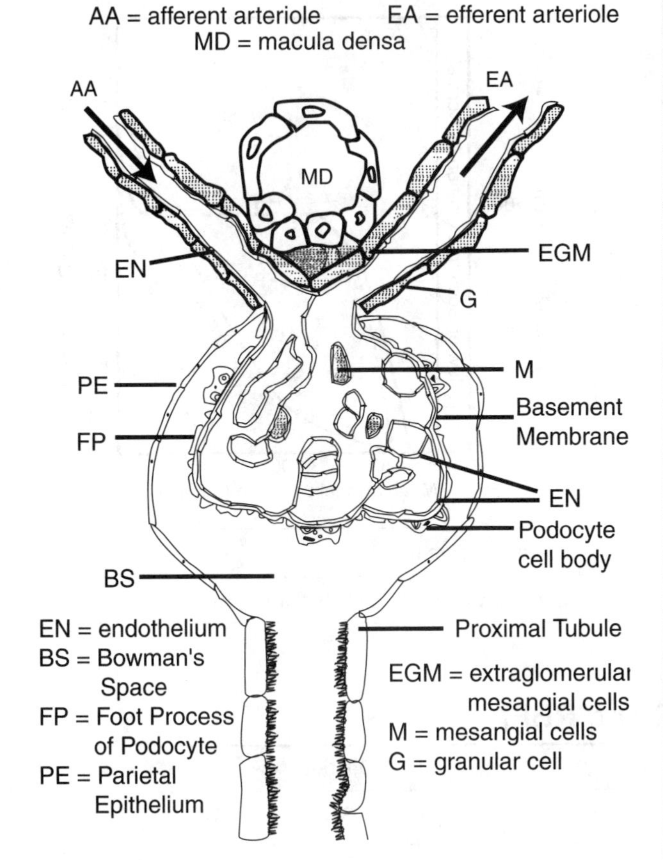

Filtration	
Bowman capsule	A pouch wrapped around capillaries
Podocytes	Specialized cells of BC, which help prevent protein filtration
Glomerular capillaries (GC)	Fenestrated, which allows rapid filtration of water and small solutes, but prevents protein filtration
Filtration = GFR = $K_f[(P_{GC} - P_{BC}) - (\varpi_{GC} - \varpi_{BC})]$ (same forces as discussed in CV)	
K_f	Filtration coefficient is high for H_2O, electrolytes, small solutes
P_{GC}	Remains high along GC due to very low GC resistance, favoring filtration
P_{BC}	Remains low unless urinary tract obstruction is present
π_{GC}	Rises along GC due to high filtration, until filtration ceases in distal capillary; reduced with low plasma protein
π_{BC}	≈ zero normally, proteinuria increases it
Filtration Fraction = GFR/RPF	
Ranges from 15–30%, usually ≈20% • Nonrenal capillary filtration fraction ≈1–2% • Nonrenal P_{cap} falls along capillary due to capillary resistance, so filtration only occurs at proximal end and reabsorption occurs distally.	

▶ Autoregulation of GFR and RBF

An isolated change in renal perfusion pressure between 75 and 175 mm Hg will not change RBF or GFR significantly because **autoregulation** maintains RBF at a constant level by adjusting afferent and efferent arteriolar resistances. Autoregulation occurs via a **myogenic response** and **tubuloglomerular feedback**.

Mechanisms of Autoregulation	
Myogenic	↑ arterial pressure stretches vessel wall, which ↑ calcium movement into smooth muscle cells, causing them to contract.
Tubuloglomerular feedback (TGF)	↓ arterial pressure causes GFR to ↓, which in turn ↓ delivery of NaCl to macula densa. This results in an ↑ in efferent arteriolar resistance and a ↓ in afferent arteriolar resistance, both of which ↑ GFR to a normal level. The ↑ in efferent arteriolar resistance occurs in response to ↑ levels of angiotensin II.

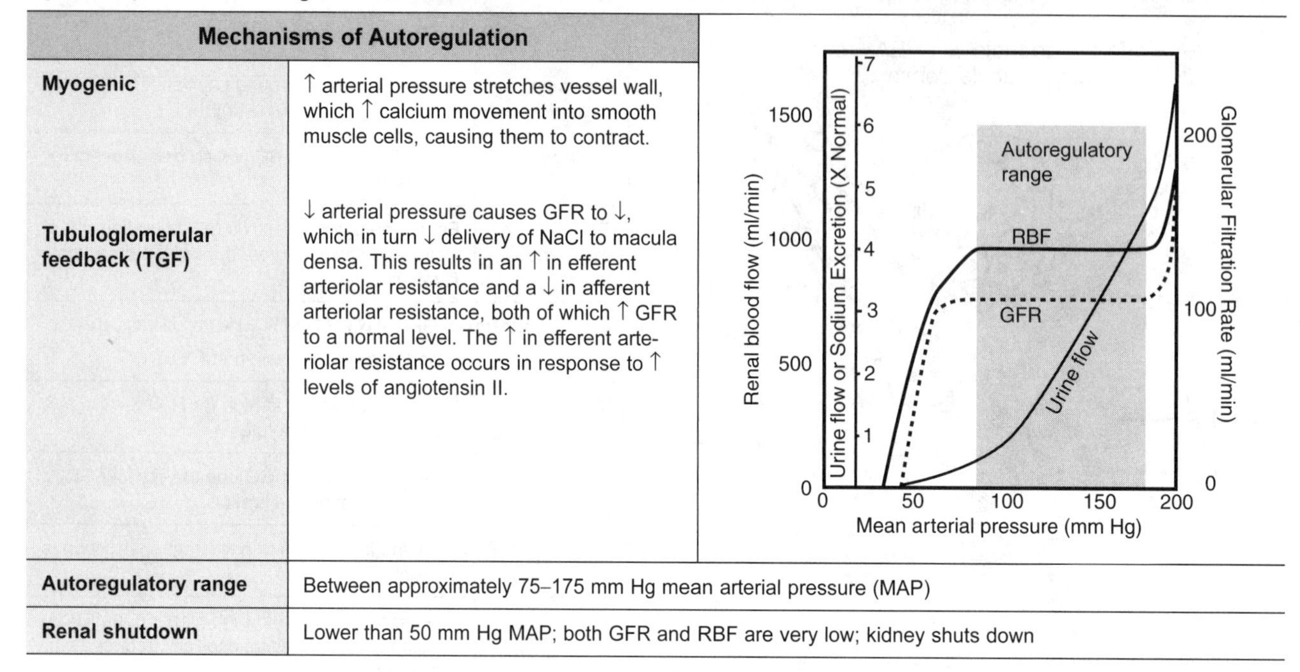

Autoregulatory range	Between approximately 75–175 mm Hg mean arterial pressure (MAP)
Renal shutdown	Lower than 50 mm Hg MAP; both GFR and RBF are very low; kidney shuts down

▶ Regulation of Filtration (GFR and RBF)

Vessel	Constriction			Dilation		
	P_{cap}	GFR	RBF	P_{cap}	GFR	RBF
Afferent arteriole	↓	↓	↓	↑	↑	↑
Efferent arteriole	↑	↑	↓	↓	↓	↑

▶ Tubular Function: Reabsorption and Secretion

Active Facilitated Transporters Display Maximum Transport (T_{max}) Rates	
Transport maximum	T_{max} = mass of solute per time transported *when carriers saturated*
Limited carrier population	Transport mediated by these carriers is saturable under pathophysiologic conditions, e.g., glucose transporters
Reabsorption < T_{max}	None of solute in urine until filtered load > T_{max}
Secretion < T_{max}	All of solute delivered in plasma appears in urine until delivery > T_{max}
Interpretation	Reduced T_{max} for glucose indicates reduced number of functioning nephrons

Estimation of T_{max} by Graphical Interpretation	
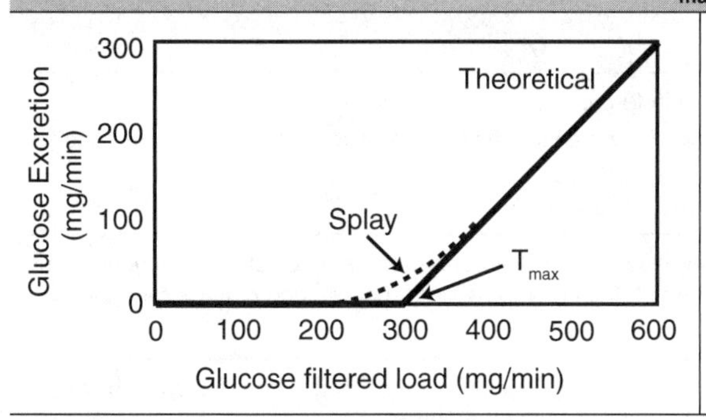	The *solid line* represents the theoretical relationship between glucose filtered load (FL) and glucose excretion (E). Actually, some glucose appears in the urine below the T_{max} due to competition for binding sites. When the filtered load > T_{max}, E = FL – T_{max}.
Select a point on linear portion: Reabsorption = excretion – filtered load T_{max} = 300 mg/min reabsorption	$FL_{glucose}$ = 500 mg/min; $E_{glucose}$ = 200 mg/min $R_{glucose}$ = 200 mg/min – 500 mg/min = –300 mg/min Therefore, $R_{glucose}$ = T_{max}, and carriers are saturated

Proximal Tubule

First Half of the Proximal Tubule

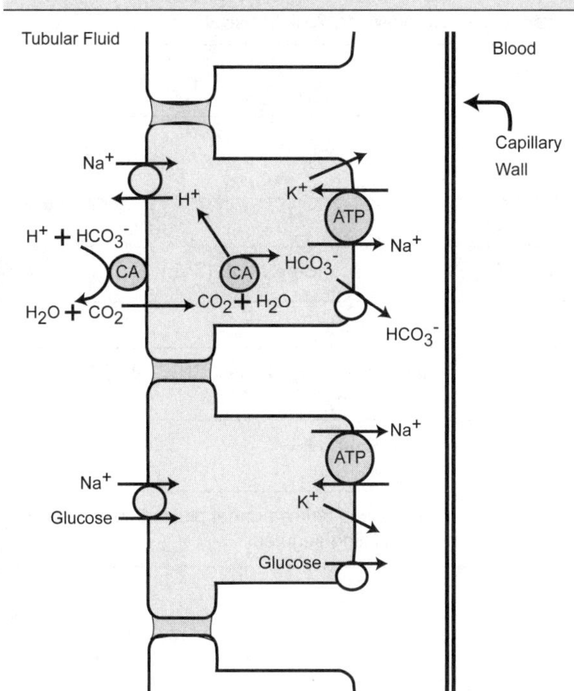

1) $NaHCO_3^-$ reabsorption is depicted at the *top of the diagram.*
2) The Na^+-glucose symporter is depicted in the *bottom cell*; other symporters reabsorb amino acids, lactate, and phosphate.

Second Half of the Proximal Tubule

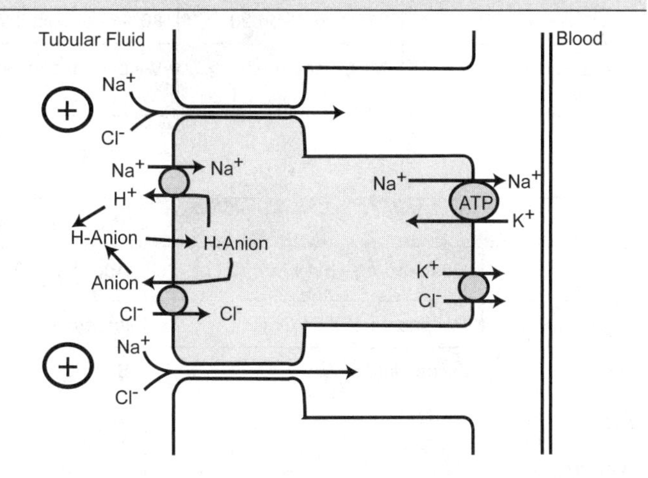

The processes in the left figure continue on in the second half of the proximal tubule and others are added:

3) Secondary active NaCl reabsorption is mediated by the parallel operation of the Na^+/H^+ antiport and Cl^-/anion antiport (e.g., OH^-, formate, oxalate, sulfate). H^+ plus anion associate into the neutral acid in the tubular fluid and diffuse passively into the cell, where they dissociate again to move through the antiports.
4) Water follows non-Cl^- solute reabsorption in the first half of the PT and increases tubular fluid [Cl^-]. Passive Cl^- diffusion pulls Na^+ electrically after it, but creates a small negative electrical potential in the tubular lumen.

(Continued)

Proximal Tubule *(Cont'd.)*		
Substance	**Action**	**Mechanism**
Water	67% reabsorbed	• Simple diffusion; filtration forces favor reabsorption • Solute reabsorption creates osmotic gradient, water follows it; paracellular and transcellular • Cell membranes permeable to water
Na^+	67% reabsorbed	• Na^+/K^+-ATPase in the basolateral membrane creates an electrochemical gradient; secondary active transport, luminal (apical) surface; co- and countertransport • Simple diffusion, paracellular
Glucose, lactate, amino acids, phosphate	≈100% reabsorbed	• Secondary active transport at the luminal (apical) surface; cotransport with Na^+; Na^+ electrochemical gradient provides the power • Transport maximum processes (*see* below)
H^+	Secreted	Primary ATPase and secondary active antiport with Na^+
HCO_3^-	≈100% reabsorbed	Reabsorbed as CO_2 at the apical surface; facilitated diffusion and exchange with Cl^- on basolateral surface; carbonic anhydrase located at the apical surface and intracellularly
Cl^-	≈67% reabsorbed	• Transcellular: at the apical surface, exchange with anions; at the basolateral surface, facilitated diffusion and symport with K^+ • Paracellular, simple diffusion
Ca^{2+}	≈70% reabsorbed	80% of proximal tubule reabsorption paracellular; 20% transcellular
Organic cations, anions	Mainly secreted	Various mechanisms

Loop of Henle: Overview			
	Descending Thin Limb (DTL)	**Ascending Thin Limb (ATL)**	**Thick Ascending Limb (TAL)**
H_2O	Reabsorbs 15% of GFR	Impermeable	Impermeable
Solutes	Diffuse into tubule	Slight active reabsorption	Reabsorbs 25% FL of NaCl, K^+; 20% FL of Ca^{2+}; 50–60% FL of Mg^{2+}
Tubular fluid volume	Decreases	No change	No change
Tubular fluid osmolarity	Increases; DTL is called the "concentrating segment"	Decreases	Decreases below normal plasma; TAL is called the "diluting segment"

Definition of abbreviation: FL, filtered load.

Thick Ascending Loop of Henle

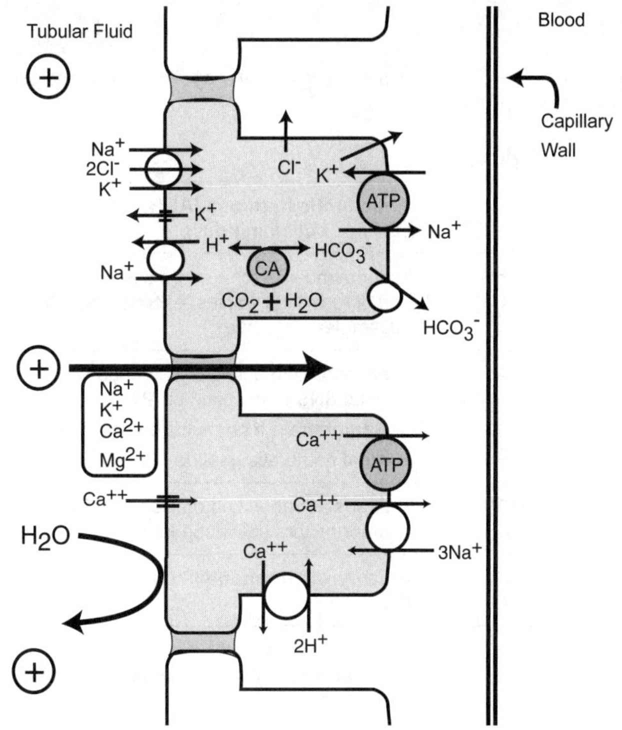

Mechanisms of Solute Transport, Thick Ascending Loop of Henle	
Na^+	**Reabsorption:** • Symport with Cl^- and K^+ • Antiport with H^+ • Paracellular due to electrical force
K^+	**Reabsorption:** • Paracellular reabsorption (electrical force) • Symport with Na^+ and Cl^-
Ca^{2+}	**Reabsorption:** • Paracellular (electrical force) • Transcellular, apical surface, simple diffusion • Basolateral Ca^{2+}-ATPase, Na^+-Ca^{2+} exchange • $2H^+/Ca^{2+}$ ATPase antiport; parathyroid hormone stimulates
Mg^{2+}	**Reabsorption:** • Paracellular (electrical force) • Transcellular, active transport
H^+	**Secretion:** • Na^+/H^+ exchange, NH_4^+

"Diluting segment": reabsorption of solutes without reabsorption of water produces hyperosmotic interstitium of medulla; needed for descending loop water reabsorption and collecting duct ability to regulate water reabsorption

Distal Tubule and Collecting Duct Overview

	Tubular Fluid Volume	Tubular Fluid Osmolarity	Tubular Fluid Solutes
Delivered from TAL	≈15% of GFR	Hypotonic at ≈1/2 P_{OSM}	• ≈10% FL of NaCl, K^+, and Ca^{2+} • ≈5% FL of HCO_3^-
Early distal convoluted tubule (EDT)	No H_2O reabsorbed	Hypotonic at ≈**1/3 P_{OSM}**	• Reabsorb ≈5% FL of NaCl via **apical Na^+-Cl^- symporter** (NCC) in secondary active transport inhibited by **thiazide** diuretics • Reabsorb ≈10% FL of Ca^{2+} via secondary active transport stimulated by **parathyroid hormone (PTH)** and indirectly by **thiazide** diuretics
Late distal tubule (LTD) and cortical collecting duct (CCD)	H_2O reabsorption regulated by **antidiuretic hormone (ADH)**	Varies primarily with H_2O reabsorption	• Reabsorb ≈4% of FL of NaCl via active transport regulated by **aldosterone** • Reabsorb ≈5% FL of HCO_3^- via active transport regulated by **aldosterone** • Secretion of K^+ determines total excretion; active secretion regulated by **aldosterone**
Medullary collecting duct (MCD)	H_2O reabsorption regulated by **ADH**	Varies primarily with H_2O reabsorption	Reabsorb ≈60% of FL of urea passively via transporter regulated by **ADH**; urea reabsorption is needed for medullary osmolarity

Definition of abbreviation: FL, filtered load.
*Some refer to the LDT as the connecting tubule.

► Renal Regulation of Excretion of Water and Solutes

Overview				
Steady-State Balance (maintained by renal regulation of excretion)		**Fractional E = (Excreted)/(Filtered)** Range	Typical	**Major Regulators of Renal Excretion**
Water	Volume liquid drunk/day = E_{H_2O}	0.5–14%	≈1%	**Antidiuretic hormone (ADH)** acting on LDT and **collecting duct**
Solute	Osmols generated as water soluble wastes = $E_{osmoles}$		≈1.5%	Major osmoles excreted are salts, acids, and nitrogenous wastes; excrete ≈600–900 mOsm/day
NaCl	Amount NaCl eaten/day = E_{NaCl}	0.1–5%	≈0.5–1%	• **Aldosterone** acting on LDT and **CCD**; renal **SNS** tone acting on **PT** • **Angiotensin II** on multiple segments • Atrial natriuretic peptide, etc.
K^+	Amount K^+ eaten/day ≈E_{K^+}	1–80%	≈15%	• **Aldosterone** acting on LDT and **CCD**; other factors described later
Ca^{2+}	Amount Ca^{2+} absorbed from GI/day = $E_{Ca^{2+}}$	0.1–3%	≈1%	**Parathyroid hormone** (PTH) acting on **DCT**
HCO_3^-	• *New* HCO_3^- added to blood/day = net acid excretion (NAE) = ($E_{ammonium}$)+ ($E_{H_2PO_4^-}$) – ($E_{bicarbonate}$) • NAE determines $[HCO_3^-]_{plasma}$	HCO_3^- $H_2PO_4^-$	≈0% ≈20%	**Aldosterone** acting on LDT and **CCD**; plasma pH and other factors described later
Urea	Amount generated/day = E_{urea}	20% – 80%	≈40%	Synthesis depends on protein metabolism; regulated by **ADH** acting on medullary **collecting duct**

Definition of abbreviations: CCD, cortical collecting duct.

Regulation of Urine Osmolarity and Urine Flow: Antidiuretic Hormone (ADH)	
Collecting Duct	**Effect of Plasma ADH on Urine Osmolarity, Urine Flow, and Total Solute Excretion**

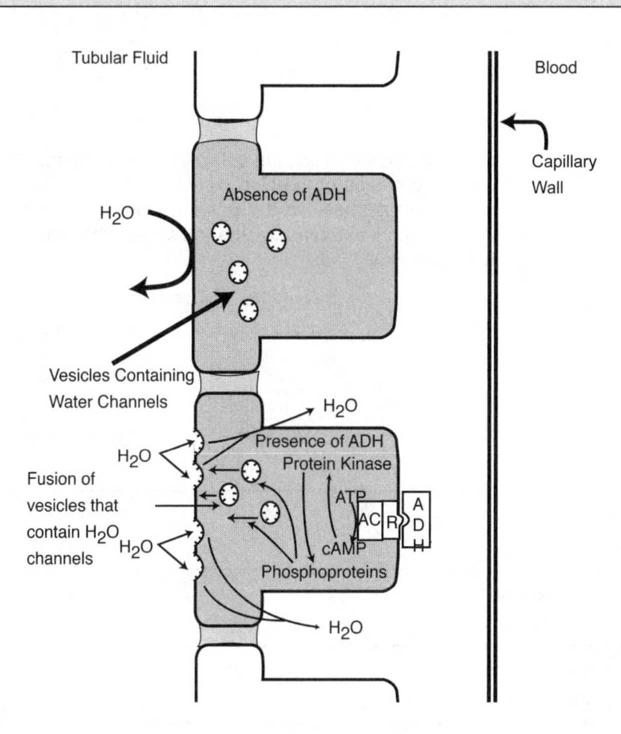

The ADH-receptor complex on the basolateral membrane activates adenylate cyclase. cAMP activates a kinase that phosphorylates proteins involved in movement of vesicles and fusion with the membrane. The resulting channels allow diffusion of water across the apical membrane.

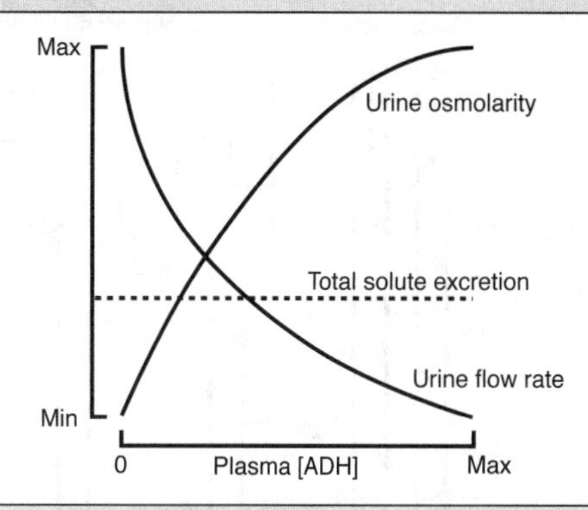

Urine Osmolarity Versus Urine Flow

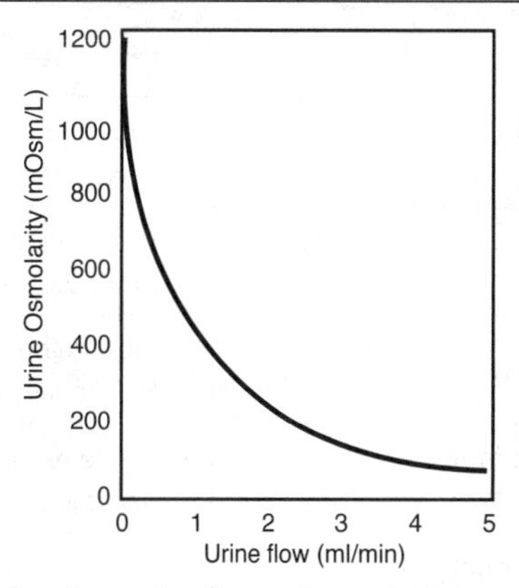

In the normal system, urine flow and osmolarity have an inverse relationship.

Loop of Henle and the Countercurrent Mechanism Antidiuresis (Presence of ADH)

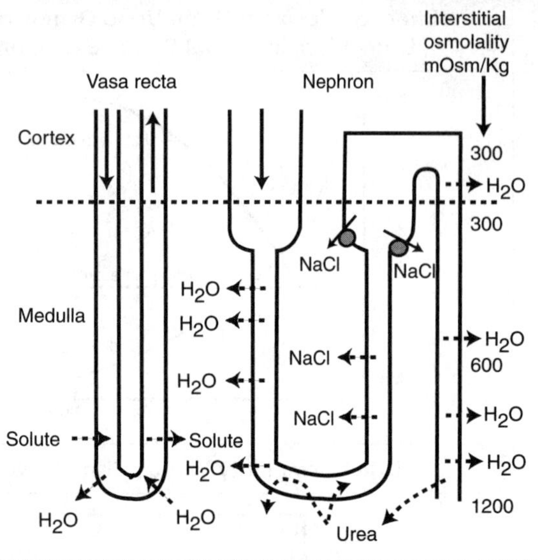

In the presence of ADH, the collecting duct is permeable to water. Because of the high osmolarity of the medulla, water is reabsorbed, so the urine volume is small and the urine concentration is the same as the medulla (hyperosmotic).

Vasa recta	• Only 1–2% of renal blood flow • Reabsorbed solutes "trapped" in medulla
Balance of solutes and water	• In total loop, more solute than H_2O reabsorbed • Produces hyperosmotic interstitium of medulla
Descending thin loop	• Fluid enters isoosmotic with normal plasma • H_2O reabsorption → ↓volume with ↑ osmolarity
Ascending loop	• Solute reabsorption only, not H_2O → hypoosmotic tubular fluid • Accumulation of solutes in interstitium of medulla
Collecting duct	• With ADH, collecting duct permeable to H_2O and urea • As duct passes through hyperosmotic interstitium, reabsorption of H_2O → small volume of concentrated urine • Negative CH_2O; dilutes plasma

Loop of Henle and the Countercurrent Mechanism Water-Induced Diuresis (Absence of ADH)	
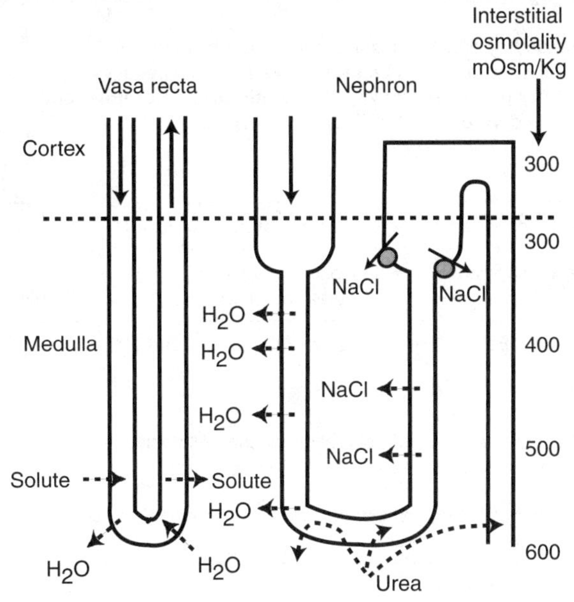	In the absence of ADH, the collecting duct is impermeable to water, so urine flow is high and the urine is dilute. Note that medullary osmolality is also lower than when ADH is present.

Vasa recta	• ↑ blood flow with lower-than-normal osmolarity • Washes out solutes, medullary osmolarity reduced
Balance of solutes and water	Medullary osmolarity still higher than normal plasma, but not as high as during antidiuresis
Descending thin loop	Less H_2O is reabsorbed because the medulla is not as concentrated as during antidiuresis
Ascending loop	Solute reabsorption continues, but a significant portion of solutes that are reabsorbed are washed into the vasa recta
Collecting duct	• Without ADH, the collecting duct is impermeable to H_2O and urea • No H_2O reabsorption because tubular fluid passes through medulla • ↑ urine flow,↓ urine osmolarity • Positive CH_2O; acts to concentrate plasma

Regulation of Plasma Osmolarity by ADH

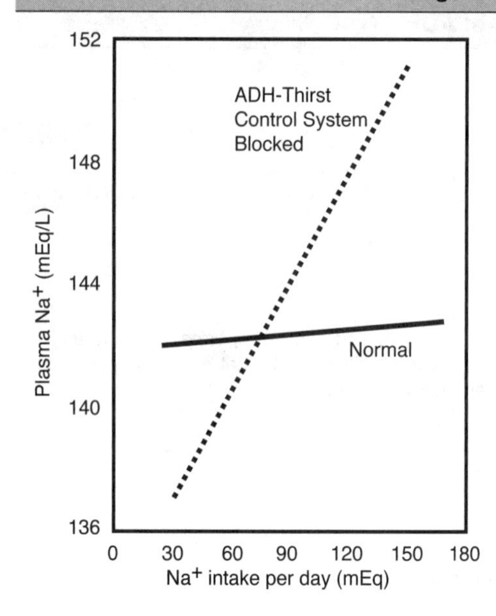

Normal function of the control system for ADH secretion and water consumption prevents large changes of plasma sodium concentration. Loss of this system causes plasma sodium concentration to increase in proportion to sodium intake.

ADH secretion is increased by elevated plasma sodium or osmolarity and decreased by high blood volume or high pressure. This acts as a negative feedback system to control plasma osmolarity. (See **Antidiuretic Hormone and Control of Osmolarity and Volume** in the Endocrine section.)

Calculation of Plasma Osmolality (mOsm/kg solution)
Plasma osmolality = 2 (plasma [Na^+]) + [glucose mg/dL]/18 + [urea mg/dL]/2.8
Example: sodium = 145 mEq/L, glucose = 180 mg/dL, urea = 28 mg/dL Plasma osmolality = 290 + 10 + 10 = 310 mOsm/kg However: plasma Na^+ dominates control of ADH because of its osmotic effect; urea and glucose usually irrelevant. *Note:* glucose may be important part of **urine** osmolality (especially in diabetes); when present, it causes osmotic diuresis.

Control Signals for Na^+ and H_2O Excretion		
System	**Action**	**Segment/Site**
Renal sympathetic nerves	↓ GFR	Afferent arteriole constriction
	↑ NaCl reabsorption	Proximal, TAL, DT, CD → ↑ H_2O reabsorption (except TAL, due to impermeability of water)
Renin–angiotensin II–aldosterone	Angio II → ↑ NaCl reabsorption	Proximal → ↑ H_2O reabsorption
	Aldosterone →↑ NaCl reabsorption	TAL, DT, CD →↑ H_2O reabsorption (except TAL, due to impermeability of water)
Atrial natriuretic peptide (ANP)	↑ GFR	Glomerulus
	↓ renin, angio II, aldosterone	JGA and adrenal cortex
	↓ NaCl and H_2O reabsorption	CD (urodilatin* assists)
	↓ ADH secretion and actions	Posterior pituitary and CD
ADH	↑ permeability to H_2O	CD → ↑ H_2O reabsorption → ↓ urine flow and ↑ urine osmolarity

*Urodilatin is a peptide produced by DT and CD when blood pressure/volume increase. Very potent inhibition of NaCl and water reabsorption, but only local action; it does not circulate.

Disorders of Solute and Water Regulation						
Disorder	**ECF Volume**	**$[Na^+]_{plasma}$**	**Blood Pressure**	**Urine Volume and Osmolarity**	**Arterial pH**	**$[K^+]_{plasma}$**
Diabetes insipidus*	↓	↑	↔ or ↓	↑ volume ↓ osmolarity	↑	↓ (↑ aldosterone and alkalosis)
SIADH	↑	↓	↑ or ↔	↓ volume ↑ osmolarity	↓	↑ (but negative balance)
Aldosterone deficiency (primary)	↓	↓	↓	↑ volume ↑ osmolarity	↓	↑ positive balance
Aldosterone excess (primary)	↑	↑	↑	↓ volume ↓ osmolarity	↑	↓ negative balance
Polydipsia	↑	↓	↔	↑ volume ↓ osmolarity	↓	↑ balance variable
Water deprivation (dehydration)	↓	↑	↓	↓ volume ↑ osmolarity	↑	↑, ↓, ↔ depends on multiple factors

Definition of abbreviation: SIADH, syndrome of inappropriate (excessive) ADH secretion.

*ADH deficiency is called primary, central, or neurogenic; ↓ renal response to ADH is nephrogenic. Distinguish by response to administration of ADH; ↑ urine osmolarity in response to ADH injection indicates primary.

Osmotic Diuresis	
Cause	**Effect**
Excessive solute in tubular fluid	Decrease reabsorption of H_2O
Diabetic ketoacidosis	• Glucose and ketones in urine → polyuria, K^+ wasting • Na^+ loss → hyponatremia
Starvation and alcoholic ketoacidosis	Ketonuria → polyuria, K^+ wasting, hyponatremia
Osmotic diuretics (mannitol, carbonic anhydrase inhibitors)	Inhibit proximal tubule H_2O reabsorption → diuresis, K^+ wasting (due to ↑ flow)
Loop diuretics (furosemide) and distal tubule diuretics (thiazides)	Inhibit NaCl reabsorption →↑ tubular solutes → diuresis and K^+ wasting

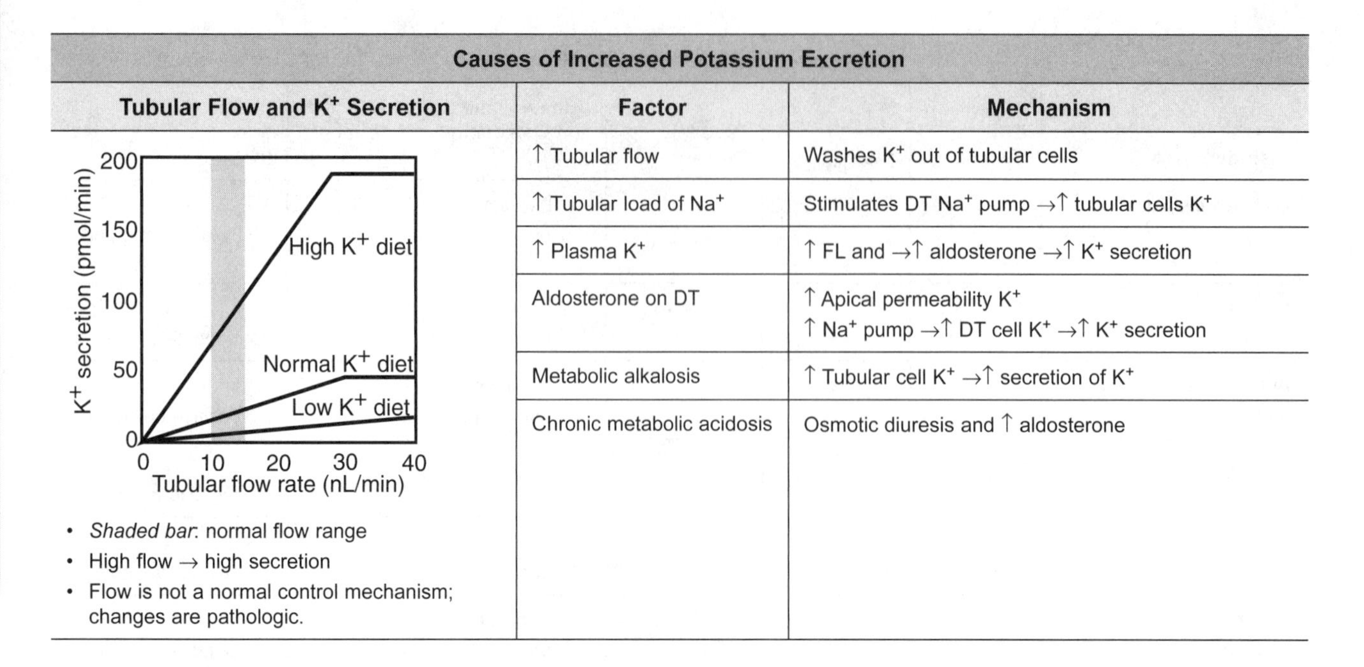

Causes of Increased Potassium Excretion

Tubular Flow and K^+ Secretion	Factor	Mechanism
(graph)	↑ Tubular flow	Washes K^+ out of tubular cells
	↑ Tubular load of Na^+	Stimulates DT Na^+ pump →↑ tubular cells K^+
	↑ Plasma K^+	↑ FL and →↑ aldosterone →↑ K^+ secretion
	Aldosterone on DT	↑ Apical permeability K^+ ↑ Na^+ pump →↑ DT cell K^+ →↑ K^+ secretion
	Metabolic alkalosis	↑ Tubular cell K^+ →↑ secretion of K^+
• *Shaded bar*: normal flow range • High flow → high secretion • Flow is not a normal control mechanism; changes are pathologic.	Chronic metabolic acidosis	Osmotic diuresis and ↑ aldosterone

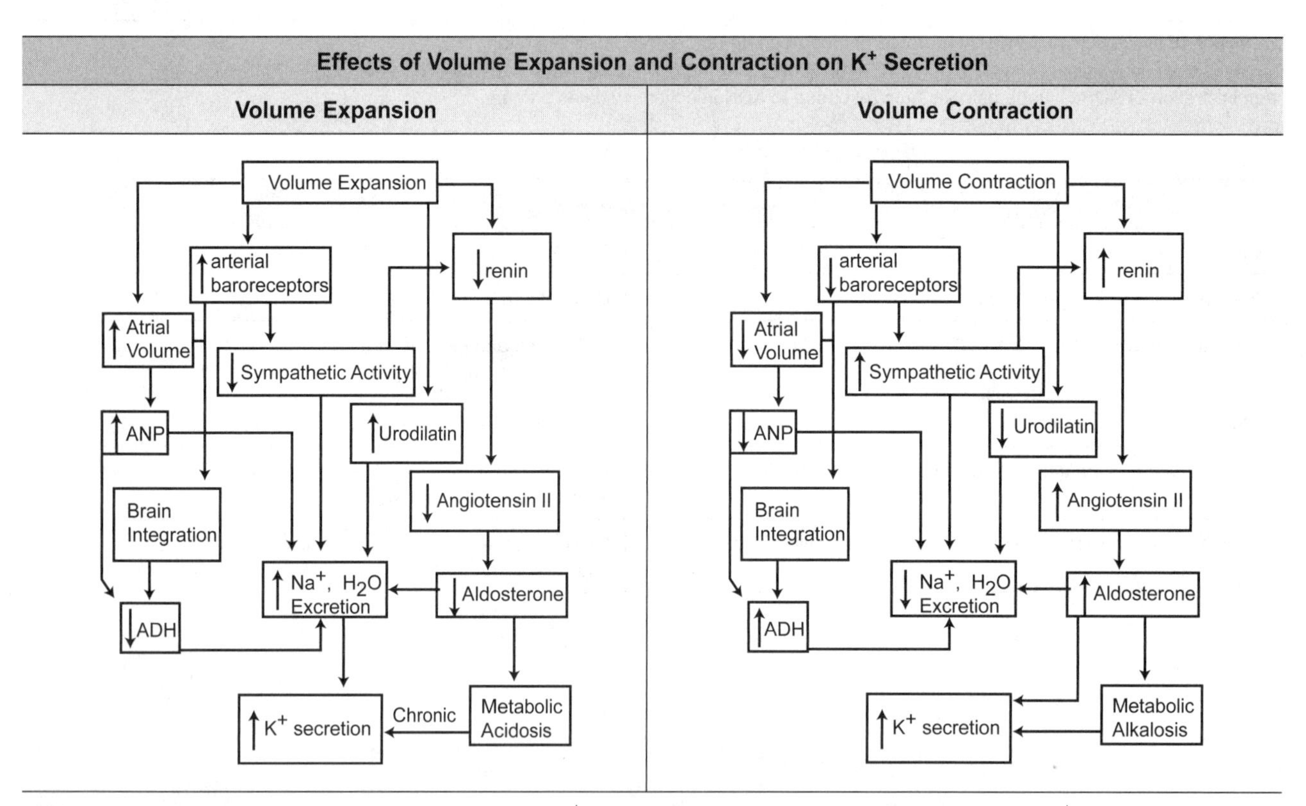

Volume expansion causes acidosis due to responses that ↓ reabsorption of bicarbonate in proximal tubule and ↓ aldosterone. Volume contraction causes alkalosis due to ↑ proximal tubular reabsorption of bicarbonate and ↑ aldosterone. Both can cause K^+ wasting because metabolic alkalosis and chronic metabolic acidosis both ↑ K^+ secretion in distal tubule and collecting duct.

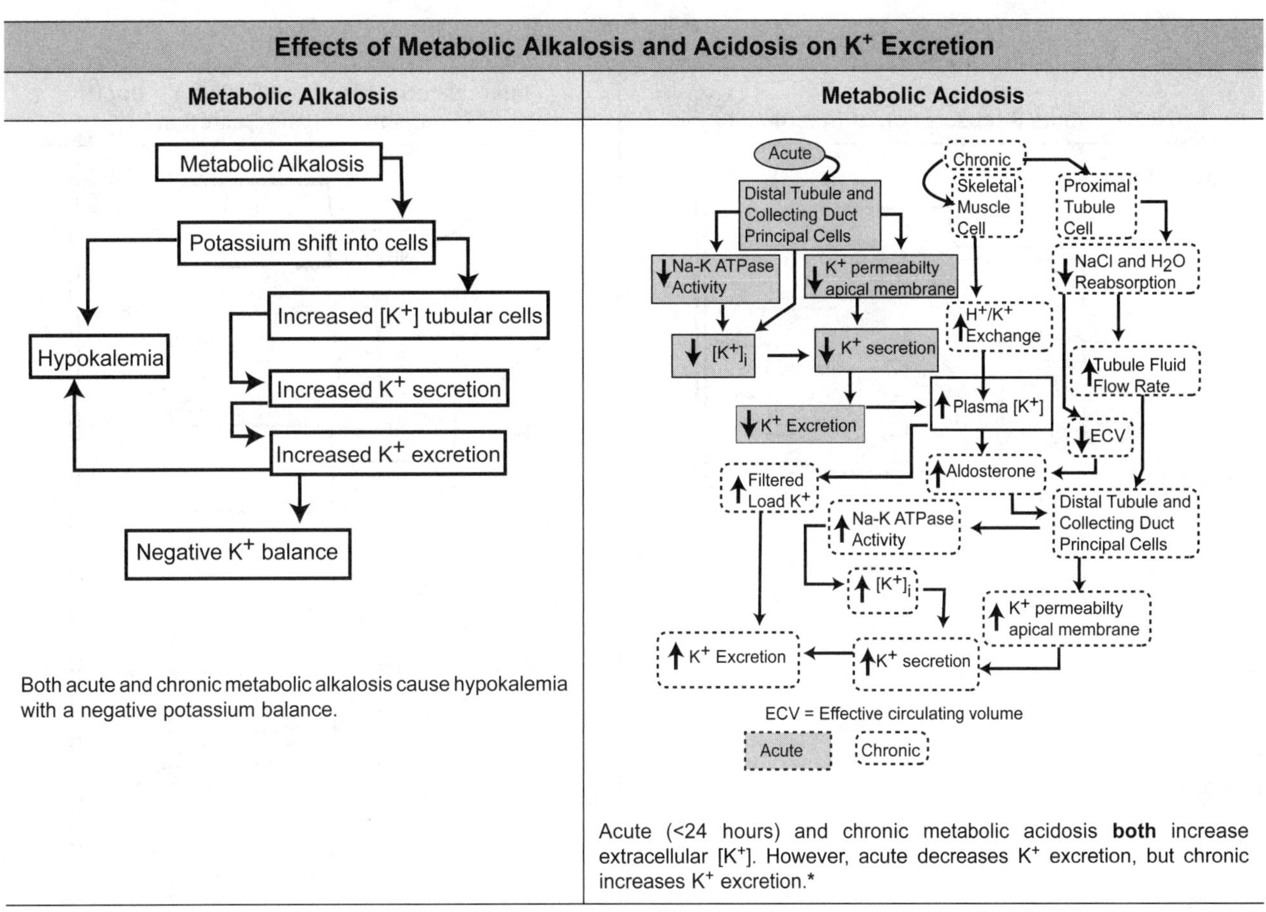

Both acute and chronic metabolic alkalosis cause hypokalemia with a negative potassium balance.

Acute (<24 hours) and chronic metabolic acidosis **both** increase extracellular [K^+]. However, acute decreases K^+ excretion, but chronic increases K^+ excretion.*

*Elevated aldosterone is the key to the reversal of the effect of acidosis on K^+ excretion. Also, increased cellular H^+ causes efflux of K^+ from cells; low cellular H^+ causes influx. Acidosis impairs metabolism; thus, ↓ solute reabsorption causes osmotic diuresis.

► Renal Mechanisms for Acid/Base Regulation

Fundamental Principles	
Tubular cell pH	• If a cell is acidotic, it will secrete acid. • If a cell is alkalotic, it will ↓ reabsorption of bicarbonate or secrete it.
Fluid volume	If ECV is decreased, ↑ Na^+ reabsorption will be accompanied by ↑ bicarbonate reabsorption.
Overall role of kidneys	This is the only significant mechanism for excretion of nonvolatile, metabolic acids.
Response time	Requires 24–72 hours for maximal compensatory response
Net acid excretion (NAE)	There is no net excretion of newly formed acid unless all bicarbonate is reabsorbed. $NAE = \dot{V} \times [NH_4^+] + \dot{V} \times [H_2PO_4^-] - \dot{V} \times [HCO_3^-]$
Free H^+ excretion	Trivial amount even at most acidic urine pH ≅ 4.4
NH_4^+ excretion	Largest quantity of excreted acid; metabolized from glutamine; "nontitratable acid"
$H_2PO_4^-$ excretion	Second largest quantity; "titratable acid"
NH_4^+ secretion	NH_4^+ secreted in the proximal tubule is reabsorbed in TAL and added to interstitium; it is then taken up by DT and CD cells and secreted.

Tubular Handling of Bicarbonate

Proximal Tubule Reabsorption of Bicarbonate Ion	Intercalated Cells of the Collecting Duct: Reabsorption and Secretion

Tubular Fluid
Blood
Na^+
$HCO_3^- + H^+$
ATP
CA
$H_2O + CO_2$
H^+
HCO_3^-
CA
$CO_2 + H_2O$
ATP
Na^+
K^+
Na^+
$3HCO_3^-$
HCO_3^-
Cl^-

Tubular Fluid
Blood
?
NH_4^+
K^+
ATP
$HCO_3^- + H^+$
ATP
CA
$H_2O + CO_2$
H^+
CA
$CO_2 + H_2O$
HCO_3^-
Cl^-
$2NH_4^+$
ATP
$3Na^+$
Tubular Fluid
Blood
HCO_3^- Secreting Cell
HCO_3^-
HCO_3^-
Cl^-
CA
$CO_2 + H_2O$
H^+
ATP
Cl^-

Tubular Handling of Ammonia

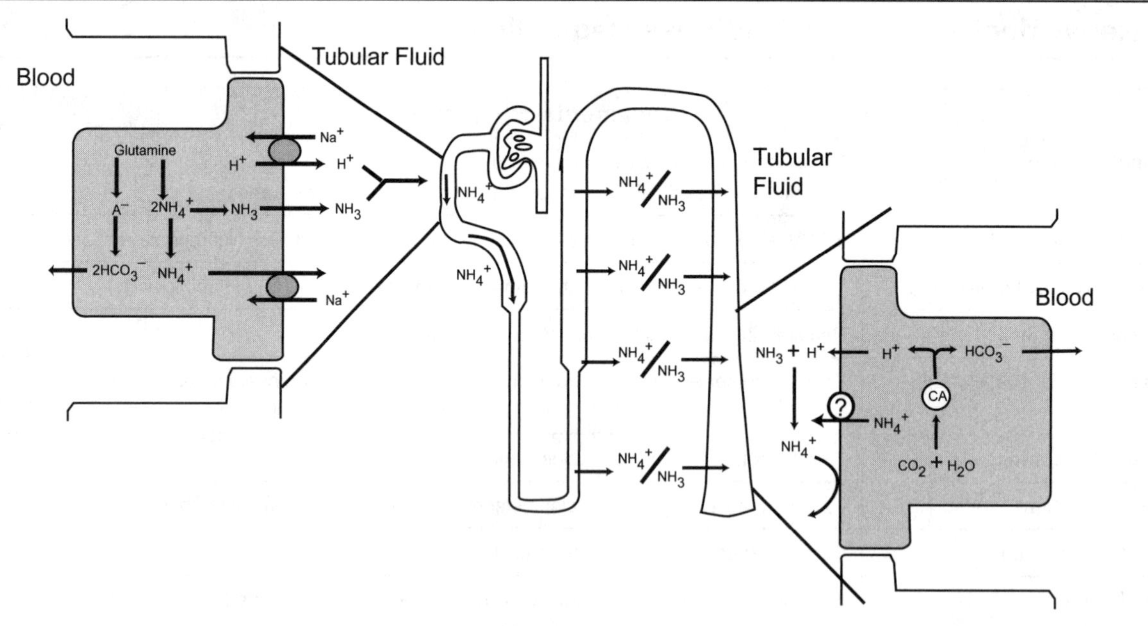

NH_4^+ can substitute for K^+ in the sodium pump. Apical secretion may be by substitution for H^+ in the H^+-K^+ exchanger or the H^+-ATPase. Production of NH_4^+ adds HCO_3^- to blood.

▶ Integrated Control of Cardiac Output and Arterial Blood Pressure

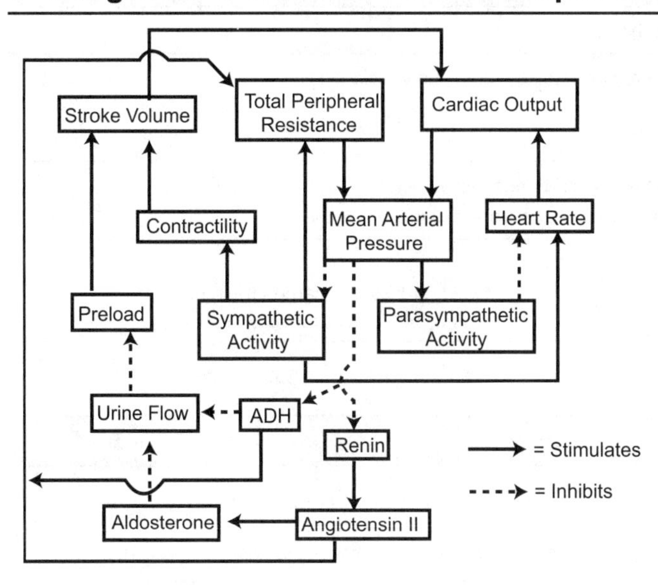

Notes on Use of Feedback Diagram
• Many intermediate steps omitted. • ADH (vasopressin) also vasoconstricts. • Diagram merges elements of cardiovascular and endocrine physiology. • For applications, make the most direct connection; e.g., hemorrhage: begin at reduced preload →↓ stroke volume, and so on. For an understanding of ACE inhibition, begin at ↓ angiotensin II.

▶ Acid/Base and Its Regulation

Fundamental Principles	
Carbonic anhydrase reaction	Produces strong acid and weak base → acidic solution
Respiratory contribution	Determined by arterial P_{CO_2}: ↑ P_{CO_2} → acidosis; ↓ P_{CO_2} → alkalosis
Metabolic contribution	Determined by arterial HCO_3^-: ↑ HCO_3^- → alkalosis ; ↓ HCO_3^- → acidosis
Simple disorder	One disorder with or without compensation
Mixed or combined disorder	Two simultaneous disorders
Compensation	• A response that tends to correct pH: compensation is never perfect. • Fully compensated means that the mechanism has come as close as possible to restoring normal pH.
Normal arterial blood gas	pH = 7.40; P_{CO_2} = 40 mm Hg; P_{O_2} = 83 – 100 mm Hg; HCO_3^- = 24 mEq/L
Henderson-Hasselbalch Eq.	Several forms; they show relationship between pH, P_{CO_2}, and HCO_3^-.

The Carbonic Anhydrase Reaction	Henderson-Hasselbalch Equation
$CO_2 + H_2O \overset{CA}{\rightleftarrows} H_2CO_3 \rightleftarrows H^+ + HCO_3^-$ CA, carbonic anhydrase	$pH = 6.1 \log \frac{[HCO_3^-]}{0.03\ P_{CO_2}}$ • ↑ P_{CO_2} → ↓ pH • ↑ HCO_3^- → ↑ pH

Examples of Calculations	
P_{CO_2} = 60 mm Hg; HCO_3^- = 18 mEq/L	pH = 6.1 + log (18/1.8) = 6.1 + log 10 = 6.1 + 1 = 7.1
P_{CO_2} = 40 mm Hg; HCO_3^- = 24 mEq/L	pH = 6.1 + log (24/1.2) = 6.1 + log 20 = 6.1 + 1.3 = 7.4
P_{CO_2} = 20 mm Hg; HCO_3^- = 24 mEq/L	pH = 6.1 + log (24/0.6) = 6.1 + log 40 = 6.10 + 1.60 = 7.70
P_{CO_2} = 60 mm Hg; HCO_3^- = 36 mEq/L	pH = 6.1 + log (36/1.8) = 6.1 + log 20 = 6.1 + 1.3 = 7.4

Note: pH is normal, but this is not normal acid/base status; both PCO_2 and HCO_3^- are abnormal.

Fundamental Mechanisms to Control pH	
Buffers	Fast, seconds: ECF bicarbonate (largest), phosphate, and proteins
Respiratory	Ventilation response fast, within a few minutes: change P_{CO_2} by changing ventilation. • ↑ ventilation →↓ P_{CO_2} → alkalosis • ↓ ventilation →↑ P_{CO_2} → acidosis
Renal	• Slow; 24–72 hours; powerful: major mechanism to excrete nonvolatile acids. • Change in renal acid excretion and HCO_3^- production is called a metabolic response.
Primary disorders	• Primary disorder ventilation = respiratory acid/base disorder • Primary disorder excess production of acid, ↓ renal excretion of acid, or loss of bicarbonate in urine or feces = metabolic disorder

The Davenport Diagram

- Above the base buffer line, there is excess metabolic base (a positive base excess).
- Below it, there is excess metabolic acid (negative base excess).
- Read pH perpendicular to the *x*-axis; HCO_3^- perpendicular to the *y*-axis, but P_{CO_2} along the curves.

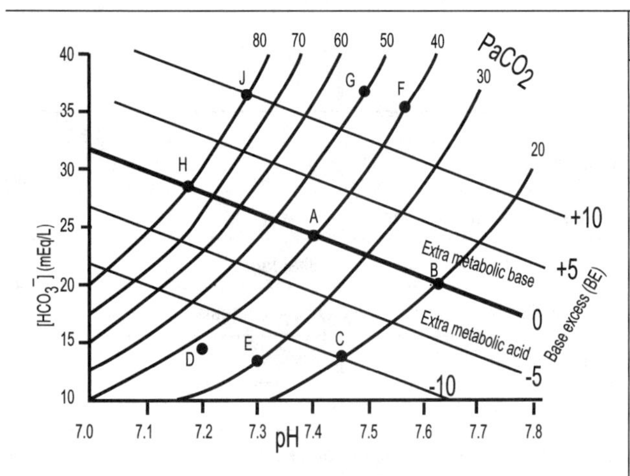

Point	Interpretation
A	Normal acid/base status
B	Uncompensated respiratory alkalosis
C	Respiratory alkalosis with compensatory metabolic acidosis
D	Uncompensated metabolic acidosis
E	Metabolic acidosis with compensatory respiratory alkalosis
F	Uncompensated metabolic alkalosis
G	Metabolic alkalosis with compensatory respiratory acidosis
H	Uncompensated respiratory acidosis
J	Respiratory acidosis with compensatory metabolic alkalosis

Patient Examples

Predicted bicarbonate concentration with changes of PCO_2

PCO_2 (mm Hg)	80	60	40	20
HCO_3^- (mEq/L)	29	27	24	21

Draw a vertical line through the PCO_2 to get the predicted HCO_3^-

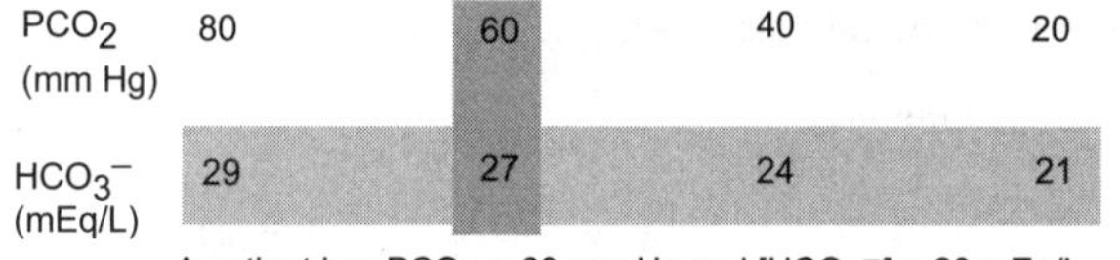

A patient has PCO_2 = 60 mm Hg and $[HCO_3^-]$ = 28 mEq/L
Has metabolism changed?
No: the bicarbonate is almost exactly as expected from the PCO_2

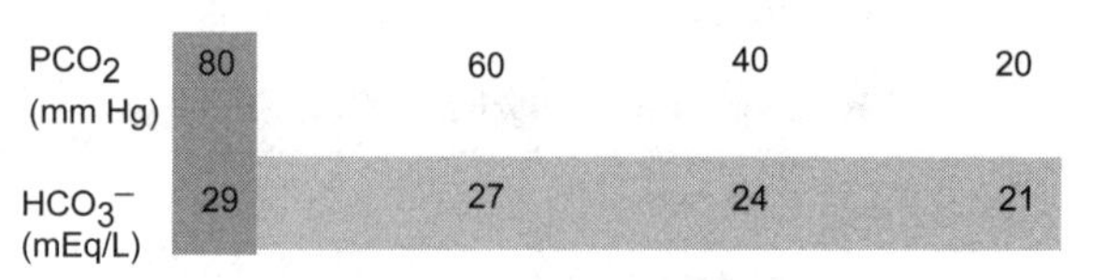

A patient has PCO_2 = 80 mm Hg and $[HCO_3^-]$ = 35 mEq/L
What is the predicted bicarbonate? 29 mEq/L
Is there more or less bicarbonate than predicted? More
Has metabolism changed? Yes, there is excess bicarbonate.
What is the metabolic contribution? Alkalosis

Simple Acid–Base Disorders			
Type of Disorder	**pH**	**$Paco_2$**	**HCO_3^-**
Metabolic acidosis	↓	↓*	↓
Metabolic alkalosis	↑	↑*	↑
Respiratory acidosis	↓	↑	↑*
Respiratory alkalosis	↑	↓	↓*

*Change due to compensation

A Step-by-Step Approach to Diagnosis of Simple Disorders

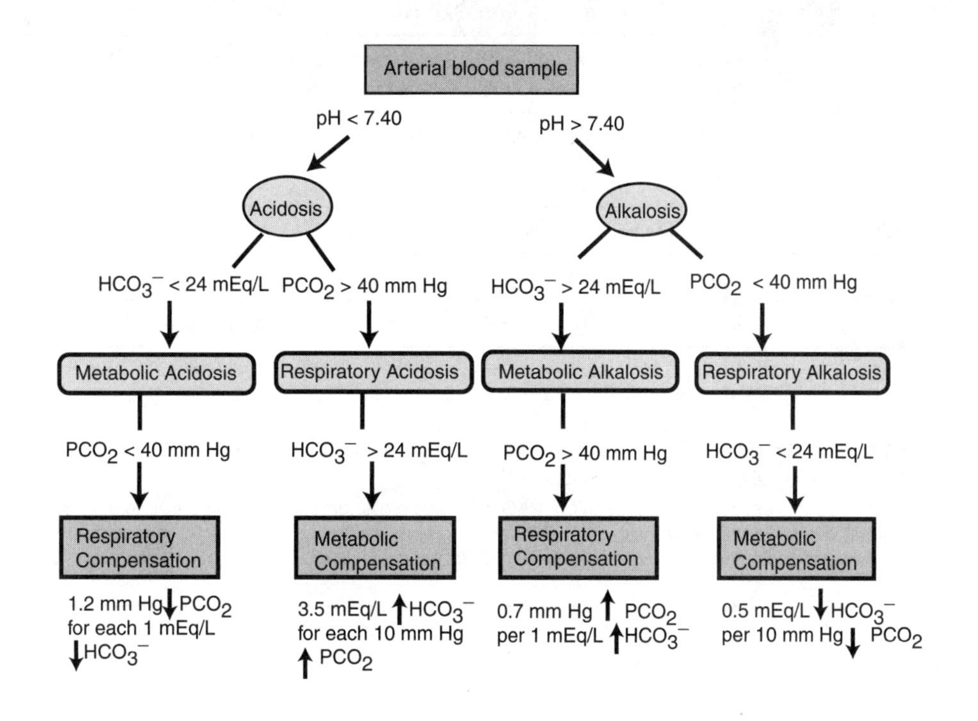

1. What is the pH? Acidotic or alkalotic?
2. What is the respiratory contribution? Acidosis, alkalosis, or no change?
3. What is the metabolic contribution? Acidosis, alkalosis, or no change?
4. What causes the acid–base disorder? *Ans:* the factor (respiratory or metabolic) that would produce the abnormal pH.
5. Is there compensation? Did the other factor change in the direction that would help offset the disorder?

Use of the Plasma Anion Gap (PAG)	
Use	To determine whether the cause of a metabolic acidosis is due to ↑ concentration of nonvolatile acid
Discriminates	Whether the primary disorder is loss of bicarbonate in urine or feces
Principle	Cations = anions in plasma, if all were measured; commonly measure Na^+, Cl^-, and HCO_3^-; metabolic acids are not measured. So there is a gap between cations and anions as measured.
Calculation	$PAG = [Na^+] - ([Cl^-] + [HCO_3^-])$; normal = 12 ± 2
Interpretation ↑ PAG	Excess molecules of acid, e.g., diabetic ketoacidosis, aspirin, lactic acidosis, etc.
Non-anion gap acidosis (normal PAG)	Acidosis due to loss of bicarbonate in urine or diarrhea results in hyperchloremic acidosis; kidneys reabsorb excess Cl^- in replacement for bicarbonate

Urinary Anion Gap (UAG)	
Principle	Hard to measure NH_4^+ in urine, but it is the major form of acid excreted Cations = anions in urine, if all are measured
Ions measured	Na^+, K^+, Cl^-: anions > cations because did not measure NH_4^+
UAG calculation	$UAG = [Na^+] + [K^+] - [Cl^-]$
Interpretation of negative	Kidneys are excreting acid
Interpretation of positive	Kidneys are excreting base
In acidosis	UAG should be negative if kidneys are compensating appropriately
In alkalosis	UAG should be positive if kidneys are compensating appropriately

Sample case:

Arterial: pH = 7.15, $Paco_2$ = 30 mm Hg, $[HCO_3^-]$ = 10 mEq/L, Cl^- = 100 mEq/L, Na^+ = 145 mEq/L

Urine: Na^+ = 100 mEq/L, K^+ = 90 mEq/L, Cl^- = 140 mEq/L

Diagnosis: metabolic acidosis with respiratory compensation; PAG = 35; therefore, grossly excessive acid in the body. Diagnosis is either non-renal metabolic acidosis or renal failure with low GFR (e.g., uremia). UAG is only reliable for measuring urine acid excretion in normal anion gap metabolic acidosis, such as to discriminate renal tubular acidosis (RTA) vs. non-renal causes. Avoid in high PAG acidosis.

Renal Pharmacology

► Diuretics: Mechanisms

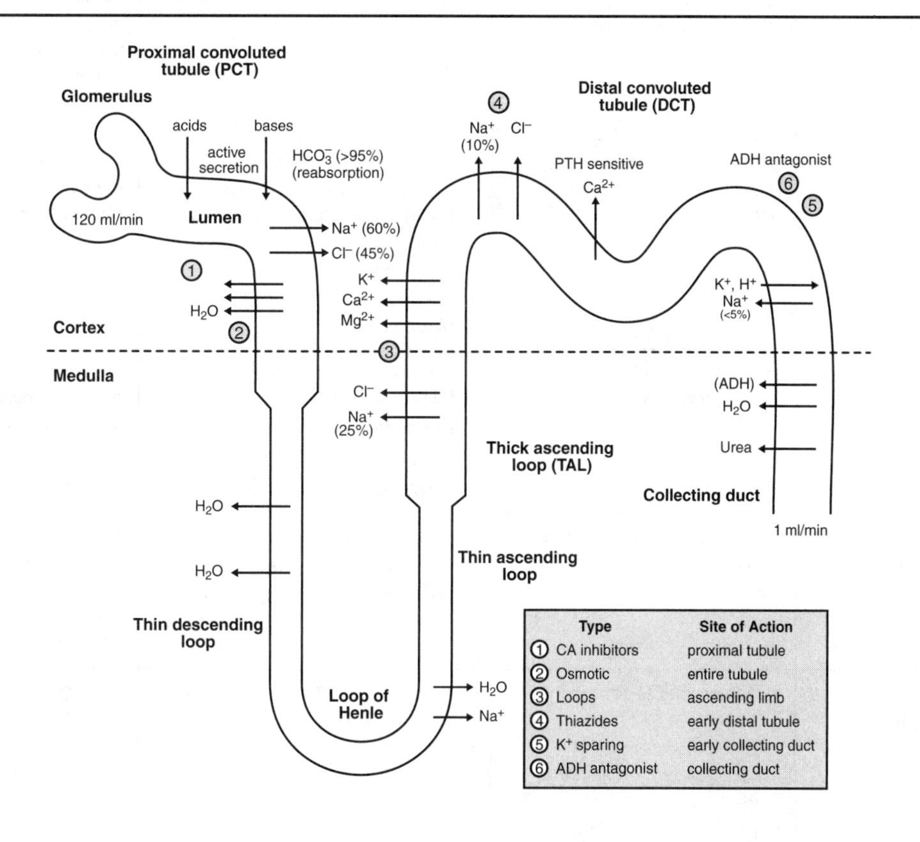

CA Inhibitor	Loop	Thiazides	K^+ Sparing*
Luminal Membrane Basolateral Membrane Proximal Tubule Acetazolamide K^+ Na^+ HCO_3^- Na^+ Na^+ H^+ H^+ HCO_3^- H_2CO_3 CA CA HCO_3^- $CO_2 + H_2O$ $CO_2 + H_2O$ $CO_2 + H_2O$	Luminal Membrane Basolateral Membrane Thick Ascending Loop K^+ Na^+ 2 Cl^- Cl^- Na^+ Na^+, K^+ K^+ Na^+ Cl^- loops (+)potential Mg^{2+}, Ca^{2+}	Luminal Membrane Basolateral Membrane Distal Tubule thiazides K^+ Na^+ Na^+ Cl^- Cl^- Na^+ Ca^{2+} Ca^{2+}	Luminal Membrane Basolateral Membrane Collecting Duct Principal Cell Aldosterone Na^+ AR Na^+ K^+ or H^+ K^+ K^+ Na^+ Cl^- H^+ $H^+ + HCO_3^-$ Cl^- HCO_3^- CA $CO_2 + H_2O$ Intercalated Cell

**Note:* K^+ sparing includes aldosterone blockers and Na^+ channel blockers.

▶ Diuretics

Diuretics	Mechanism	Clinical Uses	Side Effects
1. CA inhibitors (acetazolamide, dorzolamide*)	Inhibits carbonic anhydrase in PCT (brush border and intracellular) and other tissues	• **Glaucoma** (↓ aqueous humor secretion) • Acute mountain sickness • Metabolic alkalosis • Urinary alkalinization (eliminates acidic drugs)	Acidosis, hypokalemia, hyperchloremia, renal stones, sulfa allergy, paresthesias, possible ammonia toxicity in patients with hepatic failure
2. Osmotic (mannitol)	Is freely filtered and poorly reabsorbed, so ↑ tubular fluid osmolarity, preventing H_2O reabsorption in PCT† via an osmotic effect	• ↑ urine flow in solute overload (hemolysis, rhabdomyolysis) and if renal blood flow is ↓ (e.g., shock) • ↓ intracranial pressure • ↓ intraocular pressure (acute glaucoma)	Pulmonary edema, hypovolemia, hypernatremia, nausea/vomiting
3. Loop (furosemide, ethacrynic acid, torsemide)	Inhibit $Na^+/K^+/2Cl^-$ cotransporter in TAL	• Edematous states (e.g., heart failure, ascites) • Pulmonary edema • Hypertension • Hypercalcemia	Hypokalemic metabolic alkalosis, hypovolemia, ototoxicity, sulfa allergy (furosemide)
4. Thiazides (hydrochlorothiazide, metolazone)	Inhibit Na^+/Cl^- cotransporter in DCT	• Hypertension • Edematous states (e.g., CHF) • Lose effectiveness when GFR is very low, in which case, switch to loops	Hyponatremia, hypokalemic metabolic alkalosis, hyperglycemia (esp. diabetics), hyperuricemia, hypercalcemia, hyperlipidemia, sulfa allergies
5. K^+ sparing (spironolactone, eplerenone, amiloride, triamterene)	• Act in cortical collecting tubules • Block aldosterone receptor (spironolactone, eplerenone) • Block sodium channels (amiloride, triamterene)	• Adjunct with other diuretics to prevent K^+ loss (HTN, CHF) • Hyperaldosteronism (spironolactone) • Antiandrogen (spironolactone)	Hyperkalemia, endocrine effects, such as gynecomastia, antiandrogen effects (spironolactone)
6. ADH antagonists (demeclocycline, lithium)	V_2 antagonists	• SIADH (demeclocycline)	Bone/teeth abnormalities (demeclocycline); nephrogenic diabetes insipidus (lithium)

Definition of abbreviations: DCT, distal convoluting tubule; PCT, proximal convoluting tubule; SIADH, syndrome of inappropriate antidiuretic hormone; TAL, thick ascending limb.

*Topical for eye

†More minor effects in descending loop of Henle, collecting tubule

▶ Electrolyte Changes by Diuretics

	Urinary NaCl	Urinary $NaHCO_3$	Urinary K^+	Urinary Ca^{2+}	Body pH
CA inhibitors	↑	↑↑	↑	↑	↓
Loop	↑↑↑	↓, –	↑	↑	↑
Thiazides	↑↑	–	↑	↓	↑
K^+ sparing	↑	–	↓	–	↓

Renal Pathology

► Cystic Disease of the Kidney

Childhood polycystic disease	• Rare **autosomal recessive** disease, presenting in infancy with renal (and often hepatic) cysts and progressive renal failure • *Gross:* bilaterally enlarged kidneys with smooth surfaces. *Cut section:* sponge-like appearance with multiple small cysts in the cortex and medulla
Adult polycystic disease	• **Autosomal dominant**, usually have normal renal function until middle age • Present with renal insufficiency, hematuria, flank pain, and hypertension • *Extrarenal manifestations:* liver cysts, **circle of Willis berry aneurysms**, mitral valve prolapse • *Gross:* marked bilateral enlargement with large cysts bulging through the surface • *Micro:* cysts involve <10% of nephrons
Simple cysts	Common, can be single or multiple; have little clinical significance
Medullary sponge kidney	• Multiple **cystic dilatations of collecting ducts in medulla** • Most are asymptomatic
Acquired cystic disease	• Multiple cortical and medullary cysts may result from **prolonged renal dialysis** • ↑ risk for development into renal cell carcinoma

► Glomerular Diseases

Type	Clinical Presentation	Mechanism	Light Microscopy	Electron Microscopy	Immuno-fluorescence
Poststreptococcal glomerulonephritis (acute proliferative)	Nephritis; **elevated ASO**; low complement; children > adults Usually recover completely; occasionally progress to RPGN	Immunologic (type III hyper-sensitivity)	Polymorpho-nuclear neutrophil leukocyte infiltration; proliferation	**Subepithelial humps**	**Granular pattern**; GBM and mesangium contain IgG and C3
Minimal change disease (lipid nephrosis)	**Nephrotic syndrome; children > adults** Usually normal renal function; may respond to steroids	Unknown	Normal	No deposits; **loss of epithelial foot processes**	Negative
Membranous glomerulonephritis	**Nephrotic syndrome; adults > children** May respond to steroids	Immunologic	Capillary wall thickening	**Subepithelial spikes**; loss of epithelial foot processes	Granular pattern of IgG and C3
Membranoproliferative glomerulonephritis	Variable: mild proteinuria, mixed nephritic/nephrotic, or frank nephrotic syndrome Poor response to steroids	*Type I:* immune complex and both classic and alternate complement pathways *Type II:* immune complex and alternate com-plement pathway	Basement membrane thick and split; mesangial proliferation	*Type I:* **sub-endothelial deposits** *Type II:* **dense deposit disease**	*Type I:* IgG and C3, C1q, and C4 *Type II:* C3 (IgG, C1q, and C4 usually absent) "C3 nephritic factor"
Focal segmental glomerulosclerosis	Nephrotic syndrome (most common cause in adults) Poor prognosis; rarely responds to steroids	Immunologic; aggressive variant of lipoid nephrosis; **IV drug use; HIV nephropathy**	Focal and segmental sclerosis and hyalinization	Epithelial damage; loss of foot processes	IgM and C3 focal deposits

Definition of abbreviations: ASO, antistreptolysin O.

(Continued)

▶ Glomerular Diseases (*Cont'd.*)

Type	Clinical Presentation	Mechanism	Light Microscopy	Electron Microscopy	Immuno-fluorescence
Goodpasture syndrome	RPGN + pulmonary hemorrhage Often poor prognosis; may respond to steroids, plasmapherisis, cytotoxic agents	**Anti-GBM antibodies** (type II hyper-sensitivity)	**Crescents**; mesangial proliferation in early cases	GBM disruption; no deposits	**Linear IgG and C3**
Idiopathic RPGN	RPGN; may follow flu-like syndrome Extremely poor prognosis	Immunologic	**Crescents**	Variable, ± deposits; all have GBM ruptures	Granular or linear
Focal proliferative glomerulonephritis	Primary focal glomerulonephritis or part of multisystem disease; may be subclinical or present with hematuria, proteinuria, nephrotic syndrome Variable prognosis	Immunologic	Proliferation limited to certain segments of particular glomeruli	Variable; may show mesangial deposits	Variable; may show mesangial deposits
IgA nephropathy (Berger disease)	Variable: recurrent hematuria, mild proteinuria, nephrotic syndrome; children and young adults Usually slowly progressive course	Unknown	Variable: normal or segmental/ mesangial proliferation or crescentic	Mesangial deposits	**Mesangial IgA deposition**
Diabetic glomerulopathy	Microscopic proteinuria; or can eventually cause nephrotic syndrome Prognosis variable, depends on diabetic control	Nonenzymatic glycosylation causes glomerular basement membrane thickening and mesangial matrix expansion	Capillary basement membrane thickening; diffuse and nodular glomerular sclerosis (Kimmelstiel-Wilson)	Thickened glomerular basement membrane and well-demarcated, roughly round nodules in the glomeruli	Not really helpful; may show non-specific immunoglobulin G deposition along the basement membrane
Chronic glomerulonephritis	Chronic renal failure; may follow a variety of acute glomerulopathies Poor prognosis	Variable	**Hyalinized glomeruli**	Not specific	Negative or granular
Amyloidosis	Nephrotic syndrome Prognosis variable, depends in underlying condition	Amyloid deposition, often accompanying diseases e.g., multiple myeloma, chronic infections, tuberculosis, and rheumatoid arthritis	Eosinophilic amorphous deposits in glomeruli and interstitium that show apple-green birefringence when stained with Congo Red	Fibrillar depositions of amyloid material	Immuno-fluorescence is not usually helpful
Alport syndrome	Hematuria, proteinuria, which slowly progress to renal failure; **deafness**; ocular disorders Renal failure common by age 50	**X-linked** disorder of collagen	Segmental and focal glomerulo-sclerosis, tubular atrophy, interstitial fibrosis, chronic inflammation	**Thickening** (sometimes thinning) and **splitting the basement membrane**	**Immuno-fluorescence for individual chains of type IV collagen** can be diagnostic

Definition of abbreviations: ASLO, antistreptolysin O; GBM, glomerular basement membrane; RPGN, rapidly progressive glomerulonephritis.

▶ Tubular Diseases of the Kidney

Acute Tubular Necrosis (ATN)	**Most common cause of acute renal failure**; associated with reversible injury to the tubular epithelium; excellent prognosis if patient survives disease responsible for the ATN
Ischemic ATN	• **Most common cause of ATN** • ↓ blood flow caused by severe renal vasoconstriction, hypotension, or shock
Nephrotoxic ATN	Caused by heavy metals such as mercury, drugs, and myoglobin
Four Phases of ATN	
Initial phase	*36 hours*: after precipitating event occurs
Oliguric phase	*10 days:* ↓ urine output; uremia, fluid overload, and hyperkalemia may occur
Diuretic phase	*2–3 weeks:* gradual ↑ in urine volume (up to 3 L/day); hypokalemia, electrolyte imbalances, and infection may occur
Recovery phase	*3 weeks:* improved concentrating ability, restoration of tubular function; normalization of BUN and creatinine

▶ Tubulointerstitial Diseases of the Kidney

Pyelonephritis	• Infection of the renal pelvis, tubules, and interstitium • **Ascending infection is the most common route** with organisms from the patient's fecal flora; hematogenous infection is much less common. • Etiologic agents usually gram-negative bacilli (e.g., *E. coli*, *Proteus,* and *Klebsiella*). *E. coli* pili mediate adherence, motility aids movement against flow of urine. *Proteus* (urease, alkaline urine, struvite stones), *Klebsiella* (large capsule)
Acute pyelonephritis	• Risk factors: urinary obstruction, vesicoureteral reflux, pregnancy, instrumentation, diabetes mellitus • Under 40 more common in women (shorter urethra); over 40, ↑ incidence in men due to benign prostatic hypertrophy • **Symptoms:** fever, malaise, dysuria, frequency, urgency, and costovertebral angle (CVA) tenderness. **Fever, CVA tenderness,** and **WBC casts** distinguish pyelonephritis from cystitis. • **Urine:** many WBCs and WBC casts • **Gross:** scattered yellow microabscesses on the renal surface • **Micro:** foci of interstitial suppurative necrosis and tubular necrosis • Blunting of the calyces may be seen on intravenous pyelogram
Chronic pyelonephritis	• **Reflux nephropathy** most common cause • Interstitial parenchymal scarring deforms the calyces and pelvis • **Symptoms:** onset can be insidious or acute; present with renal failure and hypertension • **Gross:** irregular scarring and deformed calyces with overlying corticomedullary scarring • **Micro:** chronic inflammation with tubular atrophy • Pyelogram diagnostic
Acute allergic interstitial nephritis	• **Hypersensitivity reaction** to infection or drugs (e.g., NSAIDs, penicillin, methicillin) • Leads to interstitial edema with a mononuclear infiltrate • Presents 2 weeks after exposure with hematuria, pyuria, eosinophilia, and azotemia
Analgesic nephritis	Interstitial nephritis and renal papillary necrosis, induced by large doses of analgesics
Gouty nephropathy	• **Urate crystals** in tubules, inducing tophus formation and a chronic inflammatory reaction • *Note:* Urate crystals appear as **birefringent**, **needle-shaped crystals** on light microscopy.
Acute urate nephropathy	• Precipitation of crystals in the collecting ducts, causing obstruction • Seen in **lymphoma and leukemia**, especially after chemotherapy
Multiple myeloma	**Bence-Jones proteins** are directly toxic to tubular epithelium
Diffuse cortical necrosis	• Generalized infarction of both kidneys; preferentially involves the renal cortex • Seen in settings of obstetric catastrophes (abruptio placentae) and shock • Mechanism thought to be a combination of vasospasm and DIC

► Vascular Diseases of the Kidney

Ischemia	• Caused by embolization of mural thrombi usually left side of heart or aorta • **Gross:** sharply demarcated, wedge-shaped pale regions, which undergo necrosis with subsequent scarring • **Symptoms:** infarcts may be asymptomatic or may cause pain, hematuria, and hypertension
Renal vein thrombosis	• Thrombosis of one or both renal veins may occur • Associated with the nephrotic syndrome, particularly membranous glomerulonephritis • Renal cell carcinoma may also provoke renal vein thrombosis as a result of direct invasion by tumor • Presents with hematuria, flank pain, and renal failure • **Gross:** kidney enlarged • **Micro:** hemorrhagic infarction of renal tissue

► Urolithiasis

- Affects 6% of the population; men > women
- Renal colic may occur if small stones pass into the ureter, where they may also cause hematuria and urinary obstruction and predispose to infection

Calcium stones	• **75% of stones**; most patients have hypercalciuria without hypercalcemia • Calcium stones are **radiopaque**; they are the only ones that can be seen on x-ray
Magnesium-ammonium phosphate stones	• 15% of stones; occur after infection by urease-producing bacteria, such as ***Proteus*** • Urine becomes alkaline, resulting in precipitation of **magnesium-ammonium phosphate salts**; may form large stones (e.g., **staghorn calculi**)
Uric acid stones	Seen in **gout**, **leukemia**, and in patients with acidic urine
Cystine stones	• Very rare • Associated with an autosomal recessive amino acid transport disorder, leading to **cystinuria** • Most stones are unilateral and formed in calyx, pelvis, bladder

► Obstructive Uropathy and Hydronephrosis

Hydronephrosis	• Multiple etiologies, including stones, benign prostatic hypertrophy, pregnancy, neurogenic bladder, tumor, inflammation, and posterior urethral valves • Persistence of glomerular filtration despite urinary obstruction, causing dilation of calyces and pelvis. High pressure in the collecting system causes atrophy and ischemia. • **Gross:** dilatation of the pelvis and calyces with blunting of renal pyramids • **Symptoms:** – *Unilateral:* may remain asymptomatic as the kidney atrophies – *Bilateral incomplete:* loses concentrating ability, causing urinary frequency, polyuria and nocturia – *Bilateral complete:* causes anuria, uremia, and death if untreated

► Tumors of the Kidney

Benign	
Cortical adenomas	• Common finding at autopsy • **Gross:** yellow, encapsulated cortical nodules • **Micro:** may be identical to renal cell carcinoma, distinguished by size
Angiomyolipomas	• Hamartomas, composed of fat, smooth muscle, and blood vessels • Particularly common in patients with tuberous sclerosis
Malignant	
Renal cell carcinomas	• 90% of all renal cancers in adults; seen in ages 50–70 with no sex predilection • Moderate association with smoking and a familial predisposition • Occurs in 2/3 of patients with von Hippel-Lindau disease • **Symptoms:** "classic" triad of hematuria, palpable mass, and costovertebral pain (10% of cases); hematuria in middle-aged patient should raise concern • **Gross:** most common in the upper pole; usually solitary, with areas of necrosis and hemorrhage; often invades the renal vein and extends into the vena cava and heart • **Micro:** polygonal clear cells with abundant clear cytoplasm • **Genetics:** often associated with loss of function of VHL gene and activation of the MET oncogene • Paraneoplastic syndrome: polycythemia, hypercalcemia, Cushing syndrome, etc. • High incidence of metastasis on initial presentation • 5-year survival depends on stage, especially poor (25–50%) if tumor extends into the renal vein
Wilms tumor (nephroblastoma)	• Common childhood malignancy with peak incidence at age 2 • **Symptoms**: abdominal mass and hypertension, nausea, hematuria, intestinal obstruction • May be associated with other congenital anomalies • **Gross**: very large, demarcated masses; most are unilateral, but may be bilateral if familial • **Micro:** embryonic glomerular and tubular structures surrounded by mesenchymal spindle cells • 90% survival rate when patients are treated with surgery, chemotherapy, radiotherapy
Transitional cell carcinoma	• Can involve the epithelium of the renal pelvis • Histology similar to transitional cell carcinoma of the bladder, but is less common • Can present with hematuria

► Anomalies of the Ureters

Double ureters	• Form when two same-sided ureters join at some point before the junction to the bladder or enter the bladder separately • Associated with double renal pelvises or an abnormally large kidney
Ureteral Obstruction in Hydroureter and Hydronephrosis	
Internal obstruction	• **Renal calculi** are most common cause; usually impact at the ureteropelvic junction, entrance to the bladder, and where they cross iliac vessels • Other causes: strictures, tumors
External obstruction	• Pelvic tumors may compress or invade the ureteral wall; sclerosing retroperitonitis, a fibrosis of retroperitoneal structures, can cause obstruction. • Pregnancy may cause obstruction and may also cause dilation (secondary to progesterone).

▶ Pathology of the Bladder

Diverticula	• **Pouch-like evaginations** of the bladder wall • Occur in older men and women and may lead to urinary stasis and therefore infection
Exstrophy of bladder	• Caused by **absence of the anterior musculature** of the bladder and abdominal wall; developmental failure of downgrowth of mesoderm over the anterior bladder • Site of severe chronic infections, with ↑ incidence of adenocarcinoma
Patent urachus	• Fistula that **connects the bladder with the umbilicus** • Isolated persistence of the central urachus termed a urachal cyst • Carcinomas may develop in these cysts
Infectious cystitis	• Cystitis causes frequency, urgency, dysuria, and suprapubic pain • **Causative organisms:** *E. coli, Staphylococcus saprophyticus* (associated with intercourse), *Proteus, Klebsiella* (esp. in diabetics), *Pseudomonas* (capsule is antiphagocytic, exotoxin A inhibits EF-2, esp. in patients with structural abnormalities and antibiotic usage), *Enterococcus* (esp. in males with prostate problems) • No WBC casts in urine (as compared with pyelonephritis) • Systemic signs, such as fever and chills, are also uncommon with lower urinary tract infections; CVA tenderness usually absent
Hemorrhagic cystitis	Marked mucosal hemorrhage secondary to viruses, radiation, or chemotherapy (**cyclophosphamide**, protect with **mesna**)
Cystitis emphysematosa	• Submucosal gas bubbles • Occurs mostly in diabetics
Bladder obstruction	• **Men:** benign prostatic hyperplasia or carcinoma most common cause • **Women:** cystocele of the bladder most common cause • **Gross:** thickening, hypertrophy, and trabeculation of the smooth muscle bladder wall
Carcinoma of the bladder	• **Transitional cell carcinoma:** 90% of primary bladder neoplasms • **Risk factors: Smoking,** occupational exposure **(e.g., naphthylamine),** infection with *Schistosoma haematobium* **(more commonly associated with squamous cell carcinoma)** • 3% of all cancer deaths in the United States - peak incidence between 40 and 60 years of age. • Usually presents with painless hematuria, may also cause dysuria, urgency, frequency, hydronephrosis, and pyelonephritis. • **Prognosis:** Determined by grade and stage at the time of diagnosis; high incidence of recurrence at multiple locations.

The Gastrointestinal System

Chapter 5

Embryology of the Gastrointestinal System

► Development of the Gastrointestinal Tract

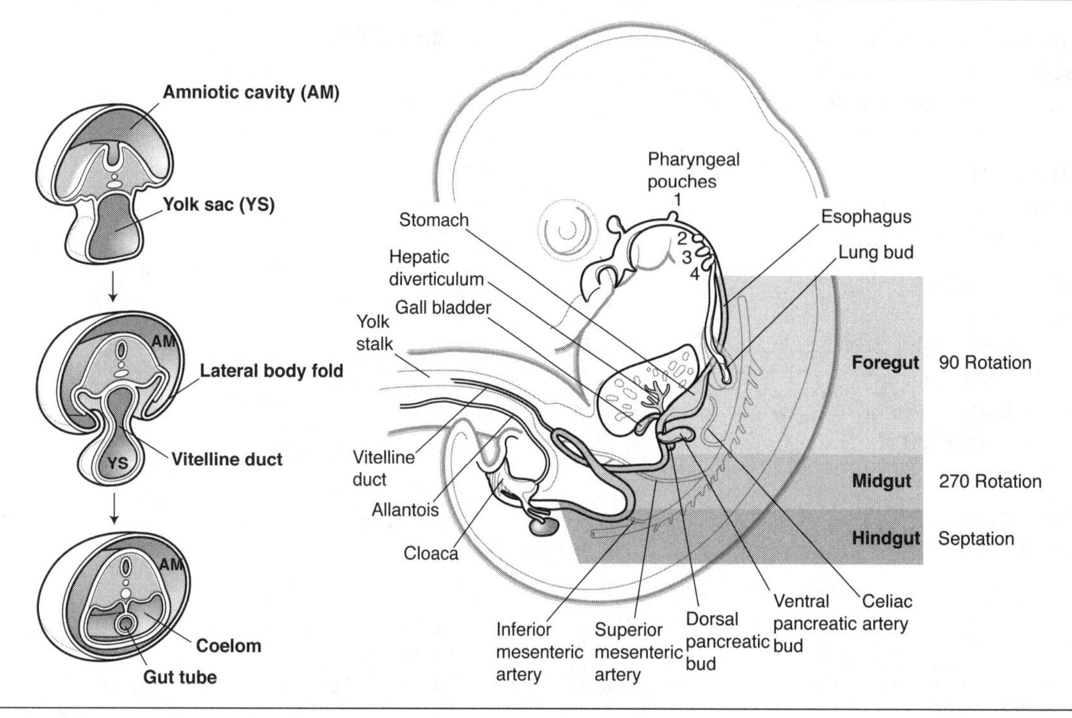

The **primitive gut tube** is formed by incorporation of the yolk sac into the embryo during cranial–caudal and lateral folding.

- The epithelial lining and glands of the mucosa are derived from **endoderm**. The epithelial lining of the gut tube proliferates rapidly and obliterates the lumen, followed by recanalization.
- The lamina propria, muscularis mucosa, submucosa, muscularis externa, and adventitia/serosa are derived from **mesoderm**.

The primitive gut tube is divided into the **foregut**, **midgut**, and **hindgut**, each supplied by a specific artery, each receiving a slightly different autonomic innervation, and each having slightly different relationships to a mesentery.

► Adult Structures Derived from Each of the Three Divisions of the Primitive Gut Tube

Foregut	Midgut	Hindgut
Artery: celiac **Parasympathetic innervation:** vagus nerves **Sympathetic innervation:** greater splanchnic nerves, T5–T9*	**Artery:** superior mesenteric **Parasympathetic innervation:** vagus nerves **Sympathetic innervation:** lesser and lowest splanchnic nerves, T9–T12*	**Artery:** inferior mesenteric **Parasympathetic innervation:** pelvic splanchnic nerves **Sympathetic innervation:** lumbar splanchnic nerves L1–L2*
Foregut Derivatives	**Midgut Derivatives**	**Hindgut Derivatives**
Esophagus Stomach Duodenum (first and second parts) Liver Pancreas Biliary apparatus Gall bladder Pharyngeal pouches† Lungs† Thyroid† Spleen‡	Duodenum (second, third, and fourth parts) Jejunum Ileum Cecum Appendix Ascending colon Transverse colon (proximal two thirds)	Transverse colon (distal third) Descending colon Sigmoid colon Rectum Anal canal (above pectinate line)

***Referred pain**—stimulation of visceral pain fibers that innervate a gastrointestinal structure results in a dull, aching, poorly localized pain that is referred over the T5 through L1 dermatomes. The **sites of referred pain** generally correspond to the spinal cord segments that provide the sympathetic innervation to the affected gastrointestinal structure.

†Derivatives of endoderm, but not supplied by the celiac artery or innervated as above.

‡Spleen is not a foregut derivative, but is supplied by the celiac artery.

Gastrointestinal Histology

► Layers of the Digestive Tract

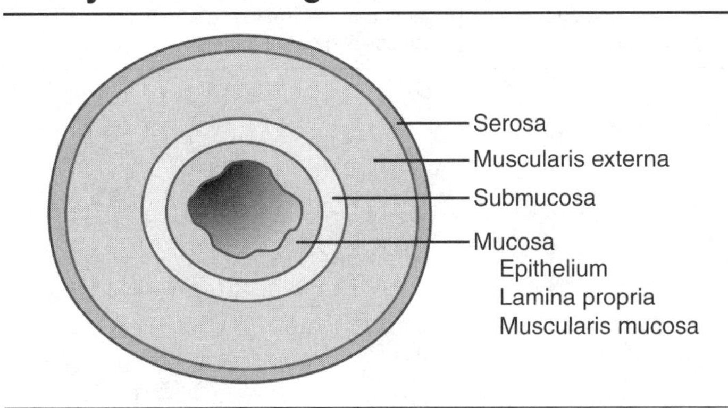

Submucosal plexus *(Meissner plexus):* a collection of ganglia and interneurons of the enteric nervous system (ENS), predominantly responsible for regulating epithelial function and some circular smooth muscle function

Myenteric plexus *(Auerbach plexus):* a collection of ganglia and interneurons of the enteric nervous system (ENS), predominantly responsible for regulating longitudinal and some circular smooth muscle function

Note that the first third of esophageal muscle is skeletal, the middle third is mixed skeletal and smooth, and the final third is smooth muscle. Also, the stomach smooth muscle contains an oblique layer between the submucosa and circular layer of smooth muscle.

▶ Histology of Specific Regions

Region	Major Characteristics	Mucosal Cell Types at Surface	Function of Surface Mucosal Cells
Esophagus	• Nonkeratinized stratified squamous epithelium • Skeletal muscle in muscularis externa (upper 1/3) • Smooth muscle (lower 1/3)	—	—
Stomach (body and fundus)	*Rugae:* shallow pits; deep glands	Mucous cells	Secrete mucus; form protective layer against acid; tight junctions between these cells probably contribute to the acid barrier of the epithelium.
		Chief cells	Secrete pepsinogen and lipase precursor
		Parietal cells	Secrete HCl and intrinsic factor
		Enteroendocrine (EE) cells	Secrete a variety of peptide hormones
Pylorus	Deep pits; shallow branched glands	Mucous cells	Same as above
		Parietal cells	Same as above
		EE cells	High concentration of gastrin
Small intestine	Villi, plicae, and crypts	Columnar absorptive cells	Contain numerous microvilli that greatly increase the luminal surface area, facilitating absorption
Duodenum	Brunner glands, which discharge alkaline secretion	Goblet cells	Secrete acid glycoproteins that protect mucosal linings
		Paneth cells	Contains granules that contain lysozyme. May play a role in regulating intestinal flora
		EE cells	High concentration of cells that secrete cholecystokinin and secretin
Jejunum	Villi, well developed plica, crypts	Same cell types as found in the duodenal epithelium	Same as above
Ileum	Aggregations of lymph nodules called Peyer patches	M cells found over lymphatic nodules and Peyer patches	Endocytose and transport antigen from the lumen to lymphoid cells
Large intestine	Lacks villi, crypts	Mainly mucus-secreting and absorptive cells	Transports Na^+ (actively) and water (passively) out of lumen

Gross Anatomy

▶ Abdominal Viscera

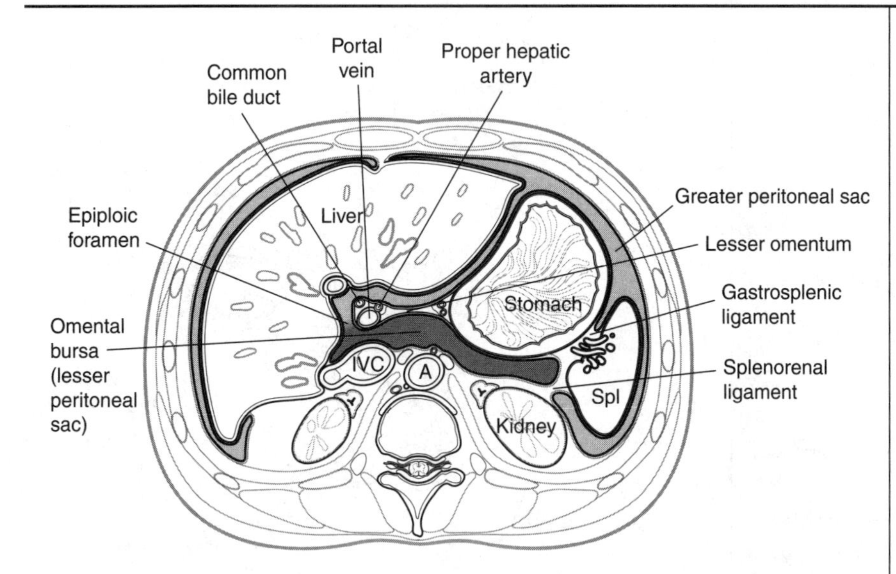

Viscera are classified as:

- **Peritoneal organs**—have a mesentery and are almost completely enclosed in peritoneum. These organs are mobile.

- **Retroperitoneal organs**—are partially covered with peritoneum and are immobile or fixed organs.

This figure is a cross-section of the abdomen that shows the greater and lesser peritoneal sacs and associated abdominal viscera.

Major Peritoneal Organs (suspended by a mesentery)	**Major Secondary Retroperitoneal Organs** (lost a mesentery during development)	**Major Primary Retroperitoneal Organs** (never had a mesentery)
Stomach	Midgut duodenum	Kidneys
Liver and gallbladder	Head, neck, and body of pancreas	Adrenal glands
Spleen	Ascending colon	Ureter
Foregut duodenum	Descending colon	Aorta
Tail of pancreas	Upper rectum	Inferior vena cava
Jejunum		Lower rectum
Ileum		Anal canal
Appendix		
Transverse colon		

▶ The Abdominal Aorta and Celiac Circulation

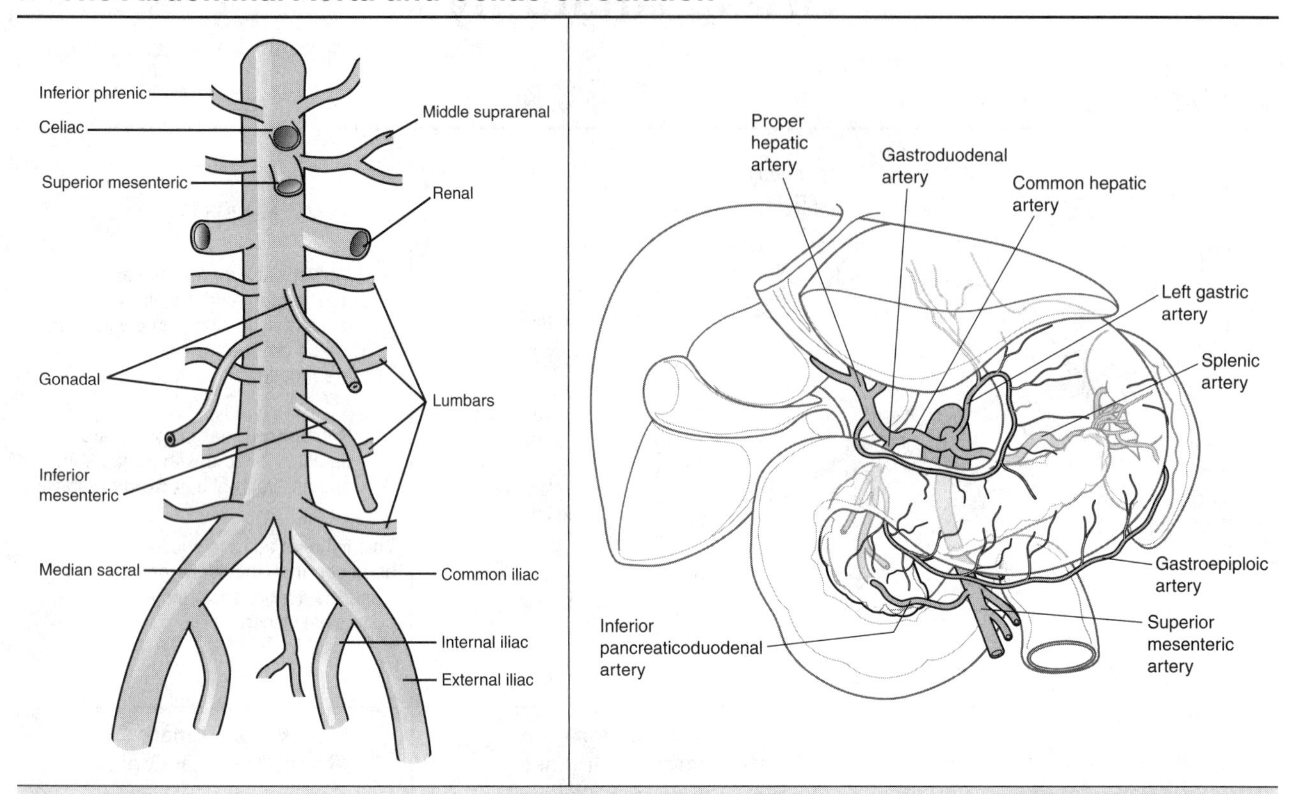

Clinical Correlations

In an **occlusion of the celiac artery** at its origin from the abdominal aorta, collateral circulation may develop in the **head of the pancreas** by way of anastomoses between the **pancreaticoduodenal branches of both the superior mesenteric** and the gastroduodenal arteries.

Branches of the **celiac circulation** may be subject to **erosion** if an ulcer penetrates the posterior wall of the stomach or the posterior wall of the duodenum.

- The **splenic artery** may be subject to erosion by a penetrating ulcer of the **posterior wall** of the stomach.
- The **left gastric artery** may be subject to erosion by a penetrating ulcer of the **lesser curvature** of the stomach.
- The **gastroduodenal artery** may be subject to erosion by a penetrating ulcer of the **posterior wall** of the first part of the duodenum.

Patients with a penetrating ulcer may have **pain referred to the shoulder**, which occurs when air escapes through the ulcer and stimulates the peritoneum covering the inferior aspect of the diaphragm. The contents of a **penetrating ulcer** of the posterior wall of the stomach or the duodenum may enter the **omental bursa**.

Hematemesis may result from bleeding into the lumen of the esophagus, stomach, or duodenum proximal to the ligament of Treitz. Hematemesis is commonly caused by a duodenal ulcer, a gastric ulcer, or esophageal varices.

▶ Superior and Inferior Mesenteric Arteries

Superior Mesenteric Artery Distribution	Inferior Mesenteric Artery Distribution

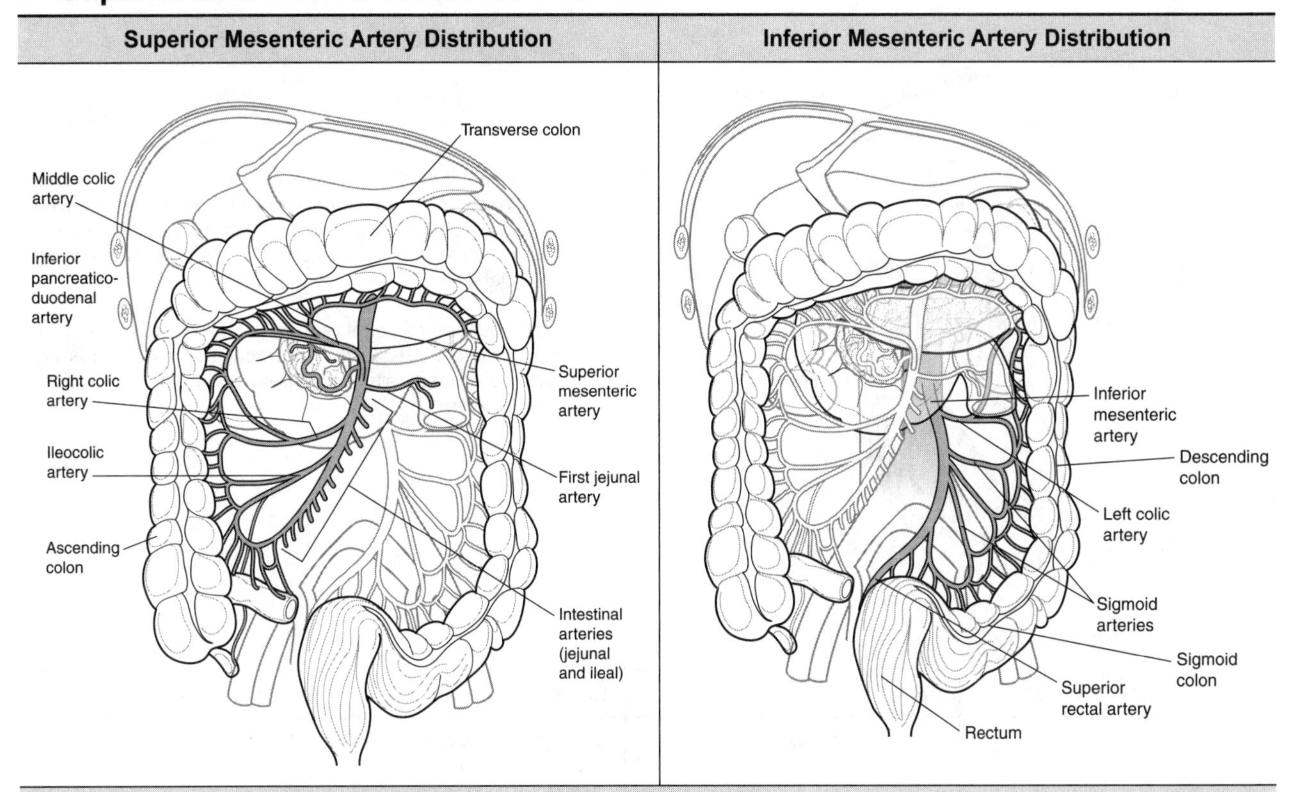

Clinical Correlation

- Common sites of **ischemic bowel infarction** are in the transverse colon near the splenic flexure and in the rectum.
- **Infarction of the transverse colon** occurs between the distal parts of the middle colic branches of the superior mesenteric and left colic branches of the inferior mesenteric arteries.
- **Infarction of the rectum** occurs between the distal parts of the superior rectal branches of the inferior mesenteric artery and the middle rectal branches of the internal iliac artery.

► Hepatic Portal Circulation

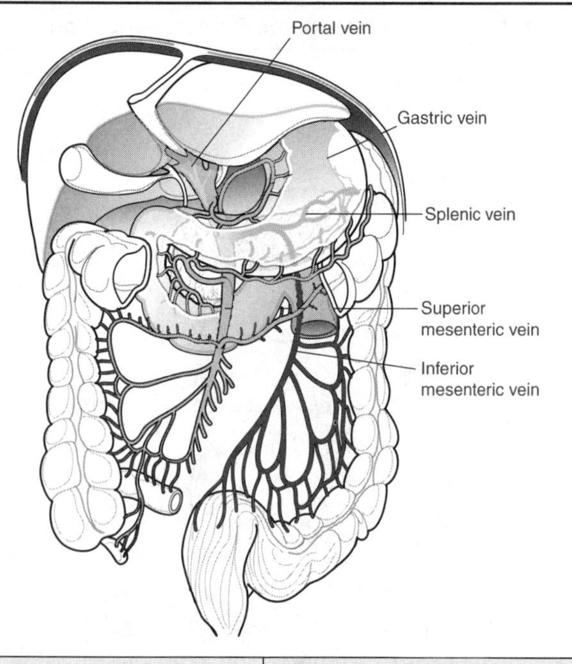

Clinical Correlation

Patients with cirrhosis of the liver may develop **portal hypertension**, in which venous blood from gastrointestinal structures, which normally enters the liver by way of the portal vein, is forced to flow in the retrograde direction in tributaries of the portal vein.

Retrograde flow forces portal venous blood into tributaries of the superior or inferior vena cava; **portacaval anastomoses** are established at these sites, permitting portal venous blood to bypass the liver.

Sites of Anastomoses	Portal	Caval	Clinical Signs
1. Umbilicus	Paraumbilical veins	Superficial veins of the anterior abdominal wall	Caput medusa
2. Rectum	Superior rectal veins (inferior mesenteric vein)	Middle and inferior rectal veins (internal iliac vein)	Internal hemorrhoids
3. Esophagus	Gastric veins	Veins of the lower esophagus, which drain into the azygos system	Esophageal varices
4. Retroperitoneal organs	Tributaries of the superior and inferior mesenteric veins	Veins of the posterior abdominal wall	Not clinically relevant

► Inferior Vena Cava (IVC) and Tributaries

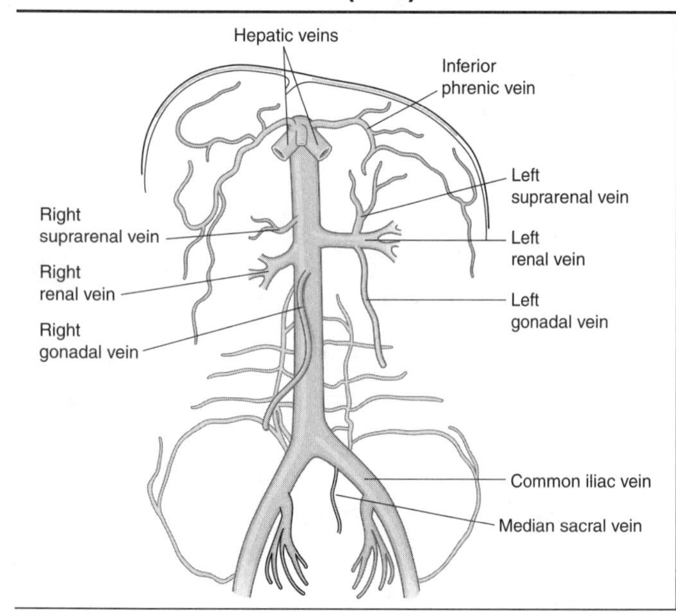

The **inferior vena cava** is formed at about the level of the L5 vertebra by the union of the common iliac veins. It ascends just to the right of the midline.

On the right, the renal, adrenal, and gonadal veins drain directly into the inferior vena cava.

On the left, only the **left renal vein** drains directly into the inferior vena cava; the left gonadal and the left adrenal veins drain into the left renal vein. The left renal vein crosses the anterior aspect of the aorta just inferior to the origin of the superior mesenteric artery.

Clinical Correlation

The **left renal vein may be compressed** by an **aneurysm of the superior mesenteric artery** as the vein crosses anterior to the aorta. Patients with compression of the left renal vein may have renal and adrenal hypertension on the left, and, in males, a varicocele on the left.

► Cross-Sectional Anatomy

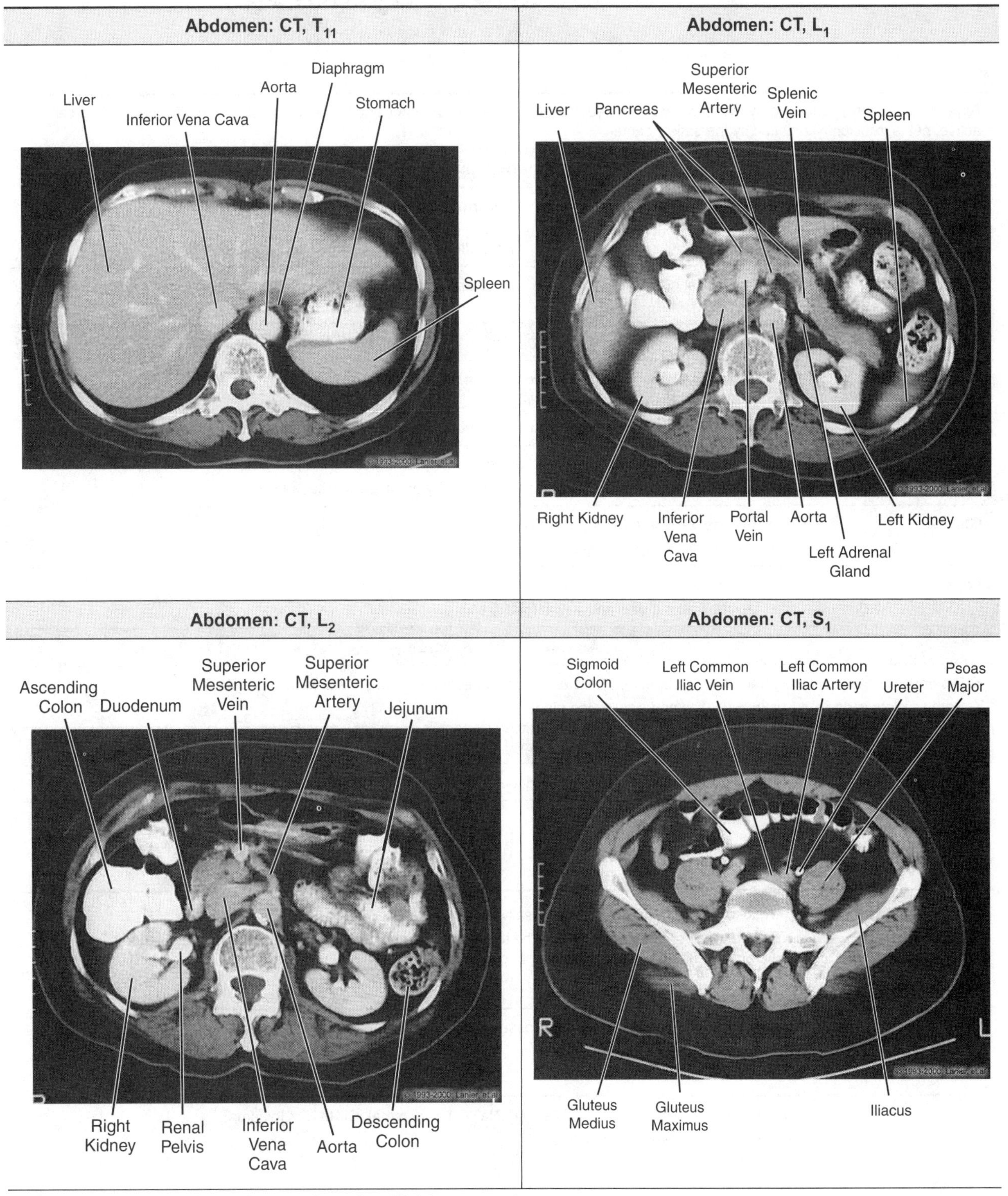

Images copyright 2005 DxR Development Group Inc. All rights reserved.

Gastrointestinal Physiology

► Appetite

Appetite is primarily regulated by two regions of the hypothalamus: a feeding center and a satiety center. Normally, the feeding center is active, but is transiently inhibited by the satiety center.

Hypothalamus			
	Location	**Stimulation**	**Destruction**
Feeding center	Lateral hypothalamic area	Feeding	Anorexia
Satiety center	Ventromedial nucleus of hypothalamus	Cessation of feeding	Hypothalamic obesity syndrome

Hormones That May Affect Appetite	
Cholecystokinin (CCK)	• Released from I-cells in the mucosa of the small intestine • CCK-A receptors are in the periphery • CCK-B receptors are in the brain—both reduce appetite when stimulated
Calcitonin	• Released mainly from the thyroid gland • Has also been reported to decrease appetite by an unknown mechanism
Ghrelin	• ↑ levels provoke hunger • Levels rise during fasting
Leptin	• Promotes satiety • Levels ↑ after meals and ↓ with fasting

Mechanical Distention

- Distention of the alimentary tract inhibits appetite, whereas the contractions of an empty stomach stimulate it.
- Some satiety is derived from mastication and swallowing alone.

Miscellaneous

Other factors that help to determine appetite and body weight include body levels of fat and genetic factors.

► Saliva

Salivary glands Submandibular Parotid Sublingual	• Produce approximately 1.5 L/day of saliva • The presence of food in the mouth, the taste, smell, sight, or thought of food, or the stimulation of vagal afferents at the distal end of the esophagus increase the production of saliva.	
Functions	• Initial triglyceride digestion (lingual lipase) • Initial starch digestion (α-amylase) • Lubrication	
Composition	**Ions:** HCO_3^- 3× [plasma]; K^+ 7 × [plasma]; Na^+ 0.1 × [plasma]; Cl^- 0.15 × [plasma] **Enzymes:** α-amylase, lingual lipase **Hypotonic** **pH:** 7–8 **Flow rate:** alters the composition **Antibacterial:** lysozyme, lactoferrin, defensins, IgA	
Regulation	**Parasympathetic**	↑ synthesis and secretion of **watery** saliva via muscarinic receptor stimulation (anticholinergics → dry mouth)
	Sympathetic	↑ synthesis and secretion of **viscous** saliva via β-adrenergic receptor stimulation

► Swallowing

Swallowing is a reflex action coordinated in the **swallowing center** in the medulla. Afferents are carried by the glossopharyngeal (CN IX) and vagus (CN X) nerves. Food is moved to the esophagus by the movement of tongue (hypoglossal nerve, CN XII) and the palatal and pharyngeal muscles (CNs IX and X).

1. **Initiation of swallowing** occurs voluntarily when the mouth is closed on a bolus of food and the tongue propels it from the oral cavity into the pharynx.
2. **Involuntary contraction** of the **pharynx** advances the bolus into the esophagus.
3. Automatic **closure of the glottis** during swallowing inhibits breathing and prevents aspiration.
4. Relaxation of the **upper esophageal sphincter (UES)** allows food to enter the esophagus.
5. **Peristaltic contraction of the esophagus** propels food toward the **lower esophageal sphincter (LES)**, the muscle at the gastroesophageal junction.
6. The LES is tonically contracted, relaxing on swallowing. **Relaxation of the LES** is mediated via the vagus nerve; **VIP** (vasoactive inhibitory peptide) is the major neurotransmitter causing LES relaxation.

Clinical Correlation	
Achalasia	**Pathologic inability of the LES to relax** during swallowing. Food accumulates in the esophagus, sometimes causing **megaesophagus**.
Gastric reflux	**LES tone** is low, allowing acid reflux into the esophagus; can lead to gastroesophageal reflux disease (**GERD**).

► The Stomach

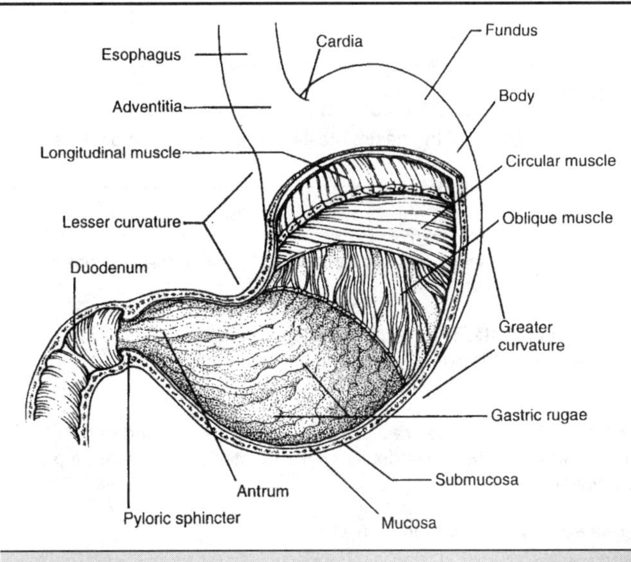

- The stomach has **three layers of smooth muscle**: longitudinal (outer, for peristalsis), circular (for mixing), and oblique (inner).
- The stomach is composed of the **fundus**, **body**, and **antrum**.
- **Receptive relaxation** mediated by VIP occurs in the fundus. As the stomach fills, a vagovagal-mediated receptive relaxation occurs, allowing storage.
- The bolus of swallowed food received by the stomach is further macerated and mixed with HCl, mucus, and pepsin. The food **(chyme)** is then discharged at a controlled rate into the duodenum. Only a small amount of chemical digestion actually occurs in the stomach.

Gastric Motility

- A pacemaker within the greater curvature produces a **basal electric rhythm (BER)** of 3 to 5 waves/min.
- The **magnitude** of the gastric contractions are **increased by parasympathetic** and **decreased by sympathetic stimulation**.
- **Migrating motor complexes (MMC)** are propulsive contractions initiated during fasting that begin in the stomach and move undigested material from the stomach and small intestine into the colon. They repeat every 90 to 120 minutes and are mediated by **motilin**. This housekeeping function lowers the bacterial count in the gut.

Gastric Emptying

- The **pylorus** is continuous with the circular muscle layer and acts as a "valve" to control gastric emptying.
- The contractions of the stomach (peristalsis) propel chyme through the pylorus at a regulated rate.
- Pyloric sphincter contraction at the time of antral contraction limits the movement of chyme into the duodenum and promotes mixing by forceful regurgitation of antral contents back into the fundus (**retropulsion**).
- **Gastric emptying is delayed by:**
  - **Fat/protein in the duodenum stimulating CCK release**, which increases gastric distensibility
  - **H^+ in the duodenum** via neural reflexes
  - Stomach contents that are hypertonic or hypotonic

Definition of abbreviation: VIP, vasoactive intestinal peptide.

▶ The Small Intestine

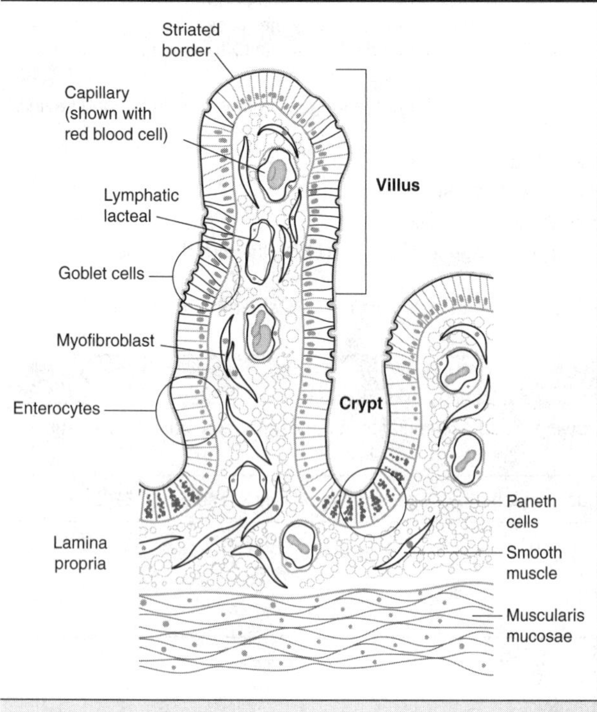

- The small intestine participates in the **digestion and absorption** of nutrients. It has specialized villi on the epithelial surface to aid in this function.
- The **duodenum** is the proximal pyloric end of the small intestine. Distal to the duodenum is the **jejunum**, and then the **ileum**.
- In the small intestine, the chyme from the stomach is **mixed** with mucosal cell secretions, exocrine pancreatic juice, and bile.
- **Mucus production** occurs in surface epithelial cells throughout the gastrointestinal tract, **Brunner glands** in the duodenum, and **goblet cells** in the mucosa throughout the intestine.
- **Mucus functions** include lubrication of the gastrointestinal tract, binding bacteria, and trapping immunoglobulins where they have access to pathogens.
- The **rate of mucus** secretion is increased by cholinergic stimulation, chemical irritation, and physical irritation.

Clinical Correlate

Any compromise of the mucous protection can lead to significant damage and irritation of the gastrointestinal tract, leading to gastritis, duodenitis, or even peptic ulcer disease.

Intestinal Motility

- **Small bowel slow waves** move caudally in the circular smooth muscle. The rate slows from approximately 12/min in the jejunum to approximately 9/min in the ileum.
- **Segmentation contractions** are ring-like contractions that **mix intestinal contents**. They occur at random "nodes" along the intestine. These relax, and then new nodes are formed at the former internodes. This action moves the chyme back and forth, increasing mucosal exposure to the chyme.
- **Peristalsis** is a reflex response initiated by stretching of the lumen of the gut. There is contraction of muscle at the oral end and relaxation of muscle at the caudal end, thus **propelling the contents caudally**. Although peristalsis is modulated by autonomic input, it can occur even in isolated loops of small bowel with no extrinsic innervation.
  - The intrinsic control system senses stretch with calcitonin gene-related polypeptide neurons (CGRP).
  - The contractile wave is initiated by acetylcholine (ACh) and substance P.
  - The relaxation caudal to the stimulus is initiated by nitric oxide (NO) and VIP.
- **Parasympathetic stimulation ↑ contractions** and **sympathetic stimulation ↓ contractions.**
- The **gastroileal reflex** is caused by food in the stomach, which stimulates peristalsis in the ileum and relaxes the ileocecal valve. This delivers intestinal contents to the large intestine.
- Small intestinal **secretions** are generally **alkaline**, serving to neutralize the acidic nature of the chyme entering from the pylorus.

Clinical Correlation

Peristalsis is activated by the **parasympathetic system**. For those suffering from decreased intestinal motility manifesting as constipation (paralytic ileus, diabetic gastroparesis), dopaminergic and cholinergic agents are often used (e.g., metoclopramide).

► The Large Intestine (Colon)

General Features

- The colon is larger in diameter and shorter in length than is the small intestine. Fecal material moves from the **cecum**, through the colon (**ascending**, **transverse**, **descending**, and **sigmoid colons**), rectum, and anal canal.
- Three longitudinal bands of muscle, the **teniae coli**, constitute the outer layer. Because the colon is longer than these bands, pouching occurs, creating **haustra** between the teniae and giving the colon its characteristic "caterpillar" appearance.
- The mucosa has **no villi**, and mucus is secreted by short, inward-projecting colonic glands.
- Abundant lymphoid follicles are found in the cecum and appendix and more sparsely elsewhere.
- The major functions of the colon are **reabsorption of fluid and electrolytes** and **temporary storage of feces**.

Colonic Motility

- **Peristaltic waves** briefly open the normally closed ileocecal valve, passing a small amount of chyme into the cecum. Peristalsis also advances the chyme in the colon. Slow waves, approximately 2/min, are initiated at the ileocecal valve and increase to approximately 6/min at the sigmoid colon.
- **Segmentation contractions** mix the contents of the colon back and forth.
- **Mass movement contractions** are found only in the colon. Constriction of long lengths of colon propels large amounts of chyme distally toward the anus. Mass movements propel feces into the rectum. Distention of the rectum with feces initiates the **defecation reflex**.

Absorption

The mucosa of the colon has great absorptive capability. **Na^+ is actively transported with water following**, and **K^+ and HCO_3^- are secreted into the colon**.

Defecation	
Feces	Contains undigested plant fibers, bacteria, inorganic matter, and water. Nondietary material (e.g., sloughed-off mucosa) constitutes a large portion of the feces. In normal feces, 30% of the solids may be bacteria. Bacteria synthesize **vitamin K**, B-complex vitamins, and folic acid, split urea to NH_3, and produce small organic acids from unabsorbed fat and carbohydrate.
Defecation	Rectal distention with feces activates intrinsic and cord reflexes that cause relaxation of the internal anal sphincter (smooth muscle) and produce the urge to defecate. If the external anal sphincter (skeletal muscle innervated by the **pudendal nerve**) is then voluntarily relaxed, and intra-abdominal pressure is increased via the **Valsalva maneuver**, defecation occurs. If the external sphincter is held contracted, the urge to defecate temporarily diminishes.
Gastrocolic reflex	Distention of the stomach by food **increases the frequency of mass movements** and produces the urge to defecate. This reflex is mediated by **parasympathetic** nerves.

► Vomiting

Vomiting occurs in three phases: **nausea**, **retching**, and **vomiting**.

- **Nausea**—hypersalivation, decreased gastric tone, increased duodenal and proximal jejunal tone → reflux of contents into stomach
- **Retching**—Gastric contents travel to the esophagus. Retching occurs if upper esophageal sphincter (UES) remains closed.
- **Vomiting**—If pressure increases enough to open the UES, vomiting occurs; vomiting can be triggered by oropharyngeal stimulation, gastric overdistention and gastroparesis, vestibular stimulation, or input from the **chemoreceptor trigger zone**, located in the **area postrema** in the floor of the fourth ventricle, which stimulates the medullary vomiting center.

▶ Antiemetics

Drug Class	Agents	Comments
$5HT_3$ antagonists	**Ondansetron** Granisetron, dolasetron	May act in chemoreceptor trigger zone and in peripheral sites
DA antagonists	**Phenothiazine**, metoclopramide*	Block D_2 receptors in chemoreceptor trigger zone
Cannabinoids	**Dronabinol**	Active ingredient in marijuana

▶ Emetics

Ipecac	• Locally irritates the GI tract and stimulates the chemoreceptor trigger zone • If emesis does not occur in 15-20 min, lavage must be used to remove ipecac
Apomorphine	• Dopamine-receptor agonist that stimulates the chemoreceptor trigger zone • Vomiting should occur within 5 min

*Also a prokinetic agent

▶ Gastrointestinal Hormones

Gastrointestinal hormones are released into the systemic circulation after physiologic stimulation (e.g., by food in gut), can exert their effects independent of the nervous system when administered exogenously, and have been chemically identified and synthesized. The five gastrointestinal hormones include **secretin**, **gastrin**, **cholecystokinin (CCK)**, **gastric inhibitory peptide (GIP)**, and **motilin**.

Hormone	Source	Stimulus	Actions
Gastrin*,‡	**G cells** of gastric antrum	• **Small peptides, amino acids, Ca^{2+}** in lumen of stomach • Vagus (via **GRP**) • **Stomach distension** • **Inhibited by: H^+** in lumen of antrum	• **↑ HCl secretion by parietal cells** • **Trophic effects on GI mucosa** • ↑ pepsinogen secretion by chief cells • ↑ histamine secretion by ECL cells
CCK*	**I cells** of duodenum and jejunum	• **Fatty acids, monoglycerides** • **Small peptides and amino acids**	• **Stimulates gallbladder contraction** and **relaxes sphincter of Oddi** • **↑ pancreatic enzyme secretion** • Augments secretin-induced stimulation of pancreatic HCO_3^- • **Inhibits gastric emptying** • Trophic effect on exocrine pancreas/gallbladder
Secretin†	**S cells** of duodenum	• **↓ pH** in duodenal lumen • **Fatty acids** in duodenal lumen	• **↑ pancreatic HCO_3^- secretion** (neutralizes H^+) • Trophic effect on exocrine pancreas • **↑ bile production** • **↓ gastric H^+ secretion**
GIP†	K cells of duodenum and jejunum	**Glucose, fatty acids, amino acids**	• **↑ insulin release** • **↓ gastric H^+ secretion**
Motilin	Enterochromaffin cells in duodenum and jejunum	Absence of food for >2 hours	Initiates MMC motility pattern in stomach and small intestine

(Continued)

► Gastrointestinal Hormones *(Cont'd.)*

Paracrines/ Neurocrines	Source	Stimulus	Actions
Somatostatin	D cells throughout GI tract	↓ pH in lumen	• ↓ gallbladder contraction, pancreatic secretion • ↓ gastric acid and pepsinogen secretion • ↓ small intestinal fluid secretion • ↓ ACh release from the myenteric plexus and decreases motility • ↓ α-cell release of glucagon, and β-cell release of insulin in pancreatic islet cells
Histamine	Enterochromaffin cells	• Gastrin • ACh	↑ **gastric acid secretion** (directly, and potentiates gastrin and vagal stimulation)
VIP[†, ¶]	Neurons in GI tract	• Vagal stimulation • Intestinal distention	• **Relaxation of intestinal smooth muscle, including sphincters** • ↑ **Pancreatic HCO_3^- secretion** • **Stimulates intestinal secretion of electrolytes and H_2O**
GRP	Vagal nerve endings	Cephalic stimulation, gastric distension	**Stimulates gastrin release** from G cells
Pancreatic polypeptide	F cells of pancreas, small intestine	Protein, fat, glucose in lumen	↓ pancreatic secretion
Enteroglucagon	L cells of intestine	—	• ↓ gastric, pancreatic secretions • ↑ insulin release

Definition of abbreviations: ECL, enterochromaffin-like cells; GIP, gastric inhibitory peptide; GRP, gastrin-releasing peptide.
*Member of gastrin-CCK family
†Member of secretin-glucagon family
Clinical Correlates:
‡**Zollinger-Ellison syndrome (gastrinoma)**—non-β islet-cell pancreatic tumor that produces gastrin, leading to ↑ in gastric acid secretion and development of peptic ulcer disease
¶**VIPoma**—tumor of non-α, non-β islet cells of the pancreas that secretes VIP, causing watery diarrhea

▶ Gastric Secretions

Secretion Product	Cell Type	Region of Stomach	Stimulus for Secretion	Inhibitors of Secretion	Action of Secretory Product
HCl	Parietal (oxyntic) cells	Body/fundus	• Gastrin • ACh (from vagus) • Histamine	• **Low pH** inhibits (by inhibiting gastrin) • Prostaglandins • Chyme in duodenum (via GIP and secretin)	• Kills pathogens • Activates pepsinogen to pepsin
Intrinsic factor					Necessary for **vitamin B_{12}** absorption by the ileum; forms complex with vitamin B_{12}
Pepsinogen (zymogen, precursor of **pepsin**)	Chief cells	Body/fundus	• ACh (from vagus) • Gastrin • HCl	H^+ (via somatostatin)	• Converted to pepsin by ↓ pH and pepsin (autocatalytic) • **Digests up to 20% of proteins**
Gastrin	G cells	Antrum	• Small peptides/aa • Vagus (via GRP) • Stomach distention	H^+ (via somatostatin)	• **↑ HCl secretion (parietal cells)** • ↑ pepsinogen secretion (chief cells) • ↑ histamine secretion by ECL cells
Mucus	Mucous cells	Entire stomach	ACh (from vagus)		Forms gel on mucosa to protect mucosa from HCl and pepsin; traps HCO_3^- to help neutralize acid

Definition of abbreviations: aa, amino acids; ECL, enterochromaffin-like cells.

▶ Mechanism of Gastric H^+ Secretion

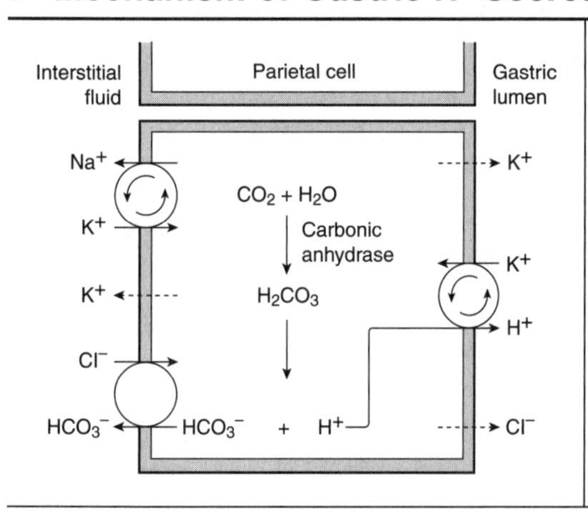

- In the parietal cell, CO_2 and H_2O are converted by carbonic anhydrase to H^+ and HCO_3^-.
- H^+ is secreted into the lumen of the stomach by H^+-K^+ pump (**H^+/K^+–ATPase**). Cl^- is secreted with H^+.
- HCO_3^- is absorbed into the bloodstream in exchange for Cl^-.
- After a meal, enough H^+ may be secreted to raise the pH of systemic blood and turn the urine alkaline **("alkaline tide")**.

The three primary triggers of H^+ secretion are:

- **ACh** (from vagus), via the M_3 muscarinic receptor
- **Histamine**, via the H_2 histamine receptor
- **Gastrin**, via unidentified receptor

▶ Drugs for Peptic Ulcer Disease

Drug Class	Agents	Comments
Antacids	Magnesium hydroxide, aluminum hydroxide, calcium carbonate	*Magnesium:* **laxative** effect *Aluminum hydroxide:* **constipating** effect
H_2 antagonists	**Cimetidine**, ranitidine, famotidine, nizatidine	• Useful in PUD, GERD, Zollinger-Ellison syndrome, but not as effective as proton pump inhibitors • **Cimetidine inhibits hepatic drug metabolizing enzymes** and has **antiandrogen effects**
Proton pump inhibitors	**Omeprazole,** lansoprazole, esomeprazole, pantoprazole, rabeprazole	• Irreversibly **inactivate H^+/K^+-ATPase**, thus blocking H^+ secretion. • Work very well—useful in PUD, Zollinger-Ellison syndrome, and GERD
Mucosal protective agents	**Sucralfate** Misoprostol	Polymerizes in the stomach and forms protective coating over ulcer beds. PGE_1 derivative used for peptic ulcers caused by NSAIDs
Antibiotics	Macrolides, metronidazole, tetracyclines (various combinations)	To treat ***H. pylori***

Definition of abbreviations: GERD, gastroesophageal reflux disorder; PGE_1, prostaglandin E_1; NSAIDs, nonsteroidal antiinflammatory drugs; PUD, peptic ulcer disease.

▶ Phases of Gastric Secretion

Cephalic phase	The smell, sight, thought, or chewing of food can increase gastric secretion via parasympathetic (vagal) pathways. Responsible for approximately 30% of acid secreted.
Gastric phase	Food in the stomach ↑ secretion. The greatest effects occur with proteins and peptides, leading to **gastrin** release (alcohol and caffeine also exert a strong effect). Gastric distention initiates vagovagal reflexes. Accounts for approximately 60% of acid secreted.
Intestinal phase	Protein digestion products in the duodenum stimulate duodenal gastrin secretion. In addition, absorbed amino acids act to stimulate H^+ secretion by parietal cells. The intestinal phase accounts for less than 10% of the gastric secretory response to a meal.

► Pancreatic Secretions

The exocrine secretions of the pancreas are produced by the **acinar cells**, which contain numerous enzyme-containing granules in their cytoplasm, and by the **ductal cells**, which secrete HCO_3^-. The secretions reach the duodenum via the **pancreatic duct**.

Bicarbonate (HCO_3^-)	• HCO_3^- in the duodenum **neutralizes HCl in chyme** entering from the stomach. This also deactivates pepsin. • When H^+ enters the duodenum, S cells secrete **secretin**, which acts on pancreatic ductal cells to increase HCO_3^- production. • HCO_3^- is produced by the action of **carbonic anhydrase** on CO_2 and H_2O in the pancreatic ductal cells. HCO_3^- is secreted into the lumen of the duct in exchange for Cl^-.
Pancreatic enzymes	• Approximately 15 enzymes are produced by the pancreas, which are responsible for **digesting proteins, carbohydrates, lipids, and nucleic acids**. • When small peptides, amino acids, and fatty acids enter the duodenum, **CCK** is released by I cells, stimulating pancreatic enzyme secretion. • **ACh** (via vagovagal reflexes) also stimulates enzyme secretion and potentiates the action of secretin. • **Protection of pancreatic acinar cells against self-digestion:** – **Proteolytic enzymes are secreted as inactive precursors**, which are activated in the gut lumen. For example, the duodenal brush border enzyme, **enterokinase**, converts trypsinogen to the active enzyme, trypsin. Trypsin then catalyzes the formation of more trypsin and activates chymotrypsinogen, procarboxypeptidase, and prophospholipases A and B. Ribonucleases, amylase, and lipase do not exist as proenzymes. – Produce **enzyme inhibitors** to inactivate trace amounts of active enzyme formed within.

Enzyme	Reaction Catalyzed
Proteases:	
Trypsin	Proteins → peptides
Chymotrypsin	Proteins → peptides
Carboxypeptidase	Peptides → amino acids
Polysaccharidase:	
Amylase	Starch and glycogen → maltose, maltotriose, and α-limit dextrins
Lipases:	
Phospholipases A and B	Phospholipids → phosphate, fatty acids, and glycerol
Esterases	Cholesterol esters → free cholesterol and fatty acids
Triacylglycerol lipases	Triglycerides → fatty acids and monoglycerides
Nucleases:	
Ribonuclease	RNA → ribonucleotides
Deoxyribonuclease	DNA → deoxyribonucleotides

► Hepatic Excretion

Physiologic roles	• Excretion of **bilirubin**, **cholesterol**, drugs, and **toxins** • Promotion of **intestinal lipid absorption** • Delivery of **IgA** to small intestine
Components of bile	• Bile—composed of **bile salts, phospholipids, cholesterol, bilirubin (bile pigments), water, and electrolytes**

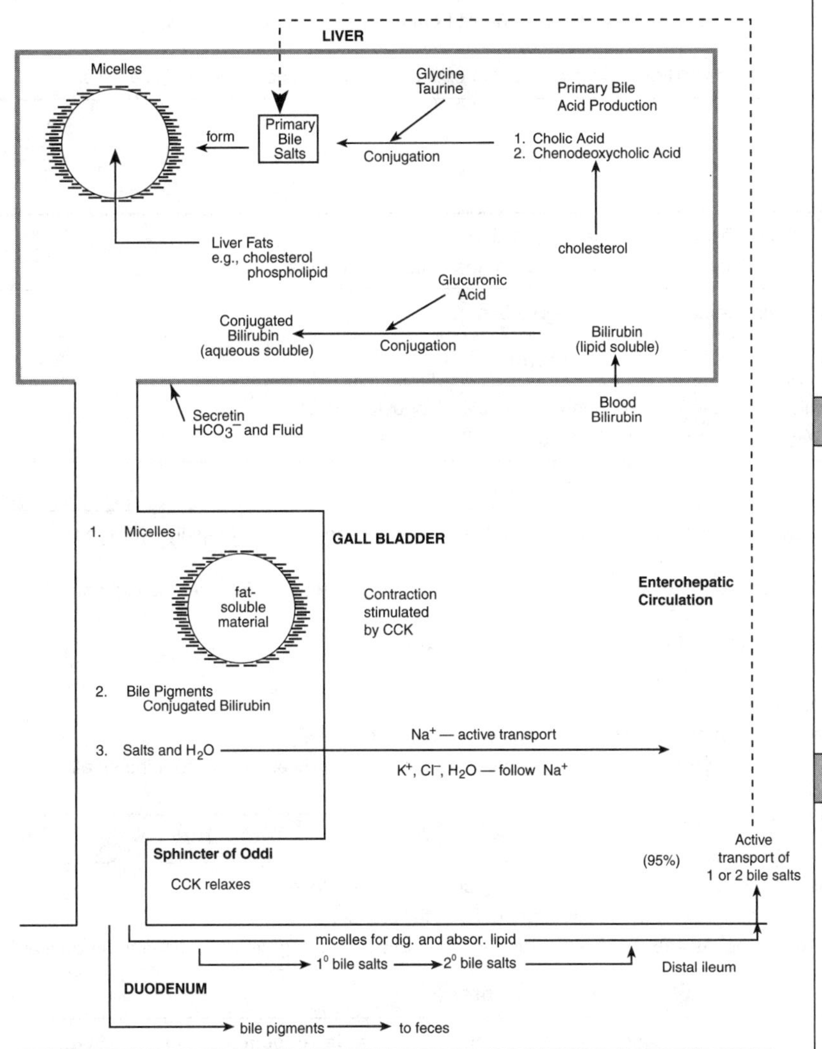

Formation of Bile

- Bile, produced by hepatocytes, drains into **hepatic ducts** and is stored in the **gallbladder** for later release.
- **Primary bile acids** (cholic and chenodeoxycholic acids) are made from cholesterol in the liver. **Secondary bile acids** (deoxycholic and lithocholic) are products of bacterial metabolism of primary bile acids in the gut.
- All bile acids must be conjugated with **glycine** or **taurine** to form their **bile salt** before being secreted into bile.
- Above a **critical micellar concentration**, bile salts form **micelles**. Electrolytes and H_2O are also added to the bile.

Micelles

- Micelles are water soluble-spheres with a lipid-soluble interior.
- Micelles are vital in the digestion, transport, and absorption of lipid-soluble substances from the duodenum to the distal ileum, where bile salts are actively reabsorbed and recycled (**enterohepatic circulation**).
- A lack of reabsorbing mechanisms or distal ileal disease can lead to deficiency of bile salts. This can lead to malabsorption, gallstones, and **steatorrhea**.

Gallbladder

- The gallbladder concentrates and stores bile for release during meals.
- During the interdigestive period, the **sphincter of Oddi** is closed and the gallbladder is relaxed, allowing it to fill with bile.
- Bile is **concentrated** in the gallbladder by water and electrolyte absorption.
- Small peptides and fatty acids in the duodenum cause **CCK** secretion, which causes **gallbladder contraction** and **relaxation of the sphincter of Oddi**. ACh also aids in this process.

(Continued)

► Hepatic Excretion *(Cont'd.)*

Bilirubin	• Bilirubin is a product of **heme metabolism**. • It is taken up by hepatocytes and **conjugated with glucuronic acid** prior to secretion into bile. This gives bile a golden yellow color. In the large intestine, bilirubin is deconjugated and metabolized by bacteria to form **urobilinogens** (colorless). Some of the urobilinogens are reabsorbed; most of the reabsorbed urobilinogens are secreted into bile, with the remainder excreted in the urine. • Most urobilinogen remains in the gut and is further reduced to pigmented compounds (stercobilins and urobilins) and excreted in feces. **Stercobilins** give a brown color to feces. • **Jaundice** (yellowing of the skin and whites of the eyes) is a result of **elevated bilirubin**.
Regulation of bile secretion	• **Secretin** stimulates the secretion of bile **high in HCO_3^- content** from the biliary ductules, but does not alter bile salt output. • Secretion of bile salts by hepatocytes is directly proportional to hepatic portal vein concentration of bile salts.

► Digestion and Absorption

- Carbohydrates, lipids, and proteins are digested and absorbed in the small intestine.
- The brush border of the small intestine increases surface area, greatly facilitating absorption of nutrients.

Carbohydrate digestion	Must be converted to **monosaccharides** in order to be absorbed
	Mouth
	• **Salivary amylase** normally hydrolyzes approximately 10 to 20% of ingested starch. • It **hydrolyzes** only **α-(1:4)-glycosidic linkages** to maltose, maltotriose, and α-limit dextrins.
	Intestine
	• **Pancreatic amylase** is found in the highest concentration in the duodenal lumen, where it rapidly hydrolyzes starch to oligosaccharides, maltose, maltotriose, and α-limit dextrins. • **Maltase**, **α-dextrinase, lactase**, **sucrase**, and **isomaltase** are found in the **brush border**, with the highest concentrations in the mid-jejunum and proximal ileum. – **α-dextrinase:** cleaves terminal α-1,4 bonds, producing free glucose – **Lactase:** converts lactose to glucose and galactose – **Sucrase:** converts sucrose to glucose and fructose – **Maltase:** converts maltose and maltotriose to 2 and 3 glucose units, respectively. • The monosaccharide end products (**glucose**, **galactose**, and **fructose**) are readily absorbed from the small intestine, primarily in the jejunum.
	Clinical Correlation
	• **Lactase deficiency** causes an inability to digest lactose into glucose and galactose. • *Consequence:* ↑ osmotic load, giving rise to **osmotic diarrhea** and **flatulence**. • Very common in African Americans, Asians, and Mediterraneans, and to a lesser degree, in Europeans/Americans.
Carbohydrate absorption	**Luminal Membrane**
	• **Glucose and galactose** compete for transport across the brush border by a **Na^+-dependent coporter (SGLT-1)**. Na^+ moves down its gradient into the enterocyte and the sugars move up their concentration gradients into the enterocyte (secondarily active). A Na^+-K^+ ATPase in the basolateral membrane helps maintain the Na^+ gradient (by keeping intracellular Na^+ low). • **Fructose:** facilitated diffusion (down its concentration gradient) via GLUT-5 transporter
	Basolateral Membrane
	Glucose, galactose, and **fructose** are transported across the basolateral membrane via facilitated diffusion (GLUT-2 transporter).

(*Continued*)

▶ Digestion and Absorption (*Cont'd.*)

Lipid digestion	**Stomach**
	• Fatty materials are **pulverized** to decrease particle size and increase surface area. • **CCK** slows gastric emptying to allow enough time for digestion and absorption in the small intestine.
	Small Intestine
	• **Bile acid micelles** emulsify fat. • **Pancreatic lipases** digest fat. • Fats are hydrolyzed by pancreatic lipases to **free fatty acids**, **monoacylglycerols**, and other lipids (e.g., **cholesterol**, and **fat-soluble vitamins A, D, E, K**), which collect in micelles.
Lipid absorption	• **Micelles** carry products of fat digestion in the aqueous fluid of the gut lumen to the brush border, where they can diffuse into the enterocyte. • Enterocytes **re-esterify the fatty acids** to form triacylglycerols, phospholipids, and cholesteryl esters, which are incorporated with **apoproteins** into **chylomicrons**. • Chylomicrons are **released by exocytosis** into the intercellular spaces, where they **enter the lacteals** of the lymphatic system. They then enter the venous circulation via the **thoracic duct**. • Glycerol diffuses into portal blood and is either oxidized for energy or stored as glycogen. • **Triacylglycerols** with **medium- and short-chain fatty acids** are hydrolyzed quickly and do not require micelle formation for absorption. They undergo little re-esterification and are absorbed directly into the portal venous system.
	Clinical Correlation
	Abetalipoproteinemia results from a deficiency of apoprotein B, causing an inability to transport chylomicrons out of intestinal cells.
Protein digestion	**Stomach**
	Pepsin begins protein digestion in the stomach. It functions best at pH 2 and is irreversibly deactivated above pH 5; therefore, it will be inactivated in the duodenum. Pepsin is not an essential enzyme.
	Small Intestine
	Protein digestion continues with **pancreatic proteases** (trypsin, chymotrypsin, elastase, carboxypeptidases A and B) activated by brush border peptidases. These are essential enzymes.
Protein absorption	**Luminal Membrane**
	• Protein products can be absorbed as **amino acids**, **dipeptides**, and **tripeptides**. • Amino acids are absorbed via **Na^+-dependent amino acid cotransport**. Many different transport systems have been identified, e.g., carriers for neutral, basic, acidic, and imino amino acids. • Dipeptides and tripeptides are absorbed via an **H^+-dependent cotransport** mechanism.
	Basal Membrane
	• Dipeptides and tripeptides are hydrolyzed to amino acids intracellularly. • Amino acids are transported to the blood by facilitated diffusion.
	Clinical Correlation
	Hartnup disease is a disorder in which neutral amino acids cannot be absorbed.
Water and electrolyte absorption	The absorption of water and electrolytes occurs mainly in the **small intestine**. Approximately 5–10 liters of fluid must be absorbed daily (intake and secretion), with 80–90% being absorbed in the small intestine at a maximal rate of 700 mL/h.

(Continued)

► Digestion and Absorption *(Cont'd.)*

NaCl	• In the **proximal intestine**, there is **Na^+-H^+ exchange**, **Na^+-glucose or Na^+-amino acid cotransport**, **Na^+-Cl^- cotransport,** and **passive diffusion** through Na^+ channels. • In the colon, passive diffusion through Na^+ channels is more important and is stimulated by **aldosterone**. • **Cl^-** is absorbed via Na^+-Cl^- cotransport, Cl^--HCO_3^- exchange, and passive diffusion.
K^+	• K^+ absorption occurs in the small intestine by passive diffusion. • K^+ is secreted in the colon (stimulated by aldosterone).
Ca^{2+}	• Absorption in the small intestine is via a vitamin D-dependent carrier. • Vitamin D deficiency → ↓ Ca^{2+} absorption → **osteomalacia** (adults) and **rickets** (children).
H_2O	• Secondary to solute absorption • Isoosmotic absorption in gallbladder and small intestine; permeability is lower in colon.
Iron	• Absorbed as **free Fe^{2+}** or as **heme** iron, primarily in the duodenum • **Fe^{2+}** is bound to transferrin in the blood.
Vitamins	• **Fat-soluble (A, D, E, K)**—incorporated into micelles and absorbed • **Water soluble**—usually via Na^+-dependent cotransporters • **Vitamin B_{12}**—absorption occurs in ileum and is transported while bound to intrinsic factor; ↓ intrinsic factor (gastrectomy) → **pernicious anemia**
H_2O, electrolyte secretion	Secretion occurs in **crypts**. **Cl^-** is the main ion secreted, via cAMP-regulated channels in the luminal membrane. **Clinical Correlation** **Cholera toxin** stimulates adenylate cyclase → ↑ cAMP → open Cl^- channels; Na^+ and H_2O follow → secretory diarrhea.

Gastrointestinal Pathology

▶ Lesions of the Oral Cavity

Leukoplakia	• White plaques on oral mucosa, produced by hyperkeratosis of the epithelium • 10% have epithelial dysplasia, a precancerous lesion • *Predisposing factors:* smoking, smokeless tobacco, alcohol abuse, chronic friction, and irritants
Erythroplakia	• Flat, smooth, and red. • Significant numbers of atypical epithelial cells • High risk of malignant transformation
Hairy leukoplakia	• Wrinkled patches on side of tongue • Epstein-Barr virus associated • No malignant transformation
Lichen planus	White reticulate lesions on the buccal mucosa and tongue
Tumors of the Oral Cavity	
Benign tumors	Hemangiomas, hamartomas, fibromas, lipomas, adenomas, papillomas, neurofibromas, and nevi
Malignant tumors	• **Squamous carcinoma** most common. Peak incidence from ages 40–70 years. • Associated with tobacco and alcohol use, particularly when used together • **Lower lip** most common site, but may affect floor of mouth and tongue

▶ Esophageal Pathology

Achalasia	• **Lack of relaxation of the LES** secondary to loss of myenteric plexus • *Most common ages:* 30–50. • *Symptoms:* dysphagia, regurgitation, aspiration, chest pain • Can be idiopathic, secondary to Chagas disease (*Trypanosoma cruzi*), or malignancy
Barrett esophagus	• Gastric or intestinal **columnar epithelium** replaces normal squamous epithelium • Occurs with chronic insult, usually reflux (**increases risk of adenocarcinoma** 30–40 times)
Boerhaave syndrome	• Violent retching causes potentially fatal esophageal rupture
Diverticula	• Sac-like protrusions of one or more layers of the pharyngeal or esophageal wall • **Zenker diverticula:** – Occur at the junction of the pharynx and esophagus in elderly men – *Symptoms:* dysphagia and regurgitation of undigested food soon after ingestion • **Traction diverticula:** true diverticula in mid-esophagus; usually asymptomatic
Esophageal carcinoma	• Most are **adenocarcinomas** occurring after 50 and have male:female ratio of 4:1 • Incidence higher in northern Iran, Central Asia • Associated with smoking, alcohol, nitrosamines, achalasia, Barrett esophagus, and vitamin A deficiency • Presents with dysphagia (first to solids) • Liver and lung most common sites of metastasis; poor prognosis
Esophageal strictures	• Narrowing of the esophagus, often as a result of fibrosis after severe inflammation. Caused by reflux, Herpes virus, Cytomegalovirus, *Candida,* chemical burns (e.g., lye ingestion). • Carcinoma should be ruled out
Esophageal varices	• Dilated tortuous vessels of the esophageal venous plexus resulting from **portal hypertension** • Esophageal varices are prone to **bleeding**; may be life-threatening

(Continued)

► Esophageal Pathology *(Cont'd.)*

Esophagitis	• Most common cause is reflux; other causes include infections (Herpes virus, Cytomegalovirus, *Candida*) and eosinophilic esophagitis	
Hernia	**Sliding**	90% of cases, gastroesophageal junction above diaphragm, associated with **reflux**
	Paraesophageal	Gastric cardia above diaphragm, gastroesophageal junction remains in the abdomen; herniated organ at risk for strangulation and infarction
Mallory-Weiss tears	Occur at gastroesophageal junction secondary to recurrent **forceful vomiting**, usually seen in **alcoholics**	
Schatzki rings	Mucosal rings at the squamocolumnar junction below the aortic arch	
Tracheoesophageal fistula	• Usually esophageal blind pouch with a fistula between the lower segment of the esophagus and trachea • Associated with congenital heart disease and other gastrointestinal malformations	
Webs	• Mucosal folds in the upper esophagus above the aortic arch • **Plummer-Vinson syndrome:** dysphagia, glossitis, iron-deficiency anemia, and esophageal webs	

Definition of abbreviation: LES, lower esophageal sphincter.

► Stomach Pathology

Acute gastritis (erosive)	Can be caused by alcohol, aspirin, smoking, shock, steroids, and uremia Patients experience heartburn, epigastric pain, nausea, vomiting, and hematemesis
Chronic gastritis	
Fundal (type A)	Autoimmune; associated with **pernicious anemia**, achlorhydria, and intrinsic factor deficiency
Antral (type B)	Caused by ***Helicobacter pylori*** and is most common form of chronic gastritis in U.S.
Carcinoma	• *Risk factors:* genetic predisposition, diet, hypochlorhydria, pernicious anemia, and nitrosamines • Usually asymptomatic until late, then presents with anorexia, weight loss, anemia, epigastric pain. Virchow node (left supraclavicular lymph node) common site of metastasis • *Pathology:* 50% arise in the antrum and pylorus • *Linitis plastica:* infiltrating gastric carcinoma with a diffuse fibrous response • *Histology:* signet ring cells characteristic of gastric carcinoma
Hypertrophic gastropathy	• **Menetrier disease:** markedly thickened rugae due to hyperplastic superficial mucus glands with atrophy of deeper glands • **Hypertrophic-hypersecretory gastropathy:** hyperplasia of parietal and chief cells in gastric glands • **Excessive gastrin secretion (e.g., gastrinoma, Zollinger-Ellison syndrome):** produces gastric gland hyperplasia. Risk of peptic ulcer disease.
Peptic ulcers	• *Common locations:* proximal duodenum, stomach, and esophagus • ***H. pylori*** infection important etiologic factor. Modification of acid secretion coupled with antibiotic therapy that eradicates *H. pylori* is apparently curative in most patients. • *Symptoms:* episodic epigastric pain; *complications:* hemorrhage, perforation • Duodenal ulcers do not become malignant. Gastric ulcers only rarely • *Stress ulcers:* burns → Curling ulcers; CNS trauma → Cushing ulcers
Pyloric stenosis	• Congenital hypertrophy of pyloric muscle • *Classic case:* firstborn boy, presenting with **projectile vomiting** 3–4 weeks after birth; associated with a palpable **"olive" mass** in epigastric region • *Treatment:* surgical

► Small Intestine Pathology

Celiac sprue	• Allergic reaction to the **gliadin** component of gluten; genetic predisposition • Predisposes to neoplasm, especially lymphoma • *Pathology:* atrophy of villi in the jejunum; affects only **proximal** small bowel
Congenital anomalies	• **Meckel diverticulum:** persistent omphalomesenteric vitelline duct. Located near ileocecal valve. May contain ectopic gastric, pancreatic, or endometrial tissue, which may produce ulceration • **Vitelline fistula:** direct connection between the intestinal lumen and the outside of the body at the umbilicus due to persistence of the vitelline duct. Associated with drainage of meconium from the umbilicus • Atresia: congenital absence of a region of bowel (e.g., **duodenal atresia**); polyhydramnios, obstruction, and bile-stained vomiting in neonate • Stenosis: narrowing that may cause obstruction • **Omphalocele:** when the midgut loop fails to return to the abdominal cavity, forming a light gray shiny sac at the base of the umbilical cord filled with loops of small intestine • **Gastroschisis:** A failure of the lateral body folds to fuse causes extrusion of the intestines through a open defect in the abdominal wall that is usually to the right of the umbilicus; unlike omphalocele, the intestines are exposed to the open air rather than being covered with a membrane.
Hernias	• Cause 15% of small intestinal obstruction, most commonly at the inguinal and femoral canals • **Inguinal hernias** – **Indirect inguinal hernia:** the intestinal loop goes through the internal (deep) inguinal ring, external (superficial) inguinal ring, and into the scrotum. Much more common in males; may present in infancy if there is a failure of the processus vaginalis to close. – **Direct inguinal hernia:** the intestinal loop protrudes through the inguinal (Hesselbach's) triangle to cause a bulge in the abdominal wall medial to the inferior epigastric artery. Hesselbach's triangle is defined by the inferior epigastric artery, the lateral border of the rectus abdominus, and the inguinal ligament. The typical patient is an older man, and the hernia is covered by the external spermatic fascia. • **Femoral hernia:** the intestinal loop protrudes below the inguinal ligament through the femoral canal below and lateral to the pubic tubercle. More common in women. This type of hernia is particularly likely to produce a dangerous bowel incarceration. • **Diaphragmatic hernia:** infants with defective development of pleuroperitoneal membrane may have intestines and other abdominal structures herniated into the thorax; potentially fatal because of impairment of lung expansion
Ischemic bowel disease	• Thrombosis or embolism of the **superior mesenteric artery** accounts for approximately 50% of cases; venous thrombosis for 25% of cases • Internal hernias can strangulate entrapped loops of bowel • Usually after age 60 and presents with abdominal pain, nausea, and vomiting
Intussusception	• Telescoping of one segment of bowel into another • More common in infants and children; may be reduced with a diagnostic barium enema • In adults, lead point usually an intraluminal mass; usually requires surgery
Lymphoma	• Usually non-Hodgkin, large cell, diffuse type • In immunosuppressed patients, the incidence of primary lymphomas of small intestine is increasing. • **MALToma:** often follicular and follow a more benign course; associated with *H. pylori* infection; may regress after antibiotic therapy.
Necrotizing enterocolitis	• Life-threatening acute, necrotizing inflammation of the small and large intestines • Usually in premature or low birth weight neonates • Peak incidence when babies start oral foods at 2 to 4 days, but can occur any time in first three months of life • If surgical treatment is required, it may cause short bowel syndrome
Tropical sprue	• Unknown etiology; high incidence in the tropics; especially Vietnam, Puerto Rico • *Pathology:* similar to changes in celiac disease, but affects **entire** length of small bowel
Volvulus	• Twisting of the bowel about its mesenteric base; may cause obstruction and infarction • May be associated with malrotation of the midgut • Typical patient is elderly
Whipple disease	• Rare, periodic acid-Schiff (PAS)–positive **macrophages in the lamina propria** of intestines • Caused by small bacilli ***(Tropheryma whippelii)***; more common in men (10:1)

► Appendix Pathology

Appendicitis	• The vermiform appendix may become inflamed as a result of either an obstruction by stool, which forms a **fecalith** (common in adults), or **hyperplasia** of its lymphatic tissue (common in children). • An inflamed appendix may stimulate visceral pain fibers, which course back in the lesser splanchnic nerves and result in colicky pain referred over the umbilical region.

► Large Intestine Pathology

Angiodysplasia	• Dilated tortuous vessels of the right colon → lower gastrointestinal bleeding in elderly • Highest incidence in the cecum
Diverticular disease	• Multiple outpouchings of colon present in 30–50% of adults; higher incidence with ↑ age • Presents with pain and fever
Hirschsprung disease	• **Absence of ganglion cells** of Meissner and Auerbach plexus in distal colon • Produces markedly distended colon, proximal to aganglionic portion • **Failure to pass meconium**, with constipation, vomiting, and abdominal distention
Imperforate anus	Failure of perforation of the membrane that separates endodermal hindgut from ectodermal anal dimple
Polyps	
Tubular adenomas	• Pedunculated polyps; 75% of adenomatous polyps • Sporadic or familial; average age of onset is 60; most occur in left colon • Cancer occurs in approximately 4% of patients
Villous adenomas	• Largest, least common polyps; usually sessile • **1/3 cancerous**
Tubulovillous adenomas	• Combined tubular and villous elements • **Increased villous elements →↑ likelihood of malignant transformation**
Polyposis Syndromes	
Peutz-Jeghers syndrome	• **Autosomal dominant**; involves entire gastrointestinal tract; **melanin pigmentation** of the buccal mucosa • Polyps—**hamartomas**; not premalignant
Turcot syndrome	Colonic polyps associated with brain tumors
Familial multiple polyposis	• **Autosomal dominant**; appearance of polyps during adolescence • Start in rectosigmoid area and spread to cover entire colon • **Virtually all patients develop cancers**; prophylactic total colectomy recommended
Gardner syndrome	• Colonic polyps associated with **desmoid tumors** • **Risk of colon cancer nearly 100%**
Malignant Tumors	
Adenocarcinoma	• **98% of all colonic cancers;** third most common tumor in both women and men; peak incidence in 60s • *Symptoms:* rectal bleeding, change in bowel habits, weakness, malaise, and weight loss • Tumor spreads by direct extension and **metastasis to nodes, liver, lungs, bones** • **Carcinoembryonic antigen (CEA) tumor marker** helps to monitor tumor recurrence after surgery • 75% of tumors in rectum, sigmoid colon • **Left-sided lesions:** annular constriction, infiltration of the wall, **obstruction** • **Right-sided lesions:** often bulky, polypoid, protuberant masses; **rarely obstruct** because fecal stream is liquid on right side
Squamous cell carcinoma	• Occur in anal region, associated with **papilloma viruses** • Incidence rising in homosexual men with AIDS

▶ Inflammatory Bowel Disease: Crohn Disease Versus Ulcerative Colitis

	Crohn Disease	Ulcerative Colitis
Most common site	Terminal ileum	Rectum
Distribution	Mouth to anus	Rectum → colon; "backwash" ileitis
Spread	Discontinuous/"skip"	Continuous
Gross features	Focal ulceration with intervening normal mucosa, linear fissures, cobblestone appearance, thickened bowel wall, "creeping fat"	Extensive ulceration, pseudopolyps
Micro	Noncaseating granulomas	Crypt abscesses
Inflammation	Transmural	Limited to mucosa and submucosa
Complications	Strictures, "string sign" on barium studies, obstruction, abscesses, fistulas, sinus tracts	Toxic megacolon
Genetic association	Family history of any type of inflammatory bowel disease is associated with increased risk.	
Extraintestinal manifestations	Less common	Common (e.g., arthritis, spondylitis [HLA B27 positive], primary sclerosing cholangitis, erythema nodosum, pyoderma gangrenosum)
Cancer risk	Slight 1–3%	5–25%

▶ Exocrine Pancreas

Acute hemorrhagic pancreatitis	• Diffuse necrosis of the pancreas by release of activated enzymes • Most often associated with **alcoholism** and **biliary tract disease** • *Symptoms:* sudden onset of acute, continuous, and intense abdominal pain, often radiating to back; accompanied by nausea, vomiting, and fever → frequently results in shock • *Lab values:* **high amylase**, **high lipase** (elevated after 3–4 days), leukocytosis • *Gross:* gray areas of enzymatic destruction, white areas of fat necrosis, red areas of hemorrhage
Chronic pancreatitis	• Remitting and relapsing episodes of mild pancreatitis → progressive pancreatic damage • X-rays reveal **calcifications** in pancreas • Chronic pancreatitis may result in **pseudocyst formation**, **diabetes**, steatorrhea
Pseudocysts	• Possible sequelae of pancreatitis or trauma • Up to 10 cm in diameter with a fibrous capsule; no epithelial lining or direct communication with ducts
Carcinoma	• *Risk factors:* smoking, high-fat diet, chemical exposure • Commonly develop in **head of the pancreas**, may result in compression of bile duct and main pancreatic duct → obstructive jaundice • **Asymptomatic until late** in course, then weight loss, abdominal pain (classically, epigastric pain radiating to back), jaundice, weakness, anorexia; **Trousseau syndrome** (**migratory thrombophlebitis**) often seen • Very poor prognosis
Cystic fibrosis	• Autosomal recessive; ***CFTR*** (cystic fibrosis transmembrane conductance regulator protein) gene located on chromosome 7; **Δ508** is a common mutation • **Defective chloride channel**: secretion of very thick mucus and **high sodium and chloride levels in sweat** • 10% present with **meconium ileus** (most present during first year with steatorrhea, **pulmonary infections**, and obstructive pulmonary disease) • ***Pseudomonas aeruginosa*** is most common etiologic agent; **blue-green sputum** • Median survival age 37; mortality most often due to pulmonary infections
Annular pancreas	Occurs when the ventral and dorsal pancreatic buds form a ring around the duodenum → obstruction of the duodenum

► Congenital Hepatic Diseases

Extrahepatic biliary atresia	• Incomplete recanalization → cholestasis, cirrhosis, portal hypertension • **Within first weeks of life:** jaundice, dark urine, light stools, hepatosplenomegaly
Intrahepatic biliary atresia	• Diminished number of bile ducts; sometimes associated with **α-1-antitrypsin deficiency** • **Presents in infancy** with cholestasis, pruritus, growth retardation, ↑ serum lipids • Icterus visible when serum bilirubin exceeds 2 mg/dL (true in any case of jaundice)
Conjugated hyperbilirubinemia	• **Dubin-Johnson syndrome:** benign conjugated hyperbilirubinemia due to impaired transport; liver grossly **black** • **Rotor syndrome:** asymptomatic, similar to Dubin-Johnson, but the liver **not pigmented**
Unconjugated hyperbilirubinemia	• Can be due to hemolysis, diffuse hepatocellular damage, enzymatic defect • **Gilbert syndrome:** autosomal recessive disease; deficiency of glucuronyl transferase; benign • **Crigler-Najjar syndrome:** – *Type 1:* autosomal recessive with **complete absence of glucuronyl transferase**, marked unconjugated hyperbilirubinemia, severe kernicterus, death – *Type 2:* autosomal dominant with **mild deficiency of glucuronyl transferase**; no kernicterus

► Common Patterns of Liver Disease Presentation

Cholestasis	• Impaired excretion of conjugated bilirubin; can have chalky stool • *Intrahepatic:* viral hepatitis, cirrhosis, drug toxicity • *Extrahepatic:* gallstones, carcinoma of bile ducts, ampulla of Vater or head of pancreas
Hepatic failure	Causes jaundice, encephalopathy, renal failure, palmar erythema, spider angiomas, gynecomastia, testicular atrophy, prolonged prothrombin time, hypoalbuminemia
Chronic passive congestion	• Associated with **right heart failure** • *Pathology:* congestion of central veins and centrilobular hepatic sinusoids (known as **"nutmeg liver"**)

► Acquired Hepatic Diseases

<table>
<tr><td rowspan="4">Alcoholic liver disease</td><td colspan="2">• Three major forms: 1) fatty liver, 2) alcoholic hepatitis, 3) alcoholic cirrhosis</td></tr>
<tr><td>Fatty liver</td><td>• Yellow, greasy liver
• Accumulation of lipids within hepatocytes
• Potentially reversible</td></tr>
<tr><td>Alcoholic hepatitis</td><td>• Alcoholic hepatitis usually associated with fatty change; occasionally seen with cirrhosis
• Results from a heavy drinking binge
• Note: Mallory bodies may be seen, but may also be seen in Wilson disease, hepatocellular carcinoma, and primary biliary cirrhosis
• AST/ALT > 2.0 indicates alcoholic liver disease</td></tr>
<tr><td>Cirrhosis</td><td>• Third leading cause of death in the 25- to 65-year-old age group
• Leading etiologies: alcoholism and hepatitis C</td></tr>
<tr><td>Alpha-1-antitrypsin deficiency</td><td colspan="2">• Autosomal recessive; characterized by deficiency of a protease inhibitor
• Results in pulmonary emphysema and hepatic damage (cirrhosis)</td></tr>
<tr><td>Budd-Chiari syndrome</td><td colspan="2">• Congestive liver disease secondary to thrombosis of the inferior vena cava or hepatic veins
• Causes “nutmeg liver” (also seen in right heart failure) with centrilobular congestion and necrosis
• May develop rapidly or slowly; more likely to be fatal if it develops rapidly</td></tr>
<tr><td>Hemochromatosis</td><td colspan="2">• Primary form autosomal recessive inheritance; secondary form usually related to multiple blood transfusions
• Deposits of iron in the liver, pancreas, heart, adrenal, skin “bronze diabetes”
• Also seen: cardiac arrhythmias, gonadal insufficiency, arthropathy
• High incidence of hepatocellular carcinoma</td></tr>
<tr><td>Portal hypertension</td><td colspan="2">• Intrahepatic: most common cause and usually secondary to cirrhosis of the liver; other causes: schistosomiasis, sarcoid
• Posthepatic: right-sided heart failure, Budd-Chiari syndrome
• Prehepatic: portal vein obstruction
• Clinical: ascites, portosystemic shunts that form hemorrhoids, esophageal varices, periumbilical varices (caput medusae), encephalopathy, splenomegaly
• Additionally, impaired estrogen metabolism: gynecomastia, gonadal atrophy, amenorrhea in females, spider angiomata, palmar erythema</td></tr>
<tr><td>Primary biliary cirrhosis</td><td colspan="2">• Autoimmune etiology; causes sclerosing cholangitis, cholangiolitis
• Associated with other autoimmune diseases; primarily affects middle-aged women
• Presents with fatigue and pruritus; elevated alkaline phosphatase
• Antimitochondrial antibody in over 90% of patients</td></tr>
<tr><td>Secondary biliary cirrhosis</td><td colspan="2">• Longstanding large bile duct obstruction, stasis of bile, inflammation, secondary infection, and scarring
• Usually presents with jaundice</td></tr>
<tr><td>Reye syndrome</td><td colspan="2">• Usually affects children between 6 months and 15 years of age
• Characterized by fatty change in the liver, edematous encephalopathy
• Etiology: unclear; frequently preceded by a mild upper respiratory infection, varicella, influenza A or B infection
• Also associated with aspirin administration at levels not ordinarily toxic</td></tr>
<tr><td>Sclerosing cholangitis</td><td colspan="2">• Chronic fibrosing inflammatory disease of the extrahepatic and larger intrahepatic bile ducts
• Associated with inflammatory bowel disease; predisposition for cholangiocarcinoma</td></tr>
<tr><td>Wilson disease (hepatolenticular degeneration)</td><td colspan="2">• Autosomal recessive—inadequate excretion of copper
• Clinical: rarely manifests before age 6, then presents with weakness, jaundice, fever, angiomas, and eventually portal hypertension; CNS manifestations: tremor, rigidity, disorders of affect and thought
• Labs: low serum ceruloplasmin; ↑ urinary copper excretion
• Pathology: macronodular cirrhosis, degenerative changes in the lenticular nuclei of brain, pathognomonic Kayser-Fleischer rings, a deposition of copper in Descemet membrane of the corneal limbus</td></tr>
</table>

► Hepatic Tumors

Liver cell adenoma (benign)	• ↑ incidence with **anabolic steroid** and **oral contraceptive use** • Forms a mass, which may be mistaken for carcinoma, or may rupture (especially during pregnancy)
Nodular hyperplasia (benign)	• Appears as solitary nodule that often has a fibrous capsule and bile ductules • Stellate fibrous core usually present • Nodular regenerative hyperplasia—multiple nodules composed of normal hepatocytes with loss of normal architecture
Cholangiocarcinomas	• 10% of primary liver neoplasms; **associated with primary sclerosing cholangitis** • In developing countries, also associated with **infection with *Clonorchis sinensis*** (liver fluke) • *Clinical:* weight loss, jaundice, pruritus • 50% metastasize to lungs, bones, adrenals, and brain, exhibiting both hematogenous and lymphatic spread
Hepatoblastoma	• Rare, malignant neoplasm of children • Hepatomegaly, vomiting, diarrhea, weight loss, elevated serum levels of AFP
Hepatocellular carcinoma	• **90% of primary liver neoplasms; strongly associated with cirrhosis, HCV and HBV** infections • *Clinical:* tender hepatomegaly, ascites, weight loss, fever, polycythemia, hypoglycemia • **Alpha-fetoprotein is present in 50–90%** of patients' serum (AFP also found with other forms of liver disease, pregnancy, fetal neural tube defects, germ-cell carcinomas of the ovaries and testes) • Death due to gastrointestinal bleed and liver failure; generally, metastases first occur in lungs

► Hepatic Infections

Acute viral hepatitis	• Can be icteric or anicteric • *Symptoms:* malaise, anorexia, fever, nausea, upper abdominal pain, hepatomegaly • *Labs:* elevated transaminases
Chronic hepatitis	• 5–10% of HBV infections and **well over 50% of HCV**; *other etiologies:* drug toxicity, Wilson disease, alcohol, α-1-antitrypsin deficiency, autoimmune hepatitis • *Histology:* chronic inflammation with hepatocyte destruction, cirrhosis, liver failure
Fulminant hepatitis	• **Massive hepatic necrosis** and progressive hepatic dysfunction; mortality of 25–90% • *Etiologies:* HBV, HCV, delta virus (HDV) superinfection, HEV, chloroform, carbon tetrachloride, certain mushrooms, acetaminophen overdose • *Pathology:* progressive shrinkage of liver as parenchyma is destroyed
Liver abscesses	Pyogenic: • *E. coli, Klebsiella, Streptococcus, Staphylococcus*; ascending cholangitis most common cause • Seeding of liver due to bacteremia another potential cause Parasitic: • *Entamoeba histolytica:* especially in men over age 40 following intestinal disease; thick, brown abscess fluid • *Ascaris lumbricoides:* can cause blockage of bile ducts, eosinophilia, verminous abscesses
Parasitic infections	• **Schistosomiasis:** splenomegaly, portal hypertension, ascites • **Amebiasis:** *Entamoeba histolytica*, bloody diarrhea, pain, fever, jaundice, hepatomegaly

Definition of abbreviations: HBV, hepatitis B virus; HCV, hepatitis C virus.

► Characteristics of Viral Hepatitides

	Hepatitis A	Hepatitis B	Hepatitis C	Hepatitis D	Hepatitis E
Nucleic acid	RNA (Picornavirus)	DNA (Hepadnavirus)	RNA (Flavivirus)	RNA	RNA (Hepevirus)
Characteristics	• 50% seropositivity in people >50 • *Clinical disease:* mild or asymptomatic; rare after childhood	• Worldwide carrier rate 300 million • 300,000 new infections/year in U.S.	• 150,000 new cases/year in U.S. • **Most important cause of transfusion-related hepatitis**	• Replication defective • **Dependent on HBV coinfection** for multiplication	**Fulminant** hepatitis 0.3–3%; **20% in pregnant women**
Transmission	• **Fecal-oral**, raw shellfish (concentrate virus) • Not shed in semen, saliva, urine • Shed in stool 2 weeks before onset of jaundice and 1 week after	• **Parenteral**, close personal contact • Transfusion • Dialysis • Needle-sticks • IV drug use • Male homosexual activity	• **Parenteral**, close personal contact • **Route of transmission undetermined in 40–50%** of cases	**Parenteral**, close personal contact	• **Waterborne** • Young adults
Incubation	2–6 weeks	4–26 weeks	2–26 weeks	4–7 weeks in superinfection	2–8 weeks
Carrier state	**None**	1% blood donors	1%	1–10% in drug addicts	Unknown
Progression to chronic hepatitis	None	5–10% acute infections, 90% in infants	**>80%**	• <5% in coinfection* • 80% superinfection†	None
Increased risk of hepatocellular carcinoma	No	**Yes**	**Yes**	**Yes**, same as for B	Unknown, although not likely
Diagnosis	IgM against HAV	• **HBcAb is the 1st antibody, HBsAg indicates current infection** • **HBeAg indicates infectivity**	ELISA for HCV Abs	Ab to Delta Agent plus HBsAg	ELISA for HEV Abs

*Coinfection: hepatitis B and delta agent acquired at the same time
†Superinfection: delta agent acquired during chronic hepatitis B infection

▶ Serology of Hepatitis B Infection

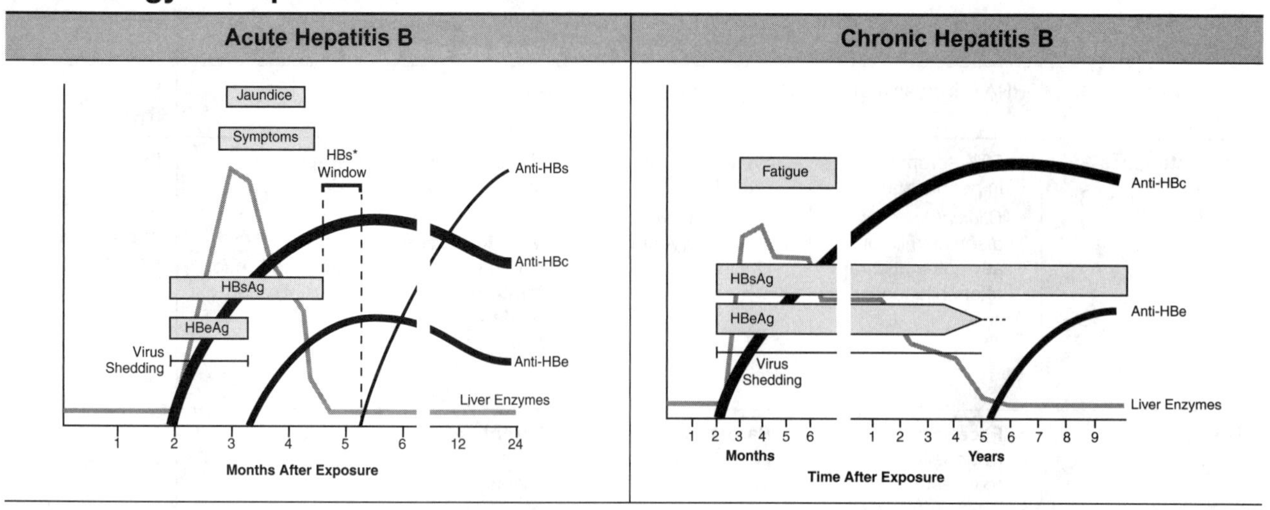

*The window is the time between the disappearance of the HBsAg and before antibody to the surface antigen is detected.

▶ Biliary Disease

Cholelithiasis (gallstones)	• 20% of women and 8% of men in U.S.; rare before age 20, but seen in 25% of persons >60 years • Most stones remain in gallbladder and are asymptomatic • Famous **"4 Fs": fat, female, fertile (multiparous), older than 40 years**
	Three Types of Stones
	Cholesterol Stones
	Pure cholesterol stones are radiolucent, solitary, 1–5 cm (diameter), yellow, more common in Northern Europeans
	Pigment Stones
	• Small, black, multiple, and radiolucent; high incidence in Asians • Associated with **hemolytic disease**, e.g., hereditary spherocytosis • Cholelithiasis occurs in the young; think of hereditary spherocytosis, sickle cell disease, or other chronic hemolytic process
	Mixed Stones
	• 80% of all stones and **associated with chronic cholecystitis** • Composed of cholesterol and calcium bilirubinate
Carcinoma of gallbladder	• Disease asymptomatic until late • *Symptoms:* dull abdominal pain, mass, weight loss, anorexia • *Pathology:* typically involves fundus and neck; 90% differentiated or undifferentiated adenocarcinomas • Poor prognosis, with 3% 5-year survival rate • *Risk factors:* cholelithiasis and cholecystitis (in up to 90% of patients), porcelain gallbladder (due to calcium deposition in gallbladder wall); occurs predominantly in elderly
Carcinoma of bile ducts (cholangiocarcinoma)	• Not associated with gallstones • Men are affected more frequently; usually elderly • *Symptoms:* obstructive jaundice • *Risk factors:* chronic inflammation, infections, (e.g., liver flukes), ulcerative colitis

Gastrointestinal Microbiology

▶ Microbial Diarrhea: Organisms Causing Inflammatory Diarrhea/Dysentery

(Invasive Organisms Eliciting Blood, Pus In Stool, Fever)

Organism	Most Common Sources	Common Age Group Infected	Incubation Period	Pathogenesis/ Vignette Clues	Diagnosis	Treatment
Campylobacter jejuni	**Poultry, domestic animals, water**	All	3–5 days	**Invades epithelium**, RBC and WBC in stools (most common bacterial diarrhea in U.S.)	Oxidase ⊕, gram ⊖, curved rod, seagull-wings shape; grows at 42°C; microaerophile	• Treatment for severe cases only • Erythromycin for invasive disease
***Salmonella* spp.**	**Poultry, domestic animals**, water	All	8–48 hours	Penetrates to lamina propria of ileocecal region → **PMN response and prostaglandin synthesis, which stimulates** cAMP	Gram ⊖, motile rods; encapsulated, oxidase ⊖	• Severe cases only • Sensitivity testing required
***Shigella* spp.**	Water, no animal reservoirs, fecal-oral transmission	All	1–7 days	**Shallow mucosal ulcerations and dysentery; septicemia rare**	Gram ⊖ rod; nonlactose fermenting; nonmotile	• Severe cases only • Fluoroquinolones, trimethoprim-sulfamethoxazole
Yersinia enterocolitica	Milk, wild and **domestic animals**, fecal-oral	All	2–7 days	Cold-climate **pseudoappendicitis**; heat-stable enterotoxin; arthritis may occur	Gram ⊖ , motile rod; nonencapsulated, oxidase ⊖; urease ⊕; bipolar staining; best growth at 25°C	• Severe cases only • Aminoglycosides, trimethoprim-sulfamethoxazole
Clostridium difficile	Associated with **antibiotic use**	Pt. on antibiotics	NA	Pt. on antibiotic (clindamycin)	Gram ⊕ rod; anaerobic spore former	Switch antibiotic; metronidazole
Enteroinvasive *E. coli*	Food, water, fecal-oral	Adults	2–3 days	Similar to *Shigella*	Gram ⊖ rod; motile, lactose fermenter; serotyping compares O, H, K antigens	Sensitivity testing required
Entamoeba histolytica	Food, water, fecal-oral	All	2–4 weeks	Trophozoites invade colon; **flask-like lesions, extraintestinal abscesses (liver)**; travelers to Mexico	Motile trophozoites or quadrinucleate cysts	Metronidazole

Definition of abbreviation: Pt., patient.

▶ Microbial Diarrhea: Organisms Causing Noninflammatory Diarrhea

(Noninvasive Organisms: No Blood, Pus In Stool)

Organism	Most Common Sources	Common Age Group Infected	Incubation Period	Pathogenesis/ Vignette Clues	Diagnosis	Treatment
Rotaviruses	Day care, water, fecal-oral	**Infants** and toddlers	1–3 days	Microvilli of small intestine blunted; dehydration	Diagnosis by exclusion: dsRNA naked, double-shelled, icosahedral (Reovirus family)	Supportive
Norwalk virus Norovirus (Norwalk-like)	Water, food, fecal-oral	**Older kids and adults**	18–48 hours	Blunting of microvilli; "cruise ship" diarrhea	Diagnosis by exclusion: ⊕ ssRNA, naked, icosahedral (Calicivirus family)	Supportive
Adenovirus 40/41	Nosocomial	Young kids, immunocompromised	7–8 days	Death of enteric cells causes diarrhea	Diagnosis by exclusion: naked, dsDNA, icosahedral	No specific therapy
Clostridium perfringens	**Beef, poultry, gravies**, Mexican food	All	8–24 hours	**Enterotoxin**	Anaerobic, gram ⊕ rods, spore-forming, Nagler reaction	Not indicated
Vibrio cholerae	Water, food, fecal-oral	All ages	9–72 hours	• Toxin stimulates adenylate cyclase • Rice water stools	Curved, gram ⊖ rod; oxidase ⊕; "shooting-star" motility	Oral rehydration therapy; tetracycline shortens symptoms
Vibrio parahaemolyticus	Raw or **undercooked shellfish**	Anyone eating raw shellfish	5–92 hours	Self-limited gastroenteritis mimicking cholera	Curved, gram ⊖ rod; oxidase ⊕; "shooting-star" mobility	Not indicated
Enterotoxigenic *E. coli* (ETEC)	Water, uncooked fruits and vegetables	All ages	12–72 hours	**Heat labile toxin (LT) stimulates adenylate cyclase**; stable toxin stimulates guanylate cyclase	Gram ⊖ rod; motile; lactose fermenter; serotyping compares O, H, K antigens	Sensitivity testing required
Enteropathogenic *E. coli* (EPEC)	Food, water, fecal-oral	Infants in developing countries	2–6 days	**Adherence to enterocytes through pili → damage to** adjoining microvilli	Gram ⊖ rod; motile; lactose fermenter; serotyping compares O, H, K antigens	Sensitivity testing required
Enterohemorrhagic *E. coli* (EHEC)	Food, fecal-oral (**hamburger**)	50% <10 years, all	3–5 days	**Verotoxin, which inhibits 60S ribosomal subunit, causes bloody diarrhea, no fever**	• Gram ⊖ rod; motile; lactose fermenter; serotyping compares O, H, K antigens • O157H7 most common • Does not ferment sorbitol	Antibiotics may increase risk of hemolytic-uremic syndrome

(Continued)

► Microbial Diarrhea: Organisms Causing Noninflammatory Diarrhea (CONT'D)
(Noninvasive Organisms: No Blood, Pus In Stool)

Organism	Most Common Sources	Common Age Group Infected	Incubation Period	Pathogenesis/ Vignette Clues	Diagnosis	Treatment
Giardia lamblia	Water, day care, camping, beavers, dogs, etc.	All, children	5–25 days	Cysts ingested; trophozoites; **multiply and attach to small intestinal villi by sucking disk**, cause fat malabsorption → steatorrhea	Flagellated binucleate trophozoites; "falling-leaf" motility; quadrinucleate cysts	Metronidazole
Cryptosporidium parvum	Day care, fecal-oral, animals, homosexuals	Children, AIDS patients	2–4 weeks	Parasites intracellular in brush border	Acid-fast oocytes in stool	Nitazoxanide, puromycin, azithromycin in immunocompromised

► Diarrhea by Intoxication

Organism	Most Common Sources	Common Age Group Infected	Incubation Period	Pathogenesis/ Vignette Clues	Symptoms	Diagnosis	Treatment
Staphylococcus aureus	Ham, potato salad, cream pastries	All	**1–6 hours**	Heat-stable enterotoxin is produced in food (contamination by food handler with skin lesions); food sits at room temperature	Abdominal cramps, vomiting, diarrhea; sweating and headache may occur; no fever	Symptoms, time of onset, food source	Recovery without treatment
***Bacillus cereus:* emetic form**	**Fried rice**	All	**<6 hours**	Heat-stable toxin causes vomiting	Vomiting 1–6 hours; diarrhea 18 hours	Symptoms, time of onset, food source	Recovery without treatment
Bacillus cereus: diarrheal form	Meat, vegetables	All	>6 hours	Heat-labile toxin causes diarrhea (similar to *E. coli* LT)	Nausea, abdominal cramps, diarrhea	Symptoms, time of onset, food source	Recovery without treatment
Clostridium perfringens	Meat, vegetables	All	18–24 hours	Enterotoxin	Nausea, abdominal cramps, diarrhea	Symptoms, time of onset, food source	Recovery without treatment

Definition of abbreviation: LT, labile toxin.

Additional Pharmacology

► Antidiarrheals

Drug Class	Agents	Mechanism
Opioid derivatives	Diphenoxylate Loperamide	*Diphenoxylate:* in combination with atropine to prevent abuse *Loperamide:* available over-the-counter

► Laxatives

Drug Class	Agents	Mechanism
Stimulant	Castor oil, phenolphthalein, senna	Have stimulant or irritant actions on bowel
Bulk-forming	Psyllium, methylcellulose	Indigestible and hydrophilic; absorb H_2O to form bulky stools → reflex bowel contraction
Osmotic	$Mg[OH]_2$ (milk of magnesia), lactulose	↑ fecal liquidity
Stool-softener/lubricant	Docusate, glycerin, mineral oil	Softens stool, enabling easier passage; glycerin and mineral oil lubricate

► Drugs That Stimulate Gastrointestinal Motility

Agent	Mechanism	Notes
Metoclopramide, cisapride*	Cause cholinergic stimulation; metoclopramide—also a DA antagonist, cisapride—also a $5HT_4$ agonist	Used for gastroparesis (e.g., from diabetes), GERD

*Cisapride has been taken off the market due to fatal arrhythmias; it is now available only on a limited basis.

► Miscellaneous Gastrointestinal Agents

Ursodiol, chenodiol	Used for dissolution of small gallstones
Pancrelipase	Used for steatorrhea, which is caused by pancreatic enzyme insufficiency
Orlistat	Used for obesity; inhibits gastric and pancreatic lipases, preventing absorption of triglycerides; available by prescription and OTC; causes steatorrhea, diarrhea, ↓ absorption of lipid-soluble vitamins

The Endocrine System

Chapter 6

Endocrine System

► General Characteristics of Hormones

	Peptides and Proteins (Water Soluble)	Steroids and Thyroid Hormones (Lipid Soluble)
Receptors	Membrane surface	Cytoplasm and/or nucleus
Mechanism	Second messenger	mRNA transcription
Storage	Yes	No (except thyroid as thyroglobulin)
Plasma protein binding	No (except somatomedins)	Yes; acts as pool and prolongs effective half-life
Synthesis	Rough endoplasmic reticulum	Smooth endoplasmic reticulum

Hypothalamus and Pituitary Overview

► Hypothalamus-Anterior Pituitary System

Hypothalamus and Anterior Pituitary Vascular System

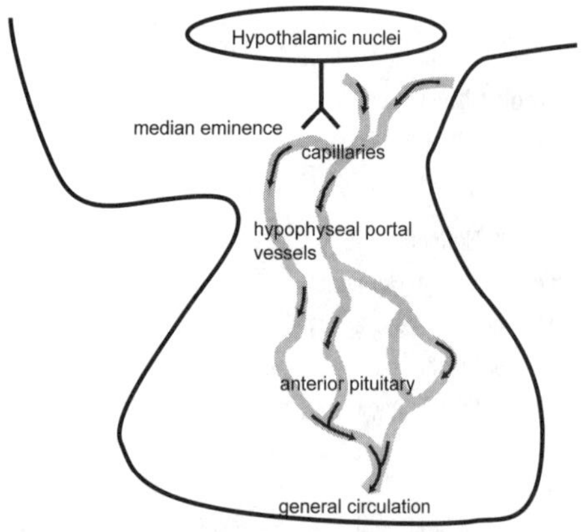

General Characteristics of the Pituitary

	Anterior Pituitary	Posterior Pituitary
Tissue	Glandular	Neuronal
Vascular	Indirect (portal via hypothalamus)	Direct
Control	Neurohormones	Neural
Hormones secreted	TSH, ACTH, LH FSH, GH, prolactin	ADH Oxytocin

Occlusion or lesion of pituitary stalk reduces secretion of all anterior and posterior pituitary hormones, *except* prolactin, which is controlled by inhibitory effects of dopamine.

Hypothalamus	Anterior Pituitary	Peripheral Target
Thyrotropin-releasing hormone (TRH)	Increases secretion of **thyroid-stimulating hormone (TSH)**	TSH stimulates synthesis and secretion of thyroid hormones; hypertrophy of thyroid gland
Corticotropin-releasing hormone (CRH)	Increases secretion of **adrenocorticotropic hormone (ACTH)**	ACTH stimulates synthesis and secretion of cortisol; hypertrophy of adrenal cortex
Gonadotropin-releasing hormone (GnRH)	Increases secretion of **luteinizing hormone (LH)** and **follicle-stimulating hormone (FSH)**	• LH stimulates gonadal steroids • FSH stimulates follicular development (females) and spermatogenesis (males)
Growth hormone–releasing hormone (GHRH) (the dominant control of GH)	Increases secretion of **growth hormone (GH)**	GH actions: causes liver to produce somatomedins; metabolic and growth effects other tissues
Somatostatin (also known as growth hormone–inhibiting hormone [GHIH])	Inhibits secretion of **GH**	—
Prolactin-inhibiting factor (dopamine [PIH])	Inhibits secretion of **prolactin**	Prolactin stimulates lactation and inhibits GnRH, LH, and FSH

► Anterior Pituitary Hyperfunction

Condition	Type	Features
Hyperprolactinemia		• Elevated serum prolactin associated with prolactinoma (chromophobic); **most common pituitary tumor** • Women: amenorrhea and galactorrhea; men: galactorrhea and infertility • **Treatment:** dopamine agonists (e.g., pergolide, bromocriptine)
Excess GH	**Gigantism**	• Results from excess GH secretion **before fusion of growth plates** • Excessive skeletal growth may result in heights close to 9 feet • Eosinophilic granuloma
	Acromegaly	• Results from excess GH secretion **after fusion of growth plates** • Circumferential deposition of bones—enlargement of the hands and feet with frontal bossing • Classic case—hat does not fit anymore
Cushing disease		• ACTH-secreting tumors in the anterior pituitary (compare with Cushing syndrome); rarely cause mass effect (see page 373)
Pathology		• May be micro- or macroadenomas; generally, if active compound is released, lesion is noticed when small • Large lesions can cause mass effect on the optic chiasm; very large masses (10 cm) may invade surrounding structures

► Pituitary Hypofunction

Condition	Type	Features
Anterior pituitary hypofunction		• **Sheehan syndrome:** postpartum hemorrhagic infarction of pituitary associated with excessive bleeding; presents as failure to lactate • **Empty sella syndrome:** atrophy of the pituitary; sella is enlarged on skull x-ray and may mimic neoplasm
Posterior pituitary hypofunction	**Diabetes insipidus (DI)**	• Insufficient or absent antidiuretic hormone • *Clinical:* polydipsia, polyuria, **hypotonic urine, high serum osmolality, hypernatremia** • Central DI responds to exogenous ADH (desmopressin) therapy; nephrogenic DI does not
Posterior pituitary hyperfunction		• **Syndrome of inappropriate ADH secretion (SIADH):** as name suggests, inappropriate, excessive ADH secretion unrelated to serum osmolality • *Causes:* May be paraneoplastic (**small cell lung cancer**), CNS damage, drugs, or infections (TB) • *Clinical:* fluid retention, weight gain, and lethargy, **low serum osmolality, hypertonic urine, hyponatremia**

Definition of abbreviation: GH, growth hormone.

Adrenal Gland

▶ Adrenal Hormones

Adrenal Gland: Cortex and Medulla

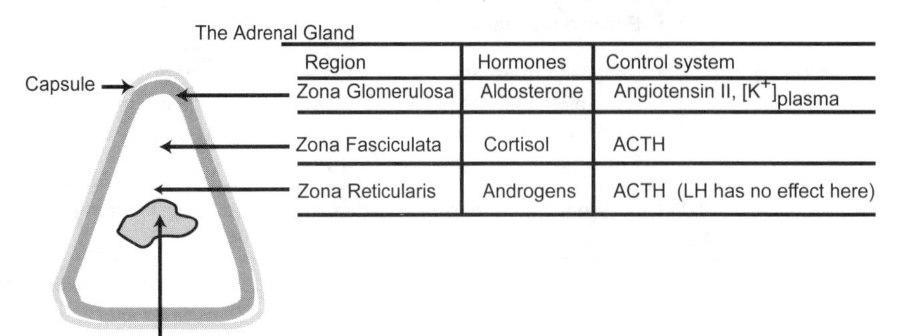

Region	Hormones	Control system
Zona Glomerulosa	Aldosterone	Angiotensin II, $[K^+]_{plasma}$
Zona Fasciculata	Cortisol	ACTH
Zona Reticularis	Androgens	ACTH (LH has no effect here)

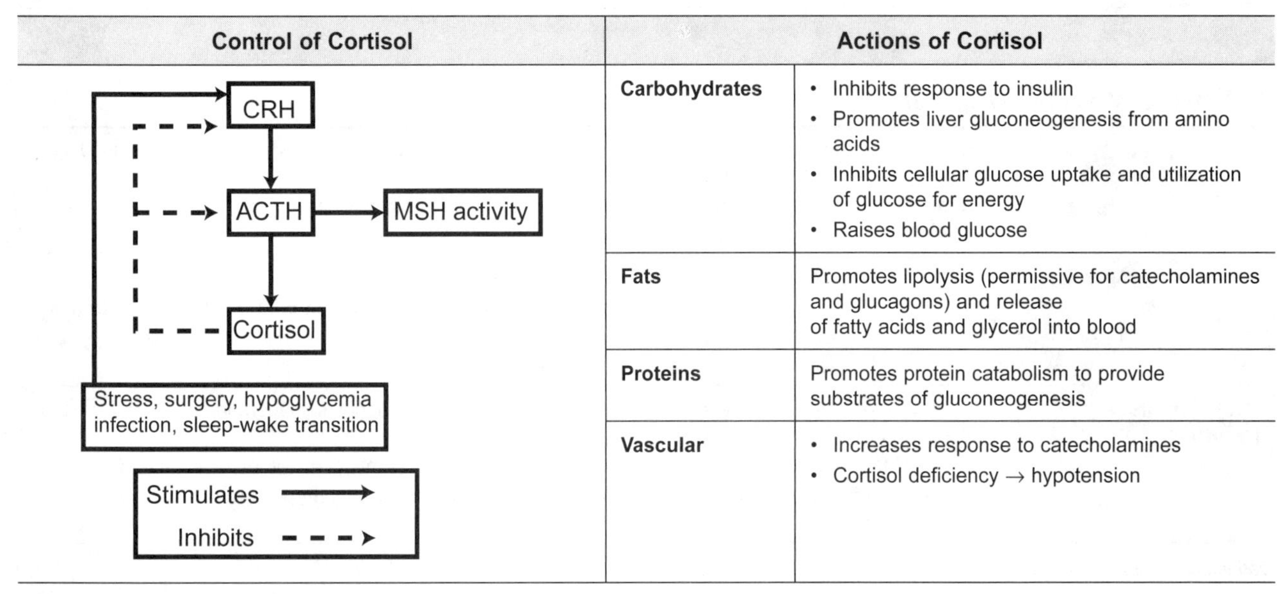

Actions of Cortisol

Carbohydrates	• Inhibits response to insulin • Promotes liver gluconeogenesis from amino acids • Inhibits cellular glucose uptake and utilization of glucose for energy • Raises blood glucose
Fats	Promotes lipolysis (permissive for catecholamines and glucagons) and release of fatty acids and glycerol into blood
Proteins	Promotes protein catabolism to provide substrates of gluconeogenesis
Vascular	• Increases response to catecholamines • Cortisol deficiency → hypotension

Note: Fragments of ACTH precursor proopiomelanocortin (POMC) are also released and have biologic effects, notably melanocyte-stimulating hormone (MSH) activity.

► Cushing Syndrome

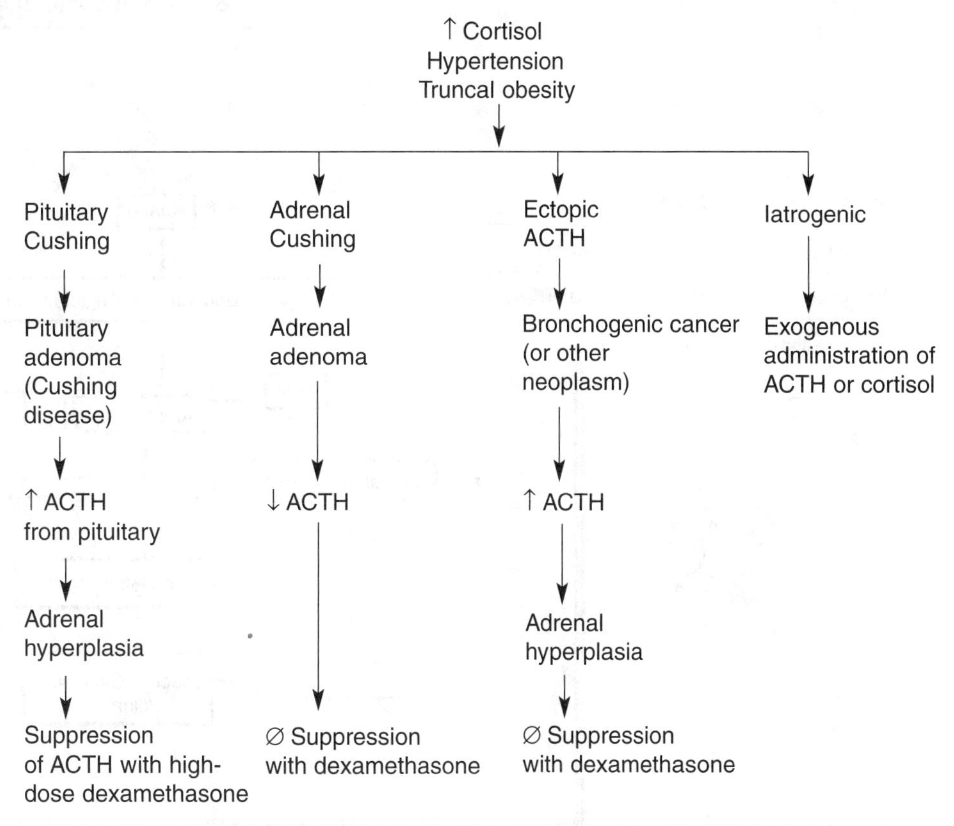

Cushing (pituitary)	↑ cortisol ↑ ACTH
Cushing (adrenal)	↑ cortisol ↓ ACTH ↑ aldosterone (often)
Clinical	
Truncal obesity, buffalo hump, moon facies, facial plethora, hirsutism, menstrual disorders, hypertension, muscle weakness, back pain (osteoporosis), striae, acne, psychological disorders, bruising	

► Differential Diagnosis of Cortisol Excess and Deficiency Based on Feedback Control

Disorder	Plasma Cortisol	Plasma CRH	Plasma ACTH	Hyperpigmentation
Primary (adrenal) excess	↑	↓	↓	No
Primary deficiency	↓	↑	↑	Yes
Secondary (pituitary) excess	↑	↓	↑	Yes
Secondary deficiency	↓	↑	↓	No
Steroid administration (synthetics other than cortisol)	↓ (but symptoms of excess)	↓	↓	No

► Aldosterone

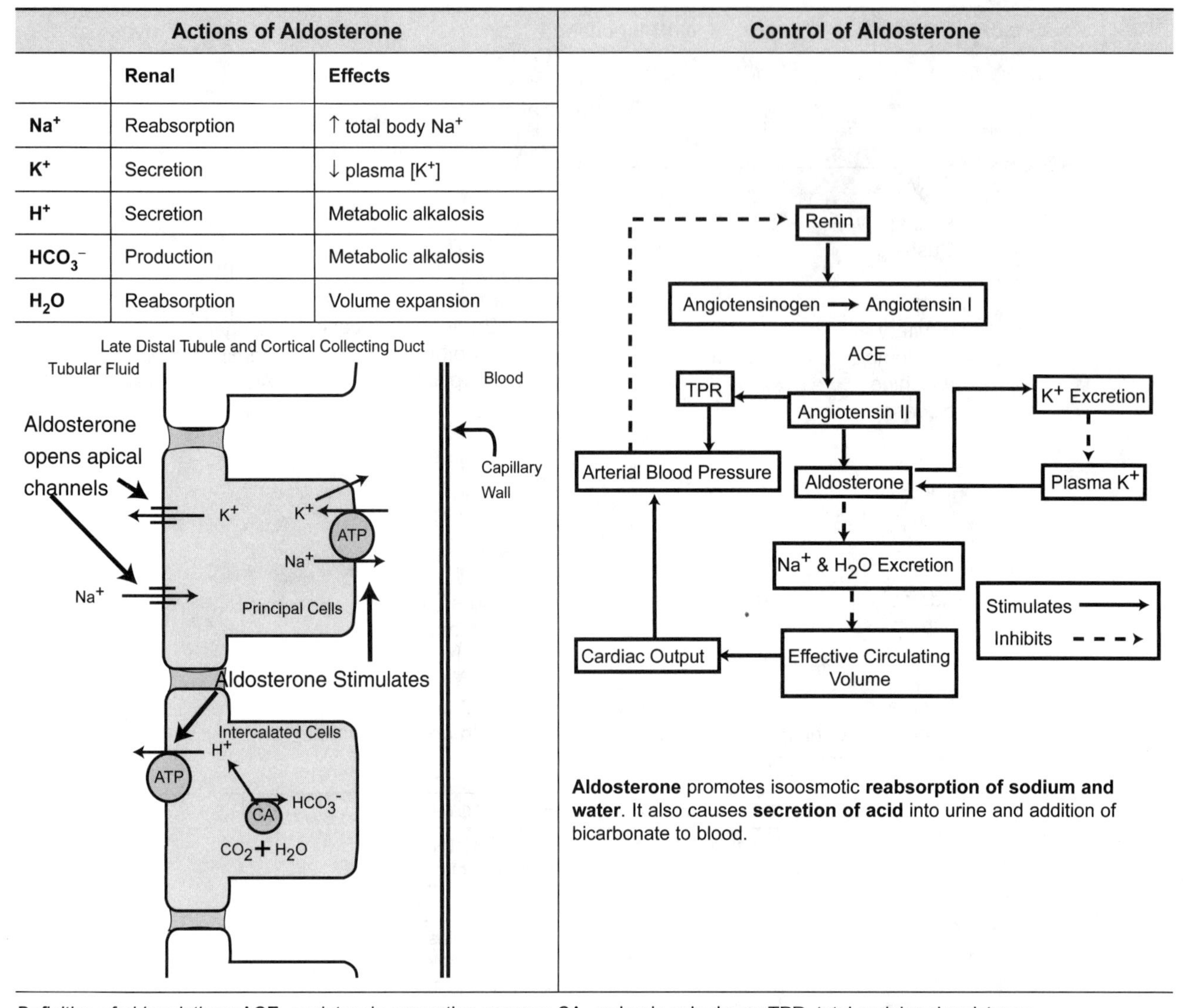

Actions of Aldosterone		
	Renal	**Effects**
Na^+	Reabsorption	↑ total body Na^+
K^+	Secretion	↓ plasma $[K^+]$
H^+	Secretion	Metabolic alkalosis
HCO_3^-	Production	Metabolic alkalosis
H_2O	Reabsorption	Volume expansion

Aldosterone promotes isoosmotic **reabsorption of sodium and water**. It also causes **secretion of acid** into urine and addition of bicarbonate to blood.

Definition of abbreviations: ACE, angiotensin-converting enzyme; CA, carbonic anhydrase; TPR, total peripheral resistance.

► Long-Term Control of Blood Pressure by Renin-Angiotensin-Aldosterone

Event	Effects	Compensation
Volume expansion Saline infusion Polydipsia	↑ preload, cardiac output, and BP	• ↓ renin →↓ AII and ALD • ↓ AII, ↓ ALD →↓ Na^+ and H_2O reabsorption →↑ urine flow →↓ EFV →↓ preload and cardiac output →↓ BP to normal
Volume loss Hemorrhage Dehydration	↓ preload, cardiac output, and BP	↑ renin →↑ AII and ALD →↑ Na^+ and H_2O reabsorption → ↓ urine flow →↑ EFV →↑ preload and cardiac output toward normal →↑ BP to normal
Heart failure	↓ cardiac output and BP	↑ renin →↑ AII and ALD → ↑ Na^+ and H_2O reabsorption →↓ urine flow →↑ EFV →↑ preload and cardiac output toward normal

Definition of abbreviations: AII, angiotensin II; ALD, aldosterone; BP, blood pressure; EFV, extracellular fluid volume.

▶ Hyperaldosteronism

Primary hyperaldosteronism (Conn syndrome)	• ↑ aldosterone secretion • Adrenal adenoma most common cause • **Clinical:** diastolic hypertension, weakness, fatigue, polyuria, polydipsia, headache, no edema • **Lab values: hypokalemia, low renin levels, metabolic alkalosis, hypernatremia**, failure to suppress aldosterone with salt loading • **Pathology:** single well-circumscribed adenoma with lipid-laden clear cells
Secondary hyperaldosteronism	• Etiologies: congestive heart failure, decreased renal blood flow (increased renin), renin-producing neoplasms, and Bartter syndrome (juxtaglomerular cell hyperplasia, hyperreninemia, hyperaldosteronism) • **Clinical:** Same as for primary, but edema may be present • **Lab values: high renin levels, hypernatremia, hypokalemia**

▶ Steroid Synthetic Pathways Most Commonly Involved in Pathology

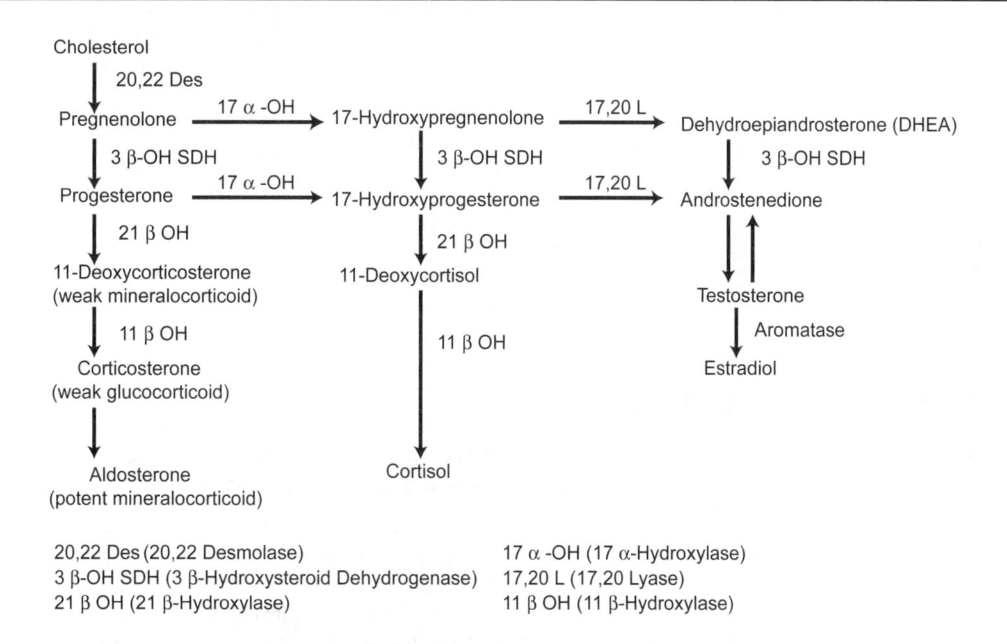

Simplified scheme for understanding enzyme deficiencies; a vertical cut (e.g., 17 α -OH deficiency) all products to the right are deficient (cortisol, androgens) and products to the left are in excess (mineralocorticoids); a horizontal cut (e.g., 11 β OH) all products above are excess (11-Deoxycorticosterone, 11-Deoxycortisol, androgens) and below are deficient (Aldosterone, Cortisol).

Congenital Adrenal Hyperplasia	• Usually due to a congenital enzyme deficiency characterized by **cortisol deficiency and enlargement of adrenal glands** • The 3 most common forms **all cause virilism** and are called **adrenogenital syndromes**: – **Partial 21-hydroxylase deficiency:** normal aldosterone function and impaired cortisol production. – **Salt-losing syndrome: near total 21-hydroxylase deficiency** and aldosterone deficiency; infants present with vomiting, dehydration, hyponatremia, hyperkalemia – **11-hydroxylase deficiency:** leads to excessive androgen production and buildup of 11-deoxycorticosterone (a weak mineralocorticoid, which, in excess, has a strong mineralocorticoid effect), causing virilization, hypertension, hypokalemia

► Congenital Enzyme Deficiency Syndromes

Enzyme Deficiency	Glucocorticoids	Mineralocorticoids	Androgens	Other Effects
20,22 desmolase	↓	↓	↓	Lethal if complete
3 β-OH SDH	↓	↓	↑ DHEA	• Masculinization of females in utero • Incomplete precocious puberty (males) • Adrenal hyperplasia • Hyponatremia and hypovolemia
21 β-OH	↓	↓	↑	• Masculinization of females in utero • Incomplete precocious puberty (males) • Adrenal hyperplasia • Hyponatremia and hypovolemia
11 β-OH	↓	↓ aldosterone ↑↑↑ DOC*	↑	• Mineralocorticoid excess (DOC) • Hypervolemia and hypernatremia • Sexual effects as above
17 α-OH	↓	↓ aldosterone ↑↑↑ DOC*	↓↓	• DOC excess • Absent secondary sexual aspects both sexes • Amenorrhea
17, 20 L	Normal	Normal	↓↓	• Absent secondary sexual aspects both sexes • Amenorrhea

*11-deoxycorticosterone, a weak mineralocorticoid that causes symptoms of mineralocorticoid overload when excessive amounts are secreted

► Adrenal Cortical Hypofunction

Acute adrenocortical insufficiency	• Rapid withdrawal of exogenous steroids in patients with chronic adrenal suppression • Adrenal apoplexy seen in **Waterhouse-Friderichsen syndrome** (adrenal hemorrhage associated with meningococcal septicemia)
Primary adrenocortical insufficiency (Addison disease)	• Etiology: most common cause autoimmune adrenalitis; other causes: tuberculosis, other infections, iatrogenic, metastases, adrenal hemorrhage, pituitary insufficiency • For clinically apparent insufficiency, 90% of the adrenal gland must be nonfunctional • **Clinical:** weakness, weight loss, anorexia, nausea, vomiting, hypotension, skin pigmentation, hypoglycemia with prolonged fasting, inability to tolerate stress, abdominal pain • **Lab values:** hyponatremia, hypochloremia, hyperkalemia, metabolic acidosis; ACTH levels high, cortisol and ALD levels low • **Pathology:** bilateral atrophied adrenal glands • ACTH and MSH share amino acid sequences; in cases of high ACTH → skin pigmentation
Secondary (pituitary) adrenocortical insufficiency	• **Etiology:** metastases, irradiation, infection, infarction, affecting the hypophysial-pituitary axis • Results in decreased ACTH (less skin pigmentation)
Hypoaldosteronism	Hyponatremia, hypovolemia, hypotension, metabolic acidosis, hyperkalemia
Primary	• ↓ aldosterone, ↑ renin and AII
Secondary	• ↓ aldosterone, ↓ renin and AII, ↓ total peripheral resistance

Definition of abbreviations: AII, angiotensin II; ACTH, adrenocorticotropic hormone; ALD, aldosterone; MSH, melanocyte-stimulating hormone.

▶ Adrenal Cortical Neoplasms

Adrenal adenomas	• Mostly asymptomatic and not steroid-producing • Steroid-producing adenomas may produce Conn syndrome, Cushing syndrome, or virilization in women • **Pathology:** small and unilateral nodule, yellow-orange on cut section, poorly encapsulated
Adrenal carcinomas	• Relatively rare and usually very malignant • Greater than 90% are steroid-producing • Pathology: tumors often large and yellow with areas of hemorrhage and necrosis

▶ Adrenal Medulla

Tissue	Hormones	Control	Actions
Neural, chromaffin	Epinephrine (80%) Norepinephrine (20%)	Sympathetic nervous system	Glycogenolysis, lipolysis, ↑ blood glucose, ↑ metabolic rate (requires cortisol and thyroid)

Disorders of Adrenal Medulla	
Pheochromocytoma	• Neoplasm of **neural crest-derived chromaffin cells** that secrete catecholamines (norepinephrine and epinephrine) → hypertension • Highest incidence in children and adults age 30–50 • **Clinical:** paroxysmal or constant hypertension is most classic symptom; also, sweating, headache, arrhythmias, palpitations • **Lab values:** elevated urinary homovanillic acid (HVA) and vanillylmandelic acid (VMA) • **The Rule of 10s for pheochromocytoma:** – 10% extra-adrenal – 10% bilateral – 10% malignant – 10% affect children – 10% familial
Neuroblastoma	• **Most common malignant extracranial solid tumor of childhood** • Occurs most frequently in the adrenal medulla, but may arise in sympathetic chain • Amplification of the **N-*myc*** oncogene—more copies = more aggressive • **Clinical:** tumors grow rapidly, metastasize widely (especially to bone); prognosis in younger patients (less than 1 year old) better than for older children • **Pathology:** lobulated with areas of necrosis, hemorrhage, calcification • **Microscopic appearance:** rosette pattern of small cells

▶ Treatment for Adrenocortical Disease

Class	Mechanism	Indications
Glucocorticoids (hydrocortisone)	Replacement therapy	For adrenocortical insufficiency (Addison disease, acute adrenal insufficiency from other causes)
Mineralocorticoids (fludrocortisone)	Replacement therapy	For chronic treatment of Addison disease in patients requiring mineralocorticoids
Glucocorticoid synthesis inhibitors (aminoglutethimide metyrapone ketoconazole)	Inhibits glucocorticoid synthesis via different mechanisms	To suppress adrenocortical steroid production in variety of disorders, e.g., Cushing disease, Cushingoid states, congenital adrenal hyperplasia

▶ Antidiuretic Hormone and Control of Osmolarity and Volume

Normal function of antidiuretic hormone (ADH, vasopressin): prevents changes of plasma osmolarity, restores normal blood volume and blood pressure, ↑ **permeability of renal collecting duct to water,** ↑ **reabsorption of water, causes** ↓ **urine flow, and** ↑ **urine osmolarity**

ADH Control of Plasma Osmolarity and ECF Volume	Release of ADH from the Pituitary

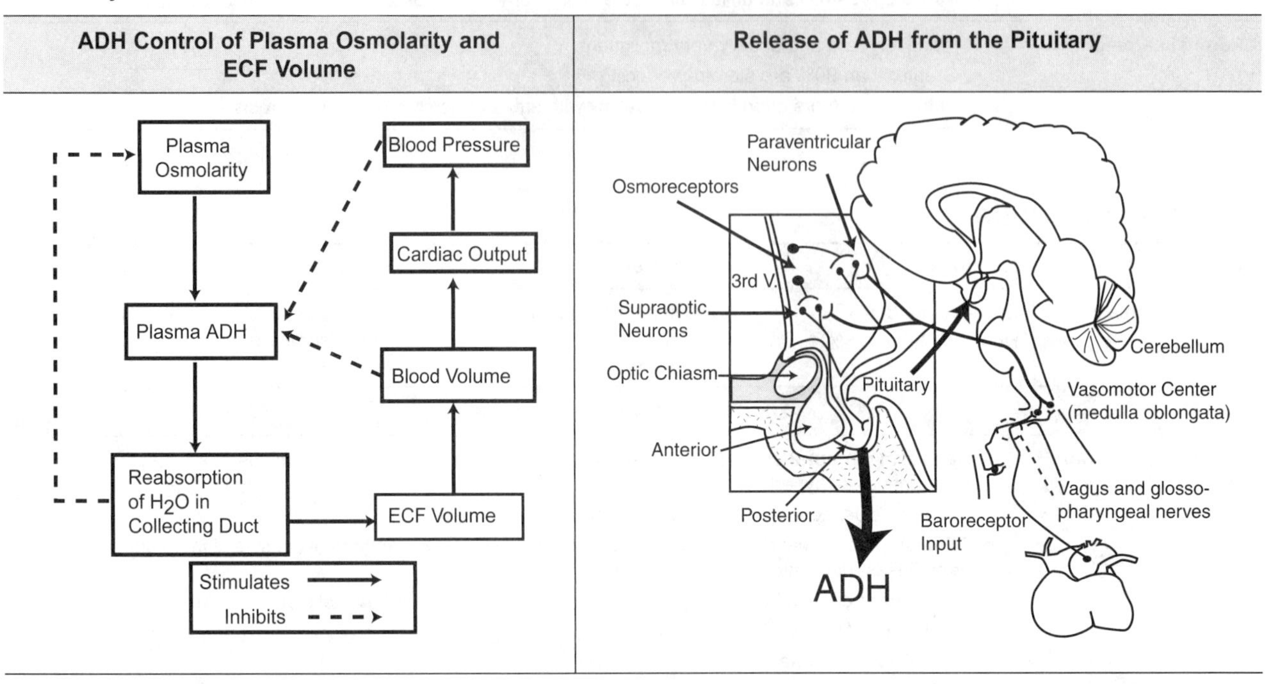

Secretion of ADH by neurons with cell bodies located in the **supraoptic and paraventricular nuclei** of the hypothalamus is **under the control of osmoreceptors and baroreceptors.** The osmoreceptors are in the AV3V region and are influenced by blood osmolarity. Atrial stretch receptors (detecting blood volume) are the largest cardiovascular influence, and arterial baroreceptors, including the carotid sinus (moderate effect) and the aortic arch (minor effect), are processed in medullary centers that regulate cardiovascular function, such as the nucleus tractus solitarius (NTS). The NTS sends efferents to the ADH-secreting neurons. The axon terminals of these neurons are in the posterior pituitary and release the hormone into the blood. Normally, osmolarity dominates control of ADH secretion; extreme changes of blood volume or cardiac output can overrride osmolarity.

▶ Disorders of Antidiuretic Hormone (Vasopressin)

Diabetes insipidus ↓ ADH (central, neurogenic) ↓ response (nephrogenic)	• Hypovolemia, hypernatremia, metabolic alkalosis, tendency toward hypotension, large volume of dilute urine (distinguishes from dehydration) • Distinguish central versus nephrogenic by testing response to ADH injection (nephrogenic → no response; central → response = concentration of urine)
Dehydration (water deprived) ↑ ADH to compensate	Hypovolemia, hypernatremia, metabolic alkalosis, tendency toward hypotension, small volume of concentrated urine (distinguishes from diabetes insipidus)
Syndrome of inappropriate ADH (SIADH; ↑ ADH)	Hypervolemia, hyponatremia, metabolic acidosis, small volume of concentrated urine (distinguishes from polydipsia)
Primary polydipsia	Hypervolemia, hyponatremia, large volume of dilute urine

▶ Atrial Natriuretic Peptide

Secretion	• Secreted by **heart (right atrium)** • Stimulated by **stretch (blood volume)**
Actions	• ↑ renal sodium excretion due to ↑ GFR and ↓ reabsorption of sodium (late distal tubule and collecting duct) • ↓ reabsorption of water due to ↓ reabsorption of sodium • Inhibits response to ADH
Clinical	• No known disorder caused by deficiency • Elevated levels have clinical use as index of the severity of congestive heart failure

Pancreas

► Endocrine Pancreas

Hormones of the Islets of Langerhans

Hormone	Control of Secretion	Target Tissues	Actions
Insulin (β cells)	• Glucose, amino acids • Effect of glucose via ↑ ATP, closes ATP-dependent K^+ channels, producing depolarization and exocytosis of insulin • Gastric inhibitory peptide (GIP) stimulates release	Liver	↑ glucose uptake (enzymatic effect) ↑ glucose utilization ↑ triglyceride synthesis ↑ protein synthesis ↑ glycogen synthesis ↓ gluconeogenesis ↓ glycogenolysis ↓ lipolysis ↓ protein catabolism ↓ ureagenesis, ketogenesis ↓ **blood glucose concentration**
		Muscle	↑ glucose uptake (GLUT4 transport) ↑ glucose utilization ↑ protein synthesis ↓ glycogenolysis ↓ protein catabolism ↓ **blood glucose concentration**
		Adipose	↑ glucose uptake (GLUT4 transport) ↑ triglyceride synthesis (↑ lipoprotein lipase to ↑ uptake of fatty acids) ↓ lipolysis (↓ hormone-sensitive lipase) ↓ **blood glucose concentration**
Glucagon (α cells)	• Glucose inhibits • Hypoglycemia stimulates • Amino acids stimulate	Liver	↑ gluconeogenesis ↑ glycogenolysis ↑ lipolysis ↑ protein catabolism ↑ ureagenesis, ketogenesis ↑ **blood glucose concentration**
Somatostatin (δ cells)	Stimulated by glucose, amino acids, fatty acids	Pancreas, GI tract	Inhibits secretion of insulin and glucagon; role is disputed

▶ Integrated Control of Plasma Glucose by Insulin and Glucagon

Use the fact that a carbohydrate meal increases plasma glucose as a sample application. Tracking through the feedback diagram predicts effects on carbohydrate, fat, and protein metabolism.

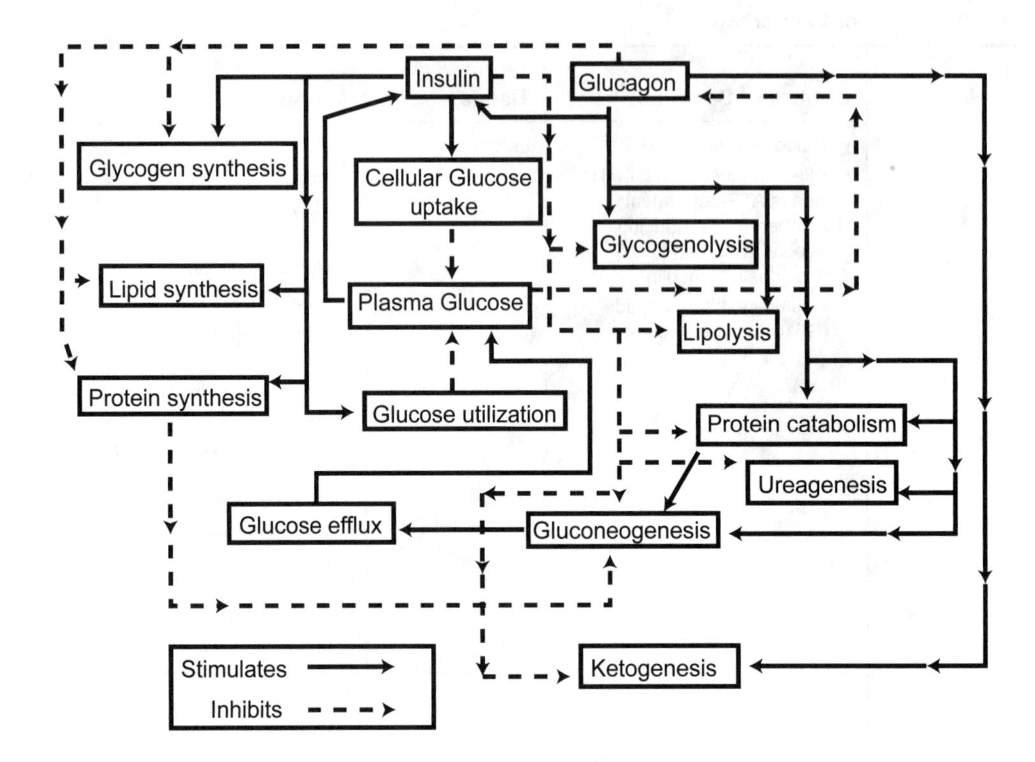

▶ Summary of Glucose Counter-Regulation

Hormone	Stimulus	Actions
Glucagon	• Fasting, hypoglycemia, stress • **Insulin/glucagon ratio is key to metabolism balance**	• Inhibits response to insulin (mainly in liver) • Promotes catabolism, gluconeogenesis, glycogenolysis
Cortisol	Stress, fasting, hypoglycemia	• Inhibits response to insulin • Reduces cellular glucose uptake • Permissive for lipolysis • Increases gluconeogenesis and protein catabolism • Inhibits cellular glucose uptake
Epinephrine	Stress, severe hypoglycemia	• Increases glycogenolysis and lipolysis
Growth hormone	Fasting, sleep, stress	• Inhibits cellular uptake of glucose • Stimulates lipolysis
Thyroid hormone	Cold, stress	• Permissive for epinephrine's effects • Required for production of GH

▶ Insulin-Related Pathophysiology

	Glucose	Insulin	C-peptide	Ketoacidosis	Other Features
Diabetes type 2	↑	↑ or ↔	↑ or ↔	Uncommon	Familial, often obese
Diabetes type 1	↑	↓	↓	Yes	Often islet antibodies
Insulinoma	↓	↑	↑	No	Tachycardia (epinephrine)
Insulin overdose	↓	↑	↓	No	Tachycardia (epinephrine)
Fasting hypoglycemia	↓	↑	↑	No	Insulin remains at postprandial level in fasting
Reactive hypoglycemia	↓	↑	↑	No	Excessive secretion with oral glucose tolerance test

▶ Disorders of the Endocrine Pancreas

Diabetes mellitus (DM) **Type 1**	• Often, abrupt onset with **ketoacidosis** • Marked, absolute insulin deficiency, resulting from **diminished β-cell mass** • Characterized by **low serum insulin** levels
Type 2	• Constitutes most cases of idiopathic diabetics, characterized by **peripheral insulin resistance** • Most patients have **central obesity**; onset of disease usually after age 40 • These patients are **not prone to ketoacidosis**
Acute metabolic complications of diabetes	• **Diabetic ketoacidosis** (DKA) may occur in type 1 diabetics, rarely in type 2 • Metabolic acidosis results from accumulation of ketones • High blood glucose → dehydration via an **osmotic diuresis** • Treatment with insulin normalizes the metabolism of carbohydrate, protein, fat • Fluids given to correct the dehydration • **Hyperosmolar nonketotic coma** in patients with mild type 2; blood glucose can be elevated; treatment similar to treatment of DKA
Late complications of diabetes	• Series of long-term complications, including **atherosclerosis** (leading to strokes, myocardial infarcts, gangrene), **nephropathy, neuropathy** (distal, symmetric polyneuropathy with “stocking-glove” distribution), **retinopathy** that may lead to blindness • Patients with DM are also at high-risk for: – ***Klebsiella pneumoniae*** – **Sinus mucormycosis** – **Malignant otitis externa (*Pseudomonas aeruginosa*)**
ß-cell tumors	• Insulinomas most commonly occur between ages of 30 and 60 • **Pathogenesis:** β-cell tumors produce hyperinsulinemia → hypoglycemia • **Clinical features:** patients experience episodes of altered sensorium (i.e., disorientation, dizziness, diaphoresis, nausea, tremulousness, coma), which are relieved by glucose intake • **Pathology:** most tumors solitary, well-encapsulated, well-differentiated adenomas of various sizes.; 10% are malignant carcinomas
Zollinger-Ellison syndrome	• Due to gastrinoma and often associated with **MEN type I** • **Pathogenesis:** tumors of pancreatic islet cells secrete gastrin → gastric hypersecretion of acid • **Clinical:** includes intractable peptic ulcer disease and severe diarrhea • **Pathology:** 60% malignant; most tumors located in pancreas (10% in duodenum)
MEN I	• Tumors of parathyroids, adrenal cortex, pituitary gland, pancreas • Associated with peptic ulcers and Zollinger-Ellison syndrome
MEN IIa	Tumors of adrenal medulla (pheochromocytoma), medullary carcinoma of thyroid, parathyroid hyperplasia or adenoma
MEN IIb/III	Medullary carcinoma of the thyroid, pheochromocytoma, and mucosal neuromas.

▶ Treatment of Diabetes

Insulin				
Classes	**Insulin Types**	**Onset (hours)**	**Peak (hours)**	**Duration (hours)**
Rapid-acting	Lispro, aspart, glulisine	0.25	0.5–1.5	3–5
Short-acting	Regular	0.5–1	2–4	5–8
Intermediate-acting	NPH	1–3	8	12–16
Long-acting	Glargine, detemir	1	No peak	20–26

Noninsulin Antidiabetic Agents		
Drug	**Mechanism**	**Notes**
Insulin secretagogues • **Sulfonylureas** – **First generation (tolbutamide, chlorpropamide)** – **Second generation (glipizide, glyburide, glimepiride)** • **Meglitinides (repaglinide, nateglinide)**	• Block K^+ channels → depolarization of pancreatic β cells → ↑Ca^{2+} influx → insulin release	• Requires functional β cells • Repaglinide: faster onset and shorter duration than sulfonylureas; taken before meals to control postprandial glucose levels • Side effects: hypoglycemia, weight gain
Biguanides (metformin)	May ↑ tissue sensitivity to insulin and/or ↓ hepatic gluconeogenesis	• Does not require functional β cells • Will not cause hypoglycemia • Most serious side effect: lactic acidosis
Thiazolidinediones (rosiglitazone, pioglitazone)	• Bind nuclear peroxisome proliferator-activated receptor-γ (PPAR-γ receptor) • ↑ target tissue sensitivity to insulin, inhibits hepatic glucose output, ↑ glucose uptake	• Hypoglycemia rare • Weight gain, edema
α-Glucosidase inhibitors (acarbose, miglitol)	Inhibits intestinal brush border α-glucosidase →↓ glucose absorption →↓ postprandial glucose →↓ demand for insulin	• GI distress, flatulence, diarrhea
Drugs affecting glucagon-like peptide-1 (GLP-1) (exenatide*, sitagliptin)	• GPL-1, an incretin released from the small intestine in response to food, augments glucose-dependent insulin release, inhibits glucagon secretion, slows gastric emptying, and increases feelings of satiety • Exenatide: long-acting GLP-1 receptor agonist • Sitagliptin: oral DDP-4 inhibitor, prevents the breakdown of GLP-1	• Extenatide: nausea, hypoglycemia when combined with sulfonylureas • Sitagliptin: headache, URI, nasopharyngitis
Pramlintide*	Synthetic analog of amylin, which decreases glucagon release, slows gastric emptying, and promotes satiety	• Used in type 1 and type 2 DM • Hypoglycemia, GI disturbances
Hyperglycemic Drugs		
Glucagon	↑ hepatic glycogenolysis and gluconeogenesis, ↑ heart rate and force of contraction, relaxes smooth muscle	• Administered via intramuscular injection • Major side effect: hyperglycemia • Glucagon receptors stimulate adenylate cyclase and ↑ cAMP; this is basis for its use in beta-blocker overdose

* Exanatide and pramlintide are injectables, the rest are given orally

Definition of abbreviations: DDP-4, dipeptidyl peptidase 4; URI, upper respiratory infection

Endocrine Regulation of Calcium and Phosphate

► Overview of Hormonal Regulation of Calcium and Phosphate

Hormone	Site Produced	Stimuli to Secretion or Production	Effect on Plasma Free Ca^{2+}	Effect on Plasma Phosphate
Parathyroid Hormone (PTH)	Parathyroid gland	Low plasma Ca^{2+}	↑	↓
1,25-$(OH)_2$-vitamin D_3	Skin → liver → kidney	Sunlight, PTH, dietary intake	↑	↑
Calcitonin (not essential for control)	Parafollicular cells of thyroid	High plasma Ca^{2+}	↓	Little net effect

► Parathyroid Hormone

Effects on kidneys	• Calcium ↑ reabsorption • Phosphate ↓ reabsorption • Vitamin D ↑ production of active form 1,25 $(OH)_2$ vitamin D_3 from precursor 25-OH D_3 formed in liver
Effects on bone	• Receptors on osteo*b*lasts, not on osteo*c*lasts • Rapid mobilization of Ca^{2+} and phosphate from bone fluid • ↑ osteoclast activity via mediators released from osteoblasts • Causes bone resorption, release of Ca^{2+}, phosphate into plasma
Effects on GI tract	Indirectly stimulates Ca^{2+} and phosphate absorption in small intestine through its effect to produce active form of vitamin D_3

► Vitamin D (Calcitriol)

Effects on kidneys	↓ Ca^{2+} and phosphate excretion (by ↑ reabsorption of both)
Effects on bone	↑ resorption (with PTH); releases Ca^{2+} and phosphate into plasma (but normal growth and maintenance also requires both vitamin D and PTH)
Effects on GI tract	↑ Ca^{2+} and phosphate absorption in intestine, ↑ both in plasma

► Calcitonin

Effects on kidneys	↑ phosphate excretion; ↓ Ca^{2+} excretion (minor effect)
Effects on bone	↓ resorption; ↑ deposition; ↓ plasma Ca^{2+} (major effect)
Effects on GI tract	↑ Ca^{2+} and phosphate absorption in intestine (minor effect)

► Regulation of Calcium and Phosphate

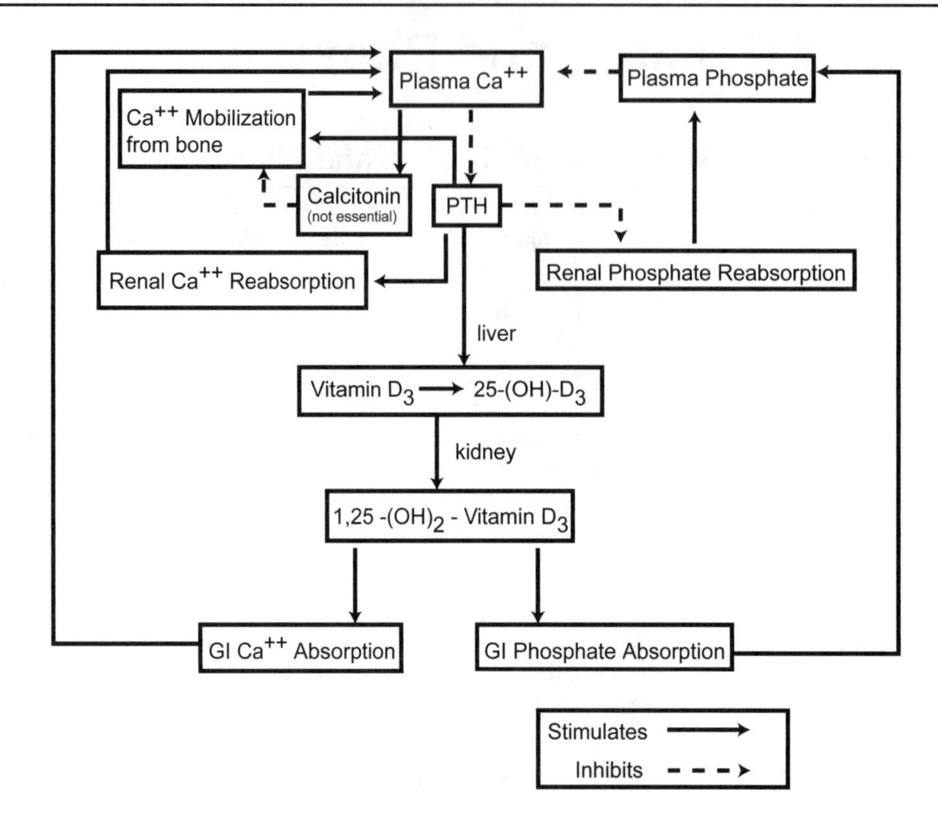

► Disorders of Calcium and Phosphate Regulation

	Plasma (Ca^{2+})	Plasma (Phosphate)	Effects on Bone
Primary hyperparathyroidism	↑	↓	Demineralization and osteopenia
Primary hypoparathyroidism	↓	↑	Malformation
Deficient vitamin D (secondary hyperparathyroidism)	↓	↓	• Osteomalacia in adults • Rickets in children
Excess vitamin D (secondary hypoparathyroidism)	↑	↑	Osteoporosis
Renal failure with high plasma phosphate (secondary hyperparathyroidism)	↓ (precipitation)	↑	Osteomalacia and osteosclerosis

Note: Clinical presentation of hypocalcemia often includes muscle spasms and tetany; hypercalcemia presents with weakness and flaccid paralysis.

► Pathophysiology of Calcium Homeostasis

Primary hyperparathyroidism	• Parathyroid adenoma is the most common cause; can see in MEN I and MEN IIa • **Clinical:** elevated serum calcium often asymptomatic • **Lab values: ↑ PTH and alkaline phosphatase, hypercalcemia, hypophosphatemia**
Osteitis fibrosa cystica (von Recklinghausen disease of bone)	• Occurs in **chronic primary hyperparathyroidism** • Cystic changes in bone occur due to osteoclastic resorption • Fibrous replacement of resorbed bone may lead to a non-neoplastic "brown tumor"
Secondary hyperparathyroidism	• Usually caused by **chronic renal failure** and decreased Ca^{2+} absorption, stimulating PTH • Hyperphosphatemia due to decreased phosphate excretion in renal failure causes hypocalcemia; this provokes increased PTH secretion. Treatment to lower plasma phosphate is often effective. • Vitamin D deficiency and malabsorption less common causes • May show soft tissue calcification and osteosclerosis • **Lab values: ↑ PTH and alkaline phosphatase, hypocalcemia, hyperphosphatemia**
Hypoparathyroidism	• Common causes are accidental removal during thyroidectomy, idiopathic, and DiGeorge syndrome • **Clinical:** irritability, anxiety, tetany, intracranial and lens calcifications • **Lab values: hypocalcemia, hyperphosphatemia**
Pseudohypoparathyroidism	• Autosomal dominant disorder, resulting in kidney unresponsive to circulating PTH • Skeletal abnormalities: short stature, shortened fourth and fifth carpals and metacarpals
Hypercalcemia	• **Mnemonic** "MISHAP": Malignancy, Intoxication (vitamin D), Sarcoidosis, Hyperparathyroidism, Alkali syndrome (Milk-Alkali), and Paget disease • **Clinical:** renal stones, abdominal pain, drowsiness, metastatic calcification

► Drugs in Bone and Mineral Disorders

Drug	Mechanism	Notes
Bisphosphonates (alendronate, etidronate, pamidronate, risedronate)	↓ bone resorption	• Osteoporosis • Paget disease • Esophageal ulceration may occur

Thyroid

► Physiologic Actions of Thyroid Hormone

Metabolic rate	↑ metabolic rate: high O_2 consumption, mitochondrial growth, ↑ Na^+/K^+-ATPase activity, ↑ food intake, thermogenesis, sweating, ↑ ventilation
Energy substrates	Mobilization of carbohydrates, fat and protein; ↑ ureagenesis; ↓ muscle and adipose mass
Growth	Required after birth for normal brain development and bodily growth (protein anabolic)
Circulation	↑ cardiac output (linked to metabolism), ↑ β-adrenergic receptors on heart

► Control of Thyroid Hormone

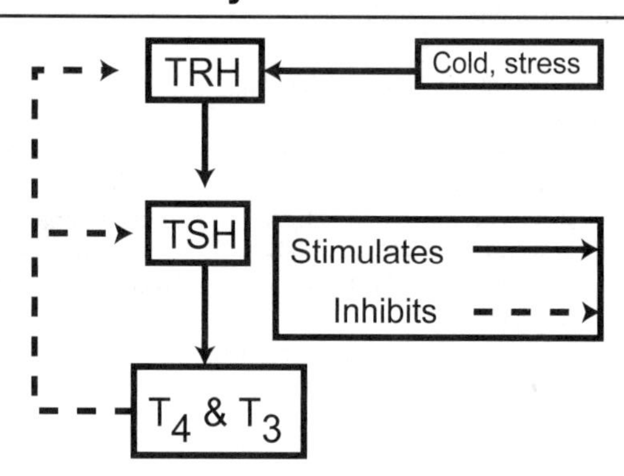

TRH = thyrotropin-releasing hormone (**hypothalamus**)

TSH = Thyroid-stimulating hormone (**anterior pituitary**); is also trophic (increases gland mass)

T_4 = thyroxine (tetraiodothyronine); 90% of production, but low biologic activity (**thyroid**)

T_3 = triiodothyronine; 10% of production, very potent (**thyroid**)

► Synthesis and Storage

Process	Mechanism	Result
Uptake of I^-	Active transport	↑ follicular cell iodide
Oxidation	Peroxidase	Produce oxidized iodine
Iodination	Peroxidase	Produces MIT and DIT within thyroglobulin
Coupling	Peroxidase	Links MITs and DITs to form T_4 and T_3
Exocytosis	Exocytosis into lumen	Storage of thyroglobulin

Definition of abbreviations: DIT, diiodotyrosine; MIT, monoiodotyrosine.

► Secretion

Process	Mechanism	Result
Transport	Endocytosis	Colloid taken up by follicular cells
Fusion	Lysosomes and colloid fuse	Incorporation in lysosomes
Proteolysis	Lysosomal enzymes	Cleave thyroglobulin into T_3, T_4, MITs, and DITs
Secretion	Simple diffusion	Lipid-soluble T_3 and T_4 diffuse into extracellular fluid to plasma
Deiodination	Deiodinase	Removes iodide from MITs and DITs to recycle it

► Regulation of Thyroid Activity

Transport	99% bound to thyroid binding globulin, 1% free; represents a pool to prevent rapid change of thyroid level
Peripheral Conversion	T_4 converted to more active T_3 by 5′-monodeiodinase or to inactive reverse T_3 by 5-monodeiodinase; occurs in most tissues giving local control of hormone action
Mechanism of action	Binding to nuclear receptors: T_4 has low affinity, T_3 has high affinity, so is responsible for most thyroid hormone effects
Degradation	Successive deiodination steps to thyronine, also sulfates and glucuronides

► Thyroid Disorders

Disorder	T_4	TSH	TRH	Gland Mass
Primary hypothyroidism	↓	↑	↑	↑, goiter possible
Pituitary hypothyroidism	↓	↓	↑	↓, due to low TSH
Hypothalamic hypothyroidism	↓	↓	↓	↓, due to low TSH
Iodine deficiency (prolonged, severe)	↓	↑	↑	↑, goiter likely
Pituitary hyperthyroidism	↑	↑	↓	↑, goiter possible
Primary hyperthyroidism (tumor)	↑	↓	↓	↓, due to low TSH
Graves disease (autoimmune production of thyroid-stimulating immunoglobulins [TSIs])	↑	↓	↓	↑, goiter possible

► Disorders of the Thyroid Gland

Hyperthyroidism	• Seen most often in Graves disease, toxic multinodular goiter, toxic adenoma • Clinical: tachycardia, cardiac palpitations (β-adrenergic effect), skin warm and flushed, ↑ body temperature, heat intolerance, hyperactivity, tremor, weight loss, osteoporosis, diarrhea and oligomenorrhea, eyes show a wide stare with lid lag; **exophthalmos** seen **only in Graves disease** • Thyrotoxic storm: severe hypermetabolic state with 25% fatality • **Lab values:** low TSH and elevated T_4; low TSH is most important • Note: in pregnancy, ↑ in TBG secondary to high estrogen levels elevates total serum T_4, but not free serum T_4
Graves disease	• Peaks in the third and fourth decades; more common in women; associated with other autoimmune diseases (including Hashimoto thyroiditis) • Type II non-cytotoxic hypersensitivity • Production of TSI and TGI bind and activate TSH receptors • **Pathology:** diffuse, moderate, symmetric enlargement of gland • **Microscopic appearance:** hypercellular with small follicles and little colloid
Hypothyroidism	• Clinical features depend on age group • Lab values: elevated TSH and low T_4
Infants	• Develop **cretinism**; major effects are on skeletal and CNS development; once apparent, syndrome is irreversible; neonatal screening for elevated TSH for early detection • **Clinical:** protuberant abdomen, wide-set eyes, dry rough skin, broad nose, delayed epiphyseal closure
Older children	Short stature, retarded linear growth (GH deficiency caused by thyroid hormone deficiency), delayed onset of puberty
Adults	• Lethargy, weakness, fatigue, decreased appetite, weight gain, cold intolerance, constipation • Myxedema: associated with severe hypothyroidism; periorbital puffiness, sparse hair, cardiac enlargement, pleural effusions, anemia
Hashimoto thyroiditis	• Chronic lymphocytic thyroiditis featuring goitrous thyroid gland enlargement • Autoimmune; may be autoantibodies to the TSH receptors, T_3 and T_4; **antimicrosomal antibodies** also seen • Type IV hypersensitivity • **Most common type of thyroiditis;** highest incidence in **middle aged females** • **Pathology:** painless goiter, gland enlarged and firm • **Microscopic appearance:** lymphocytic and plasma cell infiltrate with **Hürthle cells** (follicular cells with eosinophilic granular cytoplasm), evidence of cell-mediated cytolysis
Diffuse nontoxic goiter	Diffuse enlargement of gland in euthyroid patients; high incidence in certain geographic areas with iodine-deficient diets
de Quervain granulomatous subacute thyroiditis	• Self-limited disease; seen more often in females in the second to fifth decades • Follows **viral** syndrome, lasts several weeks with a **tender** gland • May initially have mild hyperthyroidism later, usually euthyroid
Riedel thyroiditis	• Rare, chronic thyroid disease, possibly of immune origin, that causes dense fibrosis of the thyroid gland leading to hypothyroidism • Presents with a hard, fixed, painless goiter • May be associated with idiopathic fibrosis in other sites such as the retroperitoneum
Thyroglossal duct cyst	• May communicate with skin or base of tongue • Remnant of incompletely descended midline thyroid tissue
Ectopic thyroid nests	Usually at the base of tongue; prior to removal, it must be documented that patient has other functioning thyroid tissue

Definition of abbreviations: GH, growth hormone; TBG, thyroid-binding globulin; TGI, thyroid growth immunoglobulin; TSI, thyroid-stimulating immunoglobulin; TSH, thyroid-stimulating hormone.

▶ Drugs for Thyroid Gland Disorders

Class	Mechanism	Comments/Agents
Hyperthyroidism Agents	These agents are used for short-term or long-term treatment of hyperthyroid states. The most common adverse effects are related to signs and symptoms of hypothyroidism.	
Thioamides	• Inhibit synthesis of thyroid hormones • They do *not* inactivate existing T_4 and T_3 • Propylthiouracil is able to inhibit peripheral conversion of T_4 to T_3	• **Examples: PTU, methimazole** • **Indications:** long-term hyperthyroid therapy, which may lead to disease remission and short-term treatment before thyroidectomy or radioactive iodine therapy • **Side effects:** skin rash (common), hematologic effects (rare)
Iodides	• Inhibit the release of T_4 and T_3 (primary) • Inhibit the biosynthesis of T_4 and T_3 and ↓ the size and vascularity of thyroid gland	• **Examples:** Lugol's solution (iodine and potassium iodide) and potassium iodide alone • **Indications:** preparation for thyroid surgery; treatment of thyrotoxic crisis and thyroid blocking in radiation emergency • **Note:** therapeutic effects can be seen for as long as 6 weeks
Beta-Blockers	Nonselective β-receptor blockers used for palpitations, anxiety, tremor, heat intolerance; partially inhibit peripheral conversion of T_4 to T_3	• **Examples: nadolol, propranolol** • Used to treat the signs and symptoms of hyperthyroidism
Radioactive Iodine [^{131}I]	Ablation of thyroid gland	• **Indications:** first-line therapy for Graves disease; treatment of choice for recurrent thyrotoxicosis in adults and elderly
Hypothyroidism Agents	These agents are used as thyroid replacement therapy. The most common adverse effects are related to signs and symptoms of hyperthyroidism.	
Thyroid Hormones	Acts by controlling DNA transcription and protein synthesis	**Examples:** synthetic T_4, synthetic T_3, or combination of synthetic T_4:T_3 in 4:1 ratio

Definition of abbreviation: PTU, propylthiouracil.

▶ Thyroid Neoplasms

Adenomas	• Follicular adenoma is most common; may cause pressure symptoms, pain, and rarely thyrotoxicosis • **Pathology:** usually small, well-encapsulated solitary lesions
Papillary carcinoma	• Most common thyroid carcinoma • Incidence higher in women • **Pathology:** papillary branching pattern; 40% have tumors containing psammoma bodies • Spread to local nodes is common; **hematogenous spread rare** • Resection curative in most cases
Follicular carcinoma	• More malignant than papillary cancer • **Pathology:** may be encapsulated, with penetration through the capsule; colloid sparse • Local invasion and pressure → dysphagia, dyspnea, hoarseness, cough • **Hematogenous metastasis to lungs or bones common**
Medullary carcinoma	• Arises from parafollicular C cells • Secretes calcitonin • May be associated with MEN IIa and IIb
Anaplastic carcinoma	• Rapid growing, aggressive with poor prognosis; affects older patients • **Pathology:** tumors usually bulky and invasive with undifferentiated anaplastic cells • **Clinical:** early, widespread metastasis and death within 2 years

Growth Hormone

► Control of Growth Hormone

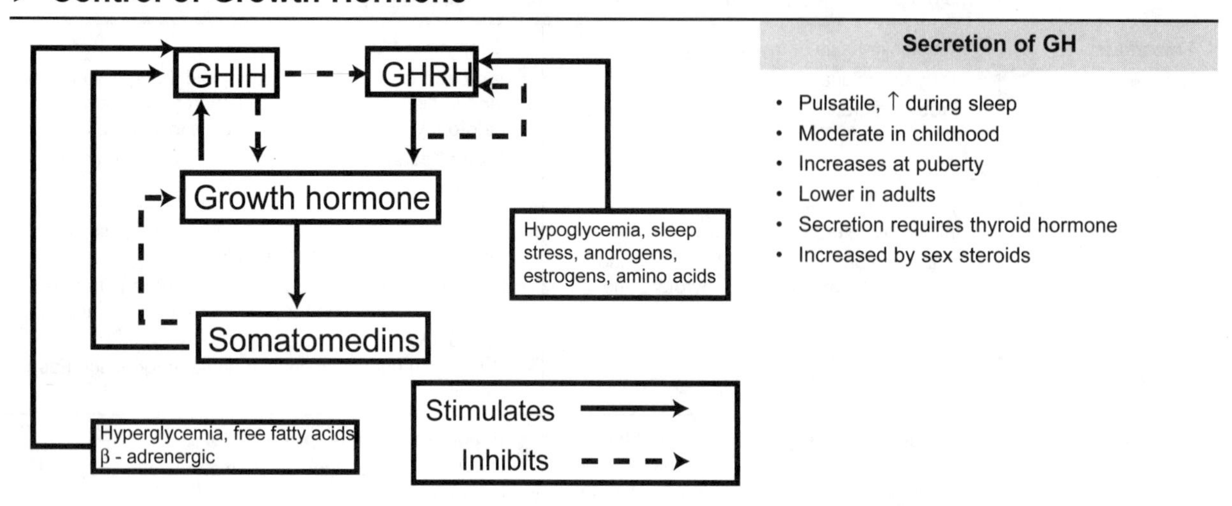

Secretion of GH

- Pulsatile, ↑ during sleep
- Moderate in childhood
- Increases at puberty
- Lower in adults
- Secretion requires thyroid hormone
- Increased by sex steroids

► Biologic Actions of Growth Hormone

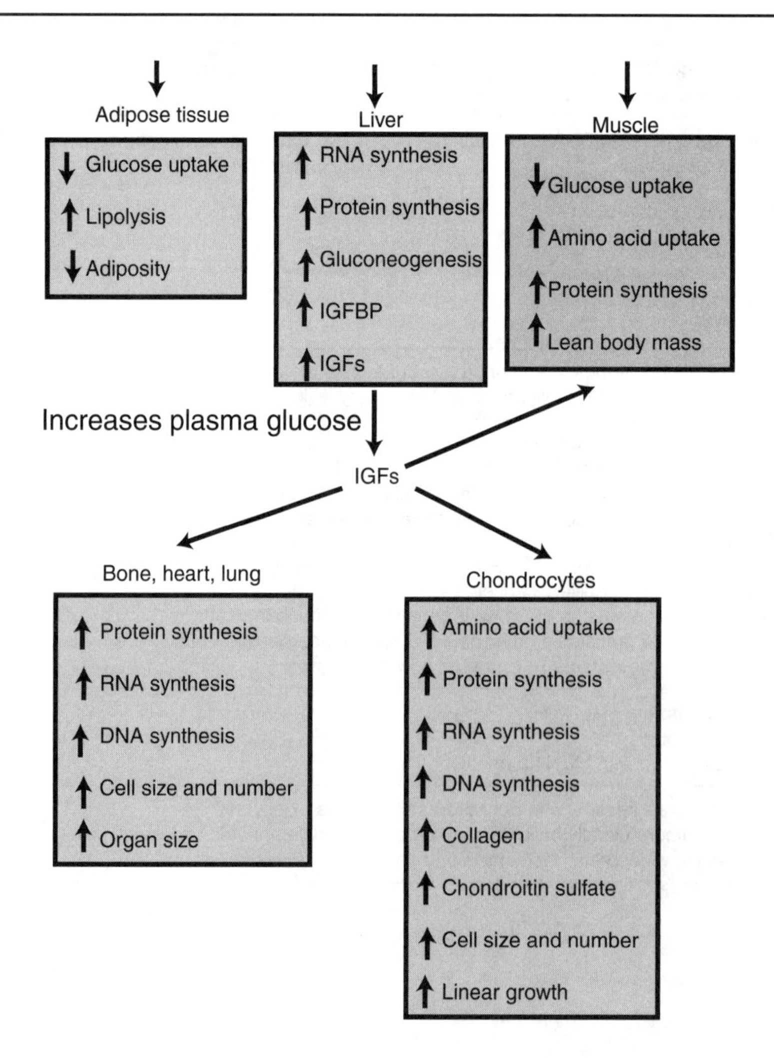

► Childhood Disorders

Dwarfism	• Caused by: – ↓ GH or ↓ liver production of IGF (Laron syndrome) – Defective GH receptors • Mental development normal • Reversible with GH treatment
Gigantism	• Caused by ↑ GH prior to epiphyseal closure • Increased height, increased body mass

Definition of abbreviation: GH, growth hormone; IGF, insulin-like growth factor.

► Adult Disorders of Growth Hormone

Acromegaly	↑ GH	• Enlargement of hands and feet • Protrusion of lower jaw, coarse facial features • ↑ lean body mass • ↓ body fat • ↑ size of visceral organs • Impaired cardiac function (related to mass) • Abnormal glucose tolerance with tendency to hyperglycemia
Deficiency	↓ GH	• Tendency to hypoglycemia when fasting • Susceptible to insulin-induced hypoglycemia • Significance of other effects disputed

Definition of abbreviation: GH, growth hormone.

► Treatment of Growth Hormone Disorders

Class	Mechanism	Comments/Agents
GH	Stimulation of linear/skeletal growth (pediatric patients only); potentiation of cell and organ growth; enhanced protein, carbohydrate and lipid metabolism	• **Somatropin** and **somatrem** (recombinant forms of human GH) • **Indications:** growth failure, Turner syndrome, cachexia, somatotropin deficiency
Octreotide	• Long-acting octapeptide that mimics somatostatin • Inhibits release of GH, glucagon, gastrin, thyrotropin, insulin	• **Indications:** acromegaly, carcinoid, glucagonoma, gastrinoma, other endocrine tumors

Definition of abbreviation: GH, growth hormone.

The Reproductive System

Chapter 7

Reproductive System

▶ Male and Female Development

Adult Female and Male Reproductive Structures Derived From Precursors of the Indifferent Embryo		
Adult Female	**Indifferent Embryo**	**Adult Male**
Ovary, follicles, rete ovarii	Gonads	Testes, seminiferous tubules, rete testes
Uterine tubes, uterus, cervix, and upper part of vagina	Paramesonephric ducts	Appendix of testes
Duct of Gartner	Mesonephric ducts	Epididymis, ductus deferens, seminal vesicle, ejaculatory duct
Clitoris	Phallus	Glans and body of penis
Labia minora	Urogenital folds	Ventral aspect of penis
Labia majora	Labioscrotal swellings	Scrotum

Congenital Reproductive Anomalies

Female Pseudointersexuality

- 46,XX genotype
- Have ovarian (but no testicular) tissue and masculinization of the female external genitalia
- Most common cause is **congenital adrenal hyperplasia**, a condition in which the fetus produces excess androgens

Male Pseudointersexuality

- 46,XY genotype
- Testicular (but no ovarian) tissue and stunted development of male external genitalia
- Most common cause is inadequate production of testosterone and müllerian-inhibiting factor (MIF) by the fetal testes; due to a 5α-reductase deficiency

5α-reductase 2 deficiency	• Caused by a mutation in the **5α-reductase 2 gene** that renders 5α-reductase 2 enzyme underactive in catalyzing the conversion of testosterone to dihydrotestosterone • *Clinical findings:* underdevelopment of the penis and scrotum (microphallus, hypospadias, and bifid scrotum) and prostate gland; epididymis, ductus deferens, seminal vesicle, and ejaculatory duct are normal • At puberty, they undergo virilization due to an increased **T:DHT ratio**
Complete androgen insensitivity (CAIS, or testicular feminization syndrome)	• Occurs when a fetus with a 46,XY genotype develops testes and female external genitalia with a rudimentary vagina; the uterus and uterine tubes are generally absent • Testes may be found in the labia majora and are surgically removed to circumvent malignant tumor formation • Individuals present as normal-appearing females, and their psychosocial orientation is female despite their genotype • Most common cause is a mutation in the **androgen receptor (AR) gene** that renders the AR inactive

▶ Male and Female Reproductive Anatomy

Male

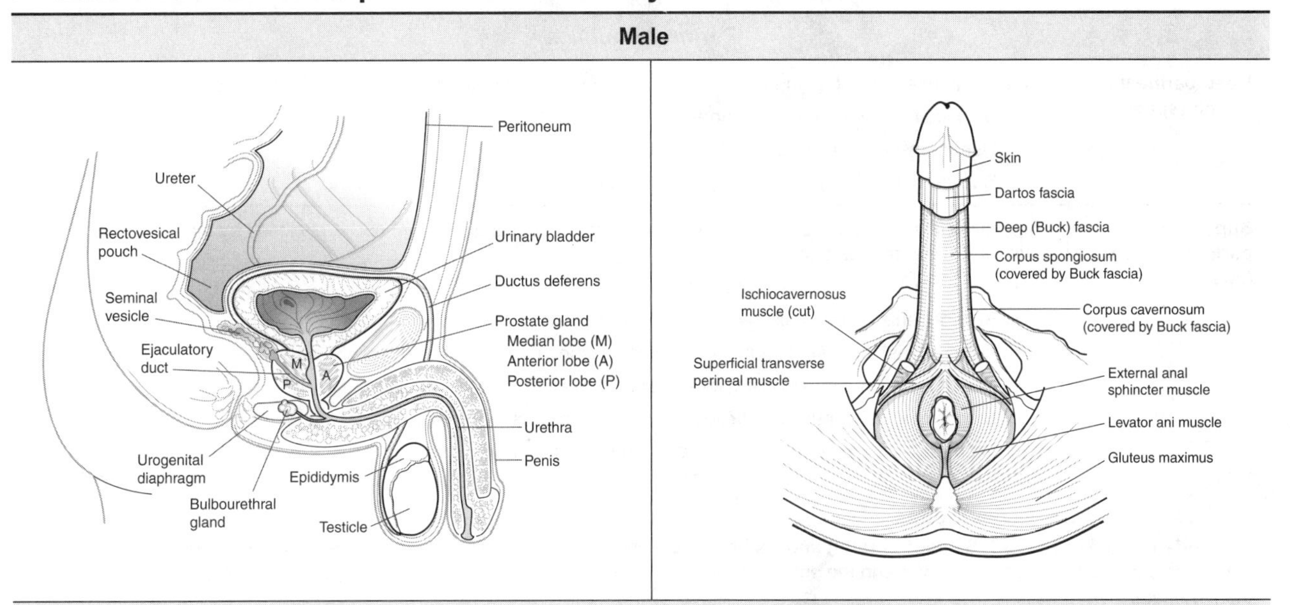

Female

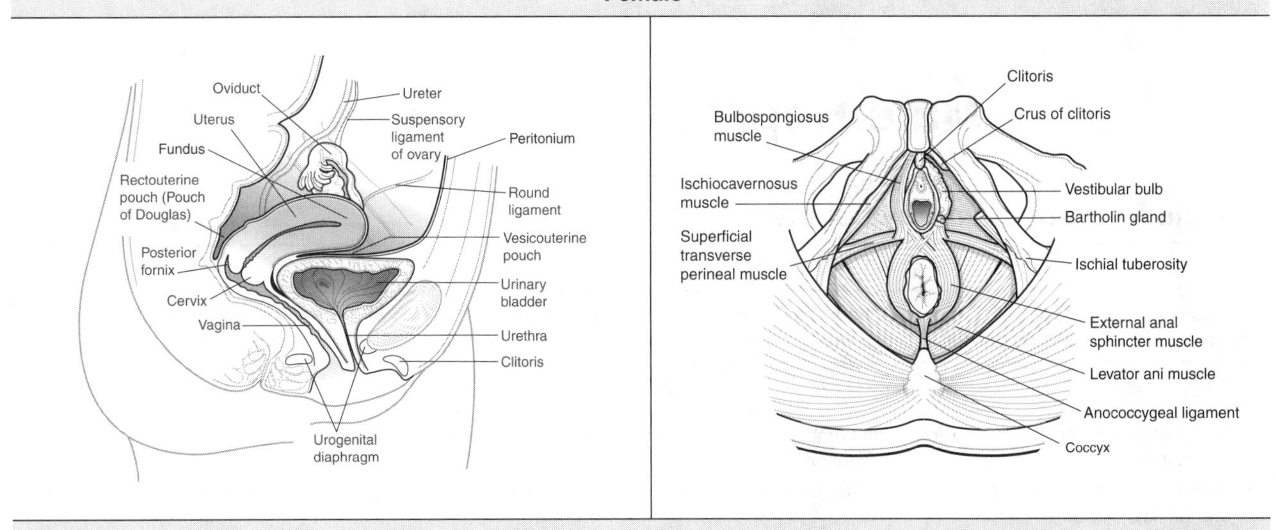

Pelvic Floor and Perineum

- The floor of the pelvis is formed by the **pelvic diaphragm** (two layers of fascia with a middle layer of skeletal muscle).
- The muscles forming the middle layer are the **levator ani** and **coccygeus**. The levator ani acts as a muscular sling for the rectum and marks the boundary between the rectum and anal canal.
- The region below the pelvic diaphragm is the **perineum**, which contains the **ischioanal fossa**, the fat-filled region below the pelvic diaphragm surrounding the anal canal.
- The **urogenital diaphragm** is in the perineum and extends between the two ischiopubic rami. The urogenital diaphragm (like the pelvic diaphragm) is composed of two layers of fascia with a middle layer of skeletal muscle.

(Continued)

► Male and Female Reproductive Anatomy (*Cont'd.*)

Perineal Pouches	
Deep perineal pouch (space)	The deep perineal pouch is the middle (muscle) layer of the urogenital diaphragm. It contains: • Sphincter urethrae muscle—serves as external sphincter of the urethra • Deep transverse perineal muscle • Bulbourethral (Cowper) gland (in the male only)—duct enters bulbar urethra
Superficial perineal pouch (space)	The superficial perineal pouch is the region below the urogenital diaphragm and is enclosed by the superficial perineal (Colles) fascia. It contains: • Crura of penis or clitoris—erectile tissue • Bulb of penis (in the male)—erectile tissue; contains urethra • Bulbs of vestibule (in the female)—erectile tissue; in lateral walls of vestibule • Ischiocavernosus muscle—skeletal muscle that covers crura of penis or clitoris • Bulbospongiosus muscle—skeletal muscle that covers bulb of penis or bulbs of vestibule • Greater vestibular (Bartholin) gland (in female only)—homologous to Cowper gland

Pelvic Innervation
The **pudendal nerve (S2, S3, S4 ventral rami) and its branches** innervate the skeletal muscles in the pelvic and urogenital diaphragms, the external anal sphincter and the sphincter urethrae, skeletal muscles in both perineal pouches, and the skin that overlies the perineum.

Male Reproductive System

► Male Reproductive Physiology

Penile Erection
Erection occurs in response to **parasympathetic** stimulation (pelvic splanchnic nerves). **Nitric oxide** is released, causing relaxation of the corpus cavernosum and corpus spongiosum, which allows blood to accumulate in the trabeculae of erectile tissue.

Ejaculation
• **Sympathetic** nervous system stimulation (lumbar splanchnic nerves) mediates movement of mature spermatozoa from the epididymis and vas deferens into the ejaculatory duct. • Accessory glands, such as the bulbourethral (Cowper) glands, prostate, and seminal vesicles, secrete fluids that aid in sperm survival and fertility. • **Somatic motor efferents** (pudendal nerve) that innervate the bulbospongiosus and ischiocavernous muscles at the base of the penis stimulate the rapid ejection of semen out the urethra during ejaculation. Peristaltic waves in the vas deferens aid in a more complete ejection of semen through the urethra.

Clinical Correlation
• Injury to the bulb of the penis may result in extravasation of urine from the urethra into the superficial perineal space. From this space, urine may pass into the scrotum, into the penis, and onto the anterior abdominal wall. • Accumulation of fluid in the scrotum, penis, and anterolateral abdominal wall is indicative of a laceration of either the membranous or penile urethra (deep to Scarpa fascia). This can be caused by trauma to the perineal region (saddle injury) or laceration of the urethra during catheterization.

► Agents for Erectile Dysfunction

Drug	Mechanism	Comments
Selective phosphodiesterase (PDE) 5 inhibitors **(sildenafil, vardenafil, tadalafil)**	Inhibits the enzyme phosphodiesterase (PDE) 5, which inactivates cGMP, leading to ↑ cGMP → vasodilation, more inflow of blood → better erection	PDE 5 inhibitors + **nitrates** (which ↑ cGMP production) → **severe hypotension**
Synthetic prostaglandin E_1 (PGE_1) agents **(alprostadil)**	↑ cAMP (via adenylate cyclase) → smooth muscle relaxation	Administered via transurethral or intracavernosal injection *Contraindications:* intercourse with pregnant women (can stimulate uterine contractions unless used with a condom); conditions that might predispose a patient to priapism (e.g., sickle cell anemia, multiple myeloma, leukemia)
Testosterone	Replacement/supplementation for males whose serum androgen concentrations are below normal	Used if diminished libido is a significant patient complaint

► Diseases of the Penis and Prostate

Noninfectious Disorders of the Penis	
Hypospadias and epispadias	With hypospadias, urethra opens onto the ventral surface of penis Often associated with a poorly developed penis that curves ventrally, known as **chordee** With epispadias, urethra opens onto dorsal surface; often associated with exstrophy of bladder Either of these malformations may cause infertility
Phimosis	Prepuce orifice too small to be retracted normally Interferes with hygiene; can also predispose to bacterial infections; if foreskin retracted over the glans, it may lead to urethral constriction → paraphimosis *Treatment:* circumcision
Penile carcinoma	
Bowen disease	Carcinoma in situ; can be associated with visceral malignancy Men >35 years; tends to involve shaft of the penis and scrotum *Gross:* thick, ulcerated plaque *Micro:* squamous cell carcinoma in situ
Squamous cell carcinoma	1% of cancers in men in the United States, usually age 40–70. Usually slow growing and non-painful; patients often delay seeking medical attention. Circumcision decreases the incidence. Human Papilloma virus (types 16 and 18) infection is closely associated. Gross: Plaque progressing to an ulcerated papule or fungating growth. Metastases can go to local lymph nodes.
Peyronie disease	Curved penis due to fibrosis of the tunica albuginea

(Continued)

▶ Diseases of the Penis and Prostate (*Cont'd.*)

Diseases of the Prostate	
Prostatic carcinoma	• Most common cancer in men; usually occurs after age 50, and the incidence increases with age • Associated with race (more common in African Americans than in Caucasians, relatively rare in Asians) • May present with urinary problems or a palpable mass on rectal examination • Prostate cancer more common than lung cancer, but lung cancer is bigger killer • Metastases may occur via the lymphatic or hematogenous route • Bone commonly involved with osteoblastic metastases, typically in the pelvis and lower vertebrae • Elevated PSA, together with an enlarged prostate on digital rectal exam, highly suggestive of carcinoma • Most patients present with advanced disease and have a 10-year survival rate of <30% • *Treatment:* surgery, radiation, and hormonal modalities (orchiectomy and androgen blockade).
Benign prostatic hyperplasia	• Formation of large nodules in the periurethral region (median lobe) of the prostate • May narrow the urethral canal to produce varying degrees of urinary obstruction and difficulty urinating • It is increasingly common after age 45; incidence increases steadily with age • Can follow an asymptomatic pattern, or can result in urinary symptoms and urinary retention
Prostatitis	
Acute	• Results from a bacterial infection of the prostate • Pathogens are often organisms that cause urinary tract infection • *Escherichia coli* most common • Bacteria spread by direct extension from the posterior urethra or the bladder; lymphatic or hematogenous spread can also occur
Chronic	• Common cause of recurrent urinary tract infections in men • Two types: bacterial and nonbacterial • Both forms may be asymptomatic or may present with lower back pain and urinary symptoms

▶ Antiandrogens

Drug	Mechanism	Clinical Uses
Flutamide	Androgen receptor antagonist	Prostate cancer
Spironolactone	Androgen receptor antagonist (also a potassium-sparing diuretic)	Hirsutism (also used for primary hyperaldosteronism, edema, hypertension)
Leuprolide	GnRH analog	Depot form is used for prostate cancer
Finasteride	5α-reductase inhibitor (prevents conversion of testosterone to DHT)	Benign prostatic hypertrophy (BPH), male pattern baldness
Ketoconazole	Inhibits steroid synthesis (also an antifungal agent)	Androgen receptor–positive prostate cancer

Definition of abbreviation: DHT, dihydrotestosterone.

▶ The Testes

Descent of the Testes

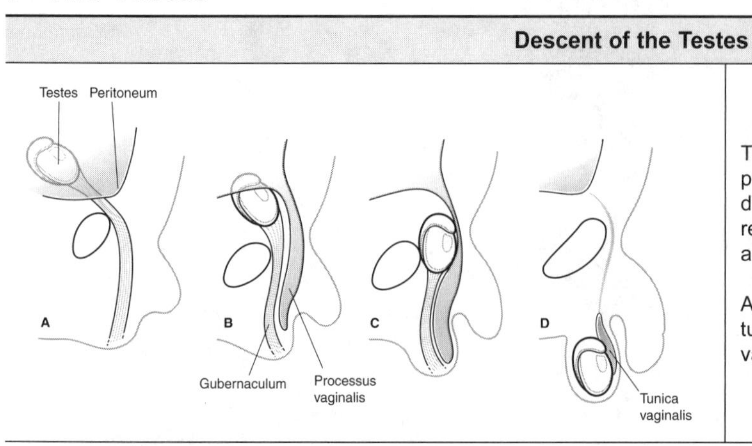

The **processus vaginalis** is an evagination of parietal peritoneum that descends through the inguinal canal during fetal life. The **tunica vaginalis** is a patent remnant of the processus vaginalis that covers the anterior and lateral parts of the testis.

A **hydrocele** is an accumulation of serous fluid in the tunica vaginalis or in a persistent part of the processus vaginalis in the cord.

(Continued)

▶ The Testes (*Cont'd.*)

Normal Anatomy and Anatomic Abnormalities

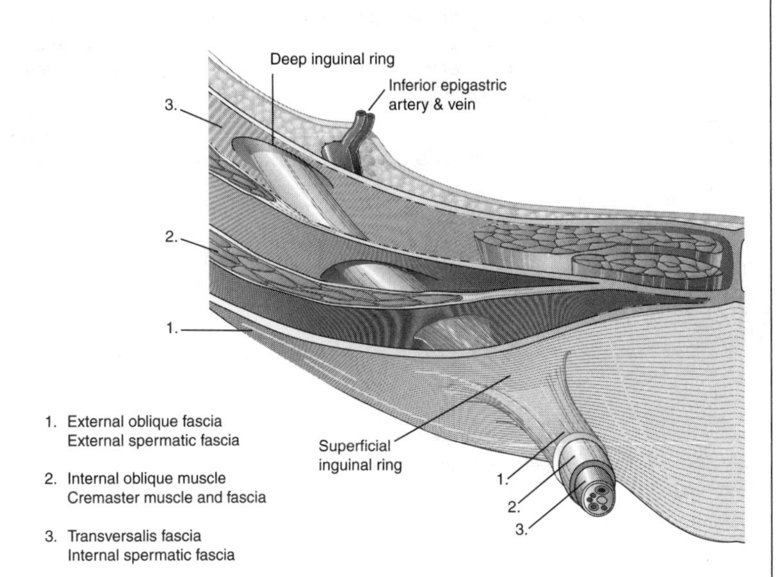

The **inguinal canal** is formed by four of the eight tissue layers of the anterior abdominal wall; outpocketings of three of these layers give rise to **spermatic fasciae**, which cover the testis and structures in the **spermatic cord**.

The spermatic cord contains the:

- **Ductus deferens**, which conveys sperm from the epididymis to the ejaculatory duct in the male pelvis
- **Testicular artery**, which arises from the abdominal aorta between the L2 and the L3 vertebrae
- **Artery to the ductus deferens**, which arises from a branch of the internal iliac artery
- **Pampiniform plexus of the testicular vein**
- **Right testicular vein**, which drains into the inferior vena cava
- **Left testicular vein**, which drains into the left renal vein.
- **Lymphatic vessels** that drain the testis. Testicular lymphatic vessels pass through the inguinal canal and drain directly to **lumbar nodes** in the posterior abdominal wall.

Cremasteric Reflex

The **cremasteric reflex** utilizes sensory and motor fibers in the ventral ramus of the L1 spinal nerve. Stroking the skin of the superior and medial thigh stimulates sensory fibers of the **ilioinguinal nerve**. **Motor fibers** from the genital branch of the **genitofemoral nerve** cause the cremaster muscle to contract, elevating the testis.

Abnormalities

Cryptorchidism	• Failure of normal descent of intra-abdominal testes into the scrotum • Most common location of a cryptorchid testis is in the inguinal canal • Unilateral or bilateral, more often on right side • Bilateral cryptorchidism can cause infertility • Maldescended testes are associated with a greatly increased incidence of testicular cancer, even once repositioned within scrotum
Torsion	• Precipitated by sudden movement, trauma, and congenital anomalies • Twisting of spermatic cord may compromise both arterial supply and venous drainage • Sudden onset of testicular pain and a loss of the cremasteric reflex are characteristic • If not surgically corrected early, may result in testicular infarction
Hydrocele	• Congenital hydrocele occurs when a small patency of the processus vaginalis remains so that peritoneal fluid can flow into the processus vaginalis; may occur later in life, often inflammatory causes (e.g., epididymitis) • Results in fluid-filled cyst near testes
Varicocele	• Results from dilatations of tributaries of the testicular vein in the pampiniform plexus • Varicosities of the pampiniform plexus are observed when the patient is standing and disappear when the patient is lying down
Spermatocele	• Retension cyst containing sperm in the rete testes or head of the epididymis

▶ Spermatogenesis

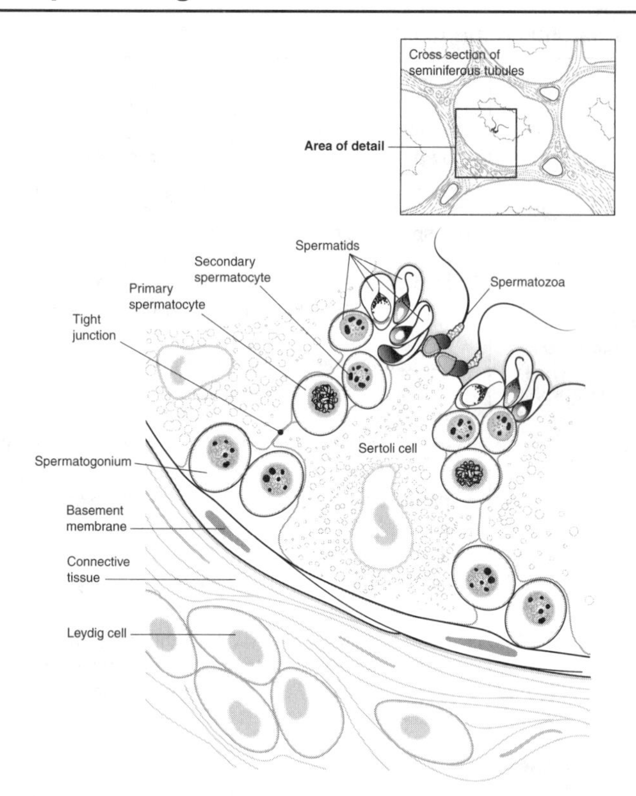

Spermatogenesis occurs in the seminiferous tubules between the **Sertoli cells**, which extend from the seminiferous tubule basement membrane to the lumen and are separated by tight junctions (blood–testis barrier) and germ cells in varying stages of spermatogenesis. The **blood–testis barrier** protects the spermatocytes and spermatids from the immune system

Three Stages of Spermatogenesis

1. **Spermatocytogenesis** begins at puberty adjacent to the basement membrane of the Sertoli cell. Spermatogonia first undergo spermatocytogenesis, during which mitosis divides the spermatogonia into spermatocytes.
2. **Meiosis** reduces the diploid spermatocytes into haploid spermatids.
3. **Spermiogenesis** is the maturation of spermatids into mature spermatozoa.

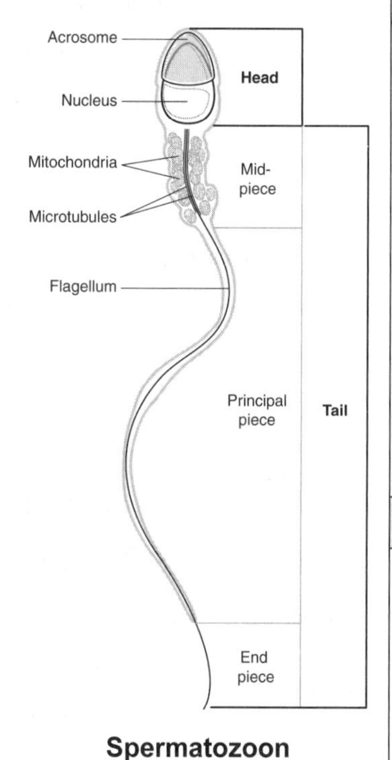

Spermatozoon

Spermiogenesis

During **spermiogenesis**, the spermatids undergo chromatin condensation and nuclear elongation, which forms the head of the spermatozoa.

The **acrosome**, a hydrolytic enzyme-containing region on the sperm cell head, also forms during this time.

The midpiece of the sperm has as a mitochondrial sheath that contains much of the ATP necessary for sperm movement. The **flagellum** contains an array of **9 + 2 microtubular pairs** linked by **dynein** for sperm motility. Patients with defective or absent dynein in **Kartagener syndrome** have reduced sperm motility, as well as reduced mucus clearance in the respiratory pathways, leading to bronchiectasis.

Once formed, spermatozoa detach from the Sertoli cells and combine with a fluid that aids in the movement of spermatozoa into the epididymis. In the epididymis, the fluid is reabsorbed, thereby concentrating sperm, and sperm interact with **forward motility factor**, a protein that aids in sperm motility. Ejaculated sperm must undergo **capacitation** in the uterus before fertilization can occur.

Fertilization

Fertilization is a three-step process:

1. **Acrosome reaction:** Sperm close to the corona radiata release hyaluronidase, which dissolves material between corona radiata cells, allowing sperm to reach the zona pellucida.
2. **Zonal reaction:** Sperm bind to a glycoprotein of the zona and release acrosin, which facilitates penetration of the zona by the sperm head.
3. **Cortical reaction:** The first sperm to penetrate the zona fuses with the plasma membrane of the ovum and induces a calcium-dependent release of cortical granules that prevents polyspermy.

► Hormonal Control of Steroidogenesis and Spermatogenesis

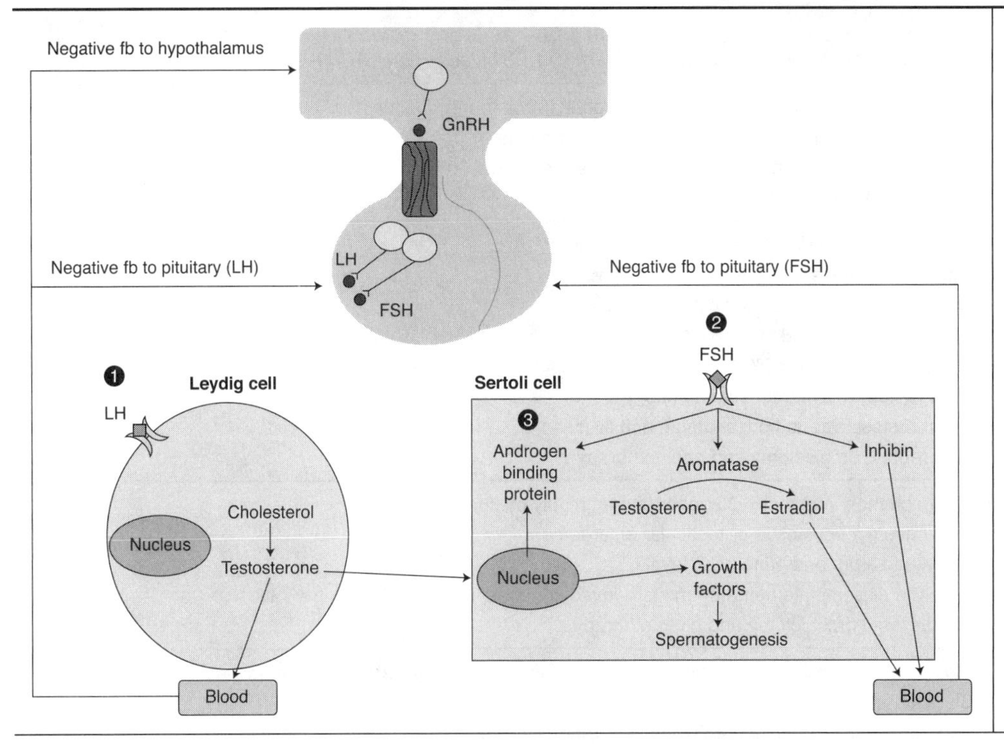

GnRH is synthesized in the **preoptic nucleus** and is released in a pulsatile manner into the hypophysial portal system. Binding of GnRH to its receptor on the anterior pituitary stimulates the release of LH and FSH.

Note: If the gonadotrophs are subjected to constant GnRH stimulation, the receptors will undergo downregulation.

❶ **LH** binds to the **Leydig cell** and stimulates the conversion of cholesterol into testosterone. Testosterone diffuses into blood and feeds back to inhibit hypothalamic GnRH and pituitary LH. Testosterone diffuses into the Sertoli cell and increases transcription of androgen-binding protein and growth factors that mediate spermatogenesis.

❷ **FSH** binds to the **Sertoli cell** and stimulates transcription of androgen-binding protein, the conversion of testosterone to estradiol, and the secretion of inhibin. Inhibin feeds back to inhibit further production of pituitary FSH.

❸ Sertoli cell-derived **androgen binding protein** provides an important reserve of testosterone in the testes. Because spermatogenesis is dependent on intratesticular testosterone rather than systemic testosterone, these reserves are important to normal spermatogenesis.

► Inflammatory Lesions

Mumps	• Orchitis develops in approximately 25% of patients over age 10, but is less common in patients under 10 • Rarely leads to sterility
Gonorrhea	• Neglected urethral gonococcal infection may spread to prostate, seminal vesicles, and epididymis, but rarely to testes
Syphilis	• Acquired or congenital syphilis may involve the testes • Two forms: gummas or a diffuse interstitial/lymphocytic plasma cell infiltrate • Can lead to sterility
Tuberculosis	• TB usually spreads from epididymis; this is almost always associated with foci of TB elsewhere

▶ Testicular Neoplasms

Germ Cell Tumors Most common malignancy in men 15 to 34 years of age	
Seminoma	• Rare in infants, incidence increases to a peak in the fourth decade • 10% are anaplastic seminomas; show nuclear atypia • *Prognosis:* with treatment; 5-year survival rate exceeds 90% • Highly radiosensitive; metastases rare
Embryonal carcinoma	• Most commonly in the 20–30-year-age group • Aggressive, present with testicular enlargement • 30% metastatic disease at time of diagnosis • Serum AFP: elevated • 5-year mortality rate 65% • Less radiosensitive than seminomas • Often metastasize to nodes, lungs, and liver • May require orchiectomy and chemotherapy
Choriocarcinoma	• Most common in men 15–25 years of age, highly malignant • May have gynecomastia or testicular enlargement • Elevated serum and urine hCG levels • Tends to disseminate hematogenously, invading lungs, liver, and brain • Treated with orchiectomy and chemotherapy
Yolk sac tumor	• Most common testicular tumor in children and infants, although rare overall in adults • Elevated alpha fetoprotein (AFP) • Very aggressive; exhibiting a 50% 5-year mortality rate
Teratoma	• Can occur at any age, but are most common in infants and children • Appears as a testicular mass • Exhibit a variety of tissues, such as nerve, muscle, cartilage, and hair • Benign behavior during childhood, malignant in adults • 2-year mortality is 30% • *Treatment:* orchiectomy, followed by chemotherapy and radiation
Non–Germ Cell Tumors	
Leydig cell tumor	• Usually unilateral • Can produce androgens or estrogens • *Children:* present with masculinization or feminization; *adults:* gynecomastia • Usually benign and only 10% are invasive; surgery may be curative
Sertoli cell	• Usually unilateral • Can produce small amounts of androgens or estrogens, usually not enough to cause endocrinologic changes • Present with testicular enlargement • Over 90% are benign
Lymphoma	• Lymphomas are the most common testicular cancer in elderly men • The tumors are rarely confined to the testes (often disseminated)

Note: Testicular neoplasms tend to metastasize to the **lumbar nodes**, whereas scrotal disease affects **superficial inguinal nodes**.

Female Reproductive System

► Hormonal Control of Steroidogenesis

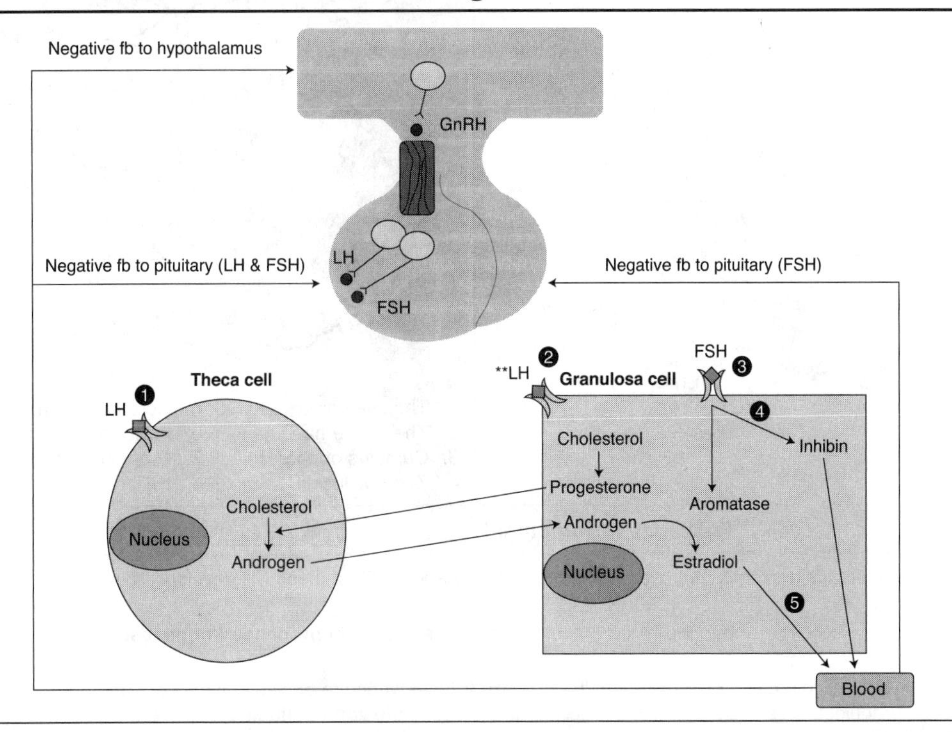

GnRH is synthesized in the **preoptic nucleus** and is released in a **pulsatile** manner into the hypophysial portal system. Binding of GnRH to its receptor on the anterior pituitary stimulates the release of LH and FSH.

Note: If the gonadotrophs are subjected to constant GnRH stimulation, the receptors will undergo downregulation.

1. **LH** binds receptors on **theca cells**, resulting in production of androstenedione or testosterone. Androgens produced by the theca cells enter the granulosa cells to be converted into estrogens.

2. **LH** also binds receptors on **granulosa cells** during the luteal phase and stimulates production of progesterone, which enters the theca cells. During the luteal phase, progesterone is required to maintain pregnancy if fertilization/implantation occur.

3. **FSH** binds receptors on **granulosa cells**, resulting in aromatization of androgens to estradiol and synthesis of new LH receptors on the granulosa cells. Estradiol can be released into the blood or can act locally to increase granulosa cell proliferation and sensitivity to FSH.

4. **FSH** also stimulates the production of **inhibin**, which negatively feeds back to inhibit further FSH secretion.

5. **Estradiol** secreted into the blood negatively feeds back to inhibit hypothalamic and pituitary secretion of GnRH, and LH and FSH, respectively. This action does not occur near the ovulatory period.

Definition of abbreviations: LH, luteinizing hormone; FSH, follicle-stimulating hormone; GnRH, gonadotropin-releasing hormone.

► Folliculogenesis and Ovulation

Follicular Development	Graafian follicle

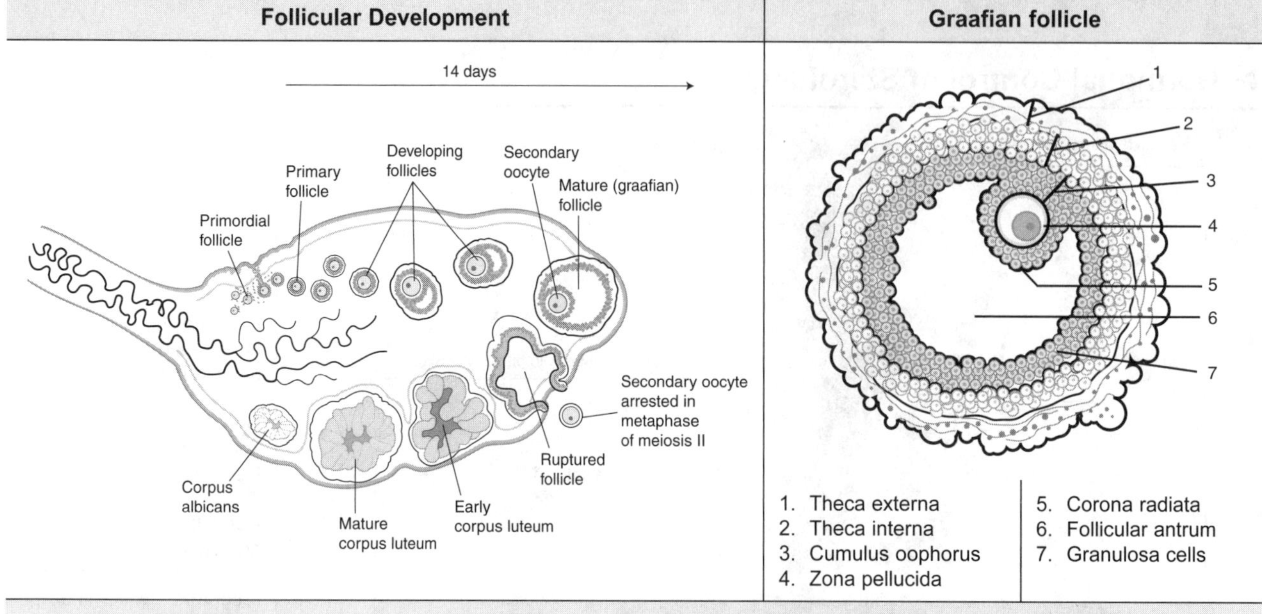

1. Theca externa
2. Theca interna
3. Cumulus oophorus
4. Zona pellucida
5. Corona radiata
6. Follicular antrum
7. Granulosa cells

Follicular Development

- At puberty, there are about 400,000 follicles present in the ovarian stroma, but only about 450 of these will develop (remaining follicles undergo atresia).
- The immature or **primordial follicle** (an oocyte surrounded by pregranulosa cells) is **arrested in prophase I of meiosis** until maturation.
- Starting at puberty, during each cycle, a primordial follicle becomes the **primary follicle** when the oocyte enlarges and the granulosa cells mature and proliferate. The granulosa cells secrete mucopolysaccharides, creating the **zona pellucida**, which protects the oocyte and provides an avenue for the oocyte to receive nutrients and chemical signals from the granulosa cells.
- As the follicle matures, additional granulosa cell layers are added and a layer of androgen-producing theca cells known as the **theca interna** surround the now **secondary follicle**. The secondary follicle continues to grow, and a fibrous theca externa surrounds the follicle. A follicular cavity (**antrum**) forms from granulosa cell secretions.
- The mature follicle, the **graafian follicle**, is now ready for ovulation. At the time of the LH surge, the oocyte resumes meiosis, **completes the first meiotic division, and is arrested in metaphase of meiosis II prior to ovulation**. Meiosis produces a nonfunctional first polar body, which degenerates, and a larger secondary haploid oocyte.
- If it is fertilized, the secondary oocyte completes meiosis II to form a **mature oocyte** and **polar body**.

Ovulation

- As the antral fluid increases, the pressure becomes greater until the follicle ruptures and the oocyte is extruded.
- Following ovulation, the theca cells enlarge and begin secreting estrogen and the granulosa cells enlarge and secrete progesterone. This new endocrine organ is called the **corpus luteum** and reaches maximal development about 7 days after ovulation.
- If fertilization does not occur, the corpus luteum degenerates.
- If fertilization does occur, the corpus luteum continues to grow for about 3 months and is maintained by **human chorionic gonadotropin (hCG)** from the embryo. Once the placenta is functional, the corpus luteum is no longer necessary to maintain pregnancy.

► Uterine Cycle

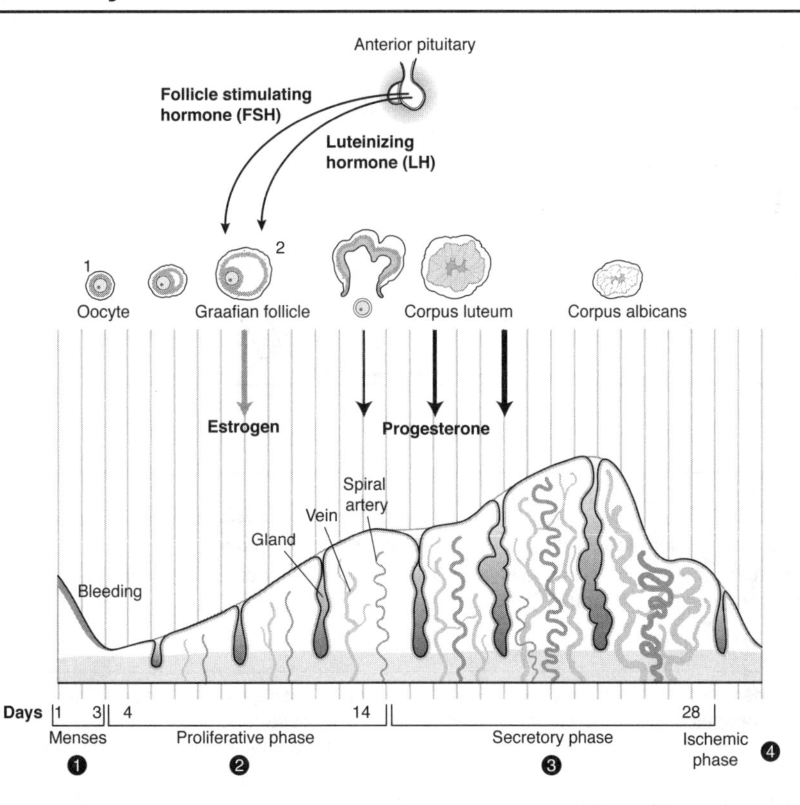

The **uterine cycle** can be divided into four phases:

❶ **menstruation**

❷ **proliferative phase**

❸ **secretory phase**

❹ **premenstruation**

As with the menstrual cycle, Day 1 of the uterine cycle begins at the onset of menses. This time period corresponds with the degeneration of the corpus luteum.

Proliferative Phase ❷

- The **proliferative phase** follows menses, corresponding to the latter portion of the follicular phase of the menstrual cycle and ending near ovulation.
- During this time, elevated estrogen levels stimulate the **proliferation of endometrial cells**, an **increase in length and number of endometrial glands**, and **increased blood flow** to the uterus.
- The endometrium increases in thickness sixfold and becomes contractile. Estrogen also stimulates a marked increase in progesterone receptors in the endometrium to prepare it for fertilization. **Edema** develops in the uterus toward the end of the proliferative phase and continues to develop during the secretory phase.

Secretory Phase ❸

- The **secretory phase** corresponds with the luteal phase of the menstrual cycle and is characterized by **endometrial cell hypertrophy, increased vascularity, and edema**.
- **Progesterone levels are elevated** during this phase and lead to a thick secretion consisting of glycoprotein, sugars, and amino acids. Like estrogen, progesterone increases cell proliferation and vascularization, but unlike estrogen, progesterone depresses uterine contractility.

Premenstrual Phase ❹

- The **premenstrual phase** consists of constriction of arteries, causing ischemia and anoxia. The **superficial layer of the endometrium degenerates**, and blood and tissue appear in the uterine lumen.

▶ Menstrual Cycle

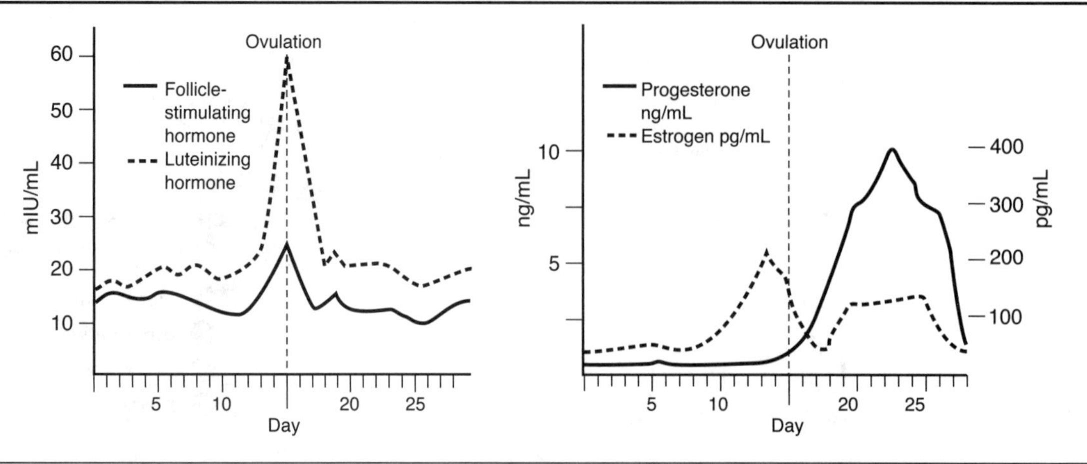

Overview

- There are **four phases** in the menstrual cycle: **menses**, **follicular phase**, **ovulation**, and **luteal phase**.
- The **average cycle length** is **28 days**, but can vary widely. The period from ovulation to the onset of menses is always 14 days, so any variation from 28 days occurs during the **follicular** phase.

Menses

- The onset of menses marks **Day 1** of the menstrual cycle and is triggered by a **decrease in estrogen** (decreased estrogen synthesis by granulosa cells and decreased LH) and **progesterone**. The decrease in estrogen and progesterone support for the endometrium results in tissue necrosis and arterial rupture, leading to sloughing of the superficial layer of the endometrium and bleeding.
- The elevated levels of progesterone in the luteal phase act to negatively feed back and decrease LH production. Also, luteal cells become less responsive to LH about 1 week following ovulation.

Follicular Phase

- The **follicular phase** begins on about **Day 5** of the menstrual cycle and lasts an average of 9 days.
- When progesterone and estrogen levels decrease, the negative feedback on the hypothalamus and pituitary is removed, allowing an **increase in the frequency of GnRH pulses**. This, in turn, **stimulates FSH secretion**, thereby stimulating follicular growth (and estrogen production from proliferating granulosa cells).
- One follicle will secrete more estradiol than the others and will become the **dominant follicle** while the others undergo atresia. **Estrogen** levels continue to increase until they reach a critical point, at which estrogen changes from **negative to positive feedback** to increase GnRH pulse frequency and LH and FSH secretion. This rapid rise in GnRH pulse frequency results in a surge of both LH and FSH.

Ovulation

- The **LH surge** and **high estrogen levels trigger ovulation** on about **Day 14** of the cycle.
- The **follicle ruptures about 24 to 36 hours after the LH surge**, during which the oocyte resumes meiosis and the first polar body is extruded.

Luteal Phase

- The **luteal phase** begins after ovulation around Day 14 and extends to about Day 28.
- During this time the follicular cells form the **corpus luteum** and secrete **high levels of progesterone** and a lower level of estrogen (even without LH stimulation). Progesterone negatively feeds back to slow the frequency of GnRH pulses, so LH and FSH levels remain low.
- In the absence of fertilization, the corpus luteum undergoes luteolysis, causing progesterone and estrogen levels to decrease until the hormonal support for the endometrial lining declines and the cells undergo apoptosis. Menses occurs and the cycle is back at the beginning.

► Fertilization

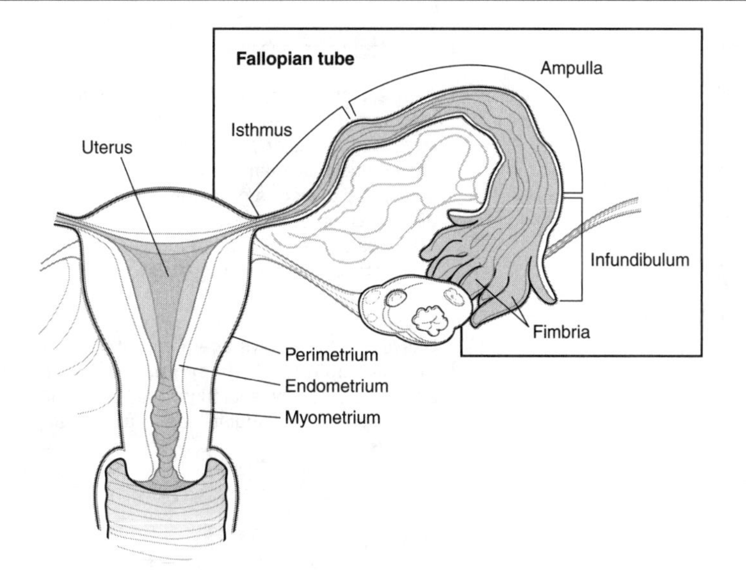

The ovulated oocyte is picked up from the intraperitoneal cavity by fimbria of the fallopian tube. Peristaltic contractions of the fallopian tube move the oocyte into the ampulla, where fertilization usually occurs 8 to 25 hours after ovulation.

Fertilization

- Semen ejaculated into the vagina quickly coagulates and **neutralizes** the **acidic vaginal fluids** to permit sperm survival. About 100,000 of the approximately 60 million sperm that enter the vagina will make it through the cervix. **Elevated estrogen** before ovulation **thins the cervical mucus**, allowing for easier transit of sperm to the uterus.
- Uterine fluid solubilizes the glycoproteins coating the sperm in a process called **capacitation**. Capacitation aids in fertilization by increasing energy metabolism, enhancing motility, and allowing the **acrosome reaction** that occurs at the zona pellucida. Sperm movement through the uterus is primarily accomplished by contraction of the female reproductive tract, primarily the uterus.

- Sperm are capable of fertilizing for as long as 72 hours after ejaculation. Once sperm reach the oocyte, they **bind to the zona pellucida** and **undergo the acrosome reaction**. This reaction releases hydrolytic enzymes stored in the acrosome cap, which dissolve the zona pellucida. Sperm motility is important to push the sperm head toward the oocyte. When the sperm reach the oocyte, the two membranes fuse and the contents of the sperm cell enter the oocyte. At this point, a **cortical reaction** occurs, during which the zona pellucida hardens and prevents additional sperm from entering the oocyte.
- Prior to fusing of the male and female pronucleus, the oocyte undergoes a second meiotic division, producing the second polar body and the female pronucleus. **The contents of the sperm form the male pronucleus, which fuses with the female pronucleus, forming the embryo**.

Early Embryogenesis

- The embryo remains in the ampulla several days, during which time rising levels of progesterone relax the uterine and fallopian tube musculature, making it easier for the embryo to pass into the uterus.
- The **embryo** usually **arrives in the uterus** by about the **third day following fertilization**, but **does not implant in the uterus for about 3 more days.** During the latter 3 days, the embryo develops a vascular system that aids in taking up nutrients it receives from uterine secretions.

► Implantation and Pregnancy

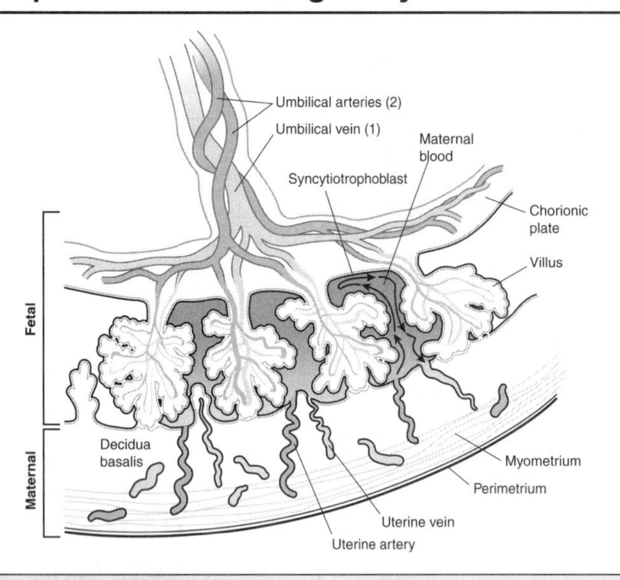

Implantation

At the time of implantation, the trophectoderm cells of the embryo contact the maternal epithelium, resulting in increased vascular permeability in the embryo, edema in the intracellular matrix, swelling of the stromal cells with addition of glycogen granules, and sprouting and ingrowth of capillaries. This reaction is called **decidualization** and prepares the decidua for the embryo. Within about 12 days after fertilization, the embryo is completely embedded in the decidua.

Between 8 to 12 days after fertilization, **human chorionic gonadotropin (hCG) is synthesized by the blastocyst** and is structurally and functionally similar to LH. It acts on the ovary to stimulate luteal growth and to suppress luteolysis.

Placenta

- The **placenta** allows for the exchange of nutrients and waste products between the maternal and fetal circulations. By about 5 weeks postfertilization, the placenta is developed and functional, although not fully mature.
- The placenta consists of **villi** from cell columns of chorionic **syncytiotrophoblast**, which have fetal blood vessels throughout. The villi branch and penetrate the maternal stroma, forming a mass of terminal villi that is separated from fetal capillaries by the thin layer of the villi.

Lactation

- During pregnancy, **estrogen and progesterone** stimulate the development of the mammary glands for **lactation** while preventing milk production.
- Following delivery, estrogen and progesterone levels decline and milk production is permitted.
- The production of milk requires **prolactin**, a hormone produced by the anterior pituitary. Prolactin is normally inhibited by hypothalamic dopamine, but suckling decreases dopamine release.
- The ejection of milk requires **oxytocin**, a hormone produced in the hypothalamus and released from the posterior pituitary. **Oxytocin is released in response to suckling or baby crying** and acts on the myoepithelial cells to stimulate contraction (milk let-down). Milk is then ejected from the nipple.

▶ Diseases of Pregnancy

Pregnancy-induced Hypertension	
Preeclampsia and eclampsia	**Preeclampsia:** Hypertension, proteinuria, and edema Symptoms can include headache, blurred vision, mental changes, facial and extremity edema, and abdominal pain Preeclampsia is common (>5% of pregnancies) from 20 weeks' gestation onward Can cause Hemolysis, Elevated Liver function tests, and Low Platelets (HELLP syndrome) Best treatment is delivery of fetus if possible; or manage with bed rest, salt restriction, IV magnesium sulfate, antihypertensives, and diazepam if needed for treatment and/or prevention of seizures. **Eclampsia** = Preeclampsia and **seizures** Rare because of aggressive management of preeclampsia
Placental Abnormalities	
Abruptio placentae	Placenta detaches prematurely from the endometrium, with risk of death for the fetus and DIC in the mother
Placenta previa	Placenta overlies the cervical os; the baby must be delivered by Caesarian section to prevent life-threatening maternal or fetal hemorrhage due to tearing of the placenta during delivery
Placenta accreta	Placenta implants directly in the myometrium rather than the endometrium; following delivery, hysterectomy is usually performed to be sure all of the placenta was removed
Amniotic Fluid Abnormalities	
Polyhydramnios	Very large amount of amniotic fluid; usually due to severe abnormalities in the fetus such as anencephaly, esophageal atresia, or duodenal atresia
Oligohydramnios	Very small amount of amniotic fluid; usually due to severe abnormalities in the fetus such as bilateral renal agenesis or posterior urethral valves (males) that prevent urination
Miscellaneous	
Ectopic pregnancy	Embryo lodges in an abnormal site (most commonly in a fallopian tube, but can also be on pelvic organs or in the abdomen) Typically because of fallopian tube pathology that prevents the egg from reaching the uterine cavity Risk of potentially fatal hemorrhage to the mother

▶ Infectous Agents That Cross the Placenta

(Mnemonic: TORCH)	Miscellaneous:
Toxoplasma **O**ther (Syphilis) **R**ubella **C**MV **H**erpes and **H**IV	*Listeria monocytogenes* Parvovirus B19 Coxsackie B Polio

► Female Reproductive Pharmacology

Class	Mechanism	Comments/Agents
Contraception		
The most common methods of reversible contraception include oral contraceptives, long-acting injectable or implantable progestins, condoms, spermicides, withdrawal, diaphragm and intrauterine devices, and timely abstinence.		
Estrogens and progestins	Suppresses production of FSH and LH, which leads to inhibition of ovulation and alteration of cervical mucus and the endometrium.	• Commonly used estrogens: ethinyl estradiol and mestranol • Commonly used progestins: norgestrel, norethindrone, and medroxyprogesterone • Are available orally as monophasic, biphasic, and triphasic combinations; also available as progestin-only preparations • Are available in many other forms, including transdermal patches, vaginal rings, IUDs, and long-acting injections • Can cause nausea, breast tenderness, headache, depression, thromboembolism, and weight gain • Absolute contraindications include thrombophlebitis, thromboembolic disorders, cerebral vascular disease, coronary occlusion, known or suspected pregnancy, smokers over the age of 35; dramatically impaired liver function, and suspected breast cancer • Other uses: female hypogonadism, HRT, dysmenorrhea, uterine bleeding, and acne
Postcoital contraceptives	Prevents pregnancy if used within 72 hours of unprotected intercourse	Different types include estrogens alone, progestins alone, combination pills, mifepristone (RU486)
Intrauterine devices (IUD)	"Devices" that create a hostile environment in the endometrium through low-grade intrauterine inflammation and increased prostaglandin formation; therefore interfere with the implantation of the fertilized ovum	Examples: Copper-T 380 (IUD) and Progesterone T (IUD)
Hormone Replacement Therapy (HRT)		
• Used in the treatment of menopause, which is defined as a permanent cessation of menstruation secondary to a loss of ovarian follicular activity. HRT is used to prevent hot flashes, atrophic changes in the urogenital tract, and osteoporosis. • When ERT is administered alone, it may induce endometrial growth and cancer; concomitant progesterone use prevents this. HRT has been associated with an increased breast cancer and stroke risk and is no longer as widely used.		

(Continued)

► Female Reproductive Pharmacology *(Cont'd.)*

Class	Mechanism	Comments/Agents
Selective Estrogen Receptor Modulator (SERM) These drugs act as estrogen agonists, partial agonists, or antagonists, depending on the target tissue.		
Tamoxifen	• Estrogen antagonist in breast • Estrogen agonist in endometrium • Estrogen agonist in bone	• Used in hormone-responsive breast CA; reduces risk of breast CA in very high risk women • Increases risk of endometrial CA • Prevents osteoporosis in woman using it for breast CA • Causes hot flashes and increases risk of venous thrombosis
Raloxifene	• Partial estrogen agonist in bone • Estrogen antagonist in breast • Estrogen antagonist in uterus	• Prevents osteoporosis in postmenopausal women • Reduces risk of breast CA in very high risk women • No increased endometrial CA risk • Causes hot flashes and increases risk of venous thrombosis
Miscellaneous Agents		
Clomiphene	Fertility agent; nonsteroidal agent that selectively blocks estrogen receptors in the pituitary, reducing negative feedback mechanism and thereby increasing FSH and LH and stimulation of ovulation	Most common side effect: multiple birth pregnancy
Danazol	Inhibits ovarian steroid synthesis	Used in endometriosis and fibrocystic breast disease
Anastrozole **Exemestane**	Aromatase inhibitor (decrease in estrogen synthesis)	Used in breast CA in postmenopausal women
Mifepristone (RU 486)	Progesterone and glucocorticoid antagonist	Used as postcoital contraceptive and abortifacient
Dinoprostone	PGE_2 analog	Used to induce labor, causes cervical ripening and uterine contractions; also an abortifacient
Terbutaline	β_2 agonists	Relax uterus, have been used to suppress premature labor

Definition of abbreviations: CA, cancer; ERT, estrogen replacement therapy; FSH, follicle-stimulating hormone; HRT, hormone-replacement therapy; LH, luteinizing hormone.

Female Reproductive System Pathology

▶ Diseases of the Vulva

Disease	Description	Distribution	Etiology/Comments
Condyloma acuminatum	Verrucous, wartlike lesions Koilocytosis, acanthosis, hyperkeratosis, and parakeratosis	Vulva, perineum, vagina, and cervix	Associated with human papillomavirus (HPV) serotypes 6 and 11 Greatly increased risk of cervical carcinoma
Papillary hidradenoma	Benign tumor similar to an intraductal papilloma of the breast	Occur along the milk line	
Extramammary Paget disease of the vulva	Erythematous, crusted rash Intraepidermal malignant cells with pagetoid spread	Labia majora	Not associated with underlying tumor
***Candida* vulvovaginitis**	Erythema, thick white discharge	Vulva and vagina	Extremely common, especially in diabetics and after antibiotic use

Note: See also Sexually Transmitted Diseases, pages 419–421.

▶ Diseases of the Vagina

Vaginal Adenosis and Clear Cell Adenocarcinoma

- Rare in the general population, but greatly increased risk in females exposed to diethylstilbestrol (DES) in utero (1940–1970)
- Vaginal adenosis—benign condition thought to be a precursor of clear cell carcinoma

Embryonal Rhabdomyosarcoma (Sarcoma Botryoides)

- Rare tumor affecting female infants and young children (age <4)
- Polypoid, "grapelike," soft tissue mass protruding from the vagina
- Spindle-cell tumor, may show cross-striations, positive for desmin, indicating skeletal muscle origin

▶ Diseases of the Cervix/Fallopian Tubes

Disease	Description	Etiologies	Disease Manifestations	Clinical
Pelvic inflammatory disease	Ascending infection from cervix to endometrium, fallopian tubes, and pelvic cavity	***Neisseria gonorrhoeae*** (Gram ⊖ diplococcus) ***Chlamydia trachomatis*** (intracytoplasmic inclusions in mucosal cells)	Cervicitis, endometritis, salpingitis, peritonitis, pelvic abscess, perihepatitis (Fitz-Hugh-Curtis syndrome), chandelier sign *Complications:* tubo-ovarian abscess, tubal scarring, infertility, ectopic pregnancy, intestinal obstruction	• Vaginal discharge/ bleeding • Midline abdominal pain, bilateral lower abdominal and pelvic pain • Abdominal tenderness and peritoneal signs • Fever
Cervical carcinoma	Third most common malignant tumor of the female genital tract in United States; peak incidence in the 40s	Associated with early first intercourse, multiple sexual partners, infection by **HPV types 16, 18,** 31 and 33, smoking, and immunosuppression	**Begins as cervical intraepithelial neoplasia (CIN)** → carcinoma in situ → invasive squamous cell cancer	May be asymptomatic, or may have postcoital bleeding, dyspareunia, discharge Early detection possible with **Papanicolaou (Pap) smear – koilocytic cells**

▶ Diseases of the Uterus

Disease	Description	Location	Pathology	Clinical
Endometritis	Ascending infection from the cervix	Endometrium and decidua	*Ureaplasma*, *Peptostreptococcus*, *Gardnerella*, *Bacteroides*, Group B *Streptococcus*, *Chlamydia trachomatis*, *Actinomyces* (yellow, granular filaments on IUD)	Associated with pregnancy or abortions (acute) Associated with PID and intrauterine devices (IUDs) (chronic)
Endometriosis	• Presence of endometrial glands and stroma outside the uterus • Most commonly affects women of reproductive age	Ovary Ovarian and uterine ligaments Pouch of Douglas Serosa of bowel and bladder Peritoneal cavity	Red-brown serosal nodules **("powder burns")** Endometrioma: ovarian **"chocolate" cyst** Adenomyosis = endometrial glands in the myometrium	Chronic pelvic pain linked to menses Dysmenorrhea and dyspareunia Rectal pain and constipation Infertility
Leiomyoma	• **Benign smooth muscle tumor that grows in response to estrogen.** • Higher incidence in African Americans Malignant variant: leiomyosarcoma	May occur in subserosal, intramural, or submucosal locations in the myometrium	Well-circumscribed, rubbery, white-tan "whorled" masses Often multiple	Menorrhagia Abdominal mass Pelvic pain, back pain, or suprapubic discomfort Infertility
Leiomyosarcoma	Smooth muscle sarcoma of the uterus	Myometrium	**Gross:** Bulky tumor with necrosis and hemorrhage **Micro:** Malignant smooth muscle cells, often with nuclear pleomorphism and increased mitotic rate	Increased incidence in blacks Aggressive tumor that tends to recur May present with cervical bleeding
Endometrial adenocarcinoma	Most common malignant tumor of the female genital tract Most commonly affects postmenopausal women	Begins in endometrium and may invade myometrium	**Gross:** Tan polypoid endometrial mass Invasion of myometrium is prognostically important **Micro:** endometrioid adenocarcinoma (most common type)	Postmenopausal vaginal bleeding **RISK FACTORS:** **Early menarche and late menopause** **Nulliparity** Hypertension and diabetes Obesity **Chronic anovulation** **Estrogen-producing ovarian tumors, estrogen replacement therapy and tamoxifen** Endometrial hyperplasia Lynch syndrome (colon, endometrial, and ovarian cancers = HNPCC)

Definition of abbreviations: HNPCC, hereditary nonpolyposis colorectal cancer.

► Diseases of the Ovary

Disease	Presentation	Laboratory/Pathology	Etiology	Treatment
Polycystic ovary disease (Stein-Leventhal syndrome)	• Young, **obese**, **hirsute** females of reproductive age • Oligomenorrhea or secondary amenorrhea • **Infertility**	• **Elevated luteinizing hormone** (LH) • Low follicle stimulating hormone (FSH) • **Elevated testosterone** Bilaterally enlarged ovaries with multiple follicular cysts	Increased LH stimulation leads to increased androgen synthesis and anovulatory cycles	Oral contraceptives or medroxyprogesterone, metformin, surgical wedge resection
		Epithelial Tumors		
Cystadenoma	Most common benign ovarian tumor Pathology: Unilocular cyst with simple serous or mucinous lining			
Cystadeno-carcinoma	Most common malignant ovarian tumor. Often asymptomatic until far advanced (presenting symptoms may be increased abdominal girth due to ascites, bowel or bladder problems) Can produce **pseudomyxoma peritonei**	CA-125- marker for cystadenocarcinoma of ovary. Used to monitor recurrence, measure response to therapy. Complex multiloculated cyst with solid areas • Serous (serous cystadenocarcinoma) or mucinous (mucinous cystadenocarcinoma) lining with tufting, papillary structures with **psammoma bodies** • Spreads by seeding pelvic cavity	**Genetic risk factors:** • BRCA-1: breast and ovarian cancers • Lynch syndrome	Surgery, antineoplastic drugs
Borderline tumor	Tumors of low malignant potential			
Brenner tumor	Rare tumor that resembles transitional carcinoma; can be benign or malignant			
Less common tumors	• Yolk sac tumor: may have structures resembling primitive glomeruli • Choriocarcinoma (see table below) and embryonal carcinoma: very aggressive tumors			
		Germ Cell Tumors		
Teratoma	• Most are benign • Occur in younger women • Contain elements from **all three germ layers (ectoderm, mesoderm, endoderm)** • **Immature teratoma**- contains primitive cells **– higher malignant potential**	Ovarian cyst containing hair, teeth, and sebaceous material	May be due to abnormal differentiation of fetal germ cells that arise from the fetal yolk sac	Surgical
Dysgerminoma	• Malignant • Affects mainly young adults	Similar to seminoma in appearance	**Risk factors:** Turner syndrome, pseudohermaphroditism	Radiosensitive, so good prognosis

(Continued)

► Diseases of the Ovary (*Cont'd.*)

Sex Cord-Stromal Tumors	
Ovarian fibroma	Common tumor. Associated with Meigs syndrome = fibroma + ascites + pleural effusion
Granulosa cell tumor	Potentially malignant, **produces estrogen** and can produce precocious puberty, irregular menses, or dysfunctional uterine bleeding. Microscopic: made of polygonal tumor cells with formation of follicle-like structures **(Call-Exner bodies)** Complications: endometrial hyperplasia and cancer
Sertoli-Leydig cell tumor (androblastoma)	• Androgen producing tumor, presents with virilization

► Gestational Trophoblastic Disease

Hydatidiform Mole (Molar Pregnancy)—tumor of placental trophoblast		
Incidence	1:1,000 pregnancies	
Clinical	"Size greater than dates," vaginal bleeding, passage of edematous, grape-like tissue, elevated β-hCG, invasive moles invade myometrium	
Treatment	Curettage, follow β-hCG levels	
Types		
Complete mole	Results from fertilization of an ovum that lost all its chromosomal material; all chromosomal material is derived from sperm	90% 46,XX; 10% contain a Y chromosome
Partial mole	Results from fertilization of an ovum by two sperm, one 23,X and one 23,Y	Partial moles are triploid = 69, XXY (23,X [maternal] + 23X [one sperm] +23Y [the other sperm])
Choriocarcinoma	• **Malignant** germ cell tumor derived from trophoblast • Gross: necrotic and hemorrhagic mass • Micro: proliferation of cytotrophoblasts, intermediate trophoblasts, and syncytiotrophoblasts • Hematogenous spread to lungs, brain, liver, etc. • Responsive to chemotherapy	

► Partial Moles Versus Complete Moles

Properties	Partial Mole	Complete Mole
Ploidy	Triploid	Diploid
Number of chromosomes	69	46 (All paternal)
β-hCG	Elevated (+)	Elevated (+++)
Chorionic villi	Some are hydropic	All are hydropic
Trophoblast proliferation	Focal	Marked
Fetal tissue	Present	Absent
Invasive mole	10%	10%
Choriocarcinoma	Rare	2%

Breast Pathology

▶ Fibrocystic Disease

- Most common breast disorder, affecting approximately 10% of women; may be mistaken for CA
- Develops during reproductive life, distortion of the normal breast changes associated with the menstrual cycle
- Patients often have lumpy, tender breasts
- *Pathogenesis:* possibly due to high estrogen levels, coupled with progesterone deficiency
- *Pathology:* several morphologic patterns recognized

Fibrosis	• Women 35 to 49 years of age; not premalignant • *Gross:* dense, rubbery mass; usually unilateral, most often in the upper outer quadrant • *Histology:* increase in stromal connective tissue; cysts are rare
Cystic disease	• Women 45 to 55 years of age; not premalignant • *Gross:* serous cysts, firm to palpation, may be hemorrhagic; usually multifocal, often bilateral • *Histology:* cysts lined by cuboidal epithelium, may have papillary projections • May be an accompanying stromal lymphocytic infiltrate (chronic cystic mastitis)
Sclerosing adenosis	• Women 35 to 45 years of age; mild increased risk of CA • *Gross:* palpable, ill-defined, firm area most often in upper outer quadrant; usually unilateral. • *Histology:* glandular patterns of cells in a fibrous stroma; may be difficult to distinguish from cancer
Epithelial hyperplasia	• Women over 30 years of age, mild increased CA risk • *Gross:* variable with ill-defined masses • *Histology:* ductal epithelium is multilayered and produces glandular or papillary configurations • Atypical hyperplasia has a moderate risk of CA

▶ Tumors

Fibroadenoma	• Most common benign breast tumor • Single movable breast nodule, often in the upper outer quadrant; not fixed to skin • Occurs in reproductive years, generally before age 30, possibly related to increased estrogen sensitivity • May show menstrual variation and increased growth during pregnancy; postmenopausal regression usual • *Gross:* round and encapsulated with a gray-white cut surface • *Histology:* glandular epithelial-lined spaces with a fibroblastic stroma • Surgery required for definitive diagnosis
Phyllodes tumor (cystosarcoma phyllodes)	• Fibroadenoma-like tumors that have become large, cystic, and lobulated • Distinguished by the nature of the stromal component • Malignant fibrous, cartilaginous; bony elements may be present • *Gross:* irregular mass; often fungating or ulcerated • *Histology:* myxoid stroma with increased cellularity, anaplasia, and increased mitoses • Tumor initially localized but may spread later, usually to distant sites but not to local lymph nodes
Intraductal papilloma	• Most common in women 20–50 years of age; solitary lesion within a duct • May present with nipple discharge (serous or bloody), nipple retraction, or small subareolar mass • *Gross:* small, sessile or pedunculated, usually close to the nipple in major ducts • *Histology:* multiple papillae • Single intraductal papillomas may be benign, but multiple papillomas associated with an increased risk of CA

► Carcinoma of the Breast

General features	• Most common cause of CA in women • Lung CA causes more deaths • Rare in women under age 25 • Lifetime risk of breast CA for the average woman with no family history: 8 to 10%	
Risk factors	• Increasing age (40+ years) • Nulliparity • Family history • Early menarche • Late menopause	• Fibrocystic disease • Previous history of breast cancer • Obesity • High-fat diet
Clinical features	• 50% in the upper outer quadrant • Ninety percent arise in ductal epithelium • Slightly more common in the left breast; bilateral or sequential in 4% of cases • Breast mass usually discovered after self-examination or on routine physical	
Tumor suppressor genes: *BRCA1* and *BRCA2*	Mutated *BRCA1*— • Almost 100% lifetime risk for breast CA, often in the third and fourth decades of life • Also at increased risk for ovarian CA (men may be at increased risk for prostate CA) Mutated *BRCA2*— • Increased incidence of breast CA in both women and men • Smaller risk of ovarian CA compared with *BRCA 1*	
Invasion	• May grow into the thoracic fascia to become fixed to the chest wall • May extend into the skin, causing dimpling and retraction • May cause obstruction of subcutaneous lymphatics, causing an orange-peel consistency to skin called "peau d'orange" • May invade Cooper ligaments within ducts to cause nipple retraction	
Metastases	• Most breast CAs disseminate via lymphatic or hematogenous routes • Involve axillary, supraclavicular, and internal thoracic nodes • Can also involve nodes of the contralateral breast	

▶ Breast Carcinoma Types

Noninfiltrating intraductal carcinoma	• *Gross:* focus of increased consistency in breast tissue • *Histology:* typical duct epithelial cells proliferate and fill ducts, leading to ductal dilatation • Often called "comedocarcinomas" because cheesy, necrotic tumor tissue may be expressed from ducts. • Rarely have a papillary pattern
Infiltrating ductal carcinoma	• Most common breast CA • *Gross:* rock hard, usually 2 to 5 cm in diameter, foci of necrosis and calcification common (may be seen on mammography) • *Histology:* malignant duct epithelial cells appear in masses or ducts, invading the stroma • Fibrous reaction responsible for the hard, palpable mass
Paget disease of the breast	• Older women; poor prognosis • Form of intraductal carcinoma involving areolar skin and nipple • *Gross:* skin of the nipple and areola ulcerated and oozing • *Histology:* ductal carcinoma, as well as large, anaplastic, hyperchromatic "Paget cells"
Medullary carcinoma	• Better prognosis than infiltrating ductal carcinoma • *Gross:* fleshy masses, often 5 to 10 cm in diameter, little fibrous tissue, although foci of hemorrhage and necrosis common • *Histology:* sheets of large, pleomorphic cells with increased mitotic activity and a lymphocytic infiltrate
Colloid (mucinous) carcinoma	• Older women, slow growing, has a better prognosis than infiltrating ductal carcinoma • *Gross:* soft, large, gelatinous tumors • *Histology:* islands of tumor cells with copious mucin
Lobular carcinoma	• Multicentric; usually have estrogen receptors, arise from terminal ductules • *Gross:* rubbery and ill-defined (result of their multicentric nature) • *Histology:* tumor cells small and may be arranged in rings

▶ Fibrocystic Disease versus Breast Cancer

Fibrocystic Disease	Breast Cancer
• Often bilateral • May have multiple nodules • Menstrual variation • Cyclic pain and engorgement • May regress during pregnancy	• Often unilateral • Usually single nodule • No menstrual variation • No cyclic pain and engorgement • Does not regress during pregnancy

▶ Miscellaneous Breast Conditions

Acute mastitis	• Fissures in nipples during **early nursing** predispose to bacterial infection; usually unilateral with pus in ducts; necrosis may occur • Usual pathogens: *Staphylococcus aureus* and *Streptococcus* • Antibiotics and surgical drainage may be adequate therapy
Mammary duct ectasia (plasma cell mastitis)	• Occurs in fifth decade in multiparous women • Presents with pain, redness, and induration around the areola with thick secretions; usually unilateral • Skin fixation, nipple retraction, and axillary lymphadenopathy may occur—must be distinguished from malignancy
Gynecomastia	• Enlargement of the male breasts; most often unilateral, but may be bilateral • Secondary to Klinefelter syndrome, testicular tumors, puberty, or old age • Associated with hepatic cirrhosis (cirrhotic liver cannot degrade estrogens) • May be important signal that patient has high-estrogen state

Genitourinary System Disease

► Infections and Sexually Transmitted Diseases (STDs)

Type Infection	Case Vignette/ Key Clues	Most Common Causative Agent(s)	Pathogenesis	Diagnosis	Treatment
Urethritis	Gram ⊖ diplococci in PMNs in urethral exudate	*Neisseria gonorrhoeae*	Invasive; pili assist adherence; have antigenic variation; are antiphagocytic; IgA protease	Growth on Thayer-Martin agar, DNA probes	Ceftriaxone
	Culture ⊖, glycogen inclusion bodies in cytoplasm, tissue culture	*Chlamydia trachomatis*	Obligate intracellular in epithelial cells; CMI and DTH cause scarring	Tissue culture; glycogen-containing inclusion bodies in cytoplasm	Tetracyclines, macrolides
	Urease ⊕, no cell wall, ↑ urine pH, non-Gram staining	*Ureaplasma urealyticum*	Urease raises pH of urine → struvite stones	Not gram staining; diagnosed by exclusion, urinary pH	Tetracyclines, macrolides
	Flagellated protozoan with corkscrew motility	*Trichomonas vaginalis*	Unknown, PMN filtrate	Flagellated protozoan, corkscrew motility	Metronidazole
Cystitis	Painful urination, hematuria, fever,	*E. coli #1,* other gram ⊖ enterics	Pili, adhesins, motility, many are β hemolytic	Culture of urine $\geq 10^5$ CFU/ml of gram ⊖ rods in urine	Fluoroquinolones, sulfonamide
	As above, in young, newly-sexually active female (honeymoon cystitis)	*Staphylococcus saprophyticus*	Sexual intercourse introduces normal flora organisms into urethra	Culture of urine, gram ⊕ cocci	Fluoroquinolones
	As above with increased urinary pH	*Proteus spp.*	Urease raises urinary pH, predisposes to struvite stones	Culture of urine, lactose non-fermenting gram ⊖ bacilli with swarming motility	Fluoroquinolones, TMP-SMX
Pyelonephritis	As above with flank pain and fever	*E. coli, Staphylococcus*	Strictures, urinary stasis allow colonization	Culture of urine	Fluoroquinolones, 3rd gen cephalosporin, ampicillin-sulbactam
Cervicitis	Friable, inflamed cervix with mucopurulent discharge	*Neisseria gonorrhoeae*	Invades mucosa, PMN infiltration, pili, IgA protease	Gram ⊖ diplococci, Thayer-Martin agar	Ceftriaxone
		Chlamydia trachomatis	Obligate intracellular, CMI and DTH → scarring	Tissue culture, cytoplasmic inclusions	Tetracyclines, macrolides
		Herpes simplex virus	Vesicular lesions, painful	dsDNA, nuclear envelope, icosahedral; Tzanck smear, intranuclear inclusions	Acyclovir, valacyclovir, famciclovir

(Continued)

► Infections and Sexually Transmitted Diseases (STDs; *Cont'd.*)

Type Infection	Case Vignette/ Key Clues	Most Common Causative Agent(s)	Pathogenesis	Diagnosis	Treatment
Vulvovaginitis	Adherent yellowish discharge, pH >5, fishy amine odor in KOH, clue cells, gram ⊖ cells dominate	Bacterial vaginosis	Overgrowth of *Gardnerella vaginalis*, anaerobes	Clue cells, gram ⊖ rods	Metronidazole
	Vulvovaginitis, pruritus, erythema, discharge with consistency of cottage cheese	*Candida* spp.	Antibiotic use → overgrowth, immunocompromised	Germ tube test, gram ⊕ yeasts in vaginal fluids	Nystatin, miconazole
	"Strawberry cervix," foamy, purulent discharge; many PMNs and motile trophozoites microscopically (corkscrew motility)	*Trichomonas vaginalis*	Vaginitis with discharge	Pear-shaped trophozoites with corkscrew motility	Metronidazole
Pelvic inflammatory disease	Adnexal tenderness, bleeding, dyspareunia, vaginal discharge, fever, chandelier sign, onset often follows menses	*Neisseria gonorrhoeae*	Pili and IgA protease production	Gram ⊖ diplococci in PMNs or culture on Thayer-Martin	Ceftriaxone + doxycycline (doxycycline given for presumed coinfection with *Chlamydia*)
		Chlamydia trachomatis	Intracellular in mucosal epithelia; causes type IV hypersensitivity damage	Tissue culture, intracytoplasmic inclusions in mucosal cells	Doxycycline, macrolides
Condyloma acuminatum (genital warts)	Lesions are papillary/wart-like, may be sessile or pedunculated, koilocytotic atypia is present, anogenital	Human papilloma virus (HPV; most common U.S. STD)	HPV proteins E6 and E7 inactivate cellular antioncogene Associated with cervical CA	dsDNA, naked, icosahedral, intranuclear inclusion bodies	Podophyllin, imiquimod
Genital herpes	Multiple, painful, vesicular, coalescing, recurring	Herpes simplex	Latent virus in sensory ganglia reactivates	Virus culture, intranuclear inclusions, syncytia (Tzanck smear), dsDNA enveloped (nuclear), icosahedral	Acyclovir, valacyclovir, famciclovir

(Continued)

► Infections and Sexually Transmitted Diseases (STDs; *Cont'd.*)

Type Infection	Case Vignette/ Key Clues	Most Common Causative Agent(s)	Pathogenesis	Diagnosis	Treatment
Syphilis **Primary**	Painless chancre forms on glans, penis (or vulva/ cervix) and heals within 1 to 3 months	*Treponema pallidum*	3-week incubation during which spirochetes spread throughout the body	Biopsy/scraping viewed with dark-field microscopy shows spirillar organisms	Penicillin, doxycycline is an alternative
Secondary	Local or generalized rash lasting 1 to 3 months, can involve the palms and soles		Develops 1 to 2 months after primary stage	Serology—VDRL ⊕ (nonspecific); FTA-ABS (specific)	
Tertiary	Affects central nervous system, heart, and skin; characteristic lesion is gumma, may be single or multiple; most common in the liver, testes, and bone		Develops in one-third of untreated patients; *neurosyphilis:* including meningovascular, tabes dorsalis, and general paresis; obliterative endarteritis of vasa vasorum of the aorta can lead thoracic aneurysm	Serology – FTA-Abs, non-specific tests may be negative	
Chancroid	Nonindurated, painful ulcer, suppurative with adenopathy; slow to heal	*Haemophilus ducreyi*	Unknown	Gram ⊖ rods, chocolate agar (requires NAD and hemin)	Cefotaxime, ceftriaxone
Lymphogranuloma venereum	Soft, painless papule heals, lymph nodes enlarge and develop fistulas, genital elephantiasis may develop	*Chlamydia trachomatis* serotypes L1–3	Obligate intracellular	Cell culture, glycogen-containing inclusions	Tetracyclines, erythromycin

Definition of abbreviations: CA, cancer; CMI, cell-mediated immunity; ds, double-stranded; DTH, delayed type hypersensitivity; PMNs, polymorphonuclear leukocytes.

The Musculoskeletal System, Skin, and Connective Tissue

Chapter 8

Structure, Function, and Pharmacology of Muscle

Head and Neck Embryology and Anatomy

Upper Extremities and Back

Musculoskeletal Disorders

Skin

Structure, Function, and Pharmacology of Muscle

► Features of Skeletal, Cardiac, and Smooth Muscle

Characteristics	Skeletal	Cardiac	Smooth
Appearance	Striated, unbranched fibers Z lines Multinucleated	Striated, branched fibers Z lines Single nucleus	Nonstriated, fusiform fibers No Z lines; have dense bodies Single nucleus
T tubules	Form triadic contacts with SR at A-I junction	Form dyadic contacts with SR near Z line	Absent; have limited SR
Cell junctions	Absent	Junctional complexes between fibers (intercalated discs), including gap junctions	Gap junctions
Innervation	Each fiber innervated	Electrical syncytium	Electrical syncytium
Action potential			
Upstroke	Inward Na^+ current	• Inward Ca^{2+} current (SA node) • Inward Na^+ current (atria, ventricles, Purkinje fibers)	Inward Na^+ current
Plateau	No plateau	• No plateau (SA node) • Plateau present (atria, ventricles, Purkinje fibers)	No plateau
Excitation-contraction coupling	AP → T tubules → Ca^{2+} released from SR	• Inward Ca^{2+} current during plateau → Ca^{2+} release from SR • cAMP increases Ca^{2+} and force in myocytes	• AP → opens voltage-gated Ca^{2+} channels in sarcolemma; hormones and neurotransmitters → open IP_3-gated Ca^{2+} channels in SR • cAMP and cGMP inhibit smooth muscle contraction
Calcium binding	Troponin	Troponin	Calmodulin

Definition of abbreviations: AP, action potential; IP_3, inositol triphosphate; SR, sarcoplasmic reticulum.

► Skeletal Muscle Fiber Morphology and Function

Skeletal muscle connective tissue (*see* right):

- **Epimysium:** dense connective tissue that surrounds the entire muscles
- **Perimysium:** thin septa of connective tissue that extends inward from the epimysium and surrounds a bundle (fascicle) of muscle fibers
- **Endomysium:** delicate connective tissue that surrounds each muscle fiber

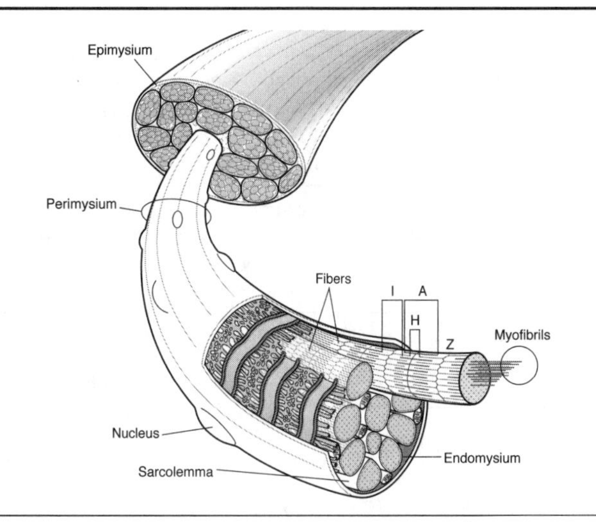

Subcellular components (*see* below):

- **Myofibrils:** long, cylindrical bundles that fill the sarcoplasm of each fiber
- **Myofilaments: actin** and **myosin**; are within each myofibril and organize into units called **sarcomeres**

During contraction:

A band: no change	**I band:** shortens
H band: shortens	**Z lines:** move closer together

The Crossbridge Cycle

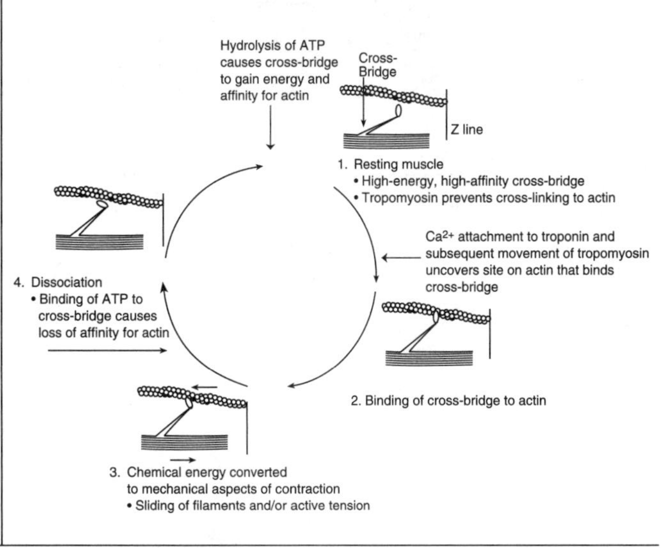

► Red versus White Skeletal Muscle Fibers

Red Fibers (Type I)	White Fibers (Type II)
Slow contraction	Fast contraction
↓ ATPase activity	↑ ATPase activity
↑ Capacity for aerobic metabolism	↑ Capacity for anaerobic glycolysis
↑ Mitochondrial content	↓ Mitochondrial content
↑ Myoglobin (imparts red color)	↓ Myoglobin
Best for slow, posture-maintaining muscles, e.g., back (think chicken drumstick/thigh)	Best for fast, short-termed, skilled motions, e.g., extraocular muscles of eye, sprinter's legs, hands (think chicken breast meat and wings)

▶ Smooth Muscle Function

Types of Smooth Muscle	
Multiunit	• Acts as individual motor unit • Little or no electrical coupling • Is densely innervated; contraction controlled by autonomic nervous system • In iris, ciliary muscle of lens, and vas deferens
Unitary (single unit)	• Extensive electrical coupling, allowing coordinated contraction • Has a resting tone; spontaneously active (slow waves), has pacemaker activity; activity is modulated by neurotransmitters and neurohormones • Found mainly in the walls of hollow viscera, e.g., GI tract, uterus, bladder, ureters
Vascular	• Has properties of both multiunit and single-unit smooth muscle

Smooth Muscle Contraction

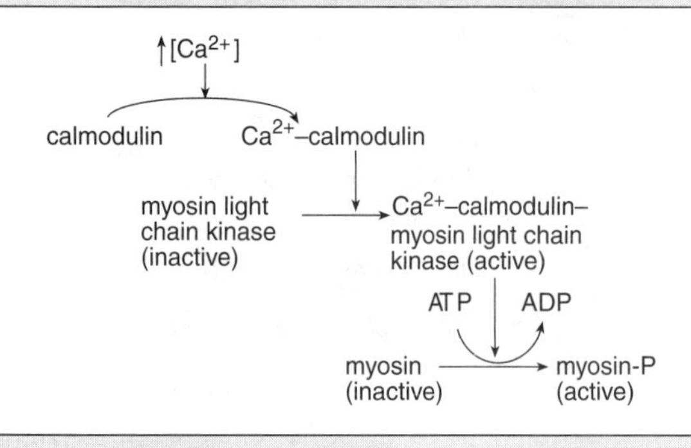

1. ↑ intracellular Ca^{2+}
2. Ca^{2+} binds calmodulin
3. Ca^{2+}-calmodulin binds to and activates **myosin light chain kinase (MLCK)**
4. Myosin is phosphorylated
5. Myosin-P binds actin and shortening occurs
6. Dephosphorylation of myosin → relaxation

Bridge to Pharmacology

α_1 and M_3 stimulation: ↑ IP_3 → ↑ intracellular Ca^{2+} → smooth muscle contraction

β_2 stimulation: ↑ cAMP → inhibits MLCK → smooth muscle relaxation

► Skeletal Muscle Relaxants

There are two general classes of skeletal muscle relaxants: **neuromuscular blockers** and **spasmolytics**. Neuromuscular blockers are used during surgical procedures and act on **skeletal muscle nicotinic cholinergic receptors**. There are two classes of neuromuscular blockers: **nondepolarizing (competitive)** and **depolarizing**. **Spasmolytics** are used for CNS disorders and acute muscle spasm and have varying mechanisms of action.

Neuromuscular Blockers

Nondepolarizing Blockers

These agents are competitive antagonists at the nAChR (N_M) on skeletal muscle and can be therefore be reversed by acetylcholinesterase inhibitors (e.g., neostigmine, pyridostigmine).

Drug	Duration	Elimination	Notes
Atracurium	Intermediate	Spontaneous	Safer in renal and hepatic disease
Cisatracurium	Intermediate	Spontaneous	Safer in renal and hepatic disease
Doxacurium	Long	Renal	—
d-Tubocurarine	Long	Renal	Also blocks autonomic ganglia and causes histamine release
Mivacurium	Short	Plasma ChE	—
Pancuronium	Long	Renal	Blocks muscarinic receptors
Rocuronium	Intermediate	Hepatic	—
Vecuronium	Intermediate	Hepatic	—

Depolarizing Blockers:

These agents act as nicotinic agonists and **depolarize** skeletal muscle. Patients often initially have **fasciculations**. Continuous depolarization of the motor end-plate leads to flaccid paralysis. When given continuously, two phases occur:

Phase I block (depolarizing)—fasciculations, flaccid paralysis; this phase is **augmented by AChE inhibitors**
Phase II block (desensitizing)—the end-plate repolarizes, but is unresponsive to ACh; **reversed by AChE inhibitors**

Succinylcholine	Ultrashort	Plasma ChE	Stimulates autonomic ganglia and muscarinic receptors; can → hyperkalemia; postoperative muscle pain

Spasmolytics

The spasmolytics reduce excessive muscle tone in CNS disorders (e.g., cerebral palsy, multiple sclerosis, stroke, spinal cord injury) or in acute muscle injury.

Drug	Mechanism of Action	Location of Action	Clinical Uses
Baclofen	$GABA_B$ receptor agonist	CNS	Spasticity of central or spinal origin
Diazepam	Benzodiazepine, potentiates $GABA_A$ receptors	CNS	Spasticity and acute muscle spasm
Tizanidine	α_2-receptor agonist	CNS	Acute muscle spasm
Botulinum toxin	Blocks ACh release	Muscle	Locally injected to relieve spasticity, e.g. in CP; used for cosmetic purposes
Dantrolene	Blocks ryanodine receptors on SR to prevent Ca^{2+} release	Muscle	Malignant hyperthermia; neuroleptic malignant syndrome
Cyclobenzaprine Carisoprodol Metaxalone	—	CNS	Acute muscle spasm

Definition of abbreviation: CP, cerebral palsy; SR, sarcoplasmic reticulum.

Head and Neck Embryology and Anatomy

► Skeletal Muscles Innervated by Cranial Nerves

Muscles Derived from a Pharyngeal Arch	Cranial Nerve	Muscles	Skeletal Elements (from neural crest)
First arch—mandibular (Mandibular hypoplasia is seen in **Treacher Collins syndrome** and in the **Robin sequence**. Both involve neural crest cells.)	**Trigeminal mandibular nerve (V3)**	Four muscles of mastication: • Masseter • Temporalis • Lateral pterygoid • Medial pterygoid Plus: • Digastric (anterior belly) • Mylohyoid • Tensor tympani • Tensor veli palatini	Mandibular process Maxillary process Malleus Incus
Second arch—hyoid	**Facial (VII)**	Muscles of facial expression: • Orbicularis oculi • Orbicularis oris • Buccinator and others Plus: • Digastric (posterior belly) • Stylohyoid • Stapedius	Hyoid (superior part) Styloid process Stapes
Third arch	**Glossopharyngeal (IX)**	Stylopharyngeus	Hyoid (inferior part)
Fourth arch	**Vagus (X) superior laryngeal (external branch)** **Vagus (X) pharyngeal branches**	Cricothyroid Levator veli palatini Uvular muscle Pharyngeal constrictors Salpingopharyngeus Palatoglossus Palatopharyngeus	Thyroid cartilage
Fifth arch	**Lost**	—	—
Sixth arch	**Vagus (X) recurrent laryngeal**	Lateral cricoarytenoid Posterior cricoarytenoid Transverse arytenoid Oblique arytenoid Thyroarytenoid (vocalis)	Cricoid, arytenoid, corniculate, cuneiform cartilages
Muscles of myotome origin	**Accessory (XI)**	Trapezius Sternocleidomastoid	
	Hypoglossal (XII)	Genioglossus Hyoglossus Styloglossus	
	Oculomotor (III)	Superior, inferior, and medial rectus; inferior oblique, levator palpebrae superioris	
	Trochlear (IV)	Superior oblique	
	Abducens (VI)	Lateral rectus	

► Pharyngeal Pouches

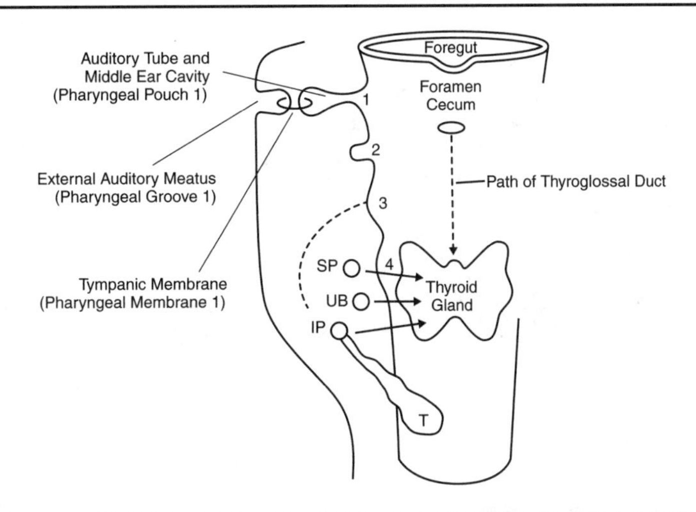

Adult Structures Derived from the Fetal Pharyngeal Pouches		
Pouch	**Adult Derivatives**	**Clinical Correlate**
1	Epithelial lining of auditory tube and middle ear cavity	The **DiGeorge sequence** occurs when pharyngeal pouches 3 and 4 fail to differentiate into the parathyroid glands and thymus. Patients have immunologic problems, hypocalcemia, and may have cardiovascular defects (persistent truncus arteriosus), abnormal ears, and micrognathia.
2	Epithelial lining of crypts of palatine tonsil	
3	Inferior parathyroid gland (IP) Thymus (T)	
4	Superior parathyroid gland (SP) Ultimobranchial body (UB)	

The thyroid gland does not develop in a pharyngeal pouch; it develops from midline endoderm of the oropharynx and migrates inferiorly along the path of thyroglossal duct. Neural crest cells migrate into the UB to form parafollicular C cells of the thyroid.

The external auditory meatus is the only postnatal remnant of a pharyngeal groove or cleft.

► Palate and Face Development

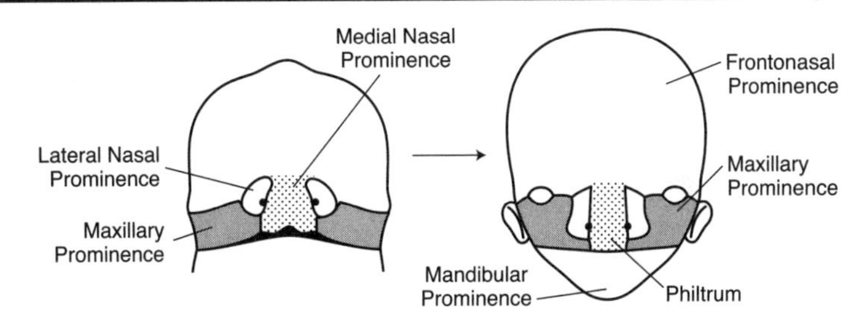

The **face** develops from the frontonasal prominence, the pair of maxillary prominences, and the pair of mandibular prominences.

The **intermaxillary segment** forms when the two medial nasal prominences fuse together at the midline and → the **philtrum of the lip**, **four incisor teeth**, and the **primary palate** of the adult.

The **secondary palate** forms from palatine shelves, which fuse in the midline, posterior to the incisive foramen.

The primary and secondary palates fuse at the **incisive foramen** to form the definitive palate.

Clinical Correlation

Cleft lip occurs when the maxillary prominence fails to fuse with the medial nasal prominence.

Cleft palate occurs when the palatine shelves fail to fuse with each other or the primary palate.

▶ Cavernous Sinuses

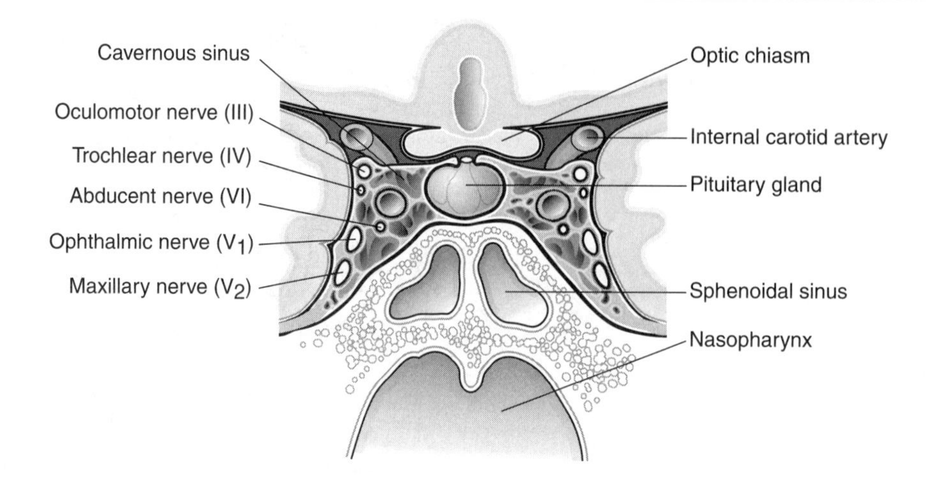

The cavernous sinuses are located on either side of the body of the sphenoid bone. Each sinus receives blood from some of the cerebral veins, ophthalmic veins, and the sphenoparietal sinus. Each cavernous sinus drains into a transverse sinus via the superior petrosal sinus and into the internal jugular vein via the inferior petrosal sinus.

Cavernous Sinus Thrombosis

Infection can spread from veins of the face into the cavernous sinuses, producing a thrombosis that may involve the cranial nerves that course through the cavernous sinuses. Cranial nerves III, IV, and VI and the ophthalmic and maxillary divisions of CN V, as well as the internal carotid artery and its periarterial plexus of postganglionic sympathetic fibers, traverse the cavernous sinuses. All of these cranial nerves course in the lateral wall of each sinus, except for CN VI, which courses through the middle of the sinus. Initially, patients have an internal strabismus. Later, all eye movements are affected, along with with altered sensation in skin of the upper face and scalp.

Upper Extremities and Back

▶ Brachial Plexus

The **brachial plexus** is formed by an intermingling of ventral rami from the C5 through T1 spinal nerves.

The ventral rami of the brachial plexus exhibit a proximal to distal gradient of innervation. Nerves that contain fibers from the superior rami of the plexus (C5 and C6) innervate proximal muscles in the upper limb (shoulder muscles). Nerves that contain fibers from the inferior rami of the plexus (C8 and T1) innervate distal muscles (hand muscles).

Five major nerves arise from the brachial plexus: the **musculocutaneous, median**, and **ulnar** nerves contain anterior division fibers and innervate muscles in the anterior arm, anterior forearm, and hand that act mainly as flexors. The **axillary** and **radial** nerves contain posterior division fibers, and innervate muscles in the posterior arm and posterior forearm that act mainly as extensors.

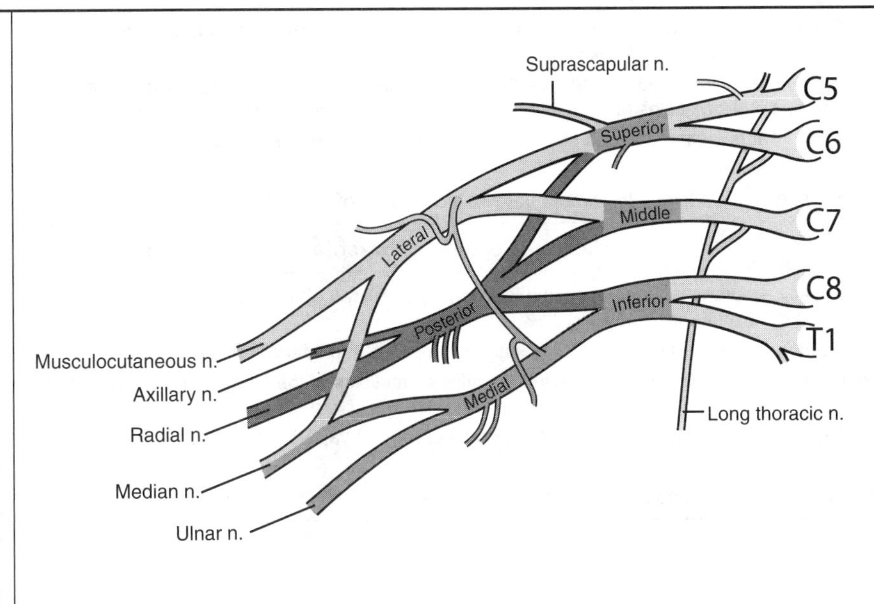

▶ Lesions of Roots of the Brachial Plexus

Lesioned Root	C5	C6	C7	C8	T1
Dermatome paresthesia	Lateral border of upper arm	Lateral forearm to thumb	Over triceps, midforearm, middle finger	Medial forearm to little finger	Medial arm to elbow
Muscles affected	Deltoid Rotator cuff Serratus anterior Biceps Brachioradialis	Biceps Brachioradialis Brachialis Supinator	Latissimus dorsi Pectoralis major Triceps Wrist extensors	Finger flexors Wrist flexors Hand muscles	Hand muscles
Reflex test	—	Biceps tendon	Triceps tendon	—	—
Causes of lesions	Upper trunk compression	Upper trunk compression	Cervical spondylosis Herniation of C6/C7 disk	Lower trunk compression	Lower trunk compression

▶ Upper and Lower Brachial Plexus Lesions

Upper (C5 and C6) Brachial Plexus Lesion: Erb-Duchenne Palsy
• Usually occurs when the head and shoulder are forcibly separated (e.g., accident, birth injury, or herniation of disk) • Trauma will damage **C5** and **C6** roots of the upper trunk • Primarily affects the **axillary, suprascapular,** and **musculocutaneous nerves** with loss of function of the intrinsic muscles of the shoulder and muscles of the anterior arm • Arm is medially rotated and adducted at the shoulder: loss of **axillary** and **suprascapular** nerves. The unopposed latissimus dorsi and pectoralis muscles pull the limb into adduction at the shoulder • The forearm is extended and pronated: loss of **musculocutaneous** nerve • **Sign is "waiter's tip"** • Sensory loss on lateral forearm to base of thumb: loss of musculocutaneous nerve
Lower (C8 and T1) Brachial Plexus Lesion: Klumpke's Paralysis
• Usually occurs when the upper limb is forcefully abducted above the head (e.g., grabbing an object when falling, thoracic outlet syndrome, or birth injury) • Trauma will injure the **C8** and **T1** spinal nerve roots of the inferior trunk • Primarily affects the ulnar nerve and the intrinsic muscles of the hand with weakness of the median-innervated muscles of the hand • Sign is combination of **"claw hand" (ulnar nerve)** and **"ape hand" (median nerve)**. • May include a Horner syndrome • Sensory loss on medial forearm and medial 1½ digits

▶ Lesions of Nerves of the Brachial Plexus

Radial Nerve (C5, C6, C7, C8)		
Axilla: (Saturday night palsy or using crutches)	**Mid-shaft of humerus at radial groove or lateral elbow (lateral epicondyle)**	**Wrist: (laceration)**
• Loss of extension at the elbow, wrist and MP joints • Weakened supination • Sensory loss on posterior arm, forearm, and dorsum of thumb • Distal sign is **"wrist drop"**	• Loss of forearm extensors of the wrist and MP joints • Weakened supination • Sensory loss on the posterior forearm and dorsum of thumb • Distal sign is **"wrist drop"**	• No **motor loss** • Sensory loss only on dorsal aspect of thumb (first dorsal web space)

Median Nerve (C6, C7, C8, T1)	
Elbow: (Supracondylar fracture of humerus)	**Wrist: (carpal tunnel or laceration)**
• Weakened wrist flexion (with ulnar deviation) • Loss of pronation • Loss of flexion of lateral 3 digits, resulting in the inability to make a complete fist; sign is **"hand of benediction"** • Loss of thumb opposition (opponens pollicis muscle); sign is **ape (simian) hand** • Loss of first two lumbricals • Thenar atrophy • Sensory loss on palmar surface of the lateral hand and the palmar surfaces of the lateral 3½ digits ***Note:* A lesion of median nerve at the elbow results in the "hand of benediction" and "ape hand."**	• Loss of thumb opposition (opponens pollicis muscle); sign is **ape or simian hand** • Loss of first two lumbricals • Thenar atrophy • Sensory loss on the palmar surfaces of lateral 3½ digits. Note sensation on lateral palm may be spared (see figure on next page). ***Note:* Lesions of median nerve at the wrist present without hand of benediction and with normal wrist flexion, digital flexion, and pronation.**

(Continued)

► Lesions of Nerves of the Brachial Plexus (*Cont'd.*)

Ulnar Nerve (C8, T1)
Elbow (medial epicondyle), wrist (lacerations), fracture of hook of hamate, midshaft clavicle fracture
• Loss of hypothenar muscles, third and fourth lumbricals, all interossei and adductor pollicis • With elbow lesion, there is minimal weakening of wrist flexion with radial deviation • Loss of **abduction** and **adduction of digits 2–5** (interosseus muscles) • Weakened IP extension of digits 2–5 (more pronounced in digits 4 and 5) • Loss of thumb adduction • Atrophy of the hypothenar eminence • Sign is **"claw hand"** (note that clawing is greater with a wrist lesion) • Sensory loss on medial 1½ digits
Axillary Nerve (C5, C6)
Fracture of the surgical neck of the humerus or inferior dislocation of the shoulder
• Loss of abduction of the arm to the horizon • Sensory loss over the deltoid muscle
Musculocutaneus Nerve (C5, C6, C7)
• Loss of elbow flexion and weakness in supination • Loss of sensation on lateral aspect of the forearm
Long Thoracic Nerve (C5, C6, C7)
• Often damaged during a radical mastectomy or a stab wound to the lateral chest (nerve lies on superficial surface of serratus anterior muscle) • Loss of abduction of the arm above the horizon to above the head • Sign of **"winged scapula"**; patient unable to hold the scapula against the posterior thoracic wall
Suprascapular Nerve (C5, C6)
• Loss of shoulder abduction between 0 and 15 degrees (supraspinatus muscle) • Weakness of lateral rotation of shoulder (infraspinatus muscle)

► Cutaneous Innervation of the Hand

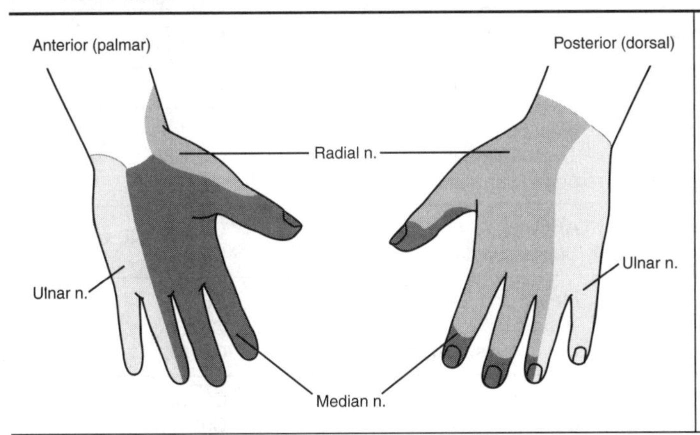

The palm is supplied mainly by the median and ulnar nerves. The median supplies the lateral 3½ digits and the adjacent area of the lateral palm and the thenar eminence. The ulnar supplies the medial 1½ digits and skin of the hypothenar eminence. The radial nerve supplies skin of the dorsum of the hand in the area of the first dorsal webbed space, including the skin over the anatomic snuffbox.

▶ Back Muscles

Action	Muscles Involved	Innervation
Extend/Rotate vertebrae	1. Splenius capitis, splenius cervicis	Dorsal rami of spinal nerves
	2. Erector spinae: • Iliocostalis • Longissimus • Spinalis	Dorsal rami of spinal nerves
	3. Transversospinalis: • Semispinalis • Multifidus • Rotatores	Dorsal rami of spinal nerves

▶ Movements of the Pectoral (Shoulder) Girdle on the Trunk

Action	Muscles Involved	Innervation	Major Segments of Innervation
Elevation	Levator scapulae Trapezius, upper part	Dorsal scapular Accessory	C4, C5 C1–C5
Depression	Pectoralis minor Trapezius, lower part	Medial pectoral Accessory	C7, C8
Protraction	Serratus anterior	Long thoracic	C5–C7
Retraction	Rhomboid major and minor Trapezius, middle fibers	Dorsal scapular Accessory	C5 C1–C5
Lateral (upward) rotation of scapula (in abduction)	Serratus anterior, lower half Trapezius, upper and lower parts	Long thoracic Accessory	C5–C7 C1–C5
Medial (downward) rotation of scapula (in adduction)	Rhomboid major and minor Levator scapulae	Dorsal scapular Dorsal scapular	C5 C4, C5

▶ Movements at the Shoulder (Glenohumeral) Joint

Action	Muscles Involved	Innervation	Major Segments of Innervation
Flexion	Pectoralis major, clavicular head Deltoid clavicular part Biceps short head	Lateral pectoral Axillary Musculocutaneous	C5–C7 C5, C6 C5, C6
Extension	Deltoid, posterior fibers Latissimus dorsi Teres major	Axillary Thoracodorsal Lower subscapular	C5, C6 C6–C8 C6
Abduction	Deltoid, middle fibers Supraspinatus	Axillary Suprascapular	C5, C6 C5
Adduction	Pectoralis major, sternocostal part Latissimus dorsi Teres major	Medial and lateral pectoral Thoracodorsal Lower subscapular	C6–T1 C6–C8 C5, C6
Lateral rotation	Deltoid, posterior fibers Infraspinatus Teres minor	Axillary Suprascapular Axillary	C5, C6 C5, C6 C6
Medial rotation	Pectoralis major Latissimus dorsi Deltoid, clavicular part Teres major Subscapularis	Medial and lateral pectoral Thoracodorsal Axillary Lower subscapular Upper and lower subscapular	C5–T1 C6–C8 C5–C7 C5, C6 C5, C6

► Rotator Cuff

The tendons of rotator cuff muscles strengthen the glenohumeral joint and include the **s**upraspinatus, **i**nfraspinatus, **t**eres minor, and **s**ubscapularis (the **SITS** muscles). The tendons of the muscles of the rotator cuff may become torn or inflamed. The tendon of the **supraspinatus** is most commonly affected. Patients with rotator cuff tears experience pain anteriorly and superiorly to the glenohumeral joint during abduction.

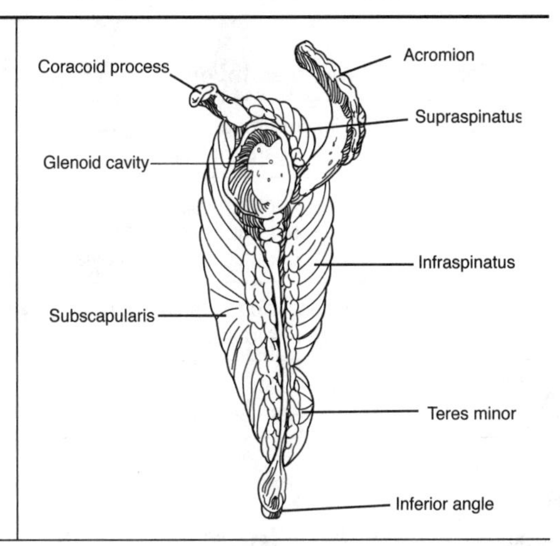

► Movements at the Elbow Joint

Action	Muscles Involved	Innervation	Major Segments of Innervation
Flexion	Brachialis Biceps brachii Brachioradialis	Musculocutaneous Musculocutaneous Radial	C5, C6 C5, C6 C5, C6
Extension	Triceps	Radial	C7, C8

► Movements at the Radioulnar Joints

Action	Muscles Involved	Innervation	Major Segments of Innervation
Pronation	Pronator teres Pronator quadratus	Median Median (anterior interosseous nerve)	C6, C7 C8, T1
Supination	Supinator Biceps brachii	Radial (deep branch) Musculocutaneous	C6–C8 C5, C6

▶ Movements at the Wrist Joint

Action	Muscles Involved	Innervation	Major Segments of Innervation
Flexion	Flexor carpi ulnaris Flexor carpi radialis	Ulnar Median	C8 C6, C7
Extension	Extensor carpi ulnaris Extensor carpi radialis longus Extensor carpi radialis brevis	Radial (deep br.) Radial Radial (deep br.)	C7, C8 C6, C7 C6, C7
Abduction	Extensor carpi radialis longus/brevis Flexor carpi radialis	Radial (deep br.) Median	C6, C7 C6, C7
Adduction	Flexor carpi ulnaris Extensor carpi ulnaris	Ulnar Radial (deep br.)	C8 C7, C8

▶ Movements of the Fingers

Action	Muscles Involved	Innervation	Major Segments of Innervation
Flexion, All Fingers			
All joints and DIP	Flexor digitorum profundus	To index and middle: **median** (ant. interosseous n.)	C8
		To ring and little: **ulnar**	T1
MP and PIP, all	Flexor digitorum superficialis Lumbricals	Median Median—index and middle Ulnar—ring and little	C7, C8, T1 C8, T1 C8, T1
MP little finger	Flexor digiti minimi	Ulnar	C8, T1
Extension—MP Joints			
Index, middle, ring	Extensor digitorum	Radial (deep br.)	C6–C8
Index only	Extensor indicis	Radial (deep br.)	C7, C8
Little finger only	Extensor digiti minimi	Radial (deep br.)	C7, C8
Extension—IP Joints			
Through extensor expansion	Lumbricals	Median: index and middle	C8, T1
		Ulnar: ring and little	C8, T1
	Interossei	Ulnar	C8, T1
Abduction—MP Joints			
All fingers, except little	Dorsal Interossei	Ulnar	C8, T1
	Abductor digiti minimi	Ulnar	C8, T1
Adduction—MP Joints			
All fingers, except middle	Palmar Interossei	Ulnar	C8, T1
Opposition little finger	Opponens digiti minimi	Ulnar	C8, T1

Definition of abbreviations: ant., anterior; br., branch; IP, interphalangeal joints; DIP, distal interphalangeal joints; MP, metacarpophalangeal joints; n, nerve; PIP, proximal interphalangeal joints.

▶ Movements of the Thumb

Action	Muscles Involved	Innervation	Major Segments of Innervation
Flexion			
All joints, especially IP	Flexor pollicis longus	Median (ant. interosseous n.)	C8, T1
MP	Flexor pollicis brevis	Median (recurrent br.)	C8, T1
Extension			
All joints	Extensor pollicis longus	Radial (deep br.)	C7, C8
MP	Extensor pollicis brevis	Radial (deep br.)	C7, C8
Abduction			
	Abductor pollicis longus	Radial (deep br.)	C7, C8
	Abductor pollicis brevis	Median (recurrent br.)	C6–T1
Adduction			
	Adductor pollicis	Ulnar	C8, T1
Opposition			
	Opponens pollicis	Median (recurrent br.)	C6–T1

Definition of abbreviations: ant., anterior; br, branch; MP, metacarpophalangeal joints; IP, interphalangeal joints.

► Effects of Lesions to Roots and Nerves of the Lumbosacral Plexus

The **lumbosacral plexus** is formed by an intermingling of the ventral rami of the L2 through S3 spinal nerves. The ventral rami of the lumbosacral plexus exhibit a proximal to distal gradient of innervation. Nerves that contain fibers from the superior rami of the plexus (L2 through L4) innervate muscles in the anterior and medial thigh that act at the hip and knee joints. Nerves that contain fibers from the inferior rami of the plexus (S1 through S3) innervate muscles of the leg that act at the joints of the ankle and foot.

Four major nerves arise from the lumbosacral plexus:

The **obturator** and **tibial** nerves contain anterior division fibers and innervate muscles in the medial and posterior compartments of the thigh, the posterior compartment of the leg, and in the sole of the foot.

The **femoral** and **common fibular** nerves contain posterior division fibers and innervate muscles in the anterior compartment of the thigh, in the anterior and lateral compartments of the leg, and in the dorsum of the foot.

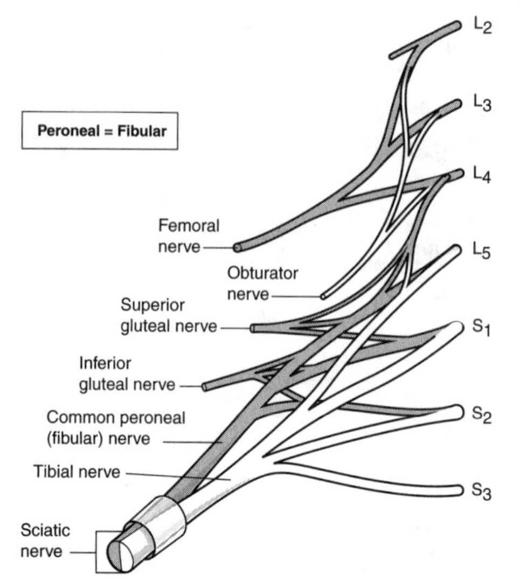

Lesioned Root	L3	L4	L5	S1
Dermatome paresthesia	Anterior thigh	Medial leg	Anterior leg, dorsum of foot	Lateral foot, sole
Reflex test	—	Patellar tendon	—	Achilles tendon
Muscles affected	Hip flexors Hip adductors	Knee extensors Hip adductors	Dorsiflexors Toe extensors	Plantar flexors Toe flexors
Causes of lesions	Osteoarthritis	Osteoarthritis	Herniation of L4/L5 disc	Herniation of L5/S1 disc
Lesioned Nerve	**Obturator (L2–L4)**	**Femoral (L2–L4)**	**Common Fibular (L4, L5, S1, S2)**	**Tibial (L4, L5, S1–S3)**
Altered sensation	Medial thigh	Anterior thigh, medial leg to medial malleolus	Anterior leg, dorsum of foot	Posterior leg, sole, and lateral border of foot
Reflex test	—	Patellar tendon	—	Achilles tendon
Motor weakness	Adduction of thigh	Extension of knee	Dorsiflexion, eversion of the foot	Plantar flexion Toe flexion
Common sign of lesion	—	—	Footdrop	—
Causes of lesions	Pelvic neoplasm Pregnancy	Diabetes Posterior abdominal neoplasm Psoas abscess	Compression at fibula neck Hip fracture/dislocation Misplaced gluteal injection Piriformis syndrome	Hip fracture/dislocation Penetrating trauma to buttock

► Sensory Innervation of the Foot and Leg

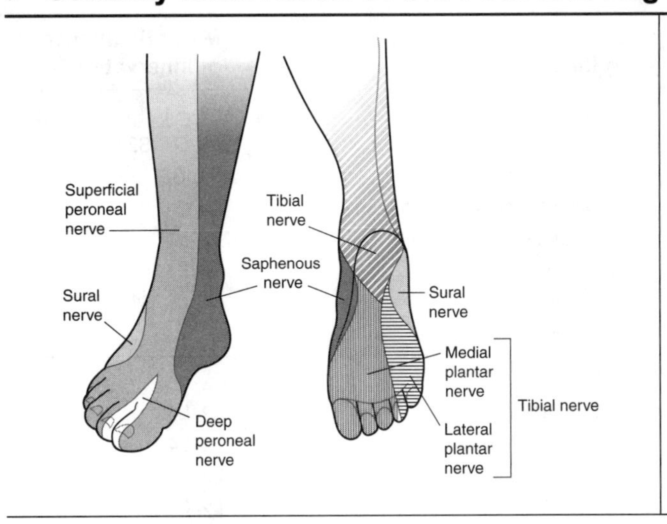

- The lateral leg and the dorsum of the foot are supplied mainly by the **superficial peroneal nerve**, with the exception of the first dorsal webbed space, which is supplied by the **deep peroneal nerve**.
- The sole of the foot is supplied by the lateral and medial plantar branches of the **tibial nerve**.
- The sural nerve (a combination of both peroneal and tibial branches) supplies the posterior leg and lateral side of the foot.
- The **saphenous nerve** (a branch of the femoral nerve) supplies the medial leg and medial foot.

► Movements at the Hip Joint

Action	Muscles Involved	Innervation	Major Segments of Innervation
Flexion	Iliacus and psoas major	Lumbar ventral rami	L2, L3
	Rectus femoris	Femoral	L2–L4
	Sartorius	Femoral	L2, L3
	Tensor fasciae latae	Superior gluteal	L4, L5, S1
Extension	Gluteus maximus	Inferior gluteal	L5, S1, S2
	Semimembranosus	Sciatic (tibial)	L5, S1
	Semitendinosus	Sciatic (tibial)	L5, S1, S2
	Biceps femoris, long head	Sciatic (tibial)	S1, S2
	Adductor magnus, ischial part	Obturator	L3, L4
Adduction	Adductors longus, brevis, and magnus	Obturator	L2–L4
	Gracilis	Obturator	L2–L4
	Pectineus	Femoral	L2, L3
Abduction	Gluteus medius and minimus	Superior gluteal	L4, L5, S1
	Tensor fasciae latae	Superior gluteal	L4, L5, S1
Medial rotation	Gluteus minimus	Superior gluteal	L4, L5, S1
	Gluteus medius, anterior fibers	Superior gluteal	L4, L5, S1
	Tensor fasciae latae	Superior gluteal	L4, L5, S1
Lateral rotation	Gluteus maximus	Inferior gluteal	L5, S1, S2
	Sartorius	Femoral	L2–L4
	Obturator internus and superior gemellus	Nerve to obturator internus	L5, S1, S2
	Obturator externus	Obturator	L3, L4
	Quadratus femoris and inferior gemellus	Nerve to quadratus femoris	L4, L5, S1
	Piriformis	Nerve to piriformis	L5, S1, S2

► Movements at the Knee Joint

Action	Muscles Involved	Innervation	Major Segments of Innervation
Flexion	Semimembranosus	Sciatic (tibial)	L5, S1
	Semitendinosus	Sciatic (tibial)	L5, S1, S2
	Biceps femoris	Sciatic (tibial)	S1, S2
	—	Sciatic (common peroneal)	L5, S1, S2
	Gracilis	Obturator	L2–L4
	Sartorius	Femoral	L2, L3
	Popliteus	Tibial	L4, L5, S1
	Gastrocnemius	Tibial	S1, S2
Extension	Quadriceps femoris:		
	• Vastus medialis	Femoral	L2–L4
	• Vastus lateralis	Femoral	L2–L4
	• Vastus intermedius	Femoral	L2–L4
	• Rectus femoris	Femoral	L2–L4
Lateral rotation of leg	Gluteus maximus	Inferior gluteal	L5, S1, S2
	Biceps femoris	Sciatic (tibial and common peroneal)	L5, S1, S2
Medial rotation of leg	Popliteus ("unlocks" extended knee)	Tibial	L4, L5, S1
	Semimembranosus	Sciatic (tibial)	L5, S1
	Semitendinosus	Sciatic (tibial)	L5, S1, S2
	Gracilis	Obturator	L2–L4
	Sartorius	Femoral	L2, L3

► Common Knee Injuries

The three **most commonly injured structures at the knee** are the tibial collateral ligament, the medial meniscus, and the ACL (**the terrible triad**).

A blow to the lateral aspect of the knee may sprain the tibial collateral ligament. The attached medial meniscus may also be torn.

Patients with a **medial meniscus tear** have pain when the leg is medially rotated at the knee. **ACL tears** may occur when the tibial collateral ligament and medial meniscus are injured. Patients with a torn ACL have an anterior drawer sign, in which the tibia may be displaced anteriorly from the femur in the flexed knee.

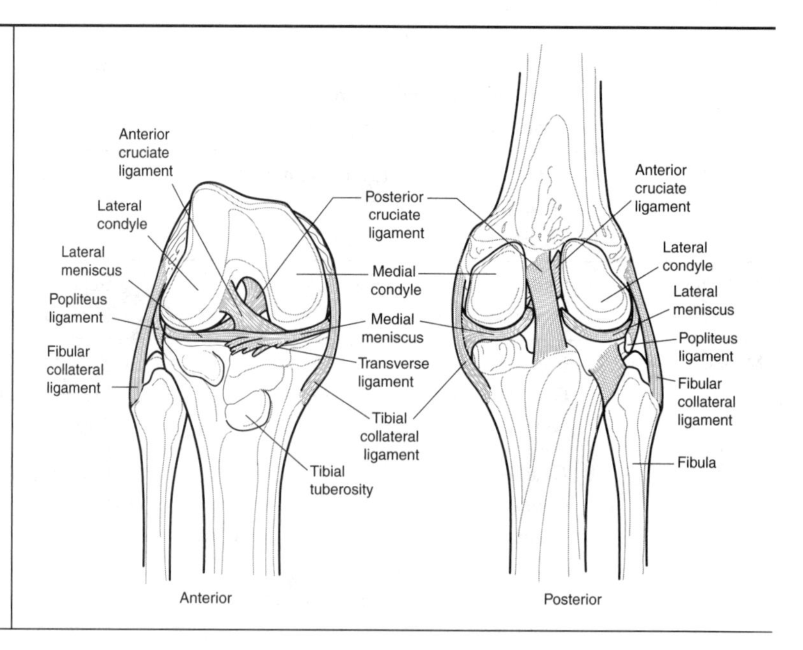

► Movements at the Ankle Joint

Action	Muscles Involved	Innervation	Major Segments of Innervation
Plantar flexion	Gastrocnemius	Tibial	S1, S2
	Soleus	Tibial	S1, S2
	Plantaris	Tibial	L5, S1
Dorsiflexion	Tibialis anterior	Deep peroneal	L4, L5
	Extensor hallucis longus	Deep peroneal	L5, S1

► Movements at the Tarsal (Transverse Tarsal and Subtalar) Joints

Action	Muscles Involved	Innervation	Major Segments of Innervation
Inversion	Tibialis anterior	Deep peroneal	L4, L5
	Extensor hallucis longus	Deep peroneal	L5, S1
	Tibialis posterior	Tibial	L5, S1
Eversion	Peroneus longus	Superficial peroneal	L5, S1
	Peroneus brevis	Superficial peroneal	L5, S1

► Movements of the Toes

Action	Muscles Involved	Innervation	Major Segments of Innervation
Flexion			
All joints	Flexor hallucis longus	Tibial	L5, S1
	Flexor digitorum longus	Tibial	L5, S1
	Quadratus plantae (flexor accessorius)	Lateral plantar	S1, S2
Extension			
All joints	Extensor digitorum longus	Deep peroneal	L4, L5, S1
Toes 2–4	Extensor digitorum brevis	Deep peroneal	L5, S1
Hallux	Extensor hallucis longus	Deep peroneal	L5, S1
IP only toes 2–5	Lumbricals	Lateral plantar, toes 3–5	L5, S1
		Medial plantar, toe 2	S1, S2

Definition of abbreviations: DIP, distal interphalangeal joints; MP, metatarsophalangeal joints; PIP, proximal interphalangeal joints.

Musculoskeletal Disorders

► Skeletal Disorders

Achondroplasia	• Autosomal dominant form of dwarfism; mutation of the fibroblast growth factor receptor 3 (FGFR3) • Abnormal cartilage synthesis • Decreased epiphyseal bone formation • Short limbs, proportionately large body and head (disease spares the cranium and vertebral bones), "saddle nose"
Enchondromatosis	• Multiple cartilaginous masses (enchondromas) within the medullary cavity of bone • **Ollier disease:** nonhereditary, multiple, most commonly hands and feet • Patients present with pain and fractures; may undergo malignant transformation • **Maffucci syndrome:** Familial; enchondromas and hemangiomas of the skin
Fibrous dysplasia	• Trabeculae of woven bone in a background of fibrous tissue • Incidence higher in male teenagers • Usually monostotic; often asymptomatic or may lead to pathologic fracture • Affects long bones, ribs, skull, and facial bones • Fibrosis starts within the medullary cavity and remains encased in cortical bone • **McCune-Albright syndrome:** Association of polyostotic fibrous dysplasia, café-au-lait spots, and sexual precocity in women
Hyperparathyroidism (osteitis fibrosa cystica)	• Osteoclasts resorb bone • Kidney wastes calcium • Osteitis fibrosa cystica more common in primary hyperparathyroidism • Bone pain and fractures • Fibrous replacement of marrow, causing cystic spaces in bone and "brown tumors"
Hypertrophic osteoarthropathy	• Idiopathic painful swelling of wrists, fingers, ankles, knees, or elbows • Periosteal inflammation, new bone forms at the ends of long bones, metacarpals, and metatarsals • Arthritis commonly seen, often with digital clubbing • Etiology: Intrathoracic carcinoma, cyanotic congenital heart disease, and inflammatory bowel disease • Regresses when underlying disease treated
Osteochondromatosis	• Bony metaphyseal projections capped with cartilage • Gardner syndrome: Exostoses and colonic polyps—may become carcinomas
Osteogenesis imperfecta	Defect in type I collagen characterized by fragile bones, blue sclera, and lax ligaments • *Type I*: autosomal dominant mild-to-moderate disease • *Type II*: autosomal recessive stillborn infant; generalized crumpled bones • *Type III*: autosomal recessive progressive severe deformity; white sclera • *Type IV:* autosomal dominant variable severity; normal sclera

(Continued)

► Skeletal Disorders *(Cont'd.)*

Osteomalacia and rickets	• **Vitamin D deficiency** due to chronic renal insufficiency, intestinal malabsorption, or dietary deficiency; osteoid produced in normal amounts but not calcified properly (diffuse radiolucency on bone films); low calcium and phosphorus and high alkaline phosphatase • *Rickets:* children, prior to closure of the epiphyses. Bone deformities, "rachitic rosary" (deformity of the chest wall), bowing of legs, and fractures • *Osteomalacia:* impaired mineralization of normal osteoid matrix; fractures, deformities
Osteomyelitis	• Spread by direct inoculation of bone or hematogenous seeding • *Staphylococcus aureus, Streptococcus, Haemophilus influenzae*, *Salmonella* (sickle cell disease), *Pseudomonas* (intravenous drug users and diabetics), *Mycobacterium tuberculosis* (Pott disease): tuberculous osteomyelitis of spine • Patients present with fever, localized pain, erythema, and swelling • X-ray may be normal for up to 2 weeks, then may show periosteal elevation • **Specific findings** include: – Sequestrum, a necrotic bone fragment – Involucrum, new bone that surrounds the area of inflammation – Brodie abscess, localized abscess formation in the bone
Osteopetrosis	• Osteoclasts unable to resorb bone • Increased density of cortex with narrowing of erythropoietic medullary cavities • Brittle bones, anemia, blindness, deafness, hydrocephalus, cranial nerve palsies • *Autosomal recessive:* affects children, causing early death due to anemia and infections (no bone marrow) • *Autosomal dominant*: adults—fractures
Osteoporosis	• Decrease in bone mass; postmenopausal women • Estrogen deficiency, low density of original bone, lack of exercise • Bone formed normally but in decreased amounts (thinned cortical bone, enlarged medullary cavity) • All bones are affected; x-ray shows generalized radiolucency • Weight-bearing bones predisposed to fractures
Paget disease	• Excessive bone resorption with replacement by soft, poorly mineralized matrix (woven appearance microscopically); x-ray: enlarged, radiolucent bones • Patients present with pain, deformity, fractures • Laboratory tests: extremely elevated alkaline phosphatase • Polyostotic: skull, pelvis, femur, and vertebrae • Progresses from an osteolytic to an osteoblastic phase • May cause bone hypervascularity with increased warmth of the overlying skin and eventually leading to high output cardiac failure

► Microbiology of Osteomyelitis

Type Infection	Case Vignette/Key	Most Common Causative Agent	Mechanism of Pathogenesis	Diagnosis	Treatment
Fever, bone pain with erythema and swelling; some patients (particularly those with diabetes) may have associated cellulitis	Adults, children, and infants without major trauma or special conditions	*Staphylococcus aureus*	Hematogenous spread → lytic bone lesions, lytic toxins	Blood culture or bone biopsy, Gram ⊕, catalase ⊕, coagulase ⊕, coccus	Nafcillin, 3rd gen cephalosporin, IV vancomycin
	Sickle cell anemia	*Salmonella* spp.	HbS patients are functionally asplenic and cannot kill bloodborne pathogens	Gram ⊖, oxidase ⊖, nonlactose fermenting	3rd gen cephalosporin, fluoroquinolones, chloramphenicol
	Trauma	*Pseudomonas aeruginosa*	Capsule protects against phagocytosis	Gram ⊖, oxidase ⊕, blue-green pigments; grape odor	Antipseudomonal beta-lactam + aminoglycoside, or carbapenem + antipseudomonal fluoroquinilone + aminoglycoside
	Spine, hip, knee, hands. Immigrants—Indian subcontinents	*Mycobacterium tuberculosis*	Tuberculous granuloma erodes into bone	Acid-fast bacilli or auramine stain	Multiple drug therapy

▶ Musculoskeletal Tumors

Osteoblastic Tumors	
Osteoblastoma	Similar to an osteoid osteoma, but is large and painless; often involves vertebrae; may be malignant
Osteoid osteoma	Benign; affects diaphysis of long bones; often tibia or femur Causes pain that is worse at night and relieved by aspirin X-ray findings—central radiolucency surrounded by a sclerotic rim Pathology: brown nodule surrounded by dense sclerotic cortical bone
Osteoma	Benign; frequently involves skull *Hyperostosis frontalis interna*: osteoma that extends into the orbit or sinuses Pathology: dense normal bone
Osteosarcoma	**Most common malignant bone tumor** Produces osteoid and bone Men are affected more often than women; usually second and third decade of life Associated with Paget disease in older patients Present with localized pain and swelling, weight loss, and anemia Classic x-ray findings—Codman triangle (periosteal elevation) and bone destruction Pathology: large, necrotic, and hemorrhagic mass Poor prognosis; patients are treated with amputation and chemotherapy Metastasis to the lungs common
Chondromatous Tumors	
Chondromyxoid fibroma	Benign, rare; affects young men Firm mass within the metaphyseal marrow cavity of the tibia or femur Contains fibrous and myxomatous tissue; must be differentiated from a malignant lesion
Chondrosarcoma	Malignant tumor; age range 30–60; men affected more than women May arise de novo or secondarily from pre-existing enchondroma Slower growing than osteosarcomas Typically presents with pain and swelling Involves the spine, pelvic bones, and upper extremities
Enchondroma	Solitary cartilaginous growth within the spongiosa of bone Solitary growths are similar to those in the multiple form (Ollier disease)
Osteochondroma	Benign metaphyseal growth; may be solitary; lesions identical to those in multiple form
Miscellaneous Tumors	
Ewing sarcoma	Malignant, rare; usually affects adolescents; often males Arises from mesenchymal cells Presents as pain, tenderness, and early widespread dissemination Commonly affects the pelvis and metaphysis of long tubular bones
Giant cell tumor	Usually benign, locally aggressive; uncommon; affects ages 20–50; arises in the epiphyseal region of long bones Presents as a bulky mass with pain and tenderness X-ray findings—expanding area of radiolucency without a sclerotic rim
Lipoma	Very common, soft, yellow, benign fatty tumor Often found subcutaneously; simple excision is curative
Liposarcoma	Malignant tumor of fatty tissue typically seen in middle-aged or older adults Typically found in deep subcutaneous tissues, deep fatty tissue of legs, or retroperitoneal areas

▶ Joint Pathology

Ankylosing spondylitis	Occurs predominantly in young men with HLA-B27 Also associated with inflammatory bowel disease Involves the sacroiliac joints and spine
Gout	**Hyperuricemia** leads to deposition of monosodium urate crystals (**needle-shaped and negatively birefringent**) in joints, leading to recurrent bouts of acute arthritis • Caused by overproduction of uric acid (under 10%) or underexcretion of uric acid (over 90%) • Joints are affected asymmetrically; **great toe** (first metatarsal joint) is classically affected (podagra) • Later stages → chronic arthritis and **tophi** in affected joints • Uric acid kidney stones develop in up to 25% of patients **Primary gout** (90% of cases): due to inborn error of purine metabolism, usually from an unknown enzyme deficiency **Secondary gout:** hyperuricemia unrelated to purine metabolism (Lesch-Nyhan syndrome is a rare cause due to HGPRT deficiency)
Juvenile rheumatoid arthritis (Still disease)	Peak incidence from 1–3 years; girls affected more frequently Often preceded by acute febrile illness Periarticular swelling, lymphadenopathy, hepatosplenomegaly, and absence of rheumatoid factor Variable course; resolution may occur
Osteoarthritis (degenerative joint disease)	Incidence ↑ with age; women more affected than men Affects 80% of people over 70 years old in at least one joint Aging or wear and tear (biomechanical) most important mechanism Insidious onset with joint stiffness, ↓ range of motion, and effusions X-ray findings—narrowing of the joint space due to loss of cartilage and osteosclerosis Joint fluid—few cells and normal mucin Most commonly affected joints—vertebrae, hips, knees, and distal interphalangeal (DIP) joints of fingers
Pseudogout (chondrocalcinosis)	Calcium pyrophosphate crystal deposition Associated with multiple diseases (e.g., Wilson disease, hypothyroidism, diabetes mellitus)
Psoriatic arthritis	Similar to rheumatoid arthritis, but absence of rheumatoid factor Associated with HLA-B27
Reactive arthritis (Reiter syndrome)	Triad of arthritis, conjunctivitis, urethritis possibly triggered by infections (e.g., *Chlamydia* and *Shigella*) Typically affects young men who are HLA-B27 positive
Rheumatoid arthritis	Progressive arthritis, associated with HLA-DR4 More common in women, ages 20–60 years Autoimmune reaction with the formation of IgM against Fc of IgG, anti-cyclic citrullinated peptide (CCP) antibodies (rheumatoid factor) Type IV hypersensitivity Symptoms—low-grade fever, malaise, fatigue, and morning stiffness Physical examination—joint swelling, redness, and warmth Synovial fluid—increased cells (usually neutrophils) and poor mucin Elevated sedimentation rate and hypergammaglobulinemia X-ray findings—erosions and osteoporosis Starts in the small joints of the hands and feet but may involve any joint; usually symmetric involvement **Felty syndrome:** polyarticular rheumatoid arthritis, splenomegaly, leukopenia
Suppurative arthritis	Tender, red, swollen joint (e.g., "a hot knee") Monoarticular; high neutrophil count in joint fluid Due to *Staphylococcus*, *Streptococcus*, and gonococci

► Infectious Arthritis

Presentation	Case Vignette/ Key Clues	Most Common Causative Agent	Mechanism of Pathogenesis	Diagnosis	Treatment
Pain, redness, low-grade fever, tenderness, swelling, reduced joint mobility	#1 overall, except in the 15–40 age group, where gonococcal is more prevalent	*Staphylococcus aureus*	Coagulase inhibits phagocytosis	Gram ⊕, coagulase ⊕ cocci, catalase ⊕	Nafcillin, 3rd gen cephalosporin, IV vancomycin
	15–40 years; mono- or polyarticular	*Neisseria gonorrhoeae*	Pili mediate adherence and inhibit phagocytosis	Gram ⊖ diplococcus; ferments glucose but not maltose	Ceftriaxone
	Prosthetic joint	Coagulase-negative staphylococci	Biofilm allows adherence to Teflon®	Gram ⊕, catalase ⊕ cocci	Nafcillin, 3rd gen cephalosporin
	Viral, polyarteritis nodosa	Rubella and hepatitis B, parvovirus	Immune complex mediated (type III hypersensitivity)	Detect immune complexes	Immunosuppressive therapy
	Chronic onset, monoarticular, weightbearing joints	*M. tuberculosis* or fungal	Granulomas erode into bone	Acid-fast bacillus or auramine stain, fungus stain	Multiple drug therapy

► Rheumatoid Arthritis Drugs

The rheumatoid arthritis (RA) **medications** can be divided into **three major classes**: anti-inflammatory drugs, bridging therapy, and disease-modifying antirheumatic drugs (DMARDs). Pharmacologic goals of therapy are to decrease pain, maintain "normal" functional status, reduce inflammation, decrease disease progression, and facilitate healing.

Drug	Mechanism/Other Uses	Adverse Effects
	Anti-inflammatory Drugs	
Salicylates (aspirin)	**Irreversibly** inhibit COX-1 and -2, decreasing PG synthesis • Low dose: ↓ platelet aggregation • Intermediate dose: antipyretic, analgesic • High dose: anti-inflammatory	• Chronic use associated with gastric ulcers, upper GI bleeding, acute renal failure, and interstitial nephritis • Large doses can produce tinnitus, vertigo, respiratory alkalosis • Overdose → metabolic acidosis, hyperthermia, dehydration, coma, death
NSAIDs (ibuprofen, naproxen, diclofenac, ketoprofen, indomethacin)	**Reversibly** inhibit COX-1 and -2, leading to decreased production of PGs Anti-inflammatory, analgesic, antipyretic; indomethacin used to close PDA	• Abdominal distress, bleeding, ulceration • Renal damage (especially in patients with renal disease) due to clearance by kidney
COX-2 inhibitors (celecoxib)	Selectively inhibit COX-2 **COX-1 pathway:** produces PGs that protect the GI lining, maintain renal blood flow, and aid in blood clotting **COX-2 pathway:** produces PGs involved in inflammation and pain	• Beneficial because of reduced GI side effects • This drug class is under scrutiny due to ↑ incidence of stroke and MI.
DMARDs	• May slow or reverse joint damage • Indicated for the treatment of RA when anti-inflammatory therapy insufficient to control patient's symptomatology • DMARDs usually do not show benefit for 6–8 weeks or longer; so, **bridging therapy** (corticosteroids) may be used until a full therapeutic effect is obtained. • DMARDs have severe and potentially fatal side effects.	
Hydroxychloroquine	• Stabilizes lysosomes, ↓ chemotaxis • Also an antimalarial	Ophthalmic abnormalities, dermatologic reactions, hematotoxicity, GI reactions
Methotrexate	• Inhibits dihydrofolate reductase, immunosuppressant • Also an antineoplastic	Hemotoxicity, ulcerative stomatitis, renal toxicity, elevated LFTs
Sulfasalazine	• Metabolized to 5-aminosalicylic acid (5-ASA) and sulfapyridine; parent drug and/or metabolites have anti-inflammatory and/or immunomodulatory properties • Used in inflammatory bowel disease	Rash, GI distress, headache, hematotoxicity
Azathioprine	• ↓ purine metabolism and nucleic acid synthesis; immunosuppressant	Hematologic, GI disturbance, secondary infection, increased risk of neoplasia
Etanercept	Binds TNF	Injection site reactions
Leflunomide	Inhibits cell proliferation and antiinflammatory	Hepatotoxicity, immunosuppression, GI disturbance, alopecia, rash, teratogen
Infliximab	• Monoclonal antibody to TNF-α • Also used in inflammatory bowel disease	Infusion reactions, infections, activation of latent TB

Definition of abbreviations: COX, cyclooxygenases; DMARDs, disease-modifying, slow-acting antirheumatic drugs; LFTs, liver function tests; PGs, prostaglandins; TNF, tumor necrosis factor; TB, tuberculosis.

▶ Drugs Used in the Treatment of Gout

There are **three** primary ways of treating gout: *1*) decreasing inflammation in acute attacks (NSAIDs, colchicine, intra-articular steroids), *2*) using uricosuric drugs to increase renal acceleration of uric acid, and *3*) decreasing conversion of purines to uric acid by inhibiting xanthine oxidase (allopurinol).

Agents	Mechanism	Comments
Acute treatment of gout	Acute treatment measures include NSAIDs (primarily indomethacin), colchicine, and corticosteroids (used only in resistant cases).	
NSAIDs	(*See* Rheumatoid Arthritis Drugs table.)	**Indomethacin** is drug of choice for acute gouty arthritis, although other NSAIDs also effective
Colchicine	**Binds tubulin and prevents microtubule assembly** Reduces the inflammatory response by decreasing leukocyte migration and phagocytosis	Previously considered drug of choice for gouty arthritis; however, its side effect profile limits its usage **Adverse effects:** severe diarrhea and abdominal distress, hematologic (bone marrow suppression, aplastic anemia, or thrombocytopenia), renal failure, hepatic failure, peripheral neuropathy, alopecia
Chronic/prophylactic treatment of gout	These agents are used for the treatment and/or prevention of hyperuricemia with gout and gouty arthritis.	
Uricosuric agents (probenecid, sulfinpyrazone)	**Inhibits reabsorption of uric acid**, thus ↑ its excretion	Inhibits the secretion of many weak acids (e.g., penicillin, methotrexate) **Adverse reactions:** may precipitate acute gouty arthritis, which can be prevented by concurrent NSAIDs; cross-allergenicity with other sulfonamides
Allopurinol	**Inhibits xanthine oxidase**, the enzyme responsible for conversion of hypoxanthine and xanthine to uric acid	Also used as an adjunct to cancer chemotherapy Inhibits metabolism of 6-mercaptopurine and azathioprine (which are metabolized by xanthine oxidase) **Adverse reactions:** GI distress, rash

▶ Tumors Involving Joint Space

Malignant fibrous histiocytoma	Relatively common, affecting adult men > women Arise in soft tissue or bones Lower extremities > upper; also abdominal cavity
Pigmented villonodular synovitis	Villous proliferation of synovium colored brown by hemosiderin deposition Probably a reactive response to recurrent trauma or possibly a neoplastic process that does not metastasize
Synoviosarcoma	Rare tumor, early adulthood; affects males and females equally Slow growing, painless masses Aggressive, early metastases to the lung and pleura Two-thirds occur in lower extremities and one-third in upper extremities

▶ Muscle Disorders

Muscular Dystrophies	
Becker muscular dystrophy	X-linked recessive inheritance or spontaneous Milder; patients may walk until age 20 or 25 Cardiac lesions are mild
Duchenne muscular dystrophy	Severe, **X-linked**, abnormal **dystrophin** protein, loss of muscle cell membrane stability Elevation of creatine kinase and histologic degeneration precedes clinical features Pelvic girdle weakness and ataxia Course is progressive; children unable to walk by the age of 10 Pseudohypertrophy of the calves characteristic Myocardial muscle involvement accompanies other muscle degeneration; may cause death Heterozygous female carriers have subclinical degeneration of muscle fibers
Facioscapulohumeral muscular dystrophy	Autosomal dominant, but spontaneous mutation relatively common Usually involves the face, neck, and shoulder muscles; pelvic muscles in later stages
Limb-girdle muscular dystrophy	Autosomal recessive Weakness begins in pelvic or shoulder girdle; may retain ambulation for 25 years
Myotonic dystrophy	Autosomal dominant pattern or spontaneous mutations Trinucleotide repeat (CTG) in a protein kinase Clinically unique: weakness, atrophy, and myotonia (tonic contractions) Head and neck muscles frequently weak and atrophic
Additional Muscular Disorders	
Myasthenia gravis	**Autoimmune** disease; antibodies against **neuromuscular junction acetylcholine receptors** (nicotinic AChR) Typically affects young women with **fluctuating weakness** but **no sensory** abnormalities; worsens with increased use of muscles Diagnosis—decremental response on EMG or improvement with **edrophonium** May have **thymic abnormalities**, including thymoma (10–20%) or thymic hyperplasia (70–80%)
Lambert-Eaton myasthenic syndrome	• Closely related to myasthenia gravis with similar symptoms • Autoantibodies are directed against NMJ Ca^{2+} channels • May occur in a paraneoplastic setting
Myositides	**Polymyositis** and **dermatomyositis**; autoimmune or collagen vascular diseases Polymyositis more common in females Neck and proximal limb muscle weakness, dysphagia, and muscle pain Dermatomyositis: purple discoloration of the eyelids (heliotrope rash) and ↑ risk of internal malignancies
Myositis ossificans	Ossification at the site of traumatic hemorrhage Pain, swelling, and tenderness
Rhabdomyosarcoma	Most common soft tissue sarcoma in children 40% have metastases at the time of diagnosis **Embryonal rhabdomyosarcoma:** • Most often located in head and neck tissues • Sarcoma botryoides: embryonal rhabdomyosarcoma with a grape-like, soft, polypoid appearance usually located in the genitourinary or upper respiratory tract

Definition of abbreviation: EMG, electromyogram; NMJ, neuromuscular junction.

Skin

▶ Skin and Skin Appendages

The integument consists of the skin (epidermis and dermis) and associated appendages (sweat and sebaceous glands, hairs, and nails). The epidermis is devoid of blood vessels and contains a stratified squamous epithelium derived primarily from ectoderm.

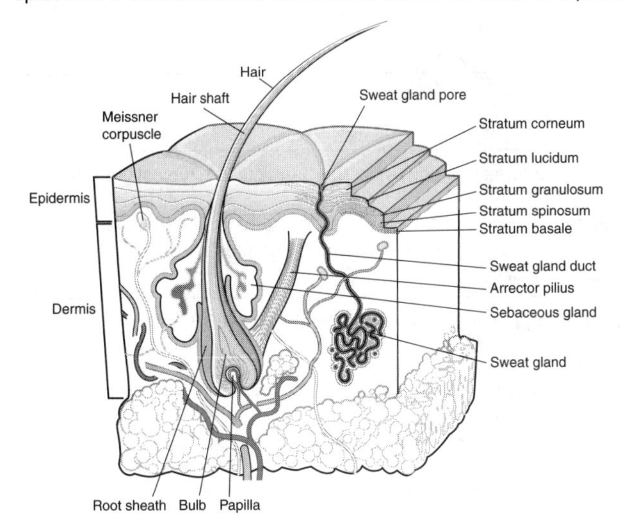

The **epidermis** is composed of five layers in thick skin:

- **Stratum basale:** a proliferative layer of columnar/cuboidal cells
- **Stratum spinosum:** a multilaminar layer of cuboidal/polygonal cells
- **Stratum granulosum:** has more flattened polygonal cells containing basophilic granules
- **Stratum lucidum:** a thin, eosinophilic layer of squamous cells
- **Stratum corneum:** a thick layer containing anucleate keratinized cells

- The **epidermis** contains four cell types: **keratinocytes**, which produce keratin; **melanocytes** (derived from neural crest) that produce melanin; **Langerhans cells**, which are antigen-presenting cells; and **Merkel cells**, associated with nerve fibers.
- The **dermis** is a connective tissue layer mainly of mesodermal origin.
- **Sweat glands** may be eccrine or apocrine.
- **Sebaceous glands** are branched, holocrine acinar glands that discharge their secretions onto hair shafts within hair follicles. They are absent in the palms and soles.
- **Hair** is composed of keratinized epidermal cells. Hair follicles and the associated sebaceous glands are known as pilosebaceous units.

▶ Skin Pathology

Acanthosis nigricans	Epidermal hyperplasia producing velvety hyperpigmentation of body folds in areas such as axilla, neck, and groin Can be associated with hyperlipidemia, diabetes, and visceral malignancies.
Actinic keratosis	Premalignant and may develop into squamous cell carcinoma Fair-skinned people of middle age associated with chronic sun exposure
Atopic dermatitis (eczema)	Itchy eruption that often preferentially involves the flexural areas Typically diagnosed at about 6 months of age — often a lifetime illness with a remitting/flaring course; thought to be due to an immediate hypersensitivity (Type I) reaction with elevated serum IgE May be associated with asthma, allergic rhinitis, and food allergy
Basal cell carcinoma	Pearly papules with telangiectasias Invasive, but rarely metastasizes Occurs on sun-exposed areas in middle-aged or elderly individuals with fair complexions Complete excision is usually curative; 50% recurrence rate
Bullous pemphigoid	Autoantibodies to dermoepidermal junction antigens Pemphigus vulgaris is due to autoantibodies to keratinocyte intercellular junction antigens
Capillary hemangiomas	Arise within the first weeks of life and usually resolve spontaneously; starting at 1–3 years of age; most completely gone by age 5 Soft, red, lobulated mass, 1–6 cm in diameter, composed of thick-walled capillaries

(Continued)

► Skin Pathology *(Cont'd.)*

Contact dermatitis	Localized skin rash caused by contact with a foreign substance Rash is typically bright red, sometimes with blisters, wheals, or urticaria; may have intense itchiness and burning sensation Can be due to chemical irritation (with or without light exposure), true allergic reaction, or physical irritation Often resolves spontaneously; treated with oral antihistamines or hydrocortisone cream; systemic corticosteroids can be used in severe, persistent cases
Dermatitis herpetiformis	Vesicular, pruritic disease often associated with celiac sprue IgA is found at the dermoepidermal junction
Erythema multiforme	Hypersensitivity reaction to drugs Stevens-Johnson syndrome is the severe form.
Erythema nodosum	Hypersensitivity reaction that causes inflammatory nodules in adipose tissue (panniculitis), typically on the anterior shins Peak age in the 20s to 30s; female predominance Can be associated with diseases that stimulate the immune system, including tuberculosis, coccidiomycosis, histoplasmosis, leprosy, sarcoidosis, and streptococcal infections
Kaposi sarcoma	Malignant mesenchymal tumor characterized by an aggressive course in patients with AIDS and by a slower course in elderly men without HIV Caused by human herpes virus type 8 (HHV8)
Lichen planus	Small, itchy, purple, polygonal papules of the skin and oral mucosa Dense infiltrate of lymphocytes at the dermal epidermal junction accompanied by irregular, "saw-tooth" extensions of the dermis into the epidermis
Malignant melanoma	Peaks by ages 40–60 Melanoma grows horizontally first (radial growth), followed by vertical dermal invasion (nodular growth) and forms a large, brown-black patch best prognosis of all forms of melanoma **Lentigo maligna** best prognosis of all forms of melanoma **Nodular** melanoma shows extensive dermal invasion and rapid growth Raised brown-black lesions may be found anywhere on skin or mucosa Worst prognosis of the melanomas Prognosis related to **depth of invasion** (Breslow thickness)
Melasma	Hyperpigmentation occurring in pregnancy, sometimes called "mask of pregnancy" because it often involves the face
Nevocellular nevus	Common benign mole containing nevocytes derived from melanocytes • Junctional nevus: nevocytes are just at the dermoepidermal junction • Dermal nevus: nevocytes just in the dermis • Compound nevus: nevocytes are both at the junction and in the dermis
Nevus flammeus (port wine stain)	Common congenital lesion; composed of telangiectatic vessels Usually located on the neck or face as a large, flat, irregular pink patch that tends to resolve spontaneously
Pemphigus vulgaris	Potentially life-threatening blistering disease of skin and mucous membranes Autoimmune etiology involving antibodies directed against desmoglein 3 Blisters form within the epidermis
Pityriasis rosea	Self-limited skin eruption that characteristically starts with a single, isolated lesion called the "herald patch" and then 1 or 2 weeks later progresses to a generalized rash with large numbers of large, flaky, reddish skin lesions
Psoriasis	Silvery, scaly plaque that primarily affects knees, elbows, and the scalp
Seborrheic keratosis	Common benign (but often biopsied) skin lesion of elderly people - flat, greasy papules that appear "stuck on" on the head, trunk, and extremities Microscopically shows a thickened, benign epidermis composed of basaloid cells with horn cyst formation within the epidermis

(Continued)

► Skin Pathology *(Cont'd.)*

Squamous cell carcinoma	Malignant tumor; most frequently in sun-exposed areas with peak at 60 years of age Preponderance among women When on sun-exposed regions, rarely metastasizes When on unexposed skin, up to 50% metastasize
Toxic epidermal necrolysis (TENS; Lyell's syndrome)	Life-threatening dermatologic condition in which the epidermis detaches from the dermis all over the body, particularly involving the mucous membranes Considered to be a very severe form of Stevens-Johnson syndrome Typically occurs as a severe reaction to medications, notably including sulfonamides, non-steroidal anti-inflammatory drugs, allopurinol, anti-HIV drugs, corticosteroids, and anticonvulsants Patients are treated similarly to those with severe burns
Urticaria	Urticaria (hives) due to localized areas of allergic-based mast cell degranulation leading to dermal edema with intense itchiness
Vitiligo	Vitiligo is an acquired pigmentation disorder characterized by localized areas of complete skin depigmentation

► Infectious Diseases of the Skin, Mucous Membranes, and Underlying Tissues

Case Vignette/ Key Clues	Diagnosis	Common Causative Agents	Mechanism of Pathogenesis	Treatment
		Furuncles, carbuncles		
Neck, face, axillae, buttocks	Catalase ⊕, coagulase ⊕, gram ⊕ cocci in grape-like clusters	*Staphylococcus aureus*	Coagulase breaks fibrin clot Neutrophils + bacteria → pus	Nafcillin for methicillin-sensitive *S. aureus* (MSSA); vancomycin for methicillin-resistant *S. aureus* (MRSA)
Inflamed follicles from neck down	Oxidase ⊕, gram ⊖ rod, blue-green pigment, grape odor	*Pseudomonas aeruginosa* (hot tub folliculitis)	Capsule inhibits phagocytosis	Antipseudomonal beta-lactam + aminoglycoside
		Acne vulgaris		
Inflammation of follicles and sebaceous glands; adolescent	Gram ⊕ rod, identified by clinical clues	*Propionibacterium acnes*	Fatty acids and peptides produced from sebum cause inflammation	Tetracycline, macrolide
		Cutaneous lesions (scratching mosquito bites, cat scratches, etc.)		
Initially vesicular; skin erosion; honey-crusted lesions	Catalase ⊖, gram ⊕ cocci, bacitracin sensitive	*Streptococcus pyogenes*	Streptokinase A and B DNAse, hyaluronidase	Penicillin, macrolide
Initially vesicular but with longer-lasting bullae	Catalase ⊕, coagulase ⊕, gram ⊕ cocci, in clusters	*Staphylococcus aureus*	Exfoliatins produce bullae	Nafcillin for MSSA; vancomycin for MRSA
		Red, raised, butterfly-wing facial rash		
Dermal pain, edema, rapid spread	Catalase ⊖, gram ⊕ cocci, bacitracin-sensitive	*Streptococcus pyrogenes* (erysipelas)	M protein inhibits phagocytosis, erythrogenic exotoxins, hyaluronidase	Penicillin, macrolide

(Continued)

► Infectious Diseases of the Skin, Mucous Membranes, and Underlying Tissues *(Cont'd.)*

Case Vignette/ Key Clues	Diagnosis	Common Causative Agents	Mechanism of Pathogenesis	Treatment
Jaw area swelling with pain, sinus tract formation, yellow granules in exudate				
Carious teeth, dental extraction or trauma	Gram ⊕, anaerobic, filamentous branching rods, non-acid fast	*Actinomyces israelii*; "lumpy jaw"; actinomycosis	Unknown	Penicillin
Vesicular lesions				
Sometimes preceded by neurologic pain	Cell culture, intranuclear inclusion, multinucleated cells	Herpes	Virus-rich vesicles ulcerate dsDNA, enveloped (nuclear membranes)	Acyclovir, valacyclovir, famciclovir
Sometimes large	Catalase ⊕, coagulase ⊕, gram ⊕ cocci, in clusters	*Staphylococcus aureus*	Exfolatins produce bullae	Nafcillin for MSSA; vancomycin for MRSA
Subcutaneous granulomas/ulcers/cellulitis				
Tropical fish enthusiasts; granulomatous lesion (most commonly freshwater)	Biopsy, slow growing acid-fast bacilli	*Mycobacterium marinum*	Trauma + water exposure → granulomas form	Clarithromycin initially, then antimycobacterial therapy
Cellulitis following contact with saltwater or oysters	Green colonies on TCBS agar (alkaline), gram ⊖ rod, oxidase ⊕	*Vibrio vulnificus*	Cytolytic compounds, antiphagocytic polysaccharides	Tetracycline, aminoglycosides
Solitary or lymphocutaneous lesions; rose gardeners or florists; sphagnum moss	Cigar-shaped yeast in pus	*Sporothrix schenckii* (rose gardener's disease)	Ulceration or abscess	Potassium iodide, ketoconazole
Subcutaneous swelling (extremities, shoulders), sinus tract formation, granules (mycetoma)	*Actinomyces*—Gram ⊕, anaerobic, filamentous branching non-acid fast rods *Nocardia*—partially acid-fast, branching filaments, aerobic	*Actinomyces*, *Nocardia*	Unknown; granules are microcolonies	*Actinomyces*—penicillin, doxycycline, clindamycin *Nocardia*—sulfonamide, sulfa antibiotics; carbapenem for resistant cases
Malignant pustule				
Pustule → dark-red, fluid-filled, tumor-like lesion → necrosis → black eschar surrounded by red margin; postal worker or wool handler/importer	Gram ⊕, spore-forming, encapsulated rods	*Bacillus anthracis*	Poly-D-glutamate capsule, exotoxin causes edema, cell death Three-component toxin	Ciprofloxacin, penicillin
As above, with pseudomonal septicemia	Blood culture, gram ⊖, oxidase ⊕, produces blue-green pigments, fruity odor	*Pseudomonas aeruginosa*, (**ecthyma gangrenosum**)	Endotoxin	Susceptibility testing necessary

(Continued)

▶ Infectious Diseases of the Skin, Mucous Membranes, and Underlying Tissues *(Cont'd.)*

Case Vignette/ Key Clues	Diagnosis	Common Causative Agents	Mechanism of Pathogenesis	Treatment
Burns, cellulitis				
Blue-green pus; grape-like odor	Oxidase ⊕, gram ⊖ rod, blue-green pigments, grape odor	*Pseudomonas aeruginosa*	Capsule inhibits phagocytosis	Antipseudomonal beta-lactam + aminoglycoside
Wounds				
Surgical wounds (clean)	Same as above for *S. aureus*	*Staphylococcus aureus*	Same as above for *S. aureus*	Nafcillin for MSSA, vancomycin for MRSA
Surgical wounds (dirty)	Gram ⊖ facultative anaerobes	Enterobacteriaceae, anaerobes	Contamination from fecal flora	3rd gen cephalosporin
Trauma with damage to blood supply	Nagler reaction, anaerobic, gram ⊕ rod, spore forming	*Clostridium perfringens* and others	Alpha toxin (lecithinase) gas production, edema, cytotoxicity	Debridement, clindamycin, chloramphenicol, tetracycline
Animal bites (various)	Gram ⊖ rods (bite wounds are not generally cultured)	*Pasteurella multocida*	Capsule	Amoxicillin/clavulanate
Human bites, fist fights	Gram ⊖ oral floral	*Eikenella corrodens*	Pili, phase variation	3rd generation cephalosporins, fluoroquinolones
Dog bites	Gram ⊖ fusiform	*Capnocytophaga canimorsus*	Sialidase allows adherence to host cells	As above
Rat bites	Gram ⊖ pleomorphic rod	*Streptobacillus moniliformis and Spirillum minus*	Endotoxin	Penicillin G or V
Cat scratches, resulting in lymphadenopathy with stellate granulomas	Gram ⊖ envelope	*Bartonella henselae*	Obligate intracellular	Various antibiotics (rifampin, ciprofloxacin, gentamicin, TMP-sulfamethoxazole)
Shallow puncture wound through tennis shoe sole	Oxidase ⊕, gram ⊖ rod, blue-green pigments	*Pseudomonas aeruginosa*	Capsule inhibits phagocytosis	Antipseudomonal beta-lactam + aminoglycoside
Leprosy				
Blotchy, red lesions with anesthesia; facial and cooler areas of skin	Acid-fast, intracellular bacilli in punch biopsy, ⊕ lepromin test	*Mycobacterium leprae* (tuberculoid form)	Cell-mediated immunity kills intracellular organisms, damages nerves	Dapsone, clofazimine +/- rifampin
Numerous nodular lesions; leonine facies	Acid-fast, intracellular bacilli in punch biopsy, ⊖ lepromin test	*Mycobacterium leprae* (lepromatous form)	Humoral immunity elicited does not stop growth of organisms	Dapsone, clofazimine +/- rifampin

(Continued)

► Infectious Diseases of the Skin, Mucous Membranes, and Underlying Tissues *(Cont'd.)*

Case Vignette/ Key Clues	Diagnosis	Common Causative Agents	Mechanism of Pathogenesis	Treatment
Keratinized area of skin (ringworm)				
Reddened skin lesion in growing ring shape, raised margin; infection nail bed or hair shaft	Wood's lamp (fluoresce), skin scraping and KOH; arthroconidia	*Trichophyton* spp. (skin, hair, nails) *Epidermophyton* spp. (skin, nails) *Microsporum* spp. (skin, hair)	Fungi germinate in moist areas, invade	Topical miconazole or oral imidazoles if hair shaft or nails infected
Dermatitis				
Itching skin rash after swimming in fresh water lakes (swimmer's itch)	Clinical signs and history	Bird schistosomes	Skin penetration by cercariae → death in skin → hypersensitivity	Topical anti-inflammatory
Snake-like tracks on bare skin exposed to dog/cat feces (plumber's itch, cutaneous larva migrans)	Clinical signs and history	Dog and cat hookworms (*Ancylostoma* spp.)	Skin penetration by larvae → death in skin → hypersensitivity (type 1)	Topical anti-inflammatory, thiabendazole
Warts				
Plantar surfaces	Intranuclear inclusion bodies	HPV 1 (dsDNA, naked icosahedral)	Virus infects basal layers of skin, stimulates cells to divide	Cryotherapy
Common warts		HPV serotypes 2, 4		
Umbilicated warts; wrestling teams; may be anogenital	Intracytoplasmic inclusions	*Molluscum contagiosum* (pox family, dsDNA, enveloped complex)	Infects epidermal cells to form fleshy lesion	Cryotherapy
Anogenital warts	Intranuclear inclusion bodies	HPV 6 and 11 (most common) HPV 16 and 18 (premalignant)	Virus stimulates cell division Cervical intraepithelial neoplasia Tumor suppressor gene inactivation	Imiquimod, interferon-α, cidofovir
Mucocutaneous erosive lesions				
Foreign immigrant or military stationed in the Middle East, ulcers; chronic facial disfigurement; sandfly vector	Finding amastigotes with flagellar pocket inside phagocytic cells in biopsy	*Leishmania* spp.	Amastigotes intracellular in macrophages, proliferate and spread	Antimonials pentamidine

► Selected Rashes

Type of Rash	Progression	Other Symptoms	Causative Agent(s)	Pathogenesis	Diagnosis	Treatment
Scarlet fever						
Erythematous maculopapular (sandpaper-like)	Trunk and neck → extremities (spares palms and soles)	Sore throat, fever, nausea	*Streptococcus pyogenes*	Exotoxins A–C (superantigens)	Gram ⊕, catalase ⊖ cocci	Penicillin, clindamycin
Toxic shock syndrome						
Diffuse, erythematous, macular sunburn-like	Trunk and neck → extremities with desquamation on palms and soles	Acute onset, fever >102 F, myalgia, pharyngitis, vomiting, diarrhea; hypotension leading to multiorgan failure	*Staphylococcus aureus*	TSST-1 (superantigen)	Gram ⊕, catalase ⊕, coagulase ⊕ cocci	Nafcillin , oxacillin; vancomycin in penicillin allergic patients
Staphylococcal skin disease: scalded skin disease and scarlatina						
Perioral erythema, bullae, vesicles, desquamation	Trunk and neck → extremities, except tongue and palate; large bullae and vesicles precede exfoliation	Abscess or some site of infection	*Staphylococcus aureus*	Exotoxin	Gram ⊕, catalase ⊕, coagulase ⊕ cocci	Nafcillin , oxacillin; vancomycin in penicillin allergic patients
Lyme disease						
Erythematous concentric rings (Bull's eye)	Originates at site of tick bite	Fever, headache, myalgia, Bell's palsy	*Borrelia burgdorferi* (#1 vector-borne disease in U.S.)	Invades skin and spreads to involve heart, joints and CNS. Arthritis is type III hypersensitivity	Serology	Doxycycline, ceftriaxone
Epidemic typhus						
Petechiae → purpura	Trunk → extremities; spares palms, soles, and face	Fever, rash, headache, myalgias, and respiratory symptoms	*Rickettsia prowazekii*	Endotoxin	Serology, Weil-Felix	Doxycycline, chloramphen-icol
Rocky Mountain spotted fever (most common on East Coast)						
Petechiae	Ankles and wrists → generalized with palms and soles	Fever, rash, headache, myalgias, and respiratory symptoms	*Rickettsia rickettsii*	Endotoxin (overproduces outer membrane fragments)	Serology, Weil-Felix	Doxycycline, chloramphen-icol
Early meningococcemia						
Petechiae → purpura	Generalized	Abrupt onset, fever, chills, malaise, prostration, exanthem → shock	*Neisseria meningitidis*	Endotoxin	Gram ⊖ diplococcus on chocolate agar; LPA for capsular antigens	Ceftriaxone

(Continued)

► Selected Rashes *(Cont'd.)*

Type of Rash	Progression	Other Symptoms	Causative Agent(s)	Pathogenesis	Diagnosis	Treatment
Secondary syphilis						
Skin: maculopapular; mucous membrane: condylomata lata	Generalized bronze rash involving the palms and soles	Fever, lymphadenopathy, malaise, sore throat, splenomegaly, headache, arthralgias	*Treponema pallidum*	Endotoxin	Serology: VDRL (nonspecific), FTA-ABS (specific)	Penicillin, doxycycline, erythromycin
Measles						
Confluent, erythematous, maculopapular rash, unvaccinated child	Head → entire body Koplik's spots	Cough, coryza, conjunctivitis, and fever (prodrome); oral lesions, exanthem, bronchopneumonia, ear infections (unvaccinated individual)	Rubeola virus; negative sense RNA virus, non-segmented = Paramyxovirus	T-cell destruction of virus-infected cells in capillaries causes rash	Virus cultures, serology	Supportive
Chickenpox/Shingles						
Asynchronous rash, unvaccinated child	Generalized with involvement of mucous membranes	Fever, pharyngitis, malaise, rhinitis, exanthem	Varicella zoster virus (Herpesviridae, dsDNA)	Virus replicates in mucosa and is latent in dorsal root ganglia	Tzanck smear (find syncytia), Cowdry type A intranuclear inclusions, PCR	Supportive, avoid aspirin due to Reye syndrome
Unilateral vesicular rash following a dermatome, 50-60 year old patient	Restricted to one dermatome	Fever, severe nerve pain, pruritus	Varicella zoster virus (Herpesviridae, dsDNA)	Reactivation of latent infection	Tzanck smear (find syncytia), Cowdry type A intranuclear inclusions, PCR	Acyclovir, famciclovir, valacyclovir

The Hematologic and Lymphoreticular System

Chapter 9

Hematopoiesis

All of the different blood cells are derived from stem cells in the bone marrow, as shown below.

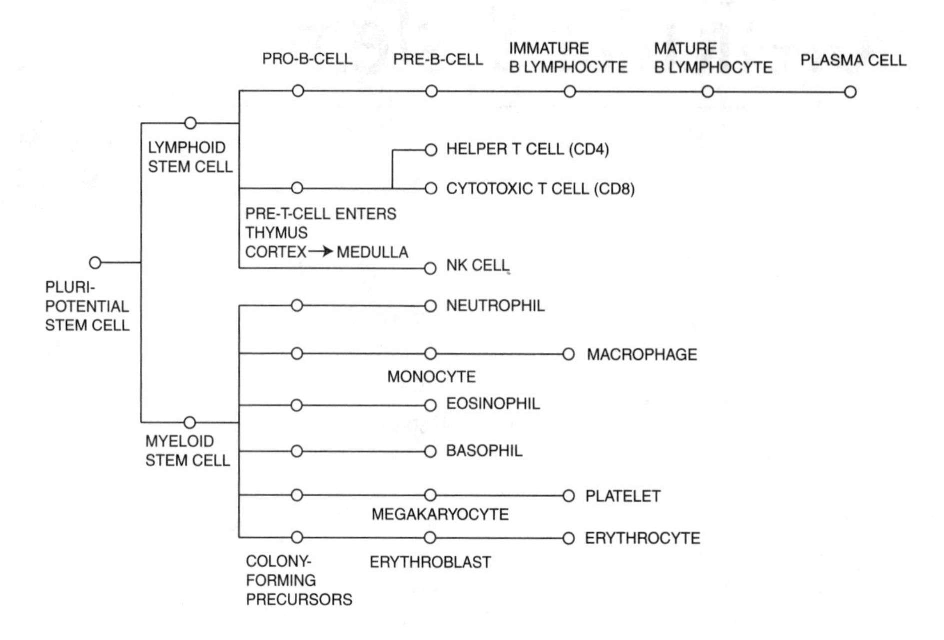

Hemostasis

Hemostasis is the sequence of events leading to the cessation of bleeding by the formation of a stable fibrin-platelet plug. It involves the vascular wall, platelets, and the coagulation system.

► Platelet Function and Dysfunction

Platelets are anuclear, membrane-bound cellular fragments derived from megakaryocytes in the bone marrow. They have a short lifespan of approximately 10 days. There are normally 150,000–400,000 platelets per mm^3 of blood. Platelet activity is measured by **template bleeding time**. Clinically, dysfunction is seen as **petechiae**. The platelet reaction consists of three steps. Dysfunction of each step is associated with different diseases:

- Adhesion (e.g., von Willebrand disease, Bernard-Soulier syndrome)
- Primary aggregation (e.g., thrombasthenia)
- Secondary aggregation and release (e.g., aspirin, storage pool disease)

► Formation of the Platelet Plug

Vascular wall injury	• Injury causes exposure of subendothelial extracellular collagen • Arteriolar contraction due to reflex neurogenic mechanisms, and the local release of **endothelin** occurs
Adhesion	• **von Willebrand factor (vWF)** binds exposed collagen fibers in the basement membrane • Platelets adhere to vWF via **glycoprotein Ib** and become **activated** (shape change, degranulation, synthesis of **thromboxane A_2, TxA_2**) • Deficiency of vWF → **von Willebrand disease**; deficiency of glycoprotein Ib receptor → **Bernard-Soulier syndrome**
Release reaction	• Release contents of platelet dense bodies (e.g., **ADP**, **calcium**, serotonin, histamine, epinephrine) and alpha granules (fibrinogen, fibronectin, factor V, vWF, platelet-derived growth factor) • Membrane expression of phospholipid complexes (important for coagulation cascade)
Aggregation	• **ADP** and **thromboxane A_2 (TxA_2)** is released by **platelets** and **promote aggregation** (TXA_2 production is inhibited by **aspirin**) • Cross-linking of platelets by **fibrinogen** requires the **GpIIb/IIIa receptor**, which is deficient in **Glanzmann thrombasthenia** • Decreased endothelial synthesis of antithrombogenic substances (e.g., prostacyclin, nitric oxide, tissue plasminogen activator, thrombomodulin)

► Disorders of Platelet Numbers

Thrombocytopenia	• **Decrease** in the platelet count (normal = 150,000–400,000/mm^3) • **Clinical features:** bleeding from small vessels, often skin, GI/GU tracts; **petechiae** and **purpura** are seen • **Classification:** decreased production (aplastic anemia, drugs, vitamin B_{12} or folate deficiency); increased destruction, (e.g., DIC, TTP, ITP, drugs, malignancy); abnormal sequestration
Idiopathic thrombocytopenic purpura (ITP)	• Spleen makes antibodies against platelet antigens (e.g., GpIIb–IIIa, GpIb–IX); platelets destroyed in the spleen by macrophages • Acute form (children): self-limited, postviral • Chronic form (adults): ITP may be primary or secondary to another disorder (e.g., HIV, SLE) • Smear shows enlarged, immature platelets; normal PT and PTT • Treatment: corticosteroids, immunoglobulin therapy, splenectomy
Thrombotic thrombocytopenic purpura (TTP)	• **Clinical features:** pentad (thrombocytopenic purpura, fever, renal failure, neurologic changes, microangiopathic hemolytic anemia); usually in young women • Inherited or acquired deficiency of **ADAMTSI3** • Smear shows few platelets, schistocytes, and helmet cells • **Hemolytic uremic syndrome (HUS):** mostly in children after gastroenteritis with bloody diarrhea; organism: verotoxin-producing *E. coli* O157:H7; similar clinical triad
Thrombocytosis (reactive)	Increase in count due to bleeding, hemolysis, inflammation, malignancy, iron deficiency, stress, or postsplenectomy
Essential thrombocythemia	Increase in count due to primary myeloproliferative disorder

► Disorders of Platelet Function Leading to Increased Bleeding

Bernard-Soulier disease	Defective platelet plug formation secondary to decreased Gp1b, which causes impaired platelet-to-subendothelial collagen aggregation
Glanzmann thrombasthenia	Defective platelet plug formation secondary to decreased GpIIb/IIIa, which causes impaired platelet-to-subendothelial collagen aggregation
von Willebrand disease	Defective platelet plug formation due to an autosomal dominant defect in quantity or quality of von Willebrand factor (vWF); increased bleeding time and increased PTT (because vWF stabilizes factor VIII)

▶ Coagulation

Coagulation begins anywhere from a few seconds to 1–2 minutes after an injury.

Intrinsic Pathway

- **Factor XII** (Hageman factor) is activated on contact with the collagen
- "a" indicates activated form

Extrinsic Pathway

Initiated by exposure to tissue thromboplastin

Common Pathway

Factors IXa, VIIa, VIIIa, platelet phospholipids, and calcium activate factor X

Thrombin (IIa):

- Able to catalyze own activation
- Increases platelet aggregation
- Activates **factor XIII** (fibrin-stabilizing factor) and potentiates binding of factors V and VIII to phospholipid/Ca^{2+} complex

Fibrinogen (I):

- Split into self-polymerizing fibrin monomers
- Initially bind via loose hydrogen and hydrophobic bonds
- **Factor XIII** catalyzes formation of strong covalent bonds

Vitamin K–Dependent Factors

Factors II (prothrombin), VII, IX, X, protein C, and protein S

INTRINSIC PATHWAY

EXTRINSIC PATHWAY

Surface contact (collagen, platelets) Prekallikrein HMW-kininogen

XII → XIIa

XI → XIa

Tissue thromboplastin

VIIa ← VII

IX → IXa

VIII → VIIIa (IIa)

Ca^{2+}

X → Xa

V → Va (IIa)

Ca^{2+}

Prothrombin (II) → Thrombin (IIa)

Fibrinogen → Fibrin (monomers)

XIII

Fibrin (polymers)

COMMON PATHWAY

▶ Final Step: Clot Retraction and Dissolution

- After the clot is formed, it begins to shrink.
- Edges of small wounds are pulled together by platelet actinomyosin.
- Fibrinolysis (dissolution) requires activation of plasminogen.
- Clot releases plasminogen activator, which converts plasminogen to plasmin, which in turn proteolyses fibrinogen and fibrin.
- Urokinase and streptokinase are exogenous sources of plasminogen activation. Tissue plasminogen activator (t-PA) is endogenously produced, but can also be administered as a drug in the setting of acute myocardial infarction.

▶ Laboratory Tests of Coagulation System

Test	Measures	Specific Coagulation Factors Involved
Prothrombin time (PT)	Extrinsic and common coagulation pathways	VII, X, V, prothrombin, fibrinogen
Partial thromboplastin time (PTT)	Intrinsic and common coagulation pathways	XII, XI, IX, VIII, X, V, prothrombin, fibrinogen
Thrombin time (TT)	Fibrinogen levels	Fibrinogen
Fibrin degradation products (FDP)	Fibrinolytic system	—

► Clotting Cascade

Physiologic Regulation of the Clotting Cascade	
Fibrin	Adsorbs most of the thrombin, keeping it from spreading
Antithrombin III	α-globulin that binds to and inactivates thrombin
Laminar Flow	Keeps platelets away from blood vessel walls
Heparin	Polysaccharide produced by many cell types, promotes antithrombin inhibition of thrombin
Protein C and Protein S	Vitamin K–dependent Endogenous anticoagulants Protein C is activated by thrombin bound to thrombomodulin In the presence of protein S, activated protein C promotes anticoagulation by inactivating factors V and VIII Protein C also promotes fibrinolysis by inactivating t-PA inhibitor, thereby promoting t-PA fibrinolytic activity
Causes of Failure to Clot	
Factor VIII deficiency (hemophilia A)	X-linked Severe cases bleed in infancy at circumcision or have multiple hemarthrosis Moderate cases have occasional hemarthrosis Mild cases may be missed until dental or surgical procedures Bleeding may require treatment with cryoprecipitate or lyophilized factor VIII
Factor IX deficiency (Christmas disease, hemophilia B)	X-linked recessive Signs and symptoms same as hemophilia A
Vitamin K deficiency	Fat-soluble, produced by gut flora Essential in the posttranslational modification of factors II, VII, IX, and X, as well as proteins C and S Vitamin K deficiency may result from fat malabsorption, diarrhea, antibiotics
Liver disease	Factors II, V, VII, IX, X, XI, and XII synthesized in the liver
Causes of Excessive Thrombosis	
Protein C or S deficiency	Deficiency of these factors decreases the ability to inactivate factors V and VIII, leading to increased risk of deep vein thrombosis and pulmonary embolism, cerebral venous thrombosis, and warfarin-induced skin necrosis.
Factor V Leiden deficiency	A mutant factor V cannot be degraded by protein C, leading to increased risks of deep vein thrombosis with pulmonary embolism, and possibly increased risk of miscarriage.
Prothrombin gene mutation	Prothrombin gene mutation in the 3' untranslated region causes increased circulating thrombin and venous clots.
ATIII deficiency	Antithrombin III is a potent inhibitor of the clotting cascade, and its deficiency leads to increased venous clots.

► Disseminated Intravascular Coagulation (DIC)

- Massive, persistent activation of both coagulation system and fibrinolytic system
- Consumption deficiency of clotting factors and platelets
- Morbidity/mortality from DIC may be related to either thrombosis or hemorrhage
- **Etiologies:** amniotic fluid embolism, infections (particularly gram-negative sepsis), malignancy, and major traumas, particularly head injury
- **Diagnosis:** low platelets, low fibrinogen, increased PT, increased PTT, and presence of fibrin degradation products

Anticlotting Agents

Anticlotting agents are used in the treatment of ischemic stroke, deep venous thrombosis, myocardial infarction, and atrial fibrillation. Anticoagulants and thrombolytics work in arterial and venous circulations; antiplatelets act in the arterial circulation. Bleeding is the most important adverse effect of these agents.

Class	Mechanism	Comments/Agents
Anticoagulants	Decrease fibrin clot formation. Differ in pharmacokinetics/pharmacodynamics. **Heparin** is used when immediate anticoagulation is necessary (acute MI, DVT, pulmonary embolism, stroke, beginning therapy); **warfarin** is used chronically. LMWHs have a longer half-life than does heparin.	
Heparin (IV, SC) LMWHs	Binds **AT-III**; this complex inactivates **thrombin**, factors IXa, **Xa**, and XIIa	Acts in **seconds**; used acutely (days) aPTT used to monitor heparin, not LMWHs **Protamine** reverses heparin and LMWHs Used in pregnancy **LMWHs** (ardeparin, dalteparin, enoxaparin) inhibit **factor Xa** more and thrombin less than heparin LMWH preferred for long-term use due to risk of HIT
Fondaparinux	Binds ATIII, potentiating inactivation of factor Xa	Indications, warnings, and precautions similar to LMWH's
Warfarin **(PO)**	Interferes with the synthesis of the vitamin K–dependent clotting factors (II, VII, IX, X)	Takes 2–5 **days** to fully work; chronic use **PT** or **INR** used to monitor **Vitamin K** reverses effect **Contraindicated** in pregnancy **Cytochrome P450**–inducing drugs ↓ effect; cytochrome P450 inhibitors ↑ effect
Direct thrombin inhibitors	Bind directly to thrombin substrates and/or thrombin (ATIII not required) Bind to soluble thrombin and clot-bound thrombin	**Lepirudin**, bivalirudin, argatroban, hirudin
Antiplatelets	Platelets adhere to site of vascular injury, where they are activated by various factors to express a glycoprotein to which fibrogen binds, resulting in platelet aggregation and formation of a platelet plug. Antiplatelet drugs inhibit this process, thus reducing the chances of thrombi formation.	
COX inhibitors	Block COX-1 and COX-2, thereby inhibiting **thromboxane A_2**–mediated platelet aggregation	**Aspirin**—also antipyretic, antiinflammatory, analgesic Affected platelets are impaired for their lifespan (9–12 days) Side effects—tinnitus, ↓ renal function, GI ulceration/bleeding, Reye syndrome (in children with viral syndromes)
ADP antagonists	Irreversibly inhibit ADP-mediated platelet aggregation	**Ticlopidine, clopidogrel**
Glycoprotein IIb/IIIa inhibitors	Reversibly inhibit binding of fibrin to platelet glycoprotein IIb/IIIa, preventing platelet cross-linking	**Abciximab**, eptifibatide, tirofiban
PDE/adenosine uptake inhibitors	Inhibit phosphodiesterase 2, thereby ↑ cAMP → inhibits platelet aggregation Block adenosine uptake → less adenosine A2 stimulation → ↑ cAMP	**Dipyridamole**, cilostazol
Thrombolytics	**Convert plasminogen → plasmin**, leading to fibrinolysis and breakdown of clots. Used IV for short-term emergency management of coronary thromboses in MI, DVT, pulmonary embolism, and ischemic stroke (t-PA).	
	t-PAs are thought to convert fibrin-bound plasminogen, thus targeting clots; others are not clot specific	Alteplase, reteplase (**t-PA derivatives**) Streptokinase, urokinase

Definition of abbreviations: ADP, adenosine diphosphate; aPTT, activated partial thromboplastic time; ATIII, antithrombin III; COX, cyclo-oxygenase; DVT, deep venous thrombosis; GI, gastrointestinal; HIT, heparin-induced thrombocytopenia; INR, international normalized ratio; IV, intravenous(ly); LMWH, low molecular weight heparin; MI, myocardial infarction; PO, by mouth; PT, prothrombin time; SC, subcutaneously; t-PA, tissue plasminogen activator.

► Drugs Used in Bleeding Disorders

Antifibrinolytic (aminocaproic acid, tranexamic acid)	Inhibits fibrinolysis by inhibiting plasminogen-activating substances	Prevents and treats bleeding in hemophiliacs during dental and surgical procedures
Vitamin K	Plays a role in coagulation by acting as a cofactor for clotting factors II, VII, IX, and X	Indicated for treatment of warfarin overdose, as well as vitamin deficiency
Oprelvekin (IL-11)	Stimulates bone marrow platelet production via stimulation of IL-11 receptor	Used to prevent thrombocytopenia after antineoplastic therapy Edema commonly occurs
Desmopressin (DDAVP)	A synthetic analog of arginine vasopressin; causes a dose-related release of factor VIII and vWF from storage sites	Hemophilia A and von Willebrand disease

Erythropoiesis

Erythropoiesis is the process of RBC formation. Bone marrow stem cells (colony-forming units, CFUs) differentiate into proerythroblasts under the influence of the glycoprotein **erythropoietin**, which is produced by the kidney.

Proerythroblasts → basophilic erythroblasts → normoblasts (nucleus extruded) → reticulocyte (still contains some ribosomes) → erythrocyte (remain in the circulation approximately 120 days and are then recycled by the spleen, liver, and bone marrow)

Disorders of Red Blood Cells

► Polycythemia (Increase in Red Blood Cell Mass)

Polycythemia vera (primary)	• Myeloproliferative syndrome • Males age 40–60 • Vessels distended with viscous blood, congestive hepatosplenomegaly, and diffuse hemorrhages • Management is generally with therapeutic phlebotomy
Secondary polycythemia	• Increased erythropoietin levels • Etiologies: high altitude, cigarette smoking, respiratory, renal and cardiac disease and malignancies (e.g., renal cell carcinoma, hepatoma, leiomyoma, adrenal adenoma, cerebellar hemangioblastoma)
Relative polycythemia	Fluid loss with stable RBC mass (vomiting, diarrhea, burns)

▶ Anemia

Secondary to decreased production, increased destruction, sometimes both
Symptoms: palpitations, high-output heart failure, pallor, fatigue, dizziness, syncope, and angina

Decreased Production (Low Reticulocyte Count)	
Decreased production (low reticulocyte count) **Iron deficiency** (smear: hypochromic, microcytic)	• An important differential feature between the thalassemia traits and iron deficiency is that thalassemia traits result in an elevated number of microcytes, whereas iron deficiency results in a decreased number of microcytes. • Serum iron, total iron-binding capacity (TIBC), and ferritin confirm the diagnosis.
Megaloblastic—B_{12}/folate (smear: macrocytic, hypersegmented neutrophils)	• Impaired DNA synthesis • Vitamin B_{12} deficiency neurologic (subacute combined degeneration) and hematologic sequelae • **Vitamin B_{12}:** copious body stores, years to develop deficiency *Causes:* dietary deficiency, malabsorption, tapeworm, bacterial overgrowth, deficiency of intrinsic factor (pernicious anemia) • **Folate:** deficiency develops much more quickly (months) *Causes:* deficient intake (poor diet, alcoholism, malabsorption), increased need (pregnancy, malignancy, increased hematopoiesis), or impaired use (antimetabolite drugs) • Must treat patient with both folate and B_{12}. Folate may reverse anemia in a B_{12} deficiency but not neurologic complications
Aplastic	• Pancytopenia • *Multiple etiologies:* idiopathic, drugs, including alkylating agents, chloramphenicol, radiation, infections, and congenital anomalies (i.e., Fanconi anemia) • Prognosis is poor • Bone marrow transplant may be curative
Myelophthisic	Displacement of hematopoietic bone marrow by infiltrating tumor
Myeloid metaplasia with myelofibrosis	• Chronic myeloproliferative disorder with small numbers of neoplastic myeloid stem cells • Resultant bone marrow fibrosis leads to pancytopenia

Thalassemias		
Types	**Key Points**	**Clinical Picture**
ALPHA	Secondary to gene deletion: **four** genes can be deleted: **1 deleted:** silent carrier **2 deleted:** trait **3 deleted:** HbH disease **4 deleted:** hydrops fetalis, Bart Hb	• Variable clinical severity • Non α-chain aggregates less toxic • Mild hemolysis and anemia tend to be milder; Bart's in the neonate leads to anoxia and intrauterine death
BETA	• Defects in mRNA processing. • Homozygous: β-thalassemia major • Heterozygotes: β-thalassemia minor	• Mediterranean countries, Africa, and Southeast Asia • Relative excess of α chains; Hb aggregates and becomes insoluble • Intra- and extramedullary hemolysis • Extramedullary hematopoiesis • Secondary hemochromatosis
HB ELECTROPHORESIS	• α-Thalassemia: normal HbA_2 and HbF • β-Thalassemia minor (trait): elevated HbA_2, HbF	

Increased Destruction (Normal–High Reticulocyte Count)	
Blood loss	Loss rather than destruction Clinical features depend on rate and severity of blood loss Chronic loss better tolerated, regenerate by increasing erythropoiesis Acute blood loss: possible hypovolemia may lead to shock and death Hematocrit may be initially normal because of equal plasma and RBC loss; will decrease as interstitial fluid equilibrates
	Extravascular—premature RBC destruction, hemoglobin (Hb) breakdown, and a compensatory increase in erythropoiesis *Intravascular*—elevated serum and urinary Hb, jaundice, urinary hemosiderin, and decreased circulating haptoglobin. Bile pigment gallstones arise from chronic, not acute hemolysis
Warm hemolytic anemia	IgG Secondary to drugs, malignancy, and SLE
Cold hemolytic anemia	IgM Functions below body temperature in the periphery Associated with mononucleosis, *Mycoplasma* infection, idiopathic hemolytic anemia, and hemolytic anemia associated with lymphoma
Paroxysmal hemolytic anemia	IgG Functions in the periphery
Hereditary spherocytosis *Presplenectomy smear:* spherical cells lacking central pallor and reticulocytosis *Postsplenectomy smear:* more spherocytes and Howell-Jolly bodies	Autosomal dominant defect in spectrin Less pliable; vulnerable to destruction in the spleen Anemia, jaundice, splenomegaly, cholelithiasis Exhibit characteristically increased osmotic fragility Treatment: splenectomy
G6PD deficiency *Smear:* reticulocytosis and Heinz bodies (Hb degradation products)	X-linked deficiency of the enzyme (hexose monophosphate shunt) Decreased regeneration of NADPH, therefore glutathione Older cells unable to tolerate oxidative stress Associated with drugs (e.g., sulfa, quinine, nitrofurantoin), infections (particularly viral), or certain foods (fava beans)
Paroxysmal nocturnal hemoglobinuria	Acquired deficiency of membrane proteins: decay accelerating factor (CD55) and membrane inhibitor of reactive lysis (CD59) Chronic intravascular hemolysis Predisposes to stem cell disorders (e.g., aplastic anemia, acute leukemia) Most frequently die of infection or venous thrombosis Diagnosis: flow cytometry shows absence of CD55 and CD59

* See **Appendix D** for Abnormal Erythryocytes on Peripheral Smear

Sickle Cell Disease		
Incidence	**Key Points**	**Clinical Picture**
• 0.2% of the U.S. African-American population has disease • 8% carry trait	• Substitution of valine for glutamic acid at position 6 of the beta chain • Sickle trait: 40% HbS-sickle in extreme conditions • "Sickle prep" is a blood sample treated with a reducing agent, such as metabisulfite; sickled cells may be seen • Definitive diagnosis made by Hb electrophoresis • HbS aggregates at low oxygen tension; leads to sickling • Heterozygote is protected from *Plasmodium falciparum* malaria	• Microvascular occlusion and hemolysis • Recurrent splenic thrombosis and infarction; autosplenectomy usually by age 5 • Also affects liver, brain, kidney, bones, penis (painful prolonged erection—priapism) • Vaso-occlusive crises ("painful crises") may be triggered by infection, dehydration, acidosis • Aplastic crises: Parvovirus • Functional asplenia: vulnerable to *Salmonella* osteomyelitis and infections with encapsulated organisms, such as *Pneumococcus* • Most patients die before age 30

Agents Used to Treat Anemia		
Class	**Mechanism**	**Indications**
Iron	Needed to form heme, the oxygen-carrying component of hemoglobin	Iron deficiency (microcytic hypochromic) anemia
Vitamin B_{12} (cyanocobalamin, hydroxocobalamin)	Required for DNA synthesis, RBC production, and nervous system function	Pernicious anemia and anemia resulting from gastric resection
Folate	Essential for DNA synthesis and maintenance of normal erythropoiesis	Folic acid deficiency secondary to malabsorption syndrome and dietary insufficiency; macrocytic/megaloblastic anemias
Hydroxyurea (HU)	An antimetabolite that inhibits ribonucleotide reductase HU reactivates HbF synthesis and increases the number of reticulocytes containing HbF in sickle cell patients	Sickle cell anemia, polycythemia vera, and chronic myelogenous leukemia
Erythropoietin (EPO) **Darbepoetin alpha**	EPO is normally produced by the kidney Stimulates RBC production	Used for a variety of anemias, including anemia of renal failure Hypertension a common and severe side effect Erythropoiesis-stimulating agents increase the risk of tumor progression or recurrence; severe caution when used in patients with cancer

White Blood Cells

Leukocytes can be divided into **granulocytes** and **agranulocytes** (mononuclears) based on the presence of cytoplasmic granules.

▶ Granulocytes

Granulocyte Type	Features	Functional Role	Relative Abundance
Neutrophils	• 3–5 nuclear lobes • Contain **azurophilic granules** (lysosomes) • Specific granules contain bactericidal enzymes (e.g., lysozyme)	First cells in **acute inflammation**	54–62% of leukocytes Normal value: 1800–7800/µl
Eosinophils	• Bilobed nucleus • Acidophilic granules contain hydrolytic enzymes and peroxidase	More numerous in the blood during **parasitic infections and allergic diseases**	1–3% of leukocytes Normal value: 0–450/µl
Basophils	Large basophilic and metachromatic granules, which contain proteoglycans, heparin, and histamine. Note that **mast cells are essentially tissue basophils.**	Degranulate in **type I hypersensitivity**, releasing granule contents and producing slow-reacting substance (SRS-A) = leukotrienes LTC_4, LTD_4, LTE_4	1% of leukocytes Normal value: 0–200/µl

▶ Granulopoiesis

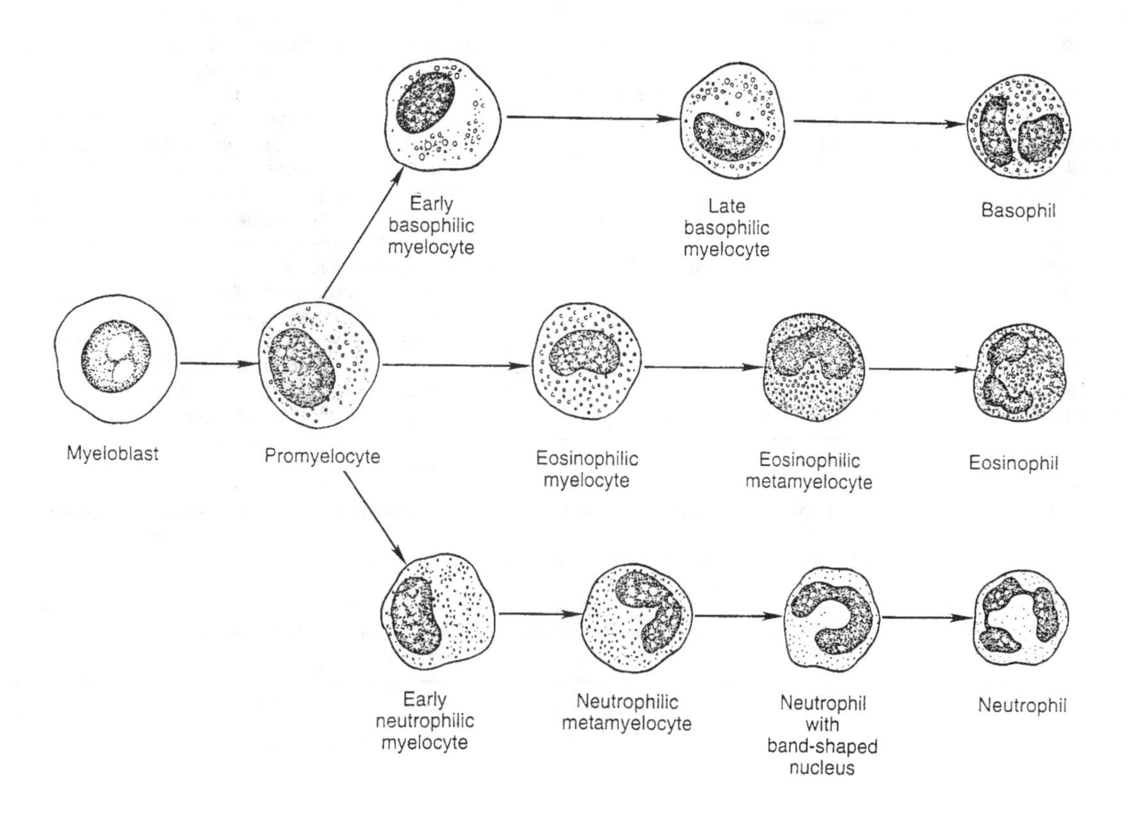

► Agranulocytes (Mononuclears)

Agranulocyte Type	Features	Functional Role	Relative Abundance
Lymphocytes	Dark blue, round nuclei; scant cytoplasm	T cells, B cells, null cells in immune system	25–33% of leukocytes Normal value: 1000–4000/µl
T cells	Differentiate in thymus	Helper and suppressor cells modulate the immune response	—
B cells	Differentiate in bone marrow	Humoral immunity—antibodies produced by plasma cells	—
NK (Natural Killer) cell	Produced in bone marrow	Destroy some tumor cells and some virus-infected cells on which MHC class I antigens are not expressed	10% of total lymphocyte count
Monocytes/macrophages	• Largest peripheral blood cells • Kidney-shaped nuclei, stain lighter than lymphocytes	Monocytes are precursors of tissue macrophages (histiocytes), osteoclasts, alveolar macrophages, and Kupffer cells of the liver	3–7% of leukocytes Normal value: 0–900/µl

Disorders of Leukocytes

► Nonneoplastic White Blood Cell Disorders

Neutropenias: decreased production (most common)	• Megaloblastic anemia • Aplastic anemia • Leukemia/lymphoma • Autoimmune destruction of stem cells	• Lack of innate immune defense • Constitutional symptoms and a high susceptibility to infection, particularly gram-negative septicemia • Poor prognosis: death from overwhelming infection • Infected, necrotic ulcers may occur in mucosa (oral cavity, skin, vagina, anus, gastrointestinal tract) • Granulocyte-macrophage colony-stimulating factor (GM-CSF) and granulocyte-stimulating factor (GSF) are now used to treat postchemotherapy neutropenias
Neutropenias: increased destruction	Splenic sequestration, often immune-mediated (e.g., Felty syndrome)	
Drug-induced neutropenia	Alkylating agents, chloramphenicol, sulfonamides, chlorpromazine, and phenylbutazone	• Usually reversible • Chloramphenicol: dose-related marrow suppression in all patients; aplastic anemia in rare individuals
Polymorphonuclear leukocytosis (most common)	• Acute bacterial infection, tissue necrosis, and "stress" • Increased bands and left shift	• Döhle bodies (round, blue, cytoplasmic inclusions, product of rough endoplasmic reticulum) • Toxic granulations: coarse, dark, granules (lysosomes)
Monocytosis	Tuberculosis, endocarditis, malaria, brucellosis, rickettsiosis	
Lymphocytosis	Tuberculosis, brucellosis, pertussis, viral hepatitis, cytomegalovirus infections, infectious mononucleosis	
Mononucleosis	• Increase in mononuclear cells: lymphocytes and monocytes • Viral infections, EBV, CMV	
Eosinophilic leukocytosis	Neoplasms, allergy, asthma, collagen vascular diseases, parasitic infections, skin rashes	

▶ Hematopoietic Growth Factors

Granulocyte colony-stimulating factor (G-CSF; filgrastim)	A glycoprotein that stimulates the bone marrow to produce granulocytes while promoting their survival and differentiation	Accelerates neutrophil recovery following chemotherapy; used for primary and secondary neutropenia
Granulocyte-monocyte colony-stimulating factor (GM-CSF, sargramostim)	A glycoprotein growth factor for erythroid megakaryocyte and eosinophil precursors. It also enhances the survival and function of circulating granulocytes, monocytes, and eosinophils	

▶ Nonneoplastic Lymph Node Disorders

Nonspecific lymphadenitis	Drugs, toxins, or infection	• In neck following dental or tonsillar infection • In axillary/inguinal regions after infections of the extremities • Enlarged abdominal lymph nodes (mesenteric adenitis) may cause abdominal pain resembling acute appendicitis
Generalized lymphadenopathy	Systemic viral or bacterial infections	• May be a precursor to AIDS • Associated with hyperglobulinemia and normal CD4 lymphocyte counts
Acute lymphadenopathy	• Swollen, red-gray nodes with prominent lymphoid follicles • Older patients have fewer germinal centers than children	
Chronic lymphadenopathy	Most common in axillary and inguinal nodes; nodes large and nontender	
	Microscopic Findings: Three Basic Patterns	
Follicular hyperplasia **B-cell antibody response**	• Large germinal centers, containing mostly B cells, helper T cells, and histiocytes • Seen in bacterial infections or with exposure to new antigens	
Paracortical hyperplasia **T-cell reaction**	• Reactive changes in cortex • Seen with phenytoin use, viral infections, or secondary immune responses	
Sinus histiocytosis	• Lymphatic sinusoids prominent and distended with macrophages • Seen in nodes draining carcinomas or any chronic inflammation	

▶ Neoplastic Leukocyte Disorders

Hodgkin Disease
• Contiguous spread (from one node to the next) with spleen involved before liver • High cure rate, rarely has leukemic component • **Reed-Sternberg (RS)** cells (large, containing large "owl-eyed" nucleoli with a clear halo; abundant cytoplasm) are necessary, but not sufficient, to make diagnosis • Bimodal age distribution (high peak: 15–35; low peak: 50+) in both men and women • Clinical: painless cervical adenopathy +/– constitutional symptoms

Four Variants of Hodgkin Disease	
Lymphocyte predominance	Sea of lymphocytes, few RS cells, variable number of histiocytes, little fibrosis, and no necrosis
Nodular sclerosis	• More common in women • Mediastinal, supraclavicular, and lower cervical nodes • Mixture of lymphocytes, histiocytes, a few eosinophils, plasma cells, and RS cells. Collagen bands create nodular pattern; RS cells called lacunar cells
Mixed cellularity	• Mixture of neutrophils, lymphocytes, eosinophils, plasma cells, and histiocytes • Large number of RS cells
Lymphocyte depletion	• Rare lymphocytes, many RS cells with variable eosinophils, plasma cells, and histiocytes • Diffuse fibrosis may be seen

Worsening Prognosis ↓

► Non-Hodgkin Lymphomas (NHL)

- Lymphadenopathy and hepatosplenomegaly. In 30% of cases, initial involvement extranodal
- Usually discovered in only one chain of nodes—usually cervical, axillary, inguinal, femoral, iliac, or mediastinal
- Patients present with local or generalized lymphadenopathy, abdominal or pharyngeal mass, abdominal pain, or GI bleeding
- Involve lymph nodes or lymphoid tissue in the gut, oropharynx, liver, spleen, and thymus
- **Do not produce Reed-Sternberg cells**, do not spread in contiguity, and frequently have a leukemic or blood-borne phase
- Occur in late 50s (rare and more aggressive in children and young adults)
- Weight loss common sign of disseminated disease
- Common in immunosuppressed patients

Two main categories: **nodular** (better prognosis) and **diffuse**

Staging similar to Hodgkin disease but staging less clinically significant in NHL because prognosis is more affected by histology, and the disease is often disseminated at time of diagnosis

Disease	Characteristics	Pathology
Follicular lymphoma	Median age 60–65; median survival 8–10 years Painless adenopathy t(14;18) translocation brings bcl-2 close to heavy chain immunoglobulin gene May be associated with immunodeficiency states	Follicular or nodular pattern of growth with areas resembling germinal centers but lacking normal germinal center architecture
Well-differentiated lymphocytic lymphoma (diffuse)	Older patients Generalized lymphadenopathy, hepatosplenomegaly Often seeds the blood late in the disease similar to CLL Bone marrow almost always involved Survival: 5–7 years	Lymph nodes replaced by small round lymphocytes with scant cytoplasm, dark nuclei, and rare mitoses
Mixed lymphocytic-histiocytic lymphoma (nodular)	Prognosis fair; remission may be achieved with combination chemotherapy	Cells with atypical lymphocytes and large histiocytes
Lymphoblastic lymphoma (diffuse) Cells similar to ALL	Bimodal—high peak: adolescents/young adults, low peak: 70s; male:female ratio, 2.5:1 Associated with a mediastinal mass, particularly in boys Often express T-cell markers Prognosis uniformly poor; T-cell lymphomas worse	Uniform size, scant cytoplasm, delicate chromatin, and absent nucleoli Nuclear membrane is loculated or convoluted Frequent mitoses
Burkitt undifferentiated lymphoma	Endemic in Africa (mandible or maxilla) and sporadic in the United States (abdomen) Children or young adults Tied to **Epstein-Barr virus (EBV),** especially the African form Leukemic phase is rare; prognosis is fair African Burkitt: translocations (8;14, 2;8, or 8;22) bring c-myc gene close to enhancers of heavy or light chain synthesis in B cells	Sea of intermediate size lymphocytes with lipid-containing vacuoles interspersed with macrophages to produce **"starry sky pattern."**
Mantle cell lymphoma	Median age 60; median survival 3 years t(11;14) translocation links immunoglobin heavy chain to bcl-1 Can present with lymphadenopathy, fever, night sweats, massive splenomegaly, or hepatomegaly	Expansion of the mantle zone surrounding germinal centers with small to medium atypical lymphocytes

▶ Cutaneous T-Cell Lymphomas

Mycosis fungoides	Three phases of skin lesions: inflammation, plaque, and tumor. Epidermal and dermal infiltrates by neoplastic T (CD4) cells with cerebriform nuclei. Nodules and fungating tumors may develop later in the disease. Nodal and visceral dissemination can occur.
Sézary syndrome	Rare chronic disease with progressive, pruritic erythroderma, exfoliation, and lymphadenopathy. "Sézary cells," T cells with cerebriform nuclei (similar to those seen in mycosis fungoides) infiltrate the peripheral blood. May be considered a preterminal phase of mycosis fungoides.

▶ Leukemias

Disease	Characteristics	Pathology
Acute lymphocytic leukemia (ALL)	• 60–70% of cases occur in childhood; peak age 4; rare over 50 • 75% of children are cured; prognosis for adults is very poor • Fatigue, fever, epistaxis, gingival petechiae, ecchymoses 2° to thrombocytopenia; may have subarachnoid or cerebral hemorrhage • Present with lymphadenopathy, bone pain, hepatosplenomegaly • Most likely leukemia to involve CNS • Prognosis: death often from infection or bleed • Most cells pre-B cells; T-cell variants occur, usually affecting boys and causing a thymic mass that may compress the trachea	Smear: lymphoblasts are prominent; mature WBCs rare **CD10 (CALLA)** is the diagnostic surface marker; terminal deoxynucleotidyl transferase (TDT) positive in both B-cell and T-cell ALL and negative in AML
Acute myelogenous leukemia (AML)	20% of acute leukemia in children, most common acute leukemia in adults Signs and symptoms resemble ALL, except usually also present with lymphadenopathy or splenomegaly AML: t(15;17); acute promyelocytic leukemia: t(1;12)	Primary cell type variable; see the French, American, and British (FAB) Classification of Myelogenous Leukemias, page 475.
Chronic myelogenous leukemia (CML)	• Middle age but may occur in children/young adults • Fatigue, fever, night sweats, and weight loss • Splenomegaly (up to 5 kg) giving abdominal discomfort • Variable remission period, may develop blast crisis • Two-thirds convert to AML; one-third to B-cell ALL • **Philadelphia chromosome (Ph1), t(9;22):** ***bcr:abl*** translocation is pathognomonic; present in 95% of cases • Prognosis in CML is worse in Ph1-negative patients	• Marked leukocytosis • Low-to-absent leukocyte alkaline phosphatase • Elevated serum vitamin B_{12} and vitamin B_{12}–binding proteins • High uric acid levels (due to rapid cell turnover)
Chronic lymphocytic leukemia (CLL)	• Over 60 years of age • Asymptomatic or fatigue and weight loss; lymphadenopathy and hepatosplenomegaly later findings • Higher incidence of visceral malignancy • Median survival with treatment is 5 years but varies widely; prognostic factor is extent of disease	Lymph node histology indistinguishable from diffuse, well-differentiated lymphocytic lymphoma Classic cell: CD5 B cell Cells do not undergo apoptosis
Hairy cell leukemia	• Rare disease; cells express tartrate-resistant acid phosphatase • Present with hepatosplenomegaly; pancytopenia common • Prognosis: may now be cured with 2-chloro-deoxyadenosine (2CdA), an apoptosis inducer	Leukemic cells have "hair-like" cytoplasmic projections visible on phase-contrast microscopy Cells express some B-cell antigens
Adult T-cell leukemia/ lymphoma (CD4 T cell)	• Endemic in Japan • Lymphadenopathy, hepatosplenomegaly, skin involvement, and hypercalcemia • Poor prognosis; however, many infected patients do not progress to disease	Caused by human T-cell leukemia/lymphoma virus (HTLV1); exposure to the virus may be decades earlier
Myelodysplastic syndromes	Proliferative stem cell disorders—maturation defect Gray zone between benign proliferation and frank acute leukemias One-third of these patients later develop frank acute myelocytic leukemia	Presents as pancytopenia in elderly patients

* See **Appendix D** for Abnormalities of White Blood Cells and Platelets on Peripheral Smear

Leukemia Clues	
Children	ALL
Myeloblasts	AML
Auer rods	AML, promyelocytic
DIC	Promyelocytic
Elderly	CLL
Splenomegaly	CML
Philadelphia chromosome	CML
Tartrate-resistant acid phosphatase	Hairy cell
HTLV-1	Adult T cell

▶ French, American, and British (FAB) Classification of Myelogenous Leukemias

M0	Undifferentiated	—
M1	Myeloblasts without maturation	Myeloblasts have round-oval nuclei Auer rods
M2	Granulocyte maturation	—
M3	Promyelocytic	Auer rods DIC
M4	Mixed myeloid and monocytic	Features of both myelocytes and monocytes
M5	Monoblastic or monocytic	—
M6	Erythroid differentiation	Di Guglielmo disease Atypical multinucleated RBC precursors Usually converts to AML
M7	Megakaryocytic differentiation	—

▶ Plasma Cell Dyscrasias

Polyclonal hypergammaglobulinemia	1–2 weeks after an antigen stimulus (e.g., bacterial infection); also associated with granulomatous disease, connective tissue disorders, and liver failure Elevated serum globulins, elevated ESR Polyclonal Bence-Jones proteins in serum or urine Hyperviscosity of blood may lead to sludging and rouleaux formation with subsequent thrombosis, hemorrhage, renal impairment, and right-sided congestive heart failure
Lymphoplasmacytic lymphoma (Waldenström macroglobulinemia)	Age 60–70 years in both men and women Monoclonal IgM resembles lymphocytic lymphoma with M-protein spike on serum protein electrophoresis Symptoms due to hypergammaglobulinemia and tumorous infiltration Hepatosplenomegaly, lymphadenopathy, bone pain, and hyperviscosity Blindness and priapism due to hyperviscosity may be seen 2–5-year survival rate with chemotherapy
Monoclonal gammopathy of undetermined significance (MGUS)	Asymptomatic M-protein spike on serum electrophoresis Prognosis: initially thought benign, but approximately 2% may later develop myeloma, lymphoma, amyloidosis, or Waldenström macroglobulinemia
Multiple myeloma	Peak incidence is 50–60 years old; male = female Multifocal plasma cell neoplasms in the bone marrow, occasionally soft tissues Monoclonal immunoglobulin (IgG) Signs and symptoms result from excess abnormal immunoglobulins (causing hyperviscosity) and from infiltration of various organs by neoplastic plasma cells Proteinuria may contribute to progressive renal failure Infiltration of bone with plasma cell neoplasms may lead to bone pain and hypercalcemia Over 99% of patients have elevated levels of serum immunoglobulins or urine Bence-Jones proteins, or both Serum protein electrophoresis (SPEP) shows homogeneous peak or "spike" Marrow is infiltrated with plasma cells (usually over 30%) in various stages of maturation, called "myeloma cells"; contain cytoplasmic inclusions (acidophilic aggregates of immunoglobulin) called Russell bodies Multiple osteolytic lesions throughout the skeleton; appear as "punched-out" defects on x-ray Kidney: protein casts in distal tubules Prognosis: less than 2-year survival without therapy; death usually results from infection, bleeding, or renal failure **(Bence-Jones proteins)**

Hematologic Changes Associated with Infectious Disease

► Changes in Blood Cells

Signs and Symptoms	Case Vignette/ Key Clues	Most Common Causative Agents	Pathogenesis	Diagnosis	Treatment
Anemia	Megaloblastic Ingestion of raw fish	*Diphyllobothrium latum*	Parasite absorbs B_{12}	Operculated eggs in stool	Niclosamide
	Normocytic	Chronic infections	Bacteria chelate iron	Culture, Gram stain	Depends on agent
	Microcytic and hypochromic (iron-deficiency anemia)	*Ancylostoma*, *Necator* *Trichuris*	Hookworms suck blood; trichuris damages mucosa	Golden brown, oval eggs; eggs with bipolar plugs	Mebendazole
Patient with cyclic or irregular fever, ↓ hemoglobin and hematocrit	Travel to tropics, parasites in RBCs	*Plasmodium* spp.	Parasite lyses RBC Autoimmune RBC destruction	Rings/trophozoites in blood film	Chloroquine, etc. (considerable drug resistance), followed by primaquine if *P. vivax* or *P. ovale*
↓ CD4 cell count	Lymphadenopathy Opportunistic infections	HIV	Virus infects and destroys CD4 ⊕ T cells, and macrophages	ELISA, Western blot	NRTIs, NNRTIs, protease inhibitors, fusion inhibitors, CCR5 antagonists, integrase inhibitors
↑ PMNs (neutrophilia)	—	Generally found in many extracellular bacterial infections	*N*-formyl methionyl peptides are chemotactic for PMNs	Culture, Gram stain	Depends on agent
↑ eosinophils (eosinophilia)	—	Allergy	ECF-A released by mast cells attracts eosinophils	Skin testing: wheal and flare	Antihistamines
		Helminths during migrations	Parasites release allergens	Depends on agent	Depends on agent
↑ monocytes and/or lymphocytes	—	Intracellular organisms: viruses, *Listeria*, *Legionella*, *Leishmania*, *Toxoplasma*	Intracellular organisms elicit TH1 cells and CMI	Depends on agent	Depends on agent
Above plus fever, fatigue, lymphadenopathy, myalgia, headache	Infectious mononucleosis Downey type II cells (reactive T cells), sore throat, lymphadenopathy, young adult	Epstein-Barr virus	Virus infects B lymphocytes via CD21; CTLs respond to kill virus-infected cells	Monospot ⊕ Complete blood count, 70% lymphocytosis	Supportive
	Heterophile ⊖	CMV	Virus infects fibroblasts; CTLs respond to kill virus-infected cells	Monospot ⊖ Virus culture	Ganciclovir (severe cases)
Lymphocytosis with hacking cough	Unvaccinated child, hypoglycemic	*Bordetella pertussis*	Tracheal cytotoxin, fimbrial antigen, endotoxin, ↓ chemokine receptors	Gram ⊖ rod Culture Bordet-Gengou agar or serology	Erythromycin, antitoxin

Section III

Appendices

Appendix A

Essential Equations

► Biostatistics

<table>
<tr><th>Name</th><th>Equations</th><th>Reasons for Use</th><th>Page</th></tr>
<tr><td>Sample 2×2 table: test versus disease</td><td><table><tr><td></td><td></td><td colspan="2">Disease</td></tr><tr><td></td><td></td><td>Present</td><td>Absent</td></tr><tr><td rowspan="2">Test Results</td><td>Positive</td><td>TP a</td><td>FP b</td></tr><tr><td>Negative</td><td>FN c</td><td>TN d</td></tr></table></td><td>Note: You should know the meanings of a, b, c, d. Data may be rotated in presentation.
TP: true positive
FP: false positive
TN: true negative
FN: false negative</td><td>34</td></tr>
<tr><td>Sensitivity</td><td>$\text{Sensitivity} = \frac{a}{a + c} = \frac{TP}{TP + FN}$</td><td>Probability that test correctly identifies people with the disease</td><td>34</td></tr>
<tr><td>Specificity</td><td>$\text{Specificity} = \frac{d}{d + b} = \frac{TN}{TN + FP}$</td><td>Probability that test correctly identifies people without disease</td><td>34</td></tr>
<tr><td>Positive predictive value</td><td>$PPV = \frac{a}{a + b} = \frac{TP}{TP + FP}$</td><td>Probability that a person with a positive test is a true positive</td><td>34</td></tr>
<tr><td>Negative predictive value</td><td>$NPV = \frac{d}{d + c} = \frac{TN}{TN + FN}$</td><td>Probability that a person with a negative test is a true negative</td><td>34</td></tr>
<tr><td>Accuracy</td><td>$A = \frac{a + d}{a + d + b + c} = \frac{TP + TN}{TP + TN + FP + FN}$</td><td>Represents the true value of the measured attribute</td><td>34</td></tr>
<tr><td>Sample 2×2 table: risk factor versus disease</td><td><table><tr><td></td><td></td><td colspan="2">Disease</td></tr><tr><td></td><td></td><td>Present</td><td>Absent</td></tr><tr><td rowspan="2">Risk Factor</td><td>Positive</td><td>a</td><td>b</td></tr><tr><td>Negative</td><td>c</td><td>d</td></tr></table></td><td>Note: You should know the meanings of a, b, c, d. The data may be rotated in presentation.</td><td>34</td></tr>
<tr><td>Odds ratio</td><td>$\text{Odds ratio} = \frac{a/c}{b/d} = \frac{ad}{bc}$</td><td>Looks at odds of getting a disease with exposure to a risk factor versus nonexposure to risk factor</td><td>35</td></tr>
<tr><td>Relative risk</td><td>$RR = \frac{a/(a + b)}{c/(c + d)}$</td><td>"How much more likely?"
Incidence rate of exposed group ÷ incidence rate of unexposed group</td><td>35</td></tr>
<tr><td>Attributable risk</td><td>$AR = \frac{a}{a + b} = \frac{c}{c + d}$</td><td>"How many more cases in one group?" Incidence rate in exposed group – incidence rate in unexposed group</td><td>35</td></tr>
</table>

► Genetics

Name	Equation	Reason for Use	Page
Hardy-Weinberg equilibrium	Allele frequencies: $p + q = 1$ Genotypic frequencies: $p^2 + 2pq + q^2 = 1$ p^2 = frequency of AA $2pq$ = frequency of Aa q^2 = frequency of aa	Used in population genetics. Assume no mutations, no selection against genotype, no migration or immigration of the population, random mating	77

► Cell Biology and Neurophysiology

Name	Equation	Reason for Use	Page
Nernst equation	$E_x = \frac{60\text{ mV}}{Z} \log_{10} \frac{[X]_o}{[X]_i}$ Z = electrical charge of ion; $[X]_o$ = extracellular ion concentration; $[X]_i$ = intracellular ion concentration	Calculation of equilibrium potential; effect of ion concentration on membrane potential	151
Estimated fluid volumes in liters	$TBW = 0.6 \times BW$ $ICF = 0.4 \times BW$ $ECF = 0.2 \times BW$ TBW = total body water; ICF = intracellular fluid volume; ECF = extracellular fluid volume; BW = body weight in kg	Assumes normal subject who is normally hydrated; starting point for evaluating changes of body fluids for diagnostic purposes or effects of administration of fluids	149

► Pharmacokinetics

Name	Equation	Reason for Use	Page
Volume of distribution	$V_d = \frac{\text{Dose}}{C^0}$ C^0 = plasma concentration at time 0	Used to estimate the fluid volume into which a drug has distributed For 70 kg person: plasma vol. = 3 L; blood = 5 L; ECF = 12–14 L; TBW = 40–42 L	166
Clearance	$Cl = \frac{\text{Rate of drug elimination}}{\text{Plasma drug concentration}}$ $Cl = k_e \times V_d$ k_e = elimination constant	Used to measure the efficiency with which a drug is removed from the body	166
Half-life	$t_{1/2} = \frac{0.7 \times Vd}{Cl}$	Used to determine the time to decrease drug plasma concentration by 1/2	166
Bioavailability (F)	$F = \frac{AUC_{PO}}{AUC_{IV}}$ AUC = area under the curve	Used to estimate the fraction of administered drug to reach the systemic circulation	166
Maintenance dose	$MD = \frac{Cl \times C_P}{F}$	Used to calculate the dose to maintain a relatively constant plasma concentration	167

(Continued)

► Pharmacokinetics *(Cont'd.)*

Name	Equation	Reason for Use	Page
Loading dose	$LD = \frac{V_d \times C_P}{F}$	Used to calculate drug dose to quickly achieve therapeutic levels	167
Therapeutic index	$TI = \frac{TD_{50}}{ED_{50}}$ or $\frac{LD_{50}}{ED_{50}}$ ED_{50}, TD_{50}, and LD_{50} are the effective, toxic, and lethal doses in 50% of the studied population	Used to determine the safety of a drug ↑ TI: safe drug ↓ TI: unsafe drug	168

► Cardiovascular

Name	Equation	Reason for Use	Page
Stroke volume	SV = EDV – ESV SV = stroke volume; EDV = end diastolic ventricular volume; ESV = end systolic ventricular volume	Basic formula for calculation of stroke volume from cardiac volumes	242
Ejection fraction	EF = SV/EDV EF = ejection fraction (multiply by 100% to turn into percentage)	Ejection fraction is a common measure of cardiac contractility.	242
Cardiac output	CO = SV × HR CO = cardiac output; HR = heart rate	Calculation of cardiac output; used to explain or predict effects of changes of pumping ability or heart rate	242
Cardiac output (Fick method)	$CO = \dot{V}O_2/(Ca - Cv)$ $\dot{V}O_2$ = oxygen consumption, Ca = arterial oxygen content, Cv = venous oxygen content	Used to measure cardiac output; most accurate if Ca is pulmonary venous and Cv is pulmonary arterial	249
Resistance	$R = 8\,\eta l/\pi r^4$ η = viscosity, l = length, r = radius	Hydraulic resistance equation shows that dominant control of resistance is vessel radius.	248
Pressure gradient	$P_i - P_o = Q \times R$ P_i = input pressure P_o = output P; Q = flow, R = resistance	Predicts pressure gradient that results from fluid flow; can be rearranged to solve for R or Q	248
Total peripheral resistance (systemic vascular resistance)	TPR = (MAP – RAP)/CO TPR = total peripheral resistance = systemic vascular resistance (Can ignore RAP in a normal person because RAP is usually small; it is important in disease, e.g., CHF)	Rearrangement of above; use to predict effects of changing input pressure or resistance on flow through a tissue; can also solve for resistance to evaluate its contribution to pathology or to compensatory mechanisms	248
Mean arterial pressure	MAP = diastolic P + 1/3(pulse pressure) Pulse pressure = systolic P – diastolic P MAP = 1/3 systolic P + 2/3 diastolic P	Common method to calculate approximate mean arterial pressure	248
Series circuit	$R_T = R_1 + R_2 + R_3 \ldots + R_n$ R_T = total resistance	Adding resistor in series increases total resistance	248
Parallel circuit	$1/R_T = 1/R_1 + 1/R_2 + 1/R_3 \ldots + 1/R_n$	Adding in parallel decreases total resistance	248

(Continued)

▶ Cardiovascular *(Cont'd.)*

Name	Equation	Reason for Use	Page
Compliance and elasticity	$C = \Delta V/\Delta P$ C = compliance; ΔV = change of volume; ΔP= change of pressure $E = \Delta P/\Delta V$ E = elasticity	Compliance measures flexibility of chamber; elasticity, its inverse, increases if it is stiff. Applied to heart in cardiomyopathy, to vessels with decreased compliance leading to isolated systolic hypertension (↑ P_S but ↓ P_D)	248
Wall tension (LaPlace law)	$T = P \times r$ T = wall tension P = transmural pressure; r = radius	LaPlace relationship for thin-walled chamber; shows that increased radius or transmural pressure increases wall tension and risk of rupture	251
Net filtration pressure (Starling equation)	Filtration = $K_f [(P_{cap} + \pi_{tissue}) - (\pi_{cap} + P_{tissue})]$ K_f = ultrafiltration coefficient P_{cap} = capillary blood pressure P_{tissue} = interstitial fluid pressure π_{cap} = plasma (capillary) protein oncotic pressure π_{tissue} = interstitial protein oncotic pressure	Used to evaluate balance of forces between filtration and reabsorption. Positive value is net filtration; negative is net reabsorption.	250

▶ Respiratory

Name	Equation	Reason for Use	Page
Compliance and elasticity	$C = \Delta V/\Delta P$ $E = \Delta P/\Delta V$ Specific compliance = C/V (corrects for body size)	Lung compliance increases in some diseases, such as emphysema; decreases in fibrosis, premature newborns, etc.	278
Total (minute) ventilation	$\dot{V} = V_T \times n$ $\dot{V}$ = minute ventilation; V_T = tidal volume; n = frequency	Measures total air movement into and out of lungs per minute	277
Alveolar ventilation	$\dot{V}_A = (V_T - V_D) \times n$ $\dot{V}_A$ = alveolar ventilation; VT = tidal volume; V_D = dead space	Volume of air per minute that participates in gas exchange; the functional ventilation	278
Alveolar ventilation (Bohr method)	$\dot{V}_A = \frac{\dot{V}_{CO_2}}{P_{CO_2}} \times K$ $\dot{V}_{CO_2}$ = CO_2 production (mL/min) P_{CO_2} = partial pressure of CO_2 (mm Hg)	Bohr method to measure alveolar ventilation ↑ V_A → ↓ $Paco_2$ ↓ V_A → ↑ $Paco_2$	278
Physiologic dead space ratio	$\frac{V_D}{V_T} = \frac{Pa_{CO_2} - PE_{CO_2}}{Pa_{CO_2}}$ Pa_{CO_2} = arterial CO_2, PE_{CO_2} = CO_2 in mixed expired gas	Measurement of physiologic dead space takes into account anatomic dead space and units with poor gas exchange.	278
Alveolar gas equation	$PAO_2 = PIO_2 - \frac{PA_{CO_2}}{R}$ *Note:* $PIO_2 = FIO_2 \times (P_B - PH_2O)$; FIO_2 = 0.21 on room air and 1.0 on pure O_2; P_B = 760 mm Hg, PH_2O = 47 mm Hg; PIO_2 is 150 mm Hg at sea level	Calculation of alveolar oxygen for evaluation of alveolar-arterial gradient; allows differential diagnosis of arterial hypoxemia, e.g., diffusion barrier versus hypoventilation	282

(Continued)

► Respiratory *(Cont'd.)*

Name	Equation	Reason for Use	Page
Forced vital capacity test	FEV_1/FVC FEV_1 = volume of air expelled in first second of forced expiration during forced vital capacity (FVC) maneuver	Clinical measure of airway resistance; reduced value indicates obstructive pulmonary disease; no change or increase in restrictive disease	281
Airway resistance	$R = \Delta P/Q = 8\,\eta l/\pi r^4$ Note: same factors as in blood flow, above.	Increased airway resistance is the hallmark of obstructive pulmonary disease.	279
LaPlace law for surface tension	$P = \frac{2T}{r}$ P = pressure due to surface tension; T = surface tension; r = radius	Shows that collapsing pressure is higher in smaller alveoli; surfactant ↓ tension	279
Fick's Law of Diffusion	$V_{gas} \propto D(P_1 - P_2) \times A/T$ D = diffusion constant; A = surface area; T = thickness	Identifies factors that change physiologically and pathologically to alter diffusion	283

► Renal and Acid/Base

Name	Equation	Reason for Use	Page
Filtered load (FL)	$FL = GFR \times [solute]_{plasma}$ $[solute]_{plasma}$ = free, not bound solute	Filtered load is the critical parameter in formation of urine; comparison with excretion allows interpretation of renal transport.	304
Excretion	$E_x = \dot{V} \times [X]_{urine}$ = FL + transport $\dot{V}$ = urine flow; $[X]_{urine}$ = solute concentration in urine; transport is either secretion or reabsorption. If excretion > FL, there is net secretion. If excretion < FL, there is net reabsorption. If excretion = FL, there is no net transport.	Mass balance concept—urine formation involves up to three processes: filtration, secretion, and reabsorption. Excretion is the final result.	304
Clearance	$Clearance = \dot{V} \times \frac{[X]_{urine}}{[X]_{plasma}}$ • To calculate **GFR**: use **inulin**, **creatinine**, mannitol • To calculate **RPF**: use renal clearance of **para-aminohippuric acid (PAH)**	Relates excretion to plasma solute concentration; essential formula used in many ways	305
Renal blood flow	$RBF = \frac{RPF}{1 - \text{hematocrit}}$	Used to determine the volume of blood/plasma delivered to the kidneys per unit time	305
Free water clearance	$CH_2O = \dot{V} - C_{osm}$ C_{osm} = osmolar clearance; use urine osmolarity and plasma osmolarity in clearance equation	Evaluates balance between excretion of water and solutes; determines if kidneys are appropriately controlling solutes and water. Positive C_{osm}: hypotonic urine; negative C_{osm}: hypertonic urine.	305
Glomerular filtration rate (GFR)	$GFR = K_f[(P_{cap} + \pi_{BC}) - (\pi_{cap} + P_{BC})]$ (Same forces as in cardiovascular section, but specific to Bowman's capsule and glomerular capillaries).	Calculation of GFR from Starling forces: illustrates importance of glomerular capillary pressure and oncotic pressures.	307

(Continued)

▶ Renal and Acid/Base *(Cont'd.)*

Name	Equation	Reason for Use	Page
Filtration fraction (FF)	$FF = \frac{GFR}{RPF}$	Use for differential diagnosis of reduced GFR states.	307
Net acid excretion	$NAE = \dot{V} \times [NH_4^+]_u + \dot{V} \times [TA]_u - \dot{V} \times [HCO_3^-]$ TA = titratable acid (usually H_2PO_4)	Net acid excretion calculates the amount of acid excreted in the urine; more accurate but more difficult to measure than urinary anion gap.	319
Henderson-Hasselbalch equation	$pH = pKa + \log \frac{[base]}{[acid]}$ pKa = dissociation constant The main form used is: $pH = 6.1 \log \frac{Hco_3^-}{0.03\ Pco_2}$	Calculates pH of solutions. Major physiologic use is to calculate pH in response to CO_2 and bicarbonate ion.	321
Plasma anion gap	$PAG = [Na^+] - ([Cl^-] + [HCO_3^-])$ Normal = 12 ± 2 (Concentrations are generally in plasma, but for bicarbonate specifically refers to arterial blood sample.) High PAG means excess acid in body.	Plasma anion gap measures amount of nonvolatile acid in blood to determine cause of (not whether they have) metabolic acidosis. Commonly called "anion gap" acidosis.	324
Urinary anion gap	$UAG = [Na^+] + [K^+] - [Cl^-]$ Negative UAG: kidneys are excreting acid Positive UAG: kidneys are excreting base	Urinary anion gap calculates amount of acid excreted; used to determine contribution of kidneys to acid/base status. Only use UAG in normal PAG, hyperchloremic disorders.	324

Appendix B

	REFERENCE RANGE	SI REFERENCE INTERVALS
BLOOD, PLASMA, SERUM		
* Alanine aminotransferase (ALT, GPT at 30°C)	8-20 U/L	8-20 U/L
Amylase, serum	25-125 U/L	25-125 U/L
* Aspartate aminotransferase (AST, GOT at 30°C)	8-20 U/L	8-20 U/L
Bilirubin, serum (adult) Total // Direct	0.1-1.0 mg/dL // 0.0-0.3 mg/dL	2-17 µmol/L // 0-5 µmol/L
* Calcium, serum (Total)	8.4-10.2 mg/dL	2.1-2.8 mmol/L
* Cholesterol, serum	140-250 mg/dL	3.6-6.5 mmol/L
Cortisol, serum	0800 h: 5-23 µg/dL // 1600 h: 3-15 µg/dL	138-635 nmol/L // 82-413 nmol/L
	2000 h: 50% of 0800 h	Fraction of 0800 h: ≤ 0.50
Creatine kinase, serum (at 30°C) ambulatory	Male: 25-90 U/L	25-90 U/L
	Female: 10-70 U/L	10-70 U/L
* Creatinine, serum	0.6-1.2 mg/dL	53-106 µmol/L
Electrolytes, serum		
Sodium	135-147 mEq/L	135-147 mmol/L
Chloride	95-105 mEq/L	95-105 mmol/L
* Potassium	3.5-5.0 mEq/L	3.5-5.0 mmol/L
Bicarbonate	22-28 mEq/L	22-28 mmol/L
Estriol (E_3) total, serum (in pregnancy)		
24-28 weeks // 32-36 weeks	30-170 ng/mL // 60-280 ng/mL	104-590 // 208-970 nmol/L
28-32 weeks // 36-40 weeks	40-220 ng/mL // 80-350 ng/mL	140-760 // 280-1210 nmol/L
Ferritin, serum	Male: 15-200 ng/mL	15-200 µg/L
	Female: 12-150 ng/mL	12-150 µg/L
Follicle-stimulating hormone, serum/plasma	Male: 4-25 mIU/mL	4-25 U/L
	Female: premenopause 4-30 mIU/mL	4-30 U/L
	midcycle peak 10-90 mIU/mL	10-90 U/L
	ostmenopause 40-250 mIU/mL	40-250 U/L
Gases, arterial blood (room air)		
pO_2	75-105 mm Hg	10.0-14.0 kPa
pCO_2	33-44 mm Hg	4.4-5.9 kPa
pH	7.35-7.45	$[H^+]$ 36-44 nmol/L
Glucose, serum	Fasting: 70-110 mg/dL	3.8-6.1 mmol/L
	2-h postprandial: < 120 mg/dL	< 6.6 mmol/L
Growth hormone – arginine stimulation	Fasting: < 5 ng/mL	< 5 µg/L
	provocative stimuli: > 7 ng/mL	> 7 µg/L
Immunoglobulins, serum		
IgA	76-390 mg/dL	0.76-3.90 g/L
IgE	0-380 IU/mL	0-380 kIU/mL
IgG	650-1500 mg/dL	6.5-15 g/L
IgM	40-345 mg/dL	0.4-3.45 g/L
Iron	50-170 µg/dL	9-30 µmol/L
Lactate dehydrogenase (L → P, 30°C)	45-90 U/L	45-90 U/L
Luteinizing hormone, serum/plasma	Male: 6-23 mIU/mL	6-23 U/L
	Female: follicular phase 5-30 mIU/mL	5-30 U/L
	midcycle 75-150 mIU/mL	75-150 U/L
	postmenopause 30-200 mIU/mL	30-200 U/L
Osmolality, serum	275-295 mOsmol/kg	275-295 mOsmol/kg
Parathyroid hormone, serum, N-terminal	230-630 pg/mL	230-630 ng/L
* Phosphatase (alkaline), serum (p-NPP at 30°C)	20-70 U/L	20-70 U/L
* Phosphorus (inorganic), serum	3.0-4.5 mg/dL	1.0-1.5 mmol/L
Prolactin, serum (hPRL)	< 20 ng/mL	< 20 µg/L
* Proteins, serum		
Total (recumbent)	6.0-7.8 g/dL	60-78 g/L
Albumin	3.5-5.5 g/dL	35-55 g/L
Globulins	2.3-3.5 g/dL	23-35 g/L
Thyroid-stimulating hormone, serum or plasma	0.5-5.0 µU/mL	0.5-5.0 mU/L
Thyroidal iodine (^{123}I) uptake	8-30% of administered dose/24 h	0.08-0.30/24 h
Thyroxine (T_4), serum	5-12 µg/dL	64-155 nmol/L
Triglycerides, serum	35-160 mg/dL	0.4-1.81 mmol/L
Triiodothyronine (T_3), serum (RIA)	115-190 ng/dL	1.8-2.9 nmol/L
Triiodothyronine (T_3), resin uptake	25-35%	0.25-0.35
* Urea nitrogen, serum (BUN)	7-18 mg/dL	1.2-3.0 mmol urea/L
* Uric acid, serum	3.0-8.2 mg/dL	0.18-0.48 mmol/L

(*) Included in the Biochemical Profile (SMA-12)

	REFERENCE RANGE	SI REFERENCE INTERVALS
CEREBROSPINAL FLUID		
Cell count	0-5 cells/mm^3	0-5 x 10^6/L
Chloride	118-132 mmol/L	118-132 mmol/L
Gamma globulin	3-12% total proteins	0.03-0.12
Glucose	40-70 mg/dL	2.2-3.9 mmol/L
Pressure	70-180 mm H_2O	70-180 mm H_2O
Proteins, total	< 40 mg/dL	< 0.40 g/L
HEMATOLOGIC		
Bleeding time (template)	2-7 minutes	2-7 minutes
Erythrocyte count	Male: 4.3-5.9 million/mm^3	4.3-5.9 x 10^{12}/L
	Female: 3.5-5.5 million/mm^3	3.5-5.5 x 10^{12}/L
Hematocrit	Male: 41-53%	0.41-0.53
	Female: 36-46%	0.36-0.46
Hemoglobin, blood	Male: 13.5-17.5 g/dL	2.09-2.71 mmol/L
	Female: 12.0-16.0 g/dL	1.86-2.48 mmol/L
Hemoglobin, plasma	1-4 mg/dL	0.16-0.62 mol/L
Leukocyte count and differential		
Leukocyte count	4500-11,000/mm^3	4.5-11.0 x 10^9/L
Segmented neutrophils	54-62%	0.54-0.62
Band forms	3-5%	0.03-0.05
Eosinophils	1-3%	0.01-0.03
Basophils	0-0.75%	0-0.0075
Lymphocytes	25-33%	0.25-0.33
Monocytes	3-7%	0.03-0.07
Mean corpuscular hemoglobin	25.4-34.6 pg/cell	0.39-0.54 fmol/cell
Mean corpuscular hemoglobin concentration	31-36% Hb/cell	4.81-5.58 mmol Hb/L
Mean corpuscular volume	80-100 m^3	80-100 fl
Partial thromboplastin time (nonactivated)	60-85 seconds	60-85 seconds
Platelet count	150,000-400,000/mm^3	150-400 x 10^9/L
Prothrombin time	11-15 seconds	11-15 seconds
Reticulocyte count	0.5-1.5% of red cells	0.005-0.015
Sedimentation rate, erythrocyte (Westergren)	Male: 0-15 mm/h	0-15 mm/h
	Female: 0-20 mm/h	0-20 mm/h
Thrombin time	< 2 seconds deviation from control	< 2 seconds deviation from control
Volume		
Plasma	Male: 25-43 mL/kg	0.025-0.043 L/kg
	Female: 28-45 mL/kg	0.028-0.045 L/kg
Red cell	Male: 20-36 mL/kg	0.020-0.036 L/kg
	Female: 19-31 mL/kg	0.019-0.031 L/kg
SWEAT		
Chloride	0-35 mmol/L	0-35 mmol/L
URINE		
Calcium	100-300 mg/24 h	2.5-7.5 mmol/24 h
Chloride	Varies with intake	Varies with intake
Creatinine clearance	Male: 97-137 mL/min	
	Female: 88-128 mL/min	
Estriol, total (in pregnancy)		
30 weeks	6-18 mg/24 h	21-62 mol/24 h
35 weeks	9-28 mg/24 h	31-97 mol/24 h
40 weeks	13-42 mg/24 h	45-146 mol/24 h
17-Hydroxycorticosteroids	Male: 3.0-10.0 mg/24 h	8.2-27.6 mol/24 h
	Female: 2.0-8.0 mg/24 h	5.5-22.0 mol/24 h
17-Ketosteroids, total	Male: 8-20 mg/24 h	28-70 mol/24 h
	Female: 6-15 mg/24 h	21-52 mol/24 h
Osmolality	50-1400 mOsmol/kg	
Oxalate	8-40 g/mL	90-445 mol/L
Potassium	Varies with diet	Varies with diet
Proteins, total	< 150 mg/24 h	< 0.15 g/24 h
Sodium	Varies with diet	Varies with diet
Uric acid	Varies with diet	Varies with diet

Essential Diseases and Findings

► Essential Eponyms (Diseases and Findings)

Name	Description
Addison disease	Primary adrenocortical insufficiency
Albright syndrome	Young girls with short stature, polyostotic fibrous dysplasia, precocious puberty, café-au-lait spots
Alport syndrome	Progressive hereditary nephritis with sensorineural deafness
Argyll-Robertson pupil	Small, irregular pupils that react poorly to light in neurosyphilis (accommodation is preserved)
Arnold-Chiari malformation	Congenital herniation of cerebellar tonsils and vermis through the foramen magnum; may compress medulla or cervical cord
Aschoff bodies	Painless nodules in rheumatic fever
Auer rods	Intracytoplasmic inclusions in acute myelogenous leukemia
Babinski sign	Upward moving great toe when sole stroked; indicates upper motor neuron lesion
Baker's cyst	Popliteal fossa cyst in rheumatoid arthritis
Bartter syndrome	Hypokalemia, metabolic alkalosis, elevated renin and aldosterone, normal to low blood pressure
Becker muscular dystrophy	Less severe than Duchenne, also due to defective dystrophin
Bell's palsy	Facial paralysis due to lower motor neuron CN VII palsy
Bence Jones protein	Kappa or lambda immunoglobin light chains in urine of patients with multiple myeloma or Waldenström macroglobulinemia
Berger disease	IgA nephropathy; most common form of primary glomerulonephritis
Bernard-Soulier disease	Thrombocytopenia, large platelets; defect in platelet adhesion
Birbeck granules	Intracellular "tennis racket"–shaped structures in histiocytosis X (eosinophilic granuloma)
Bouchard's nodes	PIP swelling in osteoarthritis secondary to osteophytes
Brushfield spots	Ring of iris spots in Down syndrome
Bruton disease	X-linked agammaglobulinemia; mature B cells absent
Budd-Chiari syndrome	Posthepatic venous thrombosis causing occlusion of hepatic vein or inferior vena cava
Buerger disease	Small/medium artery vasculitis, especially in young male smokers
Burkitt lymphoma	EBV-associated lymphoma with 8:14 translocation (starry sky appearance)
Burton's lines	Blue discoloration of gums in lead poisoning

(Continued)

► Essential Eponyms (Diseases and Findings) *(Cont'd.)*

Name	Description
Caisson disease	Gas emboli in divers
Call-Exner bodies	Small spaces with eosinophilic material in granulosa-theca cell tumor of ovary
Chagas disease	Infection with *Trypanosoma cruzi* (Central and South America)
Charcot's triad #1	Nystagmus, intention tremor, and scanning speech; suggests multiple sclerosis
Charcot's triad #2	Jaundice, RUQ pain, and fever; suggests cholangitis
Charcot-Leyden crystals	Crystals in sputum made of eosinophil membranes; suggests bronchial asthma
Chediak-Higashi disease	Phagocyte deficiency related to abnormally large granules in neutrophils
Cheyne-Stokes respirations	Terminal pattern of respirations with increasing breaths followed by apnea; indicates central apnea in coronary heart disease and increased intracranial pressure
Chvostek's sign	Facial musical spasm on tapping; indicates hypocalcemia
Codman's triangle on x-ray	Subperiosteal new bone formation; suggests osteosarcoma
Cori disease	Liver and muscle glycogen storage disease due to debranching enzyme deficiency
Councilman bodies	Eosinophilic intracytoplasmic balls in hepatocytes; suggests toxic or viral hepatitis
Cowdry type A bodies	Intranuclear inclusions; suggests herpesvirus infection
Crigler-Najjar syndrome	Mild (type 2) to life-threatening (type 1) congenital unconjugated hyperbilirubinemia
Curling ulcer	Acute gastric ulcer secondary to severe burns
Curschmann's spirals	Coiled mucinous fibrils found in sputum in bronchial asthma
Cushing ulcer	Gastric ulcer produced by increased intracranial pressure
Donovan bodies	Intracellular bacteria in granuloma inguinale
Dressler syndrome	Fibrinous pericarditis developing after myocardial infarction
Dubin-Johnson syndrome	Benign black liver secondary to congenital conjugated hyperbilirubinemia
Duchenne muscular dystrophy	X-linked recessive muscle dysfunction secondary to deleted dystrophin gene
Edwards syndrome	Trisomy 18; causes "rocker bottom" feet, low-set ears, and heart disease
Eisenmenger's complex	Uncorrected left-to-right cardiac shunt causes late right-to-left shunt with late cyanosis
Erb-Duchenne palsy	"Waiter's tip" hand secondary to superior trunk brachial plexus injury
Fanconi syndrome	Kidney dysfunction secondary to proximal tubular reabsorption defect
Gardner syndrome	Constellation of colon polyps with osteomas and soft tissue tumors
Gaucher disease	Glucocerebrosidase deficiency leading to potentially fatal glucocerebroside accumulation in multiple organs, notably spleen, liver, marrow, and brain
Ghon focus	Small lung lesion of early tuberculosis
Gilbert syndrome	Benign congenital unconjugated bilirubinemia (mostly just scares doctors)
Goodpasture syndrome	Anti-basement membrane antibodies; causes pulmonary and kidney bleeding

(Continued)

► Essential Eponyms (Diseases and Findings) *(Cont'd.)*

Name	Description
Gower's maneuver	Child using arms to help with leg weakness when trying to stand; suggests Duchenne muscular dystrophy
Guillain-Barré syndrome	Autoimmune peripheral nerve damage causing life-threatening paralysis
Hand-Schüller-Christian disease	Chronic, progressive, potentially fatal histiocytosis in which macrophages attack a child's body
Heberden's nodes	Osteophytes at DIP; suggests osteoarthritis
Heinz bodies	Red cell inclusions in G6PD deficiency
Henoch-Schönlein purpura	Hypersensitivity vasculitis causing hemorrhagic urticaria and arthritis
Homer-Wright rosette	Microscopic finding of a ring of neural cells suggesting neuroblastoma
Horner syndrome	Dysfunction of oculosympathetic pathway; ptosis, miosis, hemianhidrosis, apparent enophthalmos; causes include Pancoast tumor, lateral medullary syndrome
Howell-Jolly bodies	Red cell inclusions of DNA suggesting hyposplenism
Huntington disease	Autosomal-dominant caudate degeneration causing chorea and psychiatric problems
Janeway lesions	Hemorrhagic nodules in palms or soles; suggest endocarditis
Jarisch-Herxheimer reaction	Overaggressive treatment of infection causing endotoxin release with possible shock; classic example is syphilis
Job syndrome	Poor delayed hypersensitivity with neutrophil chemotaxis abnormality causing hyper-IgE with skin abscesses and other infections
Kaposi sarcoma	HHV-8 infection in AIDS patients causing vascular sarcoma
Kartagener syndrome	Dynein defect causes defective cilia, leading to bronchiectasis
Kayser-Fleischer rings	Green to golden copper deposits in iris around pupil; suggest Wilson disease
Kimmelstiel-Wilson nodules	Acellular glomerular nodules; suggest diabetic nephropathy
Klüver-Bucy syndrome	Bilateral amygdala lesions causing bizarre behavior with tendency to put anything in the mouth
Koplik spots	Minute white specks in buccal mucosa that may be first sign of measles
Krukenberg tumor	Gastric adenocarcinoma with ovarian metastases
Kussmaul ventilation	Diabetic ketoacidosis causes rapid, deep breathing to blow off CO_2
Lesch-Nyhan syndrome	X-linked HGPRT deficiency causing high uric acid levels with risk of brain damage
Lewy bodies	Round intracytoplasmic inclusions in neurons; seen in Parkinson disease
Libman-Sacks disease	Noninfectious endocarditis in SLE
Lines of Zahn	White streaks in arterial thrombus
Lisch nodules	Brown iris lesions in neurofibromatosis
Mallory bodies	Ropy cytoplasmic inclusions in hepatocytes in alcoholic liver disease

(Continued)

► Essential Eponyms (Diseases and Findings) *(Cont'd.)*

Name	Description
Mallory-Weiss syndrome	Esophagogastric lacerations with profuse bleeding secondary to heavy vomiting, forcing part of stomach into esophagus
McArdle disease	Muscle phosphorylase deficiency causing glycogen storage disease with prominent muscular symptoms
McBurney's point	Appendicitis is suggested by tenderness on palpation on a line between the anterior superior spine of the ilium and the umbilicus
Negri bodies	Neuron inclusions on electron microscopy in rabies
Niemann-Pick disease	Potentially fatal sphingomyelinase deficiency causing sphingomyelin deposition in brain and other organs, cherry-red macula spot, and neurologic problems
Osler's nodes	Pea-sized nodules on palms and soles suggesting endocarditis
Pancoast tumor	Apical lung cancer causing Horner syndrome
Parinaud syndrome	Dorsal midbrain syndrome often caused by compression by pineal gland; paralysis of upward gaze, may compress cerebral aqueduct → noncommunicating hydrocephalus
Parkinson disease	Motor disorder (resting tremor, rigidity) secondary to nigrostriatal dopamine depletion
Peutz-Jeghers syndrome	Benign autosomal-dominant colon polyposis syndrome
Peyronie disease	Penis deviates on erection secondary to fibrosis
Pick bodies	Round, silver-staining cytoplasmic structures in neurons in Pick disease; contain tau protein
Pick cells	Swollen (balloon) cells found in Pick disease; may contain Pick bodies
Pick disease	Frontal and temporal atrophy; progressive dementia; similar to Alzheimer disease but has a shorter course
Plummer-Vinson syndrome	Esophageal webs with iron deficiency anemia
Pompe disease	Lysosomal glucosidase deficiency causing cardiomegaly
Pott disease	Tuberculosis of the vertebrae
Raynaud syndrome	Recurrent vasospasm in extremities causing hand or foot color changes
Reed-Sternberg cells	Large binucleate tumor cells in Hodgkin disease
Reid index	Increased Reid index means thick mucous glands in bronchus and suggests chronic bronchitis
Reinke crystals	Crystals seen in Leydig cell tumors on microscopy
Reiter syndrome	Nongonococcal urethritis causes immune response, leading to conjunctivitis and arthritis
Roth spots	Retinal hemorrhages; suggest endocarditis
Rotor syndrome	Fairly benign congenital conjugated hyperbilirubinemia
Russell bodies	Round plasma cell inclusions that suggest multiple myeloma
Schiller-Duval bodies	Glomerulus-like microscopic structures in yolk sac tumors

(Continued)

▶ Essential Eponyms (Diseases and Findings) *(Cont'd.)*

Name	Description
Sézary syndrome	Cutaneous form of T-cell lymphoma with marked generalized erythema
Sheehan syndrome	Postpartum pituitary necrosis leading to massive hormonal deficits
Sipple syndrome	MEN type IIa; medullary thyroid carcinoma, pheochromocytoma, and parathyroid disease
Sjögren syndrome	Autoimmune attack on salivary glands with dry eyes, dry mouth, and arthritis
Spitz nevus	Childhood spindle cell lesion that looks like melanoma but has better prognosis
Trousseau's sign of hypocalcemia	Carpal spasm
Trousseau's sign of malignancy	Migratory thrombophlebitis suggesting visceral (pancreatic) carcinoma
Virchow's node	Left supraclavicular node enlargement suggesting metastatic gastric carcinoma
Virchow's triad	Combination of blood stasis, endothelial damage, and hypercoagulation causes venous clots with risk of pulmonary embolism
von Recklinghausen neurologic disease	Neurofibromatosis
von Recklinghausen bone disease	Osteitis fibrosa cystica
Wallenberg syndrome	Lateral medullary syndrome caused by PICA occlusion; causes contralateral pain/temperature deficits in body, ipsilateral pain/temperature deficits in face, dysphagia, vestibular dysfunction, ipsilateral Horner syndrome
Waterhouse-Friderichsen syndrome	Adrenal hemorrhage complicating meningococcemia
Weber syndrome	Medial midbrain syndrome; ipsilateral oculomotor paralysis, contralateral spastic paralysis, contralateral lower facial weakness
Wermer syndrome	MEN type I; parathyroid tumors, endocrine pancreatic tumors, and pituitary gland tumors
Whipple disease	*Tropheryma whippelii* causes malabsorption syndrome
Wilson disease	Altered copper metabolism causes damage to liver and brain; Kayser-Fleischer rings
Zenker's diverticulum	Lower esophageal diverticulum
Zollinger-Ellison syndrome	Gastrin-secreting tumor causing peptic ulcers

▶ Eye Findings on Physical Examination

Finding	Classic Disease Association (Notes)
Argyll Robertson pupil	Tertiary (neuro) syphilis; loss of light reflex constriction; accommodation is preserved; classic form bilateral
Blue sclera	Osteogenesis imperfecta, types I and II (fatal) Also may be seen in Ehlers-Danlos syndrome, pseudoxanthoma elasticum, Marfan syndrome
Brushfield spots	Down syndrome (ring of white spots around periphery of iris; trisomy 21)
Charcot's triad #1	Multiple sclerosis (nystagmus, intention tremor, scanning speech; triad #2 is for cholangitis: jaundice, fever, rigors, pain)

(Continued)

► Eye Findings on Physical Examination *(Cont'd.)*

Finding	Classic Disease Association (Notes)
Cherry-red spot	Tay-Sachs, Niemann-Pick, central retinal artery occlusion (retinal pallor contrasting with strikingly red macular spot)
Cotton-wool spots	Chronic hypertension (small areas of yellowish-white discoloration in the retina)
Horner syndrome	Impaired sympathetic innervation to eye (ptosis, miosis, anhidrosis, and apparent enophthalmos; numerous causes, including vascular, traumatic, congenital, Pancoast tumor, other tumors)
Internuclear ophthalmoplegia (INO)	Multiple sclerosis (disorder of lateral conjugate gaze; affected eye cannot adduct and nystagmus occurs in the abducting eye; convergence is intact)
Kayser-Fleischer rings	Wilson disease (greenish or golden copper deposits in crescent or ring in Descemet's membrane)
Lens dislocation	Marfan syndrome (can be accompanied by aortic dissection and joint hyperflexibility)
Lisch nodules	Neurofibromatosis type I (tan hamartomas on the iris)
Roth spots	Bacterial endocarditis (hemorrhage in retina with a white center; also seen in leukemia, diabetes, collagen-vascular diseases)

► Skin Findings

Finding	Classic Disease Association (Notes)
Adenoma sebaceum	Tuberous sclerosis (raised, erythematous papules on the face, especially around the nose)
Anesthesia	Leprosy (skin may be blotchy, red, or thickened)
Bullae (tense)	Bullous pemphigoid
Bullae (flaccid, rupturing)	Pemphigus
Brown-black lesion with fuzzy edge	Melanoma (depth of lesion most important prognostic indicator)
Butterfly rash	Systemic lupus erythematosus (nose and cheeks)
Café-au-lait spots	Neurofibromatosis (light brown spots, often over 1 cm)
Chancre	Primary syphilis (pain**less** ulcer, usually on genitalia)
Chancroid	*Haemophilus ducreyi* (pain**ful** ulcer, usually on genitalia)
Condylomata lata	Secondary syphilis (smooth, flat, painless genital lesions; scrapings may show spirochetes with darkfield microscopy)
Dermatitis, dementia, diarrhea	Pellagra caused by niacin deficiency
Dog or cat bite	*Pasteurella multocida*
Elastic skin	Ehlers-Danlos syndrome
Erythema chronicum migrans	Lyme disease (expanding red ring with central clearing at tick bite site)
Generalized hyperpigmentation	Addison disease (primary adrenal insufficiency)
Kaposi sarcoma	AIDS (usually slightly raised violaceous papules or plaques)

(Continued)

► Skin Findings *(Cont'd.)*

Finding	Classic Disease Association (Notes)
Port wine stain	Hemangioma (large, purplish lesion on face)
Rash on palms and soles	Secondary syphilis, Rocky Mountain spotted fever
Silvery, scaly plaques	Psoriasis (knees, elbows, scalp)
Slapped cheeks	Erythema infectiosum (fifth disease, parvovirus B19)
Vesicles, small painful	Herpes, dermatitis herpetiformis

► Extremity Findings on Physical Examination

Finding	Classic Disease Association (Notes)
Arachnodactyly	Marfan syndrome (very long fingers and toes)
Babinski sign	Upper motor neuron lesion (stimulation of sole of foot → upgoing great toe)
Baker's cyst	Rheumatoid arthritis (cyst in popliteal fossa)
Bouchard's node	Osteoarthritis (PIP osteophytes)
Boutonniere deformity	Rheumatoid arthritis (finger flexed at PIP and hyperextended at DIP)
Calf pseudohypertrophy	Duchenne muscular dystrophy (replacement of muscle with fat and connective tissue)
Heberden's nodes	Osteoarthritis (DIP enlargement because of osteophytes)
Janeway lesions	Endocarditis (hemorrhagic nodules in palms or soles)
Osler's nodes	Endocarditis (tender nodules on palms and soles)
Palpable purpura	Henoch-Schönlein purpura (legs and buttocks)
Rash affecting palms and soles	Secondary syphilis, Rocky Mountain spotted fever
Raynaud syndrome	Recurrent vasospasm (pale to blue to red on hands or feet)
Simian crease	Down syndrome (single long crease across palm; trisomy 21)
Splinter hemorrhage	Infective endocarditis, trauma (found under fingernails)
Tendon xanthomas	Familial hypercholesterolemia (classically Achilles tendon)
Tophi	Gout (hard nodules composed of uric acid)

► Radiologic Findings

Finding	Classic Disease Association (Notes)
Bamboo spine	Ankylosing spondylitis (rigid spine with fused joints)
Boot-shaped heart	Right ventricular hypertrophy; tetralogy of Fallot (upturned ventricular apex and large pulmonary artery make the "boot")
Codman's triangle	Osteosarcoma (new subperiosteal bone lifts periosteum)

(Continued)

▶ Radiologic Findings *(Cont'd.)*

Finding	Classic Disease Association (Notes)
Double-bubble sign	Duodenal atresia, also duodenal stenosis, duodenal webs, annular pancreas, malrotation of the gut (two air-filled structures in upper abdomen, with little or no air distally)
"Hair on end" or "crew-cut"	Beta thalassemia, sickle cell anemia (extramedullary hematopoiesis below periosteum leads to formation of bony spicules = "hair" on outside of bone)
Mammillary body atrophy	Wernicke encephalopathy (memory loss)
Periosteal elevation	Pyogenic osteomyelitis (elevation due to subperiosteal inflammation; this may be the earliest radiologic sign of osteomyelitis)
"Punched out" (lytic) lesions of bone	Multiple myeloma
Rib notching	Coarctation of aorta (dilated aorta before coarctation puts chronic pressure on ribs)
Soap bubble	Giant cell tumor of bone (lytic expansile lesion)
String sign	Crohn disease (small bowel follow-through shows very narrow lumen, typically in terminal ileum)

▶ Auscultation Findings

Sound	Possible Causes
Systolic Murmurs	
Soft systolic ejection murmurs	May be normal in infants, children, pregnancy
Systolic ejection murmur (right 2nd interspace)	Aortic stenosis
Systolic ejection murmur (mid to lower left sternal border)	Hypertrophic obstructive cardiomyopathy
Systolic ejection murmur (left 2nd interspace)	Pulmonic stenosis
Systolic ejection murmur (apex, can increase through systole)	Mitral regurgitation
Systolic ejection murmur (lower left sternal border, increases with inspiration)	Tricuspid regurgitation
Holosystolic ejection murmur (left fourth interspace)	Ventricular septal defect
Diastolic Murmurs	
Diastolic murmur (apex)	Mitral stenosis
Diastolic murmur (left 4th interspace)	Tricuspid stenosis
Decrescendo diastolic murmur (left 4th interspace)	Aortic regurgitation (see also Austin-Flint murmur)
Austin-Flint murmur (mid-to-late-diastolic rumble/low-frequency murmur over apex)	Severe aortic regurgitation
Decrescendo diastolic murmur (right sternal edge and left 2nd interspace)	Pulmonic regurgitation

(Continued)

▶ Auscultation Findings *(Cont'd.)*

Sound	Possible Causes
Continuous Murmurs	
Continuous murmur (left 2nd interspace below median end of clavicle)	Patent ductus arteriosus
Continuous murmur (centrally at 3rd interspace level)	Aorticopulmonary window defect
Continuous murmur (peripheral body sites)	Systemic arteriovenous connections
Miscellaneous Findings	
Loud S_1	Mitral stenosis
Soft or absent S_1	Mitral regurgitation if valve is stiff
Late aortic valve closure in S_2	Left bundle branch block, aortic stenosis
Late pulmonic valve closure in S_2	Atrial septal defect, right bundle branch block
Fixed split S_2 during respiration	Atrial septal defect
Paradoxical splitting of S_2	Left bundle branch block (also some cases of aortic stenosis and patent ductus)
Single S_2	Badly damaged aortic valve (regurgitation, stenosis, or atresia)
Early systolic click	Congenital aortic or pulmonic valve stenosis, severe pulmonary hypertension
Changing systolic clicks with position	Myxomatous degeneration of mitral or tricuspid valves
S_3 (pericardial knock)	Dilated and noncompliant left (strongest on expiration) or right (strongest on inspiration) ventricle, normal in kids
S_4	Right (strongest on inspiration) or left (strongest on expiration) ventricular dysfunction (myocardial ischemia or early myocardial infarction)
Summation gallop (combined S_3 and S_4)	Tachycardic patient with right or left ventricular dysfunction
Diastolic knock	Constricting pericardium
Mitral opening snap	Mitral stenosis

▶ Genetic Associations

Finding	Classic Disease Association (Notes)
5p–	Cri-du-chat syndrome (cat-like cry, feeding problems, abnormal mental development)
45,XO	Turner syndrome (infertile female, webbed neck, coarctation of aorta)
47,XXY	Klinefelter syndrome (male with small testes and eunuchoid habitus)
CFTR	Cystic fibrosis (chloride channel gene, chromosome 7, recurrent pneumonia, pancreatic exocrine insufficiency)
FBN1 gene (codes for fibrillin)	Marfan syndrome (chromosome 15, tall stature, hyperextensible joints, dissecting aortic aneurysm)

(Continued)

▶ Genetic Associations *(Cont'd.)*

Finding	Classic Disease Association (Notes)
NF1	Neurofibromatosis type I (von Recklinghausen disease, chromosome 17, neurofibromas, café-au-lait spots)
NF2	Neurofibromatosis type II (bilateral acoustic neurofibromatosis, chromosome 22)
t(8;14)	Burkitt lymphoma *(c-myc)*
t(9;22)	CML and occasionally AML (Philadelphia chromosome, *bcr-abl* hybrid)
t(14;18)	Many follicular lymphomas *(bcl-2)*
Trisomy 13	Patau syndrome (microcephaly, mental retardation, cleft palate, polydactyly, heart malformations)
Trisomy 18	Edwards syndrome (rocker bottom feet, microcephaly, mental retardation, multiple organ defects)
Trisomy 21	Down syndrome (most common chromosomal disorder, older maternal age, mental retardation, early Alzheimer disease)
VHL	von Hippel-Lindau (chromosome 3, hemangioblastomas, renal cell carcinoma)
XYY	XYY syndrome (very tall male with increased risk of behavior problems)

▶ Microscopic Findings

Finding	Classic Disease Association (Notes)
Auer rods	Acute myelogenous leukemia, particularly promyelocytic (rods in white blood cell cytoplasm)
Basophilic stippling	Lead poisoning (dots in erythrocytes)
Birbeck granules on EM	Histiocytosis X (eosinophilic granuloma)
Call-Exner bodies	Granulosa-theca cell tumor of ovary (ring of cells with pink fluid in center)
Cerebriform nuclei	Mycosis fungoides (cutaneous T-cell lymphoma)
Clue cells	*Gardnerella vaginitis* (bacteria on epithelial cells)
Councilman bodies	Toxic or viral hepatitis (pink, round cytoplasmic inclusion in hepatocytes)
Cowdry type A bodies	Herpes (intranuclear eosinophilic inclusions)
Crescents in Bowman's capsule	Rapidly progressive crescentic glomerulonephritis
Curschmann's spirals	Bronchial asthma (coiled mucinous fibrils found in sputum)
Depigmentation of neurons in substantia nigra	Parkinson disease (degeneration of dopaminergic nigrostriatal neurons)
Donovan bodies	Granuloma inguinale (oval, rod-shaped organisms in cells)
Ferruginous bodies	Asbestosis (rod-shaped structures with crystals on them)
Heinz bodies	G6PD deficiency (red cell inclusions)
Homer Wright rosettes	Neuroblastoma (ring of neural cells)
Howell-Jowell bodies	Splenectomy or nonfunctioning spleen (blue-black erythrocyte inclusions)
Hypersegmented neutrophils	Macrocytic anemia (vitamin B_{12} or folate deficiency)

(Continued)

▶ Microscopic Findings *(Cont'd.)*

Finding	Classic Disease Association (Notes)
Hypochromic microcytosis	Iron deficiency anemia, lead poisoning
Keratin pearls	Squamous cell carcinoma (concentric layers of keratin)
Kimmelsteil-Wilson nodules	Diabetic nephropathy (acellular nodules in glomerulus)
Koilocytes	HPV infections such as condyloma, cervical dysplasia (look for perinuclear halo)
Lewy bodies	Parkinson disease (round, pink nodules in neuronal cytoplasm)
Mallory bodies	Alcoholic liver disease (ropy, pink cytoplasmic structures in hepatocytes)
Needle-shaped, negatively birefringent crystals	Gout (uric acid)
Negri bodies	Rabies (large viral inclusions in neurons, see on Emergency Medicine)
Neurofibrillary tangles	Alzheimer disease (tangles of fibers in neuron cytoplasm)
Owl's eye nuclei	Cytomegalovirus (due to virus particles in nucleus)
Pick bodies	Pick disease (silver protein deposits in neurons)
Pseudopalisading tumor cell arrangement	Glioblastoma multiforme (foci of necrosis surrounded by intact tumor cells)
Pseudorosettes	Ewing sarcoma (rings of cells with central vessel)
Reed-Sternberg cells	Hodgkin lymphoma (large binucleate cell with large nucleoli)
Reinke crystals	Leydig cell tumor (rectangular crystals, ovary or testes)
Renal epithelial casts in urine	Acute toxicity/viral (epithelial casts reflect tubular damage)
Rhomboid crystals in joint fluid, positively birefringent	Pseudogout (calcium pyrophosphate crystals)
Rouleaux	Multiple myeloma (stacked erythrocytes)
Russell bodies	Multiple myeloma (hyaline spheres in plasma cells)
Schiller-Duval bodies	Yolk sac tumor (look like glomeruli)
Senile plaques	Alzheimer disease (extracellular amyloid)
Signet ring cells	Gastric carcinoma (have nucleus compressed to one side of cell)
Smudge cells	Chronic lymphocytic leukemia (smashed lymphocyte)
Spike and dome on EM	Membranous glomerulonephritis (irregular dense deposits with basement membrane material between deposits)
"Starry sky" pattern	Burkitt lymphoma (sheets of small lymphocytes with scattered histiocytes as "stars")
Subepithelial humps on Emergency Medicine	Poststreptococcal glomerulonephritis
Sulfur granules	*Actinomyces israeli* (clusters of bacteria)
Tram track appearance on light microscopy	Membranoproliferative glomerulonephritis (double contour capillary loops)
Waxy casts in urine	Chronic end-stage renal disease
WBC casts in urine	Acute pyelonephritis
WBCs in urine	Acute cystitis (heavy neutrophilic infiltrate)
"Wire loop" lesion	Lupus nephritis (thickened capillary basement membrane)

▶ Classic Antibody Findings

Finding	Classic Disease Association (Notes)
Anti-basement membrane	Goodpasture syndrome
Anticentromere	Scleroderma (CREST syndrome)
Anti-double stranded DNA (ANA antibodies)	Systemic lupus erythematosus (type III hypersensitivity-immune complexes)
Antiepithelial cell	Pemphigus vulgaris
Antigliadin	Celiac disease
Antihistone	Drug-induced SLE
Anti-IgG	Rheumatoid arthritis (rheumatoid factor)
Antimitochondrial	Primary biliary cirrhosis
Antineutrophil	Vasculitis
Antiplatelet	Idiopathic thrombocytopenic purpura
C-ANCA, P-ANCA	Wegener granulomatosis (C-ANCA), polyarteritis nodosa (mostly P-ANCA, but can have both)
CLL	*Mycoplasma pneumoniae*, mononucleosis, lymphoma, CLL

▶ Abnormal Erythrocytes on Peripheral Smear

Finding	Classic Disease Association (Notes)
Acanthocytes (spur cells)	Abetalipoproteinemia (severe burns, liver disease, hypothyroidism)
Basophilic stippling	Lead poisoning (thalassemia)
Bite cells and Heinz bodies	Glucose-6-phosphate dehydrogenase deficiency (spleen removes Heinz bodies, leading to “bitten” appearance of RBCs)
Dacrocytes (teardrop cells)	Scarring of bone marrow (myelophthisis), splenic dysfunction
Echinocytes (burr cells)	Often drying artifact, uremia
Elliptocytes (ovalocytes)	Hereditary elliptocytosis (iron deficiency, thalassemia, myelophthisis)
Howell-Jolly bodies and Cabot rings	Splenic dysfunction (thalassemia)
Macrocytes (large cells)	Vitamin B_{12} and folate deficiency (myelodysplastic syndromes, liver disease)
Microcytes (small cells)	Iron deficiency anemia (thalassemia and some cases of anemia of chronic disease)
Pappenheimer bodies	Sideroblastic anemia (splenic dysfunction)
Rouleaux formation	Multiple myeloma (RBCs stacked like coins)
Schistocytes	Intravascular hemolysis (fragmented cells)
Spherocytes	Hereditary spherocytosis (extravascular hemolysis)
Stomatocytes	Hereditary stomatocytosis (alcoholism)
Target cells	Liver disease, thalassemia (HbC, occasionally in iron deficiency)

▶ Abnormalities of White Blood Cells and Platelets on Peripheral Smear

Finding	Classic Disease Association (Notes)
Bilobed neutrophil nuclei	Pelger-Huet anomaly
Cerebriform nuclei (convoluted appearance to nucleus)	Mycosis fungoides (cutaneous T-cell lymphoma)
Dohle bodies	Sepsis, May-Hegglin anomaly (pale blue, oval cytoplasmic inclusions that can be near cytoplasmic membrane of neutrophils)
Giant platelets	Bernard-Soulier syndrome
Hypersegmented neutrophil nuclei	Megaloblastic (macrocytic) anemia
Large blue granules in cytoplasm of all white blood cells	Alder-Reilly anomaly
Large eosinophilic granules in neutrophil cytoplasm	Chediak-Higashi syndrome
Toxic granulation	Sepsis (medium-to-large sized dark blue granulations in neutrophil cytoplasm)

▶ Serum Enzymes

Enzyme	Classic Associated Conditions
Alanine aminotransferase (ALT)	Liver damage
Alkaline phosphatase (Alk phos)	Bone, biliary, and placental disease
Amylase	Pancreatic and salivary disease
Angiotensin-converting enzyme (ACE)	Sarcoidosis (also primary biliary cirrhosis, Gaucher disease, leprosy)
Aspartate aminotransferase (AST)	Acute myocardial infarction, liver disease
Creatinine kinase (CK) CK-MB	 Myocardial infarction (early 2–8 h), severe skeletal muscle injury
Elastase-1	Pancreatic disease
Lactate dehydrogenase (LDH) LD1>LD2 High LD4 and LD5 High LD1 And LD5	 Acute myocardial infarction (early), hemolysis, renal infarction Liver damage (also skeletal muscle damage) Acute myocardial infarction complicated by liver congestion; alcoholic liver disease complicated by megaloblastic anemia
Lipase	Pancreatic disease
Myoglobin	Myocardial infarction (early, but nonspecific)
Troponin I	Myocardial infarction (elevates as early as 3 h post MI, then stays elevated up to 9 days after MI)

Appendix D

Renal Pathology

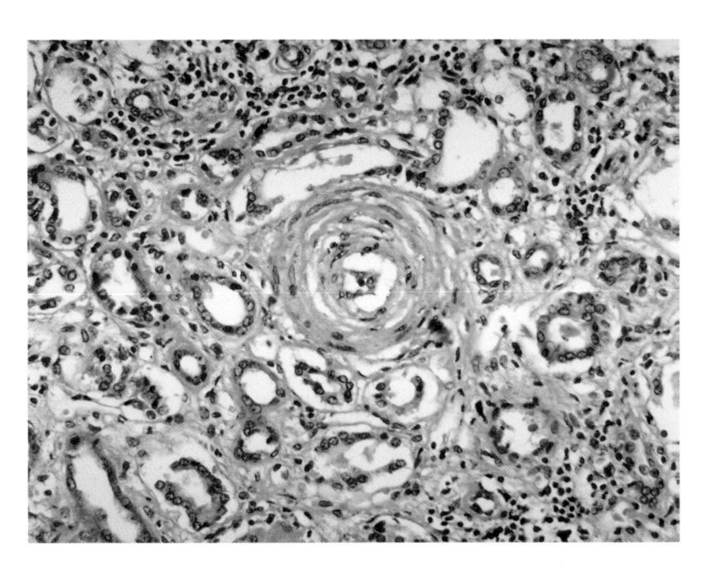

. Hyperplastic arteriolitis of kidney in malignant hypertension. Smooth nuscle cell proliferation and collagen deposition of arterioles and interlobular rtery walls result in concentric intimal thickening. This is often referred to as onion-skinning" because of the concentric appearance.

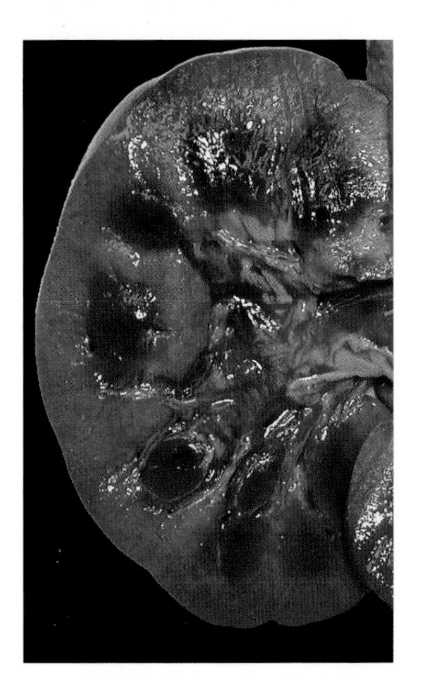

3. Color reversal of the cortex following hypotensive disease

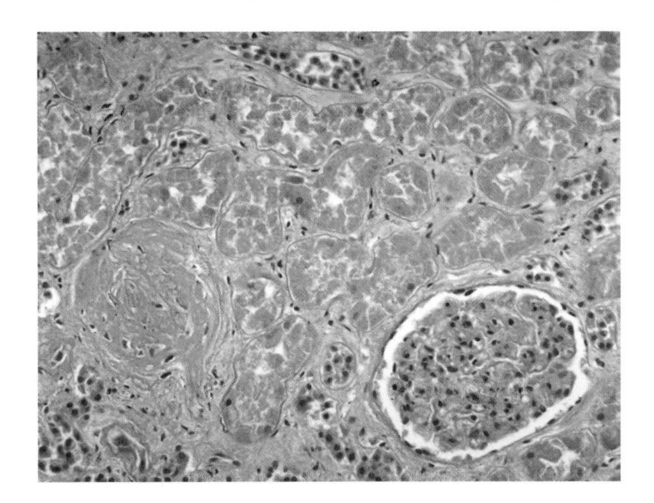

. Papillary necrosis is mostly seen in diabetic patients and in those with rinary tract obstruction. Necrotic tissue shows coagulative necrosis with reservation of the tubule outline. Tubular lumens are necrotic as a result of n extension of infection along the tubular lumens.

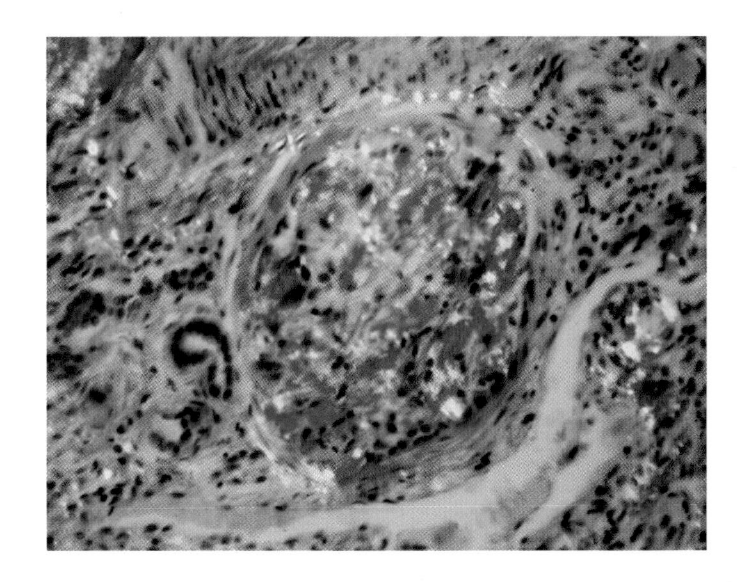

4. Kidney amyloid deposition, Congo red stain, polarized light microscopy. The photomicrograph shows a glomerulus in the kidney after staining with Congo red. When viewed under polarized light, the amyloid produces a characteristic green birefringence. This property is thought to be due to the structure of the amyloid protein. *(Source: Katsumi Miyai, M.D., Ph.D.)*

Copyright 2005 DxR Development Group Inc. All rights reserved.

Renal Pathology (*Cont'd*)

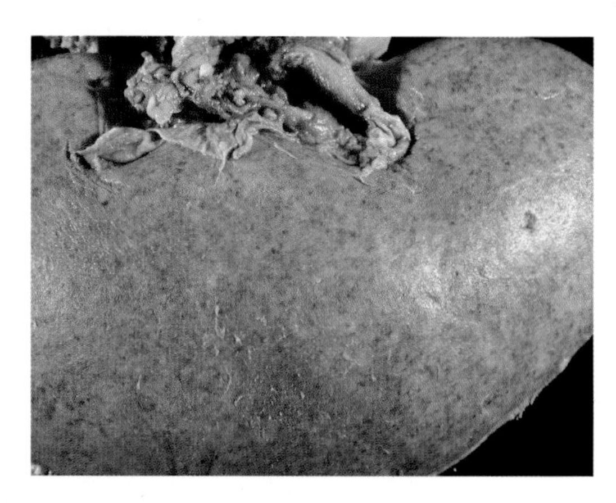

5. Nephrosclerosis of the kidney due to malignant hypertension. Pinpoint petechial hemorrhages can be seen on the surface of the kidney. This results from ruptured arterioles, giving the kidney a "flea-bitten" appearance.

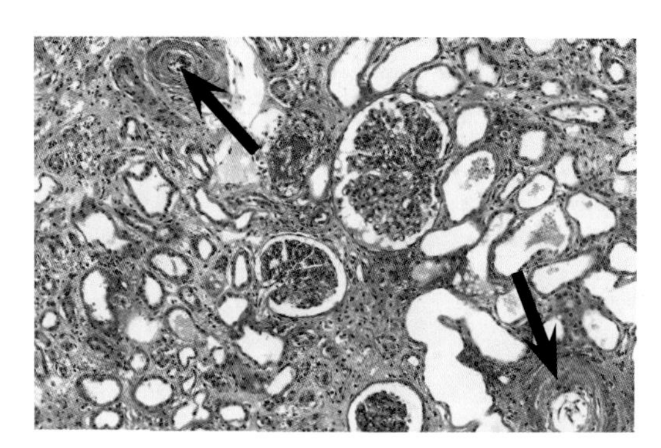

6. Fibrinoid necrosis of arterioles in malignant hypertension. Injury to the renal arterioles results in fibrinogen and platelet deposition and focal necrosis of cells in the vessel wall. This is seen as eosinophilic granular changes occuring in affected blood vessel walls.

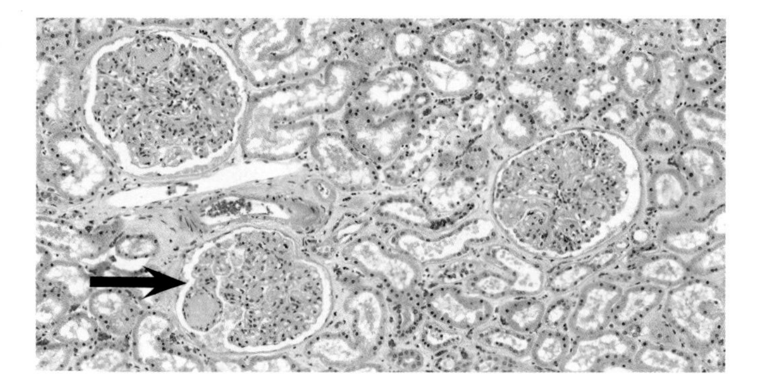

7. Nodular glomerulosclerosis in a diabetic patient. Also called Kimmelstiel-Wilson disease, this pathology results in PAS-positive glomerular nodules in the peripheral capillary loops.

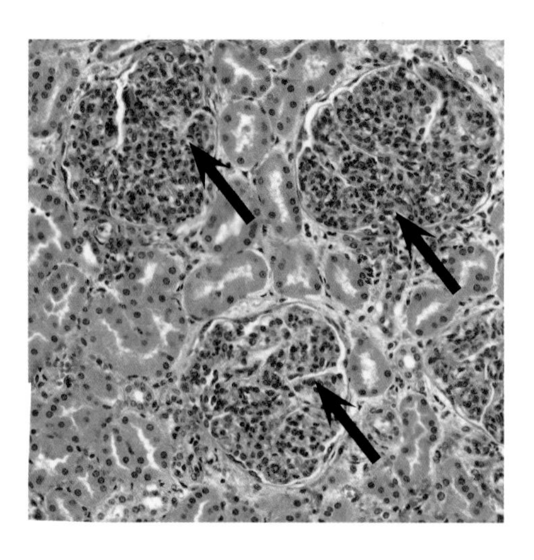

8. Poststreptococcal glomerulonephritis. Glomeruli are enlarged and hypercellular due to both infiltration of leukocytes and endothelial and mesangial proliferation. Subepithelial humps would be seen on electron microscopy (*not shown here*).

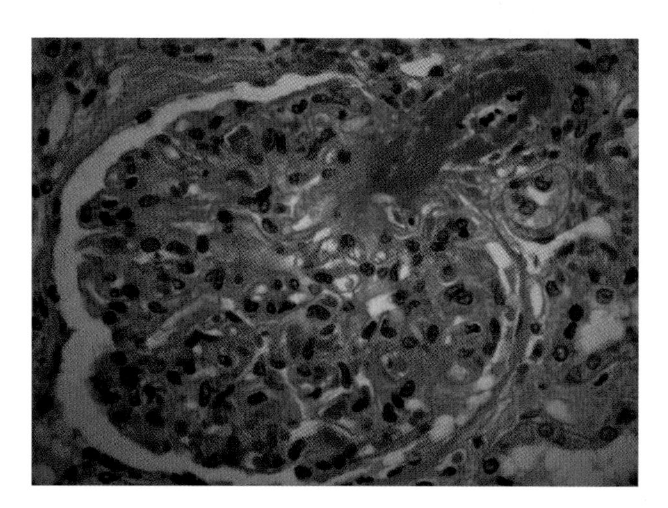

9. Fibrinoid necrosis in the kidney. This kidney photomicrograph is from a patient with malignant hypertension, and it shows the characteristic features of fibrinoid necrosis. The arteriole wall (top right) is devoid of nuclei and has an amorphous pink appearance from protein deposition.

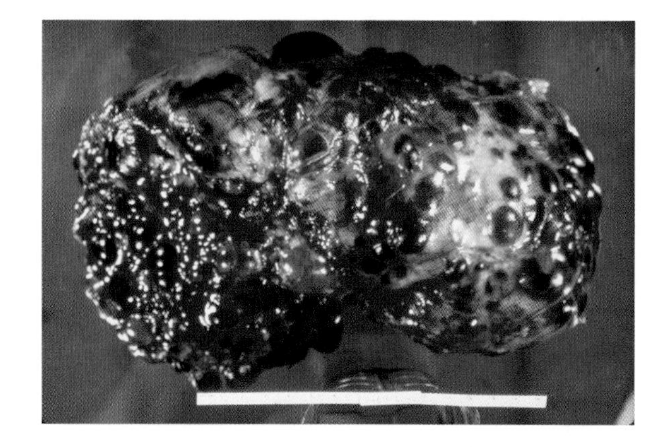

10. Autosomal dominant polycystic kidney disease. The kidney is enlarged and the parenchyma is extensively replaced by cysts of varying sizes. Some of the darker-colored cysts contain hemorrhagic fluid.

 Copyright 2005 DxR Development Group Inc. All rights reserved.

Cardiac Pathology

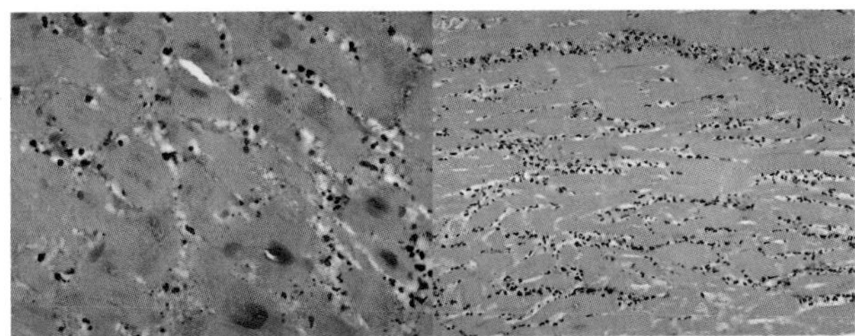

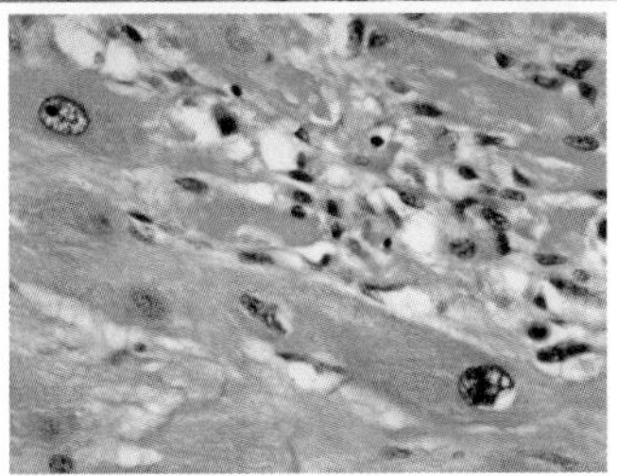

1. Microscopic features of myocardial infarction. A 1-day-old
farct (*top two images*) shows coagulative necrosis with wavy fibers,
dematous tissue, and scattered neutrophils. A 7- to 10-day-old infarct
ottom image) reveals the absence of necrotic myocytes because of
nagocytosis.

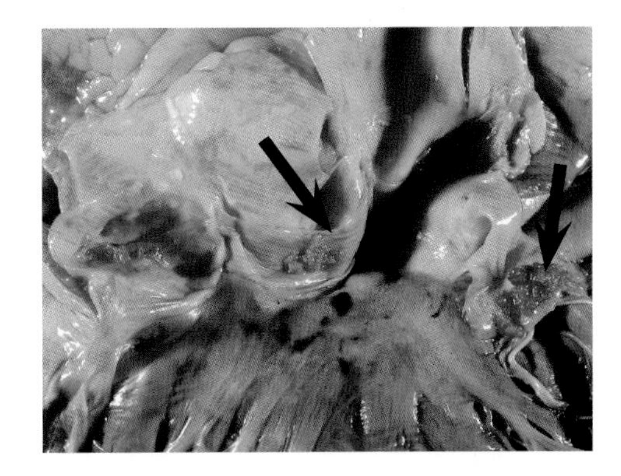

2. Infective (bacterial) endocarditis of the atrial valve. Bulky
egetations containing fibrin, inflammatory cells, and bacteria are
eposited on the heart valves. Sometimes vegetations can erode into the
yocardium and cause ring abscesses. Typically, subacute endocarditis
as less valvular destruction than is found in acute endocarditis.

3. Nonbacterial thrombotic endocarditis (NBTE) is characterized by
ne deposition of small masses of fibrin platelets on the leaflets of the
ardiac valves. In contrast to infective endocarditis, these vegetations
ontain no microorganisms. It is often related to cancer or sepsis and is
gnificant because of the potential for embolization of the deposits, with
esultant infarcts in secondary locations.

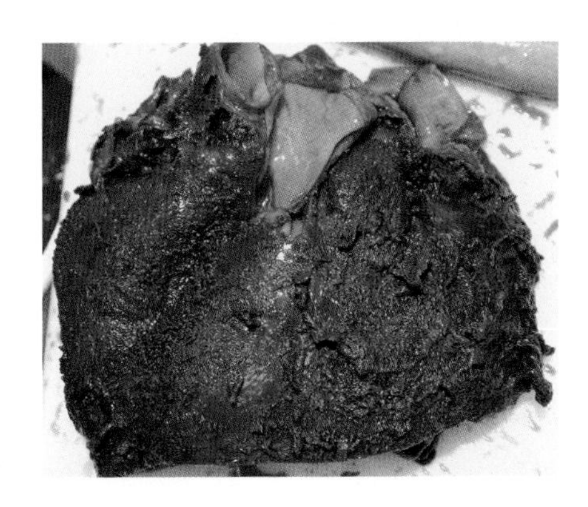

14. Fibrinous pericarditis is the most common form of pericarditis and is composed of fibrinous exudates. It is associated with myocardial infarction and uremia.

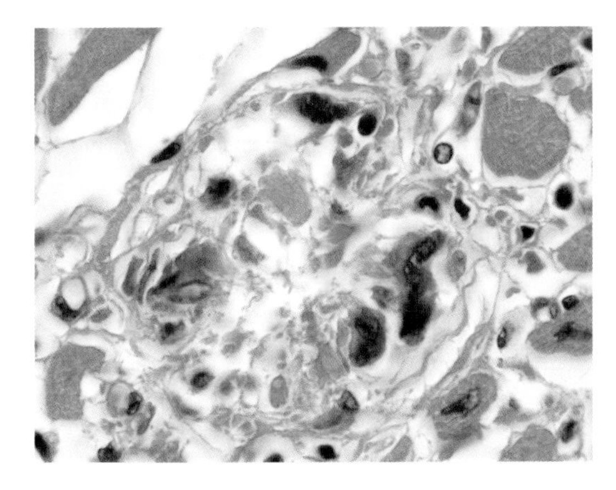

15. Rheumatic heart disease. The characteristic microscopic lesion is an Aschoff body, a focal area of eosinophilic fibrinoid necrosis surrounded by inflammatory cells.

16. Dilated cardiomyopathy (DCM). In DCM, the heart is usually enlarged and flabby with dilatation of all chambers. The myocardial wall thins as the heart enlarges, resulting in impaired contractility (systolic dysfunction).

Copyright 2005 DxR Development Group Inc. All rights reserved.

Cardiac Pathology (*Cont'd*)

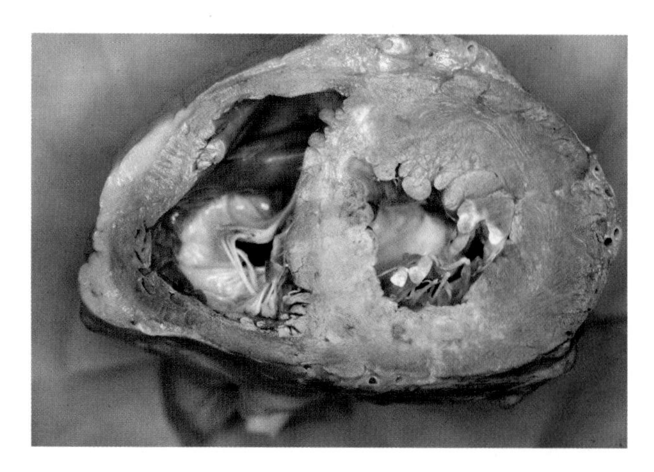

17. Cardiac hypertrophy. The heart exhibits concentric hypertrophy of the left ventricle and is from a patient with a history of hypertension. The left ventricle wall is thickened and has areas of white fibrosis consistent with prior ischemic injury.

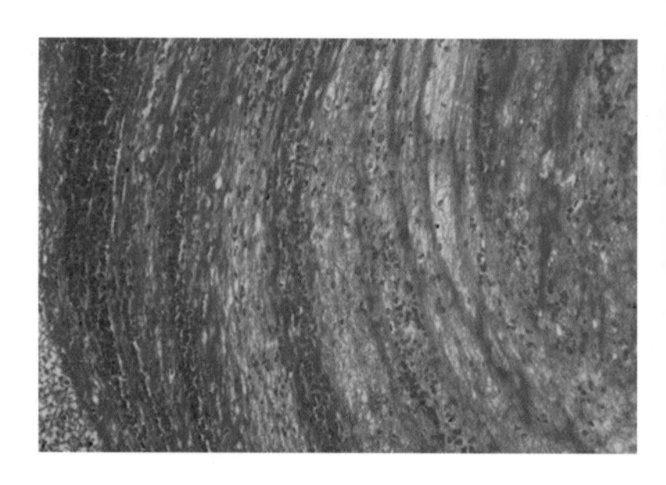

18. Lines of Zahn. Alternating pale pink bands of platelets with fibrin and red bands of red blood cells form a true thrombus.

Gastrointestinal Pathology

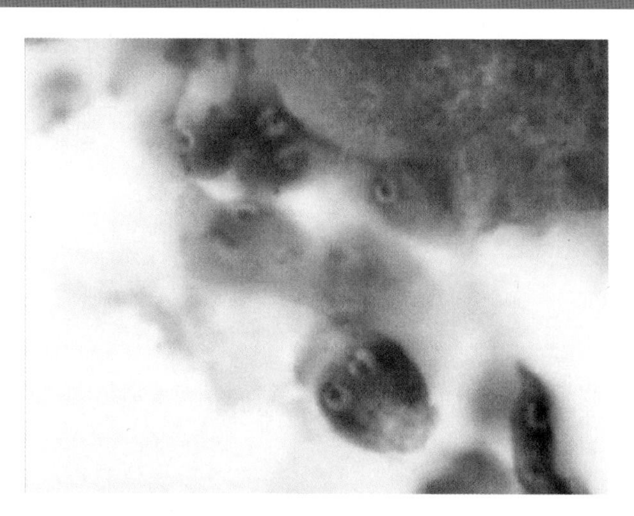

19. ***Giardia lamblia*** trophozoites are pear-shaped binucleated parasites. Trophozoites are noninvasive but can result in blunting of the intestinal villi.

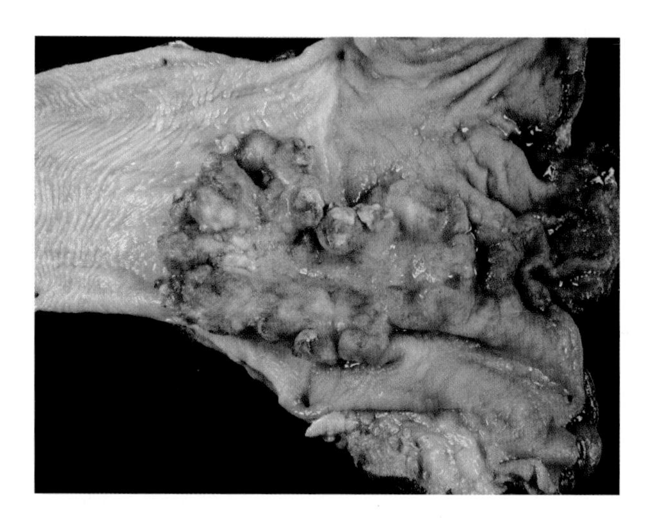

20. Adenocarcinoma of the esophagus. Adenocarcinomas arising from the distal esophagus are usually found in Barrett esophagus. Most of these tumors are mucin-producing, glandular tumors.

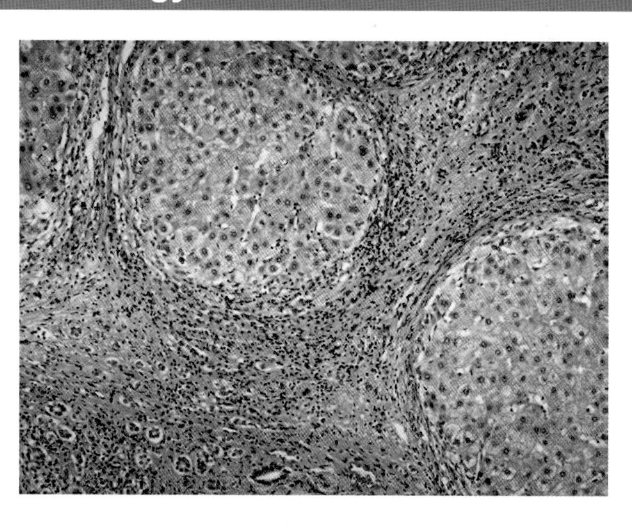

21. Chronic active hepatitis and cirrhosis from hepatitis C: inflammation in the portal tracts with predominantly lymphocytes and macrophages. There is bile duct damage and septal and bridging fibrosis.

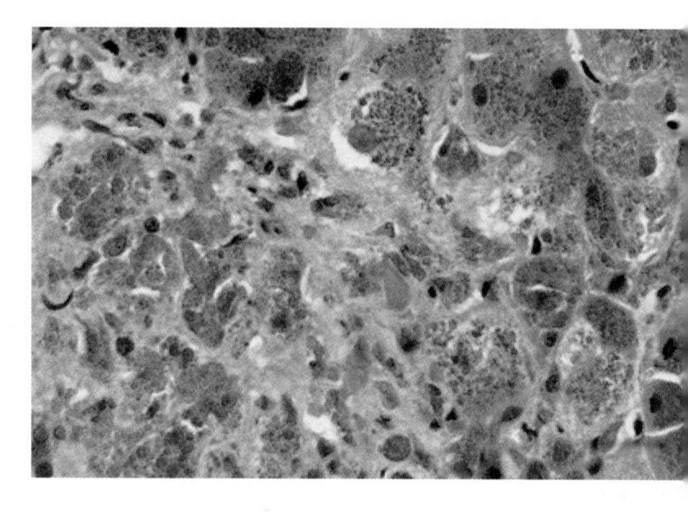

22. Hemochromatosis of the liver. Prominent hemosiderin deposition can be seen in the cytoplasm of these hepatocytes.

 Copyright 2005 DxR Development Group Inc. All rights reserved.

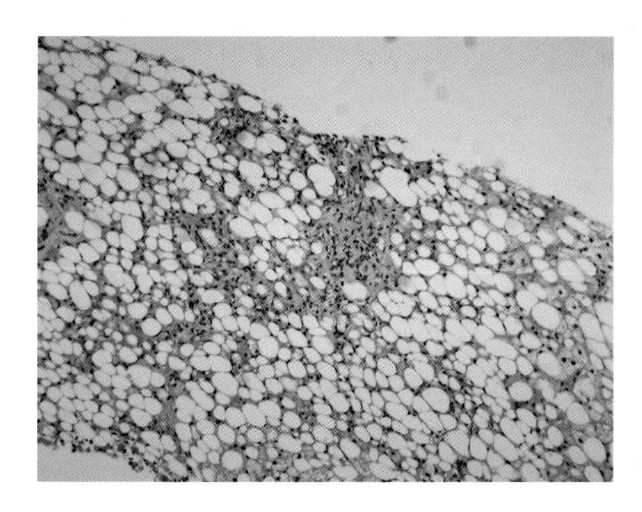

3. **Fatty liver, nonalcoholic.** Defects in any of the steps in uptake, catabolism, or secretion of fatty acids in the liver can result in steatosis. Fat vacuoles enlarge until the nucleus is displaced into the periphery of the hepatocyte.

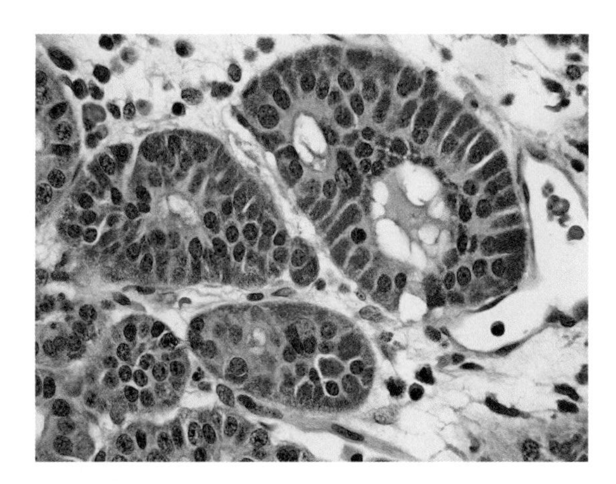

4. **Carcinoid, small intestine.** The small intestine is the most common location for a carcinoid tumor (the appendix is the second). The tumor cells are uniform in appearance and have a scant pink granular cytoplasm and a round-to-oval stippled nucleus. Electron microscopy will reveal (*not shown here*) membrane-bound secretory granules with dense cores.

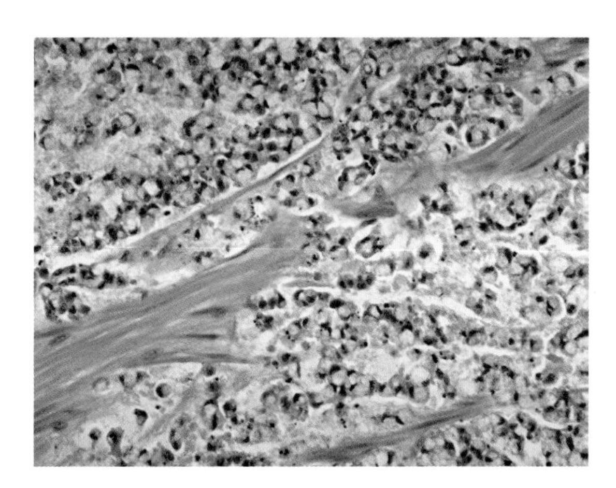

5. **Invasive adenocarcinoma of the colon.** The malignant cells have a distinctive signet ring appearance from intracellular mucin vacuoles, which displace the nucleus to the side of the cell. This subtype is considered a more aggressive form of adenocarcinoma.

Hematologic/Lymphoreticular

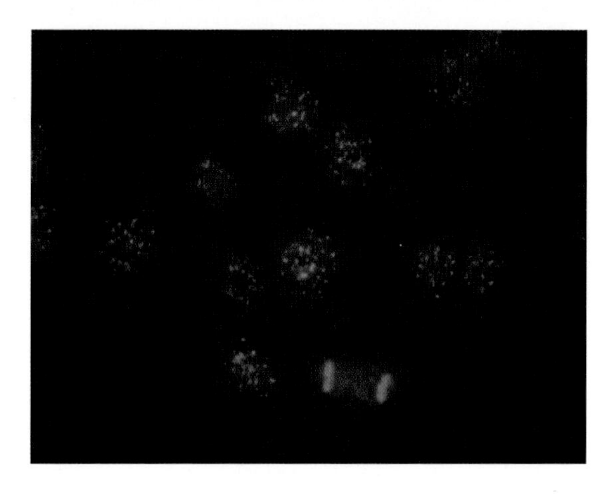

26. ANA immunofluorescence, centromere pattern. Antinuclear antibodies (ANA) are a sensitive (yet not specific) indicator of autoimmune diseases, such as SLE, scleroderma, and Sjögren syndrome. The pattern of fluorescence suggests the type of antibody in the patient's serum, as each pattern is associated with specificity to different nuclear components.

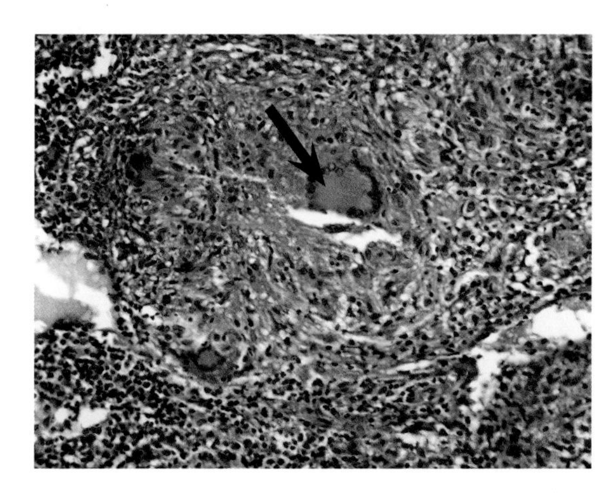

27. Lymph node, sarcoid. Classic lesions on microscopic appearance are noncaseating granulomas with epithelioid cells and occasional giant cells (Langhans cells).

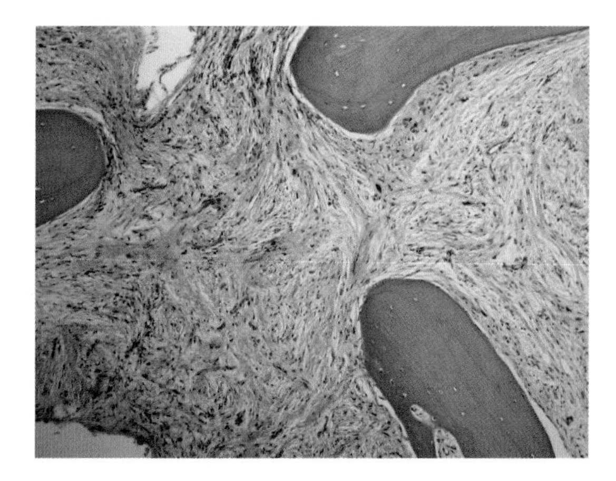

28. Bone marrow myelofibrosis. Marrow spaces become obliterated by fibrosis. This results in extensive extramedullary hematopoiesis principally in the spleen.

Copyright 2005 DxR Development Group Inc. All rights reserved.

Nervous System

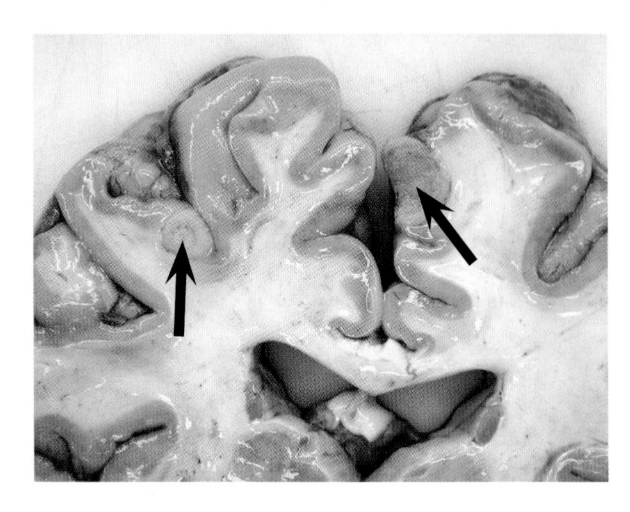

29. Metastatic disease to the brain. Metastatic lesions account for 25% of intracranial tumors. Metastases form sharply demarcated masses usually with surrounding edema. The most common location is at the gray matter–white matter junction.

30. Subarachnoid hemorrhage. This kind of hemorrhage may result from a ruptured intracranial aneurysm, trauma, or other vascular malformation. Irrespective of the etiology, there is an increased risk of injury after the hemorrhage due to vessel spasm in surrounding vessels that may involve the circle of Willis.

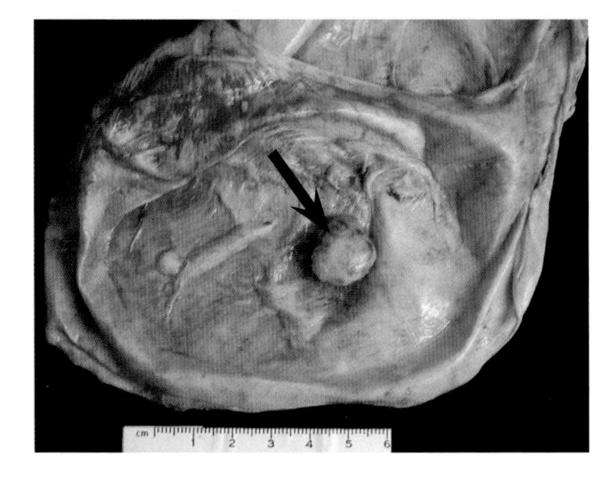

31. Dural meningioma. Meningiomas are well-defined, rounded masses that compress the underlying brain. The tumor may extend into the overlying bone. Characteristic microscopic features include a whorled pattern of cell growth and psammoma bodies.

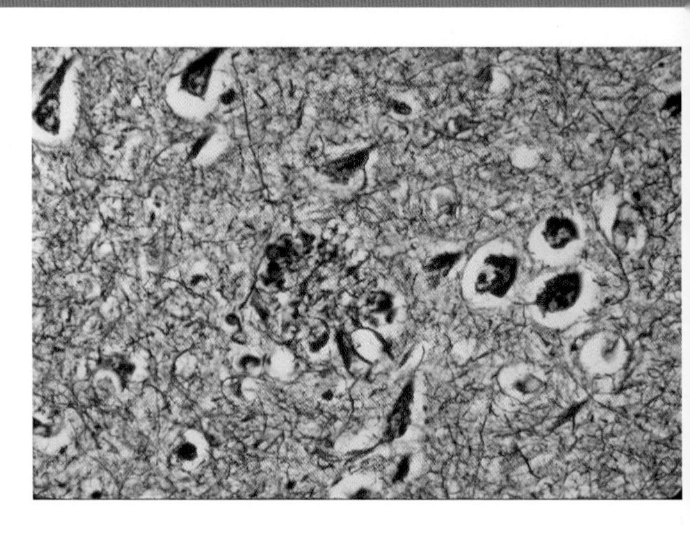

32. Neurofibrillary tangles within a neuron in Alzheimer disease. These tangles are bundles of paired filaments in the cytoplasm (abnormally hyperphosphorylated *tau* proteins). A senile plaque is seen in the center.

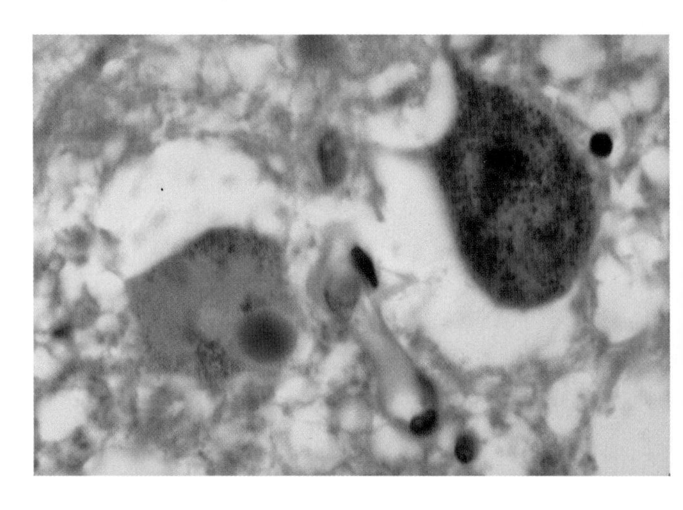

33. Lewy bodies in a substantia nigra neuron of Parkinson disease. These eosinophilic cytoplasmic filament inclusions are composed of α-synuclein. Lewy bodies may also be found in cholinergic cells of the basal nucleus of Meynert.

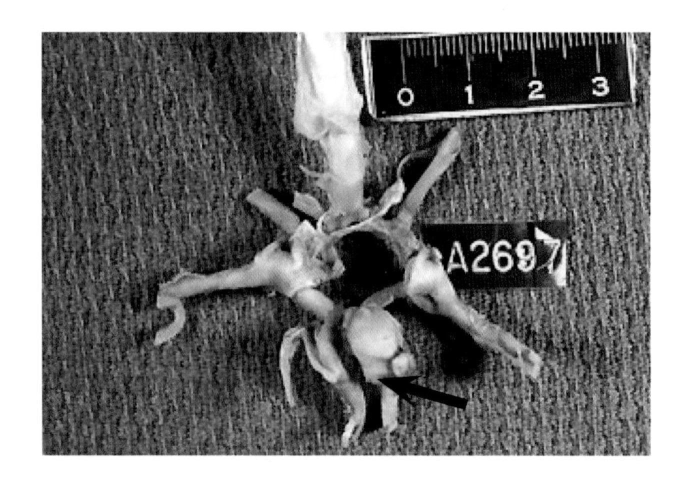

34. Berry aneurysm of the circle of Willis. Aneurysms occur on the anterior circle of Willis in 80% of cases. The anterior communicating artery and middle cerebral artery trifurcation are the most common locations. Sudden onset of severe headache is often the first symptom of aneurysm rupture.

 Copyright 2005 DxR Development Group Inc. All rights reserved.

ervous System (*Cont'd*)

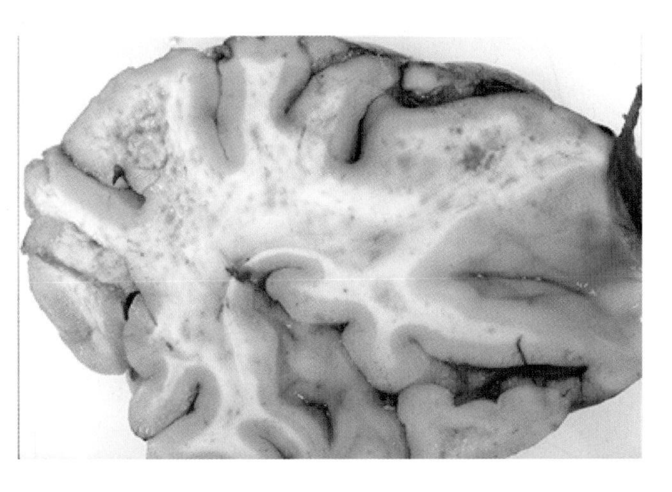

5. Progressive multifocal leukoencephalopathy is caused by the C polyomavirus, which primarily infects oligodendrocytes and results in emyelination. Lesions or patchy white matter destruction has affected e entire lobe of this brain.

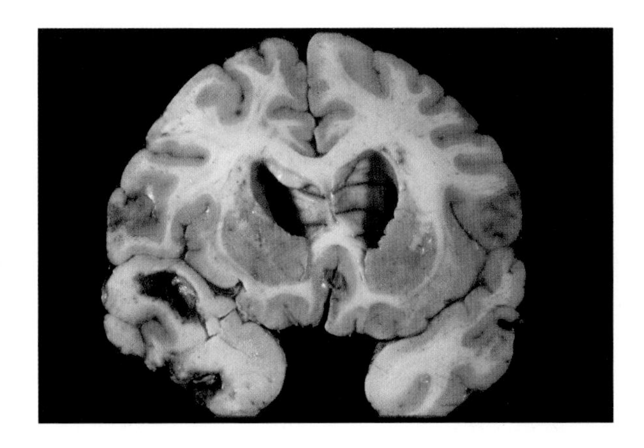

6. Temporal lobe abscess. A cerebral abscess is seen here on the right emporal lobe.

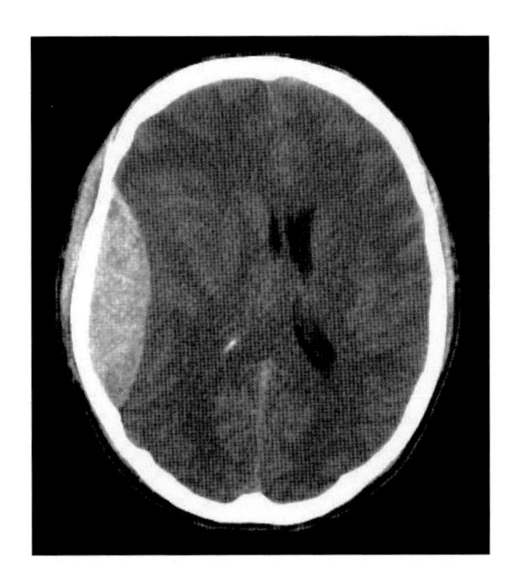

7. Epidural hematoma. This noncontrast CT shows the typical yperdense convex or lens-shaped appearance of an acute epidural ematoma. A midline shift to the left can also be seen. *(Source: Andrew ullins)*

Musculoskeletal and Skin Pathology

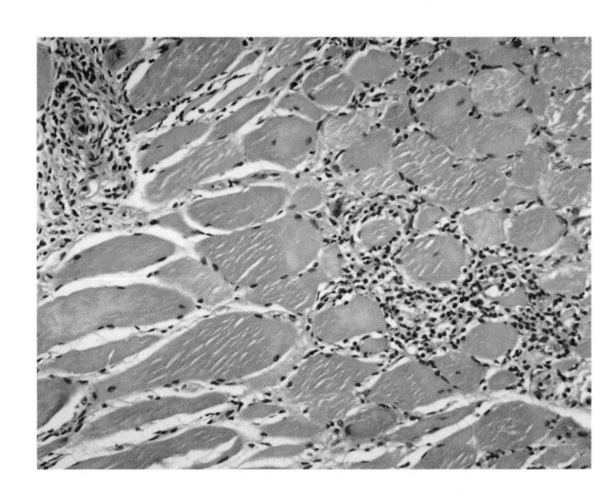

38. Muscle polymyositis. Inflammatory cells are found in the endomysium, and both necrotic and regenerating muscle fibers are scattered throughout the fascicle.

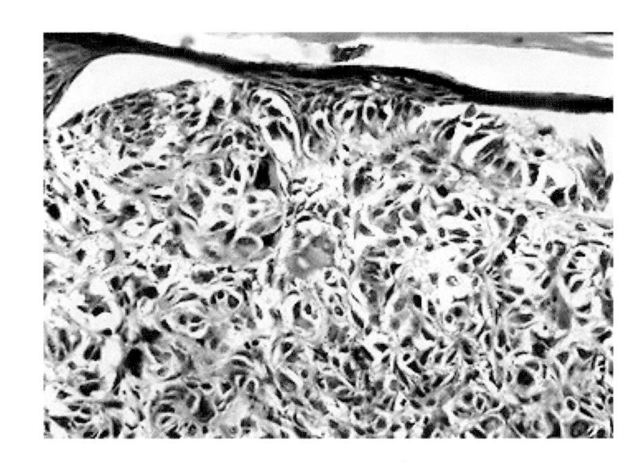

39. Melanoma cells are larger than normal nevus cells. They contain large nuclei with irregular contours, clumped chromatin, and prominent red nucleoli.

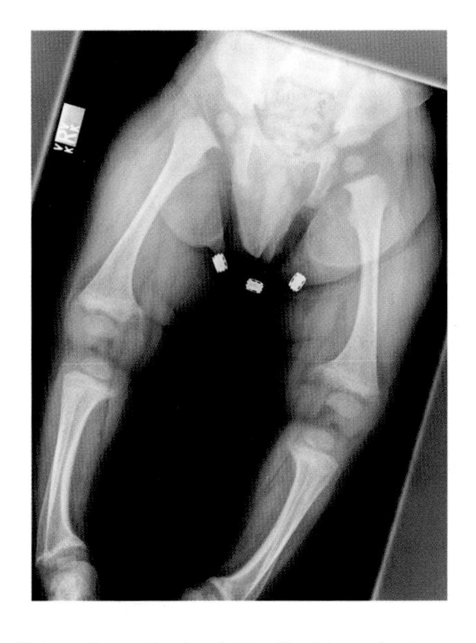

40. Rickets. This radiograph of a child with rickets depicts outward curvature (bowing) of the femur and tibia due to inadequate mineralization of bone. *(Source: Paul M. Michaud)*

Copyright 2005 DxR Development Group Inc. All rights reserved.

Pulmonary Pathology

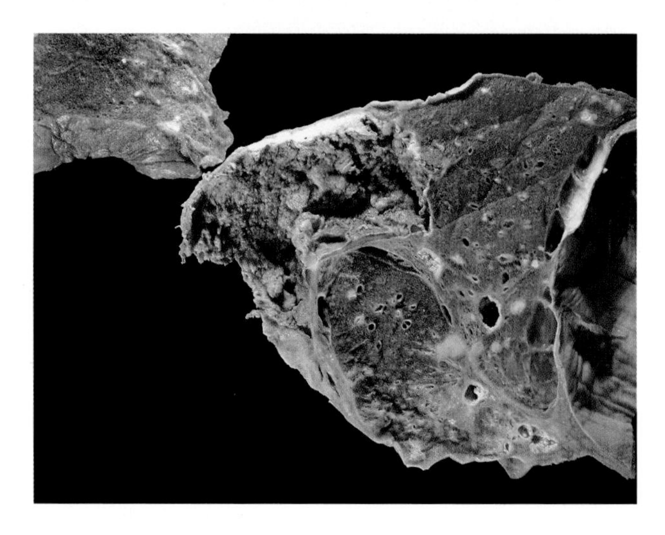

41. Cavitary lung abscess in the setting of prior tuberculosis scar tissue

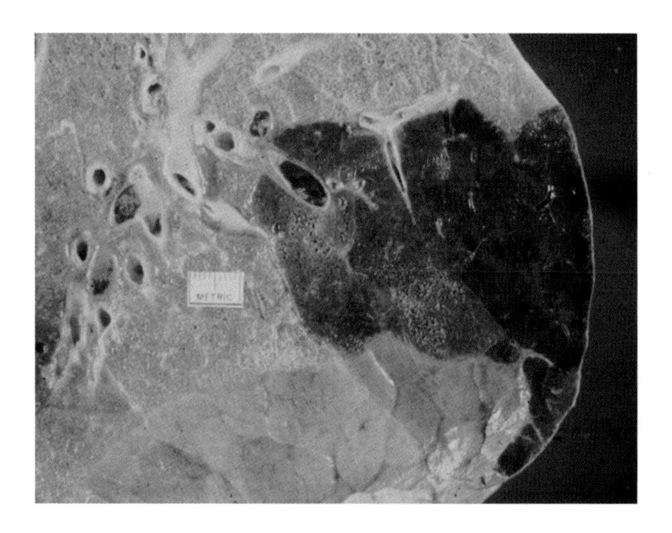

42. Pulmonary Infarction. This lung is from a patient with a history of a recent pulmonary embolus. There is a wedge-shaped hemorrhagic infarction. At the apex of the wedge are several arteries containing pulmonary emboli. *(Source: Katsumi Miyai, M.D., Ph.D.)*

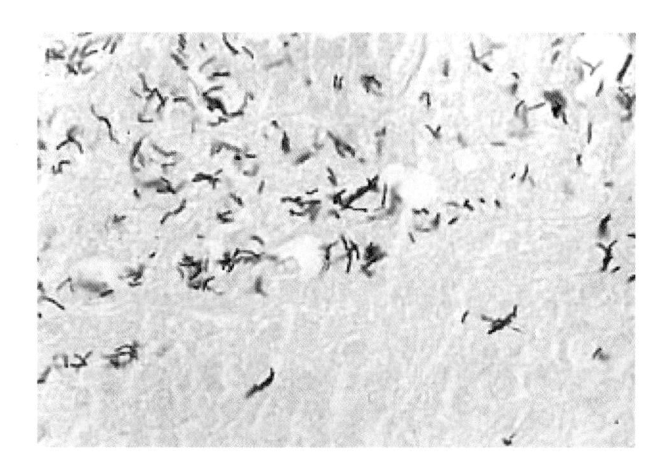

43. *Mycobacterium tuberculosis*, acid fast stain. An acid fast stain is used to diagnose the presence of mycobacteria in tissue and cytologic preparations. Many thin, red, rod-like *M. tuberculosis* organisms are shown here.

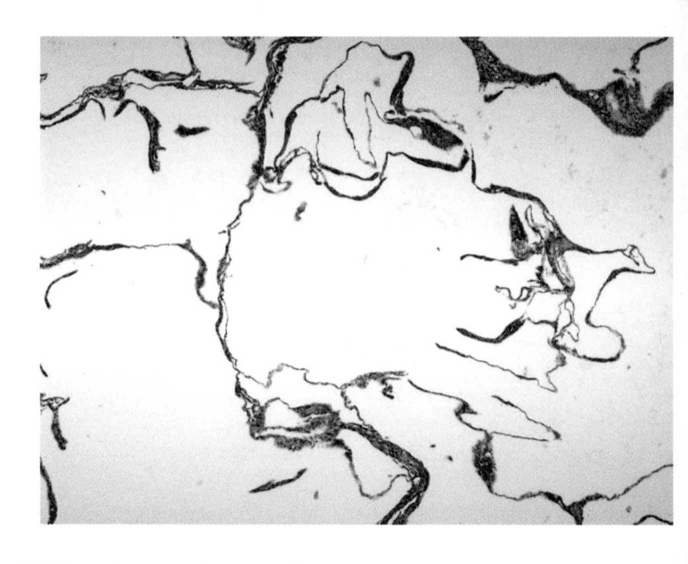

44. Panacinar emphysema. Emphysema is characterized by abnormally large alveoli separated by thin septa. These alveoli are permanently enlarged due to alveolar wall destruction. This type of emphysema more commonly affects the lower zones and anterior margins of the lung and is associated with α-1 AT deficiency.

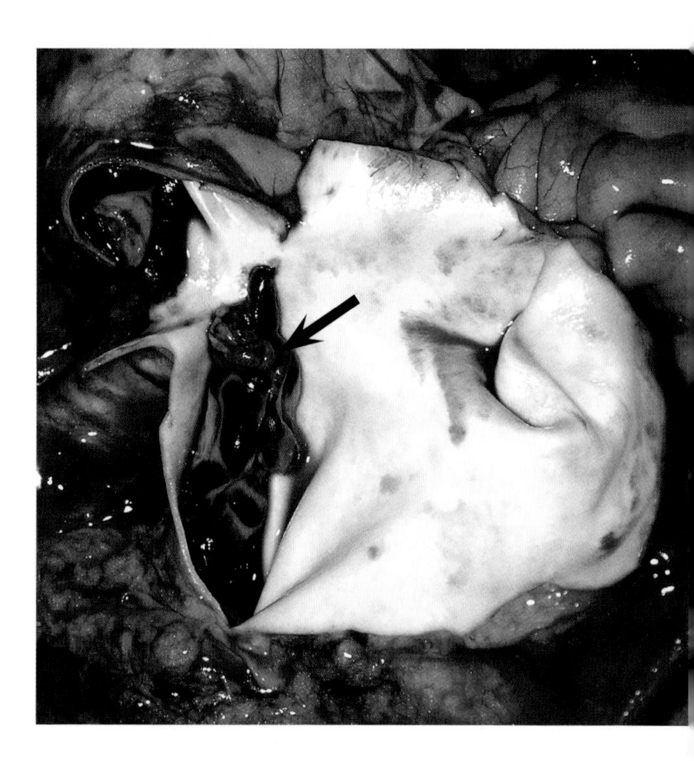

45. Pulmonary embolism, saddle embolus. A large majority of pulmonary emboli arise from deep venous leg thrombi. They pass through larger vascular channels until they lodge in smaller arteries of the lung. In this case, a large embolus occludes the bifurcation of the pulmonary artery (saddle embolus) and has caused sudden death in the patient.

 Copyright 2005 DxR Development Group Inc. All rights reserved.

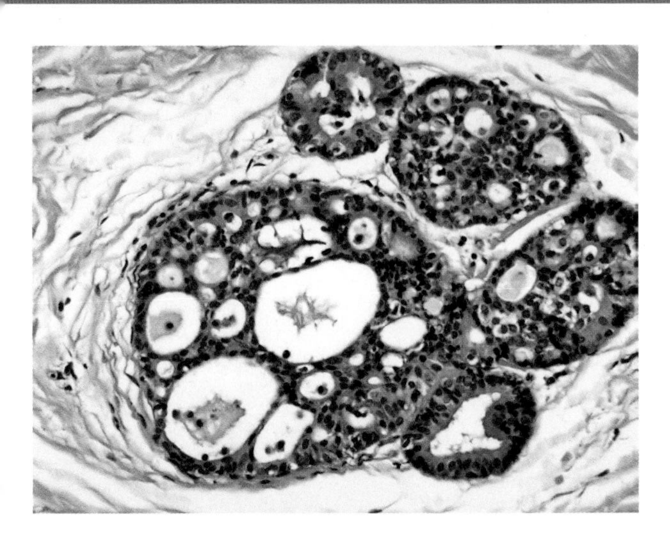

46. **Atypical ductal hyperplasia of the breast.** The duct is filled with columnar and rounded cells. The appearance is similar to lobular carcinoma in situ (LCIS), except that the atypia is limited in extent and the cells do not completely fill the ductal spaces. There is an increased risk of developing breast carcinoma when these lesions are identified.

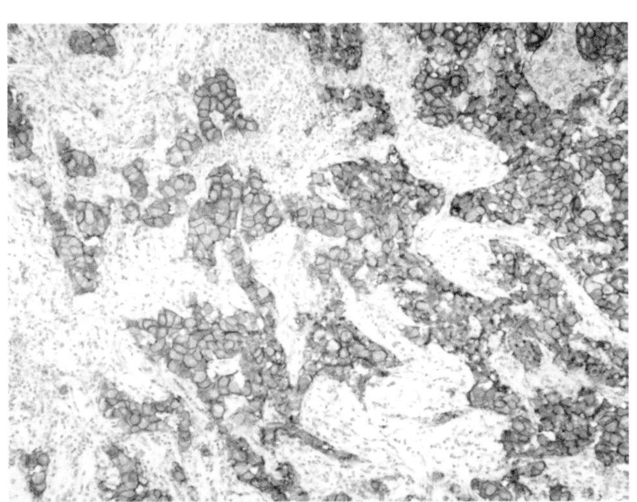

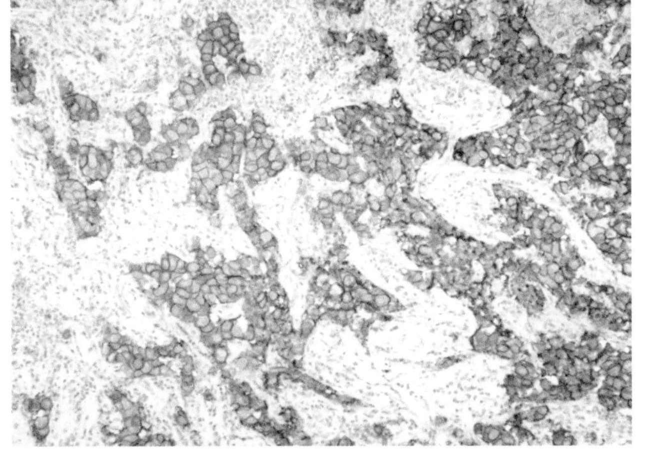

47. **HER2-positive breast carcinoma.** Breast cancers are routinely assayed for the HER2/neu gene using fluorescent in situ hybridization (FISH) and protein using immunohistochemistry (brown stain). These tests predict the clinical responses to antibodies targeted to the protein. Carcinomas that are HER2-positive tend to be poorly differentiated.

48. **Paget disease of the nipple.** Presents as an erythematous scaly crust in a unilateral nipple. Malignant ("Paget") cells extend from the ductal system into the skin of the nipple. There is an increased incidence of underlying invasive carcinoma.

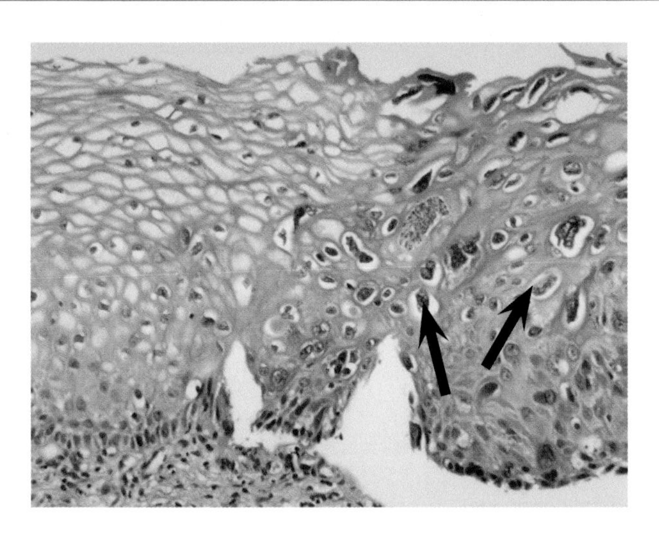

49. **Condyloma acuminata** is a benign tumor of stratified squamous epithelium caused by sexually transmitted HPV. Characteristic histologic features are acanthosis, parakeratosis, hyperkeratosis, and nuclear atypia with vacuolization (koilocytes).

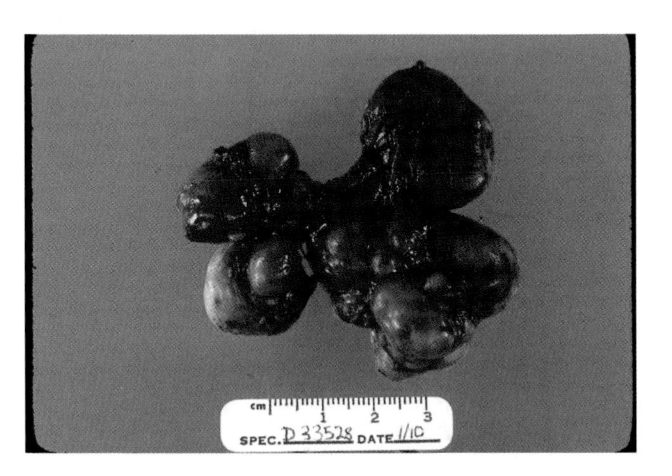

50. **Fibroadenomas** of the breast are sharply demarcated mobile masses within the breast. It is the most common benign tumor of the female breast and is more common in women under 30 years old.

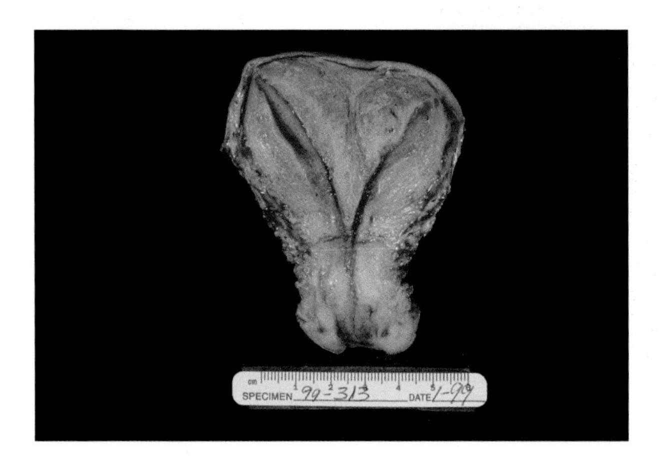

51. **Bicornuate uterus.** A heart-shaped uterus results from an incomplete fusion of the Müllerian or paramesonephric ducts. This abnormality is the most common type of uterine malformation.

Copyright 2005 DxR Development Group Inc. All rights reserved.

Reproductive Pathology (*Cont'd*)

52. Hydatidiform mole, gross specimen. Numerous swollen (hydropic) villi are shown here.

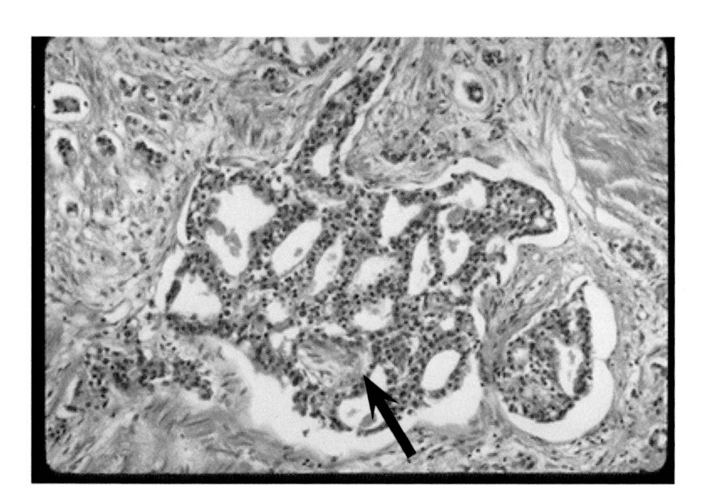

53. Prostate cancer. Small malignant cells with enlarged nuclei and dark cytoplasm make up this abnormal prostate gland. Perineural invasion by malignant cells is shown here (*arrow*).

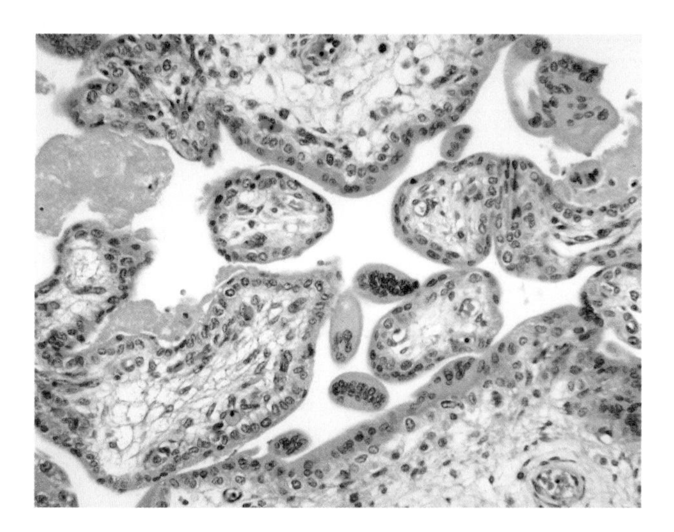

54. Normal placental villi

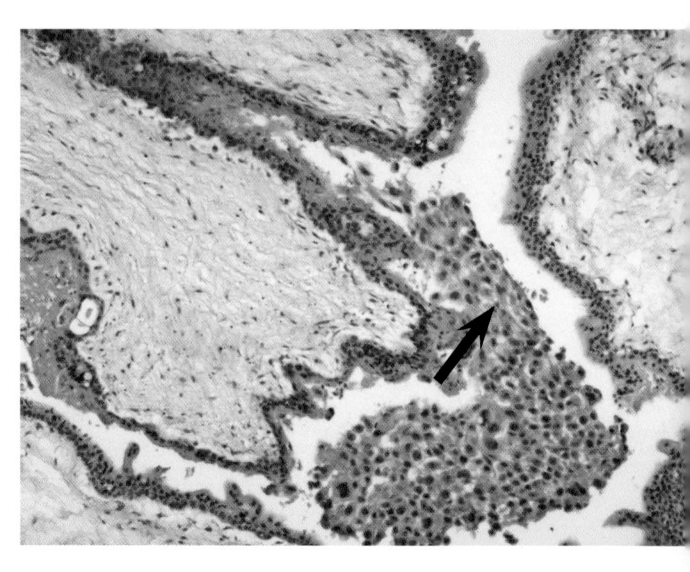

55. Complete hydatidiform mole. A complete hydatidiform mole shows hydropic swelling of most chorionic villi, trophoblast hyperplasia (*arrow*), and little-to-no vascularization of the villi. The majority of complete moles result from the fertilization by a single sperm of an egg that has lost its chromosomes.

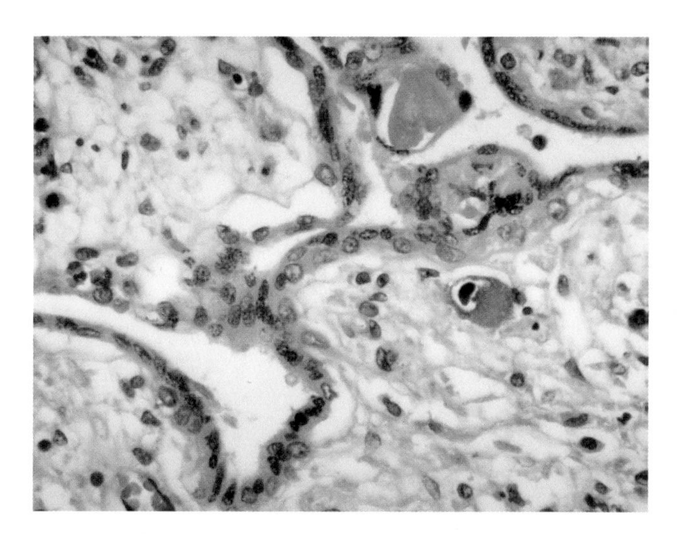

56. CMV infection of the placenta. Placental villitis occurs with hydropic change within the placenta with congenital cytomegalovirus (CMV) infection. An enlarged cell with mauve intranuclear inclusions is seen here, which is typical for congenital cytomegalovirus infection. Congenital infection with CMV is a common cause for hydrops fetalis.

 Copyright 2005 DxR Development Group Inc. All rights reserved.

Normal Hematology

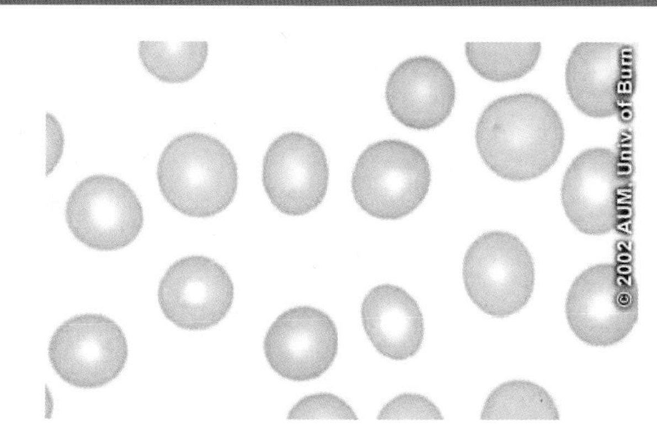

7. Normal red blood cell on peripheral smear

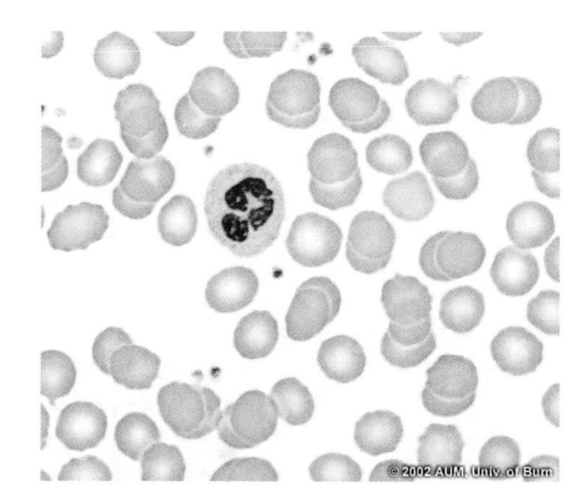

3. Segmented neutrophil. These are the first cells in acute flammation and have two types of lysosomal granules: azurophilic rimary) granules and specific (secondary). An increase of these cells und on peripheral smear indicates bacterial infections, inflammation, auma, and hemorrhage.

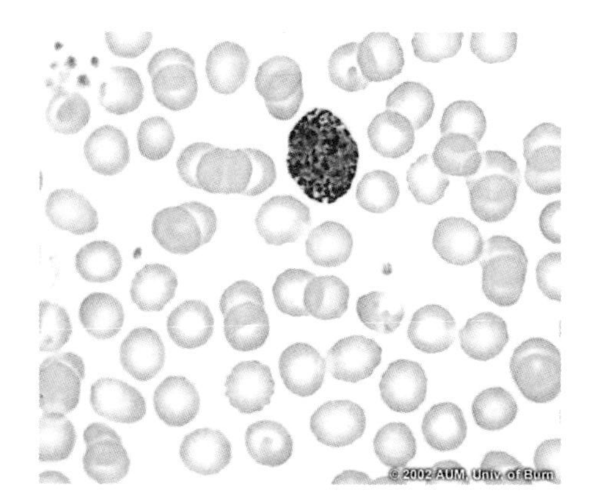

9. Basophil. These cells are involved in type I hypersensitivity and have rge basophilic and metachromatic granules. An increase in basophils on peripheral smear is seen in viral infections, urticaria, postsplenectomy, nd hematologic malignancies.

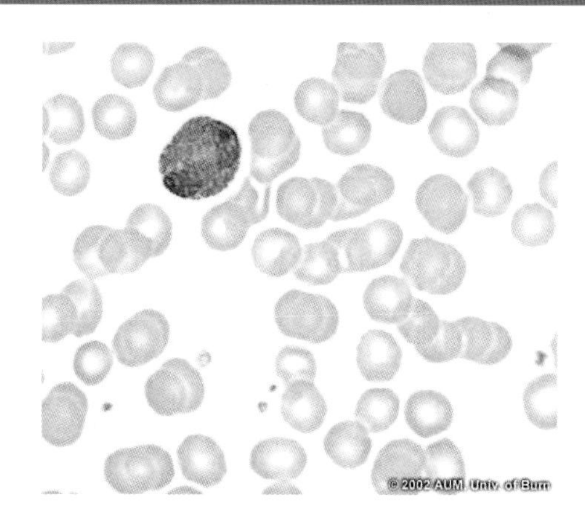

60. Eosinophil. These cells are most numerous in the blood during parasitic infections and allergic diseases.

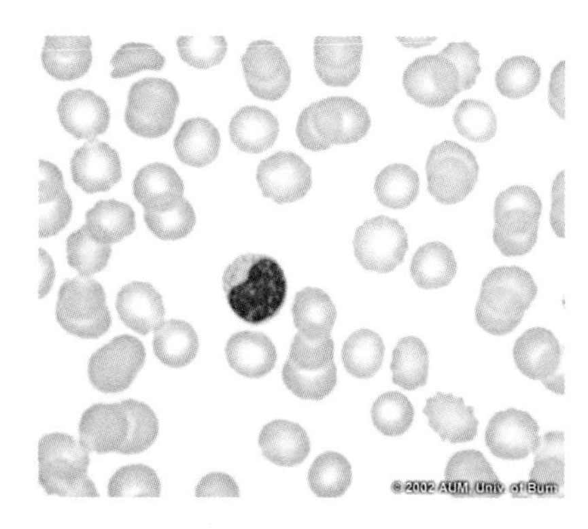

61. Lymphocyte on peripheral blood smear

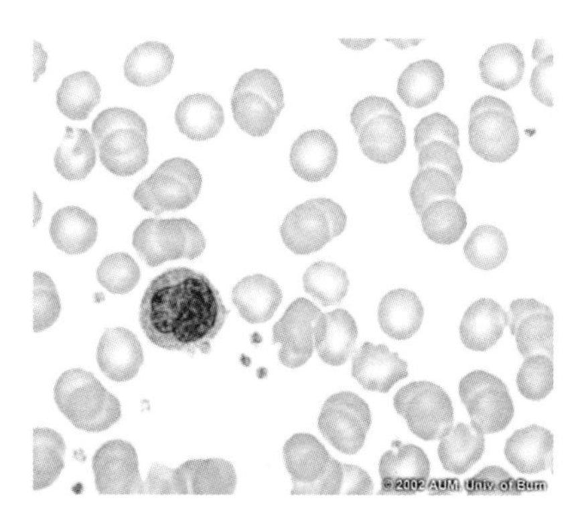

62. Monocyte. These are the precursors of phagocytic cells (macrophages, osteoclasts, Kupffer cells).

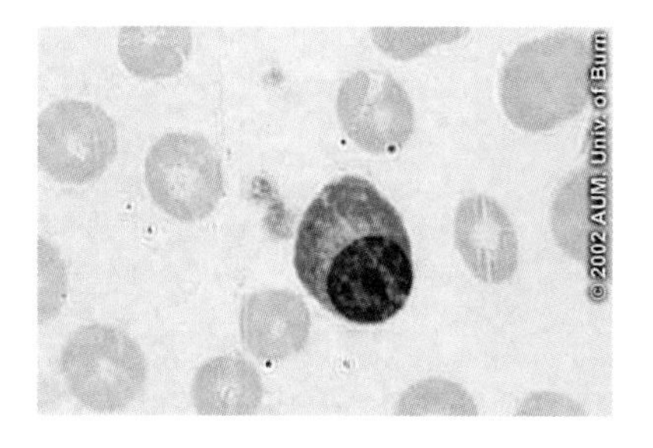

63. Plasma cell on peripheral blood smear

Copyright 2005 DxR Development Group Inc. All rights reserved.

Abnormal Hematology

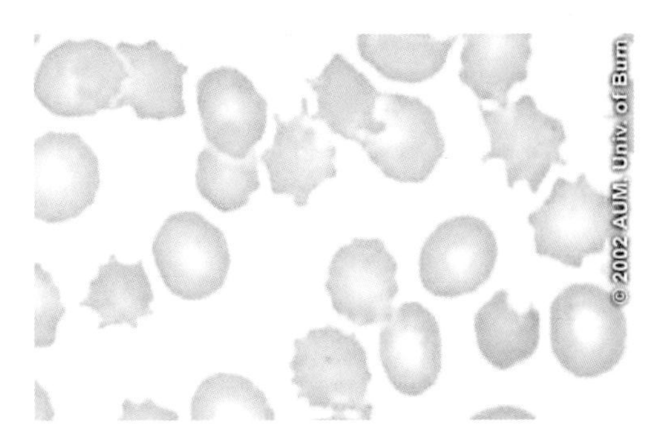

64. Acanthocytes. Red blood cells show many spicules on the surface of the cell. This can be seen in abetalipoproteinemia.

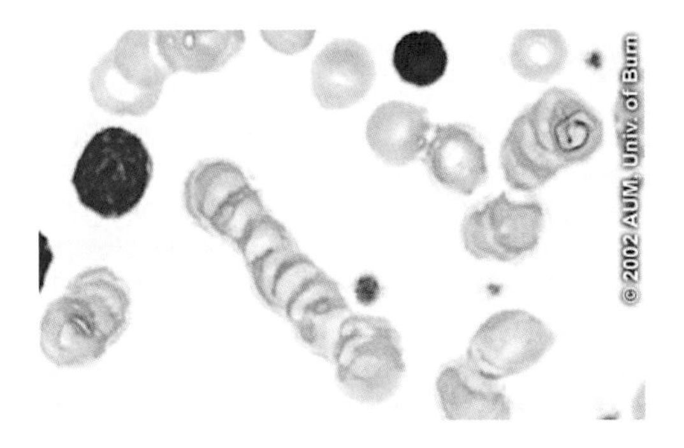

65. Rouleaux formation. Red blood cells appear to stack on each other in long chains. This results in a high erythrocyte sedimentation rate (ESR).

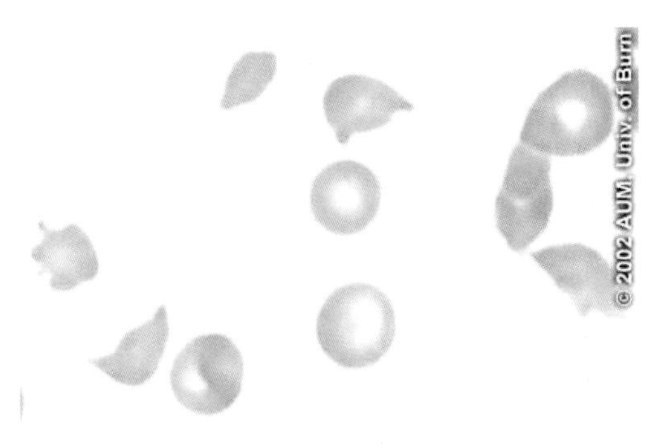

66. Schistocytes. These red blood cells are fragmented due to mechanical trauma (fibrin bands) and are seen in intravascular hemolysis.

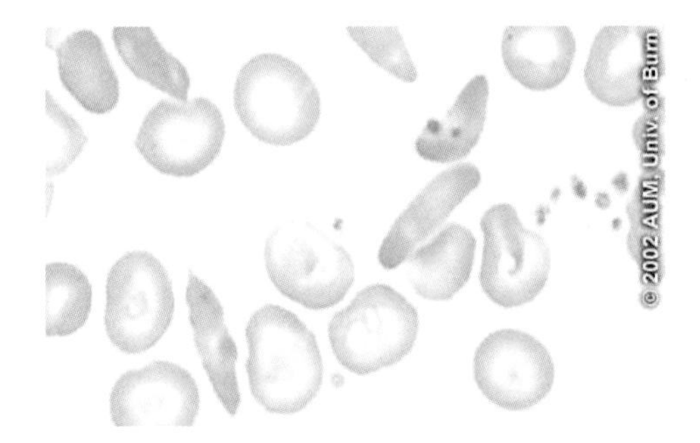

67. Sickle cells. Sickle cell anemia is an inherited blood disease in which the red blood cells produce abnormal hemoglobin. The abnormal hemoglobin causes crescent or sickle-shaped deformities.

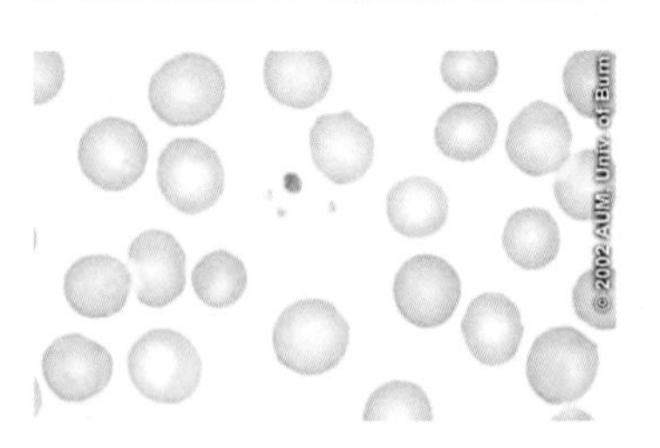

68. Spherocytes. These cells are near-spherical in shape and have no area of central pallor as seen in normal red blood cells. Spherocytes are found in hereditary spherocytosis, hemolytic anemia, and severe burns.

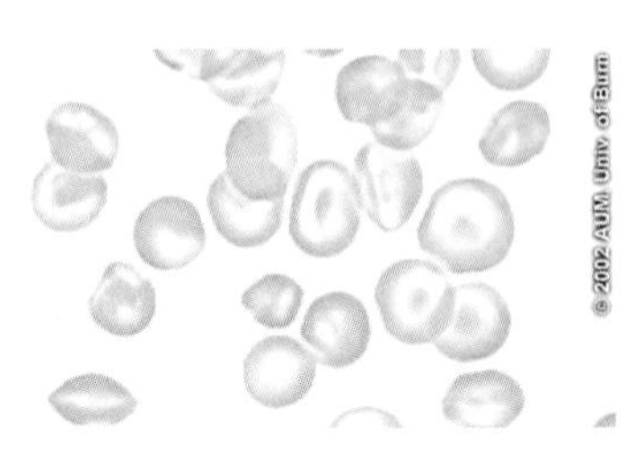

69. Target cells are erythrocytes with central staining and an inner ring and outer rim of pallor. They are seen in liver disease, thalassemia, or sickle cell disease.

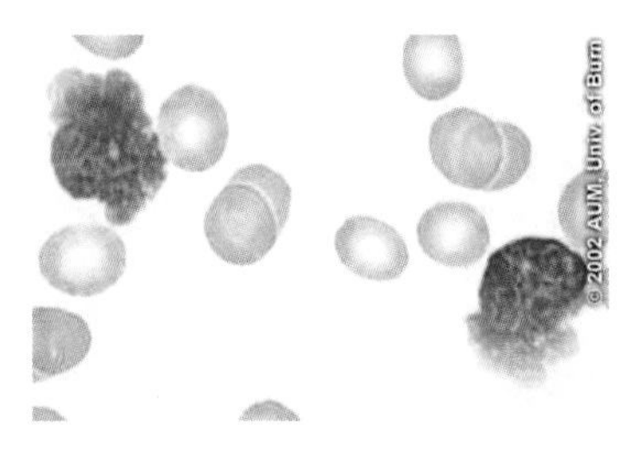

70. Smudge cells are leukocytes that have been damaged during preparation of the peripheral blood smear. This usually occurs because o the fragility of the cell and is seen in chronic lymphocytic leukemia (CLL)

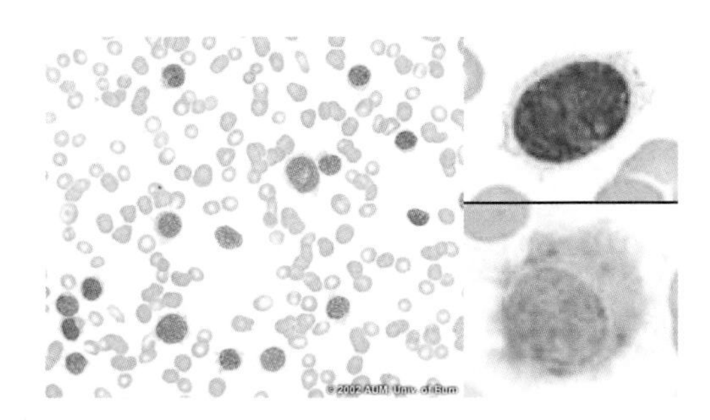

71. Hairy cells are characterized by fine, irregular pseudopods at the cell surface and immature nuclear features. These cells are found only ir hairy cell leukemia.

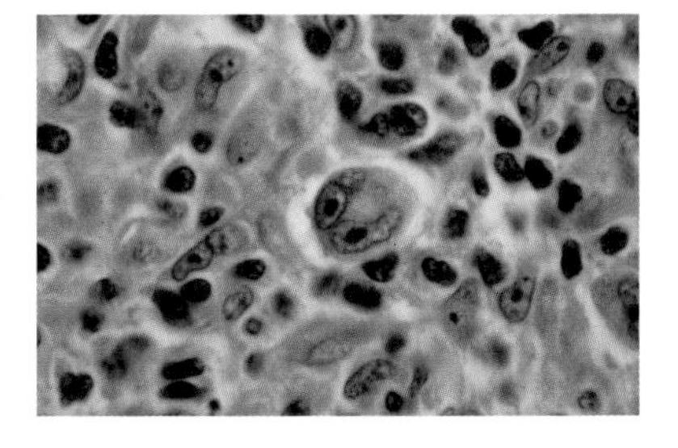

72. Hodgkin disease, Reed-Sternberg cell. A classic Reed-Sternberg cell has two nuclear lobes, large inclusion-like nucleoli, and abundant cytoplasm. These cells are essential in the histologic diagnosis of Hodgkin lymphoma.

 Copyright 2005 DxR Development Group Inc. All rights reserved.

Drug List

This drug table is designed for reference. Do not attempt to memorize it; we suggest you use this as a starting point to create your own personal drug list. Many of the most commonly used drugs are presented, but the list is not exhaustive. Also remember that most drugs have more side effects than noted in this table. We have included brand names in addition to the generic names—even though brand names are **not** tested on the USMLE—because you will be exposed to many brand names during your clinical years.

Generic Name (Brand Name) • Primary Use	Therapeutic Class and Mechanism	Most Common Adverse Effects
Acebutolol (Sectral®) • Hypertension, ventricular arrhythmias	β_1-adrenoreceptor blocker with mild intrinsic sympathomimetic activity	Hypoglycemic unawareness, bradycardia, hypotension, use with caution in asthmatics (not completely β_1 selective), impotence
Acetaminophen (Tylenol®) • Pain, fever, headache	Nonopiate, nonsalicylate, non-NSAID analgesic/antipyretic	Hepatic failure in overdose situation (treat with *N*-acetylcysteine)
Acetazolamide (Diamox®) • Glaucoma, high altitude sickness, edema if accompanied by metabolic alkalosis, urinary alkalization	Carbonic anhydrase inhibitor	Hypochloremic metabolic acidosis, renal stones, hypokalemia, paresthesias, hyperammonemia/hepatic encephalopathy in cirrhotic patients
Acyclovir (Zovirax®) • Herpes simplex	Antiviral (inhibits viral DNA polymerases)	Seizures in overdose, nephrotoxicity (crystalluria)
Albuterol (Proventil®) • Asthma	Antiasthma (β_2-receptor agonist)	Tremor, tachycardia, CNS stimulation, hypertension
Alendronate (Fosamax®) • Osteoporosis	Bisphosphonate	Esophageal ulcers, esophagitis, abdominal pain, gastric reflux, dysphagia
Allopurinol (Zyloprim®) • Gout	Xanthine oxidase inhibitor (irreversible suicide inhibitor)	Rash that can progress to Stevens-Johnson syndrome, and/or generalized vasculitis, irreversible hepatotoxicity
Alprazolam (Xanax®) • Anxiety	Benzodiazepine, sedative/hypnotic (increases frequency of $GABA_A$-receptor opening)	Drowsiness, dizziness, amnesia, decreased motor skills, dependence, respiratory depression with high doses (can be reversed with flumazenil)
Alteplase (t-PA; Activase®) • Acute myocardial infarction, stroke, pulmonary embolism	Tissue-plasminogen activator	Bleeding
Amiloride (Midamor®) • Hypertension, congestive heart failure	Potassium-sparing diuretic (blocks Na^+ channels in cortical collecting tubules)	Hyperkalemia

(Continued)

Generic Name (Brand Name) • Primary Use	Therapeutic Class and Mechanism	Most Common Adverse Effects
Amiodarone (Cordarone®) • Cardiac arrhythmias	Antiarrhythmic (class III)	Pulmonary fibrosis, deposits in skin, photosensitivity, thyroid dysfunction
Amitriptyline (Elavil®) • Depression, anxiety	Tricyclic antidepressant (blocks norepinephrine [NE] and serotonin [5HT] reuptake)	Strong antimuscarinic side effects (e.g., urinary retention, tachycardia, increased intraocular pressure), cardiac arrhythmias on overdose
Amlodipine (Norvasc®) • Hypertension, coronary artery disease	Calcium channel blocker	Dizziness, edema, somnolence, palpitations
Amoxicillin (Trimox®) • Respiratory tract infections, otic infections, urinary tract infections	Beta-lactam antibiotic (penicillin)	Diarrhea, maculopapular rash (especially in patients with mononucleosis), pseudomembranous colitis, anaphylactic reactions
Amoxicillin/potassium clavulanate (Augmentin®) • Respiratory tract infections, otic infections, urinary tract infections	Beta-lactam antibiotic (penicillin)/ beta-lactamase inhibitor	Diarrhea, maculopapular rash (especially in patients with mononucleosis), pseudomembranous colitis, anaphylactic reactions
Amphetamine mixed salts (Adderall®) • Treatment of attention deficit hyperactivity disorder	Stimulant	CNS stimulation, tachycardia, cardiac arrhythmias, dependence
Amphotericin B • Systemic mycoses	Antifungal (binds ergosterol)	Chills and fevers, hypotension, nephrotoxicity (dose-limiting)
Ampicillin (Omnipen®) • Urinary tract infections, upper respiratory tract infections	Beta-lactam antibiotic (penicillin)	Diarrhea, maculopapular rash (especially in patients with mononucleosis), pseudomembranous colitis, anaphylactic reactions
Anastrozole (Arimidex®) • Estrogen-dependent breast cancer in postmenopausal women	Aromatase inhibitor (decreases estrogen synthesis)	Hot flushes, fatigue
Aripiprazole (Abilify®) • Schizophrenia, bipolar disorder	Atypical antipsychotic	Fewer extrapyramidal side effects than typical antipsychotics
Aspirin (many name brands) • Pain/fever/headache, prevention of clotting with myocardial infarction/ transient ischemic attack	NSAID, salicylate; analgesic, antipyretic, antiplatelet, antiinflammatory (irreversibly inhibits COX-1/COX-2)	Gastric ulcers/bleeding, hypersensitivity, bronchoconstriction, nephrotoxicity, Reye syndrome; high doses: tinnitus, hyperventilation, acid/base disorders
Atenolol (Tenormin®) • Hypertension, angina pectoris due to coronary atherosclerosis, acute myocardial infarction	Antihypertensive (β_1 antagonist)	Hypoglycemic unawareness, use with caution in asthmatics (not completely β_1 selective), bradycardia, hypotension, impotence
Atomoxetine (Strattera®) • Attention deficit hyperactivity disorder	Norepinephrine (NE)–reuptake inhibitor	CNS stimulation, agitation, mood swings, potential increase in suicidal ideation
Atorvastatin (Lipitor®) • Hyperlipidemia, hypertriglyceridemia	HMG-CoA reductase inhibitor	Hepatic dysfunction, rhabdomyolysis, myalgia, myopathy

(Continued)

Generic Name (Brand Name) • Primary Use	Therapeutic Class and Mechanism	Most Common Adverse Effects
Azithromycin (Zithromax®) • Respiratory tract infections	Macrolide (azalide) antibiotic	Gastrointestinal distress, does not inhibit P450
Baclofen (Lioresal®) • Muscle spasticity secondary to multiple sclerosis, spinal cord injury	Muscle relaxant/spasmolytic ($GABA_B$ agonist)	Drowsiness, dizziness, mental confusion, incoordination
Benazepril (Lotensin®) • Hypertension, congestive heart failure, diabetic nephropathy	ACE inhibitor	Cough, angioedema, hyperkalemia, neutropenia (rare)
Benztropine (Cogentin®) • Parkinsonism, extrapyramidal disorders due to neuroleptics	Anticholinergic (muscarinic antagonist)	Strong antimuscarinic side effects (e.g., dry mouth, tachycardia, urinary retention, worsening of glaucoma, paralytic ileus, hyperthermia)
Bleomycin (Blenoxane®) • Hodgkin disease; testicular, ovarian, and bladder cancers	Antineoplastic (antibiotic)	Pulmonary fibrosis
Botulinum toxin (Botox®) • Spastic disorders, local muscle spasms, cosmetic (wrinkle reduction)	Spasmolytic (prevents ACh release)	Weakness of injected and adjacent muscles
Brimonidine (Alphagan P®) • Open-angle glaucoma, ocular hypertension	Antiglaucoma agent (α_2 agonist; decreases aqueous humor secretion)	Conjunctivitis, ocular itching
Bromocriptine (Parlodel®) • Parkinsonism, hyperprolactinemia	Dopamine agonist	Dyskinesias, nausea/vomiting, behavioral effects, postural hypotension
Budesonide (Pulmicort®) • Asthma	Corticosteroid	Oral candidiasis
Budesonide nasal spray (Rhinocort Aqua®) • Allergic rhinitis	Corticosteroid	Epistaxis (nosebleed)
Bupivacaine (Marcaine®) • Local anesthesia	Local anesthetic (amide)	Cardiovascular toxicity, various CNS symptoms (excitation or depression, dizziness, seizures), allergic reactions
Bupropion (Wellbutrin®) • Depression, smoking cessation	Antidepressant (heterocyclic)	Seizures, anxiety, insomnia, mania
Buspirone (BuSpar®) • Generalized anxiety disorder	Nonbenzodiazepine, nonbarbiturate anxiolytic; partial agonist at $5\text{-}HT_{1A}$ receptors	Dizziness, nervousness
Butalbital (with acetaminophen/caffeine; Fioricet®; with aspirin/caffine, Fiorinal®) • Tension headaches	Antiheadache; barbiturate	Drowsiness, may be habit forming
Candesartan (Atacand®) • Hypertension, heart failure	Angiotensin II–receptor (type AT_1) antagonist	Hypotension, increased potassium, contraindicated in pregnancy

(Continued)

Generic Name (Brand Name) • Primary Use	Therapeutic Class and Mechanism	Most Common Adverse Effects
Captopril (Capoten®) • Hypertension, congestive heart failure, diabetic nephropathy	ACE inhibitor (prototype; others in class include benazepril, enalapril, lisinopril, quinapril, ramipril)	Cough, angioedema, hyperkalemia, neutropenia (rare)
Carbamazepine (Tegretol®) • Seizures, trigeminal neuralgia	Anticonvulsant	Ataxia, diplopia, blood dyscrasias, P450 inducer, teratogen
Carisoprodol (Soma®) • Painful musculoskeletal conditions	Skeletal muscle relaxant	α_1 block (postural hypotension, tachycardia), mental confusion, drowsiness, incoordination
Carvedilol (Coreg®) • Heart failure, hypertension	Mixed α_1/β receptor antagonist	Heart failure, hypoglycemic unawareness, bronchospasm in asthmatics, bradycardia, hypotension, impotence
Celecoxib (Celebrex®) • Pain and inflammation secondary to a variety of conditions (osteoarthritis, rheumatoid arthritis, others)	NSAID (specific COX-2 inhibitor)	Increased risk of serious cardiovascular thrombotic events (myocardial infarction and stroke), gastrointestinal bleeding, ulceration, hepatic and renal dysfunction
First-generation cephalosporins Cefadroxil (Duricef®), cefazolin (Ancef®, Kefzol®), cephalexin (Keflex®)	Beta-lactam antibiotic (gram-positive infections [except *Enterococci* and *Listeria*], some enterics; not for CNS infections)	Anaphylaxis, serum sickness, rashes, diarrhea, pseudomembranous colitis; may potentiate renal toxicity of aminoglycosides; the *N*-methylthiotetrazole side chain found in cefamandole, cefotetan, and cefoperazone is associated with hypoprothrombinemia and intolerance to ethanol
Second-generation cephalosporins Cefaclor (Ceclor®), cefonicid (Monocid®), cefotetan (Cefotan®), cefoxitin (Mefoxin®), cefprozil (Cefzil®) and cefuroxime (Ceftin®, Zinacef®)	Beta-lactam antibiotic (less gram-positive activity and more gram-negative activity than first generation; not for CNS infections [except cefuroxime])	
Third-generation cephalosporins Cefdinir (Omnicef®), cefixime (Suprax®), cefotaxime (Claforan®), cefpodoxime (Vantin®), ceftazidime (Fortaz®, Tazidime®, Tazicef®), ceftibuten (Cedax®), ceftizoxime (Cefizox®) and ceftriaxone (Rocephin®)	Beta-lactam antibiotic (less gram-positive activity and more gram-negative activity than second generation; usually reserved for serious infections; ceftriaxone is drug of choice for gonococcal infections; also used for late-stage Lyme disease)	
Fourth-generation cephalosporins Cefepime (Maxipime®)	Beta-lactam antibiotic (more gram-negative activity and same gram-positive activity as first generation)	
Cetirizine (Zyrtec®) • Seasonal allergic rhinitis	Histamine H_1-receptor antagonist (second generation)	Fewer CNS side effects than first-generation antihistamines
Cholestyramine (Questran®) • Hyperlipidemia	Antihyperlipidemic	Constipation
Chloroquine (Aralen®) • Malaria	Antimalarial	Gastrointestinal distress, rash, visual and auditory impairment, peripheral neuropathy

(Continued)

Generic Name (Brand Name) • Primary Use	Therapeutic Class and Mechanism	Most Common Adverse Effects
Cimetidine (Tagamet®) • Peptic ulcer disease, gastroesophageal reflux disease, Zollinger-Ellison	Histamine H_2-receptor antagonist	Potent P450 inhibitor, gynecomastia, decreased libido
Ciprofloxacin (Cipro®) • Urinary tract infections, chronic bacterial prostatitis, respiratory tract infections, many others	Fluoroquinolone antibiotic (bacteriocidal inhibitor of topoisomerases)	Gastrointestinal distress, CNS dysfunction, superinfection, collagen dysfunction, pregnancy category D
Cisplatin (Platinol®) • Testicular, bladder, lung, and ovarian cancers	Antineoplastic (alkylating agent)	Nausea/vomiting, neurotoxic, nephrotoxic
Citalopram (Celexa®) • Depression	Selective serotonin reuptake inhibitor (SSRI) antidepressant	CNS stimulation, sexual dysfunction, serotonin syndrome, seizures in overdose; drug interactions with MAOIs, TCAs, meperidine
Clarithromycin (Biaxin®) • Respiratory tract infections	Macrolide antibiotic	P450 inhibitor, pseudomembranous colitis
Clomiphene (Clomid®) • Infertility	Fertility agent (blocks estrogen receptors in pituitary, induces ovulation)	Multiple births
Clomipramine (Anafranil®) • Obsessive-compulsive disorder, depression	Tricyclic antidepressant; norephinephrine (NE) and serotonin (5HT) reuptake inhibitor	Strong antimuscarinic side effects (e.g., cardiac arrhythmias, urinary retention, tachycardia, increased intraocular pressure, cognitive impairment)
Clonidine (Catapres®) • Hypertension, opioid withdrawal	Antihypertensive (α_2-agonist)	Sedation, rebound hypertension if stopped suddenly
Clonazepam (Klonopin®) • Seizures, panic disorder	Benzodiazepine, sedative/hypnotic, anticonvulsant (increases frequency of opening of $GABA_A$ receptor)	Drowsiness, dizziness, amnesia, respiratory depression
Clopidogrel (Plavix®) • Reduction of antithrombotic events	Antiplatelet agent, blocks ADP receptors on platelets	Bleeding, thrombotic thrombocytopenic purpura, neutropenia
Clozapine (Clozaril®) • Schizophrenia in patients unresponsive to other agents	Atypical antipsychotic	Agranulocytosis (requires weekly WBC count), seizures
Codeine/Acetaminophen (Tylenol® #2, #3 and #4 [least codeine in #2, most in #4]) • Moderate-to-severe pain	Opioid analgesic/acetaminophen combination	Respiratory depression, euphoria, constipation, pruritus, dependence
Colchicine (Colchicine®) • Gout	Inhibits microtubule assembly	Gastrointestinal distress, hepatic and renal damage
Cromolyn (intranasal: NasalCrom®, inhalational: Intal®, oral: Crolom®, ophthalmic: Opticrom®) • Asthma, allergies	Mast cell stabilizer	Cough, throat irritation when inhaled
Cyclobenzaprine (Flexeril®) • Treatment of muscle spasm	Skeletal muscle relaxant	Antimuscarinic side effects, drowsiness

(Continued)

Generic Name (Brand Name) • Primary Use	Therapeutic Class and Mechanism	Most Common Adverse Effects
Cyclophosphamide (Cytoxan®) • Lymphomas, ovarian and breast cancer, neuroblastoma	Antineoplastic (alkylating agent)	Bone marrow suppression, hemorrhagic cystitis (use mesna), gastrointestinal distress, alopecia
Cyclosporine (Restasis®) • Immunosuppressant for organ transplants	Immunosuppressant (binds cyclophilin)	Nephrotoxicity, peripheral neuropathy, hypertension, hirsutism, gingival hyperplasia
Dantrolene (Dantrium®) • Malignant hyperthermia, neuroleptic malignant syndrome (unlabeled use)	Skeletal muscle relaxant (blocks Ca^{2+} release from sarcoplasmic reticulum by blocking the ryanodine receptor)	Related to skeletal muscle relaxation
Desmopressin (DDAVP®) • Central diabetes insipidus, primary nocturnal enuresis	Synthetic vasopressin	Hyponatremia, decreased plasma osmolality, seizures
Dextroamphetamine (Dexedrine®) • ADHD, narcolepsy	Stimulant	CNS stimulation, tachycardia, cardiac arrhythmias, dependence
Diazepam (Valium®) • Status epilepticus, anxiety disorders, acute alcohol withdrawal, muscle spasms	Benzodiazepine; sedative/hypnotic, anticonvulsant (increases frequency of opening of $GABA_A$ receptor)	Respiratory depression, somnolence, dizziness, dependence
Digoxin (Lanoxin®) • Heart failure, atrial arrhythmias	Cardiac glycoside, antiarrhythmic, inhibits Na^+/K^+-ATPase	Arrhythmias, visual defects (green-yellow halos), nausea
Diltiazem (Cardizem®) • Hypertension, arrhythmias, angina	Calcium channel blocker	Peripheral edema, dizziness, bradycardia, AV block, hypotension
Diphenhydramine (Benadryl®) • Allergies, motion sickness	Antihistamine (first generation)	Antimuscarinic side effects, sedation
Diphenoxylate/Atropine (Lomotil®) • Diarrhea	Antidiarrheal (diphenoxylate: weak opioid; atropine: antimuscarinic)	Atropine is added to prevent abuse; side effects related to constipation, some CNS effects
Donepezil (Aricept®) • Mild-to-moderate dementia of Alzheimer disease	Anti-Alzheimer disease (reversible cholinesterase inhibitor)	Gastrointestinal (diarrhea, nausea, vomiting, increased gastric acid secretion), bradycardia or heart block
Dopamine • Shock	Sympathomimetic amine vasopressor (low dose: increases renal blood flow; moderate dose: positive inotropic effects)	Multiple (careful monitoring of patient's vitals required)
Doxazosin (Cardura®) • Benign prostatic hypertrophy, hypertension	α_1 antagonist	Orthostatic hypotension and syncope, particularly as a "first-dose effect," tachycardia
Doxepin (Sinequan®) • Depression, resistant pruritus	Tricyclic antidepressant (norephineprhine [NE] and serotonin [5HT] reuptake blocker)	Strong antimuscarinic side effects (urinary retention, tachycardia, increased intraocular pressure, cognitive impairment), cardiac arrhythmias
Doxorubicin (Doxil®) • Cancer	Antineoplastic (anthracycline antibiotic)	Cardiotoxicity (dexrazoxane may protect), myelosuppression, alopecia, gastrointestinal distress

(Continued)

Generic Name (Brand Name) • Primary Use	Therapeutic Class and Mechanism	Most Common Adverse Effects
Doxycycline (many name brands) • Prostatitis, sinusitis, *Chlamydia*, pelvic inflammatory disease, acne, prophylaxis against anthrax	Tetracycline antibiotic	Photosensitivity and skin reactions, gastrointestinal distress, dental enamel dysplasia, decreased bone growth
Duloxetine (Cymbalta®) • Major depressive disorder, diabetic peripheral neuropathic pain, generalized anxiety disorder	Serotonin and norepinephrine reuptake inhibitor (SNRI)	Nausea, sleep disorders, dizziness, dry mouth, anxiety, hypomania
Edrophonium (Enlon®) • Diagnosis of myasthenia gravis	Cholinesterase inhibitor (reversible, short-acting)	Bradycardia
Efavirenz (Sustiva®) • HIV infection	Nonnucleoside, reverse transcriptase inhibitor (NNRTI); used in combination regimens	CNS dysfunction, skin rash, elevated plasma cholesterol
Enalapril (Vasotec®) • Hypertension, congestive heart failure, diabetic nephropathy	ACE inhibitor	Cough, angioedema, hyperkalemia, neutropenia (rare)
Enoxaparin (Lovenox®) • Prevention of thrombosis	Anticoagulant; enhancer of antithrombin III activity; low molecular weight (LMW) heparin	Bleeding, much lower incidence of thrombocytopenia than heparin
Entacapone (Comtan®) • Parkinson disease	Antiparkinson (COMT inhibitor)	Exacerbates the effects of levodopa
Erythromycin (E-Mycin®) • Upper respiratory tract infections (including *Mycoplasma* and *Legionella*), skin infections, *Chlamydia*	Macrolide antibiotic	Potent P450 inhibitor, gastrointestinal distress, arrhythmia
Escitalopram (Lexapro®) • Depression, anxiety disorders, obsessive-compulsive disorder	Selective serotonin reuptake inhibitor (SSRI) antidepressant	Serotonin syndrome, somnolence, insomnia, tachycardia, postural hypotension, paresthesias, sexual dysfunction
Esomeprazole (Nexium®) • Peptic ulcer disease, gastroesophageal reflux disease, Zollinger-Ellison	Proton pump inhibitor	Gastrointestinal side effects, dizziness, headache
Estrogens (various preparations and combinations) • Contraception, hormone-replacement therapy, osteoporosis, female hypogonadism, dysmenorrhea	Estrogen-receptor agonist	Nausea, breast tenderness, endometrial hyperplasia, biliary disease, clot formation
Etanercept (Enbrel®) • Rheumatoid arthritis, ankylosing spondylitis, psoriatic arthritis	Tumor necrosis factor (TNF) inhibitor, immunosuppressant, DMARD (disease-modifiying antirheumatic drug)	Infections, injection site infections
Ethosuximide (Zarontin®) • Absence seizures	Anticonvulsant (blocks T-type Ca^{2+} channels)	Gastrointestinal distress

(Continued)

Generic Name (Brand Name) • Primary Use	Therapeutic Class and Mechanism	Most Common Adverse Effects
Ezetimibe (Zetia®) • Hyperlipidemia	Antihyperlipidemic (inhibits absorption of dietary cholesterol)	Diarrhea, myalgia, myopathy (rare)
Famciclovir (Famvir®) • Herpes simplex	Antiviral (inhibits viral DNA polymerases)	Fatigue
Famotidine (Pepcid®) • Peptic ulcer disease, gastroesophageal reflux disease, Zollinger-Ellison	Histamine H_2-receptor antagonist	
Felodipine (Plendil®) • Hypertension	Calcium channel blocker	Headache, dizziness, reflex tachycardia, gingival hyperplasia
Fentanyl (Duragesic®) • Moderate-to-severe pain	Analgesic (opioid agonist)	Respiratory depression, constipation, miosis, emesis, pruritus, dependence
Fenofibrate (TriCor®) • Hypertriglyceridemia, hypercholesterolemia	Antihyperlipidemic (fibric acid derivative; ligand for peroxisome proliferator-activated receptor-alpha [PPAR-α])	Gastrointestinal distress, gallstones, myopathy (especially in combination with statins), elevated liver enzymes
Fexofenadine (Allegra®) • Seasonal allergic rhinitis	Histamine H_1-receptor antagonist (second generation)	Fewer CNS side effects than first-generation antihistamines
Filgrastim (Neupogen®) • Neutropenia (e.g., chemotherapy, bone marrow transplant)	Granulocyte colony-stimulating factor (G-CSF), cytokine	Bone pain, splenomegaly, splenic rupture (rare)
Finasteride (Proscar®) • Benign prostatic hyperplasia, male pattern baldness	5α-reductase inhibitor	Pregnancy category X, impotence, hypotension
Fludrocortisone (Florinef®) • Hypotension	Mineralocorticoid	Hypertension, edema
Fluconazole (Diflucan®) • Esophageal and invasive candidiasis, coccidiomycosis, cryptococcal meningitis	Conazole antifungal (inhibits 14-α-demethylase, preventing the conversion of lanosterol to ergosterol)	Hepatic dysfunction, decreased steroid synthesis, inhibits CYP3A4
Fluticasone (Floven®) • Asthma	Corticosteroid; antiasthmatic	Epistaxis, pharyngitis, angioedema, upper respiratory infection
Flumazenil (Romazicon®) • Benzodiazepine overdose	Benzodiazepine-receptor antagonist	Seizures
Fluoxetine (Prozac®) • Depression, obsessive-compulsive disorder, anxiety disorders, bulimia	Selective serotonin reuptake inhibitor (SSRI) antidepressant	CNS stimulation, sexual dysfunction, gastrointestinal distress, serotonin syndrome, seizures in overdose; drug interactions with MAOIs, TCAs, meperidine
Flutamide (Eulexin®) • Prostatic carcinoma	Androgen-receptor antagonist	Hepatotoxicity
Fluvastatin (Lescol®) • Hypercholesterolemia	HMG-CoA reductase inhibitor; antihyperlipidemic	Myalgia, myopathy, rhabdomyolysis, hepatic dysfunction, elevated transaminases

(Continued)

Generic Name (Brand Name) • Primary Use	Therapeutic Class and Mechanism	Most Common Adverse Effects
Fluvoxamine (Luvox®) • Depression, obsessive-compulsive disorder	Selective serotonin reuptake inhibitor (SSRI) antidepressant	CNS stimulation, sexual dysfunction, serotonin syndrome, seizures in overdose
Fosfomycin (Monurol®) • UTIs	Broad-spectrum, bactericidal antibiotic	Diarrhea
Fosinopril (Monopril®) • Hypertension, heart failure, diabetic nephropathy	ACE inhibitor	Cough, angioedema, hyperkalemia
Furosemide (Lasix®) • Edema and hypertension	Loop diuretic (blocks $Na^+/K^+/2Cl^-$ transporter in thick ascending limb)	Hypokalemia, hypocalcemia, hyperuricemia, hyponatremia, tinnitus, and hearing loss
Gabapentin (Neurontin®) • Epilepsy, postherpetic neuralgia, diabetic peripheral neuropathy	Anticonvulsant	Dizziness, somnolence
Ganciclovir (Cytovene®) • CMV, HSV, VZV infections	Antiviral (inhibits viral DNA polymerases)	Myelosuppression, fever, rash
Gemfibrozil (Lopid®)	Antihyperlipidemic (fibric acid derivative; ligand for peroxisome proliferator-activated receptor-alpha [PPAR-α])	Gastrointestinal distress, gallstones, myopathy (especially in combination with statins), elevated liver enzymes
Gentamicin (Garamycin®) • Severe gram-negative infections	Aminoglycoside antibiotic	Ototoxicity, nephrotoxicity, muscle weakness (caused by ↓ acetylcholine release)
Glatiramer (Copaxone®) • Relapsing-remitting multiple sclerosis	Immune-modifying agent	Injection site reaction, chest pain
Glimepiride (Amaryl®) • Type 2 diabetes mellitus	Oral hypoglycemic agent, sulfonylurea	Hypoglycemia, weight gain, rash
Glipizide (Glucotrol®) • Type 2 diabetes mellitus	Oral hypoglycemic agent, sulfonylurea	Hypoglycemia, weight gain, rash
Glyburide (Micronase®, DiaBeta®, Glynase®) • Type 2 diabetes mellitus	Oral hypoglycemic agent, sulfonylurea	Hypoglycemia, weight gain, rash
Granisetron (Kytril®) • Nausea/vomiting	Antiemetic ($5HT_3$ antagonist)	Headache, dizziness
Haloperidol (Haldol®) • Schizophrenia, Tourette syndrome	Antipsychotic, butyrophenone (blocks dopamine D_2 receptors; high potency)	Extrapyramidal symptoms, tardive dyskinesia, hyperprolactinemia, neuroleptic malignant syndrome, fewer autonomic side effects than low-potency neuroleptics

(Continued)

Generic Name (Brand Name) • Primary Use	Therapeutic Class and Mechanism	Most Common Adverse Effects
Halothane (Fluothane®) • General anesthesia	General inhalational anesthetic	Cardiovascular and respiratory depression, sensitizes heart to catecholamines, hepatitis
Heparin (many brand names) • Prevention of thrombosis	Anticoagulant (enhances antithrombin III activity)	Bleeding, thrombocytopenia; antidote is protamine
Hydrochlorothiazide (HydroDIURIL®) • Edema and hypertension	Antihypertensive and thiazide diuretic (inhibits Na^+/Cl^- transporter in distal convoluted tubule)	Hypokalemia, hypercalcemia, hyperuricemia, hyperglycemia, hyperlipidemia, possible sulfonamide allergenicity
Hydralazine (Apresoline®) • Hypertension	Antihypertensive (vasodilator; releases nitric oxide from endothelial cells)	Systemic lupus erythematosus–like syndrome, tachycardia, salt/H_2O retention
Hydrocodone (Vicodin®) • Moderate-to-severe pain	Opioid analgesic	Respiratory depression, euphoria, constipation, nausea, pruritus, dependence
Ibuprofen (Motrin®) • Osteoarthritis, rheumatoid arthritis, inflammatory conditions, mild-to-moderate pain, antipyretic	NSAID (nonselective COX inhibitor)	Increased risk of serious cardiovascular thrombotic events, bleeding, gastrointestinal ulceration
Imipenem-cilastatin (Primaxin®) • For severe infections, e.g., respiratory, intraabdominal, others	Imipenem: carbapenem antibiotic; cilastatin: renal dehydropeptidase inhibitor (prevents inactivation of imipenem)	Allergy (cross-allergenicity with penicillins), gastrointestinal distress, seizures
Imipramine (Tofranil®) • Depression	Tricyclic antidepressant (norepinephrine [NE] and serotonin [5HT] reuptake blocker)	Strong anticholinergic side effects (e.g., urinary retention, tachycardia, increased intraocular pressure), cardiac arrhythmias on overdose
Indinavir (Crixivan®) • HIV infection	Protease inhibitor	Nephrolithiasis, hematologic abnormalities, inhibition of P450
Indomethacin (Indocin®) • Arthritis, acute inflammation	Antiinflammatory, NSAID (nonspecific COX inhibitor)	Gastrointestinal bleeding, increased risk of thrombotic events, renal toxicity
Interferon-α (INF-α; Roferon-A®, Intron-A®) • Hepatitis B and C, leukemias, melanoma, Kaposi sarcoma	Interferon	Flu-like symptoms, depression, bone marrow suppression
Interferon-β (INF-β; Avonex®, Refib®) • Multiple sclerosis	Interferon	Flu-like symptoms, depression, bone marrow suppression
Interferon-γ (INF-γ; Actimmune®) • Chronic granulomatous disease	Interferon	Flu-like symptoms, depression, bone marrow suppression
Ipratropium (Atrovent®) • Bronchospasm associated with chronic obstructive pulmonary disease	Anticholinergic bronchodilator (muscarinic antagonist)	Quaternary amine, so there is little systemic absorption
Irbesartan (Avapro®) • Hypertension, heart failure, diabetic neuropathy	Angiotensin II–receptor (type AT_1) antagonist	Hypotension, increased BUN and potassium, contraindicated in pregnancy

(Continued)

Generic Name (Brand Name) • Primary Use	Therapeutic Class and Mechanism	Most Common Adverse Effects
Isoniazid (INH; Nydrazid®) • Tuberculosis	Antimycobacterial	Hepatotoxicity, hemolysis (in G6PD deficiency, peripheral neuropathy; reversed by pyridoxine)
Isoproterenol (Isuprel®) • Heart block, bronchospasm	Nonspecific β agonist	Tremor, angina, arrhythmia
Isosorbide (dinitrate [Isordil®]; mononitrate [Imdur®]) • Angina pectoris	Nitrate vasodilator	Hypotension, tachycardia, headache
Isotretinoin (Accutane®) • Severe cystic acne	Retinoid	Pregnancy category X, depression, suicidal ideation, decreased night vision, dry mouth
Itraconazole (Sporanox®) • Blastomycoses, sporotrichoses, others	Conazole antifungal (inhibits 14-α-demethylase, preventing the conversion of lanosterol to ergosterol)	Hepatic dysfunction, decreased steroid synthesis, inhibits CYP3A4
Ivermectin (Stromectol®) • *Strongyloides*, onchocerciasis	Anthelmintic	Hypotension, headache, muscle aches
Ketoconazole (Nizoral®) • *Blastomyces*, *Histoplasma*, *Candida*, other	Conazole antifungal (inhibits 14-α-demethylase, preventing the conversion of lanosterol to ergosterol)	Hepatic dysfunction, decreased steroid synthesis, inhibits CYP3A4
Lamivudine (3TC) (Epivir®) • HIV infection, hepatitis B	Antiretroviral, nucleoside reverse transcriptase inhibitor (NRTI)	Least toxic of the NRTIs, some headache, gastrointestinal distress
Lamotrigine (Lamictal®) • Partial seizures, adjunctive for other seizure types, bipolar disorder	Anticonvulsant	Stevens-Johnson syndrome, life-threatening rash, sedation, ataxia
Lansoprazole (Prevacid®) • Ulcers, gastroesophageal reflux disease, Zollinger-Ellison	Proton pump inhibitor (irreversible blocker of H^+/K^+ ATPase on parietal cells)	
Latanoprost (Xalatan®) • Open-angle glaucoma, ocular hypertension	$PGF_{2\alpha}$ agonist	Eyelash changes, iris pigmentation changes
Leflunomide (Arava®) • Rheumatoid arthritis	DMARD, pyrimidine synthesis inhibitor (inhibits dihydroorotate dehydrogenase)	Diarrhea, elevated hepatic enzymes, alopecia, rash
Leuprolide (Lupron Depot®) • Advanced prostatic cancer	Gonadotropin-releasing hormone (GnRH) analog (others in class include goserelin, nafarelin)	Bone pain, gynecomastia, impotence, testicular atrophy, hematuria
Levodopa-carbidopa (Sinemet®) • Parkinsonism	Levodopa: dopamine precursor; carbidopa: peripheral dopa decarboxylase inhibitor	Dyskinesias, behavioral changes, hypotension, on-off phenomena
Levofloxacin (Levaquin®) • Urinary tract infections, chronic bacterial prostatitis, respiratory tract infections, many others	Fluoroquinolone antibiotic (bacteriocidal inhibitor of topoisomerases)	Gastrointestinal distress, CNS dysfunction, superinfection, collagen dysfunction, pregnancy category D

(Continued)

Generic Name (Brand Name) • Primary Use	Therapeutic Class and Mechanism	Most Common Adverse Effects
Levothyroxine (Synthroid®) • Hypothyroidism	Synthetic T_4	Symptoms of hyperthyroidism
Lisinopril (Prinivil® and Zestril®) • Hypertension, congestive heart failure, diabetic nephropathy	ACE inhibitor	Cough, angioedema, hyperkalemia
Lithium (Eskalith®) • Bipolar disease	Antimanic	Nephrogenic diabetes insipidus, tremor, goiter, seizures, teratogen
Loratadine (Claritin®) • Seasonal allergic rhinitis	Histamine H_1 receptor antagonist (second generation)	Fewer CNS side effects than first-generation antihistamines
Lorazepam (Ativan®) • Anxiety	Benzodiazepine, sedative/hypnotic (increases frequency of $GABA_A$ receptor opening)	Drowsiness, dizziness, amnesia, respiratory depression
Losartan (Cozaar®) • Hypertension, diabetic neuropathy, heart failure	Angiotensin II–receptor (type AT_1) antagonist (prototype; other "**-sartans**" include candesartan, irbesartan, olmesartan, telmisartan, valsartan)	Hypotension, increased BUN and potassium, contraindication in pregnancy
Lovastatin (Mevacor®) • Hypercholesterolemia	HMG-CoA reductase inhibitor, antihyperlipidemic (prototype; other "**-statins**" include atorvastatin, fluvastatin, pravastatin, rosuvastatin, simvastatin)	Myalgia, myopathy, rhabdomyolysis, hepatic dysfunction, elevated transaminases
Mannitol (Osmitrol®) • Increased intracranial pressure, to promote diuresis in renal failure, increased intraocular pressure (narrow-angle glaucoma), to promote excretion of renal toxins	Osmotic diuretic	Extracellular fluid volume expansion causing hyponatremia, nausea, headache
Mebendazole (Vermox®) • Whipworm, pinworm infections	Anthelminthic	Gastrointestinal distress
Meclizine (Antivert®) • Motion sickness	Antiemetic (H_1 antagonist)	Dizziness, drowsiness
Medroxyprogesterone (Provera®) • Contraceptive injection, hormone-replacement therapy	Progestin	Thromboembolic disorders, myocardial infarction, galactorrhea
Memantine (Namenda®) • Alzheimer disease	NMDA-receptor antagonist	Dizziness
Meperidine (Demerol®) • Moderate-to-severe pain	Opioid analgesic	Seizures, typical opioid side effects (no miosis), dangerous in combination with SSRIs and MAOIs, antimuscarinic
Mesalamine (Canasa®) • Inflammatory bowel disease	Antiinflammatory	Gastrointestinal distress, dizziness
Metaproterenol (Alupent®) • Asthma	Antiasthmatic (β_2-receptor agonist)	CNS stimulation, hypertension, tachycardia

(Continued)

Generic Name (Brand Name) • Primary Use	Therapeutic Class and Mechanism	Most Common Adverse Effects
Metformin (Glucophage®) • Type 2 diabetes mellitus	Antidiabetic, biguanide	Lactic acidosis
Methadone (Dolophine®) • Maintenance treatment and detoxification of opioid addiction, moderate-to-severe pain	Opioid analgesic	Respiratory depression, euphoria, constipation, pruritus, dependence
Methotrexate (Trexall®) • Neoplastic disease, arthritis, psoriasis	Antineoplastic, immunosuppressant (inhibits dihydrofolic reductase)	Myelosuppression, gastrointestinal distress, crystalluria (leucovorin rescue used to lower toxicity)
Methylphenidate (Ritalin®) • Attention deficit hyperactivity disorder	Stimulant	CNS stimulation, tachycardia, cardiac arrhythmias
Metoclopramide (Reglan®) • Gastroesophageal reflux disease, diabetic gastroparesis, nausea/vomiting	Antiemetic, prokinetic agent, dopamine antagonist	Extrapyramidal side effects, hyperprolactinemia
Metoprolol (Lopressor®, Toprol XL®) • Hypertension, angina pectoris, heart failure	Antihypertensive (β_1-anatgonist)	Heart failure, hypoglycemic unawareness, use with caution in asthmatics (not completely β_1 selective), bradycardia, hypotension, impotence
Metronidazole (Flagyl®) • Trichomoniasis, amebiasis, giardiasis, anaerobic bacterial infections, numerous other infections	Trichomonacide, antiprotozoal, and antibacterial agent	Disulfiram-like reaction, neuropathy, metallic taste, reversible neutropenia, seizures
Miconazole (Monistat-3® and -7®) • Vulvovaginal candidiasis, topical fungal infections	Conazole antifungal	Allergic contact dermatitis
Mifepristone (RU486®) • Abortifacient, postcoital contraceptive	Progestin and glucocorticoid antagonist/abortifacient	Vaginal bleeding, infection, sepsis
Mirtazapine (Remeron®) • Major depressive disorder	Antidepressant α_2 agonist	Weight gain, sedation
Misoprostol (Cytotec®) • Prevention of NSAID-induced ulcers	Antiulcer medication (prostaglandin E_1 agonist)	Diarrhea, miscarriage
Mometasone (Elocon®) • Inflammatory and pruritic manifestations of corticosteroid-responsive dermatoses	Topical corticosteroid, antiinflammatory	HPA-axis suppression, increased topical infection (bacterial, viral, and fungal)
Mometasone (Nasonex®) • Allergic rhinitis	Corticosteroid	Epistaxis, pharyngitis, angioedema, upper respiratory tract infection
Montelukast (Singulair®) • Asthma • Allergies	Antiasthma (for prevention, not to reverse acute attacks), selective antagonist of leukotriene D_4 (LTD_4) receptors	Gastrointestinal disturbances, hypersensitivity reactions

(Continued)

Generic Name (Brand Name) • Primary Use	Therapeutic Class and Mechanism	Most Common Adverse Effects
Morphine (MS Contin®) • Moderate-to-severe pain	Opioid analgesic/narcotic analgesic	Respiratory depression, euphoria, constipation, pruritus, dependence
Mupirocin (Bactroban®) • Impetigo, methicillin-resistant *Staphylococcus aureus* (MRSA)	Topical antibiotic	Contact dermatitis
Mycophenolate mofetil (CellCept®) • Prophylaxis of organ rejection in patients receiving allogenic renal, cardiac, or hepatic transplants	Immunosuppressant (inosine monophosphate dehydrogenase [IMPDH] inhibitor)	Myelosuppression
Nafcillin (Unipen®) • *Staphylococcal* infections	Penicillinase-resistant penicillin	Penicillin allergy
Nalbuphine (Nubain®) • Pain	Opioid, mixed agonist/antagonist (stimulates kappa, weak mu antagonist)	Sedation, CNS effects; less respiratory depression, less analgesia, and less abuse potential than strong mu agonists
Naloxone (Narcan®) • Used to reverse acute opioid overdose	Opioid antagonist	Short half-life may necessitate multiple doses
Naltrexone (ReVia®) • Decreases alcohol cravings, used in opioid dependence	Opioid antagonist	Longer half-life than naloxone, can cause abstinence symptoms
Naproxen (Naprosyn®, Naprelan®) • Osteoarthritis, inflammatory conditions, rheumatoid arthritis, pain	NSAID	Increased risk of serious cardiovascular thrombotic events, bleeding, gastrointestinal ulceration
Nedocromil (Tilade®) • Asthma	Mast cell stabilizer	Coughing, airway irritation
Nefazodone (Serzone®) • Depression	Antidepressant (heterocyclic)	Hepatotoxicity, P450 inhibitor
Nelfinavir (Viracept®) • HIV infection	Protease inhibitor	Diarrhea, P450 inhibitor
Neostigmine (Prostigmin®) • Myasthenia gravis, reversal of neuromuscular blockade	Cholinesterase inhibitor (quaternary amine)	Excess cholinomimetic effects
Nevirapine (Viramune®) • HIV infection	Antiretroviral, nonnucleoside reverse transcriptase inhibitor (NNRTI)	Fatal hepatotoxicity, Stevens-Johnson syndrome, toxic epidermal necrolysis
Niacin (Niaspan®) • Hypercholesterolemia	Antihyperlipidemic	Flushing, hepatotoxicity
Nifedipine (Procardia®, Adalat CC®) • Angina, hypertension	Dihydropyridine calcium channel blocker	Orthostatic hypotension, tachycardia, dizziness, peripheral edema, syncope, gingival hyperplasia
Nitrofurantoin (Macrodantin®, Macrobid®) • Urinary tract infections	Urinary antiseptic	Hypersensitivity pneumonitis

(Continued)

Generic Name (Brand Name) • Primary Use	Therapeutic Class and Mechanism	Most Common Adverse Effects
Nitroglycerin (Nitro-Dur®, Nitro-Bid®, Nitrostat®) • Angina	Antianginal vasodilator, nitrate	Tachycardia, hypotension, headache
Olanzapine (Zyprexa®) • Schizophrenia	Atypical antipsychotic (blocks $5HT_2$ receptors)	Increased mortality in elderly with dementia-related psychosis, postural hypotension
Omeprazole (Prilosec®) • Ulcers, gastroesophageal reflux disease, Zollinger-Ellison	Proton pump inhibitor (irreversible blocker of H^+/K^+-ATPase on parietal cells; prototype, "**-prazoles**" include esomeprazole, lansoprazole, pantoprazole, rabeprazole)	
Ondansetron (Zofran®) • Nausea/vomiting	Antiemetics, ($5HT_3$ antagonist; prototype, others in class include granisetron, dolasetron)	Headache, dizziness
Orlistat (Xenical®) • Obesity	Antiobesity, pancreatic lipase inhibitor	Steatorrhea, flatulence, bloating
Oxazepam (Serax®) • Anxiety, alcohol withdrawal	Benzodiazepine; sedative/hypnotic, anticonvulsant (increases frequency of opening of $GABA_A$ receptor)	Respiratory depression, somnolence, dizziness
Oxybutynin (Ditropan XL®) • Urinary incontinence	Genitourinary smooth muscle relaxant, antimuscarinic	Typical antimuscarinic side effects
Oxycodone (sustained-release, OxyContin®; with aspirin: Percodan®, with acetaminophen: Percocet®) • Moderate to severe pain	Opioid analgesic	Respiratory depression, euphoria, constipation, pruritus, dependence
Pantoprazole (Protonix®) • Ulcers, gastroesophageal reflux disease, Zollinger-Ellison	Proton pump inhibitor (irreversible blocker of H^+/K^+-ATPase on parietal cells)	
Paroxetine (Paxil®) • Depression, obsessive-compulsive disorder, anxiety disorders, bulimia	Selective serotonin reuptake inhibitor (SSRI) antidepressant	CNS stimulation, sexual dysfunction, gastrointestinal distress, serotonin syndrome, seizures in overdose; drug interactions with MAOIs, TCAs, meperidine
Penicillin G, penicillin V (many brand names) • Numerous bacterial infections	Beta-lactam antibiotic	Allergic reactions, anaphylaxis, drug fever, Stevens-Johnson syndrome, pseudomembranous colitis
Phenelzine (Nardil®) • Depression	Antidepressant, irreversible and nonselective monoamine oxidase (MAO) inhibitor	Hypertensive crisis with tyramine-containing foods and indirect-acting sympathomimetics, serotonin syndrome with serotonergic drugs, e.g., SSRIs
Phenobarbital (Pb®) • Seizures, preanesthetics, insomnia	Sedative hypnotic, anticonvulsant, long-acting barbiturate (increases duration of $GABA_A$-receptor opening)	Sedation, P450 induction, dependence, additive with other CNS depressants
Phenoxybenzamine (Dibenzyline®) • Pheochromocytoma	Irreversible α-adrenergic antagonist	Hypotension, gastrointestinal distress

(Continued)

Generic Name (Brand Name) • Primary Use	Therapeutic Class and Mechanism	Most Common Adverse Effects
Phenytoin (Dilantin®) • Generalized tonic-clonic and complex partial seizures	Anticonvulsant (hydantoin)	Gingival hyperplasia, sedation, diplopia, hirsutism, teratogen
Physostigmine (Antilirium®) • Anticholinergic overdose, glaucoma	Cholinesterase inhibitor (tertiary amine)	Muscarinic effects (diarrhea, urination, miosis, bronchoconstriction, bradycardia, excitation, lacrimation, salivation, sweating [DUMBBELSS])
Pilocarpine (Isopto Carpine®) • Glaucoma	Antiglaucoma (muscarinic agonist)	Muscarinic agonist effects
Pindolol (Visken®) • Hypertension	Nonselective β-adrenergic antagonist with intrinsic sympathomimetic activity	Heart failure, hypoglycemic unawareness, bronchospasm in asthmatics, bradycardia, hypotension, impotence
Pioglitazone HCl (Actos®) • Type 2 diabetes mellitus	Thiazolidinedione; stimulates peroxisome proliferator-activator receptors (PPARs)	Cardiovascular toxicity
Piroxicam (Feldene®) • Osteoarthritis, inflammatory conditions, rheumatoid arthritis	NSAID	Increased risk of serious cardiovascular thrombotic events, bleeding, gastrointestinal ulceration
Pravastatin (Pravachol®) • Hypercholesterolemia	HMG-CoA reductase inhibitor, antihyperlipidemic	Myalgia, myopathy, rhabdomyolysis, hepatic dysfunction, elevated transaminases
Prednisone (Deltasone®) • Inflammatory conditions, immunosuppressive	Antiinflammatory, glucocorticosteroid, immunosuppressant	Sodium retention, fluid retention, potassium loss, hypokalemic alkalosis, peptic ulcer disease, cushingoid state, osteoporosis
Pregabalin (Lyrica®) • Neuropathic pain associated with diabetic peripheral neuropathy, seizures	GABA analog	Somnolence
Procainamide (Procanbid®) • Ventricular arrhythmias	Class IA antiarrhythmic (Na^+ and K^+ channel blocker)	Lupus erythematosus–like syndrome, hematoxicity, hypotension, cardiovascular effects (torsades)
Prochlorperazine (Compazine®) • Nausea/vomiting	Phenothiazine antiemetic	Extrapyramidal side effects, lowers seizure threshold, neuroleptic malignant syndrome
Propranolol (Inderal®) • Hypertension, angina, arrhythmias, hyperthyroidism, migraine, benign essential tremor	Nonselective β-adrenergic antagonist	Heart failure, hypoglycemic unawareness, bronchospasm in asthmatics, bradycardia, hypotension, impotence
Propylthiouracil (PTU) (generic) • Hyperthyroidism	Antithyroid agent (blocks tyrosine iodination, inhibits coupling)	Rash, immune reactions (rare)
Quetiapine (Seroquel®) • Schizophrenia, bipolar disorder	Atypical antipsychotic	Increased suicidal risk, somnolence, hypotension, tachycardia

(Continued)

Generic Name (Brand Name) • Primary Use	Therapeutic Class and Mechanism	Most Common Adverse Effects
Quinidine (Quinaglute®, Quinidix®) • Atrial and ventricular arrythmias	Class IA antiarrythmic (Na^+ and K^+ channel blocker)	Cinchonism (Gastrointestinal symptoms, tinnitus, visual disturbances, CNS excitation), torsade
Rabeprazole (AcipHex®) • Ulcers, gastroesophageal reflux disease, Zollinger-Ellison	Proton pump inhibitor (irreversible blocker of H^+/K^+-ATPase on parietal cells)	
Raloxifene (Evista®) • Osteoporosis	Selective estrogen-receptor modulator (SERM)	Hot flashes, leg cramps, blood clots
Ramipril (Altace®) • Hypertension, congestive heart failure, diabetic nephropathy	ACE inhibitor	Cough, angioedema, hyperkalemia
Ranitidine (Zantac®) • Peptic ulcer disease, gastroesophageal reflux disease, Zollinger-Ellison	Histamine H_2-receptor antagonist	
Rifampin (Rifadin®) • Tuberculosis	Antitubercular agent (inhibits DNA-dependent RNA polymerase)	Hepatotoxicity, induces P450
Risedronate (Actonel®) • Osteoporosis	Bisphosphonate	Esophageal ulcers, esophagitis, abdominal pain, gastric reflux, dysphagia
Risperidone (Risperdal®) • Schizophrenia	Atypical antipsychotic	Fewer extrapyramidal side effects than typical neuroleptics, postural hypotension, dizziness, stroke
Ropinirole (Requip®) • Parkinson disease, restless legs syndrome	Dopamine-receptor agonist	Sedation, dyskinesias, nausea/vomiting
Rosiglitazone (Avandia®) • Type 2 diabetes mellitus	Thiazolidinedione, stimulates peroxisome proliferator-activator receptors (PPARs)	Hypoglycemia and weight gain may occur, cardiovascular toxicity
Rosuvastatin (Crestor®) • Hyperlipidemia, hypertriglyceridemia	HMG-CoA reductase inhibitor	Hepatic dysfunction, rhabdomyolysis, myalgia, myopathy
Salmeterol (Serevent®) • Asthma, chronic obstructive pulmonary disease	Bronchodilator (β_2-receptor agonist; long acting)	CNS stimulation, hypertension, tachycardia
Selegiline (Eldepryl®) • Parkinson disease	Antiparkinson (MAO_B inhibitor)	CNS stimulation, insomnia, dyskinesias
Sertraline (Zoloft®) • Depression, obsessive-compulsive disorder, anxiety disorders, bulimia	Selective serotonin reuptake inhibitor (SSRI) antidepressant	CNS stimulation, sexual dysfunction, gastrointestinal distress, serotonin syndrome, seizures in overdose; drug interactions with MAOIs, TCAs, meperidine

(Continued)

Generic Name (Brand Name) • Primary Use	Therapeutic Class and Mechanism	Most Common Adverse Effects
Sildenafil (Viagra®) • Erectile dysfunction	Cyclic guanosine monophosphate (cGMP)–specific phosphodiesterase type 5 (PDE5) inhibitor	Postural hypotension, tachycardia, myocardial infarction, priapism, vision loss
Simvastatin (Zocor®) • Hyperlipidemia, hypertriglyceridemia	HMG-CoA reductase inhibitor, antihyperlipidemic	Myalgia, myopathy, rhabdomyolysis, hepatic dysfunction, elevated transaminases
Spironolactone (Aldactone®) • Primary hyperaldosteronism, edematous conditions, CHF, hypertension, hypokalemia, female hirsutism	Potassium-sparing diuretic, aldosterone receptor antagonist, androgen receptor antagonist	Gynecomastia, hyperkalemia, dilutional hyponatremia
Stavudine (d4T) (Zerit®) • HIV Infection	Antiretroviral, nucleoside reverse transcriptase inhibitor (NRTI)	Peripheral neuropathy (dose-limiting)
Succinylcholine (Anectine®) • Muscle relaxation (adjunct to surgery, intubation)	Depolarizing muscle relaxant (short duration, metabolized by plasma cholinesterases)	May have role in malignant hyperthermia, muscle pain, hyperkalemia
Sulfasalazine (Azulfidine®) • Inflammatory bowel disease	Antiinflammatory (derivative of mesalazine)	Gastrointestinal distress, dizziness
Sumatriptan (Imitrex®) • Migraine	Antimigraine (abortive); selective 5-hydroxytryptamine$_{1D}$ (5-HT$_{1D}$)–receptor agonist	Coronary artery vasospasm, hypertension, tachycardia, chest or throat pain/pressure, myocardial infarction, stroke, cerebral hemorrhage, asthenia
Tadalafil (Cialis®) • Erectile dysfunction	Cyclic guanosine monophosphate (cGMP)–specific phosphodiesterase type 5 (PDE5) inhibitor	Postural hypotension, tachycardia, myocardial infarction, priapism, vision loss
Tamoxifen (Nolvadex®) • Metastatic breast cancer, prevention of breast cancer in high-risk patients	Selective estrogen-receptor modulator (SERM)	Hot flushes, increased risk of venous thrombosis
Tamsulosin (Flomax®) • Benign prostatic hypertrophy	α_{1A}-adrenergic antagonist (prostate selective)	Orthostatic hypotension, syncope (first-dose response), hypotension, tachycardia, fewer side effects than pure α_1-adrenergic antagonists
Temazepam (Restoril®) • Insomnia	Benzodiazepine; sedative/hypnotic (increases frequency of GABA$_A$-receptor opening)	Drowsiness, dizziness, amnesia, respiratory depression
Terazosin (Hytrin®) • Benign prostatic hypertrophy, hypertension	α_1-adrenergic antagonist	Orthostatic hypotension and syncope, particularly as a "first-dose effect," tachycardia
Terbinafine (Lamisil®) • Onychomycosis	Antifungal (inhibits squalene epoxidase)	Gastrointestinal distress, headache, hepatotoxicity
Tetracycline (Achromycin®) • Acne, *Chlamydia*, numerous sexually transmitted infections, Rocky Mountain spotted fever, many others	Tetracycline antibiotic	Teratogenicity (tooth developmental problems, bone hypoplasia, hepatic failure), photosensitivity, pseudomembranous colitis, renal toxicity, maculopapular and erythematous rashes

(Continued)

Generic Name (Brand Name) • Primary Use	Therapeutic Class and Mechanism	Most Common Adverse Effects
Theophylline (Theo-Dur®) • Chronic asthma, chronic obstructive pulmonary disease	Bronchodilator, methylxanthine, phosphodiesterase inhibitor, adenosine antagonist	Insomnia, tremor, gastrointestinal distress
Timolol (ophthalmic: Timoptic®, Betimol®; oral: Blocadren®) • Glaucoma, hypertension	Antiglaucoma, antihypertensive; nonselective β-adrenergic antagonist	Heart failure, hypoglycemic unawareness, bronchospasm in asthmatics, bradycardia, hypotension, impotence
Tolterodine tartrate (Detrol®) • Urinary incontinence	Muscarinic antagonist	Antimuscarinic side effects (e.g., tachycardia, dry mouth, urinary retention)
Topiramate (Topamax®) • Epilepsy, migraine	Anticonvulsant, antimigraine	Cognitive deficits, fatigue, renal stones, anorexia
Tramadol (Ultram®) • Moderate-to-moderately severe pain	Centrally acting synthetic opioid analgesic, inhibits serotonin (5HT) and norepinephrine reuptake	Seizures, dizziness, constipation
Trazodone (Deseryl®) • Depression	Antidepressant (heterocyclic)	Priapism, sedation, cardiac arrhythmias
Tretinoin (Retin-A®) • Acne vulgaris	Antiacne (vitamin A derivative)	Irritation, erythema, peeling, dryness, burning/stinging
Triamterene (Dyrenium®) • Edema, hypertension	Potassium-sparing diuretic (Na^+ channel blocker in collecting duct)	Hyperkalemia
Trihexyphenidyl (Artane®) • Control of extrapyramidal disorders, Parkinson disease	Anticholinergic (muscarinic antagonist)	Strong antimuscarinic side effects (e.g., dry mouth, tachycardia, urinary retention, worsening of glaucoma, paralytic ileus, hyperthermia)
Trimethoprim/sulfamethoxazole (Bactrim®, Septra®) • Urinary tract infections, *Pneumocystis jiroveci* infections, many gram-negative bacteria, CA-MRSA	Combination blocks folic acid synthesis	Toxicity primarily due to sulfonamide; hypersensitivity, hematologic disorders, kernicterus, competes for plasma proteins
Valacyclovir (Valtrex®) • Herpes simplex	Antiviral (inhibits viral DNA polymerases)	Confusion, hallucinations, seizures, thrombotic thrombocytopenic purpura/hemolytic uremic syndrome, especially with high doses in AIDS patients
Valproic acid (Depakote®) • Epilepsy, mania, migraine prophylaxis	Anticonvulsant/antimanic (inhibits T-type Ca^{2+} channels, blocks Na^+ channels)	Hepatic failure (rare but can be fatal), teratogenicity (neural tube defects), P450 inhibitor
Valsartan (Diovan®) • Hypertension, heart failure	Angiotensin II–receptor (type 1 AT_1) antagonist	Hypotension, increased BUN and potassium, contraindicated in pregnancy
Vancomycin (Vancocin®) • Severe infections caused by susceptible strains of methicillin-resistant (beta-lactam–resistant) staphylococci and other serious gram-positive infections	Antibiotic (glycopeptide bacteriocidal; inhibits cell wall synthesis)	Red man syndrome, ototoxicity, nephrotoxicity, hypersensitivity

(Continued)

Generic Name (Brand Name) • Primary Use	Therapeutic Class and Mechanism	Most Common Adverse Effects
Venlafaxine (Effexor®) • Depression	Antidepressant (heterocyclic)/inhibits norepinephrine [NE] and serotonin [5HT] uptake	Somnolence, nausea, impotence, tachycardia, CNS stimulation
Verapamil (Calan®, Verelan®) • Hypertension, arrhythmias, angina, prophylaxis of paroxysmal supraventricular tachycardia (PSVT)	Vasodilator and cardiac depressant, calcium channel blocker (blocks L-type calcium channels)	Strong negative inotropic effects, hypotension, atrioventricular block, heart failure, constipation
Vinblastine (Velban®) • Lymphomas, neuroblastoma, testicular carcinoma, Kaposi sarcoma	Antineoplastic/M phase–specific agent (inhibits mitotic spindle formation)	Myelosuppression, alopecia, gastrointestinal distress
Vincristine (Oncovin®) • Leukemias, lymphomas, Wilms tumor	Antineoplastic/M phase–specific agent (inhibits mitotic spindle formation)	Peripheral neuropathy, alopecia, gastrointestinal distress
Warfarin (Coumadin®) • Coagulation disorders, venous thrombosis, pulmonary embolism, atrial fibrillation, stroke, systemic embolism after myocardial infarction	Anticoagulant (vitamin K–dependent clotting factor inhibitor)	Bleeding, contraindicated in pregnancy, multiple drug interactions with P450 inducers or inhibitors
Zafirlukast (Accolate®) • Asthma	Antiasthma (for prevention, not to reverse acute attacks), selective antagonist of leukotriene D_4 and E_4 (LTD_4 and LTE_4) receptors	Hepatic failure
Zidovudine (ZDV; Retrovir®) • HIV infection	Nucleoside reverse transcriptase inhibitor (NRTI); formerly called azidothymidine (AZT)	Bone marrow suppression leading to anemia and neutropenia (may require transfusions), headache, asthenia, myalgia, gastrointestinal distress; all NRTIs may cause lactic acidemia and hepatomegaly with steatosis
Zileuton (Zyflo®) • Asthma (prophylaxis, chronic treatment)	5-lipoxygenase inhibitor	Dyspepsia, elevation of liver function tests
Zolmitriptan (Zomig®) • Migraine	Antimigraine (abortive)/selective 5-hydroxytryptamine$_{1D}$ (5-HT_{1D})– receptor agonist	Coronary artery vasospasm, hypertension, tachycardia, chest or throat pain/pressure, myocardial infarction, stroke, cerebral hemorrhage, asthenia
Zolpidem (Ambien®) • Insomnia	Nonbenzodiazepine hypnotic; binds BZ_1 site on the $GABA_A$ receptor; can be reversed by flumazenil	Daytime drowsiness, dizziness, abnormal behaviors

Appendix F

Bacterial Algorithms

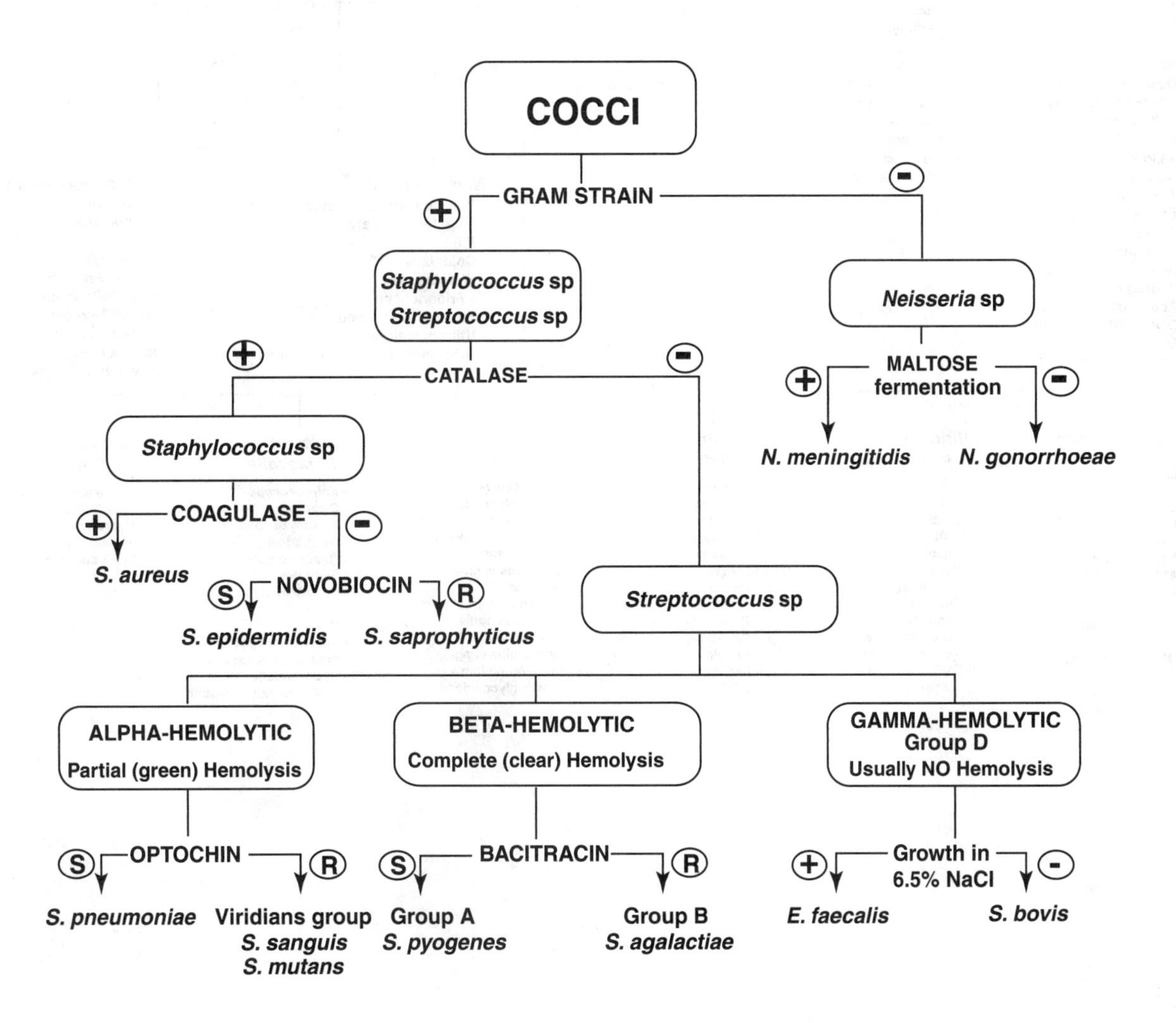

(Continued)

▶ Bacterial Algorithms *(Cont'd.)*

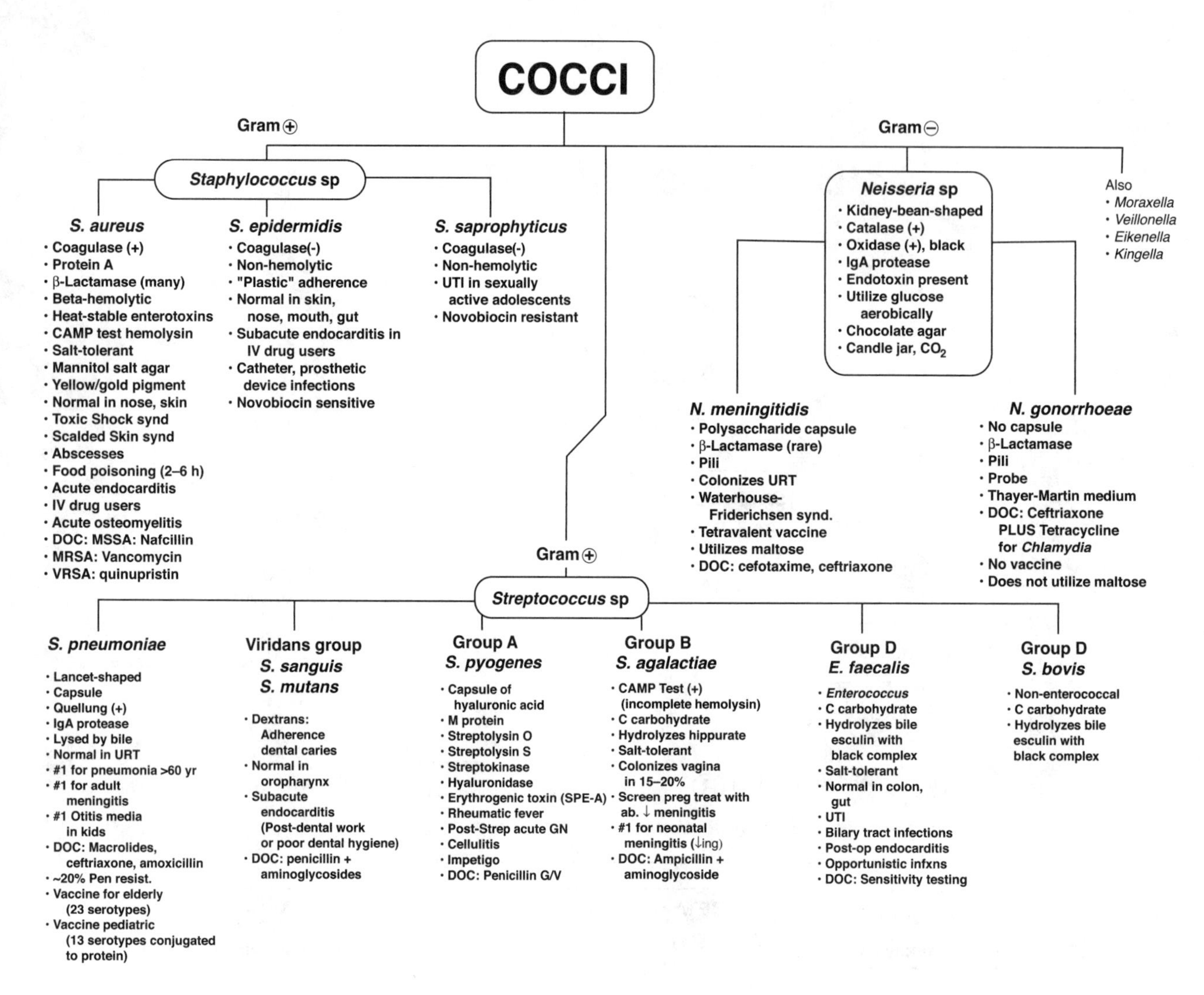

(Continued)

▶ Bacterial Algorithms *(Cont'd.)*

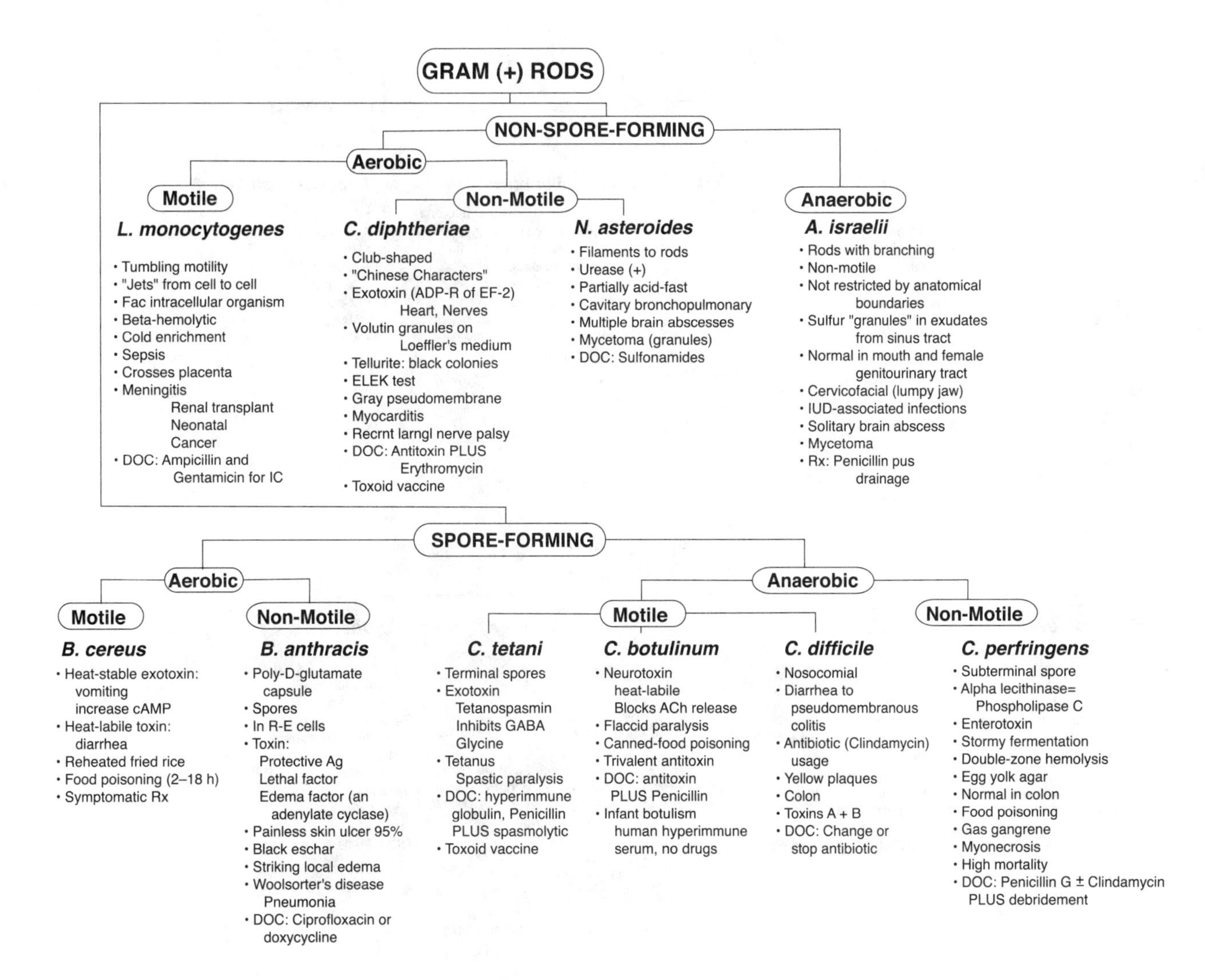

(Continued)

► Bacterial Algorithms *(Cont'd.)*

GRAM (-) RODS & SPIROCHETES

Facultative Anaerobes →

AEROBES

B. pertussis

- Adhesion to cell via hemagglutinin and pertussis toxin
- Adenylate cyclase txn (local edema)
- Tracheal toxin
- Dermanecrosis toxin
- Endotoxin - Lipid X, A ADP-R of GNBP
- Bordet-Gengou agar
- Regan-Lowe agar
- Whooping cough
- DOC: Erythromycin
- Vaccine toxoid and filamentous hemagglutinin

***Brucella* sp**

- In R-E cells
- Endotoxin
- Requires CYS, CO_2
- Unpasteurized milk
- Undulant Fever
  Bang's disease
  Malta fever
- *B. abortus*
  cattle, mild
- *B. suis*
  pigs
  suppurative, chronic
- *B. melitensis*
  goats
  severe, acute
- DOC: rifampin and doxycycline

F. tularensis

- In R-E cells
- Requires CYS
- *Dermacentor* tick bite
  Transovarian trans.
- Aerosol
- Rabbits, rodents
- Granulomatous rxn
- Tularemia - AR, MO, TX
- Live, attntd vaccine
- DOC: Streptomycin

Coxiella burnetti

- Obligate intracellular bacteria
- Q fever-atyp pneumo, hepatitis
- Urine, feces, amniotic fluid, placenta-airborne, resistant to drying
- Reservoir: cats, domestic livestock, high titers in pregnant
- Weil-Felix test negative
- No rash

L. pneumophila

- Water-loving
  air conditioning
- Requires CYS & Fe
- Buffered Charcoal
  Yeast agar
- Dieterle silver stain
- Stains poorly Gram (-)
- Atypical pneumonia
- Mental confusion
- Diarrhea
- DOC: Erythromycin
- Not contagious

P. aeruginosa

- Slime-layer
- Grape-like odor
- Exotoxin A:
  ADP-R of eEF-2
  Liver
- Oxidase (+)
- Pigments
  pyocyanin, pyoverdin
- Transient colonization
  In 10% of normal pop
- Osteomyelitis in drug abusers
- Pneumonia in cystic fibrosis
- Nosocomial infections
  Burn patients
  Neutropenic patients
- Ecthyma gangrenosum
- DOC: Penicillin
  PLUS Aminoglycoside

ANAEROBES

***Bacteroides* sp**

- *B. fragilis* - obligate
- Modified LPS, capsules
- Predominant colonic flora
- Normal in oropharynx, vagina
- Predisposing factors:
  surgery, trauma
  chronic disease (cancer)
- Septicemia, peritonitis
  aspiration pneumonia
- *Prevotella melaninogenica*
  Human oropharynx
- Fusobacterium (combined w/ *Treponema microdentium)*
  Vincent's angina
  Trench mouth
- DOC: Metronidazole OR Clindamycin OR Cefoxitin

SPIROCHETES

- Thin-Walled
- Spiral-Shaped
- Axial Filaments
- Jarisch-Herxheimer Rxn

***Treponema* sp**

- *T. pallidum* - Syphilis
  Obligate parasite
- 1°- PAINLESS chancre, infectious
- 2°- Rash infectious
- 3°- Gummas, CVS, CNS
- Congenital: stillbirths, malformed
- VDRL & RPR - Screening tests
- Reagin ab - xrxn with Cardiolipin
- FTA-ABS (immunofluorescence) specific test
- Dark-field microscopy
- DOC: Benzathine Penicillin

***Borrelia* sp**

- Microaerophillic
- Giemsa stain
- *B. burgdorferi*
  Lyme disease
  (*I. scapularis*), *I. pacificus*
  Reservoirs: mice, deer
  CT, WI, CA
  Erythema Migrans
  Target lesions
- *B. recurrentis*
  Relapsing fever
  Vector: body louse
  Antigenic variation
- DOC: Penicillin or azithromycin

***Leptospira* sp**

- Dark-field microscopy
- Contaminated water
  Animal urine
- Fever, jaundice, uremia
- Non-icteric Leptospirosis
  Meningitis - No PMN in CSF
  uveitis, rash
- Icteric Leptospirosis
  Weil's disease
  Renal failure, myocarditis
- DOC: Penicillin G or doxycycline

(Continued)

▶ Bacterial Algorithms *(Cont'd.)*

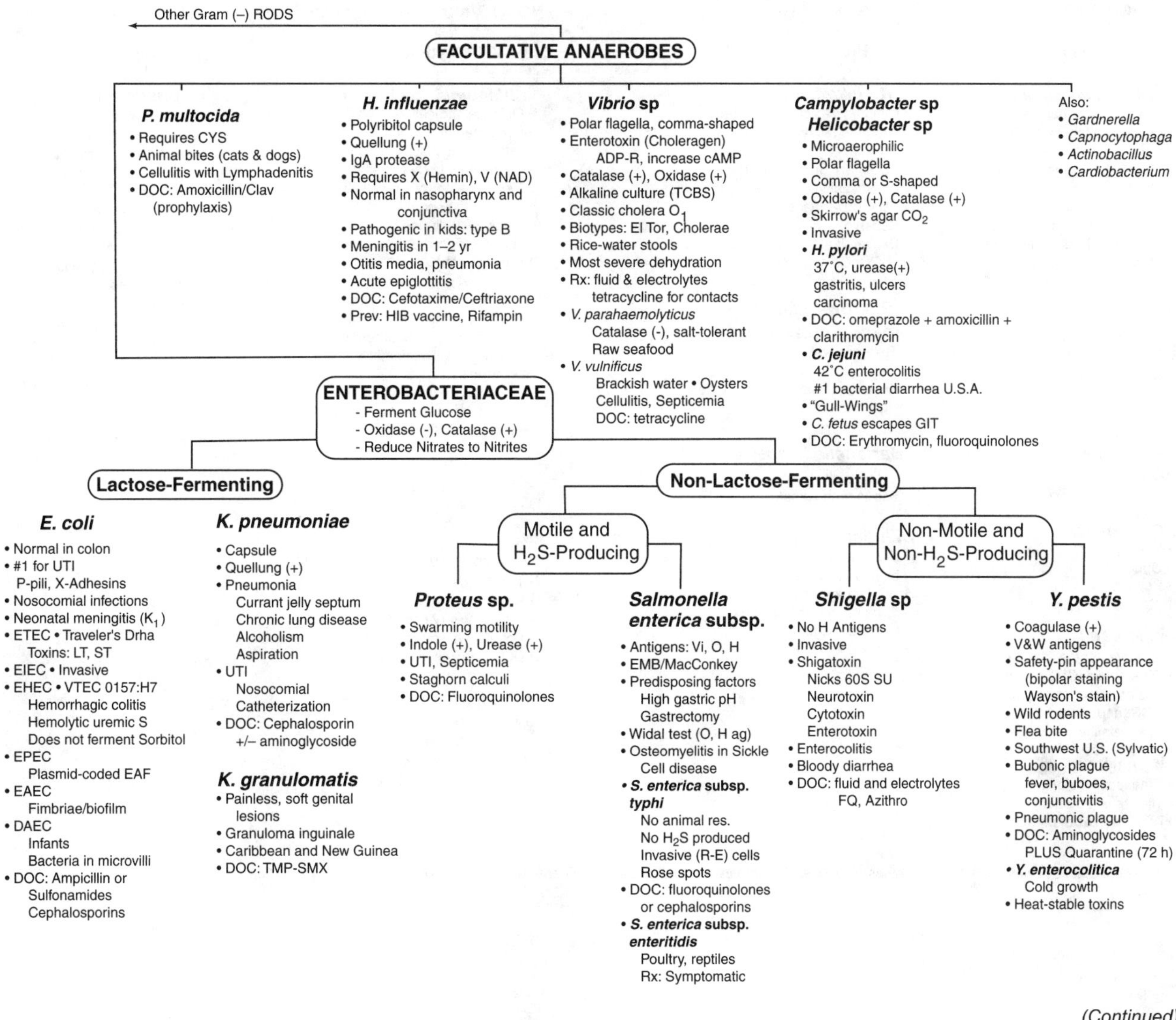

(Continued)

► Bacterial Algorithms *(Cont'd.)*

Poorly Gram-Staining Organisms*

ACID FAST → **Mycobacteria**

M. tuberculosis
- Gram (+) wall but doesn't stain due to waxy CW
- Acid fast, obligate aerobe
- Respiratory transmission
- Cord factor-trehalose mycolate-inhib. WBC migration mitoch. resp./ oxid. phosphor
- Sulfatides-inhib. phagosome-lysosome fusion
- Niacin (+), catalase (+) at 37˚C, (–) at 68˚C
- Slow growing
- Drug resistance
- Lowenstein-Jensen medium
- DOC: isoniazid + rifampin + pyrazinimide (2 mo) then isoniazid + rifampin (4 mo)

M. avium-intracellulare
- Gram (+) wall but doesn't stain due to waxy CW
- Acid fast
- Obligate aerobe
- Soil organism
- Opportunist, non-contagious
- Pulmonary → diss infections CA pts, late AIDS pts

M. leprae
- Obligate intracellular bacterium
- Tuberculoid (CMI damage)
- Lepromatous leprosy (poor CMI)
- DOC: dapsone + rifampin + clofazimine

M. marinum
- Cutaneous lesions (fish tank granuloma)
- DOC: isoniazid, rifampin, ethambutol

SOME ATP → **Rickettsias**

R. rickettsii
- Obligate intracellular bacteria
- Gram-negative envelope but stain poorly
- Rocky MT Spt'd Fever-rash on wrists/ankles → trunk, palms, soles
- Vector: *Dermacentor* tick
- Reservoirs: ticks, wild rodents
- Dx: serol: 4x incr indir Fl. Ab + Weil-Felix DOC: Doxycycline

R. prowazekii
- Obligate intracellular bacteria
- Epidemic typhus
- Vector: *Pediculus* louse
- Reservoir: humans, squirrel fleas, flying squirrels

Bartonella henselae
- Cat scratch fever
- Bacillary angiomatosis in AIDS

Ehrlichia
- Ehrlichiosis
- Morulae in WBC
- DOC: doxycycline
- *E. chafeensis*-monocytes + macrophages
- *E. phagocytophila* - PMNs
- *Ixodes* tick

NO ATP, mod. peptidoglycan → **Chlamydiaceae**

Chlamydia trachomatis
- Obligate intracellular bacteria
- Gram-negative envelope but stain poorly; lack muramic acid
- Elementary body-transmitted
- Reticulate body-intracellular
- Dx: serology or tissue culture growth confirmed by inclusion bodies (Fl Ab, Giemsa, iodine)

Serotypes D-K
- U.S.-Most common bacterial STD (HPV and HSV2 more common)
- Neonatal/adult inclus. conjunct, neonatal. pneumo; urethritis cervicitis, PID, infertility

Serotypes L1, 2, 3
- Lymphogranuloma venereum
- STD in Africa, Asia, S. America

Serotypes A, B, Ba, C
- Trachoma-follic conjunctivitis → conj. scarring, entropion → corneal scarring
- Leading infectious cause blindness
- DOC: Doxycycline or azithromycin

Chlamydophila pneumoniae
- TWAR agent
- Respiratory infections
- Probably very common
- Potential association with atherosclerosis
- DOC: macrolides and tetracycline

Chlamydophila psittaci
- Atypical pneumonia
- Birds (parrots)
- DOC: tetracycline

NO CELL WALL → **Mycoplasmas**

M. pneumoniae
- Lack cell wall peptidoglycan → non-Gram-staining
- Cholesterol (req'd) in membr.
- Atypical pneumonia in youth and young adults
- Free living (culturable, extracell.)
- Slow growth, special media: Mycoplasma, Eaton's or Hayflick's media-sterols+pur/pyrimidines: mulberry colonies
- Cold aggulutinins in 65% cases
- No Penicillins nor Cephalosporins
- DOC: erythromycin, azithromycin

Ureaplasma urealyticum
- Urethritis, prostatitis
- Urease positive
- No cell wall
- DOC: erythromycin or tetracycline

*Also note that *Legionella* and the spirochetes (*Treponema*, *Leptospira*, and *Borrelia*)—all Gram-negative—do not show up reliably with Gram stain.

Viral Algorithms

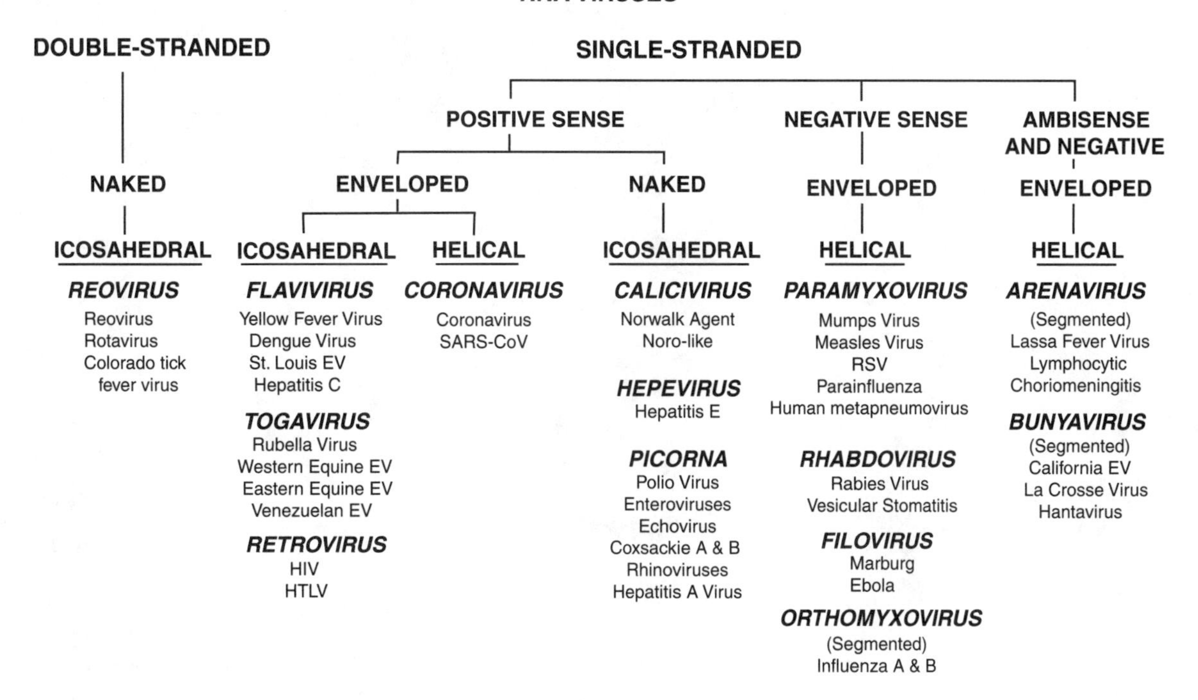

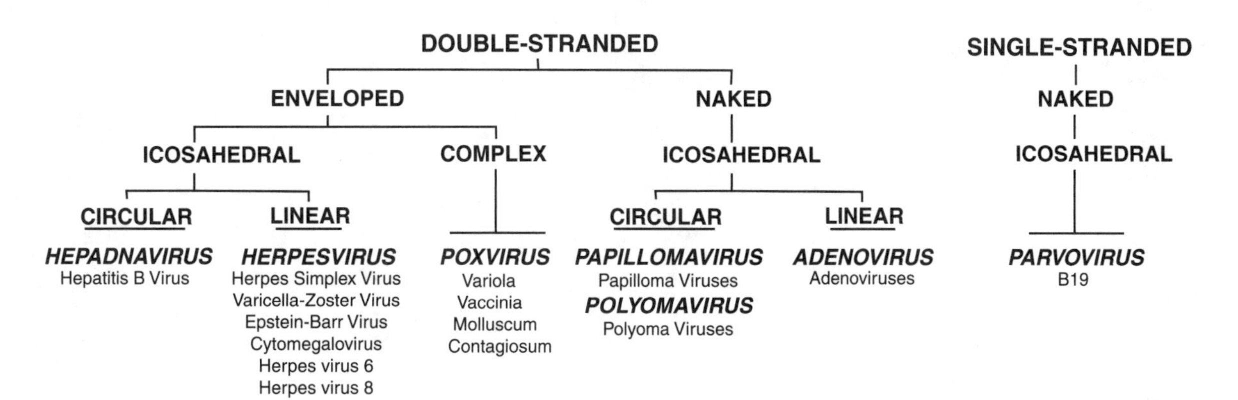

Index

B

D

E

G

H

I

J

K

L

M

N

O

Q

R

S

T

V